GEOTECHNICS FOR DEVELOPING AFRICA
LA GEOTECHNIQUE AU SERVICE DU DEVELOPPEMENT DE L'AFRIQUE

COMPTES RENDUS DU DOUZIEME CONGRES REGIONAL AFRICAIN
DE LA GEOTECHNIQUE AU SERVICE DU DEVELOPPEMENT DE L'AFRIQUE
DURBAN/L'AFRIQUE DU SUD/25-27 OCTOBRE 1999

La Géotechnique au Service du Développement de L'Afrique

Rédacteurs

G. R. Wardle
Jones & Wagener (Pty) Ltd, Rivonia, L'Afrique du Sud

G. E. Blight & A. B. Fourie
Université de Witwatersrand, Johannesburg, L'Afrique du Sud

A.A.BALKEMA/ROTTERDAM/BROOKFIELD/1999

PROCEEDINGS OF THE TWELFTH REGIONAL CONFERENCE FOR AFRICA
ON SOIL MECHANICS AND GEOTECHNICAL ENGINEERING
DURBAN/SOUTH AFRICA/25-27 OCTOBER 1999

Geotechnics for Developing Africa

Edited by

G. R. Wardle
Jones & Wagener (Pty) Ltd, Rivonia, South Africa

G. E. Blight & A. B. Fourie
University of the Witwatersrand, Johannesburg, South Africa

A. A. BALKEMA/ROTTERDAM/BROOKFIELD/1999

Cover photograph: Courtesy of Satellite Application Centre, CSIR, Pretoria

The texts of the various papers in this volume were set individually by typists under the supervision of each of the authors concerned.
Les textes des divers articles dans ce volume ont été dactylographiés individuellement sous la supervision de chacun des auteurs concernés.

Published by/Publié par
A.A.Balkema, P.O.Box 1675, 3000 BR Rotterdam, Netherlands
Fax: +31.10.413.5947; E-mail: balkema@balkema.nl; Internet site: www.balkema.nl
A.A.Balkema Publishers, Old Post Road, Brookfield, VT 05036-9704, USA
Fax: 802.276.3837; E-mail: info@ashgate.com

ISBN 90 5809 082 5
© 1999 A.A.Balkema, Rotterdam
Printed in the Netherlands/Imprimé aux Pays-Bas

Table of contents
Table des matières

2 Foundation engineering and lateral support
Conception des fondations et soutènements

3 Transport geotechnics
La géotechnique dans les moyens de transport

4 Monitoring, laboratory and field testing
Contrôle et le suivi, essais de laboratoire et in-situ

5 Methods of design and analysis
  Les méthodes d'étude et de dimensionnement

6 *Soil improvement*
  Amélioration des sols

7 *Case studies*
Etudes de cas

Geotechnics for Developing Africa, Wardle, Blight & Fourie (eds) © 1999 Balkema, Rotterdam, ISBN 90 5809 082 5

Preface
Préface

The Twelfth Regional Conference for Africa on Soil Mechanics and Geotechnical Engineering: 'Geotechnics for Developing Africa' was held in October 1999 in Durban, South Africa.

As Africa is one of the few extensive underdeveloped areas in the world, it was decided to emphasise the role played by geotechnics in development. Papers deal with the following themes, as they relate to development in Africa:
– Foundation engineering and lateral support;
– Methods of design and analysis;
– Monitoring, laboratory and field testing;
– Municipal, industrial and mining waste and environmental geotechnics;
– Soil improvement;
– Transportation geotechnics;
– Case studies of the above.
The reader will find something of use and interest under each of these headings.

The series of Regional Conferences for Africa (the first was held in Pretoria in 1955) has become an institution among the geotechnical fraternity of Africa and geotechnical engineers working in Africa. The proceedings (most of which, since 1967, have been published by A.A.Balkema) represent a valuable reference on geotechnical problems peculiar to Africa and for engineering solutions to local problems. The proceedings are also an invaluable source of data on the properties of African soils, the properties of residual and tropical soils, as well as climate-related problems.

The proceedings of the Twelfth Regional Conference follow in the mould of its forerunners and add to the store of knowledge represented by this series.

G.R.Wardle, G.E.Blight and A.B.Fourie
Editors

Geotechnics for Developing Africa, Wardle, Blight & Fourie (eds) © 1999 Balkema, Rotterdam, ISBN 90 5809 082 5

Organisation

Organising Committee
Peter Day, Chairman
Kallie Strydom, Secretary
Fritz Wagener, Treasurer
Geoff Blight, Ex-officio
John Everett
Lionel Moore
Eben Rust, SAICE Geotechnical Division
Ken Schwartz
South African Institution of Civil Engineering

Technical Committee
Geoff Blight, Chairman
Gavin Wardle, Secretary
Gavin Byrne
Peter Day, Ex-officio
Andy Fourie
Ernst Friedlaender
Eben Rust, SAICE Geotechnical Division
Ken Schwartz
Nico Vermeulen
South African Institution of Civil Engineering
Chris McKnight
South African Institute of Engineering Geologists

Referees

Geoff Blight	Philip Paige-Green
Gavin Byrne	Alan Parrock
Peter Day	Lynette Röhrs
Bill Ellis	Eben Rust
John Everett	Friedrich Scheele
Andy Fourie	Ron Scheurenberg
Ernst Friedlaender	Ken Schwartz
Michel Gambin	Jonathan Shamrock
Brian Harrison	Andrew Smith
Gerhard Heymann	Dereck Sparks
Gary Jones	Gerald Strayton
Gerhard Keyter	Kallie Strydom
Irvin Luker	Wim van Steenderen
Chris McKnight	Nico Vermeulen
Lionel Moore	Fritz Wagener
Zvi Ofer	Gavin Wardle

1 Municipal, industrial and mining waste and environmental geotechnics
   Déchets urbains, miniers et industriels géotechnique de l'environnement

Geotechnics for Developing Africa, Wardle, Blight & Fourie (eds) © 1999 Balkema, Rotterdam, ISBN 90 5809 082 5

Hydraulic stability of sanitary landfills liners

S. Bayoumi, H. El-Karamany & D. El-Monayeri
Environmental Engineering Department, Faculty of Engineering, Zagazig University, Egypt

ABSTRACT: To prevent the migration of waste from the sanitary landfills to the surroundings, a suitable barrier should be used. One of the most suitable barrier materials is the compacted sand-clay mixture of low permeability. In the present study, the efficiency of such barrier layer material (compacted sand-clay mixture) to isolate the sanitary landfills have been investigated through evaluating the impact of two inorganic anions on the hydraulic conductivity of a compacted sand-clay mixture. The two inorganic anions used in this study are ammonium chloride and ammonium sulphate. Solutions at different concentrations were prepared using each anion and then applied as permeant fluids to sand-clay mixture samples. The concentration of $S0_4^{--}$ solutions ranged from 137 ppm to 137×10^3 ppm, while the concentration of Cl^- solutions ranged from 75 ppm to 75×10^3 ppm.

1 INTRODUCTION

Waters passing through the boundaries of a sanitary landfill will leach hazardous material (organic and inorganic) from the waste disposed of in the facility to the surrounding. These contaminant leachates must not be permitted to complete their migration through the sanitary landfill boundaries to reach the environment. This can be achieved by lining the containment by a low permeable blanket, e.g., compacted low permeable soils.

The main function of low-permeability, compacted soil (barrier layers) is either to restrict infiltration of water into buried waste (in cover system), or to limit seepage of leachate from the waste (in the liner system). The advantages of compacted soil as a sanitary landfill lining or cover may include enhancement of efficiency of an overlying drainage layer, development of composite action with a flexible membrane liner, adsorption and attenuation of leachate, restriction of gas migration, and others.

For the liner system, compacted soil must also have the ability to withstand chemical degradation from the liquid to be contained (Daniel and Estortlell,1990). In addition, low-permeability compacted soil must have adequate shear strength to support itself on slope and to support the weight of overlying .materials or equipment. In case of a cover system, compacted soil must also have the ability to withstand subsidence and must be repairable if damaged by freezing, desiccation, or burrowing animals. Many organic chemicals tend to shrink the

diffuse double layer that surrounds clay particles, causing the clay particles to flocculate, the soil skeleton to shrink, and cracks, called syneresis cracks, to from (Gregory and Broderrick, 1990).

The permeability of three clays to five simulated liquid wastes have been investigated by Sanks and Gloya (1977). The aqueous waste solution tested were chosen to be representative of materials that might leak from containers of solid wastes. The various substances contained acid, base, and heavy metals. Tested clays contained large percentage of montmorillonite. Among the auther's observations are the following: (1) permeability value using d - ionized water as a permeant ranged from 4×10- 10 to 1.2×10-9 m/sec. (2) Using the acid waste as a permeant resulted in higher permeability; (3) On the other hand, lower permeability was observed when basic waste was used as a permeant; and (4) No effects were shown by using salts of heavy metals $HgCl2$ at 8100mg/l and $ZnSO4$ at 8610 mg/l.

2 MATERIALS AND METHOD:

Investigating the efficiency of a compacted sand-clay mixture as a sanitary landfill liner against certain chemical species (namely: $S0_4^{--}$ and Cl^-) is the main objective of the present study. Kaolinite was mixed with fine sand in preparing the compacted sand-clay mixture samples investigated in this study. Kaolinite used in this study was a commercially processed soil obtained from the Georgia Kaolin

Company, Elizabeth, New Jersey, and the sand was natural 1900 grain, silica sand, obtained from the Ottawa Silica Company, Ottawa, Illinois. Physical and chemical properties of the used soils were characterised using several laboratory tests, and the results are presented in Tables 1 and 2. Deionized water, in addition to the two inorganic chemical solutions, were employed in the study as permeant fluids. Among the main functions of using deio - ized water in the experimental program were: (1) to saturate the samples and establish their base values of hydraulic conductivity before switching to the chemical solution as a permeant fluid; and (2) to permeate the reference samples, used for comparison purpose, parallel to the other samples.

Ammonium chloride and ammonium sulfate solutions, used individually as permeant fluids, were prepared by dissolving the proper weight of pure in-organic chemical (either ammonium chloride NH4C1 or ammonium sulfate (NH4)2S04 in deionized water to yield specific concentrations of the chemical species (Cl-, or S04--) in the solution. The different properties of both NH4C1 and (NH4)2S04 salts are presented in Table 3 according to the manufacture's information.

The leaching and/or permeation equipment used in this study is a system of ten leaching and/or permeation cells assembled together (in addition to all the necessary fittings, measuring and controlling devices, and hydraulic system) in one board. The leaching and/or permeation cell constructed in this study is, more or less, a flexible-wall perimeter consistent in design and construction with that in literature (Daniel, et al., 1984). A schematic diagram of the cell is shown in Figure (1).

Table (1): Physical Properties of Soils used in the Present Study:

Property	Reference	Fine Sand	Kaolinite	Mixture
% of Kaolinite		0	100	33.33
% of Fine Sand		100	0	66.67
Source of Soil		Ottawa Silica Co.	Georgia Kaolin. Co.	Laboratory Prepared
Atterberg Limits	ASTM D4318-84			
LL (%)			64	17
PL (%)			34	12
Specific Gravity	ASTM D854-8	2.65	2.60	2.63

Table (2): Chemical Properties of Soils used in the Present Study

Property	Reference	Fine Sand	Kaoliniie	Mixture
PH	McLean, (1982)	5.85	4.73	5.55
Specific Surface Area (m,²lg)	Carter, et al., (1982)	N/A	50	31
CEC (meq/100 g)	Jackson, (1958)	N/A	37.73	13.88
NH4 (% w/w)	Keeney, (1982)	N/A	0.000889	0.000215
cl⁻ (%w/w)	Pfaffi (1991)	0.000495	0.001017	0.00,1107
S 04⁻⁻ (%w/w)	Pfaff, (1991)	0.000954	0.001165	0.005178

Table (3) Properties of Chemicals used in Preparing the Permeant Fluids

Chemical / Property	Ammonium chloride	Ammonium Sulphate
Formula	$(NH_4)Cl$	$(NH_4)_2S0_4$
Molecular Weight	53.5	132.14
Assay (NH₄)Cl	99,7%	N/A
Assay (NH₄)2 SO₄	N/A	99.1%
pH of 5 % Solution @ 25 c	4.9	5.2
Specific Gravity @ 20 C	1.53	1.77
Solubility in water @ 20 C	37 % (by wt.)	75.4% (by wt.)
Chloride	N/A	< 3 ppm
Sulphate	N/A	N/A
Nitrate	< 5 ppm	< 0.0005 %
Heavy Metals	< 2 ppm	< 2 ppm

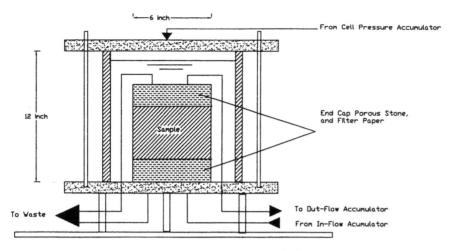

Figure 1 Schematic Diagram of the Flexible-Wall Permeameter Cells

The cell consists mainly of: (1) Cast acrylic tube that fits in recesses machined into the end plates. The cast acrylic tube is sealed between the two end plates with O-rings and has the dimensions of ¼ in. thick wall, 6 in. inner diameter, and 12 in. length. This tube with those specifications is suitable for a cofining pressure up to 80 psi which is enough in the present study; (2) Top and bottom plates are machined from acrylic sheets 1.0 in. thick, and (3) Six nuts screwed onto three stainless steel all-thread rods to clamp the two end plates together. The soil specimen is located inside the acrylic tube between a base pedestal and top cap, which are machined from PVC. PVC was selected because of its light weight, low cost, easy machineability and enough resistivity to the permeant fluid used.

3 RESULTS AND DISCUSSION:

3.1 Leaching with Chloride Anion:

Two sets of experiments were conducted in this phase of experimental program of the present study. In the first set, the investigated soil was permeated with $(NH_4)Cl$ solutions of different concentrations. Figure 2 shows the results of hydraulic conductivity investigation in terms of the change in the hydraulic conductivity, k relative to k_o which is the base value of hydraulic conductivity determined at the steady state condition using deionized water as a permeant versus the initial concentration of Cl^- in the permeant fluid.

In the second set of experiments, the investigated soil was permeated with the same permeant fluid (38,000 ppm NH_4Cl solution) until different volumes of the effluent has been collected. The col-

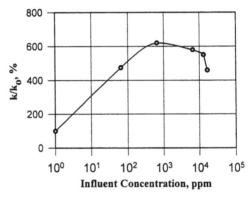

Figure 2 Effect of $(NH_4)Cl$ concentration upon the liner's hydraulic conductivity

lected effluent has different volumes ranged between 1 and 8 the volume of pores (pv) in the permeated sample. Figure (3) shows the results of hydraulic conductivity investigation in terms of the change in k versus the initial concentration of Cl^- in the permeant fluid.

3.2 Leaching with Sulphate Anion:

In this phase of the experimental program, two sets of experiments were conducted. In the first set, the investigated soil was permeated with $(NH_4)_2SO_4$ solutions of different concentrations. Figure (4) shows the results of hydraulic conductivity investigation in terms of the change in k versus the initial concentration of SO_4^- in the permeant fluid.

In the second set of experiments, the investigated soil was permeated with the same permeant fluid (69,000 ppm $(NH_4)_2SO_4$ solution) until different vol-

5

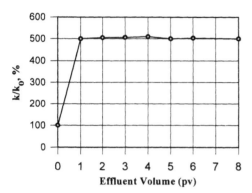

Figure 3 Effect of (NH$_4$)Cl seepage volume upon the liner's hydraulic conductivity.

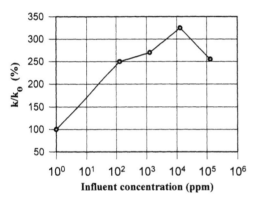

Figure 4 Effect of (NH$_4$)$_2$SO$_4$ concentration upon the liner's hydraulic conductivity.

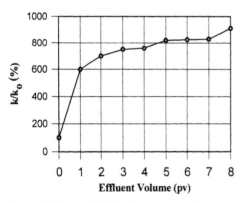

Figure 5 Effect of (NH$_4$)$_2$SO$_4$ seepage volume upon the liner's hydraulic conductivity.

umes of the effluent has been collected. Figure (5) shows the results of hydraulic conductivity investigation in terms of the change in k versus the initial concentration of SO$_4^{--}$ in the permeant fluid.

3.3 Effect of Electrolyte Concentration:

To determine the effect of the electrolyte concentration as well as seepage volume and ion type on the hydraulic conductivity of the tested material, the base line hydraulic conductivity has been determined using deionized water at the start of each permeation run before switching to the different permeant fluids.

From those results presented in Figures 2 and 4 it is observed that introducing either ammonium chloride or ammonium sulfate solutions in different concentrations to the compacted soil samples saturated with deionized water causes an increase in permeability of the tested soil.

On the one hand, this increase reached values of 325 and 600 times greater than the base line value of the hydraulic conductivity in case of ammonium sulfate and ammonium chloride respectively. That amount of increase in permeability of the liner material means a dramatic decrease in its containment effe - tiveness. On the other hand, the changes in hydraulic conductivity shown in Figures 3 and 5 may be attributed to the permeant fluid chemistry which can influence permeability apart from altering the clay fabric by its hydropliobicity or solubility in water, and by precipitation of certain chemicals or growth of micro-organisms in porous. This explanation is consistent with Pauls et al., 1988.

In other way, these changes in hydraulic conductivity upon increasing the chemical concentration in the permeant fluid may be attributed to the resulting change in double layer thickness which is inversely proportional to the electrolyte concentration. This change in double layer thickness due to changing the electrolyte concentration can affect the fabric and/or behaviour of clays (Anderson and Brown, 1981). Also, Alter et al., 1985 studied the influence of inorganic permeants on the permeability of bentonite. They found that K$^+$ and Cl$^-$ ions had a greater effect of increasing flocculation and permeability than did the Na$^+$ and CO^{2-} anions. They explained the effect of inorganic solutions on the beonite clay with the Gouy-Chapman diffuse double layer. The double layer thickness depends upon many factors including temperature, ion concentration, unit electronic charge, and the valence of the ion (Mitchell, J. K. 1993). Also, they observed that bentonite clay displayed the greatest amount of shrinkage and resulting increase in permeability when exposed to divalent cation salts such as Ca^{++} and Mg^{++} but a saturation limit of these salts is reached so that above this concentration, the salt has no greater effect.

3.4 Influence of Seepage Volume

The tested soil was permeated with the same permeant of the same concentration for different time intervals under the same conditions to assess the impact of the exposure of a clay liner to those solutions for a long-term periods. In presenting and discussing the results of that part of the experimental study in the present paper, the exposures time periods are expressed in terms of the outflow (effluent or seepage) volumes which in turn is measured in terms of the initial volume of voids in the soil sample (pv). Based on the properties of the tested materials and under the testing conditions, about one week was needed to collect one pore volume (1 pv) of the outflow.

Permeating the tested soil with ammonium chloride solution of constant concentration for different time intervals (or until different volumes of the out flow has been collected) yielded the "pv - k" relationship in Fig. 3. Within the first two weeks starting from switching the permeant fluid from deionized water to ammonium chloride solution, the permeability of the tested soil increased to five times greater than the base value established using deionized water as a permeant. Then, after about one month of perme - tion (the time required for 4pv of the out-flow to be collected) , there was no increase in the permeability. Two of replicates were tested for 1.5 and 2 months, respectively (times required for 6 and 8 pv to seep through the test sample), but there was no trend toward increased permeability during the passage of the applied permeant.

These results are consistent with the results of the test series run for nearly 30 years and described by Brandi, 1992. From that test series he found that: *"most effects of chemicals on soil do not follow a general pattern . So, soil parameters may increase or decrease depending on pollutant concentration, time, temperature, ..., etc".* Similarly, Van Ree et al., 1992 concluded that: *"liners composed of tertiary clay and sand-bentonite mixture are physically and geochemically dynamic systems. So, time-dependent behaviour for example is the consequence of consolidation process, shrinking and cracking related to changes in moisture content. Furthermore, oxidation and reduction processes as well as acid and base reactions, and precipitation and dissolution provide mechanisms which probably will influence the behaviour* and *effectiveness of a barrier system".*

A possible explanation of the first increase in permeability after two weeks permeation with ammonium chloride is the increase in electrolyte conce - tration of the permeant fluid.

Figure 5 presents the "pv - k" relationship of the tested soil when it was permeated with ammonium sulfate solution of constant concentration. Co - pared with the base value of hydraulic conductivity established using deionized water, there was an increase in the hydraulic conductivity of the tested soil upon permeation with ammonium sulfate at the tested concentration. The rate of that increase di - fers from that occurred upon permeation with ammonium chloride. Also, most of the increase occurred within the first two weeks (or during collecting the first 2 pv) and then the rate of increase decreased. Changing the rate and amount of increase in hydraulic conductivity upon changing the permeation fluid to either ammonium chloride or ammonium sulfate may be attributed to the change of type of anion of the permeant fluid from chloride to sulfate.

4 CONCLUDING DISCUSSION:

A comparison of the permeability of the samples permeated with chemical solutions and those permeated with deionized water shows significant changes in those properties upon changing the permeant fluid chemistry.

Based on the experimental program executed in this research, and limited to both the tested materials and the testing procedures employed, the following conclusions could be drawn:

(1) Introducing either ammonium chloride or ammonium sulfate solutions in different concentrations (ranged from 75 to 75,000 ppm and from 137 to 137,000 ppm respectively) to compacted soil samples, initially saturated with deionized water, causes an increase in permeability of the tested soil. This increase ranges between 325 and 600 times greater than the base line value of the hydraulic conductivity (i.e., compared with the. hydraulic conductivity of the reference samples permeated with deionized water) in case of ammonium sulfate and ammonium chloride respectively *(Figure 2& 4)*

(2) The changes in clay behaviour (regarding the hydraulic conductivity) due to changes in pore fluid composition (especially electrolyte concentration) were found to be consistent with the changes predicted by the use of Gouy-Chapman theory.

(3) Permeating the tested soil with ammonium chloride solution of constant concentration (=38,000 ppm) for different time intervals (or until different volumes of the out flow has been collected) yielded that:

(a) Within the first two weeks starting from switching the permeant fluid from deionized water to ammonium chloride solution, the permeability of the tested soil increased to five times greater than the base value established using deionized water as a permeant. (Figure 3).

(b) Then, after about one month of permeation (the time required for an amount of effluent equivalent to 4pv to be collected), there was no increase in the permeability.

7

(c) The two replicates permeated for 1.5 and 2 months, respectively (times required for 6 and 8 pvs to seep through the test sample), there was no trend toward increased permeability during the passage of the applied permeant.

(4) Permeating the tested soil with ammonium sulfate solution of constant concentration (=69,000 ppm) for different time intervals (or until different volumes of the outflow has been collected) produced:

(d) An increase in the hydraulic conductivity of the tested soil upon permeation with Ammonium sulfate at the tested concentration compared with the base value of hydraulic conductivity established using deionized water.

(e) The rate of that increase differs from that occurred upon permeation with ammonium chloride. Also, most of the increase occurred within the first two weeks (or during collecting the first 2pvs) and then the rate of increase decreased *(Figure 5)*.

REFERENCES

(1) Acar, Y. B., Hamidon, A., Field, S. D., and Scott, L. (1984a), "Organic Leachate Effects on Hydraulic Conductivity of Compacted Kaolinite," Hydraulic Barriers in Soil and Rock, ASTM STP 874, A. 1. Jolinson, R. K. Frobel, N. J. Cavalli, and C. B. Pettersson, Eds., American Society for Testing and Materials, Philadelphia, pp. 171 - 187.

(2) Acar, Y. B., Hamidon, A., Field, S. D., and Scott, L. (1984b), "The Effects of Organic Fluids on the Pore Size Distribution of Compacted Kaolinite," Hydraulic Barriers in Soil and Rock, ASTM STP 874, A. 1. Johnson, R. K. Frobel, N. J. Cavalli, and C. B. Pettersson, Eds., American Society for Testing and Materials, Philadelpilia, pp. 203-212.

(3) Acar, Y. B., Olivieri, I., and Field, S. D. (1984c), "Organic Fluid Effects on the Structural Stability of Compacted Kaolinite,"Land Disposal of Hazardous Waste: Proceedings of the Tenth Annual Research Symposium, Cited by Goldman L. J. et al., (1 990), "Clay Liners for Waste Management Facilities: Design, Construction, and Evaluation, Noys Data Corporation, NJ, USA.

(4) Anderson, D. and Brown, K.W. (1 98 1), "Organic Leachate Effects on the Permeability of Clay liners," EPA-600/9-81-0026 (PB81-173882), U.S. EPA, Washington, D.C.

(5) Brandi, H. (1992), "Mineral Liners for Hazardous Waste Containment," Geoteclinique, Vol. 42, No. 1, pp. 57-65.

(6) Buchanan, P. N. (1964), "Efrect of Temperature and Adsorbed Water on Permeability and Consolidation of Sodium- and Calcium- Montmorillonite," Ph.D. Dissertation, Texas A&M University, Colege Station, Texas , Cited by Goldman L. J. et al., (1 990), "Clay Liners,for Waste Management Facilities: Design, Construction, and Evaluation," Noys Data Corporation, NJ, USA.

(7) Daniel, D. E., and Estornell, P. M. (1990), " Compilation of Information on Alternative Barriers for Liner and Cover Systems," EPA 60012-91/002 US Environmental Protection Agency, Cincinnati, Ohio 45268.

(8) Daniel, D. E., Trautwein, S. J., Boynton, S. S., and Foreman, D. E. (1984), "permeability Testing with Flexible-Wall Permeameters," Geotechnical Testitig Journal, GTJODJ, Vol. 7, No. 3, pp. 113-122.

(9) Gregory P. B., and Daniel, D. E. (1990), " Stabilizing Compacted Clay Against Chemical Attack", Journal of Geotechnical Engineering, Vol. 1 1 6, No. 1 0.

(10) Michales, A. S., and Lin C. S. (1954), "Perm - ability of Kaolinite," Industrial and Engineering Chemistry, Vol. 46, No. 6, pp. 1239 - 1246.

(11) Mitchell, J. K. (1993), "Fundamental of Soil Behavior," 2nd. Edition, John Wiley and sons, Inc., New York, USA.

(12) Pauls, D. R., Moran, S. R., and Macyk, T. (1988), "Review of Literature Related to Clay Liners for Sump Disposal of Drilling Wastes," Alberta Land Conservation and Reclamation Council Report # RRTAC 88-10. 61 pp.

(13) Sanks, R. L., and Gloyna, E. F. (1977), "Clay Beds for Storing Solid Industrial Wastes - A survey," Proceedings of the 32nd. Industrial Waste Conference, Purdue University, Lafayette, Indiana, Ann Arbor Science, An Arbor, Michigan.

(14) Yong, R. N., Mohamed, A. 0., and Warkentin, B. P. (1992), " Principles of Contaminant Transport in Soils," Elsevier, NY, USA.

(15) Yong, R. N. (1989), "Waste Generation and Disposal: keynote address" 2nd International Symposium on Environmental Geotechnology, Shanghai, China, May 15- 18, 1989, Vol. 1, pp. 1 - 24.

Geotechnics for Developing Africa, Wardle, Blight & Fourie (eds) © 1999 Balkema, Rotterdam, ISBN 90 5809 082 5

Environmental geotechnics in the new millennium

Craig H. Benson
Department of Civil and Environmental Engineering, Geological Engineering, University of Wisconsin-Madison, Wis., USA

ABSTRACT: This paper reviews the development of environmental geotechnics and describes thrust areas for the new millenium. The discussion focuses on four key areas: site characterization, remediation, waste containment, and beneficial reuse of industrial byproducts. Thrust areas for the new millenium include innovative remediation technologies such as reactive walls and other treatment methods not constrained by mass transfer limitations, risk-based clean up, bioreactor landfills, and beneficial reuse of industrial byproducts

1 INTRODUCTION

During the last twenty years environmental geotechnics, otherwise known as geoenvironmental engineering, has grown from a novelty to a well-defined and mature engineering discipline existing somewhere between geotechnical engineering, environmental engineering, hydrogeology, and soil physics. The rise of importance of environmental geotechnics in the last 20 years in the United States (US) occurred because of the plethora of environmental regulations promulgated by federal and state governments.

Most of the early regulations [e.g., Resource Conservation and Recovery Act (RCRA) of 1976 and the Comprehensive Environmental Response, Compensation, and Liability Act (CERCLA or Superfund) of 1980] were a response to public outcry regarding poor waste management practices that led to severe soil and ground water contamination at sites such as Love Canal in New York. Adverse health effects, including cancer in children, were attributed to contamination by industrial chemicals. This outcry led to regulations that were designed as an immediate remedy to the problem, without much consideration of long-term implications or economic impacts. These regulations provided the impetus for the development of environmental geotechnics in the US. Highly engineered waste containment facilities were now needed, site investigations were to be carried out to define the nature and extent of contamination at existing waste sites, and remedial actions were to be designed to clean up soil and ground water. Similar regulations soon were promulgated in other nations, resulting in worldwide growth of environmental geotechnics.

Experience garnered over 20 years has shown that some of the initial approaches to protect soil and ground water and to restore the environment are not necessarily practical. Some approaches are unwarranted or too costly, particularly on lands that will continue to be used by heavy industry. Moreover, some of the strategies for waste containment (e.g., the dry tomb landfill) have resulted in environmental legacies requiring long-term maintenance and eventual restoration by future generations. Consequently, philosophies in environmental geotechnics are changing, and will continue to change throughout the next quarter century.

This paper provides a review of the development of the profession, and discusses where the profession is likely to go in the new millenium. The paper is divided into four key areas of environmental geotechnics: site characterization, remediation, waste containment, and beneficial reuse of industrial byproducts

2 ENVIRONMENTAL SITE CHARACTERIZATION

Environmental site characterization has evolved from traditional drilling and sampling used by geotechnical engineers for decades to sophisticated non-intrusive and intrusive exploratory techniques that provide indirect and direct measures of the properties of soil and the characteristics of contaminants. Traditional drilling and sampling still

9

remains the mainstay of environmental site characterization, but a wealth of standards have been developed by organizations such as ASTM that define specific drilling and sampling protocols that ensure quality soil and ground water samples are obtained for environmental investigations. The following ASTM standards provide an example of the type of protocols that have been developed:

- D 5781 - Dual-wall reverse-circulation drilling for geoenvironmental exploration
- D 5783 - Direct rotary drilling for geoenvironmental exploration
- D 5872 - Casing advancement in geoenvironmental drilling
- D 6169 - Soil and rock sampling devices for geoenvironmental exploration.
- D 6286 - Geoenvironmental drilling methods

Unlike previous standards that have focused on mechanical sampling disturbance, protocols in these standards ensure that the chemical characteristics of samples are undisturbed and that clean portions of the subsurface are protected. Issues such as cross-contamination, volatilization, and sample preservation are of paramount importance in these standards.

One of the key differences between traditional geotechnical exploration and geoenvironmental exploration is the need to define the stratigraphy in far greater detail. Contaminants travel along the pathway of least resistance and narrow permeable channels usually govern their movement (e.g., Benson and Rashad 1996, Webb and Anderson 1996). Defining these transport pathways is of paramount importance, but is difficult using traditional drilling and sampling because the volume explored is small, sampling is slow, and the cost of sampling is high. In addition, traditional drilling and sampling normally brings contaminated cuttings to the surface, which often need to be handled as "hazardous wastes" in the US. As a result,

exploration methods are being developed that can be used to more rapidly sample at lower cost and without waste products so that more extensive characterization can be performed.

Direct-push technologies (DPT) are one of the primary beneficiaries of the need for more rapid and less costly exploration. DPTs evolved from cone penetrometer testing (CPT), and CPT remains a common DPT used today. CPT provides a means for more rapidly and extensively describing the stratigraphy of a site without bringing contaminated cuttings to the surface or providing the opportunity for installing a monitoring well in a drill hole. Owners interested in controlling costs find the latter aspect of CPT desirable, because installation of monitoring wells at unnecessary locations can result in expensive quarterly sampling and chemical analysis that provides little benefit in terms of remediating a site.

Another key advantage of CPT is that other sensors can be added to the penetrometer to collect additional environmental data. For example, Wenner arrays have been added to penetrometers to measure continuous profiles of electrical conductivity (e.g., see Fig. 1 for the penetrometer developed by Campanella and Weemes 1990). Variations in electrical conductivity can be used to clarify the stratigraphy, but more importantly can be used to locate contaminant sources or define contaminant plumes. For example, non-aqueous phase liquids (NAPLs) generally are non-conductive whereas ground water contaminated with metals is usually more conductive than non-contaminated ground water.

An advantageous use of a penetrometer equipped with a Wenner array (herein referred to as a CPTE) is for delineating free product, as shown in Fig. 2. In this application, jet fuel from a US Air Force base had leaked from tanks and fueling spills and was floating on the ground water table. When the CPTE penetrated free product floating on the ground water table, pore pressures were measured, but the

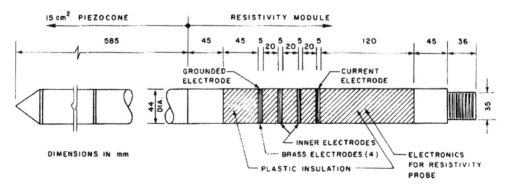

Figure. 1. Schematic of Penetrometer Equipped with a Wenner Array (from Campanella and Weemes 1990).

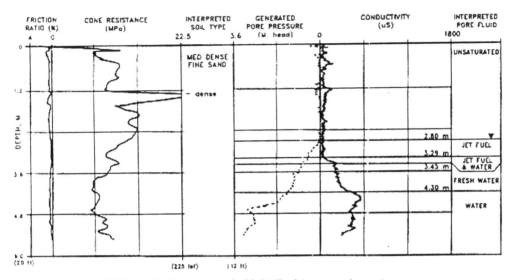

Figure. 2. Logs from CPTE at Site Contaminated with Jet Fuel (source unknown).

electrical conductivity remained very low (depths 2.8 to 3.3 m). Below 3.3 m, the CPTE was in ground water and the electrical conductivity increased. Concurrently, the pore water pressure increased in proportion to depth below the surface of the free product. Examination of both the profiles of electrical conductivity and pore water pressure permitted clear delineation of the floating layer of jet fuel.

Penetrometers have also been adapted for collecting other types of data. The US Department of Energy has funded development of a penetrometer that has a scintillation gamma sensor for detecting radioactive contaminants (Marton et al. 1988). This penetrometer can be used to readily detect and delineate radioactive contamination in deep vadose zones existing at some of the US nuclear weapons facilities, such as beneath the waste liquid storage tanks at the Hanford facility in central Washington state. Devices have also been added to collect ground water samples, measure the pH and conductivity of the pore water, and to determine water content (Bratton et al. 1995).

A popular add-on to the penetrometer is laser-induced fluorescence (LIF) technology. LIF involves directing a laser beam along a fiber optic cable routed through the rods and into a 6-mm-diameter sapphire window mounted flush along the side of the penetrometer approximately 0.5 m from the tip (Lieberman et al. 1995, Reuter et al. 1998). Focusing the laser beam on some contaminants causes them to fluoresce. The fluorescence is routed back to the surface through another fiber optic cable to analysis hardware. The hardware determines the intensity of fluorescence, which is related to the

concentration of the contaminant that was targeted, and the type of soil. An example of a fluorescence intensity profile from a site contaminated with fuel oil (Olie et al. 1993) is shown in Fig. 3. Peaks in the fluorescence profile correspond to locations of fuel oil. Recent research on LIF applications in CPT has focused on tunable lasers that can be used to target specific contaminants (Apitz et al. 1992, Bratton et al. 1995).

A disadvantage of LIF is that some of the most

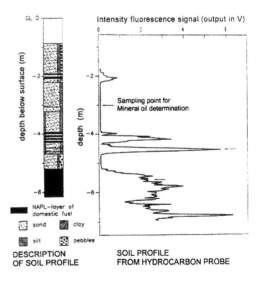

Figure. 3. Fluorescence Intensity Profile from Site Contaminated with Fuel Oil (from Olie et al. 1994).

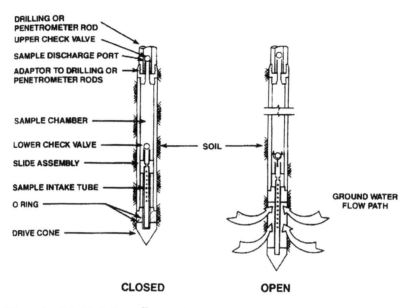

DRILLING OR
PENETROMETER ROD

UPPER CHECK VALVE

SAMPLE DISCHARGE PORT

ADAPTOR TO DRILLING OR
PENETROMETER RODS

SAMPLE CHAMBER

LOWER CHECK VALVE

SLIDE ASSEMBLY

SAMPLE INTAKE TUBE

O RING

DRIVE CONE

SOIL

GROUND WATER
FLOW PATH

CLOSED   OPEN

Figure 4.   Schematic of the HydroPunch® Ground Water Sampler (from Kuhlmeier and Sturdivant 1992).

ubiquitous organic contaminants, such as trichloroethylene (TCE) and perchloroethylene (PCE), do not fluoresce and thus cannot be detected with LIF. Penetrometers that include video-imaging systems are being used for these contaminants (e.g., Lieberman et al. 1998). Soil in contact with the sapphire window is imaged using a miniature CCD color camera located in the penetrometer. The video image is viewed at the surface using a video monitor and is recorded using a video cassette recorder. Light emitting diodes are used to illuminate the soil. Field tests conducted by Lieberman et al. (1998) at a site contaminated with chlorinated solvents and fuels showed that non-aqueous phase liquid contaminants were readily visible in the video display. Globules of NAPL were observed, as well as gas bubbles associated with the globules.

Less costly DPTs, particularly the HydroPunch® and the GeoProbe®, are being used at many sites where the cost associated with CPT is not warranted. The HydroPunch is similar to a cone penetrometer, but the sleeve retracts near the tip allowing ground water to enter into a sampling chamber (Fig. 4). Check valves keep the sample container closed during pushing and during sample retrieval. A comparison of ground water samples collected with a HydroPunch and from adjacent monitoring wells by Smolley and Kappmeyer (1991) shows that concentrations similar to those from a monitoring well can be obtained using the HydroPunch.

Sampling ground water without installing a monitoring well and the ability to penetrate with a traditional drill ring (provided deep exploration is not needed) are key advantages of the HydroPunch.

A key disadvantage is that the HydroPunch cannot be used to obtain detailed stratigraphic information as can be done with CPT. In some cases, however, CPT is conducted first to define stratigraphy, and then the HydroPunch is pushed down the same hole to collect ground water samples (Smolley and Kappmeyer 1991).

GeoProbes are similar to the HydroPunch, but they can also be used to collect soil and gas samples. GeoProbes are installed using a percussion hammer that is normally mounted on a truck or a van (Fig. 5). Thus, the GeoProbe is a lower cost and mobile exploration tool. More information on the GeoProbe can be found at www.geoprobesystems.com. Other DPTs that are seeing more frequent use are the SCAPS system (Kram and Lory 1998), the Envirocore (Einarson et al. 1998), and the Envirocone (Piccoli and Benoit 1995).

Figure. 5. Truck Equipped with Percussion Hammer for Driving the GeoProbe (from www.geoprobesystems.com)

While use of DPTs has remedied some of the shortcomings of traditional drilling and sampling, the volume of soil that is evaluated is still extremely small. The future of geoenvironmental site characterization is in methods that can be used to sample a large volume of the subsurface at reasonable cost. Geophysical methods fall into this category, especially electrical methods. The disadvantage of these methods in the past was that large data sets were difficult to handle and invert. However, the advent of high-speed desktop computers and efficient three-dimensional inversion codes has made regular use of electrical geophysics practical. Methods are now available to develop detailed images of subsurface conditions using electrical resistance methods and electromagnetic techniques. For example Paprocki and Alumbaugh (1999) have used cross-borehole ground penetrating radar (GPR) and tomographic imaging techniques to develop detailed cross-sections of water content in the vadose zone (Fig. 6). This same technique can be used to develop a detailed image of DNAPL saturation at spill sites.

Other applications include mapping plumes of jet fuel using induced polarization techniques (Morgan et al. 1999), DC resistivity mapping for delineating plumes at uranium mill tailings facilities (Buselli and Lu 1999), and electrical resistance tomography of steam stripping applications (LaBrecque et al. 1998).

3 REMEDIATION

3.1 Treatment Technologies

The original intent of US environmental regulations was to return hazardous waste sites to their original pristine condition. Attempting to meet this objective involved treatment or removal of soil, removal of sources, and treatment of contaminated ground water to very low concentrations. Most of the remediation projects initiated in the 1980s with this intent included pump and treat systems. Ground water was extracted from a series of wells, treated above ground in a conventional water treatment system, and then discharged to a surface water body. In some cases the treated water was reinjected to the aquifer. The fallacy of this approach became apparent in the late 1980s and early 1990s, particularly at sites contaminated with non-aqueous phase liquids. Contaminant concentrations decreased, but reached an asymptotic level typically

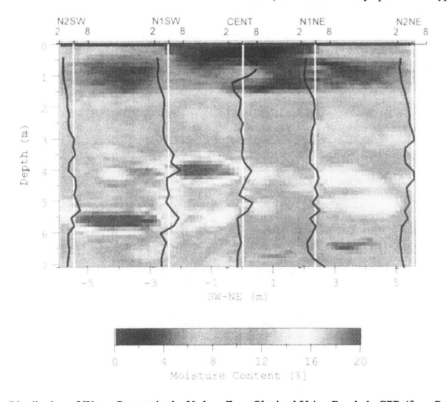

Figure 6. Distribution of Water Content in the Vadose Zone Obtained Using Borehole GPR (from Paprocki and Alumbaugh 1999).

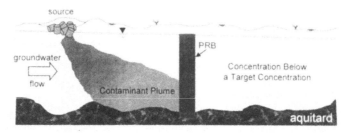

Figure 7. Schematic of a Permeable Reactive Wall (adapted from Eykholt et al. 1999).

above the maximum contaminant level (MCL), which defines the clean up target. Mass transfer limitations were found to be responsible for this asymptotic tailing, and revised predictions of clean up times were in the several hundreds or thousands of years. As a result, very few of the hazardous waste sites covered by Superfund have actually been cleaned up.

While pump and treat systems were being installed, other technologies were being developed for treating soil and ground water ex situ or for removing contaminants in situ in soils such as silts and clays where pump and treat was not feasible. Large research efforts were conducted on topics such as electrokinetic technologies, in situ bioremediation, and soil washing. However, most of these technologies were ultimately determined to be too difficult, too uncertain, or too costly to implement at all but the most contaminated and well-endowed sites. Research and development efforts are still underway on these remediation techniques in the US, but at a far lower level of effort.

While many remediation technologies have failed to deliver on expectations, several remediation techniques have emerged that appear promising and practical, and will probably see more widespread use in the next millenium. One particularly novel technology is the permeable reactive wall (PRW). A PRW is installed in the pathway of a plume, as shown in Fig. 7. As the contaminated water travels through the PRW, the contaminant reacts with the medium contained in the wall. The reaction may be sorption combined with biodegradation or decay, reductive dehalogenation, precipitation, or any other conversion technique (Pankow et al. 1993, Gillham and O'Hannesin 1994, Rael et al. 1995, Kershaw and Pamucku 1997, Carlet et al. 1998). The result is water emanating from the effluent end of the wall having contaminant concentrations below the MCL. Examples of reactive walls are described in Blowes et al. (1995), Payne et al. (1996), Benner et al. (1997), and O'Hannesin and Gillham (1998).

If the plume is narrow or small, the PRW can be installed in the aquifer alone. In other cases, ground water cut-off walls are installed to direct the plume into the PRW. The latter approach is referred to as the "funnel and gate" system (Starr and Cherry 1994).

PRWs are a desirable remediation technology because regular input of energy (e.g., as in a pump and treat system) is not required, little or no maintenance is needed if the reactive medium is carefully selected, and success is not affected by mass transfer limitations. However, because PRWs are relatively new, little is known about their long-term performance. Also, the design methodologies (e.g., Shoemaker et al. 1995, Gavaskar et al. 1998, Powell et al. 1998) are relatively simplistic and provide no rational means to account for the influence of geologic heterogeneity on performance or monitoring schemes.

One concern with PRWs is that effluent being discharged can be more toxic than the influent (Eykholt et al. 1999). For example, if TCE is being treated using iron metal filings and the residence time within the wall is too short, the effluent may contain vinyl chloride (VC) at unacceptable concentrations, rather than chloride and ethene. VC is one of the more toxic organic compounds, and has a very low MCL (2 µg/l).

Residence times that are too short can be caused by heterogeneity in hydraulic conductivity of the aquifer on either side of the PRW. Eykholt et al. (1999) used a three-dimensional ground water flow and transport model to evaluate the behavior of a PRW in a heterogeneous aquifer. The PRW was sized using conventional methods for treating a TCE plume having uniform concentration. Heterogeneity of hydraulic conductivity in the aquifer was described using a stochastic simulator. The model was used to determine residence times and concentrations of VC in the effluent.

An example of results from Eykholt et al. (1999) is shown in Fig. 8, which is a map of VC concentrations at the effluent face of the PRW. The concentrations vary over five orders of magnitude, and hot spots exist where concentrations exceed the MCL. The likelihood of hot spots at the effluent face of the wall can reduced by increasing the wall thickness. However, the wall thickness may need to be increased by a factor of five or more to preclude exceedances of MCLs for VC under typical conditions (Eykholt et al. 1999). As a result,

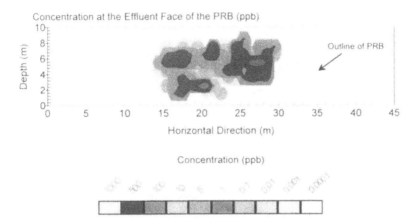

Figure 8. Distribution of VC Concentrations at the Effluent Face of a PRW (adapted from Eykholt et al. 1999).

increasing the wall thickness to ameliorate the effects of heterogeneity can be impractical.

Another approach to reduce the likelihood of hot spots in the effluent is to use "caisson" or "vertical flow" designs for the PRW (Fig. 9). These designs employ ground water cutoff walls (e.g., soil-bentonite or geomembrane walls formed into a funnel) to route contaminated ground water into a section of wall where flow passes upward through a caisson or a set of sheet piles containing the reactive media. Because ground water is forced vertically through a local region, variations in the seepage velocity are tempered and the effluent concentration is more uniform.

Another technology that will be in more widespread use is deep mixing of reagents for in situ destruction. A reactant is selected that results in

rapid and near complete destruction of the contaminant without creating toxic products. The reactant is injected while stirring the contaminated soil with deep mixing augers. Cline et al. (1997) describe two applications of this technology. Hydrogen peroxide and potassium permanganate were used as oxidizers of TCE at US Department of Energy facilities in Piketon, Ohio and Kansas City, Missouri. Both oxidants resulted in extensive destruction of TCE in the soil, which had initial concentrations as high as 950 mg/kg.

Natural attenuation will also see more widespread application at sites contaminated with fuels. This technology consists of no more than monitoring the natural breakdown of hydrocarbons, and is viable provided the plume does not extend beyond the boundaries of the site. The ability to predict the rate

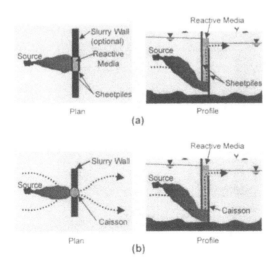

Figure 9. Schematic of Vertical Flow (a) and Caisson (b) Designs for PRWs (from Elder 1999).

of degradation is a critical aspect of natural attenuation because assurances need to be made that the plume will remain within the site boundaries while the contaminants degrade.

In many cases, natural attenuation will be combined with low cost methods of source control, such as the air-based remediation technologies soil vapor extraction (SVE) and in situ air sparging (IAS). SVE and IAS are used to strip volatile organic chemicals (VOCs) from the unsaturated zone and ground water, respectively, and use air as the carrier for removing the contaminant (Fig. 10). These air-based remediation technologies are reasonably mature. Rational methods exist that can be used to size SVE and IAS systems and to predict the rate of mass removal (e.g., Johnson et al. 1990, Elder et al. 1999). The low cost of these technologies will ensure their continued use into the next millenium.

3.2 Risk-Based Corrective Action and Brownfields

One of the most important and fundamental changes in the US approach towards hazardous waste site remediation has been to tailor the level of clean-up to future uses of the contaminated land so that the risk associated with the site after clean-up is commensurate with its use. For example, lead concentrations in soil at a site to be used for manufacturing for the foreseeable future can be higher than those in a residential area where children are likely to ingest the soil. This approach towards remediation is referred to as Risk-Based Corrective Action (RBCA) and the treated industrial lands are referred to as brownfields (Maldonado 1996). Several case histories of RBCA and brownfields are described in the ASCE publication *Risk Based Corrective Action and Brownfields Restoration* (Benson et al. 1998).

The impetus to use RBCA occurred, in part, because large tracts of industrial real estate were idle in the US. New owners for this real estate could not be found due to the extreme financial liability associated with remediating a site to pristine conditions. Implementation of RBCA has resulted in re-development of idle real estate, which frequently has prompted economic re-birth of surrounding areas.

Kay and Barton (1998) describe a typical brownfield case history. The site was an abandoned industrial facility that had been used since before the American Revolution. Industrial operations for over 200 years resulted in soil contaminated with metals, VOCs, oils, asbestos, and PCBs. The land had remained idle for about five years before it was cleaned up under a brownfields program. A new shopping mall now exists at the site and strong sales at the shopping mall have stimulated economic recovery in the surrounding area.

4 WASTE CONTAINMENT FACILITIES

4.1 Barrier Technologies

Inadequate containment of municipal and industrial wastes was the primary cause of most hazardous waste sites in the US. Prior to RCRA in 1976, few states in the US required any type of containment system for disposal of wastes. Consequently, solid and liquid wastes were disposed by placing them in unlined holes referred to as dumps. Often times

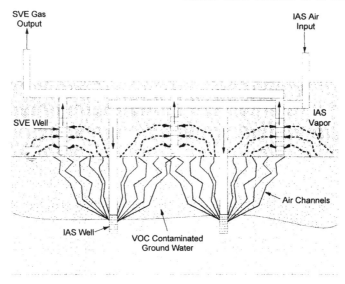

Figure 10. Schematic of a Combined SVE and IAS System Used to Treat the Vadose Zone and Ground Water Contaminated with Volatile Organic Chemicals (from Baker and Benson 1999).

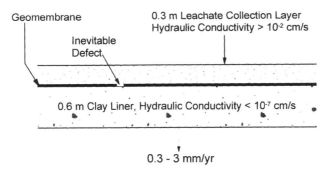

Figure 11. Schematic of a Composite Liner.

liquid wastes (particularly aqueous solutions) were intentionally placed in surface impoundments with the intention of disposing the wastes by "evaporation and infiltration." In other cases, existing fine-grained soil layers were assumed to be capable of isolating the wastes from the surrounding environment.

The fallacy of reliance on natural fine-grained soil layers was revealed by the fiasco at Love Canal. Engineers believed that clay underlying the chemical disposal pit in Love Canal would retain liquid chemical wastes. However, extensive ground water contamination ultimately occurred (Daniel 1993). One of the first projects the author worked on is a similar example. A hazardous waste landfill was sited in southern Illinois. The landfill consisted of trenches excavated in the natural clay at the site. In the late 1970s, the site received an award as a state-of-the-art hazardous waste facility. Unfortunately, dense non-aqueous phase liquids (DNAPLs) poured in the trenches moved downward to ground water through small cracks and fissures in the clay. By 1990, this award-winning facility was a hazardous waste site undergoing costly remediation.

The clear need for barrier systems in landfills led to extensive research on containment technologies. Landfills initially were lined with compacted clay layers. However, research by Anderson (1982) showed that concentrated organic chemicals can attack compacted clays, resulting in large increases

in hydraulic conductivity. As a result, emphasis shifted away from compacted clay and focused on thin plastic liners, i.e., geomembranes.

Field experience quickly showed that geomembrane liners were also vulnerable to high leakage rates because punctures were inevitable (Giroud and Bonaparte 1989). The vulnerability of geomembranes, combined with the ban on land disposal of liquid wastes promulgated by amendments to RCRA in 1984, resulted in renewed emphasis on compacted clay barriers, but now in conjunction with geomembranes, i.e., composite liners (Fig. 11).

Composite liners are now required for all new municipal and hazardous waste facilities in the US because of their very low leakage rates (~0.3 to 3 mm/yr) relative to compacted clay liners (~30 mm/yr) and geomembrane liners (~150 mm/yr). Double liners with an intermediate leak detection layer (Fig. 12) are required for hazardous waste landfills, and often both liners are composites.

Covers for landfills have evolved in a similar manner as liners. Regulations now require that new facilities include a cover that incorporates a composite barrier system (Fig. 13). This cover design is intended to prevent the "bath tub effect." That is, to prevent accumulation of contaminated liquid within the landfill cell, leakage through the cover is required to be less than or equal to leakage

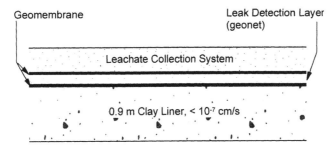

Figure 12. Double Liner Required for Hazardous Waste Landfills in the US.

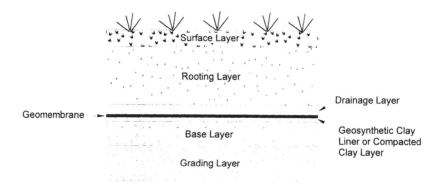

Figure 13. Cover Required for New Landfills in the US.

from the liner.

The technology of barrier systems is mature. Most of the innovations in this area include use of new materials, such as geosynthetic clay liners (GCLs) or geomembranes constructed with different polymers, improvements in construction methods, and improved monitoring techniques. However, there are two areas where significant change is likely to occur in the new millenium: performance-based design and alternative cover systems.

Performance-based design consists of selecting the barrier systems to ensure that water quality criteria are met at a compliance surface. The compliance surface usually is a prescribed distance from the edge of the facility, or may just be the property lines of the site. Ground water quality must be maintained at this boundary for infinitum. Engineers must select and design a barrier system that will ensure that the mass discharged from the containment facility is small enough so that the

water quality criteria at the compliance surface are not violated. No prescription is placed on the liner and cover system. No liner may be necessary, or a sophisticated triple composite liner may be required.

Performance-based design requires considerable more analysis than design for prescriptive barrier systems, but provides the engineer with the opportunity to be creative in selecting a cost-effective design that protects the environment. As an example, consider design of mine waste facilities in the State of Wisconsin. The state code requires that containment facilities for mine waste be designed based on performance criteria. Design of a tailings disposal facility for a new copper and zinc mine required the capability to predict the rate of influx of water and oxygen through the cap, chemical reactions within the tailings, chemical composition of the tailings leachate, and ultimately mass flux discharged from the tailings disposal facility. Sophisticated three-dimensional numerical

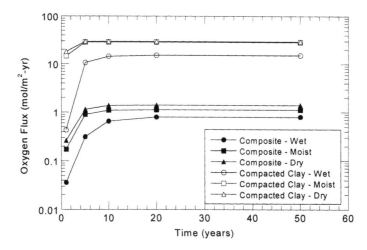

Figure 14. Oxygen Flux through Proposed Covers Over Mine Tailings (adapted from Kim and Benson 1999). Proposed Covers Included Compacted Clay and Composite Barriers. Different Moisture Conditions (Wet, Moist, and Dry) of Soils Overlying the Barrier were Considered in the Simulations.

modeling was required to predict oxygen transport and water flow through the cap, and chemical transport through the bottom liner. Only through these type of analyses could rational comparisons between designs be made.

An example of results from analyses that were required is shown in Fig. 14, which is a graph of oxygen flux across a cover placed over the tailings. Fluxes were predicted using a three-dimensional finite-element model of the cover (Kim and Benson 1999). Comparisons were made between a cover employing a compacted clay barrier and a cover employing a composite barrier. Much lower oxygen fluxes were obtained for the cover with a composite barrier. Other analyses ultimately showed that the composite cover was necessary to ensure oxidation of the tailings occurred at a slow enough rate to ensure ground water quality criteria were met.

Alternative covers is another area likely to undergo significant growth in the new millenium. An alternative cover is an all earthen cover used in lieu of a prescribed cover employing a composite barrier. The alternative cover is required to be hydrologically equivalent to the prescribed cover; i.e., percolation from the alternative cover must be less than or equal to percolation from the prescribed cover. Consequently, alternative covers are usually practical only in semi-arid and arid regions.

The high cost associated with covers employing a composite barrier (e.g., Fig. 13) has been the reason for the growing interest in alternative covers. Covers similar to the schematic shown in Fig. 13 currently cost approximately US$400,000/ha. Alternative covers having equivalent performance can be constructed for half of this cost (Dwyer 1998).

Two types of designs are being used for alternative covers: capillary barriers and monolithic covers. Capillary barriers employ a finer-textured soil overlying a coarser-textured soil (Fig. 15). Under unsaturated conditions, the low hydraulic conductivity of the coarser-textured soil forms a capillary break that precludes the flow of water from the base of the finer-textured soil (Khire et al. 1999). As a result, water is stored near the surface where it is readily removed via evapotranspiration. Monolithic covers are simply a thick layer of finer-textured soil that retains moisture during wet periods and then emits the moisture to the atmosphere via evapotranspiration during the drier periods.

A large research project called the Alternative Cover Assessment Program, or ACAP, is currently underway in the US to improve design capabilities for alternative covers. A series of instrumented alternative covers are being constructed throughout the western US during 1999 and 2000. These covers will be monitored through 2005. Water balance data collected from the covers will be used to verify and improve models used to simulate the hydrologic performance of alternative covers.

4.2 Bioreactor Landfills

Excavation of waste from closed landfills has shown that very little degradation occurs when wastes are enclosed in a "dry tomb." Thus, there is significant concern that current waste containment strategies are leaving an undesired legacy for future generations. In addition, corporations that own and operate "dry tomb" landfills are concerned about the long-term financial liability associated with their closed facilities. As a result, research is now being

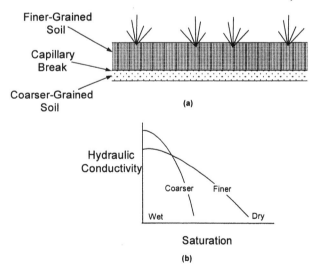

Figure 15. Schematic of Capillary Barrier (a) and Hydraulic Conductivity of Finer and Coarser Layers for Different Moisture Conditions (b).

19

conducted on how to design landfills as "bioreactors" where the waste is degraded into an innocuous compost (Reinhart and Townsend 1997). The ultimate objective is "clean closure" for landfills that successfully bioreact their waste.

Bioreactor landfills employ leachate circulation and, in some cases, addition of water to facilitate biodegradation of the waste. In some cases the waste is processed (e.g., milled) to ensure that the particle sizes are similar and that water flows through the waste uniformly. Wastes are also stirred and/or equipped with a pipe network to facilitate ingress of oxygen needed for aerobic biodegradation.

Bioreactor landfills are only in their infancy. Design, construction, and performance of assessment of bioreactor landfills will be a thrust area for research in the new millenium. Improved methods to describe water and oxygen flow through waste are needed and the hydraulic properties of waste need to be characterized. Conditions that optimize degradation rates also need to be identified.

5 BENEFICIAL REUSE OF INDUSTRIAL BYPRODUCTS

An important theme of the new millenium will be engineering for sustainable development. Sustainable development requires that fewer wastes be generated and that those wastes that are generated be reused in other applications whenever possible. When wastes are reused they are no longer "wastes." Accordingly, they are now being referred to as industrial byproducts.

Several steps are required for successful beneficial reuse of industrial byproducts (Edil and Benson 1998):

1. Identification of a creative and cost-effective application for the by-product.
2. Defining the engineering properties of the byproduct relative to the application for which it is to be used.
3. Conducting demonstrations showing that the byproduct can be successfully used in the field.
4. Identification of any special handling and construction techniques that are necessary when using the byproduct.
5. Cultivating a marketplace for the byproduct.

Each of these steps is critical; ignoring any step may result in the byproduct being considered risky or impractical.

The author has been involved in numerous projects dealing with beneficial reuse with his colleagues at the University of Wisconsin-Madison. Byproducts that have been considered include scrap plastics, discarded and remnant carpets, paper sludges, foundry sands and slags, shredded tires, fly ashes, and bottom ashes. Research on these topics is of paramount importance to sustainable development.

6 SUMMARY

Environmental geotechnics has evolved into a broad and mature discipline within geotechnical engineering during the last twenty years. Much of the early growth of the profession was fueled the promulgation of environmental regulations by federal and state governments. Future growth of the profession will rely on developing more rational and less costly solutions to remediation and containment problems and towards achieving sustainable development.

This paper has reviewed development of the profession in terms of four key areas: site characterization, remediation, waste containment, and beneficial reuse of industrial byproducts. Thrust areas in the new millenium were also described. These areas include innovative remediation technologies such as reactive walls and other treatment methods not constrained by mass transfer limitations, risk-based clean up, bioreactor landfills, and beneficial reuse of industrial byproducts

7 ACKNOWLEDGMENT

Industry and a variety of federal, state, and local government agencies have sponsored Dr. Benson's research and practice in environmental geotechnics. Although the number of sponsors is too numerous to permit them to be mentioned individually, their support is gratefully acknowledged. Dr. Benson is also indebted to his graduate students, who have done much of the work associated with his research projects, and to his co-investigators at the University of Wisconsin-Madison. Mr. Xiaodong Wang assisted in preparing some of the figures for this paper.

REFERENCES

Anderson, D. (1982), Does Landfill Leachate Make Clay Liners More Permeable?, *Civil Engineering*, 52(9), 66-69.

Apitz, S., Theriault, G., and Lieberman S. (1992), Optimization of the Optical Characteristics of a Fibre-Optic Guided Laser Fluorescence Technique for the In-Situ Evaluation of Fuels in Soils, *Proc. Environmental Process and Treatment Technologies*, International Society for Optical Engineering.

Baker, D. and Benson, C. (1999), Factors Affecting Air Sparging Plumes, *J. of Geotechnical and Geoenvironmental Engineering*, ASCE, in review.

Benner, S., Blowes, D., and Ptacek, C. (1997), A Full-Scale Porous Reactive Wall for Prevention of Acid Mine Drainage, *Ground Water Monitoring and Remediation*, 17 (3), 99-107.

Benson, C. and Rashad, S. (1996), Enhanced Subsurface Characterization for Prediction of Contaminant Transport Using Co-Kriging, *Geostatistics for Environmental and Geotechnical Applications*, STP 1283, ASTM, Rouhani, S., Srivastava, M., Desbarats, A., Cromer, M., and Johnson, A., Eds., 181-199.

Benson, C., Meegoda, J., Gilbert, R. and Clemence, S. (Eds.) (1998), *Risk-Based Corrective Action and Brownfields Restorations*, GSP No. 82, ASCE.

Blowes, D., Ptacek, C., Cherry, J., Gillham, R., and Robertson, W. (1995), "Passive Remediation of Groundwater Using In Situ Treatment Curtains," *Geoenvironment 2000: Characterization, Containment, Remediation, and Performance in Environmental Geotechnics (2)*, Daniel, D. and Acar, Y. (Eds.), ASCE, GSP 46, 1588-1607.

Bratton, W., Bratton, J., and Shin, J. (1995), Direct Penetration Technology for Geotechnical and Environmental Site Characterization, *Geoenvironment 2000: Characterization, Containment, Remediation, and Performance in Environmental Geotechnics (2)*, Daniel, D. and Acar, Y. (Eds.), ASCE, GSP 46, 105-122.

Busselli, G. and Lu, K. (1999), Applications of Some New Techniques to Detect Ground Water Contamination At Mine Tailings Dams, *Proc. Symp. on the Application of Geophysics to Environ. and Engr. Probs.*, Environmental and Engineering Geophysical Society, Wheat Ridge, CO, 507-516.

Campanella, R. and Weemes, I. (1990), Development and Use of an Electrical Resistivity Cone for Groundwater Contamination Studies, *Canadian Geotech. J.*, 27(5), 557-567.

Cantrell, K., Martin, P., and Szecsody, J. (1994), Clinoptilolite as an In Situ Permeable Barrier to Strontium Migration in Ground Water, *In Situ Remediation: Scientific Basis for Current and Future Technologies*, G. Gee and N. Wing, Eds., Battelle Press, Columbus, OH, 839-850.

Charlet, L., Liger, E., and Gerasimo, P. (1998), "Decontamination of TCE- and U-Rich Waters by Granular Iron: Role of Sorbed Fe(II)," *J. of Environmental Engineering*, 124(1), 25-30.

Dwyer, S. (1998), Alternative Landfill Covers Pass the Test, *Civil Engineering*, 68(9), 50-52.

Daniel, D. (Ed.) (1993), *Geotechnical Practice for Waste Disposal*, Chapman and Hall, London.

Edil, T. and Benson, C. (1998), Geotechnics of Industrial Byproducts, *Recycled Materials in Geotechnical Applications*, GSP No. 79, ASCE, C. Vipulanandan and D. Elton, Eds., 1-18.

Einarson, M., Casey, M., Winglewich, D., and Morkin, M. (1998), Envirocore: A Dual-Tube Direct Push System for Rapid Site Characterization, *Proc. Symp. on the Application of Geophysics to Environ. and Engr. Probs.*, Environmental and Engineering Geophysical Society, Wheat Ridge, CO, 1-10.

Elder, C. and Benson, C., and Eykholt, G. (1999), Modeling Mass Removal During In Situ Air Sparging, *J. of Geotechnical and Geoenvironmental Engineering*, ASCE, in press.

Elder, C. (1999), Effect of Heterogeneity on the Design and Performance of Permeable Reactive Barriers, PhD Dissertation proposal, University of Wisconsin-Madison.

Eykholt, G. (1997), "Uncertainty Based Scaling of Iron Reactive Barriers," *Proc. In Situ Remediation of the Geoenvironment*, GSP 71, October 5-8, Minneapolis, Minnesota, 41-55.

Eykholt, J., Elder, C., and Benson, C. (1999), "Effects of Aquifer Heterogeneity and Reaction Mechanism Uncertainty on a Reactive Barrier," *J. of Hazardous Materials*, in press.

Gavaskar, A., Gupta, N., Sass, B., Janosy, R., O'Sullivan, D. (1998), *Permeable Barriers for Groundwater Remediation: Design Construction, and Monitoring*, Battelle Press, Columbus, OH.

Gillham, R. and O'Hannesin, S. (1994), Enhanced Degradation of Halogenated Aliphatics by Zero-Valent Iron, *Ground Water*, 32(6), 958-967.

Giroud, J. and Bonaparte, R. (1989), Leakage Through Liners Constructed with Geomembranes: Parts I and II, *J. of Geotextiles and Geomembranes*, 8, 27-107.

Johnson, P., Stanley, C., Kemblowski, M., Beyers, D., and Colthart, J. (1990), A Practical Approach to the Design, Operation, and Monitoring of In Situ Soil-Venting Systems, *Ground Water Monitoring Review*, 10(2), 159-177.

Kay, W. and Barton, E. (1998), Brass Factory to Regional Mall: A Model Brownfield, *Risk-Based Corrective Action and Brownfields Restorations*, GSP No. 82, ASCE, Benson, C., Meegoda, J., Gilbert, R. and Clemence, S. (Eds.), 267-276.

Kershaw, D., and Pamucku, S. (1997), Ground Rubber: Reactive Permeable Barrier Sorption Media, *Proc. In Situ Remediation of the Geoenvironment*, Evans, J. (Ed.), ASCE, GSP 71, 26-40.

Khire, M., Benson, C., and Bosscher, P. (1999), Field Data from a Capillary Barriers and Model Predictions with UNSAT-H, *J. of Geotech. and Geoenvironmental Eng.*, ASCE, 125(6), 518-528.

Kim, H. and Benson, C. (1999), Oxygen Transport Through Multilayer Composite Caps Over Mine

Waste, Proc. Sudbury '99 - *Mining and the Environment II*, Centre in Mining and Mining Environment Research, Laurentian University, Sudbury, Ontario.

Kram, M. and Lory, E. (1998), Use of SCAPS Suite of Tools to Rapidly Delineate a Large MTBE Plume, *Proc. Symp. on the Application of Geophysics to Environ. and Engr. Probs.*, Environmental and Engineering Geophysical Society, Wheat Ridge, CO, 85-100.

Kuhlmeier, P. and Sturdivant, T. (1992), Delineation of Lithology and Groundwater Quality in a Complex Fluvial Estuarine Depositional Environment, *Current Practices in Ground Water and Vadose Zone Investigations*, STP 118, ASTM, D. Nielsen and M. Sara, Eds., 183-199.

LaBrecque, D., Bennett, J., Heath, G., Schima, S., and Sowers, H. (1998), Electrical Resistivity Tomography Monitoring for Process Control in Environmental Remediation, *Proc. Symp. on the Application of Geophysics to Environ. and Engr. Probs.*, Environmental and Engineering Geophysical Society, Wheat Ridge, CO, 613-623.

Lieberman, S., Anderson, G., and Taer, A. (1998), Use of a Cone Penetrometer Deployed Video-Imaging System for In Situ Detection for NAPLs in Subsurface Soil Environments, *Proc. 1998 Petroleum Hydrocarbons and Organic Chemicals in Ground Water: Prevention, Detection, and Remediation*, National Ground Water Assoc., Westerville, OH, 169-180.

Maldonado, M. (1996), Brownfields Boom, *Civil Engineering*, ASCE, 66(5), 36-40.

Marton, R., Taylor, L., and Wilson, K. (1988), Development of an In Situ Radioactivity Detection System - the Radcone, *Proc. Waste Management '88*, University of Arizona, Tucson.

Morgan, F., Scira-Scappuzzo, F., Shi, W., Rodi, W., Sogade, J., Vichbian, Y., and Lesmes, D. (1999), Induced Polarization Imaging of a Jet Fuel Plume, *Proc. Symp. on the Application of Geophysics to Environ. and Engr. Probs.*, Environmental and Engineering Geophysical Society, Wheat Ridge, CO, 541-548.

O'Hannesin, S. and Gillham, R. (1998), Long-Term Performance of an In Situ "Iron Wall" for Remediation of VOCs, *Ground Water*, 36(1), 164-170.

Olie, J., Van Ree, C., and Bremmer, C. (1993), Status Report on In Situ Detection of NAPL Layers of Petroleum Products with the Oil Prospecting Probe Mark 1.

Pankow, J., Johnson, R, and Cherry, J. (1993), Air Sparging in Gate Wells in Cutoff Walls and Trenches for Control of Plumes of Volatile Organic Compounds (VOCs), *Ground Water*, 31 (4), 654-662.

Paprocki, L. and Alumbaugh, D. (1999), An Investigation of Cross-Borehole Ground Penetrating Radar Measurements for Characterizing the 2D Moisture Content Distribution in the Vadose Zone, *Proc. Symp. on the Application of Geophysics to Environ. and Engr. Probs.*, Environmental and Engineering Geophysical Society, Wheat Ridge, CO, 583-592.

Payne, F., Rogers, D., and Buehlman, M. (1996), Chlorinated Solvent Reduction Induced by an Aquifer Sparge Fence, *Proceedings of the First International Symposium on In Situ Air Sparging for Site Remediation*, Oct. 24-25, Las Vegas.

Piccoli, S. and Benoit, J. (1995), Geoenvironmental Testing with the Envirocone, *Geoenvironment 2000: Characterization, Containment, Remediation, and Performance in Environmental Geotechnics (2)*, Daniel, D. and Acar, Y. (Eds.), ASCE, GSP 46, 93-104.

Powell, R., Blowes, D., Gillham, R., Schultz, D., Sivavec, T., Puls, R., Vogan, J., Powell, P., and Landis, R. (1998), Permeable Reactive Barrier Technologies for Contaminant Remediation, EPA/600/R-98/125, Washington DC.

Rael, J., Shelton, S., and Dayaye, R. (1995), Permeable Barriers to Remove Benzene: Candidate Media Evaluation, *J. of Environmental Engineering*, 121(5), 411-415.

Reinhart, D. and Townsend, T. (1997), *Landfill Bioreactor Design and Operation*, CRC Press, Boca Raton, Florida, USA.

Reuter, G., Karp, C., and Marr, T. (1998), Rapid Site Investigations Using a State of the Art Cone Penetrometer Equipped with a Fuel Fluorescence Detector, *Proc. 1998 Petroleum Hydrocarbons and Organic Chemicals in Ground Water: Prevention, Detection, and Remediation*, National Ground Water Assoc., Westerville, OH, 169-180.

Shoemaker, S., Greiner, J., and Gillham, R., (1995), Permeable Reactive Barriers, *Assessment of Barrier Containment Technologies: A Comprehensive Treatment for Environmental Remediation Application*, Rumer, R. and Mitchell, J. Eds., International Containment Technology Workshop, Baltimore, Maryland, August 29-31, 301-353.

Smolley, M. and Kappmeyer, J. (1991), Cone Penetrometer Tests and HydroPunch sampling: A Screening Technique for Plume Definition, *Ground Water Monitoring Review*, 11(2), 101-106.

Starr, R. and Cherry, J. (1994), In Situ Remediation of Contaminated Ground Water: The Funnel-and-Gate System, *Ground Water*, 32(3) 465-476.

Webb, E. and Anderson, M. (1996), Simulation of Preferential Flow in Three-Dimensional Heterogeneous Conductivity Fields with Realistic Internal Architecture, *Water Resources Research*, 32(3), 533-545.

Geotechnics for Developing Africa, Wardle, Blight & Fourie (eds) © 1999 Balkema, Rotterdam, ISBN 90 5809 082 5

The microlysimeter technique for measuring evapotranspiration losses from a soil surface

J.J. Blight & G.E. Blight
University of the Witwatersrand, Johannesburg, South Africa

ABSTRACT: The evapotranspirative loss of moisture from a soil surface (e g the surface of a landfill or waste dump) is one of the most difficult components of a water balance to estimate. The radiation balance method is a powerful, rational, though indirect technique to estimate evaporative losses, but requires specialized instrumentation and calculations to apply (e g Blight and Blight, 1998). The microlysimeter method, however, is simple and requires only rudimentary apparatus and experimental skills.
The paper will present the microlysimeter method, originally suggested by Boast and Robertson (1982) and compare results obtained by this method with those obtained by parallel measurements of the radiation balance.

1 INTRODUCTION: THE SOIL WATER BALANCE

The 'water balance' principle has been widely used to predict changes in moisture in a soil body (e g Fenn et al, 1975). It depends on the application of the principle of conservation of mass to water movement within the soil. The balance can be most simply stated as:

$$\text{water input} = \text{water output} + \text{water stored} \tag{1}$$

Considering the components of each term of equation (1), the balance may be written as:

$$P = ET + RE + RO + \Delta ST. \tag{2}$$

Where P is precipitation
ET is water lost to evapotranspiration
RE is water lost as deep recharge
RO is water lost as runoff
ΔST is the change in water stored in the soil.
RO is small for most rainfall events, and hence can usually be disregarded (Blight and Blight, 1993). This leads to a simplified equation for the water balance, from which the deep recharge may be calculated if the loss to evapotranspira-tion and the change in storage can be estimated:

$$RE = P - ET - \Delta ST \tag{3}$$

The larger ET, and the smaller P, the smaller will be the chance of recharge occuring. The value of ΔST will fluctuate, the soil acting as a sponge, storing moisture which may later be released as recharge (if field capacity is exceeded), or lost to evapotranspiration.

In arid and semi-arid areas, where ET is perennially larger than P, ΔST may remain small enough that the field capacity of the soil is not exceeded and recharge never occurs, or occurs only sporadically and in very small quantities. Over a reasonably long period (i.e. a couple of years) ET cannot exceed P even if the annual pan evaporation far exceeds P. In this situation, $\Delta ST = 0$. It is thus crucial that ET be accurately evaluated, particularly in semi-arid regions where ET may not greatly exceed P and there is therefore some chance that recharge will occur. This is particularly important when assessing the potential for a waste deposit of some type to cause pollution as a result of fugitive leachate.

2 ESTIMATION OF ET

The determination of evapotranspiration has been of concern to agriculturalists and hydrologists for over a century. Many methods of assessing *potential evapotranspiration* (evapotranspiration under non-limiting moisture conditions) have been developed.

Among these the best known is due to Penman (1963).

Lysimeters are commonly used to measure evapotranspiration - but are often constructed to a depth of only few metres, and may therefore not detect changes of soil moisture at greater depths. Most current computer programmes to compute evapotranspiration are based on lysimeter results, and will hence reproduce this underestimate. It is in fact, a common belief that significant evaporation from soil surfaces takes place to depths of less than 1m; and that transpiration effects will extend no deeper that the root zone (e g Penman, 1963, Schroeder, et al, 1983). Measurements of moisture content through entire profiles of soil as well as similar observations of waste deposits (Blight, 1997) show that seasonal wetting and drying can penetrate to depths of the order of 10m and that longer term drying can occur to 20m or more in the presence of closely-spaced trees.

However, it is sometimes sufficient or necessary to obtain a short-term estimate of evapotranspi-ration losses from close to the surface, and this is where the microlysimeter method, originally suggested by Boast and Robertson (1982) becomes both convenient and useful.

3 THE ENERGY BALANCE METHOD OF MEASURING ET

A method of measuring evapotranspiration which is not subject to assumptions of the depth to which the effects extend, and is valid under moisture-limiting conditions, is the energy 'balance method' method, proposed by Bowen (1926).

The evaporative process is a large consumer of energy. If the amount of energy consumed by evaporation can be computed, the corresponding volume of water evaporated can be deduced. The evaporative energy is calculated by considering the surface energy balance:

$$R_n = G + H + L_e \qquad (4)$$

Where: R_n is the net radiation flux for the surface (incoming solar radiation, less reflected radiation and terrestial radiation)
G is the soil heat flux
H is the sensible heat flux for the air
L_e is the latent heat flux for evaporation
R_n can be measured directly; and G can be estimated from changes in the temperature gradient in the soil

together with the specific heat capacity for soil. H, the sensible heat may be calculated from temperature and humidity gradients above the soil surface.

Data for energy balance calculations can be collected automatically by monitoring and recording the outputs of a radiometer, buried soil heat flux plates and a psychrometer, (air temperature at two levels). Readings taken at two minute intervals, and averaged over longer periods (15 - 20 minutes) are recommended for best results.

Hand held portable instruments can also be used to make recordings of the energy balance at significantly lower cost and with considerably greater flexibility. This manual method is to be preferred if it is wanted to monitor several sites with less detail.

An 'A' class evaporation pan should be monitored at all the test sites, for comparison to the quantity of evapotranspiration calculated from the energy balance. Wind speed is also measured, as advection of heat and water vapour due to gusts of wind may give rise to anomalies in the measured heat budget. (If the test site is located in the centre of a large uniform area, this effect is minimised). Precipitation must also be measured at the same time.

Figures 1a and 1b show a comparison between a set of automated (A) and a set of manual (M) energy balance measurements taken simultaneously at the same site.

β, mentioned in Figure 1, is Bowen's ratio, where

$$\beta = H/L_e \quad \dots\dots\dots\dots\dots (5a)$$
and
$$L_e = (R_n - G)/(1 + \beta) \quad \dots\dots\dots (5b)$$

Although the two sets of curves look rather different, the overall results of the two sets of measurements were very close and also agreed with measured Apan evaporation. Hence the two techniques are equivalent.

The measurements were only taken during daylight hours as there is little interchange of energy during the night.

Further details of the energy balance method for estimating evapotranspiration have been given by Blight (1997) and Blight and Blight (1998)

4 THE MICROLYSIMETER METHOD OF MEASURING ET

A microlysimeter consists of a short section of rigid plastic pipe open at both ends. Sections of 150 mm

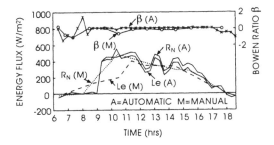

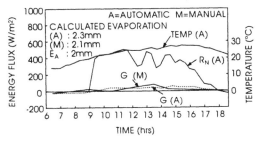

FIGURE 1: Comparison of manual (M) and automated (A) radiation balance measurements

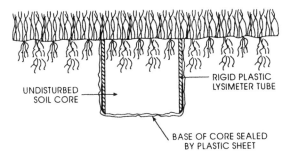

FIGURE 2   Principle of the microlysimeter

to 200 mm diameter and of a similar length are convenient, but in principle, any suitable size can be used. If the soil is soft enough and free of stones, the lysimeter can be driven into the surface, and an undisturbed core taken of the surface soil, including the surface vegetation. The core is carefully extracted and the lower end of the extracted soil is sealed by means of a plastic sheet taped over the outside of the lysimeter pipe, using adhesive plastic tape (see Figure 2). The core, in its lysimeter pipe, is weighed and then re-inserted into the hole from which it was extracted. Losses of moisture from the microlysimeter are then measured over a period of few days by re-extracting the lysimeter each day and weighing it.

If the soil is too hard or strong to drive in the lysimeter tube, it will be necessary to trim the soil

core to fit the lysimeter tube by excavating all round the cylinder, trimming away the excess soil, and easing the tube down over the soil core. Once the core has been extracted, sealed at its lower end and weighed, it is reinserted into the hole. The hole is then backfilled, tamping the loose soil around the walls of the microlysimeter to achieve a similar condition to that of a lysimeter driven into the soil. The aim is to produce a removable undisturbed soil core that has the same surface condition as the undisturbed surface that surrounds it.

5 AN EVALUATION OF THE MICROLYSIMETER METHOD

To evaluate the method, three microlysimeters measuring 150 mm in diameter by 150 mm deep were inserted into a well grass-covered soil surface adjacent to an automatic energy balance (Bowen Ratio) recording apparatus (see Figure 1). Daily measurements were then made of the lysimeters and the radiation balance over a period of 30 days. The results of the measurements are shown in Figure 3. Figure 3 records cumulative data for the following measurements:

- Apan evaporation, E_A;
- evaporation calculated from the radiation balance, E_β;
- the maximum and mean evaporation measured by means of the three microlysimeters; E_L (max) and E_L (mean); and
- the cumulative rainfall during the 30 day period.

Figure 3 also shows a summary of the water balance for the 150 mm deep microlysimeters. Initially, the top 150 mm of soil contained 30mm of water. Assuming that all of the rain infiltrated the surface where it fell, rain would have added 38mm, to bring the total water that was stored in the profile during the 30 day period to 68mm. Immediately after the 22mm precipitation on day 3 had infiltrated, the upper 150 mm of soil would have held 46mm of water, while the total void volume was 60mm (a degree of saturation S = 77%). Thereafter the water content of the soil declined progressively, with only small additions by rain. By day 30, the mean loss from the microlysimeters was 51mm, although the maximum loss recorded from one of the three lysimeters was 63mm*. Figure 3 shows that mean losses from the three lysimeters agreed closely with E_β until about day 15. From then on, E_L (mean) lagged behind

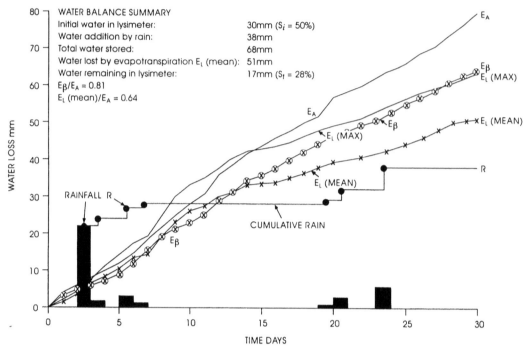

FIGURE 3   Comparison of microlysimeter and energy balance methods with evaporation pan measurements.

E_β, probably because the water held in the lysimeters was not being replenished from the soil below, whereas elsewhere replenishment could take place by upward flow. As the radiation balance measured evaporation from the general area, E_β was not affected.

The Apan evaporation exceeded that measured from the soil for most of the test period, but this was to be expected, as evaporation from a water surface will usually exceed evapotranspiration from a soil surface unless unlimited water is available in the soil (see, eg, Penman, 1963, Hillel, 1980) or if the surface vegetation is vigorously transpiring.

6 CONCLUSIONS

Microlysimeter measurements are a cheap and effective way of measuring evaporation from a soil surface, providing that only a short term estimate is required. The comparative measurements described in this paper show that measurements can be made over periods of up to 2 weeks, with results that compare well with radiation balance measurements.

One of the problems of the technique is that the behaviour of nominally identical lysimeters can be very different. This can be ameliorated, however, by using more lysimeters, and a minimum of 5 individual measurements is suggested.

REFERENCES

Blight, G E, and Blight, J J. 1993 Runoff from landfill isolation layers under simulated rainfall. In Geology and Confinement of Toxic Wastes. Balkema, Rotterdam: 313-318.

Blight, G E, 1997 Interactions between the atmosphere and the earth. Geotechnique, 42: 715-766.

Boast, C W and Robertson, T M 1982 A Microlysimeter method for determining

---

*This apparently anomalous result could have arisen because this particular lysimeter had a larger void volume than the other two, and therefore started out with a bigger stored water volume.

evaporation from bare soil: description and laboratory evaluation. Journal, Soil Science Society of America, vol 46: 689-696.

Bowen, I S. 1926 The ratio of heat losses by conduction and by evaporation from any water surface. Physical Review, 27: 779-787.

Fenn, D G, Harley, K J, De Geare, TV. 1975 Use of the water balance method for predicting leachate generation from solid waste disposal sites. US. EPA report No EPA/530/S168.

Hillel, D. 1980 Applications of Soil Physics, Academic Press, New York.

Penman, H L. 1963 Vegetation and Hydrology, Tech.Com 53, Commonwealth Agric Bur, UK.

Schroeder, P R, Morgan, J M, Walski, T M, Gibson, A C. 1983 The Hydrologic Evaluation of Landfill Performance (HELP) Model. US EPA Report No EPA/DF-83/001a.

Biochemical clogging of drainage systems in landfills

J.E.S. Boswell & H. Weber
Bohlweki Environmental, Midrand, South Africa

K.C. Fairley
Pulles, Howard & De Lange, Melville, South Africa

ABSTRACT: There is currently an extensive body of knowledge on the causes of biochemical clogging with much emphasis being placed on biofilm development within leachate collection systems within landfills. This paper gives a synopsis of the current state of knowledge of the causes of clogging phenomena, which includes biological, chemical and physical clogging, and the implications thereof for drainage systems. Recommendations are made with respect to remediating the clogging phenomena. It is suggested that the incidence of biochemical clogging can largely be controlled by the incorporation of drainage design criteria. However, the incorporation of specific criteria into the design of drainage systems is currently limited by a lack of knowledge of the incidence of clogging within the South African situation. Recommendations for future research to determine these specific criteria are thus also made.

1. INTRODUCTION

Biochemical clogging refers to the complete or partial blocking of drainage pipes and the pores of drainage systems and filters. The filling of this void space is due to an unattributed combination of physical, chemical and biological processes. The clogging results in a decrease in the hydraulic conductivity and thus the long term performance of the drainage system.

In order to increase the long term functioning of drainage systems, the design of new systems and the redesign of existing systems need to take the control of clogging phenomena into account. The management of these systems should also aim at reducing the potential for biochemical clogging phenomena. The geotechnical design engineer thus has a significant role to play in preventing and mitigating problems of biochemical clogging. The understanding of causes of clogging phenomena thus has an important role to play in drainage system design.

2. CAUSES OF BIOCHEMICAL CLOGGING

2.1 Biological Clogging

There is a growing body of evidence suggesting that microbiological processes are a major component in the clogging of drainage systems. This of particular importance in leachate collection systems within landfills, as leachate contains nutrients that encourage bacterial growth within the system. In the presence of excess nutrients certain bacterial cells produce slime capsules (McBean et al 1993, Mlynarek & Rollin 1995) which allow the cells to attach to surfaces of the drainage system. A clogging mat eventually forms due to the accumulation of the bacterial cells. The end result is the formation of a biofilm which reduces void space and results in a reduction in the hydraulic conductivity of the drainage system.

Landfill practitioners widely report the presence of a black "slime" at the bottom of the waste mass. This material has been found to block leachate and gas drains as well as form a large proportion of the total mass in the lower layers of the site (Aspinwall & Company 1995). Clogging becomes significant in landfills when the hydraulic conductivity of the drainage layer is less than the conductivity of the overlying waste (Rowe 1998). This results in the build-up of a leachate mound within the landfill and this can subsequently result in impact on the surface water by leachate seepage from the side slopes of the landfill as well as increased contaminant migration through the barrier system and into the groundwater (Manassero et al 1998). Leachate mounding on the base of the landfill system may result in a significant increase in temperature at the landfill liner (Rowe 1998). This in turn can increase both the diffusive and advective contaminant transport through the liner system and this can reduce the service life of both a geomembrane and the underlying compacted clay layer.

There is a clear relationship between the surface area available and biofilm development. The larger the surface area available the greater the incidence of bacterial attachment. This has important implications for the selection of materials used within drainage systems.

2.2 Chemical Clogging

Chemical changes leading to precipitation have the potential to lead to the clogging of drainage systems. Clogging is most often experienced when there are high concentrations of organic substances and precipitating metals such as calcium, magnesium, iron and manganese occurs present within the effluent (Reinhart 1998).

Hard incrustations composed mainly of calcium carbonate and iron sulphides in various proportions, with a subordinate organic carbon content are common phenomena within the leachate drainage systems. Although these are largely organic in nature, laboratory studies and growth experiments demonstrate that they are due to the activities of anaerobic bacteria (Aspinwall & Company 1995). Microbiological and chemical studies have shown that the clogging of drainage systems is a result of mobilisation involving "fermentative bacteria together with iron- and manganese-reducing bacteria, followed by precipitation processes involving primarily methane and sulphate-reducing bacteria." (Manassero et al 1995, Rowe 1998).

The clogging material typically has a hard and soft component. While under anaerobic conditions organic components remain in solution but when exposed to aerobic conditions they are may be oxidised to form a soft precipitate which could lead to clogging.

The oxidation of ferrous iron to ferric oxide which is insoluble in water, also leads to a decrease in the void space available within the drainage system. When water penetrates a drain, ferrous iron is partly chemically oxidised by atmospheric oxygen and biologically oxidised by autotrophic bacteria (Puig *et al* 1986).

Although most studies have implicated some form of microbiological process as being involved in the clogging phenomena (Brune *et al* 1991, Puig *et al* 1995, Rowe *et al* 1995, 1997, Rittman *et al* 1996), chemical analysis of leachate from certain

South African landfills suggests that the environment within the drainage system is too saline to allow for the survival of most bacteria. Clogging observed in these landfills can thus be attributed to non-biological events resulting in chemical precipitation and physical clogging of the leachate drainage systems.

Wastes having high concentrations of dissolved orthophosphates, free and saline ammonia and magnesium ions result in the precipitation of what is known as mineral struvite (Loewenthal 1997). Precipitation of struvite is a common phenomenon in the anaerobic treatment wine distillery, piggery waste waters and also sludge waste from excess biological phosphate removal systems (Loewenthal 1997). The trigger mechanism is a reduction in the partial pressure of carbon dioxide and the concomitant increase in pH giving rise to a state of super saturation with respect to struvite that is sufficiently high to cause precipitation of the mineral.

Recent research has indicated that clogging is also directly related to mass loading in terms of the mass Chemical Oxygen Demand (COD) and calcium per unit area (Rowe 1998). Loss of CO_2 from leachate (particularly if it has a high alkalinity), or an increase in pH can also lead to precipitation of calcium carbonate.

2.3 Physical Clogging

Particulate clogging often results due to the presence of silt and mud found in the effluent and can generally be controlled by appropriate design (e.g. see Giroud 1996). The phenomenon of particulate clogging is aggravated by the small drain or filter interstices, high sediment loads, poor selection of filter criteria and poor selection and placement of materials.

Exposure of leachate or drain flow to the atmosphere may result in significant precipitation of the sediment load at the point where oxidation takes place. If this occurs in the drain or collector, it may result in virtually total physical clogging.

The biochemical clogging then may become a self sustaining process whereby chemical or nutrient precipitation leads to accelerated biological growth which removes the precipitation. The bacteria and micro-organism population then die back (after clogging the drain) and the precipitation induced clogging then continues. The whole process would tend to follow a path of diminishing returns whereby drain flow continues to reduce.

Drains can also be crushed or compressed by

overburden pressure such as waste piled on top of leachate collection drains. In addition drainage pipes may be weakened by chemical attack by acids, solvents, oxidising agents or corrosion (Environmental Research Foundation 1998).

Sulphate-reducing bacteria are currently perceived to be the central role players in microbially influenced corrosion. These are a group of anaerobic bacteria which oxidise hydrogen and reduce sulphate and other oxidised sulphur compounds to sulphide.

3 REMEDIATION OF BLOCKED DRAINS

The following factors are considered to have the most significant contribution to the occurrence of clogging of drainage systems:
- Surface area available in drainage system
- Flow velocity
- Void space available in drainage system
- Chemical load
- Sediment load
- Drainage system design

A summary of possible remediation measures is given in Table 1.

3.1 Control of Surface Area and Void Size

By reducing the surface area available within the drainage system, the potential for biological clogging is reduced. Since the surface area of the filter systems is proportional to the square of the diameter of the particles used in the system, the incidence of biological clogging can be reduced by increasing the diameter of the grain size used in the filter system.

The large surface area provided by the interlocking fibres of geotextiles also provide a suitable environment for biofilm development. The larger the void size the less the potential of the voids becoming blocked.

In general the current design criteria against clogging is to use the largest, possible opening size and/or the largest possible porosity and lowest density of fibres in terms of geotextile selection (Manassero et al 1998).

The following basic suggestions regarding the use of geotextiles in leachate collection systems in landfills have been made (Manassero et al 1998):
- the drainage layer should cover the entire bottom of the landfill.
- a geotextile layer having a high percentage of open area should cover the drainage layer (such

as PVC coated Multi-Filament woven polyester or geomesh, or an equivalet).
- a secondary 16 to 32 mm cobble layer should cover the geotextile to provide an alternative leachate collection system on the case of geotextile clogging.
- the geotextile filter should be discontinuous allowing direct connection between the upper and lower drainage layers.
- geotextile layers surrounding pipe networks should be avoided.

The internal diameter of drainage pipes should also be at a maximum possible percentage open area. Manassero et al (1998) suggest an internal diameter of at least 250 mm for leachate pipes within leachate collection systems. It must also be possible to access, clean and inspect pipe networks. Provision should be made to allow for localised flushing of drain collector pipes. Manholes or flushing points should be included in the drainage system design to allow for backflushing and selective cleaning or rodding of drains. Drain flows should also be monitored and measured to determine the nature and extent of clogging effects.

The problem of crushing of drains due to overburden pressure can largely be alleviated by correct drainage system design. If there is limited access to the drainage systems the potential for secondary redundancy should be incorporated into the design such as the incorporation of gravel around the removal pipes to provide a secondary conduit, should the drainage pipes collapse.

Design criteria to prevent physical clogging of drainage systems are well documented and will therefore not be discussed in detail.

3.2 Maximising Flow Velocity

By maximising the flow velocity, the time that the effluent remains in the system is reduced and therefore the possibility of sedimentation and chemical precipitation is reduced. The flow velocity can be increased by increasing the gradient of the drainage system or by increasing the hydraulic conductivity of granular layers or even introducing pumping or liquid or gaseous flushing.

3.3 Management of Load

Correct mass loading can significantly decrease the occurrence of both chemical and biological phenomena. Pre-treatment of organic waste entering landfills to minimise the carbon content of the leachate can function to reduce the potential for biofilm development, by removing the energy source needed for bacterial growth.

The Chemical Oxygen Demand (COD) and the Biological Oxygen Demand (BOD) of the effluent needs to measured and reduced. The salt load could be reduced to decrease the incidence of chemical precipitation. Overliming, particularly in landfill management should be avoided to reduce the incidence of chemical clogging by calcium carbonate.

It is suggested that the silt load, pH, COD, BOD, total dissolved solids and alkalinity of the material be monitored and related to incidence of clogging phenomena.

3.4 Chemical Control of Clogging

Flushing of drainage pipes with biocides (antimicrobial agents) can be considered to prevent microbial growth. However, it is essential to apply the correct frequency. Successful application of biocide to drainage systems depends on knowledge of the micro-organisms responsible for chemical precipitation or biofilm development, selection of the correct biocide or combination of biocides, scientific determination of the dosage frequency required and the monitoring the control of micro-organisms through analysis and data processing (Cloete *et al* 1998). However, biocide treatment tends to be an expensive option for the treatment and control of biological clogging.

Since chemical precipitation occurs more readily at high pH (above 7), precipitation can be prevented by continuous injection of a small amount of acid into the drainage system. Acidification not only prevents but can also be used to remove chemical precipitates that have formed within drainage systems. However, when the effleunt contains high concentrations of metals, as is the case with certain landfill leachate, acidification may result in the mobilisation of heavy metals.

Polyphosphates are used to prevent calcium and magnesium precipitations with various cations. They also act to remove magnesium and calcium precipitates from irrigation systems. Polymaleic polymers sequester iron and magnesium and are used to prevent precipitation of various cations.

The use of material, such as certain plastic polymers, not prone to bacterial growth for the construction of drainage systems will also act to reduce the occurrence of biofilm development.

As the formation of soft organic precipitates is catalysed by the presence of iron, the phenomenon can largely be reduced by limiting the iron load within the effluent.

Keeping the material under anaerobic conditions will also reduce the formation of soft precipitates of this nature.

4 CONCLUSIONS

The long term performance of drainage systems can be significantly reduced by the incidence of biological, chemical and physical clogging. However, the indications are that clogging phenomena can be remediated by correct drainage design criteria. The control of clogging phenomena therefore needs to be included in the numerous factors requiring consideration during the design and construction of drainage systems.

However, the knowledge of improved design criteria is currently limited by a lack of knowledge regarding the occurrence of clogging phenomena under South African conditions. It is thus recommended that urgent further work be undertaken on the occurrence of biochemical clogging in order to assist in the development of improved drainage system design.

5 FURTHER WORK

Based on current trends in research on biochemical clogging of drainage systems it is suggested that future work should focus at addressing the following aspects:

- the measurement of drain flows and the rate of clogging in order to obtain a full understanding of the causes of clogging, particularly under South African conditions.
- optimal drainage design through laboratory scale experiments focussed at the determination of:
  1. the chemical formulation of appropriate geosynthetic materials.
  2. the minimum size of drainage aggregates.
  3. minimum gradients for drainage blankets, collectors and outfall pipes.
  4. maximum recommended spacing of manholes and flushing points.
  5. selection of factors of safety for inaccessible drains to cater for long term clogging build up.
  6. methods of prediction of physical, chemical and biological clogging.

Health hazards (including skin rashes and bronchial infections) are currently limiting the possibility of research involving landfill leachate in South Africa. Further work should thus aim at establishing safe conditions for carrying out laboratory research on leachate from landfills before an understanding of clogging phenomena within leachate drainage systems can be investigated further.

Table 1. Possible Causes and Solutions of Physical, Chemical and Biological Clogging of Drainage Systems

CLOGGING MECHANISM	CAUSE	SOLUTION
PHYSICAL CLOGGING:		
Particulate Clogging	• Filter and drain interstices too small • High sediment load • Poor drainage design (filter criteria) • Poor Construction (selection and placement of materials)	• Selective drain cleaning/rodding • Backflushing • Maximise flow velocity • Maximise void size
Drain Compression	• Overburden pressure • Corrosion of pipes	• Drain design • Gravel around removal pipes
Poor Material Selection	• Poor engineering supervision	• Drainage material with high permeability and porosity • Large pipe diameters
CHEMICAL CLOGGING:		
Soft Precipitates	• Oxidation of organic components	• Reduce Fe load • Keep effluent material under anaerobic conditions
Hard Precipitates	• Salts present in effluent • $Fe_2O_3 \rightarrow Fe_3O_4$ • $Ca(HCO_3) \leftrightarrow \downarrow CaCO_3 + CO_2 + H_2O$ • Precipitation of struvite	• Reduce salt load • Maximise flow velocity • Maximise void size • Acidification • Polyphosphates & Polymaleic polymers • Reduce COD load • Chemical equilibrium
BIOLOGICAL CLOGGING:		

Biofilm development	• Clogging mat of bacteria
	• Anaerobic conditions
	• Excess of nutrients
		• Decrease surface area
		• Pre-treatment to reduce carbon content
		• Biocides
		• Backflushing
		• Reduce BOD load
		• Use of material not prone to bacterial growth:
		    HDPE > polyster
		    plastic > sand

It is recommended that research on leachate drainage systems within landfills include the following:
- quantification of the effects of reduced co-disposal ratios on the capacity requirements of leachate drains.
- for long term stability and optimal performance of a landfill as an anaerobic digester, how to reduce the dependence of the landfill on the long term performance of the leachate drainage system.

REFERENCES

Aspinwall & Company (1995). Review of the Behaviour of Fluids in Landfill Sites. *Report Submitted to the Energy Technology Support Unit.*

Brözel, V.S., McLeod, E. & Dawood, Z. (1997). Old corrosive culprit newly identified. *SA Waterbulletin* 23 (6), 10.

Brune, M., Ramke, H.G., Collins, H. & Hanert, H.H. (1991). Incrustation Processes in Drainage Systems of Sanitary Landfills. *Proceedings of 3rd International Landfill Symposium, Sardinia:* 999-1035.

Cloete, T.E., Jacobs, L. & Brözel, V.S. (1998). The Chemical Control of Biofouling in Industrial Water Systems. *Report Submitted to Bohlweki Environmental.*

Environmental Research Foundation (1998). The Basics of Landfills. How they are constructed and Why they Fail. *Microsoft Internet Explorer.*

Giroud, J.P. (1996). Granular Filters and Geotextile Filters. Proceedings of Geofilters '96: 565-680.

Loewenthal, R.E. (1997). Computer program predicts and prevents mineral fouling. SA Waterbulletin 23(6), 14-15.

Manassero, M., Almeida, M.S.S., Bouazza, A., Daniel, D.E., Parker, R.J., Pasqualini, E. & Szabò, I (TC 5) (1998). Drainage and Leachate collection systems. *Report of ISSMFE Technical Committee TC 5 on Environmental Geotechnics.*

McBean, E.A., Mosher, F.R. & Rovers, F.A. (1993). Reliability-based design for leachate collection systems. *Fourth International Symposium. Int. Solid waste and Pub. Cleans. Assoc.,* 431-441.

Mlynarek, J. & Rollin, A.L. (1995). Bacterial clogging of Geotextiles - Overcoming Engineering Concerns. *Proc. Of Geosynthetics '95, Tennessee,* 177-188.

Puig, J., Gouy, J.L. & Labroue, L. (9186). Ferric Clogging of Drains. *Third International Conference on Geotextiles,* Vienna, 1179-1184.

Reinhart, D.R. (1998). Assessment of Biological Clogging on Leachate Collection Systems at Florida Municipal Solid Waste Landfills. *Project Proposal.*

Rittman, B.E., Fleming, I. & Rowe, R.K. (1996). Leachate Chemistry: Its implications for Clogging. *North American Water and Environment Congress '96, Anaheim.*

Rowe, R.K. (1998). Geosynthetics and the Minimization of Contaminant Migration through Barrier Systems Beneath Solid Waste. *Sixth International Conference on Geosynthetics:* 27-102.

Rowe, R.K., Fleming, I., Cullimore, R., Kosaric, N. & Quigley, R.M. (1995). A Research Study of Clogging and Encrustation in Leachate Collection Systems in Municipal Solid Waste Landfills. *Geotechnical Research Centre, University of Western Ontario, Report Submitted to Interim Waste Authority Limited.*

Rowe, R.K., Fleming, I.R., Armstrong, M.D., Millward, S.C. Van Gulck, J. & Cullimore, R.D. (1997). Clogging of Leachate Collection Systems: Some Preliminary Experimental Findings. Proceedings of 50th Canadian Geotechnical Conference, Ottawa: 153-160.

Geotechnics for Developing Africa, Wardle, Blight & Fourie (eds) © 1999 Balkema, Rotterdam, ISBN 90 5809 082 5

Contribution of chemical hardening to strength gain of fly ash

A. B. Fourie & G. E. Blight
University of the Witwatersrand, Johannesburg, South Africa

N. Barnard
Generation Engineering Department, Eskom, Johannesburg, South Africa

ABSTRACT: At some hydraulically built fly ash dams in South Africa it has been observed that a marked cementation of the fly-ash occurs, producing a very erosion resistant surface. The issues addressed in this paper include the characteristics of fly ash that enhance this hardening behaviour and whether atmospheric environmental factors influence the degree of hardening. Also addressed is the question of whether it is only a surface skin effect or whether the ash retains this added strength upon subsequent burial. If the chemical hardening is irreversible it will substantially improve the overall stability of the ash facilities and reduce the risks associated with these deposits.

1 INTRODUCTION

Thermal power stations contribute about 29000 MW to the South African national electricity grid. A typical power station burns 12 million tonnes of coal per year and the estimated mass of the resulting ash for all South African stations is about 21 million tonnes annually. About 80% of this ash consists of fly ash, which is collected from the flue gases by electrostatic precipitators. The remainder is bottom ash. Little of the ash is used or is usable industrially and the vast majority (± 95%) is disposed of on land. Before 1984 all of Eskom's fly ash disposal facilities were dams into which ash was placed hydraulically.

The vast majority of the power stations use coal from a dedicated coal mine. The coal used by the different power stations is thus different, and together with differences in method and degree of coal preparation, cleaning and pulverisation and type and operation of the power plant unit results in significant differences in properties of the fly ash (Toth et al, 1988). Some of these differences are explored in this paper.

2 FORMATION OF FLY ASH

The combustion process through which coal is processed has a significant effect on the chemical and physical properties of the resulting fly ash. There are two general procedures that are normally used. These are the fluidised bed combustion process, which normally takes place at about 800°C, and the pulverised coal combustion process which takes place at about 1450°C. All modern power stations in South Africa use the pulverised coal process. This process results in the melting of the major part of the inorganic coal minerals that constitute aluminosilicate clays, quartz, limestone and pyrite. When the molten and decomposed clays in the flue gases cool quickly, glass and mullite ($Al_6Si_2O_{13}$) are formed. The composition and amount of glass that is formed depends on combustion conditions such as boiler temperature. The smaller mineral particles cool most rapidly and tend to form spherical shapes, a typical example of which is shown in Figure 1. The smaller particles also tend to have a higher glass content because of their greater cooling speed and are known to be enriched with volatile trace elements at their surfaces. Larger particles can become bloated by gases such as CO, CO_2, SO_2 and H_2O vapour, forming hollow spheres during the quenching of the melt, (Roode, 1998). These spherical particles may have an important influence on the engineering properties of fly ash, as discussed later.

The glass that is thus formed is considered to be the cause of the pozzolanic character of many fly ashes (Scheetz et al, 1998). Kruger (1998) suggests that it is the reactivity of elements such as sodium, that precipitate on the surface of the glass spheres, that are primarily responsible for the pozzolanic characteristics of fly ash. However, it is also the

Figure 1: Micrograph of Lethabo fly ash showing spherical particles

amount of calcium oxide that determines reactivity.

In the American classification of fly ash, those ashes with a large amount of CaO and that demonstrate self-hardening properties are designated as class C ash. These ashes are usually derived from lignite coal. Fly ash from bituminous and sub-bituminous coals tend to not have self-hardening properties and are designated as class F ash. Although not having true cementitious properties, class F fly ash may, in finely divided form and in the presence of moisture, chemically react with calcium hydroxide ($Ca(OH)_2$) to form compounds that display cementitious properties. Class F fly ash is thus regarded as a 'pozzolan' but is only able to react at a significant rate in a highly alkaline environment.

3 POWER STATIONS INVESTIGATED AND BASIC GEOTECHNICAL PROPERTIES

Fly ash from two hydraulic fill ash dams (Duvha and Matla) was studied. Samples of ash (about 50kg

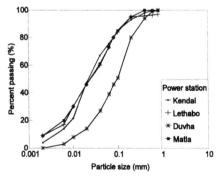

Figure 2: Particle size distribution curves for fly ash samples

from each station) were collected directly from the precipitators.

The particle size distribution curves for the two ashes are shown in Figure 2 and the results of basic geotechnical tests summarised in Table 1. Also shown on Figure 2 are grading curves for fly ash obtained from two dry ash dumps (designated as Lethabo and Kendal). The Duvha ash is significantly coarser than the other three, which are all very similar and indeed similar to much of the data previously published in the literature (Heystee et al, 1991). The numbers in brackets after the specific gravity results for Duvha and Matla samples are tests that were done on ash previously subjected to oedometer tests. There was no observed change in the specific gravity of the Matla specimens, but quite a large increase in the values for Duvha ash. This could be because the Duvha ash contained hollow spheres that crushed during the application of a vertical load. The results in Table 1 are consistent with those reported by Brackley et al (1987), although they found higher values for the coefficient of consolidation (400m²/year). The friction angles are all essentially the same and the cohesion intercept was consistently zero. As shown later, however, the development of significant cohesion is possible due to the process of carbonation.

The primary difference between the Duvha ash and the other three ashes, aside from its coarser grading, is a greater compressibility. The value for the Duvha ash is up to ten times greater than that of the Matla ash, which may have some bearing on the storage capacity of the two impoundments, ie the Duvha ash will undergo greater settlement under a particular load, thus providing additional storage capacity. This behaviour may also be due to crushing of hollow spheres. Aside from this difference, the basic geotechnical properties of fly ash appear to be well understood and well defined; it is the chemical

Table 1: Basic geotechnical properties of four fly ashes

	Kendal	Lethabo	Duvha	Matla
Specific gravity	2.12	2.17	2.21 (2.24-2.79)	2.26 (2.26)
Compression index, Cc	0.25	0.13	0.34-0.4	0.02-0.16
Coefficient of consolidation c_v (m^2/year)	20-50	20-50		
d_{50} (µm)	22	22	100	30
Angle of friction (°)	36	34	35	39
Apparent cohesion (kPa)	0	0	0	0

Table 2: Basic chemical properties of fly ashes

Mass (%)	SiO_2	TiO_2	Al_2O_3	Fe_2O_3	MgO	CaO	P_2O5	L.O.I.
Duvha	45.26	1.6	28.89	8.98	1.84	8.35	1.53	2.25
Matla	46.18	1.7	29.64	5.2	2.5	9.75	1.25	1.38

(where L.O.I. ≡ Loss on ignition)

Table 3: Chemical properties reported for fly ash from Duvha and Matla power stations (after Willis, 1987)

Mass (%)	SiO_2	Al_2O_3	Fe_2O_3	CaO	Quartz	Glass
Duvha	53	36	7.1	2.4	13	35
Matla	45	27	3.7	9.7	5.3	67.8

properties and their significance that are as yet not fully appreciated.

4 BASIC CHEMICAL PROPERTIES

Chemical constituents were determined by X-ray diffraction and X-ray fluorescence spectrometry. The results for the two samples are summarised in Table 2. The results are all fairly similar, although the iron content of Duvha ash is high. Similar tests were carried out by Willis (1987) on samples of fly ash from Duvha and Matla power stations. His results are summarised in Table 3, where only selected parameters are tabulated.

These results are all for tests on precipitator ash and are the average of four replicate tests. The results are generally similar to those in Table 2, although for the Duvha ash the values of SiO_2 and Al_2O_3 have increased while the lime content has decreased quite substantially. An interesting observation is that in the Matla ash the quartz is present as glass in a much greater percentage than is the case for Duvha ash. Willis' results also show a much smaller CaO content for Duvha ash than any

of the other results. This may be important in the discussion that follows later.

The likelihood of an ash exhibiting pozzolanic hardening properties has been correlated with the amount of ($SiO_2+Al_2O_3+Fe_2O_3$) present in the ash (ASTM Standard C 618-84). This standard suggests that if:
($SiO_2+Al_2O_3+Fe_2O_3$) > 70% by mass of the sample and LOI < 6% and SO$_3$ < 5%,
the ash is likely to exhibit pozzolanic hardening. These requirements are satisfied by both the ashes tested and pozzolanic hardening would appear to be a possibility for these samples. However, as reported by Aitcin et al (1985), the above criteria are overly simplistic and account also needs to be taken of the amount of lime and sulphate present in the ash, since these two components enhance pozzolanic hardening. The presence of available water is another prerequisite for pozzolanic hardening.

5 STRENGTH GAIN DUE TO HARDENING

At sites where ash is disposed of hydraulically it has often been noticed that overtopping of the retaining embankment does not necessarily lead to the formation of erosion gullies, which would be the case for most conventional tailings dams. There thus appears to be the formation of a competent, erosion-resistant outer layer covering the surfaces of an ash dam. Stuart et al (1997) report the formation of a very strongly cemented surface of the terraces at the Kriel ash dam that had to be ripped and broken down using heavy earthmoving equipment to facilitate rehabilitation. They also found that once the ash was broken up it remained loose and recementation did not occur.

At a visit to Matla power station, a vertical section through the ash had been exposed in a trench. The side of the trench was sprayed with phenol phthalein, from the surface to a depth of 1m. Below about 10cm the ash changed to a pinkish colour, indicating carbonation was occurring above this depth. To check the possible strength gain due to carbonation, a series of laboratory strength tests was carried out. The testing programme included unconsolidated undrained tests, confined undrained triaxial tests and consolidated confined triaxial tests. Only results of the former tests are reported in this paper.

Two air-tight 50ℓ polyethylene drums were equipped with inlet and outlet fittings through which gas could be circulated at a constant rate. One drum was continually flushed with nitrogen and the other with carbon dioxide. Specimens of Matla and Duvha ash were prepared as slurries and poured into cylindrical moulds 38mmϕ by 76mm long, and allowed to drain and consolidate. The moulds were removed and the specimens placed on trays suspended within the drums. They were then exposed to either a nitrogen or CO_2 atmosphere for different periods of time. Specimens were removed after 7, 14 and 28 days and subjected to unconfined, unconsolidated compression tests. The results are summarised in Figure 3 and each result is the average of two replicate tests. The Matla ash developed unconfined strengths that were more than three times greater than the Duvha ash specimens. These observations are consistent with the relatively greater amount of lime present in the Matla ash, which as mentioned earlier, is thought to enhance carbonation hardening by conversion of the calcium hydroxide content of the ash to calcium carbonate. Although there is a data point missing for the 28-day exposure Duvha specimen, it can be concluded that the increase in shear strength as a consequence of aging the specimens in a nitrogen environment is effectively zero for both Matla and Duvha specimens. Curing the specimens in a carbon dioxide environment, however, has a much more significant impact. The Matla ash undergoes an approximate two-fold increase in unconfined strength. There thus appear to be (at least) two factors contributing to the phenomenon of carbonation hardening. A greater amount of lime, as occurs in the Matla ash, provides some strength gain, which appears to be enhanced by exposure to a carbon dioxide environment. The reason that the Duvha ash experiences a decrease in strength from 7 to 28 days after an initial increase is more difficult to explain, particularly as the 28-day strength has decreased to a value almost equal to that of the ash exposed to nitrogen. This means that the process of carbonation has had no lasting effect on the strength for the Duvha ash, even though the calcium hydroxide content is relatively high.

When considering the implications of these findings it is important to remember that the Duvha ash is unrepresentative of typical ash (refer to Figure 2). The results of the Matla ash tests are therefore probably of more interest as they are more representative of a 'typical' ash. Since curing took place under zero effective confining stress, the measured increase in strength due to hardening is likely representative (in terms of ambient stress conditions) of what occurs on the surface of hydraulically placed ash that is exposed to the atmosphere. It is clearly not representative of air quality, since the specimens were cured under a 100% CO_2 environment. The importance of knowing whether this effect may be relied upon, is that the erosion resistance (and thus the stability of the ash dams during overtopping) may be significantly increased by carbonation hardening.

6 FIELD MEASUREMENTS OF STRENGTH VARIATIONS

In order to determine whether the laboratory observed strength gains (particularly for the Matla ash) were realistic or whether they were primarily due to the saturated carbon dioxide atmosphere in which specimens were cured, field tests were carried out at both Matla and Duvha ash dams. The tests utilised a conventional dynamic cone penetrometer and the penetration profile with depth was determined at a number of locations on the outer slope of the dams, as well as within the inner, daywall paddock and on top of newly-constructed daywalls.

Results for Matla are summarised in Figure 4. The newly placed ash, despite having been compacted, has a very low resistance to penetration.

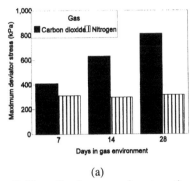
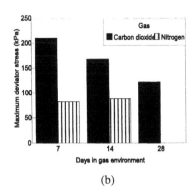

(a)   (b)

Figure 3. Unconfined compressive strengths of (a) Matla and (b) Duvha specimens

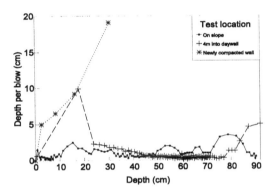

Figure 4. Penetration resistance of Matla fly ash

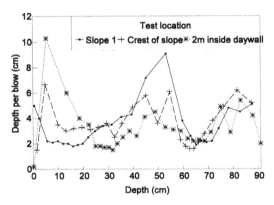

Figure 5. Penetration resistance of Duvha ash

The result illustrated for the slope of the dam was typical of eight separate tests and indeed at the base of the slope it was impossible to penetrate to a depth of more than 30cm below the surface. For the test on the slope, the penetration per blow was generally of the order of 0.5 to 1cm per blow. The penetration resistance of the ash 4m inside the daywall was also generally high, although there is evidence of a layering effect, eg at depths of 20cm and again at 90cm below surface, the resistance to penetration decreases.

Nevertheless, there is clear evidence that Matla ash left exposed to the atmosphere for a period of time undergoes a significant gain in strength and confirms the previously mentioned observation of an erosion-resistant slope. It is also important to note that the high strength ash is not restricted to the surface, but extends to a depth of at least 90cm and probably more. This is attributable to the method of deposition of the ash, whereby relatively thin layers of ash are exposed to the atmosphere and allowed to drain before placement of subsequent layers. The fly ash is thus exposed to carbon dioxide (which is present in air, albeit in a low concentration), in the presence of substantial free water. Subsequent burial by newly deposited ash does not affect this strength gain (at least for shallow depths).

The results for the Duvha ash dam are shown in Figure 5. The penetration per blow is generally much more than the Matla ash, with the greatest resistance corresponding to a depth per blow of 2cm. These results show even more pronounced evidence of layer formation, with all three tests showing very weak ash close to the surface, and at depths of about 50cm and 80cm below the surface. There is also a very good correlation between the depths at which the weaker and stronger layers occur at the three test locations, with the separation between two stronger (or two weaker) layers of about ½m corresponding to the thickness of an individual layer of hydraulically placed ash. Although there is thus some evidence of strength gain in the Duvha ash, it is not as significant as in the Matla ash and, as has been found on site, it is insufficient to prevent erosion of the slope of the ash dam, should localised overtopping occur. The variability at this site may also be due in some measure to the previously mentioned variability of the CaO content of Duvha ash (which is directly related to hardening potential).

Only two fly ash dams have been investigated in this study and it is thus not possible to generalise the findings. Previous work on South African fly ash dams by Stuart et al (1997) reported similarly erosion-resistant outer slopes to those encountered at the Matla ash dam. However, it does not at present appear possible to predict with a high degree of certainty how a particular fly ash will increase in strength with time. It does not appear to be only a matter of pozzolanic hardening, which is usually a fairly slow process, but to include some form of chemical hardening (termed carbonation in this paper) which requires the presence of carbon dioxide to facilitate the necessary reactions.

7 X-RAY DIFFRACTION TESTS ON FIELD SAMPLES

Blocks of ash were taken from the two ash dams at locations adjacent to the dynamic cone testing positions. X-ray diffraction tests were carried out on samples taken from these blocks. Among the primary conclusions of this phase of the investigation were:

- The Matla ash contained a considerable proportion of calcite ($CaCO_3$), whereas Duvha did not.

- Duvha ash exhibited a 'hump' in the diffractogram very close to that of pure silica glass, whereas Matla appears to contain some, but not much, calcium in a silica glass matrix.

- There was no hydrated lime ($Ca(OH)_2$) present in any of the samples.

These results seem consistent with the observation that the Matla ash (as obtained from the precipitators) contains a relatively large amount of CaO. Although there is some disagreement whether it is the amount of CaO or the percent glass present in a fly ash that governs its potential for chemical hardening, Lesch and Cornell (1987) presented data for fly ash from nine South African power stations that demonstrated a clear, direct correlation between these two constituents. Either indicator is thus likely equally useful. The CaO is converted into $CaCO_3$, leaving behind no hydrated lime.

Although some of the tests indicated the Duvha ash to have in excess of 4% CaO by mass, it was usually below this value. The observation that there was no hydrated lime in the Duvha sample may not necessarily indicate that the available CaO has all been converted to calcium carbonate. The coarseness of the Duvha ash makes it much more permeable to water and some of the CaO could be leached from the surface layers before it has an opportunity to react with atmospheric carbon dioxide.

8 CONCLUSIONS

Laboratory tests on fly ash from two different hydraulic fill ash dams showed relatively little variation in basic geotechnical properties, despite differences in grain size distribution. However, in-situ the two dams were found to have significantly different strength properties, with the fly ash from Matla dam forming an erosion-resistant crust whereas the Duvha ash showed relatively little gain in strength. Laboratory tests on specimens of the two fly ashes showed that curing the ash in the presence of carbon dioxide increased the unconfined undrained shear strength of the Matla ash considerably, whilst having relatively little effect on the Duvha ash. Dynamic cone penetration tests in the field confirmed this general observation, with the Matla ash exhibiting extremely high penetration resistance and the Duvha ash relatively low resistance. In both cases, the penetration resistance was not constant with depth and indeed there seemed to be the development of alternating weak and strong layers. The thickness of these layers correlated quite well with the deposition thickness used at the dam, lending further credence to the conclusion that exposure to atmospheric carbon dioxide increased the degree of strength gain due to carbonation.

ACKNOWLEDGEMENTS

The financial assistance provided by the South African Electricity Supply Commission, Eskom, is gratefully acknowledged. Ms Quirina Roode carried out the XRD tests on the samples of field ash and assisted with interpretation of results.

REFERENCES

Aitcin, P.C., Autefage, F., Carles-Gibergues, A. and Vaquier, A. (1985). Comparative study of the cementitious properties of different fly ashes. *Compositional Analysis*. Wiley and Sons, New York.

Brackley, I.J.A., Barker, M.J. and Smith, E. (1987). The stability of ash dumps and ash dams. *Proc International Conference on Mining and Industrial Waste Management*, Johannesburg 1987, pp 175-184.

Heystee, R.J., Johnston H.M. and Chan H.T. (1991). Coal fly ash properties and ash landfill design. *Proc. 1st Canadian Conference on Environmental Geotechnics*, Montreal, Canada, 65-71.

Kruger, R. (1997). Personal communication.

Lesch, W. and Cornell, D.H. (1987). The mineralogy and morphology of fly ash from South African power stations. *Proc. Conference on Ash - A valuable resource*, Pretoria 1987, 13pp.

Roode, Q. (1998). Personal communication

Scheetz, B.E., Menghini, M.J., Hornberger, R.J.,Owen,T.D., Schuek, J. and Giovannitti, E. (1998). Beneficial use of coal ash in mine reclamation and mine drainage pollution abatement in Pennsylvania. *Proc. 5th International Conference on Tailings and Mine Waste*, Fort Collins, USA, 1998, pp795-805

Stuart, R.J., Dorman, S.A., Geldenhuis, S.J.J. and Timm, R. (1997). Hydraulic ash disposal at Kriel thermal power station and its impact on the ground water regime. *Proc. 4th International Conference on Tailings and Mine Waste*, Fort Collins, USA, 1997, pp405-415.

Toth, P.S., Chan, H.T. and Cragg, C.B. (1988). Coal ash as structural fill, with special reference to Ontario experience. Canadian Geotechnical Journal, Vol.25, pp694-704.

Willis, J.P. (1987). Variations in the composition of South African fly ash. *Proc. Conference on Ash - A valuable resource*, Pretoria 1987, 12pp.

Geotechnics for Developing Africa, Wardle, Blight & Fourie (eds) © 1999 Balkema, Rotterdam, ISBN 90 5809 082 5

A unique approach to evaluate the utility of landfill monitoring boreholes

M. Levin
Africon, Pretoria, South Africa

B.Th. Verhagen
Schonland Research Centre, University of the Witwatersrand, South Africa

ABSTRACT: Environmental tritium ^{3}H is a very useful tracer of water and widely used in hydrological studies. This paper concentrates specifically on the use of tritium as a tracer to establish the hydrodynamics of boreholes as well as for tracing leachate movement. Rain water contains environmental tritium at concentrations of about 5 TU. Tritium has been used to establish the recharge to boreholes. Low or zero tritium level in ground water would indicate slow or no recharge while at the other end of the scale 5 TU would indicate recent or ongoing recharge by rainwater. Monitoring boreholes with low or zero tritium indicate very slow ground water movement and would therefore not be able to reflect pollution presently emanating from a landfill site or a leaking leachate containment pond. Anomalously high levels of artificial tritium have been discovered in leachates from landfill sites in the Republic of South Africa and elsewhere and indications are that this may be ubiquitous in such sites. The materials from which the tritium originates has not yet been established firmly. Whatever the source material, this artificial tritium in the leachate is a unique tracer. When the pre-pollution background values for recently recharged rainwater are established we can utilise the contrast to detect leachate leakage. With measured contrasts of up to 4 orders of magnitude this tracer is sensitive for picking up spillages or leakages. The origin of the tracer is unique (the leachate) and is indisputable. Environmental tritium can therefore be used to evaluate the usefulness of the monitoring boreholes and the artificial tritium in the leachate can detect movement from the leachate system to the monitoring boreholes

1. INTRODUCTION

Solid waste landfill designs are normally based on geotechnical and geohydrological information collected during site suitability investigations. These investigations are conducted according to the guidelines (Minimum Requirements, 1998) laid down by the Department of Water Affairs and Forestry for waste disposal by landfill. According to this document "a mandatory physical separation between the waste and the surface and ground water regimes, as well as an effective surface water diversion drainage system, are fundamental to all landfill designs".

Should the hydraulic properties of the underlying material be such that it cannot adequately retard the leachate after compaction or treatment, then the regulatory authority may prescribe the importation and use of suitable material for lining and/or geotextile to ensure the required physical separation.

The integrity of the leachate collection and separation system is monitored by a monitoring borehole network designed and installed during the investigative or construction period of the site. The main objective of installing a monitoring borehole is to intersect ground water moving away from a waste management facility. Base line data is obtained from the monitoring borehole(s) before commissioning of the site. The monitoring boreholes are sampled at a certain frequency to detect any change in ground water quality which could be due to leakage of the system. The suitability of a monitoring borehole to detect leakage cannot always be ensured by geophysical and/or geohydrological investigations.

This paper discuss an approach using tritium, a radioactive isotope of hydrogen to evaluate the suitability of the monitoring borehole as well as unambiguously to identify leakage of the leachate containment system.

2.   WHAT IS TRITIUM?

Environmental tritium is a very useful tracer of water and widely used in hydrological studies (IAEA 1983, Verhagen et al. 1991). Tritium is produced in nature by cosmic ray interaction with the upper atmosphere, and readily oxidised to water in which it is a conservative tracer as it is part of the water molecule. Depending on geographical location, rain water contains natural tritium at concentrations of up to some 5 TU (1 TU = $[^3H]/[^1H] = 10^{-18}$). Tritium is radioactive and decays through low-energy beta ray emission with a half-life of 12.43 years. Its radioactivity can only be measured in the laboratory. Minimum detectable values are about 5 TU by direct (screening) counting; following isotope enrichment, 0.2 TU can routinely be attained. The useful range of measurement of environmental tritium in geohydrological applications spans four to five half-lives and it is therefore measurable only in, and can act as an indicator or, recently recharged ground water.

Thermonuclear weapons testing in the latter 50's and early 60's increased tritium concentrations in rainfall by up to three orders of magnitude in the northern hemisphere and about one order in the southern hemisphere. This tritium "pulse" could be traced through hydrological systems such as ground water. At present, the widespread use of artificial tritium in e.g. Europe maintains tritium in rainfall at about one order of magnitude above the natural value. In Southern Africa, where the use of artificial tritium is much less common, tritium in rainfall has returned to close to natural values. Ground water usually has lower values, depending on residence times or recharge rates. Any water labelled with artificial tritium can therefore readily be detected and traced in the Southern African environment.

3.   TRITIUM IN GROUND WATER

The presence of tritium in ground water in amounts similar to that in present day rainfall of the area, would be an indication of rapid and active percolation. Provided it is well protected against direct inflow from surface a borehole yielding water with present day rainfall values of tritium can therefore be considered an excellent monitoring borehole. It would rapidly respond to leakages and spillages of leachate which might already be visible after the first rainfall following the incident. Such boreholes should be sampled at about two or three monthly intervals.

Low but measurable tritium in ground water points to longer residence times for ground water, due to the delay in the unsaturated zone or low recharge/storage ratio in the unsaturated zone. Monitoring boreholes displaying such values may for a number of years show very little change. However, once pollution is detected in such a borehole it is certain that, even if the source is stopped, the levels of pollution will be maintained for a number of years until it dissipates. Such boreholes are not considered good monitoring boreholes and frequent sampling is not recommended. A six monthly or annual sampling frequency is recommended depending on the measured base line tritium level.

The absence of tritium in ground water sampled from a monitoring borehole indicates very low recharge and ground water residence times well in excess of 50 years. Such boreholes cannot be considered good for early detection pollution monitoring and frequent sampling would be meaningless.

As tritium labels the water molecule itself, its concentration in e.g. ground water is, to a good approximation unaffected by processes in the sub-surface. It is therefore a conservative tracer. Sampling and handling water for tritium analysis is straightforward and requires no special precautions as compared to e.g. sampling for chemistry or dissolved gases. For direct counting a 50 ml sample suffices. For low level detection following isotope enrichment, a litre of water sample is usually required.

4.   TRITIUM IN LANDFILLS

Levels of artificial tritium well above the usual environmental levels have been detected in association with landfill sites in the Republic of South Africa during a Water Research Commission project in Bloemfontein and Johannesburg (Verhagen et al. 1995). In a polluted borehole at the Bloemfontein North site a tritium level of 26.5 TU was measured. Other boreholes gave values

Table 1: Results of the Tritium in leachate measurements in the recent study

Site No	Province	Classification	Waste Source	Tritium (TU)
1	Kwazulu-Natal	G closed	Domestic high income	1787±14
2	Kwazulu-Natal	G closed	Domestic high income	425±6
3	Kwazulu-Natal	G closed	Domestic/industrial	65,2±0,4
4	Kwazulu-Natal	H	Industrial hazardous	98288±148
5	Kwazulu-Natal	G	Domestic lower income	3064±19
6	Kwazulu-Natal	G	Domestic/industrial	2144±12
7	Western Cape	G	Domestic mixed	191±5
8	Gauteng	G	Domestic industrial	2605±17
9	Gauteng	G closed	Mixed	75±3
10	Gauteng	G	Domestic high income	607±13
11	Gauteng	G	Domestic high income	6,3±0,6
12	Gauteng	G	Domestic/industrial	10,0±2,6
13	Gauteng	G	Domestic/industrial	1771±11
14	Gauteng	G	Domestic/lower income	104±4
15	Gauteng	G	Domestic mixed	16±3
16	Gauteng	G	Domestic/industrial	89±4
17	Gauteng	G	Domestic/industrial	48±3
18	Gauteng	G	Domestic/lower income	2569±9
19	Gauteng	H	Industrial hazardous	448±12

significantly above the expected maximum environmental level. In a borehole on the Bloemfontein South site a value of 28.1 TU was measured. Seven months later this value had increased to 40.1 TU. In both cases, significant amounts of leachate must clearly have entered the aquifer/borehole. At the time the source material in these landfills was thought to be associated with medical research.

During the same project, values up to 23,2 TU were found in the culvert water draining from the Waterval landfill in Johannesburg. As in the other sites, the reason for the elevated tritium values is not fully understood at present; the more so as this landfill had been closed for some 20 years.

During 1997 a total of 19 samples of leachate for tritium analysis was collected in a survey from landfill sites in Kwazulu Natal, Gauteng and Western Cape Province (Verhagen et al. 1998, Fourie et al. 1998). The locations of the sites are indicated on the map of South Africa in Figure 1. Using the direct counting technique, it was possible to screen samples with high tritium content from those with low and background values.

The values for the various sites are shown in Table 1 and the values vary between just above background at 6,3 TU to 98288 TU. The samples are categorised according to province, with 6 from

Kwazulu Natal, one from the Western Cape and 12 from Gauteng. Only two H class sites were sampled; the rest are all G class sites.

The results presented in Table 2 highlight two important points:

- The tritium value for the total leachate is highly variable and can be influenced by rainfall, slumping in the waste etc.
- There are large differences in tritium concentration at different points in the site. Tritium source(s) may therefore be highly localised within the body of the fill.

Table 2. Detailed leachate samples from site 4 November 1997

Sample	Tritium (TU)
Total leachate	5800±42
Leachate south	405±7
Leachate centre	405±7
Leachate north	12829±38

5. TRITIUM AS A TRACER OF LEACHATE

Simultaneous with these discoveries in South Africa, rather high levels of artificial tritium were reported in leachate from English landfill sites (Robinson & Gronow 1995). The potential of these

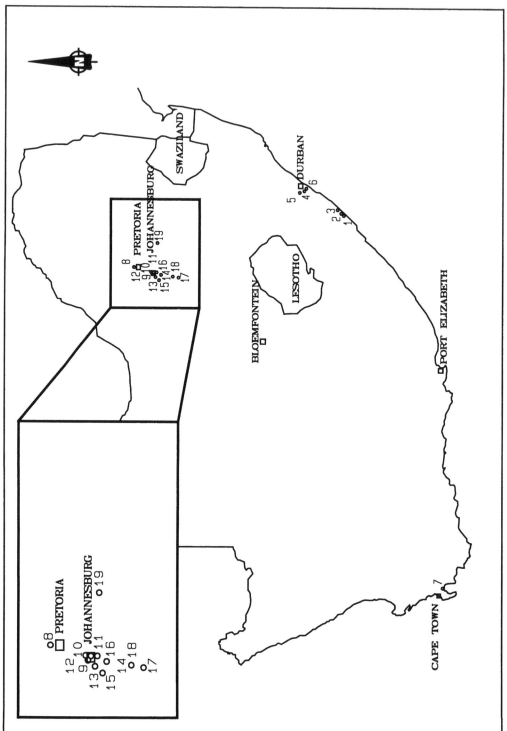

Figure 1 : Location of landfills from which leachate samples were collected.

tritium levels for tracing pollutants in water was soon recognised. However, the nature of the source materials of tritium in landfill sites are at present still open to conjecture. Artificial tritium synthetised into organic molecules is used as a tracer in medical research and diagnostics. The concentration of tritium in such compounds, although generally low in terms of biological radiation hazard, is many orders of magnitude higher than ambient levels in surface water and ground water. Even where the tracer materials themselves are conscientiously disposed of, associated contaminated materials might find their way onto landfill sites. Artificial tritium at considerably higher concentrations is used as an energy source with phosphorescent material in luminous signs and dials (Robinson & Gronow 1995). When such items are discarded and damaged, the tritium they contain can readily exchange with hydrogen-containing compounds and moisture in the landfill matrix, and the leachate which gravitates out of it.

As the use of tritiated material and items is assumed to be fairly limited in the southern African environment at present, it follows that the distribution of tritium-containing materials, both geographically in different sites and within individual landfill sites, is likely to be patchy. The method of detection in the leachate is extremely sensitive. Tritium from a single item within a landfill could therefore produce significant levels in the total leachate output.

Monitoring boreholes evaluated as tapping actively recharged ground water, which show an increase in tritium above the baseline values during routine monitoring, should be regarded as giving an early warning of pollution. Resampling and further periodic sampling will confirm tritium levels above the base values obtained before site commissioning.

6.   DISCUSSION

Evaluation of pollution at a number of South African sites (Verhagen et. al. 1998) showed that chemistry and chemical changes are not sensitive enough to confirm pollution in the monitoring system (Levin 1997). This is especially true where the background salinity of the ground water chemistry is high or where ground water movement is through clay with high ion exchange and absorption capacities for heavy metals, radionuclides or macro ions. As stated previously,

(artificial) tritium is almost unique to landfill, and a near-perfect tracer, in that it does not participate in chemical reactions, is only slightly exchanged onto geological material and is easy to sample for.

In the case of the coastal sites the contrast between environmental tritium (a few TU) in ground water and leachate (up to 10^5 TU) is several orders of magnitude. The chloride content of the ground water in these coastal areas is typically about 10^2 mg/l and in the leachate some 10^3 mg/l, a contrast of only one order of magnitude. The chloride content of water for instance need not be derived from the leachate but from sources external to the landfill. Furthermore the tritium is uniquely derived from the waste and identifies the leachate in the presence of pollutants from other sources. Tritium in landfill leachates and the high contrast this presents with existing environmental levels at many landfill sites, makes this tracer most promising in studying leachate dispersal into the environment.

In the absence of known contamination, boreholes with a tritium signal close to values in present day rainfall will be excellent monitoring boreholes. These boreholes should respond to any leakage and should be monitored frequently. Boreholes with low but measurable tritium indicate slow movement of water especially through the unsaturated zone. Such boreholes need only to be sampled 6 monthly or annually. Once pollution has been observed, one can expect it to dissipate slowly over a long period after remediation of the leak. At the lower end of the scale are boreholes with near-zero tritium indicating very low recharge and groundwater residence times well in excess of 50 years. Such boreholes cannot be considered monitoring boreholes.

After commissioning of a waste site any change in the tritium content should be carefully monitored to establish any trend. Only boreholes monitoring actively recharged ground water will respond rapidly to leakage. Monitoring boreholes indicating lower recharge will show a delayed response.

7.   CONCLUSIONS

Leakage from the leachate containment systems is monitored by boreholes located at localities where they can intersect movement of water away from the site. The dynamics of these boreholes can be evaluated by environmental tritium before commissioning of the site. Changes in tritium after the site is commissioned can uniquely be linked to leakage from the site.

From only a limited number of landfills investigated thus far, it can be concluded that artificial tritium levels useful for leachate tracing are to be found in the majority of landfill sites in South Africa. The sources of this artificial tritium in waste are still open to conjecture. A more detailed study of one site shows that such sources may be highly localised within the fill. The contrast between the levels of tritium in waste and in the natural surroundings found in this study gives it tremendous potential to be used as an environmental tracer of pollution or leakage at other landfill sites.

8.   REFERENCES

1)   Fourie A B, Verhagen B Th, Levin M & Robinson H D. 1998 Tritium as indicator of contamination from landfill leachate. *Wastecon 98*, Kemptonpark, 12 – 15 October 1998.

2)   Levin M & Verhagen B Th. 1997 The use of environmental isotopes in pollution studies. SAIEG Conference, Geology for Engineering, Urban Planning, and the Environment. Eskom Conference Centre, Midrand, Gauteng. 12-14 November 1997.

3)   International Atomic Energy Agency (IAEA) 1983 *Guidebook on Nuclear Techniques in Hydrology*. Tech.Rep. Series No. 91, Vienna

4)   Robinson, H D & Gronow, J R. 1995 Tritium levels in leachate from domestic wastes in landfill sites. In: *Procs Sardinia 95. Fifth International Landfill Symposium* CISA, Cagliari, Italy.

5)   Verhagen B Th, Levin M, Walton D G & Butler M J. 1995 Environmental isotope, hydrogeological and hydrogeochemical studies of ground water pollution associated with waste disposal. *Draft Final Report to the Water Research Commission*: Project #K5/311.

6)   Verhagen B Th, Levin M & Fourie A B. 1998 High level tritium in leachate from landfill sites in the Republic of South Africa with emphasis on its distribution and value as an environmental tracer. *WISA 98 Conference*, Cape Town, 5 – 7 May 1998.

7)   Verhagen, B Th, Geyh, M A, Froehlich, K & Wirth, K. 1991 Isotope hydrological methods for the quantitative evaluation of ground water resources in arid and semi-arid areas. Development of a methodology. *Research Reports of the Federal Ministry for Economic Cooperation*, FRG. Bonn.

Geotechnics for Developing Africa, Wardle, Blight & Fourie (eds) © 1999 Balkema, Rotterdam, ISBN 90 5809 082 5

Material selection and structural design, construction and maintenance of ramp roads in open pit mines

Phil Paige-Green & Andrew Heath
Division of Roads and Transport Technology, CSIR, Pretoria, South Africa

ABSTRACT: Traditional material and structural design of unsealed mine haul roads requires a thick pavement structure in order to dissipate stresses from heavy haul vehicles and thereby prevent structural damage to weak subgrade materials. Ramp roads in open pit mines are, however, generally constructed above solid or fractured bedrock and the placing of thick structural layers of imported material is therefore not required and expensive. The imported material is, in fact, often the weakest portion of the pavement and the thickness should be limited as far as possible. This paper evaluates the behaviour of the imported and in situ material in ramp roads under a range of typical haulage vehicles used in South Africa. Linear and non-linear material modelling was used to determine stress concentrations and areas of potential plastic failure. Recommendations for the quality of the wearing course aggregate are made and the influence of drainage, construction and maintenance of the ramp roads on vehicle operation is evaluated.

1 INTRODUCTION

Unlike traditional public roads, which are designed by pavement engineers, mine ramp roads are usually constructed by mine officials as a necessary requirement outside the core business of mining. These officials are often unacquainted with the basic principles of pavement materials and design. This often results in roads of inferior quality, although in many cases satisfactory roads have been built. Unsealed public roads on the other hand are currently mostly designed and constructed to certain norms (CSRA, 1990).

Unsealed ramp roads which are rough, slippery and dusty result in high road maintenance costs, safety problems, unnecessarily high vehicle operating costs and a general increase in vehicle downtime and mine production costs. Burton (1975) has estimated that haulage costs often represent up to 50 per cent of total mining costs and sometimes as much as 25 per cent of the overall operating, overhead and other costs of the entire mining operation.

It has been shown that the operating costs of vehicles on roads increase dramatically as the road condition deteriorates (CSRA, 1990). This has been extrapolated to traditional mine haul vehicles (Paige-Green & van Huyssteen, 1993) and can probably be extrapolated further to the new generation of very large mine haul trucks (heavier than 250 tonnes).

This paper summarises an evaluation of the materials, design, construction and maintenance of open pit mine ramp roads and investigates the expected effects on these roads of very heavy haul trucks.

It should be noted that there are apparently no industry standards or accepted guidelines for the design, construction and maintenance of mine ramp roads, either internationally or in southern Africa currently in use.

2 QUO VADIS ?

2.1 *Design*

Various types of haul roads exist in open pit mines. Although no formal classification in terms of their characteristics apparently exists, they are expected to provide different degrees of service over different design lives. The primary ramp and surface haul roads exist until the pit is enlarged (up to 20 or more years) before they are ripped up and often incorporated into the ore for processing. These roads are the critical ones, requiring all-weather passability to ensure that production is not affected during periods of extended rainfall.

Secondary ramp roads occur at lower levels in the pits and are only used for perhaps 6 to 8 years but frequently for shorter periods as the depth and width of the pit increase. Temporary haul roads are those at each working level and are utilised only during processing of the material at that level. These roads are unlikely to be used for more than a few months.

Mine ramp roads are currently not designed in the true engineering sense to any specified standard. Experience over the years has resulted in a method of constructing the necessary roads which are usually adequate for their purpose, although certain limitations are encountered. However, as the mass of the haul trucks increases, it is likely that more serious problems will be encountered.

A semi-empirical design technique for surface haul roads has recently been developed (Thompson & Visser, 1996) based on mechanistic principles and empirically-derived transfer functions. Rutting of the subgrade resulting from vertical compressive strains and shear failure of granular layers are the traditional failure criteria in mechanistic analysis. Although this design technique is considered suitable for layered structures for unsealed surface haul roads, these roads are subjected to regular and frequent maintenance. It is the authors opinion, however, that aspects such as subgrade rutting would be of minor consequence and would soon stabilise with appropriate and regular maintenance. Similarly, limited shear failure or accumulated permanent non-elastic (i.e. plastic) deformation is also manageable during the maintenance process and weak areas would gradually achieve equilibrium during routine grading and spot regravelling with appropriate material. Rectification of localised drainage deficiencies, which are generally the cause of stability problems, assists with equilibration of the road surface.

Ramp roads on the other hand are a completely different scenario. In these cases a gravel wearing course is placed on a high modulus rock "subgrade" with the purpose of achieving a maintainable road which will provide a smooth, all-weather riding surface and minimise total transportation costs (construction, maintenance and vehicle operation). The wearing course is mostly affected by the loads imposed by the traffic. Surface haul roads have a relatively balanced pavement structure with a succession of material layers which decrease in quality with depth. Ramp roads on the other hand consist of a relatively thin wearing coarse sandwiched between the high tyre loads imposed on the layer and the significantly stiffer supporting layers.

It is well known in road design that the most detrimental condition affecting road performance is excessive moisture within the structural layers. This is traditionally controlled by an adequate cross-fall on the road, side drains and, where necessary, by sub-surface drains to ensure that the moisture content of the layers does not increase to the point where their strength is adversely affected. A V-drain alongside the ramp roads between the road and the pit-wall is essential. This should be of a suitable shape and size and must be regularly maintained.

The geometric layout of ramp roads is constrained by the geometry of the pit and particularly the optimum pit-slope design angle. Widening of surface haul roads has a relatively minor cost implication. The widening of ramp roads in open-pits, however, results in a need for significantly greater excavation (often of waste material) as any widening necessarily affects the pit for the full depth to the road elevation. Road widths have thus to be carefully considered to optimise the cost of excavation and the safety aspects.

Ramp roads are generally constructed with a maximum grade of about eight per cent, a constraint imposed mostly by haul vehicle specifications with an appropriate cross-fall for drainage purposes.

2.2 Construction

The construction of ramp roads generally attempts to make use of materials that are readily available from the mining operation and requires minimal construction equipment. The available materials are usually spread to a depth of 150 to 200 mm and then compacted by the haul trucks. This is a fairly successful technique in most mines but the performance depends on the properties and thickness of the dumped material, the lateral distribution of the vehicles during compaction and the moisture content of the material during compaction. Compaction continues in the road until the maximum density possible under the specific loads applied (ie compaction effort) and moisture regime is obtained. It should be noted that, as the compaction increases, the material strength increases and the permeability of the material to water decreases.

In general, where roads are to be mined out later, material that can eventually be processed with the ore should preferably be utilised for ramp road construction.

2.3 Maintenance

Maintenance of ramp roads typically consists of five primary processes:

- Water spraying for dust suppression;
- Clearing of material which has dropped off haulers;
- Routine grader maintenance;
- Replacement of gravel to the riding surface, and
- Cleaning and shaping of drains .

These processes have been adapted and developed with time at most mines to provide successful roads, but, in certain respects, improvement is possible as discussed later.

2.4 Safety

Safety within open-cast mine pits is of primary concern. One of the aspects affecting safety is the passage of vehicles and the capability of avoiding serious consequences should the loss of control of a vehicle occur for some reason. Berms along the road on the open pit side are typically provided for safety.

The effect of dust generated from the road by traffic on the overall visibility is probably of greater consequence. Various measures to reduce the dustiness of the roads are implemented or have been evaluated. The most common technique currently employed requires the regular application of water by spray-tankers on the roads although the application of various commercial dust palliatives is practised on some mines. Many roads have been found to become excessively slippery when these are used.

2.5 Mine haul vehicles

Large open-pit mines generally make use of haulers with loaded masses of between 70 and 370 tonnes with significant moves in recent years towards the larger vehicles i.e. heavier than 250 tonnes.

The small numbers of vehicles, although moving frequently on the roads and allowing significant interactions, can generally be considered to contribute to relatively safe conditions.

The major vehicle problems associated with the roads are apparently tyre penetrations (nearly 50 per cent of tyres scrapped are the result of penetrations) and left hand final drive failure (possibly the result of a disproportionate load on the left side of the vehicle resulting from the road camber). Many of the tyre penetrations are the result of driving over aggregate which has fallen off haulers on sharp bends or rough sections of road which result in "bouncing" of the haul trucks. A problem which affects the new generation of large haulers is the fatigue failure of suspension mountings as a result of excessive bouncing on uneven roads.

The smaller haul trucks run on 24.00-35 tyres at cold inflation pressures of 720 kPa (equivalent running pressure of 750 kPa). The back axles have four of these tyres, carrying a maximum load of about 62 000 kg, resulting in a load of 15 500 kg per tyre. The larger haul trucks run on 36.00-51 tyres at cold inflation pressures of 760 kPa . The back axles have four tyres carrying a maximum load of about 184 570 kg, resulting in a load of 46 142 kg per tyre, or about three times the tyre load of the smaller existing trucks. The ratio of the approximate tyre contact area (assuming 100 per cent of the tyre is tread) of the new (0.657 m^2) to the old trucks (0.29 m^2), however, is 2.26, indicating that the increased tyre contact pressure is only effectively about 30 per cent greater for the new trucks (the tyre pressures and tread patterns also affect the contact pressure but are not taken into account in this calculation as the respective areas have not been quantified). It should be noted that recent research on contact pressures under conventional truck tyres (De Beer et al, 1997) has shown that the stress distribution is not constant beneath a tyre but has significant concentrations at the edges of the tyres and around tread patterns.

Similar research has not been carried out on the large tyres used on mine haulers but significant stress gradients can be expected between the tread rubber and the tread groove. The total load on the back axle and the stress interactions between the loaded tyres, however, differ considerably between the small and heavy vehicles.

Little detailed information on operating costs of large mine haulers and the influence of road condition on the costs is available. Observation of the condition of a number of ramp roads and the problems encountered indicated that where appropriate maintenance is applied, the condition is such that road roughness has a relatively minor consequence on the operating costs. Where construction and maintenance is poor, road roughness results in significant bouncing of the vehicles which is expected to increase the operating costs of the vehicles significantly. These roads are usually associated with excessive stoniness and significant loose material which also has a highly detrimental effect on tyre life. Significant reductions in vehicle operating costs can only be effected where the inherent roughness of the haul and ramp roads is at a relatively high level.

The loss of tyres on haul trucks as a result of penetrations by stones as well as the down-time during repair of the tyres is a significant problem. It has been indicated that stones as small as 50 mm in diameter lying on the roads result in tyre penetrations and potential destruction of the tyre. Stones such as these on the roads have generally fallen off the haul trucks and, although there are procedures for their removal, a number of vehicles may ride over them prior to their being removed. It should be noted that it is difficult to see these stones from the driver's cab (particularly at night) and probably even harder to avoid riding over them.

In addition to the haul trucks, other large plant such as tracked Rope Shovels and drag lines also use the haul roads periodically. As these shovels can have masses up to 1 200 tonnes, widths in excess of 9 metres and crawler shoes 1.8 metres wide and 11 metres long, they obviously have potential to cause considerable damage to the roads. It is, however, considered impracticable and uneconomic to design the roads for these vehicles and provision should be made in the maintenance programme to rectify any periodic damage which may be caused by these machines where necessary. In dry conditions, it is considered unlikely that significant damage to the roads will occur but shearing and displacement of material is likely when the roads are wet. It should be noted that although there are stress concentrations at points in the tracks, the average stresses are probably less than those developed under normal pneumatic tyres as a result of ther large surface areas of the tracks.

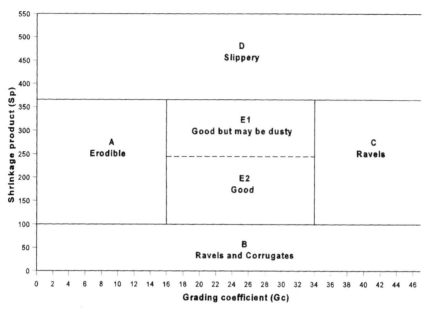

Figure 1: Relationship of performance of unsealed roads to grading coefficient and shrinkage product

Table 1: Material specifications for mine haul roads (after CSRA, 1990: Paige-Green & Bam, 1995; Jones & Paige-Green, 1996)

Maximum size (mm)	75
Oversize Index (I_o)	⊁ 10 %
Shrinkage product (S_p)	100 - 365 (max. preferably <240)
Grading coefficient (G_c)	16 - 34
Soaked CBR (%)	≥ 18 at 95 % Mod AASHTO density
Treton Impact value (%)	20 - 65

I_o = Percentage retained on 37.5 mm sieve
S_p = Bar Linear shrinkage x per cent passing 0.425 mm sieve
G_c = (Per cent passing 26.5 mm - per cent passing 2.0 mm) x per cent passing 4.75mm/100

Note: all grading analyses must be normalised to 100 per cent passing 37.5 mm.

3 MATERIALS

As for any unsealed road, the quality of the materials used in the road is critical. The materials must be such that they:
- have sufficient cohesion to resist ravelling and erosion;
- have a grading which will allow a high degree of packing and interlock of the particles;
- have sufficient bulk strength to resist shear and plastic deformation under applied wheel loads, and
- have adequate aggregate strength to resist fracture under tyre contact stresses.

Specifications for materials for mine haul roads were proposed (CSRA, 1991) and have been modified subsequently (Paige-Green & Bam, 1994; Jones & Paige-Green, 1994) following observations of haul and ramp roads in a number of open-cast mines. These are summarised in Table 1 and the plasticity/grading relationships in Figure 1.

The relationship between the Shrinkage Product and Grading Coefficient is directly related to the performance, as shown in the figure, Zones E1 and E2 being the recommended areas for best performance (Paige-Green, 1989). The figure also shows the predicted performance and the implications (potential problems) of not using material falling within the specified limits. These specifications have been found to conform generally with the requirements for surface haul roads (Thompson, 1997), although minor revisions of the recommended limits were suggested.

It should be noted that these limits were mostly derived from traditional vehicles and the significantly higher loads (mostly point contacts) of the large haul trucks on the aggregate appears to result in significantly more aggregate degradation. This may, however, not necessarily be deleterious in the mine haul road situation, as discussed in the next chapter.

Field compactions measured on ramp roads are often considerably higher than 100 per cent of Modified AASHTO maximum dry density, as one would expect from the compaction effort of the trucks. This effort is probably significantly higher than that provided during standard laboratory testing or during conventional compaction of an unsealed road. The densities have also been found to much higher than those predicted from the material properties, an indication of the packing which occurs as the materials break down under loading and that the particle size distributions vary with time.

A range of laboratory-determined soaked CBR strengths (at 100 per cent Mod AASHTO compaction) of ramp road materials between 1 and 14 per cent have been measured, although the field strengths were substantially higher. This is generally the result of the material in the field being at higher densities and also never in the soaked condition. Despite these low soaked strengths the materials in the field supported the haul vehicles without significant deformation, even after routine water spraying. Field studies indicated that a soaked CBR of 15 at 95 per cent Modified AASHTO compaction would support normal road vehicles (CSRA, 1990; Paige-Green & Bam, 1991). Subsequent investigations showed that a soaked CBR of 18 per cent would support the heavy vehicles traditionally used on haul roads (Jones & Paige-Green, 1996) for a 150 to 200 mm thick layer. It has, however, been noted that the lower CBR strength materials tend to have higher plasticities which results in unacceptably slippery conditions during periods of rainfall and under excessive water spraying for dust-suppression.

Tests conducted on materials from one mine indicate that nearly all the materials were potentially slippery, although the aggregate (mine tailings), when applied to the road fell inside the required E-Zone of Figure 1. It was clear from the grading analyses, strength tests and durability mill tests that the materials were being crushed under the trucks, resulting in generation of fines and an increase in plasticity.

4 STRUCTURAL DESIGN

Most pavement structures for ramp roads are presently not designed as such, although they result in a layer structure which can be analysed in terms of the structural capacity of the pavement. When traditional mine haul roads are constructed, the "imported" layer is placed on the subgrade to provide an even, maintainable, preferably dust-free, all-weather surface which distributes the loads applied by vehicles so that the subgrade is not overstressed. In the unique situation of ramp roads for open-pit mines, the subgrade is potentially much stronger than the imported wearing course and, if a smooth surface could be obtained it would be the best riding surface. However, because of the uneven nature of the fractured rock subgrade (usually prepared during blasting), the application of a material which can be shaped and maintained is required.

After application of a wearing course, an open-pit haul-road has a surface layer which is more likely than the stiffer rock sub-grade to be deformed by shearing, compaction or plastic squeezing. In normal structural design, the wearing course would be placed to minimise stresses on the underlying layer and the subgrade is thus the layer more likely to suffer distress under loading.

One of the simplest but most practical design methods currently used for haul roads is that developed for highway design in the United States in the 1940's and later adapted for mine haul roads (Kaufman & Ault, 1977). This empirical method defines the cover necessary to protect subgrades from overstressing as a function of the subgrade strength, wheel loads and cover material quality (Figure 2).

From this figure, the ramp road situation with a high-strength subgrade would require about 200 mm of cover material for typical smaller haul trucks. (However, the method assumes that the cover material has a CBR in excess of 80 (typically a crushed stone)). With weaker wearing course materials, similar to those tested at a number of mines (average soaked CBR values of about 15 and 30) and, on the basis of Figure 2, the roads would require 380 mm and 250 mm of cover respectively to protect them from damage. However, the roads observed in the mines are mostly structurally sound with only 150 to 200 mm of wearing course material, indicating that:

- the design criteria are too conservative for the high subgrade strengths prevailing;
- the technique is invalid for this situation; or
- the materials in the roads are not affected significantly by water (relatively impermeable with low in situ moisture contents) and that the in situ strengths (most of the in situ CBR's determined from DCP data are over 80 per cent) are adequate.

A more sophisticated method of pavement design (mechanistic analysis) makes use of the calculated stresses and strains within the pavement under applied loads which are determined from the elastic properties (Elastic modulus (E) and Poisson's ratio) of the material. Mechanistic analysis of the pavement structure takes into account the layered nature of the pavement better than the previous method, but

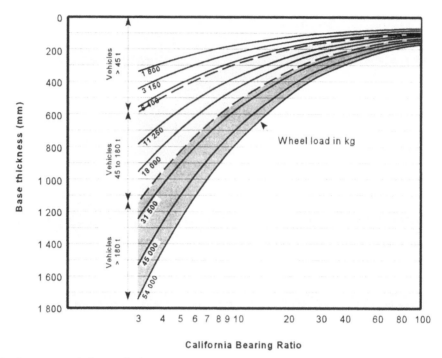

Figure 2: Cover curve design method

assumes (probably incorrectly) that the materials all behave in a linear elastic manner. Mechanistic analysis has recently been investigated for surface haul roads with soil or gravel subgrades (Thompson & Visser, 1996) but no investigations appear to have been carried out where the subgrade is in situ or fractured rock.

An example of this for two typical ramp roads is illustrated. The Effective Elastic Modulus (E_{eff}) (De Beer, 1991) was determined from the DCP data (an average of various values obtained for the ramp roads was used) and a standard value was used for Poisson's ratio (ν) for granular materials. The values for E_{eff} used in the analyses were 464 and 580 MPa for wearing course and 1 619 and 1 414 MPa for the subgrade (which are conservative values based on the lowest values determined prior to refusal of the DCP test). The values for the wearing course are higher than would be expected from typical compacted gravels but were attributed to the high densities developed under the haul trucks and the relatively dry state of the in situ material when tested.

Stress distributions under the centre, under the edge and between the tyres of the existing and new vehicles were determined for the two roads (Figures 3a and b).

Experience has shown that high stresses are developed under the tyres of the haul trucks. It is also been observed that typical ramp roads currently in service can absorb the stresses from even 100 tonne haulers without significant deformation or shearing, although "biscuiting" which has been observed within the pavement structure may result in the layers forming individual slabs under the larger haul trucks (300 to 350 tonnes). Experience has not yet shown what will happen to loose material being compacted under the larger haul trucks although, it is anticipated that after initial settlement and disintegration, the dry strength should be adequate to carry the loads.

A more sophisticated analysis technique than the linear elastic evaluation using mechanistic procedures is that in which the tyre/pavement interaction is modelled using finite element analyses. This can be modelled using either linear elastic techniques or non-linear, plastic theory which is considered to be more realistic in terms of the behaviour of a thin layer of high plasticity material sandwiched between a tyre and a stiff subgrade or a combination of both.

The road structures analysed earlier were modelled assuming non-linear, plastic behaviour of the gravel wearing course and linear elastic behaviour of the subgrade/bedrock. The stress distributions in the wearing course using the linear elastic and non-linear plastic material models, were similar. However, the results differed in that the behaviour of the wearing course was modelled as permanent plastic flow instead of recoverable elastic strain. The predicted plastic flow of the wearing course layer for the two roads analysed previously under the various vehicles masses is shown in Figure 4.

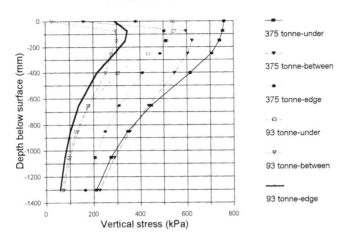

Figure 3a: Stresses generated by vehicles on a typical
haul road

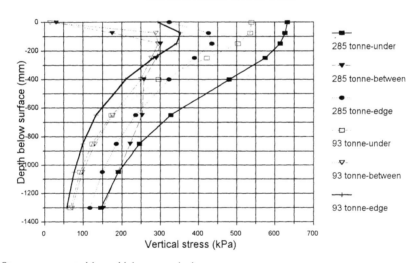

Figure 3b: Stresses generated by vehicles on typical
haul roads

It is clear from the figures that the use of the larger trucks results in slightly higher plastic flows over larger areas although they are distributed in the same zones, i.e. at the outside edges of the tyres and between the dual wheels. No transfer functions currently exist to relate these stresses, strains or plastic flow to actual failure induced by vehicles. The type of failure is such that significant lateral "squeezing" of the wearing course is unlikely to occur unless the material becomes excessively wet.

However, a study of these figures and the relevant in situ and laboratory test results indicates that major structural failure under the heavier haul trucks is unlikely, provided the layer is correctly constructed, compacted, maintained and drained. Routine maintenance is likely to rectify these strains as they occur in most instances.

5 GEOMETRICS

The geometrics of ramp roads affect the economy of mining, safety aspects and pavement performance. Only aspects related to the width of the road are discussed in this paper as the grade is pre-determined for any pit mine and complies with the recognised norms of 7 to 9 per cent (maximum 10 per cent) (Kaufman & Ault, 1977) whilst the radius of horizontal curvature, other than for the switch-backs, is determined by the geometry of the pit.

The width of the roads has a significant effect on the total dimensions of the pit and can severely affect the economics of mining if too wide a cross-section is selected. However, the width must be sufficient to allow safe manoeuvring of trucks. In accordance with traditional pavement width design, each lane of travel should have adequate clearance to the left and right of the widest vehicle making use of the lane. The number of vehicles using the roads are usually such that two lanes are considered to be sufficient without resulting in traffic delays or unsafe interactions.

Ramp roads usually consist of the compacted trafficked way, a less compacted shoulder on each side, a drain on the pit-wall side and the safety berm on the open-pit side. Based on the US guidelines (Kaufman & Ault, 1977), roads for the 100 tonne trucks should be 19.5 metres wide. To this should be added the width of the drain (3 metres) and that of the safety berm (about 6 metres) making a total width of 28.5 metres. Roads of 25 metres width including the drain and berm are considered, however, to be more than adequate for 100 tonne haulers. The larger haulers (> 250 tonnes), however, are 2.3 metres wider, resulting in a potential increase in width of 4.6 metres plus an increased safety margin, bringing the total width to about 35 metres. The extra 10 metre width would result in significant cost implications to the mine.

Based on recent work carried out at Transportek on busy urban streets, and extrapolating this to the mining environment, a proposed minimum width for the new haulers is 29.8 metres. This is illustrated in Figure 5.

A major problem exists in switch-back areas, where the larger trucks turn very sharply around curves with a maximum radius of about 25 metres. Depending on the line of travel of the vehicle, for each pair of tandem wheels, the innermost portion (with respect to the corner) travels a distance between 10 and 20 per cent less than that travelled by the outermost section (a distance of nearly two metres) of the adjacent wheel farthest from the inside of the corner, with no differential movement. As this results in significant abrasion and scuffing of the road, switchbacks should be made as wide as is practically possible to minimise this abrasion.

The cross-fall of the pavement should be maintained at 3 or 4 per cent to ensure rapid run-off of surface water without erosion of the surface layer. Problems with left-hand final drive failures of haulers have been identified. These have been tentatively attributed to the cross-fall on the roads. Evaluation of mine ramp roads has indicated that the possibility of these problems being caused by the high loads on the left side of the vehicle during negotiation of the switchbacks is considered to be of greater consequence. Stresses due to the leaning of the vehicles towards the left on the sharp corners are considerably higher than those caused by the cross-fall. In addition the rougher surface typical of the sharp bends in the switchbacks results in more vehicle bounce and thus in oscillating stresses.

6 DRAINAGE

The removal of water from groundwater seepage, rainfall and dust suppression from the surface of the roads into drains and away from the structural layers of the pavements is imperative for successful performance of the roads. Any ponding of water, with resulting seepage into the road results in significant softening of the material and deformation under load. This has been noted in localised areas along a number of ramp roads, particularly where ponding occurs in the drains adjacent to areas of low compaction in the road, e.g. on the shoulders. In these areas, deformation occurs where vehicles pass over the damp material.

Drains need to be maintained regularly and carefully. They should have no debris or eroded material from the cut face and should definitely not have road gravel and dropped material bladed into them. Maintenance operations should ensure a smooth clean profile which will allow free flow of water without ponding.

During periods of heavy rain, the flow in the drains may be too high for a single discharge at the bottom of the road and culverts beneath the pavement are necessary. For the design and spacing of these culverts an accurate knowledge of rainfall intensity, catchment area and drain dimensions is required. It should, however, be noted that loaded trucks crossing these culverts will cause large stresses around them (the culverts will be loaded by the full axle load and not by a distributed tyre load as is the pavement) and would need to be buried at some depth to prevent collapse. Placement of the drains at an angle to the road would significantly reduce the loads on the culvert.

7 CONSTRUCTION

Construction procedures of most ramp roads cannot be significantly improved on. However, before new

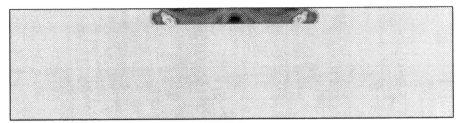

464 MPa W/C over 1619 MPa S/G 93 tonne hauler

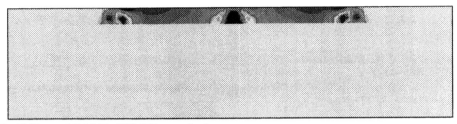

464 MPa W/C over 1619 MPa S/G 375 tonne hauler

580 MPa W/C over 1414 MPa S/G 93 tonne hauler

580 MPa W/C over 1414 MPa S/G 285 tonne hauler

Note: 150mm gravel above blasted bedrock.
Darker areas indicate higher plastic strain.

Figure 4: Horizontal plastic strain under dual rear
tyres

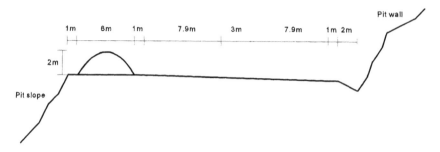

Figure 5: Minimum haulage-way section for proposed haul vehicles

layers of gravel are added to the surface, it should have some preparation prior to dumping of the gravel. The procedure of dumping material on a smooth surface and compacting it under traffic results in the formation of biscuit layers, an undesirable condition. This type of layering of the road is extremely sensitive to fracture under load and to ravelling into plate-like fragments.

The problem can be overcome by scarifying the layer with the tines on the graders to a depth of 75 to 100 mm and by spraying water onto the scarified surface prior to dumping of the gravel. The density of the material on the road may necessitate that some of the tines be removed, in order to allow more power at the remaining tines (often only 2 or 3 can be used at a time). The addition of layers in excess of 50 mm after compaction also helps to alleviate the formation of biscuit layers.

8 MAINTENANCE

Maintenance of ramp roads should concentrate on the following aspects:

8.1 Surface stones
A major cause of tyre damage on haulers is particles which have dropped off haulers onto the roads. The problem of removing this material is best solved by using wheel-dozers. However, prevention of the problem at source is possible by ensuring that loader operators distribute the loads in the haulers carefully and flatten off precariously balanced loads prior to the departure of the hauler. The time involved in this operation is easily off-set by the reduction in time lost and additional costs as a result of tyre penetration. By retaining the road in a smooth condition, vehicle bounce will be limited and the possibility of material falling off the trucks will be reduced.

The addition of a V-shaped rubber scraper in front of the front wheels to move stones to the side of the road (similar to the old "cow catchers" on trains) may be a cost-effective solution.

8.2 Dust suppression
Dust is a significant safety problem as well as a contributor to increased vehicle operating costs. Dust is usually controlled using regular water applications. This is effective but is very costly and can lead to surface deterioration problems if the water ponds in depressions and to slippery conditions if excessive quantities of water are applied. Problems are also encountered when the water tankers (only one is often available at most mines) have mechanical problems and no water can be sprayed - some contingency operation is essential, even if this only takes the form of a small towed tanker with hand spray. Consideration should be given to investigating the use of appropriate dust palliatives which are available in southern Africa. Controlled trials using ligno-sulphonates, hygroscopic salts, emulsified petroleum resins and acrylate polymers are recommended.

8.3 Routine grader maintenance
Routine grader maintenance is necessary to maintain the surface of the road as evenly as possible. In general the maintenance of the roads observed is usually of a high standard, but localised areas with undulations and depressions frequently occur, mainly associated with areas of ponding in the side-drains. In such areas, after rectification of the side-drain problem, the grading operation should smooth the road out by cutting the tops off the crests and scraping this material to waste. If the undulations are severe, the area should be scarified, additional gravel added if necessary and the road levelled before being compacted under traffic. The area should be monitored during compaction and worked with the grader if necessary to ensure that an even surface is obtained.

Waste material cut by the grader should be graded towards the side of the safety berm and not towards the side-drains. Once there is an accumulation of material alongside the berm it should be removed or added to the side of the berm if the width of the road is not affected.

8.4 Gravel replacement

Because of the nature of the pavement structure on most ramp roads (i.e. weak material sandwiched between the tyres and the rock subgrade), the thickness of the wearing course material should be limited to between 150 and 200 mm. As the thickness becomes greater the potential for shearing and plastic flow increases. Gravel replacement should only be carried out to replace material lost under traffic and by grading and not to restore the surface finish, which should be done by grading. When it is obvious that the layer of gravel is becoming too thin, gravel replacement should be carried out.

8.5 Drainage maintenance

It is imperative that all side drains are maintained properly. Effort should be put into carefully shaping them so that they have a smooth even profile at a depth of about one metre below the level of the road surface. The slope from the road to the bottom of the V-drain should be between 1:2 and 1:3. Maintenance should then be carried out regularly to ensure that there is no accumulation of loose material or areas of ponding and particularly that water does not flow into the pavement layers. Maintenance can be carried out with the graders or manually where necessary.

8.6 Safety berms

The use of safety berms on the open-pit side of the road is essential. No alternative to these is likely to be successful in open-pit mines as the mounting structures for traditional road type barriers, such as Armco, need to be founded very close to the pit-wall and are thus unlikely to resist the impact of a large truck.

The height of the berms should be not less than the radius of the wheel bearing in mind that each unit increase in height results in between two and three units increase in width. This either detracts from the road width or necessitates additional excavation of the pit wall.

The use of median berms is considered impracticable because of the extra widening of the road which would be required (at least 5 metres) and because of the lack of straight lengths of road (at least 50 metres) necessary for their effective use.

9 CONCLUSIONS

Observation and evaluation of ramp roads in various open-cast pit mines has indicated that, although these roads are seldom designed in the road engineering sense, their performance is mostly adequate. Experience on the mines has resulted in the optimum use of local materials, although better material selection would in many cases result in an improved road surface which would be easier to maintain. Recommendations for material selection are given.

The behaviour of ramp roads differs from conventional surface haul roads in the weakest part of the pavement structure being sandwiched between a strong subgrade and the applied load, resulting in stress concentrations in the wearing course on ramp roads and not at depth as is typical of surface haul roads.

Analysis using mechanistic and finite element methods has indicated that the stresses developed by large haulers are concentrated in the same areas as those generated by the haulers currently in use and are unlikely to result in significantly greater maintenance requirements.

Efficient drainage of surface water is essential to ensure satisfactory performance of ramp roads and a high degree of good maintenance is necessary to minimise the vehicle operating costs down-time as a result of tyre or mechanical failures.

REFERENCES

Burton, AK. 1975. *Off-highway trucks in the mining industry: Part 1*. Mining Engineering, Vol 27, pp 28-34.

Committee of State Road Authorities (CSRA). 1990. *The structural design, construction and maintenance of unpaved roads*. Pretoria: CSRA, Technical Recommendations for Highways (Draft TRH 20).

Kaufman, WW & Ault, JC. 1977. *Design of surface mine haul roads - a manual*. Washington, DC: US Department of the Interior, Bureau of Mines, (information Circular 8758).

De Beer, M.. 1991. *Use of the Dynamic Cone Penetrometer (DCP) in the design of road structures*. Geotechnics in the African Environment: Ed by G Blight *et al*. AA Balkema, Rotterdam, pp 167-176.

De Beer, M, Fisher, C & Jooste, FJ. 1997. *Determination of tyre/pavement surface contact stresses under moving loads and some effects on pavements with thin asphalt surfacing layers*. Pretoria: Department of Transport (Technical Report TR-96/050).

Jones, DJ & Paige-Green, P. 1996. *The development of performance-related material specifications and the role of dust palliatives in the upgrading of unpaved roads*. Christchurch, New Zealand: 18th Australian Road Research Board Conference, Part 3, pp 199-212..

Paige-Green, P. 1989. *The influence of geotechnical properties on the performance of gravel wearing course materials*. PhD thesis, University of Pretoria, Pretoria.

Paige-Green, P and Bam, A. 1994. *The hardness of gravel aggregate as an indicator of performance in unpaved roads*. Pretoria: Dept of Transport (Research Report RR 93/560).

Paige-Green, P and Bam, A. 1991. *Passability criteria for unpaved roads.* Pretoria: Dept of Transport (Research Report RR 91/172).

Paige-Green, P & Van Huyssteen. 1993. *Pavement design and life-cycle cost modelling for underground roads in coal mines.* Johannesburg: Proceedings of Conference on Cost-effective Transportation of Materials, Minerals and Men in Mines, International Executive Communications.

Thompson, RJ. 1997. *Structural design of surface haulage roads on mines.* PhD Thesis, University of Pretoria, Pretoria.

Thompson, RJ & Visser, AT. 1996. *The structural design of surface haul roads on mines.* Johannesburg: Surface Mining, South African Institute of Mining and Metallurgy, pp 59-69.

Geotechnics for Developing Africa, Wardle, Blight & Fourie (eds) © 1999 Balkema, Rotterdam, ISBN 90 5809 082 5

Static liquefaction of Merriespruit gold tailings

George Papageorgiou, Andy B. Fourie & Geoff E. Blight
University of the Witwatersrand, South Africa

ABSTRACT: Static liquefaction is a concept that describes flow failures, such as those associated with tailings dams. This mechanism of failure is characterised by a sudden and dramatic loss in shear strength, followed by a flow of the liquified material over some distance. Static liquefaction may result from an increase in pore pressure due to a rise of the phreatic surface or an increase in the *in situ* shear stress due to an increase in tailings dam height, among various other processes. However, static liquefaction cannot be triggered unless the *in situ* material is inherently susceptible to a rapid collapse of the soil skeleton, characterised by a rapid and marked increase in pore pressure. The paper introduces the concept of static liquefaction and presents current research being carried out on static liquefaction, with particular emphasis to Merriespruit gold tailings. The paper provides detailed background on static liquefaction; the concepts, testing techniques, factors affecting and influencing test results and its importance and application to flow failures, liquefaction susceptibility and tailings dam safety. Current research being performed on Merriespruit gold tailings in terms of static liquefaction susceptibility is presented, including the effect of variation in grading on the steady state line. A brief introduction to the application and interpretation of static liquefaction susceptibility of actual tailings dams is presented. This is illustrated by applying liquefaction concepts to what is known about the *in situ* state of the Merriespruit tailings dam prior to its flow failure.

1. INTRODUCTION

Liquefaction is a concept which has received much attention abroad. The concept and its application to Southern African geotechnical problems is however limited. Liquefaction is of particular relevance to hydraulic fill structures such as tailings dams, as well as natural and man made embankments. Very little work has to date been done in South Africa to develop the techniques and expertise required to carry out liquefaction testing and the evaluation of liquefaction susceptibility of soils.

It has been shown by numerous researchers Sladen et al (1985), Chillarige et al (1997), Dawson et al (1998) among others that liquefaction is responsible and describes the process occurring in flow slides. Static liquefaction may also have contributed to flow slides that occurred at Arcturus (1978), Saaiplaas (1992) and Merriespruit (1994) gold tailings dams, Bafokeng (1974) platinum tailings dam and Stava (1985) fluorite tailings dam to mention but a few.

There may have been a misconception among South African geotechnical engineers that liquefaction is only associated with seismic events and of materials of a particular grading and that most of the tailings (particularly gold tailings) encountered in Southern Africa are not susceptible to liquefaction. Cyclic triaxial tests performed on cyclone underflow tailings, full plant tailings and cyclone overflow tailings, Blight (1987), revealed that these materials were not susceptible to liquefaction. It was suggested that this was probably due to the angular and irregular shape and the harsh surface texture of the particles. As will be shown further on in the article, the observation of Blight's may have been due to the stress and void ratio state of the samples tested existing in a state/condition not susceptible to liquefaction and not due to the particles characteristics.

The article aims to provide a brief overview of the concepts of liquefaction, the testing techniques for evaluating liquefaction susceptibility, clarifying some misconceptions which may exist regarding liquefaction susceptibility and its application to tailings dams among other geotechnical structures. Liquefaction testing performed on the Merriespruit gold tailings at the University of the Witwatersrand, Civil Department is used to illustrate the application of the liquefaction concepts and liquefaction susceptibility of tailings dams.

2. LIQUEFACTION

Liquefaction is defined as *a phenomenon wherein a saturated material loses a large percentage of its shear resistance (due to monotonic or cyclic loading) and flows in a manner resembling a liquid until the shear stresses acting on the mass are as low as its reduced shear resistance*, Castro & Poulos (1977), Sladen et al (1985).

Liquefaction has been shown to occur in both saturated and unsaturated soils which exist in a metastable state i.e. a stress and void ratio condition above a critical value. For liquefaction to occur a triggering mechanism in the form of a static or dynamic loading must exist e.g. seismic activity, increase in surcharge, a rise in the phreatic surface or the sudden loss of confining stress such as the removal of the embankment of an impoundment. Fully drained conditions can exist in the material up to the point when liquefaction is triggered. Furthermore, a sufficient portion of the material must exist in this critical/metastable state for liquefaction to result in a flow failure.

Liquefaction and liquefaction susceptibility can be determined by monotonic compression or cyclic triaxial loading. For monotonic compression loading, samples can be saturated or dry (e.g. debris flows in deserts) and drainage conditions can be undrained or drained. Although the paper concentrates and presents results performed on fully saturated, consolidated, undrained, compression triaxial tests, the concepts generally apply to all of the above mentioned forms of testing.

Materials subjected to triaxial compression loading can display one of three forms of behaviour, referred to as contractive (liquefaction, strain softening), limited liquefaction (quasi steady state) and dilative (strain hardening) behaviour. These three types of behaviour are illustrated by the stress & pore pressure - strain curves and stress paths for an undrained compression triaxial sample in figure 1. Only material existing in a state which displays contractive behaviour is susceptible to liquefaction and thus of concern with respect to flow failures.

2.1. *Steady state line*

The steady state condition is defined as *that state in which the mass is continuously deforming at constant volume, constant normal effective stress, constant shear stress and constant velocity*, Poulos (1981), McRoberts & Sladen (1992). In terms of an undrained compression triaxial test, the steady state condition represents the point in the shear stress - strain curve, where the shear stress remains constant with an increase in strain, or the terminal point of the stress path, refer to figure 1 curves a.

Liquefaction and susceptibility to liquefaction is evaluated by determining the steady state line for a specific material. This line represents the mean effective confining stress and corresponding void ratio at steady state conditions for a material. The line is fairly linear when plotted in log mean effective stress - void ratio space and represents the division between dilative and contractive behaviour and is unique for a particular material. The type of behaviour displayed by a material and its potential susceptibility to liquefaction is determined by its existing state i.e. the *in situ* void ratio of the particular material and its mean effective stress in relation to its steady state line. Figures 2 & 3 show a typical steady state line together with how a material loaded in compression would behave depending on its state and drainage condition in relation to its steady state line. A material whose state exists above the steady state line will display contractive behaviour and liquefy. Limited liquefaction is displayed by material whose state exists above, but very close to the steady state line. Dilative behaviour is displayed by material whose state is below the steady state line.

The steady state line is independent of sample setup procedure, soil fabric, stress path followed to reach the steady state condition, strain rate, consolidation stresses prior to shear or the mode of loading, Sladen et al (1985), Ishihara (1993), Roberston & Fear (1995). Further it should be noted that the steady state line can only be determined by contractively behaving samples which have reached a steady state condition, McRoberts & Sladen (1992), Desrues et al (1996).

2.2. *Factors affecting the steady state line*

Some material properties influence the position and slope of the steady state line e.g. particle size distribution, shape and fines content (particles below 0.075mm) among others. Of these, the fines content is possible one of the more dominating factors.

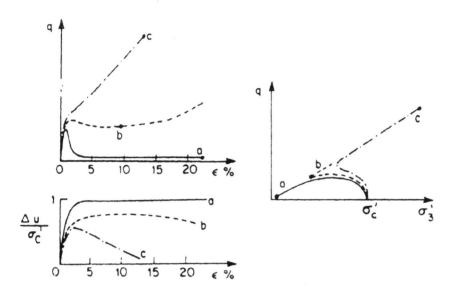

Figure 1. Different types of behaviour displayed by saturated soils during undrained consolidated compression triaxial testing (Konrad 1990), where a - liquefaction, b - limited liquefaction & c - dilative behaviour, $q = \sigma_1 - \sigma_3$, σ'_c = consolidated σ'_3, ϵ = strain and Δu = change in pore pressure

Sladen et al (1985), Pitman et al (1994), Lade & Yamamuro (1997) and Thevanayagam (1998) among others have shown that a change in the fines content of a particular material results in a shift in the position in the steady state line and its susceptibility to liquefaction. The potential for liquefaction is much lower for a clean sand than for a silty sand, all else being equal, i.e. the density required for a given steady state shear strength is much lower for a clean sand than for a silty sand. An increase in the fines content in addition reduces the permeability and increases the compressibility of the material resulting in an undrained response to any loading condition being more likely.

An increase in the fines content tends to decrease the density range achievable by a particular placement technique. This undesirable trend is further compounded by the fact that very high densities are required to produce dilative behaviour as the fines content increases. The result is that the potential for liquefaction is much lower for a clean sand than for one containing fines. Tests have shown that clean sands need to exist at a very low relative density 0%-20% for liquefaction to occur, whereas sand containing about 40% fines may liquefy at a relative density of 60% and greater.

The location of the steady state line for a specific material at a particular mean effective stress moves downwards in terms of the void ratio as the percentage fines increases in the material, but remains parallel to

the other steady state lines for the material. The slope of the line is generally steeper for angular particles and flatter for rounded particles.

Liquefaction and liquefaction susceptibility is therefore a function of a materials *in situ* state relative to its steady state line. Although factors such as particle shape and grading among other factors influence the location of a material's steady state line they do not indicate whether a material is susceptible to liquefaction.

Having provided a brief overview of liquefaction, the remaining part of the article concentrates on the application of the liquefaction concepts to the Merriespruit gold tailings. Liquefaction concepts will also be used to provide an alternative and possibly more plausible explanation for the flow failure which occurred.

3. MERRIESPRUIT FLOW FAILURE

The failure of the Merriespruit gold tailings dam on 22 February 1994, resulted in a flow failure which destroyed 80 houses, severely damaged another 200 and killed 17 people. The failure had severe consequences for the mining and tailings industry. The inquiry following the failure attributed the failure to overtopping of the embankment wall, resulting in the eroding away of the embankment consequently releasing the impounded tailings.

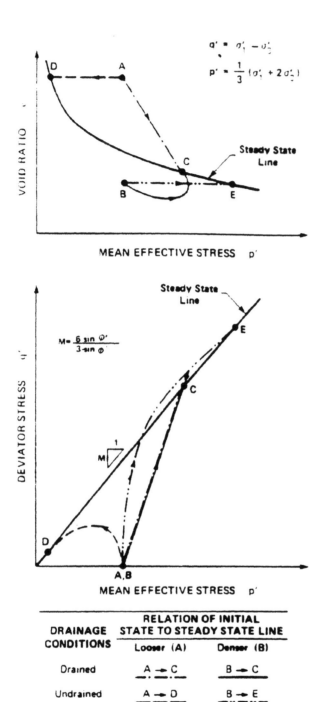

Figure 2. Behaviour of a material in relation to its steady state line and drainage condition (Sladen et al 1985)

Wagener et al (1997) described the process which resulted in the flow failure, comprising of the erosion of the embankment which resulted in steep unstable slopes which failed in a domino fashion moving retrogressively into the tailings dam.

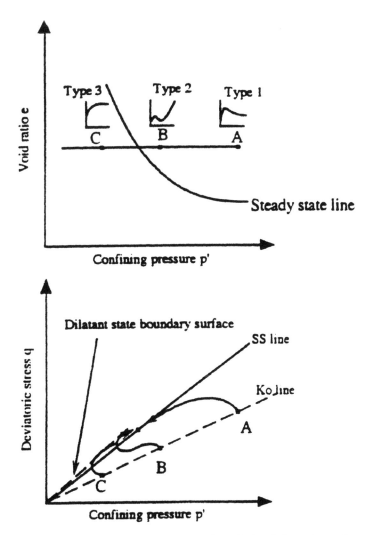

Figure 3. Behaviour of undrained compression triaxial samples in relation to the steady state line (Chillarge et al 1997)

Although at the time of the inquiry, the susceptibility of the tailings to liquefaction was considered, tests revealed that most of the tailings behaved dilatively. However, samples used to evaluate the liquefaction susceptibility during the enquiry investigation were generally in a dense state that could only produce dilative behaviour. Following the failure a research project was initiated to obtain a better understanding of the liquefaction concepts and to develop suitable testing techniques for evaluation liquefaction susceptibility.

3.1. Description of tailings

The first requirement was to establish the variability in the grading of the tailings which existed in the tailings dam. Figure 4 shows the envelope of the particle size distribution existing within the tailings dam. Figure 5 shows the distribution of the grading of the tailings in terms of percentage fines (below 0.075mm). This size was arbitrarily chosen as the boundary between sand and silt-sized particles. These distributions were established from the numerous grading curves obtained during the enquiry. 75% of the tailings graded contained fines of 60% or more. Merriespruit gold tailings contain 60% fines was obtained from the tailings dam and used in the liquefaction testing In addition, an approximation of the *in situ* void ratio and its variability within the dam was also established from field tests conducted during the enquiry into the Merriespruit flow failure, refer to figure 6.

4.   LIQUEFACTION TESTING TECHNIQUE

Static liquefaction testing by means of consolidated, undrained, compression triaxial loading was chosen to establish the steady state line of the gold tailings. As mentioned previously, the steady state line is independent of the testing technique. The moist tamping technique was used to prepare the triaxial test samples. This enables one to easily produce test samples in a state which will behave contractively and thus liquefy when loaded.

All tests were performed in triaxial cells of specimen size 38mm x 76mm. Sun dried tailings was mixed with a predetermined amount of distilled water, 10% by dry mass. Once the required setup void ratio of the test samples was determined, the total bulk mass of the test sample was calculated and divided into five equal masses. A perspex split mould with markings every 15.2mm (1/5th of the 76mm sample height) was used around the triaxial pedestal to assist in the setup of the test samples.

Each fifth of soil mass was placed into the perspex mould and tamped to the corresponding height marking. It should be noted that due to the low compaction energy required for each layer, the underlaying layers don't experience any additional densification, enabling one to use equally spaced layers. Sladen et al (1993) showed that the moist tamping technique produces samples that are uniform in terms of void ratio and moisture content throughout the sample.

Once the final layer was prepared the latex membrane, porous stone, filter paper and top cap were placed onto the test sample. The dimensions of the sample were then measured by vernier callipers.

The drainage outlet valves to the triaxial cell were closed and the triaxial chamber was carefully secured into position. A volume gauge was connected between the triaxial cell chamber valve and the cell pressure regulator. Cell water was allowed to flow under a very small head of about 5kPa into the triaxial cell until filled. The air relief valve at the top of the triaxial cell was kept open during the water filling stage. Once water flushed out of the air relief valve, the cell chamber valve was closed, after which, the air relief

valve was closed. This order of events was followed so as to avoid applying any stress to the test sample, thereby resulting in an undetected/unmeasured change in the sample volume/void ratio. During the cell chamber filling operation it was also ensured that the loading ram was locked in the fully extended position.

An initial zero volume gauge reading was taken prior to the cell pressure being increased to between 20kPa and 50kPa, while the pore pressure was increased to 10kPa less than the cell pressure. The drainage valve connected to the top of the sample was initially not connected to the pore water pressure regulator, but in a closed position. Upon applying the initial stresses the test sample consolidated/collapsed to some extent. Once this collapse had stabilized the drainage valve connected to the top cap was gradually opened to atmospheric pressure.

By opening the valve in small increments and over a period of time any additional collapse in the test sample structure due to the moving water front and escaping air through the sample was minimised. Water was allowed to flush through the sample for approximately 1-2 hours. The drainage valve to the top cap was then closed and connected to the pore water pressure source.

The pore water pressure was then maintained at 10kPa less than the cell pressure (an effective pressure of 10kPa) and saturation of the sample commenced. Every 24 hours a *B* value test was conducted until a value of 0.95 or higher was attained at which point the sample was assumed to be 100% saturated. A volume gauge reading was then taken prior to consolidating the sample to the required effective stress by increasing the cell pressure and consolidating against the back pressure. A final volume gauge reading was taken once the sample had consolidated. The pore water drainage valves were then closed and the test sample loaded.

By measuring the volume of cell water flowing in or out of the cell chamber during the flushing, saturation and consolidation phase of the test, the change in volume/void ratio of the sample during any stage of the preparation phase of the test could be determined. It should be noted that the cell chamber was also calibrated for expansion which would occur with the application of the cell pressure, and this was accounted for when determining the test samples void ratio.

5.   LIQUEFACTION TEST RESULTS

Figure 7 shows the shear stress - strain plots for the Merriespruit gold tailings triaxial tests. In all cases the

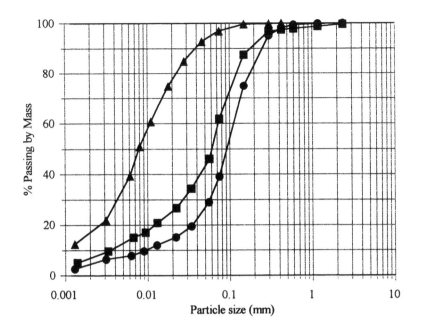

Figure 4. Merriespruit gold tailings grading curves

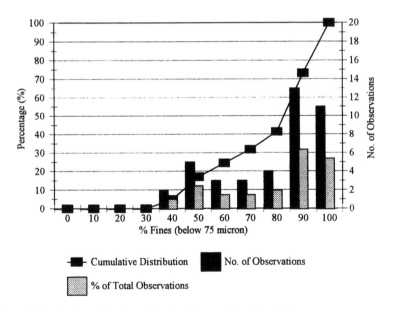

Figure 5. Distribution of Merriespruit gold tailings as a function of fines content

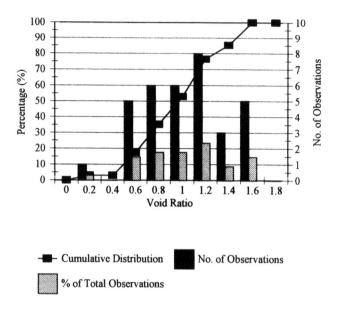

Figure 6. Distribution of Merriespruit gold tailings *in situ* void ratio

peak shear stress was reached within the first 5% strain, after which the shear stress drops to a minimum, steady state value between 12% and 17% strain. In some tests the shear stress increased slightly after the 20% strain mark. This apparent increase in shear stress with increasing strain is most likely due to the barrelling effect which is displayed by the test samples at these high strains. The pore water pressure of the relevant test samples do not decrease at these high strains which would be expected if the samples behaved dilatively, figure 8.

Figure 9 shows the stress paths for the various tests, which display typical contractive behaviour. The stress paths for all the tests reach a peak shear stress after which they decrease until they reach a terminal, steady state shear stress. The peak shear stress for the individual tests plot onto a single line referred to as the instability line. This line represents the point at which liquefaction is triggered. Any loading situation which would result in the stress path moving beyond the instability line would result in liquefaction failure. The portion of the stress path curves between the peak shear and steady state shear stress defines a boundary beyond which no stress condition for the material exists, the state boundary

The steady state line is shown in figure 10 for the Merriespruit gold tailings. In addition, the steady state line for Merriespruit gold tailings containing 20% fines (artificially graded from the Merriespruit gold tailings) is shown. The steady state line for the tailings containing 20% fines plots in a higher location in the steady state diagram relative to the tailings containing 60% fines. Both steady state lines however are parallel to each other. This highlights the effect fines content has on the location of the steady state line and the liquefaction susceptibility of a material, all else being equal. It also shows that for situations where the liquefaction susceptibility of structures such as tailings dams are to be determined one cannot use a single steady state line. Instead an upper and lower state line corresponding to the range of the coarsest and finest material found on the tailings dam should be used.

6. LIQUEFACTION SUSCEPTIBILITY OF THE MERRIESPRUIT GOLD TAILINGS DAM

Figure 5 shows the distribution of the Merriespruit gold tailings with respect to its fines content. 75% of the tailings samples contained 60% or more fines. Figure 6 shows the *in situ* void ratio distribution of the tailings, with 65% existing at a void ratio of 0.8 or higher.

By superimposing the range of *in situ* void ratios and the likely range of *in situ* confining/horizontal stresses for the Merriespruit gold tailings, its liquefaction susceptibility can be determined

The shaded area in figure 10 represents 80% of the *in situ* void ratios determined during the

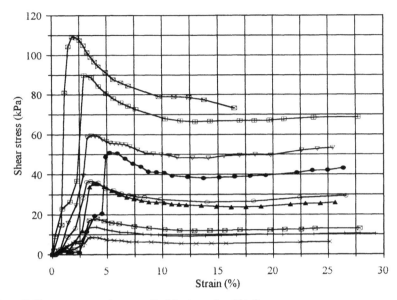

Figure 7. Shear stress - strain curves for Merriespruit gold tailings

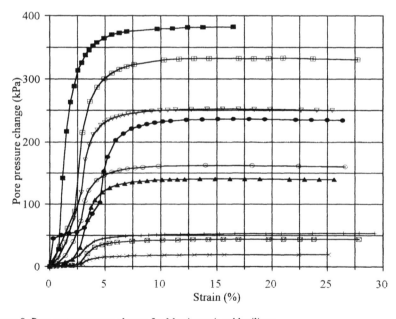

Figure 8. Pore water pressure change for Merriespruit gold tailings

Merriespruit enquiry investigation. The steady state lines shown in figure 10 indicate that the steady state line is located at a lower void ratio range as the fines content of the tailings increases. This suggests that the steady state line corresponding to 75% of the Merriespruit gold tailings (at a fines content greater than 60%) would be located somewhere below that of the 60% fines line. If one was to however use the Merriespruit 60% fines steady state line to represent the steady state condition for all the tailings containing

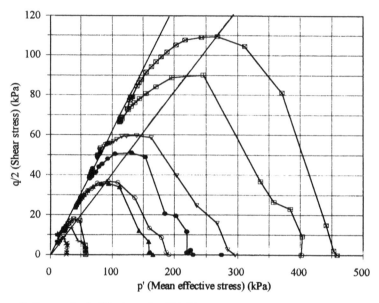

Figure 9. Stress paths for Merriespruit gold tailings

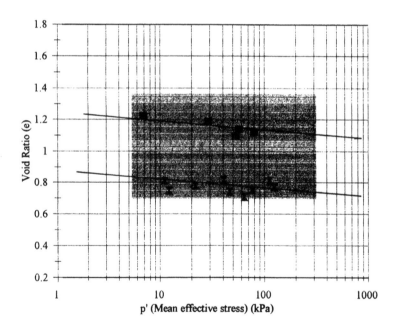

X Merriespruit 60% fines   ■ Merriespruit 20% fines

Figure 10   Steady state diagram for Merriespuit gold tailings

60% or more fines at least 50% to 60% of the *in situ* samples existed in a contractive state i.e. above the 60% fines steady state line.

Applying this to the Merriespruit tailings dam reveals that a large portion of the tailings dam, in which the *in situ* void ratios were sampled was susceptible to liquefaction. With this in mind the Merriespruit flow failure could have occurred as follows. Once the embankment wall was breached, the horizontal stress was reduced while the vertical stress remained unchanged.. This would have resulted in the potentially liquefiable, contractive behaving tailings moving toward the steady state line in an undrained manner and possibly liquefying. This would have resulted in a series of shallow retrogressively moving slips as a result of liquefaction and reduction in shear stress. The reason for the retrogressive shallow circles is the time lag which occurs once the confining stress is removed, during which the pore pressure increases, until liquefaction is triggered.. This process will continue until dilative or sufficiently less contractive tailings is encountered within the dam.

7. CONCLUSIONS

- Liquefaction is a concept which describes the process that occurs during flow failures, such as those associated with tailings dams.
- Liquefaction and liquefaction susceptibility are functions of the material's *in situ* state relative to its steady state line.
- The fines content (defined as particles below the 0.075mm sieve size) has a marked effect on the location of the steady state line of a material in the steady state diagram, all else being equal.
- The higher the fines content within a material the higher the relative density required to reduce its liquefaction susceptibility.
- The steady state line in conjunction with the *in situ* state of a material can be used to establish its potential liquefaction susceptibility.

8. ACKNOWLEDGEMENT

The authors wish to acknowledge the financial support made available for the research by the Foundation for Research and Development as well as Fraser Alexander Tailings.

9. REFERENCES

Blight G E (1987) "The effect of dynamic loading on underground fill in South African gold mines" Victor de Mello Volume1 pp 37-44.

Castro G & Poulos SJ (1977) "Factors affecting liquefaction and cyclic mobility" Journal of the Geotechnical Engineering Division, Proceedings of the ASCE, Vol 103, No GT6, June, pp 501-516.

Chillarige ARV, Morgenstern NR, Robertson PK & Christian H (1997) "Evaluation of the in situ state of Fraser river sand" Canadian Geotechnical Journal, Vol 34, pp 510-51.

Dawson RF, Morgenstern NR & Stokes AW (1998) "Liquefaction flowslides in Rocky Mountain coal mine waste dumps" Canadian Geotechnical Journal, Vol 35, pp 328-343.

Desrues J, Chambon R, Mokni M & Mazerolle F (1996) "Void ratio evaluation inside shear bands in triaxial sand specimens studied by computed tomography" Geotechnique, Vol 46, No 3, pp 529-546.

Ishihara (1993) "Liquefaction and flow failure during earthquakes" Geotechnique, Vol 43, No 3, pp 351-415.

Konrad JM (1990) "minimum undrained strength versus steady state strength of sands" Journal of Geotechnical Engineering ASCE, Vol 116, No 6, June pp 948-963.

Lade PV & Yamamura JA (1997) "Effects of nonplastic fines on the static liquefaction of sands" Canadian Geotechnical Journal, Vol 34, pp 918-928.

McRoberts EC & Sladen JA (1992) "Observations on static and cyclic sand liquefaction methodologies" Canadian Geotechnical Journal, Vol 29, pp 650-665.

Pitman TD, Robertson PK & Sergo DC (1994) "Influence of fines on the collapse of loose sands" Canadian Geotechnical Journal, Vol 31, pp 728-736.

Poulos SJ (1981) "The steady state of deformation" Journal of Geotechnical Engineering Division ASCE, Vol 107, pp 553-562.

Robertson PK & Fear CA (1995) "Liquefaction of sand and its evaluation" IS Tokyo'95, First International Conference on Earthquake Geotechnical Engineering, Keynote lecture.

Sladen JA D'Hollander RD & Krahn J (1985) "The liquefaction of sands, a collapse surface approach" Canadian Geotechnical Journal, Vol 22, pp 564-578.

Thevanayagan S (1998) "Effects of fines and confining stress on undrained shear strength of silty sands", Journal of Geotechnical and Environmental Engineering, Vol 124, No 6, June, pp 479-491.

Wagener F, Strydom K, Craig H & Blight G (1997) "The tailings dam flow failure at Merriespruit, South Africa, Causes and Consequences" Conference on Tailings and Mine Waste 1997, Colorado State University, January 1997.

Environmental geotechnics for an outfall sewer traversing a Ramsar wetland site

P.H.Oosthuizen & A.L.Parrock
ARQ Specialist Engineers, Pretoria, South Africa

A.N.Cave
Cave Klapwijk and Associates, Pretoria, South Africa

ABSTRACT: The ERWAT DD5A Regional Outfall Sewer traverses an area alongside the Blesbokspruit; an internationally proclaimed Ramsar wetland site. This paper describes analysis methods, design, risk studies, construction control and post-construction monitoring of the pipeline in accordance with an Environmental Management Plan, to ensure compliance with strict controls imposed by Authorities.

Deep percussion holes to bedrock, combined with a risk analysis of spillage were used to predict the reliability of the system.

Multi-point piezometers installed in and beneath a geotextile enclosed bedding layer, enabled monitoring and testing of ground water to be undertaken.

Laboratory testing of locally available materials which did not satisfy conventional specifications, combined with finite element analyses simulating envisaged construction sequences, enabled low cost materials to be specified with confidence. Those portions of the line installed at depths up to 10m, required detailed analysis and design to ensure safety of workpersons during pipeline installation.

1 INTRODUCTION

The proposed regional outfall sewers are required as part of the optimal regional system development to meet ongoing demand growth and will feed wastewater to a new regional water care works to be established in the vicinity.

2 BACKGROUND

2.1 ERWAT

The East Rand Water Care Company (ERWAT) was established for the purpose of co-ordinating, centralising and optimising the management, conveyance and treatment of wastewater generated within the various municipalities that comprise the East Rand. The overall objective of ERWAT is the realisation of the lowest unit cost for wastewater management. ERWAT is responsible for bulk conveyance of wastewater and subsequent treatment. Internal networks remain the responsibility of the local authority.

3 ENVIRONMENTAL APPROACH

The environmental assessment for the project was developed in a phased approach as advocated by the relevant environmental authorities. Initially a "Preliminary Site Assessment Report" was drawn up and thereafter an "Initial Environmental Examination" report was compiled in three phases:
- Stage One: A scoping exercise to identify potential issues of concern; reported in an Initial Environmental Examination (IEE) report.
- Stage Two: The preparation of an Initial Environmental Impact Report (IEIR) to identify the critical environmental impacts.
- Stage Three: The preparation of an Environmental Management Plan (EMP)

The most critical aspect identified during the IEIR was the possible deterioration of seepage water quality into the wetland as a result of a pipe leak or break. This would affect water quality in the stream/river, fauna, aquatic life and vegetation. It could ultimately have put the Ramsar site at risk of de-proclamation if the leaked volume was high and prolonged and of very low quality. Specific details which required address were:

- A strategy was to be developed for handling the increases in effluent flow resulting from rain events in the catchment. This strategy was based on a comprehensive risk assessment of spillage and overflow events.
- A strategy was to be developed for handling of the incoming effluent when the pipeline needed to be closed for whatever reason, be it a break or to clear a blockage.
- The quality of ground water was to be monitored before, during and after the project along the pipeline route. This required that measuring points be installed during construction.
- A suitably qualified person was to be appointed to monitor and report on the environmental management and controls as well as the rehabilitation procedures of the project during construction.

The above environmental rehabilitation and control specifications were to be included in contracts with the Pipeline Contractor.

4   RISK STUDY

The largest risk of environmental pollution was identified in the IEIR as the possibility of flow in excess of pipe capacity being generated as a result of infiltration to the input pipelines in the upper reaches of the catchment area during wet periods.

In addition blockages which may have occurred would also lead to spillage from the line to the Blesbokspruit and/or recharge of the dolomite areas located subsurface of the pipe with untreated waste water.

The risk potential of the above occurrences is addressed hereunder.

4.1   Background

The sizing of the DD5A Regional Outfall Sewer (ROS) pipe, was a complex and involved process, involving not only the projected future capacity but also the projected risk of capacity exceedance in wet weather conditions.   As the area is extremely sensitive to environmental pollution and as this is likely to become more so in future, a certain conservativeness was built into the process.

4.2   Risk of pipe capacity exceedance

The layout as detailed in Figure 1 pertained at the inlet works at the head of the proposed Regional Outfall Sewer (ROS).

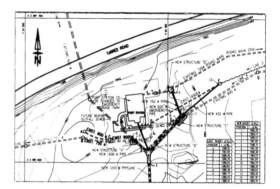

Figure 1: Layout at the head of the ROS

Waste water generated in Daveyton and Etwatwa drained under gravity through input pipelines designated Lines 1,2,3 and 4, was pumped via a rising main to the Daveyton Water Care Works (WCW) where it was treated together with the waste water from eastern Etwatwa (not included in Figure 1 but designated line 5). Once treatment was complete, discharge to the upper reaches of the Blesbokspruit was undertaken.

The capacity of the Daveyton WCW in 1997 was 16Ml/day. According to data collected, line 5 generated some 6Ml/day with the rest being pumped via the rising main to the works. Three pumps, each with a capacity of 12,5Ml/day were installed at the pump station located on the western edge of the Blesbokspruit.

From an environmental point of view it would be necessary to maintain some of the waste water purification capacity at Daveyton for later discharge to the Blesbokspruit. If the Daveyton WCW were closed and all flow diverted to the new ROS, the upper reaches of the spruit would dry out with vast negative environmental impacts.

Thus during the preliminary design of the new ROS the idea was mooted that the Daveyton WCW should remain open, processing the flow from line 5 and discharging treated water to the Blesbokspruit. The rising main was planned to be used in reverse only in an emergency situation, such that any flow from line 5 in excess of the capacity of the Daveyton WCW would by-pass the works and be diverted via the old rising main to the new DD5A ROS.

However, if the flow generated during peaks, exceeded the capacity of the new DD5A ROS, then spillage would occur and untreated waste water would be discharged to the Blesbokspruit.

To predict the risk of the above occurrence the following analysis was undertaken:

1) The expected growth of the waste water draining via line 5 to the works was calculated to increase from the 1995 average daily flow of some 6Ml/day to 8Ml/day at the end of the planning horizon in 2015.

2) The expected growth of the region whose waste water was drained via lines 1,2,3 and 4, and the construction of other ROS's which will drain to the diversion works at the head of the DD5A ROS will increase from the 1995 average daily flow of some 8Ml/day to 35Ml/day at the end of the planning horizon in 2015.

3) Complete hourly flow records, for the period 1 January 1996 to end April 1997, were obtained for the Daveyton works. This data was however only available in graphical format and also unfortunately contained certain periods when no measurements were made. Data before the latter date, ie 1 January 1996 was also not available in the same hourly format and only monthly flow records with minimum, maximum and average values were available.

4) Graphical data can not be used in a statistical analysis. This hourly data was thus digitised in a CAD package, exported to a .DXF file which was interrogated by a purpose-written program such that only x,y (hour,flow) data was obtained in ASCII format.

5) This approximately 15 months of data was then "grown" parabolically for every month of the year, by a factor of 35/14 to reflect the growth from the present flow of 14Ml/day to the predicted flow of 35Ml/day in 2015.

6) The above flows were thereafter interrogated for the number of occurrences when the predicted peak flows exceeded the pipe capacity. Table 1 details the results obtained:

The above process, conducted for predicting the number of occurrences in which the peak flow exceeded the 100% full capacity of 1 222l/sec for a 1 200mmϕ 50D concrete unit manufactured to the minimum diameter specifications of SABS 677 (1986), was also conducted at the points where the pipeline was of 1 350mmϕ and at 1 500mmϕ with the associated additional inflows through these sections. In all cases the smallest pipe generated the most critical condition.

If it is considered that the above occurrences represent the number of occasions when approximately one hours flow would exceed the capacity of the pipe, then it is obvious that the total chance of non-conformance is very small. (A year

Table 1: Predicted number of occurrences in which peak flow exceeds pipe capacity

Year	Month	Predicted number of occurrences in which peak flow exceeds pipe capacity
2000		Zero
2001		Zero
2002		Zero
2003		Zero
2004		Zero
2005		Zero
2006		Zero
0667		Zero
2008		Zero
2009		Zero
2010		Zero
2011	Jan	1
	Feb	2
2012	Jan	1
	Feb	2
	Apr	2
2013	Jan	2
	Feb	2
	Apr	6
2014	Jan	2
	Feb	3
	Apr	8
2015	Jan	2
	Feb	4
	Apr	10

contains approximately 9 000 hours and thus the 16 occasions in which overflow is predicted in the final design year of the pipeline, represents a risk of only some 1:600). This figure was considered acceptable when the option and associated cost of an increase in pipe size was evaluated .

4.3 *Risk of exceedance of storage capacity due to blockage.*

The proposal for the inlet works was that the old pump station would be converted to a storage facility capable of accommodating some 6Ml.

Due to the fact that it is envisaged that the Daveyton WCW will remain in operation to augment the natural flow to the upper reaches of the Blesbokspruit, certain personnel will man the station during normal working hours (08:00 to 17:00). Thus if a blockage in any part of the DD5A ROS were to occur, the maximum amount of time that this blockage would remain undetected is postulated at 15 hours (17:01 after the end of work on day 1 to 08:01 to the beginning of work on day 2). At the end of the planning horizon (ie 2015) the base flow in the pipe would be some 35Ml/day or some

1,5Ml/hour. Thus simplistically over a 15 hour period a total waste water volume of 22,5Ml would need to be accommodated if spillage were to be avoided.

However if a more detailed analysis is undertaken utilising the theoretical flow hydrograph then some 17,5Ml of waste water was calculated would require accommodation.

The pipe itself could theoretically generate some 8Ml of storage while the holding storage tank at the inlet works is planned for an additional 6Ml. Thus in this extreme situation theoretically some 3,5Ml would be spilled. It is however emphasised that the chances of blockage in the system itself will be extremely remote due to the size of the line and the inaccessibility to the manholes.

An additional factor bearing on the situation was that by the year 2015, the drainage systems in the upper catchment areas of DD5A will no doubt have been upgraded and improved to greatly decrease the wet weather penetration to the pipes. This small volume of spillage was considered to be acceptable.

4.4 *Risk of dolomite recharge with untreated waste water*

If the ROS were to discharge to the environment then the possibility exists that pollution of the subsurface formations may occur. If these formations were dolomitic and the water pressure differentials were such that negative pressure existed within the water in the dolomite bedrock, then an extremely pessimistic situation would occur as the pollutant would be drawn into the dolomitic water. The study as detailed below evaluates this possibility.

No traces of dolomite and/or chert were found during the geotechnical investigation for the route. However this investigation was confined to the near-surface alluvial deposits (maximum depth 4m). Available geological maps indicate the area to be underlain at depth by dolomite.

Percussion drilling was thus undertaken to confirm/establish the depth to dolomite bedrock. Three holes were drilled along the route of the ROS: one at the start near the Daveyton WCW pump station, the second near the N12 crossing and the third near Welgedacht. Four holes were also drilled at the proposed site of the new RWCW at Welgedacht. See figures 2 and 3.

The profiles of these holes may be summarised as follows for risk of pollution to the water of the dolomitic bedrock due to a break in the ROS.

Hole 1 near the start of the ROS at Daveyton

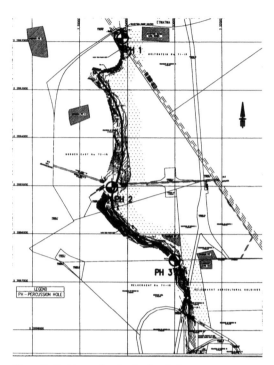

Figure 2: Positions of evaluation boreholes

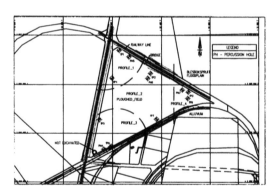

Figure 3: Positions of evaluation boreholes

was characterised by clayey alluvium to a depth of 11m, shale to 45m and thereafter chert (which is usually indicative of dolomite at depth) to 50m.

The clayey alluvium, due to its very low permeability and the shale, due to its density, would ensure practically zero chance that any pollutant would be able to migrate to depth.

Hole 2 near the N12 did not encounter any dolomite to 50m depth, with soft rock coal being present from 14 to 50m. For the same reason as in hole 1 there is practically zero chance that any pollutant would be able to migrate to depth.

Hole 3 near Welgedacht encountered alluvium to 7m depth and thereafter dolerite to 21m depth. There is little chance of a pollutant reaching dolomite bedrock under these circumstances; dolerite being a very dense material.

Three of the four holes drilled at the triangular site of the RWCW encountered dolomite at depths which varied from 4.5 to 9m.

As 1) positive water pressure gradients were found and, 2) as the area will be subject to extensive near-surface improvement due to the RWCW being located there, the risk of pollution of the subsurface water was assessed as LOW.

Notwithstanding the comments as above, it was specified in the design of the pipeline and the accompanying Environmental Management Plan (EMP) that regular monitoring of the ground water near to and below the pipe was to take place.

Piezometers as detailed in Figure 4 were designed such that the quality of water contained in the pervious material surrounding the pipe itself could be monitored. As regular monitoring is specified, any leakage would immediately be identified and remedial actions to inhibit leakage instituted. In addition the quality of ground water in the deeper sections of the profile is also monitored, providing a mechanism for assessing that if leakage had occurred, to what extent had it penetrated the insitu profile. Thus the risk of: a) overflow of the pipeline due to excessive inflow, is small, b) overflow of the pipeline occurring for prolonged periods due to a blockage is small, and c) infiltration of pollutant to the deep ground water is small. Even if c) were to occur it would be quickly detected and the leak repaired. It was thus reasoned that all reasonable precautions had been taken in the design process and in the monitoring systems specified to ensure that little or no damage to the environment would occur due to the pipeline being built to the standards specified.

5 TRENCH STABILITY

The pipe route traverses many geotechnical profiles over its total length of some 20km. Alluvial clays and high water tables were encountered in the floodplain adjacent to the Blesbokspruit, where highly slickensided surfaces in the alluvial profile lead to many small failures if trenches were excavated vertically. In contrast to this, dense silty sand profiles in other locations enabled trenches of 6m depth to be excavated vertically with a large margin of safety.

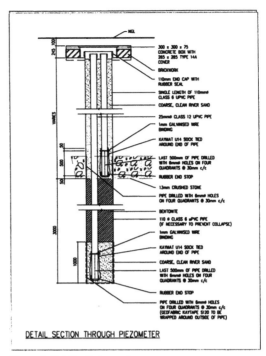

DETAIL SECTION THROUGH PIEZOMETER

Figure 4: Details of piezometers

Two of the most interesting situations however arose where very deep ($\pm$ 10m) excavations were required in an area where the profile was characterised by very soft rock.

Workperson safety is of paramount importance and in order to assist the Contractor in deciding on the excavation profile to be adopted, a finite difference analysis was performed as the first stage in an evaluation of safety.

5.1 Derivation of shear strength parameters

From a visual examination of the insitu profile and using the formulations contained in Hoek (1995) for the derivation of the Geological Strength Index (GSI), the Rock Mass Rating (RMR_{76}) as advocated by Bieniawski (1976), generated representative RMR_{76}-values as depicted in Table 1.

For RMR_{76}-values $>= 18$, the Geological Strength Index (GSI) = RMR_{76}. Using the formulations contained in Hoek (1995) and the spreadsheet HOEK95 encoded to calculate those formulations, the input parameters as contained in Tables 2 and 3 were used to derive the Hoek Brown and Mohr Coulomb shear parameters of the material as detailed in Table 4.

Using the computer program EXAMINE and the Hoek-Brown shear parameters as detailed in

Table 2: Bieniawski's (1976) Rock Mass Rating (RMR$_{76}$)

Variable	Value
Strength - UCS = 0.7-3.0MPa	0
RQD = < 25%	3
Joint spacing = < 50mm	5
Joint condition - Soft gouge	0
Water	10
TOTAL	18

Table 3: Hoek's (1995) input parameters for derivation of shear strength

Variable	Value
Geological Strength Index - GSI	10
Diamictite - m$_i$	17
UCS (MPa)	1.85

Table 4: Hoek Brown and Mohr Coulomb shear parameters

Variable	Value
m	0.909
s	0.000
a	0.560
c (kPa)	19.8
ϕ (°)	31.6
γ' (kN/m^3)	20.0

Table 4 above, the safety factors in graphical format for two trench shapes as depicted in Figures 5 and 6 were obtained.

Examination of these two figures clearly indicates the benefits of the battered slopes as opposed to vertical faces. Observations during field trials confirmed the absence of any indicators of instability. These 10m deep sections were excavated, pipes inserted, bedded and backfilled without any incident.

The sloping profile as depicted in Figure 6 contrasted very markedly with that deemed to be the most effective in excavations conducted in the alluvial deposits alongside the Blesbokspruit. Here the excavation profile as depicted in Figure 7 to a maximum depth of 4m was selected after field trials.

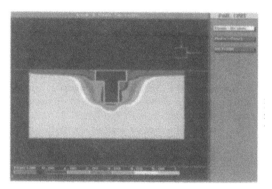

Figure 5: Contours of safety factor

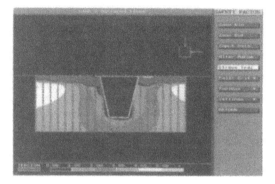

Figure 6: Contours of safety factor

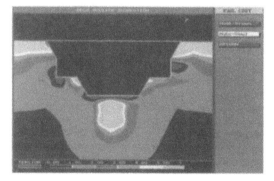

Figure 7: Safety factors for Link 1 route

6   PIPE INTEGRITY

An analysis was performed where the pipeline traverses areas where the depth of installation exceeded calculated "design" depths based on the D-load classification of the concrete pipe. Non-linear finite element techniques were used to assess and compare stresses and deflections in the pipe wall subject to different load combinations. "Load relief measures" were also assessed.

6.1 Design parameters

For the purpose of the non-linear finite element analysis the soil parameters were quantified in terms of the hyperbolic stress/strain parameters - Duncan et al (1980). The derivation of these parameters is based on calculations and results obtained from triaxial soil testing. The theory and derivation of these parameters will not be discussed in this paper. The soil tests undertaken for the derivation of the parameters is given in Table 5.

6.2 Modelling

A two-dimensional finite element grid was constructed using the programs AUTOMESH (1991) and SOILSTRUCT (1990) to model a cross section through the trench excavation, placement of bedding and pipe and subsequent backfilling. A concrete pipe with inside diameter 1 600mm and wall thickness of 150mm was used throughout the analyses discussed here.

At the base of the mesh the elements were restricted against movement in the X and the Y directions. The sides of the mesh were restricted against movement in the X direction and the mesh extended to a distance such that the deformation at the edges was practically zero.

Construction steps. As the purpose of the analysis was to determine the stresses and deformations in the concrete pipe walls induced by the placement of the backfill and NOT to determine the stability of the trench excavations, an initial step was specified consisting of the completed trench excavation. Subsequent construction steps consisted of the placement of a bedding layer, concrete pipe, a straw bale layer and backfill in layers until the trench was filled (a total of 7 construction steps).

Analyses. The concrete pipe, straw bales and concrete cradle were modelled as linear elastic materials. As such a modulus value (E), Poisson's ratio value (ν) and density (γ) were defined for these materials. All soils and bedding were modelled as non-linear, inelastic materials whose properties were characterised by the Duncan et al (1980) hyperbolic stress/strain parameters.

Different analyses were performed to quantify the effect of various parameters on the stresses and deformations in the concrete pipe. The parameters investigated included:
* different backfilling materials
* different pipe invert levels
* the placement of straw bales above the pipe to determine the effect on the pipe
* different dimensions of straw bales

Table 5: Soil tests performed for the derivation of hyperbolic parameters

Test description	To simulate
Direct shear test Slow/fast on concrete	Soil/concrete interface action
Triaxial test Unconsolidated/undrained	Short term conditions
Triaxial test Unconsolidated/undrained	Short term conditions
Triaxial test Consolidated/drained	Long term conditions
Triaxial test Consolidated/drained	Long term conditions
Triaxial test Consolidated/undrained	Used to derive the K_0-value*
Triaxial test Consolidated/undrained	Used to derive the K_0-value*

* K_0 - Coefficient of lateral earth pressure at rest

* different straw 'stiffness'
* the effect of the placement of a geotextile wrapped layer above the straw bale
* the effect of variance in the stiffness of the geotextile wrapped layer
* the effect of a concrete cradle as bedding medium
* the effect of different dimensions of concrete cradles

Output. For each analysis the predicted deflection of the pipe wall at the top crown and bottom crown were measured. These two values were used to calculate the change in inside diameter of the pipe. Furthermore the maximum and minimum, major (S1) and minor (S3) principal stresses were noted for the 16 pipe elements.

6.3 Observations

Differing backfill materials. An initial investigation concentrated on the effect of differing backfill materials as modelled by the hyperbolic parameters calculated in the cases of the Saturated Consolidated Drained (S_{CD}) triaxial test (simulating long term conditions) and the Natural moisture content, Unconsolidated, Undrained (N_{UU}) (simulating short term conditions) triaxial test. The worst case scenario as indicated by maximum pipe deflections and highest S1 and S3 stresses in the pipe elements were used throughout the rest of the analyses. As could be expected, the long-term conditions as modelled by the S_{CD} rendered the most critical case.

Straw bales. The various analyses with different straw bale configurations and dimensions suggested that the placement of straw bales above the pipe had a positive effect in that the stresses in the pipe elements were reduced with an accompanying decrease in pipe deflection. Theoretically it was concluded that the thicker the layer of straw above the pipe, the lower the stresses in the pipe. Doubling the thickness of the straw bale from 688mm to 1 350mm reduced the maximum S1 principal stress by nearly 10%. A further increase in straw thickness to 2 030 mm further reduced the S1 principal stress by an additional 10%. This phenomenon obviously has a practical limit and a standardised straw thickness of 680mm was used for further analyses.

It was noted that the "stiffness" of the straw bale (as determined by the elastic modulus value) had a marked effect on the stresses in the pipe. Merely modelling a void filled with air instead of straw reduced the maximum principal S1 stress in the pipe by 31%. Obviously the stiffness of the straw bale will be somewhere located between the stiffness of the soil and that of the air. An E-value of 5MPa was assumed for the elastic modulus of straw and was used throughout subsequent analyses.
Geotextile wrapped layer. The effect of the installation of a geotextile wrapped, stiffer layer above the straw bale was also investigated. This installation did not produce the desired results and in fact increased the stresses in the pipe. The placement of a 350mm thick "stiffened" geotextile wrapped layer above the straw bales actually increased the maximum principal S1 stress in the pipe by 4%. An increase in the stiffness of this geotextile wrapped layer (as modelled by an increase in the non-linear hyperbolic parameter K) further worsened matters.

This possibly may be explained by the fact that the placement of a stiff layer above the pipe and the straw bale inhibits the arch forming effect of the backfill material as it settles with much more direct downward force being applied to the pipe with increasing stresses.
Concrete cradle. The effect of a concrete cradle as a bedding medium for the pipe was modelled at a pipe invert level of 14,0 m. A layout with no cradle was compared to a cradle supporting the bottom 90° segment of the pipe as well as a cradle of 180°. The introduction of the 90° concrete cradle did not have a significant effect on the compressive stresses in the pipe wall but a slight increase in the tensile stresses was however predicted. The 180° concrete cradle configuration reduced the principal

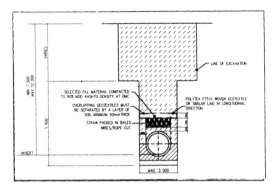

Figure 8: Layout adopted over deep sections

compressive stresses by 25% but the principal tensile stresses nearly doubled.

The layout eventually adopted to ensure acceptable stresses on the pipe was as given in Figure 8.

7   CONCLUSION

Risk studies conducted to quantify risk of overflow generated very low values which were deemed to be acceptable if certain mitigating measures were instituted.

The design and installation of monitoring piezometers at manhole positions along the entire length of the route of the pipe line will enable any leakage to be detected without delay.

Stability proposals enabled the most cost effective excavation profile to be generated with undue risk to workpersons.

A parametric study was conducted using finite element analysis techniques to determine the stresses and deflections of the ERWAT concrete sewer pipe subjected to different loading scenarios. By comparing the predicted stresses at the centroids of the pipe wall elements and the deflection of the pipe as a unit, conclusions could be drawn as to the most advantageous combinations and dimensions of "stress relieving measures" on the pipeline.

8   REFERENCES

Arranz G and Filz G. 1991. *Automesh - IBM Reference Manual and IBM Tutorial Manual, Version 1,0*. Virginia Polytechnic Institute and State University, The Charles E. Via Jr. Department of Civil Engineering.

Bieniawski ZT. 1973. Engineering classification of jointed rock masses. *The Civil Engineer in*

South Africa, December, Vol 15 No 12. pp 335-343. "Errata" published in Vol 16 No 3 (March 1974) pp 119. "Discussion" in Vol 16 No 7 (July 1974) pp 239-254.

Bieniawski ZT. 1976. Rock mass classification in rock engineering. Exploration for rock engineering, Proceedings of the Symposium, (editor ZT Bieniawski). Cape Town: Balkema, Vol 1 pp 97-106.

Duncan JM, Byrne P, Wong Kai S and Mabry P. 1980. *Strength, Stress-Strain and Bulk Modulus Parameters for finite element analyses of stresses and movements in soil masses.* Virginia Polytechnic Institute and State University, The Charles E. Via Jr. Department of Civil Engineering

EXAMINE. A 2-dimensional finite difference *computer program* version 3.0. Rock Engineering Group, Department of Civil Engineering University of Toronto.

Filz G, Clough WG and Duncan JM. 1990. *Draft user's manual for program Soilstruct (Isotropic) plane strain with beam elements.* Virginia Polytechnic Institute and State University, The Charles E. Via Jr. Department of Civil Engineering.

Hoek. 1995. Strength of rock and rock masses. Extract from a book entitled *"Support of Underground Excavations in Hard Rock"* by E Hoek, PK Kaiser and WF Bawden. pp 4-16.

Geotechnics for Developing Africa, Wardle, Blight & Fourie (eds) © 1999 Balkema, Rotterdam, ISBN 90 5809 082 5

Bearing capacity of surface footings resting on treated and reinforced fly ash

S.A. Raza, M.S. Ahmad, M.A. Khan, M. Husain & A. Sharma
Geotechnical Engineering Section, Department of Civil Engineering, Aligarh Muslim University, India

ABSTRACT: The present work deals with to study the bearing capacity aspect of surface footings(square, circular and rectangular) resting on untreated and unreinforced, treated and reinforced fly ash bed. The square footing gave better performance over circular and rectangular by enhancing load carrying capacity to (114%). Priority wise , the circular footing were judged next to the square with (105%) improvement in bearing capacity while rectangular showed (95%) improvement in bearing capacity. The findings would pave the way for reclaiming urban lands back filled by treated and reinforced fly ash for the purpose of laying foundations of low to middle order structures.

1. INTRODUCTION

Land becomes increasingly in short supply specially in urban areas and in the vicinity of thermal power plants. As demand increases the need to develop marginal sites such as fly ash dumps etc. must be examined as it becomes imperative. As a result derelict or marginal sites which is considered unutilizable can be converted into an asset. Around 75 major thermal power plants of our country produce fly ash about 100 Million Tones/Annum. For a period of 30 years of life span of these thermal power plants about 50,000 acres of land area is needed for disposal of fly ash. Besides the cost of transportation for disposal of fly ash alone is 500 million rupees annually. Unfortunately the. measures required for the rapid disposal or utilization of these wastes lagged far behind in comparison to the quantum of waste produced, resulting in substantial contribution towards environmental pollution as well.

Now in this developing era many researchers viz., Singh, Amarjit (1968). Uppal, H.L. & Dhawan, P.K. (1968). Swaminathan, C.G. & Nair, K.P. (1968). Shah, S.S. (1980). Raju, M.J. , Raju, P.A.R.K & Reddy, B.J. (1989). Raza, S.A. & Murtaza, G. (1991). Okumuro, T. & Takahashi, T. (1996), have carried out several investigations to explore the possibility of meaningful and safe utilization of this solid waste in various construction works. But inspite of their efforts hardly 3-5% of total fly ash produced in this country is being utilized for the gainful purpose. Keeping the main objective of mass scale utilization the present study has been carried out to evolve an economical strategy for specially those areas which falls in the vicinity of thermal power plants and where leveling cost of undulating lands or filling large low lying areas abandoned by brick kilns with conventional soil is very high and such areas are required for erecting low to medium order residential/commercial units. The bearing capacity of surface footings of various shapes (square, circular & rectangular) resting on plain fly ash, mixed with lime, surfactants and finally with geofibres, separately and collectively have been carried out. On the basis of experimentations and simultaneously, plotting the load settlement curves for each setup, the study reveals that out of three shapes of footings the square shape has (114%) enhanced bearing capacity while circular and rectangular shapes have (105%) and (95%) when compared with plain fly ash. On the other hand the treatment and reinforcement of fly ash also affects greatly the bearing capacity up to optimized percentages of lime, surfactant and geofibres.

2. MATERIAL AND METHOD

For laboratory test fly ash was collected from Harduagunj Thermal Power Plant, Aligarh (U.P) India. The main chemical, physical properties of fly ash, chemical composition of lime, properties of surfactants and geofibre along with model surface footing details are given in tables (1-6).

Table 1. Chemical Properties of Fly ash.

1. Silicon Dioxide   45 – 60%
2. Aluminum Oxide   21 – 29%
3. Iron Oxide   5 – 18%
4. Calcium Oxide   0.5 – 6%
5. Magnesium Oxide   0.2 – 3%
6. Sulpher Oxide   Upto 0.4%

Table 2. Physical Properties of Fly ash.

1. Specific Gravity   1.98
2. Optimum Moisture Content   44%
3. Maximum Dry Density   0.94 gm/cc
4. Percentage Finer (75µ size)   92%
5. Cohesion (undrained)   0.01 kg/cm^2
6. Angle of Internal Friction   16°

Table 3. Chemical Composition of Lime (Calcium Hydroxide).

1. Minimum Assay   95%
2. Chloride (Cl)   0.1%
3. Sulphate (So$_4$)   0.5%
4. Iron (Fe)   0.1%
5. Lead (Pb)   0.02%
6. Loss on Ignition   10%

Table 4. Properties of Surfactant (Cetyl Trimethyl Ammonium Bromide).

1. Chemical Formula $C_{19}H_{42}BrN$
2. Nature of Surfactant   Cationic
3. Molecular Wt.   364.446 a.m.u
4. Insoluble Matter   0.01%
5. Sulphated Ash   0.1%
6. Loss on drying   1%
7. Shape of Micell   Spherical

Table 5. Properties of Geofibre.

1. Mode of fabric   Felt
2. Base Material   Polyster
3. Mode of Manufacture   Non-Woven
4. Thickness of Fibre   01 mm
5. Unit Weight   70 – 80 gm/ m^2
6. Tensile Strength   KN/m in all direction
7. Permeability   250 Lit/m^2/sec.

Table 6. Model Footings Detail.

1. Material   Mild Steel
2. Shape   Circular, Square, Rectangular
3. Size   8cm Dia., 8cm.x.8cm & 8cm x 12cm

2.1 Testing Routine

The following test procedures were adopted for determining the index properties of fly ash.

1. Proctor's compaction test for determination of moisture content and dry density.
2. Specific gravity by pycnometer.
3. Particle size distribution by sieve and hydrometer analysis.
4. Determination of shear strength parameters by direct shear test.

A wooden tank of size 45cm x 45cm x 30cm having inner lining of thin aluminum sheet to make it water tight and joint stiffeners were used to protect the side walls from bulging out during the experimental work. An oven dried fly ash was mixed with OMC (44%), deposited in three equal layers in the experimental tank and each layer was tamped by 100 number of blows with the help of rectangular shaped hammer of area 15cm x 15cm having a free fall of 25cm.The surface footings were kept in middle of tank to avoid the effect of tank wall on the pressure zone. Then the whole assembly was placed under loading machine and the point load was applied axially through a ball which rested in a groove on the footings. Two dial gauges of 0.01mm sensitivity were used to measure the settlement of footings placed diagonally to each other. The testing operations were carried out in the phases (I-IV).

Phase –I Footings resting on plain fly ash

Each shape of surface footings such as square, circular and rectangular were tested on plain fly ash bed separately. On the basis of experimental data the load and settlement curves were drawn and hence, the ultimate bearing capacities were determined.

Phase – II Optimization of lime

Plain fly ash was mixed with the increasing percentages of lime (0.5, 1.0, 2.0 & 5.0). Each mix was tested for all three shapes of footings separately. The optimized percentage of lime was found to be 1% on the basis of load settlement curves drawn for each setup.

Phase – III Optimization of surfactant

Cetyl Trimethyl Ammonium Bromide was used in this study as stabilizing agent. In the process of optimization of it the specimen contains the optimized percentage of lime (1%) is further mixed with increasing percentage of surfactant (0.1% & 0.2%). The load settlement plot shows an increasing trend up to 0.1% of surfactant, hence 0.1% was taken as optimized value of surfactant.

Phase – IV Optimization of geofibre

For the optimized percentage of lime and surfactant 1% and 0.1% respectively, the increasing percentage of geofibre (0.1%, 0.2% and 0.3%) mixed randomly in the specimen. On the basis of load settlement curve the optimization of geofibre was done at 0.2%.

3. RESULTS AND DISCUSSION

The surface footings of three shapes viz., square, circular and rectangular were tested on bed of plain, treated and reinforced fly ash. The observations are plotted in the form of load settlement curves as shown in (figs. 1 – 4).
First of all, the square, circular and rectangular footings were tested on plain fly ash bed. The data were plotted in terms of load settlement curves for each shape as shown in (fig.1). The ultimate bearing capacity were obtained for square, circular and rectangular footings as 3.95t/m^2, 3.88t/m^2 and 3.90t/m^2 respectively by double tangent method. The results highlighted the domination of square footing over other shapes.
Further the plain fly ash was mixed with varying percentages of lime and on the basis of load settlement curves as shown in (fig.2) drawn for each setup, the lime was optimized at 1% and the corresponding values of bearing capacity for square, circular and rectangular footings were obtained as 6.71 t/m^2, 6.66 t/m^2 and 6.21 t/m^2 respectively.
Fig. 3 shows the load settlement pattern for three shapes of footings resting on fly ash mixed with 1% lime and treated with varying percentages (0.1% & 0.2%) of surfactant. The curves show that at 0.1% of surfactant the ultimate bearing capacity for square, circular and rectangular footings have rising trend and it was found as 7.58 t/m^2, 6.8 t/m^2 and 6.5 t/m^{2}respectively. From the analysis of fig. 3 it was

also observed that the treatment of lime and surfactant enhances bearing capacity of square footing up to 92%, while circular and rectangular shapes responded up to 76% and 67% respectively, as compared to plain fly ash. Finally plain fly ash was mixed with 1% lime, 0.1% surfactant and reinforced with varying percentages (0.1%, 0.25 and

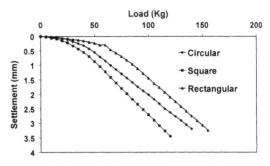

Fig. 2 Load Settlement curves for optimized percentage of lime.

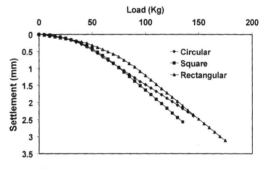

Fig. 3 Load settlement curves for optimised percentages of lime and surfactant.

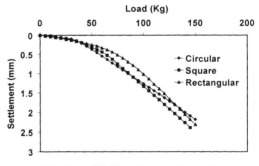

Fig. 4 Load Settlement curves for optimized percentages of lime, surfactant and geofibre.

0.3%) of geofibre mixed randomly. The data obtained were plotted in the form of load settlement curves(fig.4). On the basis of these curves the geofibre was optimized at 0.2%. and the ultimate

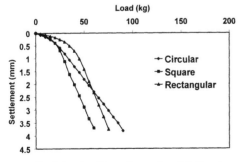

Fig. 1 Load settlement curve for plain fly ash.

Table 7. Shape effect with load criteria.

Shape of footing	Ultimate load carrying capacity (t/m^2)				Bearing capacity ratio w.r.t PFA		
	Plain fly ash	PFA+OPS	PF+OPL +OPS	PFA+OPS +OPG	OPL	OPL+OPS	OPL+OPS+OPG
Circular	3.88	6.66	6.80	7.95	1.69	1.75	2.04
Square	3.95	6.71	7.58	8.43	1.70	1.92	2.15
Rectangu lar	3.90	6.21	6.50	7.60	1.59	1.67	1.95

PFA   Plain fly ash.
OPL   Optimized percentage of lime.
OPS   Optimized percentage of surfactant.
OPG   Optimized percentage of geofibre.

Table 8. Shape effect with settlement criteria.

Shape of footing	Ultimate load carrying capacity(t/m^2) at 0.5mm & 1.0mm settlement				Bearing capacity ratio w.r.t PFA		
	Plain fly ash	PFA+OPS	PFA+OPL +OPG	PFA+OPS +OPG	OPL	OPL+OPS	OPL+OPS+OPG
Circular	4.58	9.55	10.3	12.3	2.08	2.26	2.69
Square	3.44	5.86	8.44	9.84	1.73	2.45	2.86
Rectangu lar	3.45	6.60	6.85	7.40	1.92	1.99	2.15

PFA   Plain fly ash.
OPL   Optimized percentage of lime.
OPS   Optimized percentage of surfactant.
OPG   Optimized percentage of geofibre.

load carrying capacity of square, circular and rectangular footings were obtained as 8.5 t/m^2, 8 t/m^2 and 7.6 t/m^2 respectively. Also the tremendous increase in bearing capacity of square, circular and rectangular footings were up to 114%, 105% and 95%.w.r.t plain fly ash.On the basis load and settlement criterion the experimental data were further analyzed separately for each setup and it was confirmed that the square footing has better performance over remaining two shapes of footings as shown in tables (7 & 8).

4. CONCLUSIONS

The laboratory investigations based on model testing lead to the conclusion that treated and reinforced fly ash samples have improved load settlement behavior. The following are the main conclusions drawn from the present study.

1. The effect of various treatments viz., lime, surfactant and geofibre reinforcement have significant effect on the bearing capacity of fly ash besides shape effect of footings.

2. Maximum bearing capacity as well as maximum binding effect were obtained at 0.2% non woven geofibre along with 0.1% surfactant and 1% lime.

3. On the basis of load and settlement criterion it was also confirmed that besides treatment, square footing gives better performance over circular and rectangular footings.

4. This study also suggests a ready reckoner for calculating bearing capacity of fly ash treated with lime, surfactant and reinforced with geofibre up to suggested optimized value. The bearing capacity of square, circular and rectangular surface footings can be enhanced by a factor 2.5, 2.25 and 2.0 respectively, when resting on to plain fly ash bed.

REFERENCES

Okhmura, T. & Takahashi, K. 1996. Geotechnical properties of fresh coal ashes from thermal power plants. *Environmental geotechnics, kamon (ed.) pp. 869-874.*
Ramaswamy, S.V. & Senathipathi. 1983. Design of

fabric reinforced earth walls. *Indian geotechnical society. 4:21-25*

Raza, S.A. & Murtaza, G. 1992. Lime treated fly ash reinforced with non woven geofibrics. *Indian geotechnical society, Baroda chapter. ppB.8-B.18.*

Shah, S.S. 1989. Study of the behavior of treated fly ash with non metallic reinforcement. *Indian geotechnical conference, pp. 289-293.*

Singh, A. Use of lime fly ash in a soil stabilization for roads. *JIRC,. 31: 143-153.. India*

Uppal, H.L. 1968. A resume on the use of fly ash in soil stabilization. *JIRC, 31: 379. India.*

Vidal, Henri. 1966. Mechanism of reinforced earth. *Journal of geotechnical engineering, ASCE.*

Geotechnics for Developing Africa, Wardle, Blight & Fourie (eds)© 1999 Balkema, Rotterdam, ISBN 90 5809 082 5

Evaporation from clay soils – A laboratory study

K. Dineen

Soil Mechanics Section, Department of Civil and Environmental Engineering, Imperial College of Science, Technology and Medicine, London, UK

ABSTRACT: The loss of moisture through evaporation and evapo-transpiration has been the subject of a number of research initiatives. Field studies considering the overall relationships between the atmospheric state and subsequent ground water conditions have been considered (Blight, 1997).

Laboratory studies of evaporation under a controlled environment allow consideration of the influence of particular soil parameters on the process of moisture loss. This paper considers the influence of the soil moisture characteristics on evaporation for a number of clay soils.

1. INTRODUCTION

Evaporation describes the change from a liquid to a gas. In the particular case of evaporation from a soil it describes the mechanism whereby water is removed form the soil matrix, changing from a liquid to a gas state during the process.

There are a number of approaches to the consideration of evaporation, many of which are based around the Dalton type mass balance equation, eqn 1:

$$E = f(u) (e_s - e_a) \qquad (1)$$

where:

E	=	Rate of evaporation
e_s	=	Saturated vapour pressure of the water at the clay surface
e_a	=	Partial vapour pressure of the atmosphere above the clay surface
$f(u)$	=	A function dependent on the particular atmospheric conditions at the interface between e_s and e_a.

The measurement of the relevant soil and atmospheric parameters can prove to be very difficult. In field situations there is often the added complexity of continuous environmental variation. A common approach to both field and laboratory studies is to compare the evaporative losses from the soil surface, actual evaporation (AE) to those from a free water surface, potential evaporation (PE).

In the field, continuous recording of the environment is essential for meaningful interpretation (Blight, 1997). In the laboratory the environmental conditions are more stable, and can often be controlled. This allows consideration of the influence of the soil type and condition on the evaporation process.

The soil moisture characteristics of a soil describe the relationship between soil suction, moisture condition and volumetric behaviour. In addition to being fundamental to the understanding of partly saturated soil behaviour, they provide important information for both the interpretation of soil behaviour during evaporation, and the influence of the soil moisture condition on evaporation.

2. SOIL PROPERTIES

Three plastic clay soils were chosen for this study representing a broad range of clay soil types: Speswhite Kaolin, London Clay and Bangali Clay (a highly expansive soil from northeast Kenya). The soil properties are given in table 1.

3. EXPERIMENTAL PROCEDURE

Evaporation tests were carried out on each soil type and the moisture loss from the surface compared to that from a free water surface under the same environmental conditions. In addition the soil moisture characteristics during drying were determined.

Table 1. Soil properties and mineralogy.

Soil	Soil properties					Clay mineralogy			
	Liquid limit	Plastic limit	Plasticity index	Specific gravity	Clay fraction	Chlorite	Illite	Kaolinite	Montmo-rillonite
Kaolin	64	32	32	2.61	77	4	5	82	8
London Clay	74	29	45	2.64	71	1	52	6	32
Bangali Clay	95	42	53	2.76	68	-	-	-	100

3.1 Sample preparation

All samples were initially reconstituted at 1.25 WL. Following hydration, the samples were consolidated one dimensionally to a maximum vertical stress of 200 kPa in a 105 mm diameter oedometer. Each sample was then trimmed to 100 mm diameter, 25 mm deep for testing.

3.2 Evaporation tests

A schematic diagram of the experimental set up is given in figure 1. The evaporation tests were conducted in a laboratory environment under controlled temperature conditions of 20^0C +/- 1^0C at, a relative humidity of approximately 60% and standard atmospheric pressure. Following preparation, each sample was placed on an electronic balance and allowed to dry through evaporation in the laboratory. The mass of the sample was monitored continuously as drying occurred. This allowed the actual evaporation to be defined (AE).

The initial surface area of each sample was 255 cm^2 and evaporation from the sample was compared to that from a free water surface. This was achieved by placing a pan of water, of comparable surface area on a separate electronic balance adjacent to the sample and recording the free water, potential, evaporation (PE).

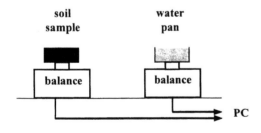

Figure 1. Experimental set-up

3.3 Soil moisture characteristics

The soil moisture characteristics of each soil were determined from initial conditions to an air dry state. Following preparation, the soil suction, moisture content and volume of each sample was measured. The initial suction was generally between 100 - 200 kPa reflecting the initial sample preparation procedure. From this point the suction was increased through controlled air drying. Further measurements of suction, moisture content and volume were made at each stage during drying until the sample reached an air dry state in the laboratory environment.

All suction measurements were made using Whatman No 42 filter papers in the in-contact mode giving matrix suction values. The calibration of Chandler et al (1992) was used for the interpretation of all measurements.

4. RESULTS AND DISCUSSION

4.1 Soil moisture characteristics

The results from the drying tests on each soil are given in figures 2 - 4. The data is presented in terms of moisture content and degree of saturation against suction. The drying behaviour is quite different for each soil type.

The kaolin sample, figure 2, de-saturates at about 1300 kPa undergoing a rapid reduction in degree of saturation over a relatively small subsequent increase in suction. The moisture content also reduces considerably once the soil de-saturates. This type of behaviour can be associated with a soil that has a relatively uniform and rigid pore structure.

The kaolin has a uniform particle size distribution, and a shrinkage limit that coincides with the point of de-saturation. The soil, once de-saturated, can therefore be considered as a rigid porous medium.

The London Clay, figure 3, is a more plastic clay soil, and remains saturated until the suction approaches 3000 kPa. Continued shrinkage

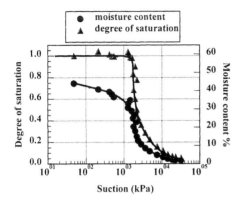

Figure 2. Soil moisture characteristics – Kaolin.

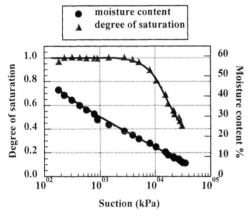

Figure 3. Soil moisture characteristics – London Clay.

following de-saturation produces a gentle curvature to the moisture content-suction relationship and a high degree of saturation at the conclusion of drying.

The Bangali clay, figure 4, displays a remarkable ability to retain moisture and remains fully saturated to very high suctions, de-saturation occurring at 9000 kPa. Under an air dry condition, at a suction approaching 30,000 kPa the degree of saturation is in excess of 80%. The moisture content - suction relationship is one of a gentle curve influenced primarily by the extensive shrinkage that the soil undergoes during drying.

4.2 *Evaporation tests*

The soil moisture characteristics for each soil were incorporated into the data analysis for the evaporation tests. This allowed the raw data (in terms of soil mass) to be considered in terms of

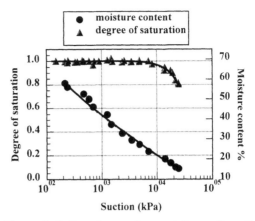

Figure 4. Soil moisture characteristics – Bangali Clay.

moisture content, matrix suction and degree of saturation. It also enabled a continuous correction to be made for the reduction in the sample volume and hence exposed surface area due to shrinkage, during the drying process. The results are given in figure 5, with evaporative losses shown as cumulative specific moisture loss against elapsed time. This enables a direct comparison to be made between the specific evaporation from the clay surface and that from a free water surface. The sample suction, degree of saturation and moisture content are given at the point where the rate of evaporation from the sample reduces below that from a free water surface. It can be seen from the data that there are two distinct phases during evaporation:

i)   A phase where the rate of evaporation from the clay is equivalent to that from a free water surface (AE/PE = 1).
ii)   A phase of reducing rate of evaporation (AE/PE <1).

Within the second phase, a transitional period can be observed as the rate of evaporation moves towards a residual value. The soil moisture condition at the point of reduction in the initial rate of evaporation (AE/PE=1) is also indicated on each plot in figure 5. A number of comments can be made with respect to soil moisture condition and evaporation:

•   De-saturation does not control the rate evaporation. The degree of saturation varies from 98% to 40%, at the end of phase (i).
•   The matrix suction does not control the rate of evaporation. The matrix suction varies between 2500 - 10000 kPa at the point of reduction in the rate of evaporation. It has

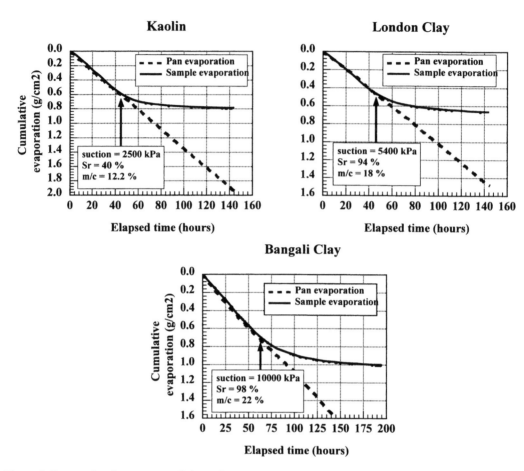

Figure 5. Evaporation from a range of clay soils

been shown that the osmotic suction component remains fairly constant over a wide range of moisture contents, (Schreiner, 1987), (Marhino, 1994). Therefore changes in matrix suction can be considered broadly equivalent to changes in total suction over this range. The osmotic suction component has been estimated to be less than 500 kPa in each soil.

A general trend can be observed in the overall soil moisture condition at the point of reduction in rate of evaporation. The Kaolin has, at this point a relatively low suction, combined with a low degree of saturation and a limited availability of moisture in the soil mass. The London Clay has a noticeably higher degree of saturation and moisture content, but also a higher suction, probably offsetting the greater availability of moisture. The Bangali Clay extends this trend with a high degree of saturation and moisture content, and a significantly higher suction.

5. CONCLUSIONS

This study provides important and useful data on both the soil moisture characteristics and evaporative behaviour of a range of plastic clay soils. A number of summary conclusions can be made:

- Clay soils have an inherent ability to retain moisture over large increases in suction
- Even in an "air dry" state, many plastic soils may be at high degrees of saturation with considerable moisture remaining within the soil mass.
- From the initial condition all soils studied lost moisture at a rate equivalent to that from a free water surface for a considerable period of time.
- Although a single soil moisture parameter cannot be identified which controls moisture loss through evaporation, a trend is observed

indicating that the overall soil moisture condition does influence evaporation.

6. ACKNOWLEDGEMENTS

The study presented here was jointly funded by the Engineering and Physical Sciences Research Council and the Transport Research Laboratory. The author gratefully acknowledges their support and the help and advice given by his colleagues Prof J.B.Burland and Dr A.M.Ridley during the research project.

7. REFERENCES

Blight G.E 1997 Interactions between the atmosphere and the earth *Geotechnique* , 47 No 4, 715 - 767

Ridley.A.M & Wray.W.K. 1995 Suction measurement: a review of current theory and practices. *1ˢᵗ Int. Conf. Unsaturated Soils*, Paris, Vol 3. 1293 1322

Marinho.F.A.M. 1994 Shrinkage behaviour of some plastic soils *PhD Thesis*, University of London

Schreiner.H.D. 1987 Measurement of solute suction in high plasticity clays *9th Regional Conf. for Africa on Soil Mechanics and Foundation Engineering,* Lagos, 163-171

Dineen 1997 The influence of suction on compressibility and swelling. *PhD Thesis*, University of London

Chandler R.J, Crilly M.S, Montgomery-Smith G. 1992 A low-cost method of assessing clay desiccation for low rise buildings *Proc. ICE*, 92, No2 pp 82-99.

Geotechnics for Developing Africa, Wardle, Blight & Fourie (eds) © 1999 Balkema, Rotterdam, ISBN 90 5809 082 5

Penstock failures on gold tailings dams in South Africa

F.von M.Wagener & S.W.Jacobsz
Jones and Wagener, Johannesburg, South Africa

ABSTRACT: Penstock failures occur at fairly regular intervals on South African gold tailings dams. This results in disruptions of the tailings deposition programme on the mine and sometimes in overloading of other tailings dams while the failed penstock is being repaired or replaced. The failed penstock is normally replaced by the installation of a "floating" penstock at considerable cost to the mine. The name implies that the penstock is floating in tailings. This paper investigates the mechanism of penstock failures, reasons for failure and possible mitigating measures.

1. TAILINGS DAM CONFIGURATION AND OPERATION

The majority of South African gold tailings dams are constructed using the upstream semi-dry paddock method. An idealised plan and section through such a tailings dam is shown in Figure 1. Dams constructed according to this method comprise of two components, viz. the "daywall" and "nightpan". The outer wall or daywall of the dam is raised by constructing a paddock with low containment walls, 10 m to 30 m wide, to form the perimeter of the dam. The tailings previously deposited in the paddock is used to raise the impounding walls with a tractor-plough combination or by labour intensive shovel packing. The paddock is then filled with approximately 150 mm to 200 mm of slurry from the delivery stations. The slurry flows from the delivery station into the paddock and distributes by gravity in the daywall zone. The slurry is allowed to settle out before the excess water is drained into the body of the dam or nightpan via decant pipes through the inner wall. The deposited material is left to dry and consolidate while deposition takes place at another delivery station. Consolidation and strength gain of the daywall is achieved by desiccation. The nature of the deposition process results in an intensely layered daywall with many desiccation cracks. The daywall is approximately 1 m to 2 m higher than the outer perimeter of the nightpan.

This wall-building operation takes place during the day when supervision is available, hence the name daywall. In order to achieve operational safety, discharge of tailings slurry at night takes place into the body of the dam, hence the basin or the pond of the dam is known as the nightpan. Water and slurry discharged into the nightpan runs down the concave surface or beach of the dam towards the penstock, which is intended to be centred in the tailings dam from where it drains excess water off the dam.

This method of tailings dam construction results in a daywall which is normally well consolidated and reasonably firm, while the material in the nightpan and especially the pool is significantly less consolidated and hence softer.

A typical penstock consists of a vertical inlet tower. An outlet pipe joins the inlet structure at the base of the tower and conveys decant water to a return water facility outside the tailings dam. The diameter of a typical inlet tower is approximately 500 mm, while the diameter of the flanged and bolted steel outlet pipe commonly varies between 250 mm and 550 mm.

During construction of a new tailings dam the penstock inlet structure and outlet pipe are constructed together with the initial earthworks. The inlet structure and outlet pipe are founded on natural material following some foundation preparation. After commissioning of the dam, tailings is gradually deposited in the dam basin until the penstock outlet pipe is covered. The inlet structure is raised as the dam gains height by adding concrete rings or steel sections, depending on the type of penstock. A cross section through a penstock arrangement in a new dam is shown in Figure 2.

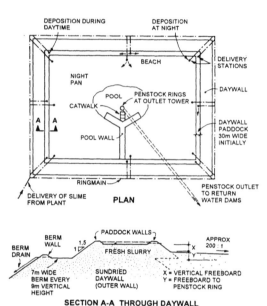

Figure 1 Plan and section of typical tailings dam.

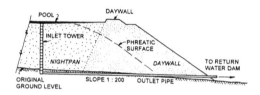

Figure 2 Typical penstock arrangement in a new tailings dam.

2. PENSTOCK FAILURES

2.1 Background

As a tailings dam gains height the pressure exerted on the penstock outlet pipe constantly increases. Although protected with coatings, cathodic protection and using corrosion resistant bolts at flanged connections, corrosion of the penstock pipes can occur. From case studies available it appears that penstock outlet pipes normally fail 10 to 15 years after commissioning when the dam reaches heights of 15 m to 25 m. Failure of the inlet towers rarely occurs.

In addition to load or corrosion related failures, failures also occur on dams built on dolomite. These failures usually occur soon after commissioning of the dam when subsurface erosion of the underlying material leads to the formation of subsidences or sinkholes. Dolomite related penstock failures are not discussed in this paper.

Penstock failures are fairly common on South African gold tailings dams, often resulting in disruptions of the tailings deposition programme on the mine. Depending on the availability of additional deposition space, a penstock failure may result in overloading of other tailings dams while the failed penstock is being repaired or replaced. During overloading, insufficient time is available for adequate consolidation of newly placed tailings in the daywall. This can result in the formation of weak layers in the daywall. As the dam gains height the weaker layer may form a preferential flow path resulting in horizontal seepage with corresponding stability problems. Overloading of a tailings dam can thus affect the long term performance and stability of a tailings dam detrimentally.

After failure the penstock is normally replaced with a "floating" penstock which is founded in the tailings and is not supported on a firm base. Depending on the nature of the failure it may also be possible to repair the penstock by inserting a lining in the form of a second pipe into the failed penstock. A cross section through a floating penstock arrangement is shown in Figure 3.

2.2 "Symptoms" of penstock failures

The most common and also most dramatic result of a penstock failure on a tailings dam is the formation of a sinkhole above the failed outlet pipe as illustrated in Photo 1. The formation of the sinkhole may be sudden and without warning or it may be preceded by the formation of "mini" sinkholes (see Photo 2) and subsidence in the area around the pipe failure long before a large sinkhole develops. The diameter of sinkholes may vary from less than 1 m to several metres. Sinkholes can pose danger to the personnel of tailings dam operators as illustrated in Photo 3, where a tractor had fallen into a sinkhole which suddenly opened up as the tractor passed above it. Fortunately the tractor driver escaped unharmed.

Penstock problems will usually also be manifested by the discharge of turbid water from the penstock outlet while clear water is decanted from the pool on the dam. In addition the outlet pipe sometimes appears to be blocked resulting in a delay to decant water off the dam.

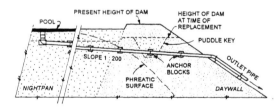

Figure 3 Typical section through a floating penstock

Photo 1 A large sinkhole opened up in line with a penstock pipe.

Photo 2 Small sinkholes in line with a penstock outlet pipe.

2.3 Mechanism of penstock failures

Detailed investigations into the causes of penstock failures in recent years shed new light on the mechanisms of penstock failures. A number of failed penstock outlet pipes were inspected with remote controlled video cameras. The investigations revealed that in most cases penstocks fail in shear at

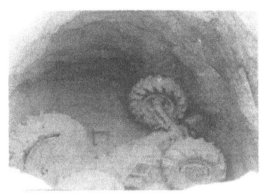

Photo 3 A tractor that had fallen into a hidden sinkhole on a tailings dam daywall.

pipe joints. Pipes were never found to be flattened by the overburden pressure.

As mentioned in Section 1 the method of tailings dam construction most commonly used in South Africa results in a well consolidated, firm daywall, while the material towards the pool on the dam is softer and less consolidated.

When a floating penstock is installed in a tailings dam the outlet pipe crosses from the soft, unconsolidated material near the centre of the dam to the firm, more consolidated material in the daywall. This results in the penstock pipe settling unevenly as the tailings dam gains height. The maximum differential settlement would be at the transition between the daywall and the nightpan and also where the penstock outlet pipe is connected to the inlet tower.

Saturated conditions occur in the nightpan of a dam, however, the same does not hold for the daywall where unsaturated conditions can be present. Under unsaturated conditions the presence of oxygen results in an increase in the potential for corrosion.

The tensile and shear stress in the bolts will increase where differential settlements occur. If in addition these bolts are weakened by corrosion they may fail in tension or shear. Where this occurs a gap will develop between the flanges resulting in tailings being drawn into the pipe. A cavity is then gradually formed above the failure zone in the pipe. As more material is washed into the pipe, the size of the cavity increases and it moves upwards until it appears as a sinkhole at the tailings surface on the dam.

2.4 Remedial action after penstock failure

When it becomes apparent that a problem is being experienced on a penstock, the outlet pipe and inlet tower should be inspected by remote controlled camera to determine the cause of the problem. Often it would not be necessary to replace the penstock and the installation of a liner in the affected area may suffice e.g. where the bolts have corroded at a joint in the penstock pipe and the pipe is slightly out of alignment and tailings is being drawn into the pipe through a gap between the flanges.

The liner should be installed to extend a substantial distance to both sides of the problem zone. After installation of the liner tailings may still be drawn into the recess between the liner pipe and the original outlet pipe. This problem will be manifested in that turbid water will still be discharged from the penstock outlet. Such an ingress of tailings into the pipe can be reduced by creating a gravel filter pack in the sinkhole. The gravel filter should consist of coarse material at the bottom, gradually becoming finer towards the top. The fine material should be selected to form a stable interface between the tailings and the underlying coarser filter material. With time the gravel filter will move downward and settle at the level of the outlet pipe as more tailings is drawn into the pipe, eventually blocking the migration of tailings into the pipe (Yssel, 1998).

If the damage to the penstock pipe is severe, the penstock will have to be replaced. In such cases it is common to install a floating penstock. This consists of an inlet structure of concrete rings or steel sections and an outlet pipe of steel.

The outlet pipe is normally installed at a slope of 1:200 towards the daywall. This commonly results in a deep excavation of up to 6 m where the pipe crosses the daywall, depending on the geometry of the dam. An adequate batter must be used where the depth of the excavation is more than 1,5 m to ensure that the sides are stable during construction. The excavation through the daywall is backfilled by puddling as shown in Figure 3 to avoid piping through loose backfilled material after recommissioning of the dam.

The new penstock outlet pipe is normally installed into the poolwall or next to the poolwall because of the availability of firm, relatively dry, consolidated material. The installation of a floating penstock is costly.

3. REDUCING THE RISK OF PENSTOCK FAILURES

3.1 Monitoring programme

When a penstock failure is manifested by the formation of a sinkhole, serious damage has usually already been caused to the outlet pipe. A method of pre-empting a failure is to compare the turbidity of

water decanting off the dam with the turbidity of water discharged by the penstock. Should the discharge water be more turbid than the water entering the penstock, tailings is entering the penstock and it is likely that a failure is developing.

The next step would be to inspect the penstock pipe by camera to determine the cause of the problem. Suitable remedial action can then be instituted.

It is recommended that penstock water should be inspected weekly as explained above. Penstock pipes should be inspected every few years by camera to ensure that problems are identified timeously.

3.2 Present design details

In order to maximise the operational life of a floating penstock, the authors have developed a number of design details, some of which are discussed below:

Anchor blocks
Reinforced concrete anchor blocks are cast around penstock pipe joints. The function of the anchor blocks are two-fold: The blocks weigh the pipe down to prevent floating once the dam is recommissioned and they also reinforce the pipe joints to protect the bolts. In the transition area between the daywall and the nightpan a special flexible tape (Densowrap) is tied around the pipe joints before casting the anchor blocks to provide some flexibility.

Puddle key
A section of the trench for the floating penstock in the daywall is backfilled with fresh tailings to prevent piping through loose material after the dam is recommissioned as shown in Figure 3. A geotextile is used to protect the interface between the dry in situ and wet tailings during construction of the puddle key.

Cathodic protection
Cathodic protection is used on steel penstock pipelines. All the pipes are linked electrically and a number of sacrificial cathodes, commonly consisting of zinc ingots connected to the pipe, are buried in the tailings near the pipe. The cathodes will corrode with time and it is essential that the integrity of the cathodic protection system be evaluated after a number of years. New cathodes will have to be installed from time to time.

Corrosion resistant coatings
The steel penstock outlet pipes are lined with a corrosion resistant coating (e.g. Copon) on the inside and with a bituminous paint on the outside to protect the pipes in the corrosive environment inside a tailings dam.

Photo 4 A large sinkhole opened up in the daywall of the Grass Dam.

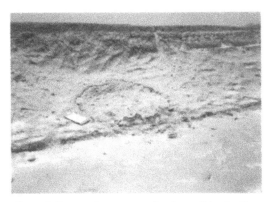

Photo 5 Depression next to the daywall in the East Dam.

Photo 6 After 24 hours the depression in the East Dam had developed into a sinkhole.

Monitoring programme
A monitoring programme as discussed in Section 3.1 should be implemented to provide early warning of possible penstock failures.

4. CASE STUDIES

4.1 Penstock failures in daywalls

Grass Dam, December 1997
A sinkhole opened up in the daywall of the dam (see Photo 4). Inspection by camera revealed that the bolts at a joint near the transition between the daywall and the nightpan failed, resulting in the pipes moving out of line by approximately 50 mm. Tailings was gradually washed into the outlet pipe resulting in the formation of the sinkhole which slowly migrated upwards.

After the sinkhole opened up a steel lining was inserted into the penstock pipe and a gravel filter pack was placed in the sinkhole and allowed to settle to the level of the pipe to block tailings from entering. The penstock has since been operating without problems for over a year. The repair costs amounted to a fraction of the cost of a new floating penstock.

This is an example of a failure that developed on the transition zone between the well consolidated daywall and less consolidated nightpan where maximum differential settlements and an increased potential for corrosion can be expected.

East Dam - East Compartment, 1998
It was reported that the penstock on this dam was discharging turbid water. A camera was used to inspect the pipe, but no problems could be identified. It was suspected that tailings was leaking through a gap in the pipe between flanges. The penstock pipe was then lined by installing an HDPE sleeve in an attempt to solve the problem. Shortly after lining, a depression was noticed on the upper berm of the tailings dam in the slope of the daywall in line with the penstock outlet pipe (see Photo 5). The day thereafter a sinkhole had opened up. The sinkhole was closed by placing a gravel filter at the bottom and backfilling with tailings. The backfilled surface continued to settle and the penstock still discharged tailings for some time. As the gravel filter gradually blocked off tailings from entering the recess between the penstock pipe and the lining the sinkhole stabilised and clear water decanted from the penstock. This is another example of a sinkhole which developed at the transition zone between the daywall and the nightpan.

5. CONCLUSIONS

Unexpected penstock failures can result in a disruption of the tailings deposition programme on a mine, resulting in temporary overloading of other tailings dams. This may affect the long term performance and stability of the affected dams. By instituting a programme of inspecting outlet pipes by camera, as well as checking the turbidity of discharge water, an early indication can be obtained of imminent penstock failure. This will enable appropriate action to be taken to reduce the impact of a failure.

A number of modifications to conventional designs, as described in this paper, can be made to reduce the risk of penstock failure and increase the operational life of a penstock. By implementing these recommendations significant cost savings with regard to tailings dam operation can be ensured over the medium to long term.

6. REFERENCES

Yssel, H.J.J. 1998. Personal communication.

Geotechnics for Developing Africa, Wardle, Blight & Fourie (eds) © 1999 Balkema, Rotterdam, ISBN 90 5809 082 5

Long-term integrity of residue deposits

J.A.Wates, G.Strayton & G.de Swardt
Wates, Meiring and Barnard, Johannesburg, South Africa

ABSTRACT: The long-term integrity of residue deposits is central to sustainable closure. Integrity is in turn a function of overall structural stability, erosional stability and hydraulic integrity. These three aspects of integrity, although inter-related, depend on different parameters.

This paper deals specifically with loss of integrity through cracking. Piping and the erosional damage, which may follow, is a function of the properties of the tailings, depositional methods and stress history. The paper examines the parameters which influence integrity from the perspective of shrinkage and cracking of the outer walls of residue deposits. The influence of slurry density, grading, mineralogy and chemical additives are examined in relation to stress history and rate-of-rise. The influence of each parameter is discussed and the conclusions are backed up with empirical test results, which provide a basis for the assessment of the potential for post-depositional and long-term cracking and the associated loss of integrity.

1 INTRODUCTION

Geotechnical Engineers usually associate the notion of loss of integrity of a residue deposit with rotational, wedge or sloughing slope failures. These conventional mechanisms, although spectacular, are the least likely to threaten long term integrity of a deposit. Loss of integrity associated with erosion is far more common and more likely to undermine long-term sustainability.

The probability of slope failure for a given slope configuration can be predicted with a high degree of reliability. The risk of occurrence of erosion, and internal erosion in particular, is on the other hand, at best estimated by rule of thumb and few engineers or scientists would be prepared to guarantee their predictions. There is thus no substantive basis for assessing the long-term risk of loss of integrity.

Loss of integrity can occur during the operational phase as well as post closure. Damage can usually be rectified quite easily whilst residue deposits are operational. After closure loss of integrity can have expensive consequences.

Two case histories, in which internal erosion has had serious consequences, have been used to illustrate the consequences of internal erosion. These case histories illustrate operational and post-closure failures. The authors propose an empirical approach for assessing the potential for cracking and demonstrate how the problem has been practically overcome on a gold and a coal tailings dam.

2 PROBLEM STATEMENT

2.1 Sustainability

The term "sustainability" has many uses and as many definitions. In the context of this paper a residue deposit design is defined as being sustainable if it will withstand the reasonably anticipated forces of nature for its design life without unforeseen maintenance.

Post-closure sustainability is an illusive objective since there is no known natural or man-made system that will endure indefinitely. Hence sustainability has here been defined within the timeframe of a design life that spans from first commissioning, through decommissioning, closure and into post-closure use. In terms of this definition design life may exceed 200 years.

Sustainability may be threatened by structural instability, by gradual erosion or by loss of utility. In most cases long term, and especially post-closure sustainability, is threatened by loss of integrity of the tailings dam structure.

Photograph 2.1: Rathole in Kloof Gold residue deposit

Photograph 2.2: Initiation of rathole in West Driefontein No.2 Dam (approximately 30 m high)

Photograph 2.3: End result of rathole in West Driefontein No. 2 Dam (loss in excess of 100 000 m³ of tailings)

2.2 Loss of integrity

Loss of integrity has, for the purposes of this paper been defined as the failure of a residue deposit to retain the tailings it was designed to hold. Loss of integrity may be caused by many mechanisms of which, overtopping and conventional slope failure are the best known. Erosion is also a common cause of loss of integrity.

Examples of loss of integrity are illustrated in Photos 2.1, 2.2 and 2.3. Photo 2.1 shows a typical "rathole" in the Leeudoorn Gold Mine residue deposit. Photo 2.2 shows the initial breach and 2.3 the extreme consequences of "ratholing" in West Driefontein Gold Mine's No. 2 tailings dam. Both of the ratholes illustrated have occurred as a result of internal erosion.

The consequences of the West Driefontein failure have been severe. In excess of 100 000 m3 of tailings was discharged and a breach about 25 m deep formed in the outer wall.

2.3 Consequences of loss of integrity

The consequences of loss of integrity may be pollution, impaired land capability or increased post-closure land management obligations. During the operational phase of the life of a residue deposit these consequences are of the initial breach seldom severe since intervention by the operator usually occurs quickly. Loss of integrity after closure can however have serious consequences for mine owners or the state. Unsustainable residue deposits furthermore represent a huge liability that will only manifest long after closure of many mining operations. As such they can also have costly consequences for future generations.

2.4 Need for understanding of mechanisms

The long-term performance of closed residue deposits has not been good. Although a measure of success has been achieved with the empirical approach to rehabilitation that has been adopted in the past, ongoing surveillance and maintenance is still required for all of the old Witwatersrand dams in order to identify failures and initiate interventions to forestall deterioration. In some instances the level of post-closure maintenance required has been substantial. Most of this has been necessary as a result of loss of integrity. If this could have been prevented in the early designs the savings would have been significant.

The causes of overtopping, of siltation of waterways and of other obstructions are well understood. Erosion by wind and water is also relatively well understood and is beyond the scope of this paper. Loss of integrity through internal erosion is however not well understood. One cannot start to design for sustainable closure of a tailings dam if the mechanisms influencing integrity are not understood. This paper specifically examines the loss of integrity associated with internal erosion.

3 INTERNAL EROSION

Internal erosion may impair the performance of waterways and hydraulic structures, create channels that concentrate stormwater and reduce freeboard. Internal erosion may open pathways for stormwater and seepage to cause further damage.

Internal erosion is usually preceded by loss of watertightness. This may occur through pathways opened up by collapsed pipes and structures by decaying roots, through insect or animal burrows or through cracks. A special form of internal erosion also occurs in dispersive materials where the colloidal fraction of the tailings matrix may be dispersed and carried away by flowing water.

Loss of integrity in fine-grained residue deposits is primarily caused by cracks. In most instances these cracks are formed during the process of shrinkage caused by desiccation. Desiccation occurs through an ongoing process of wetting and drying through the operational stage and on into the post-closure stage. Tailings dam construction techniques that have evolved over the years have been designed to overcome the consequences of shrinkage and the associated loss of integrity mainly during the operational stage.

The most common technique of tailings dam construction in South Africa is paddocking. This involves deposition of the tailings into a perimeter paddock (illustrated in Figure 3.1). The tailings is deposited in thin layers, typically no more than 150 mm deep and allowed to sun dry before the next layer is placed. The freshly placed tailings desiccate, shrink and crack (see Photo 3.1). After drying the tailings cannot form a watertight barrier and must be remoulded either by handpacking (see Photo 3.2) or machine packing. The remoulded barrier in turn retains the next layer of wet tailings.

It is important that each layer of tailings in the outer wall be allowed to desiccate to the point where no further shrinkage will occur with further loss of matrix water. If this is achieved then the cracks of previously deposited layers are filled and sealed with fresh tailings to form a watertight barrier. The remoulded outer and inner barrier walls do not shrink further and as such also provide a watertight paddock.

Figure 3.2 illustrates a typical relationship between remoulded moisture content measured by mass and linear shrinkage for gold tailings. The moisture content below which no further shrinkage occurs provides an indication of the moisture content levels that must be reached for paddock construction.

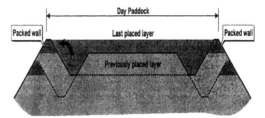

Figure 3.1: Paddock deposition technique

Figure 3.3 illustrates a typical drying curve for gold tailings derived from open air drying in bench tests of 100 mm thick samples.[1] The variation in drying rate that can be expected, given the variability in climatic conditions in the field is illustrated by the confidence limits also shown on Figure 3.3. The moisture content at the zero shrinkage limit derived from Figure 3.2 for the 95 percentile drying rate is about 10 days. Now if the tailings are deposited every 14 days in layers of 150 mm (which dry to 100 mm) then the rate of rise would be about 3,65 m/annum (= 0,1*365/10).

Provided that the rate of rise does not exceed 3,65 m/annum the tailings will not shrink further when they dry out further after closure of the residue deposit. Integrity should therefore theoretically be maintained indefinitely notwithstanding extreme variations in climatic conditions. As the rate of rise is increased ratholing may be expected to increase especially when the rate of rise is reduced substantially or when a bench is created by stepping in the outer wall.

The authors observation of behaviour of a number of gold residue deposits over recent years has lead to the conclusion that long-term post-closure shrinkage does occur in some tailings dams constructed at rates less than the theoretical maximum allowable rate of rise adopted by the gold mining industry. Drying rates and shrinkage properties of gold tailings have been observed to vary quite markedly from one mine to another and even between metallurgical processes. These observations have lead the authors to believe that an industry wide limit to maximum allowable rate of rise is not appropriate and that allowable rate of rise should be determined uniquely for each tailings.

3.1 How might tailings properties influence integrity

Tailings behaviour may differ for a variety of reasons. The most common differences giving rise to noticeably different field behaviour are:

- Particle size distribution
- Slurry water content
- Flocculent type and concentration
- Mineralogy

For a given tailings, permeability is proportional to pore size and this in turn, is proportional to particle size.[2] Since water loss rate is proportional to permeability, the time required to reach the shrinkage limit will thus increase as the particle size decreases. Finer grained material will need longer to dry before packing in order to ensure that ratholing does not occur. Allowable rate of rise is thus expected to decrease with decreasing particle size.

Tests have shown that the settled density of slurries with higher moisture content will be lower than settled density of slurries with lower moisture content.[3] Slurry water content, although not directly influencing the drying rate or shrinkage limit of tailings, therefore affects the initial moisture content at which de-saturation begins as well as the settled layer thickness. Since water content at the shrinkage limit tends to decrease as the voids ratio at the onset of de-saturation increases.[4] The slurry moisture content therefore indirectly influences the shrinkage limit. Lower allowable rates of rise should therefore abe pplicable to tailings dam constructed with high moisture content slurries.

Flocculent concentration has a significant influence on the settled density of a thin layer of tailings. Increasing flocculent concentrations (up to the optimum) will increase settling rate but will decrease the settled density. Flocculents could therefore have the same effect as slurry moisture i. e. to reduce the water content at the shrinkage limit with increasing flocculent content. It is therefore possible that flocculent could influence the potential for ratholing.

Mineralogy may have the single largest influence on the post-depositional shrinkage properties of tailings. Mineralogy influences particle shape

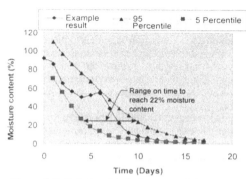

Figure 3.3: Relationship between moisture content and time for air drying under Highveld climatic conditions

Photograph 3.1: Shrinkage and cracking of residue

Photograph 3.2: Hardpacking of copper residue

and specific surface area and therefore the manner in which flocculent will react with the tailings. The presence of phyllosilicates are particularly troublesome. Phyllosilicates are a group of micaceous or platy minerals that predominantly occur in the form of sericite, chlorite and pyrophillite. Talc is also suspected to influence behaviour. In the authors experience tailings having high proportions of phyllosilicates and talc are the most susceptible to ratholing.

Severe ratholing problems have also been observed in coal tailings. Coal tailings binds water within its molecular lattice structure. This bound water is not lost under normal atmospheric wetting and drying cycles. The water is however

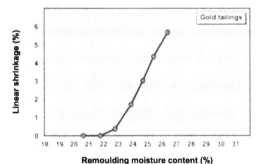

Figure 3.2: Typical relationship between remoulding moisture content and linear shrinkage

Photograph 3.3: Machine packed trench at Leeudoorn Gold Mine residue dam

driven off at an ever-increasing rate as temperature increases. High temperatures would be expected to be generated in the exposed flanks of a coal tailings dam owing to the colour of the tailings. This may explain why ratholing is so prevalent in coal tailings. Another explanation may be found in the volume loss and shrinkage associated with oxidation of the carbonaceous component of the coal. Weathering and foliation of the coal may also play a role.

3.2 How might operational practices influence integrity

Tailings layer thickness will have a significant impact on integrity. Drying time increases as layer thickness increases. Although cycle time, and hence the time available for drying will also increase as the layer thickness increases, practice has shown that an optimum exists for gold tailings at a layer thickness of between 100 and 150 mm.

The depth of trench excavated by the trenching machines used for paddock construction varies between 400 and 700 mm as shown in Figure 3.4. This trench penetrates earlier layers and is filled with each deposition. At this point in the paddock tailings up to 700 mm deep must be dried during each depositional cycle. It is here where the shrinkage limit may not be reached after each cycle and hence where a weak, albeit narrow zone is created. Photo 2.1 also illustrates ratholes that are believed to be related to the deep trenches used on this dam which are shown in Photo 3.3.

Shrinkage can also create dangerous conditions in situations where the tailings close to the outer wall is not permitted to dry out and consolidate. This may occur, for example, when the pond is allowed to stand adjacent to the wall for long periods. When the pool is eventually moved away deep shrinkage cracks, illustrated in Figure 3.5, can develop. These cracks can then fill with water and create high pore water pressures at depth close to the face of the outer wall. Where the pore pressures exceeds the total stress hydraulic fracturing may follow leading to a breach in the outer wall. This type of failure is suspected to have occurred on at least two coal fines tailings dams in South Africa.

4 PREDICTION

The design of new tailings dams could in future be influenced by the increasing propensity for post-depositional shrinkage. Allowable rate of rise may need to be reduced substantially and paddocking may even be precluded in some cases unless remedial steps are taken. It is therefore important to know, at design stage, how the tailings will behave.

There is no established technique for predicting the propensity for post-depositional shrinkage and rat-holing. The authors have therefore devised an approach that provides an indication of the possible behaviour. The method involves comparison of the remoulding moisture content below which there is o further measurable linear shrinkage with the average time required to reach that moisture content under average drying conditions. If the required rate of rise will impose a cycle time close to the time required to reach the required moisture content then further, more accurate shrinkage tests are undertaken.

More accurate relationships between moisture content and linear shrinkage can be obtained in modified shrinkage apparatus and by accurately measuring shrinkage and moisture content. A shrinkage curve for three different gold tailings samples determined by the modified shrinkage tests and from conventional linear shrinkage test method (shrinkage measured by micrometer) is shown in Figure 4.1. This figure shows that there is little difference between the results above the zero shrinkage limit derived from the conventional shrinkage test on remoulded samples (Figure 3.2) but does show that shrinkage continues after the point of zero shrinkage determined from conventional tests, albeit at a reduced rate. The shrinkage that occurs after the point of zero conventional shrinkage (finer cyclone overflow) is significant. The difference in shrinkage limits will have a significant impact on the time required between deposition cycles.

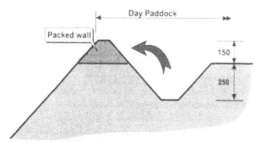

Figure 3.4(a): Conventional paddock

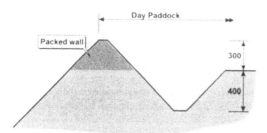

Figure 3.4(b): Deep paddock trench

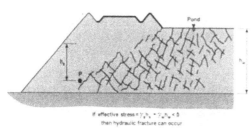

Figure 3.5: Conditions leading to hydraulic fracture

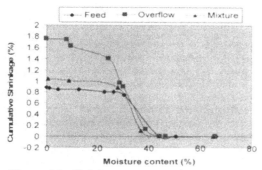

Figure 4.1: Shrinkage versus moisture content from modified shrinkage test

. The time to achieve zero shrinkage would be much longer for the overflow slime and hence allowable rate of rise would be reduced. If the 95 percentile time to achieve the required moisture were 18 days then allowable rate of rise would be 2,02 m/annum (0,1 x 365/18) i.e. reduced from 3,65 m/annum for 10 days drying time.

5 REMEDIES

5.1 Rate of rise and cycle time

Ratholing can be minimised by decreasing rate of rise. Even tailings with very high shrinkage potential can be placed by paddocking methods if the rate of rise can be reduced enough to allow the requisite drying time. Layer thickness can also be decreased, but this has the effect of decreasing the cycle time as well. There is therefore a limit to the reduction of layer thickness if a given rate of rise must be maintained.

Reducing layer thickness may not be sufficient to overcome ratholing. Reducing rate of rise is also not always practical or economical. It may therefore be necessary to consider other measures to mitigate against the consequences of ratholing.

5.2 Deposition method

Changing the deposition technique can change tailings behaviour. A change from paddlocking to a spigot distribution system can for example, allow increased rate of rise since deposition cycle times can be increased. Cyclones may also be used to decrease moisture content and increase particle size of the material placed in the outer wall. Both of these options could have the effect of eliminating ratholing but are unlikely to be workable with gold tailings.

5.3 Modification of paddocking practices

Geosynthetics have been used to overcome ratholing by modifying conventional paddocking methods of wall construction. Figure 5.1 and Photo 5.1 show how a geosynthetic fabric has been introduced into the outer wall at Leeudoorn Gold Mine to successfully overcome ratholing. The ratholing on this dam had reached alarming proportions before introduction of the geosynthetic. It has now been reduced to negligible levels over the sectors where the fabric is in use.

Although ratholing in the paddocks can be eliminated by introduction of the geosyntheic, failures initiated through hydraulic fracturing as illustrated in Figure 3.4 would not be prevented.

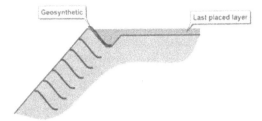

Figure 5.1: Geosynthetic introduced into outer wall eliminates ratholing

Photograph 5.1: Geosynthetic fabric introduced in outer wall

6 CONCLUSIONS

Post-depositional shrinkage can have serious consequences. Shrinkage opens pathways, which can lead to internal erosion and loss of integrity. This mechanism has, in two instances known to the authors, lead to costly failures.

Tailings with similar grading but from different reefs of the Witwatersrand gold mines have been found to behave quite differently from the norm. Certain tailings derived from reefs containing pyrophillite and talc minerals, in particular, appear to have high propensity to shrink and crack at low moisture contents.

The propensity for post-depositional shrinkage and the associated ratholing that appears to be more prevalent on gold tailings dams in recent years can be assessed on an empirical basis. An indication of the potential for excessive ratholing may be obtained by comparing the zero shrinkage moisture content of remoulded samples with the time required to reach this moisture content under average drying conditions. In those cases where the cycle time will be close to the maximum allowable calculated in this manner, further accurate measurement of the relationship between remoulding moisture content and shrinkage is advocated.

7 REFERENCES

1. Wates, J.A., Stevenson G., Purchase, A.R., The effect of relative density on gold and uranium tailings dams. Proceedings of the International Conference on Mining and Industrial Waste Management, SAICE, Johannesburg 1987
2. Lambe., T.W., and Whitman, R.V. Soil Mechanics, John Wiley & Sons Inc., 1969.
3. Robertson A., Mac G., Keynote address. International Conference on Mining and Industrial Waste Management, SAICE, Johannesburg 1987.

Geotechnics for Developing Africa, Wardle, Blight & Fourie (eds) © 1999 Balkema, Rotterdam, ISBN 90 5809 082 5

Design of soil covers to reduce infiltration on rehabilitated opencast coal mines

J.A.Wates, J.J.G.Vermaak & N.Bezuidenhout
Wates, Meiring and Barnard, Johannesburg, South Africa

ABSTRACT: Regulations regarding the standards to which opencast coal mines and coal discard dumps should be rehabilitated are becoming increasingly stringent. This has placed the coal mining industry and the design engineers of these measures in a predicament, as costs of the rehabilitation are becoming prohibitive. The paper documents a five-year pilot scale experiment to develop sufficient data to calibrate a numerical model that would accurately predict recharge through a natural soil cover. Ten test cells were constructed using different natural soil cover configurations. The cells were instrumented thoroughly to ensure a complete understanding of the performance of the soil cover. The results of the experimental data, and its relevance in terms of the design paradigms, are discussed. Further, shortcomings of existing numerical models have been identified. The paper also presents a geochemical assessment of the leachate water quality database. It concludes with design recommendations for rehabilitation using natural soil covers.

1 INTRODUCTION

The economic benefits associated with the exploitation of coal reserves by means of opencast and strip mining have resulted in large areas of land being disturbed. Rehabilitation of mined-out areas is designed for the protection of surface and groundwater resources by ensuring that the mined out areas are topsoiled and vegetated, free draining, with preventative soil erosion measures. In addition, similar measures have to be met in the rehabilitation of coal discard dumps, associated with both opencast and underground coal mines. Of all the rehabilitation aspects, the assessment of and mitigation against contamination of groundwater resources pose the greatest challenge. Because the surface and groundwater impacts associated with the opencast sections and discard dumps form part of a grater water cycle impacts are often seen down stream or gradient from these sources of pollution. Since mining activities tends to be concentrated in particular geographical regions, the combined effect of contaminant releases could contribute significantly to the total contaminant load within a particular river catchment area.

Arguably the greatest environmental impacts associated with rehabilitated opencast coal mines and discard dumps are the deterioration in groundwater quality. Experience from other countries, especially the USA, has shown that rehabilitation of groundwater resources is prohibitively expensive and technically very difficult (Travis and Doty, 1990; Stone,

1991). Based on these findings it is clear that a preventative approach is desirable.

One of the most obvious and practically sustainable methods in limiting the generation of leachate from rehabilitated opencast mines and discard dumps, is the construction of vegetated soil covers. The main objective of constructing soil cover systems is to limit infiltration of water into the discard material. Soil covers are probably the most commonly used mitigation measures to limit the generation of leachate from landfills, tailings facilities and spoils. These covers have traditionally been designed in order to support vegetation. However, the focus of design is now shifting towards the need to limit water infiltration and oxygen diffusion into the discard material.

2. PROBLEM STATEMENT

In order to quantify and qualify the groundwater impacts associated with coal discard facilities and rehabilitated opencast mines, an understanding is required of all the physical, chemical and biological processes that influence the release rate and concentration of contaminants from the discard material. These processes include: water movement, gas flux, dissolution, complexation, precipitation, adsorption/absorption, and co-precipitation.

The main processes determining the quality of leachate from coal discard facilities are pyrite oxidation and pH buffering. These processes, in general, have the effect of acidification and salinification of water percolating through the discard material. The key factors affecting pyrite oxidation and pH buffering processes are listed below in Table 1.

The main process determining solute or pollution transport in coal discard facilities is advection, implying that the quantity of leachate reaching the groundwater will mainly be a function of the recharge rate (or outflow) through the rehabilitated opencast mines or discard dumps. Once contaminants have been leached from the discard material they can impact on associated aquifer systems such as perched aquifers, which could eventually impact on surface streams.

In order to predict contaminant loads entering the groundwater regime, it is of great importance to have a thorough understanding of the complex mechanisms associated with recharge through rehabilitated opencast mines and discard dumps. In the field of groundwater research, Bredenkamp et al. (1995) stated that "reliable quantification of even average annual recharge remained problematic and the issue of annual variability of recharge was covered rather superficially". Results from various groundwater and mining related studies, including this study, have shown that the common method of expressing recharge as a percentage of rainfall are not adequate in predicting contamination loads entering the groundwater regime. In fact, this study has shown that in the case of experimental cell 9, outflow (recharge) rates varied between 0.0 to 17.6% of mean annual rainfall for the three years monitored.

Flow through the unsaturated zone is complex. Soil moisture can move downwards (or upwards) under gravitational, capillary and other forces. The rate of soil moisture flow is controlled by the physical properties of the porous medium, i.e. the unsaturated hydraulic conductivity. Which is a function of the volumetric water content with the flow rate de-

creasing with decreasing moisture content at constant hydraulic gradient.

The soil moisture content is related to the effective pore-size, hydraulic head and soil suction. This relationship is hysteretic, i.e. a different relationship exists for a drying and wetting cycle for a particular soil. The relationship between soil moisture and hydraulic head/soil suction can be obtained by means of water retention tests. The unsaturated hydraulic conductivity – soil moisture relationship can be obtained indirectly by means of the statistical approach. Many authors have confirmed the accuracy of this approach (van Genuchten, 1980, Fredlund et al., 1994, van Schalkwyk and Vermaak, 1998).

The Richards equation for one-dimensional unsaturated flow, used in most unsaturated flow models, assumes a piston-like downward flow of water through a soil profile. However, in many soils, water flows preferentially along pathways of least resistance within the soil profile, thereby nullifying the application of Darcian based equations. Many preferential flow processes have been identified including macropore channelling, fingering flow and funnelled flow. Macropore channelling refers to flow of water along fissures, cracks and other discontinuities within the soil profile. These discontinuities originate due to desiccation cracking, shrinking clays, bio-activities and other processes. Fingering flow refers to finger-like flow of water through mainly layered and water-repellent soils and occurs due to the wetting front becoming unstable. Funnelled flow refers to funnel-like flow of water along either a coarse-grained or fine-grained soil layer, depending on the moisture content. Preferential flow processes are discussed by van Schalkwyk and Vermaak (1998) and Miyazaki (1996).

The quantity of water infiltrating and evaporated from the soil surface depends on the hydrological condition occurring at the soil surface of any given time. Climate is an obvious factor contributing to the specific hydrological condition. Water infiltration occurs mainly by means of rainfall. Evaporation is not easily quantified and depends on a number of climatic conditions including temperature, net radiation, humidity and wind velocity. Transpiration is also not easily quantified and depends mainly on the amount of moisture available to plants.

Table 1: Key factors affecting pyrite oxidation and acid water neutralisation

Pyrite Oxidation	Neutralisation of acid water
Presence of bacteria	Type of base mineral
Microscopic texture and crystal form of the sulphide mineral	Particle size
Presence and availability of oxygen	Reaction rate and contact time
Temperature	Presence of secondary minerals to armour the base mineral
Particle size distribution	pH / Acidity
Presence of water	Presence of oxygen

3 SOIL COVER DESIGN

The main purpose in designing soil covers is to separate the waste material from the surface environment, to restrict infiltration of water into the waste material and to restrict ingress of oxygen into the waste material. By restricting water infiltration, and therefore limiting recharge, the volume of leachate generated from these facilities can be substantially reduced. The US Environmental Protection Agency

(USEPA) considers keeping water separated from the contaminated material to be the prime element in reducing long term water quality impacts (USEPA, 1989). In addition, by restricting influx of oxygen and water, the rate of pyrite oxidation can be substantially reduced. The subsequent anaerobic conditions will limit iron-oxidising bacteria being established in the waste material, which will further limit pyrite oxidation processes. By placing an effective soil cover, a decrease in the quantity of leachate generated can also be attained.

Cover systems have to perform efficiently for long periods of time. The design life of the cover will mainly depend on the hydrogeology of the site and the length of time that maintenance and funds are available for repair of the cover. The design is complicated by numerous environmental factors that includes (Daniel & Koerner, 1993):

1. Cyclic wetting and drying;
2. Wind and water erosion;
3. Total and differential settlement of the discard material or foundation material;
4. Penetration by plant roots and burrowing animals;
5. Temperature extremes, possibly freeze/thaw cycles;
6. Down-slope slippage or creep;
7. Vehicular movement or haul trucks traversing the cover;
8. Deformation caused by earthquakes;
9. Long term moisture changes caused by water movement into or out of the underlying waste;

3.1 Types of cover systems

The type of cover system should be cost effective but should also consider the climate prevailing at the site. The design of cover systems can be very challenging because of the number of site-specific environmental stresses that could act on the cover system. Cover systems can be grouped into soil, synthetic, and composite covers. Due to the fact that synthetic covers are generally more expensive compared to soil covers (considering the large areas of rehabilitated opencast mines) soil covers are preferred. Soil covers can further be grouped into simple and complex covers.

3.2 Simple single-layer soil covers

Since cost is always an important consideration, especially when soils material is scarce, simple single-layer soil covers are preferred. These soil covers generally comprise of a fine textured soil and high soil-moisture contents have to be maintained to limit oxygen diffusion. Large, seasonal variations in moisture content could result in the degradation of the soil cover with time. These fluctuations are most

prominent near surface and hence, thin soil covers are most susceptible. The large fluctuations could result in the formation of desiccation cracks, which results in increased permeability. Steffan, Robertson & Kirsten (SRK), (1989) reported that in British Columbia, Canada, the minimum thickness of simple, single layer soil covers to maintain saturated conditions throughout the dry season is 2m. Upward migration of salts, caused by evapotranspiration processes, could result in difficulty in establishing vegetation and cause contaminated surface run-off water. However, considering the generally coarse nature of coal discard, upward migration of salts should not be a problem. Simple soil covers may not adequately withstand water and wind erosion and as such, vegetation or rip-rap is required as erosion protection.

3.3 Complex, multiple-layer soil covers

Because of the number of limitations associated with simple soil covers, complex multiple-layered soil covers gained credibility in limiting infiltration of water and oxygen diffusion. Complex soil covers comprises of a number of soil layers, each performing a specific function. Complex soil covers could comprise of the following layers:

1. Erosion control layer
2. Moisture retention zone
3. Upper drainage layer
4. Infiltration barrier
5. Lower capillary barrier
6. Basic layer

Erosion protection can be provided either by vegetation or by a layer of coarse gravel or rip-rap. Environmental conditions generally favour vegetation. Research by the Anglo American Corporation (1986) have shown that top soil depths of 400mm, 600mm and 700mm are required for pasture, maize and sorghum production respectively. The USEPA requires soil covers sustaining vegetation growth to be at least 600mm and 150mm for hazardous and municipal waste respectively. Apart from erosion protection, the erosion protection layer could serve several important functions:

1. Protect underlying layers against excessive wetting and drying that could result in desiccation cracking;
2. Physically separate the waste from burrowing animals and plant roots;
3. Store water that has infiltrated the cover which is later removed by evapotranspiration;
4. Minimise human intrusion into the waste

A moisture retention zone could be constructed in addition to the erosion protection layer. The purpose of the moisture retention zone is firstly to maintain a constant moisture content in the underlying barrier layer, thereby preventing desiccation cracking and secondly to retain and store moisture that has infiltrated the cover. This moisture could support vegetation growth and are subsequently removed by evapotranspiration processes.

An upper drainage zone generally comprising of highly permeable sandy and gravelly material or a geotextile filter could also be installed. The upper drainage layer could serve several purposes namely:

1. To prevent seepage forces from developing within the soil cover, thereby improving slope stability;

2. Reducing the hydraulic head acting on the barrier layer thereby minimising infiltration;

3. Draining water from the overlying layers, thereby increasing the storage capacity of these zones;

4. Minimising evapotranspiration from the barrier layer thereby preventing desiccation cracking.

The infiltration barrier layer is often viewed as the most critical engineered component of the final cover system. The barrier layer limits infiltration of water into underlying waste material. In addition, the barrier layer promotes storage and drainage and ultimate, by evapotranspiration of water from overlying soil layers. Typically, the barrier layer comprises of a relatively thin (150mm-600mm) layer of soil with a design hydraulic conductivity of 1×10^{-9} m/s. Daniel and Koerner (1993) stated that many problems are associated with clay liners including:

1. Clay liners are difficult to compact properly on a soft foundation, such as coal discard;

2. Compacted clay will tend to desiccate from above and below resulting in increasing permeabilities;

3. Differential settlement of the coal discard could result in cracks developing in the liner if tensile strains become excessive;

4. Compacted clay liners are difficult to repair if damaged;

5. Compacted clay is vulnerable to damage from freezing;

Research by Montgomery and Parsons (Daniel & Koerner, 1993) has shown that neither 150mm nor 450mm of top soil is adequate in preventing desiccation of the underlying clay liner. Plant roots penetrated 250mm into the clay liner and extended up to 750mm into the clay. Similar findings have been observed from research conducted by Corser and Cranston (Daniel & Koerner, 1993). Daniel and Koerner (1993) suggested that the only practical way to protect compacted clay against desiccation is to cover the barrier layer with both a geomembrane and a layer of soil.

4. ASSESSMENT OF THE PROBLEM

Flow through the unsaturated zone needs to be well understood to predict infiltration into discard material and recharge rates. Due to the highly complex nature of recharge processes, the research focussed on predicting recharge by means of numerical unsaturated flow models. Unsaturated flow modelling is discussed in section 7.

Since moisture movement in the unsaturated zone is predominantly downward (or upward), recharge can be estimated by means of a one-dimensional flow model, taking the area of the rehabilitated opencast mine into account.

Infiltration into the soil covers will depend on the specific hydrological situation occurring at ground surface at any given time. The hydrological situation will obviously be a factor on the specific climatic situation occurring at any given time. In the case of excessive water occurring at the surface (due to rainfall or irrigation), infiltration will result, otherwise, evapotranspiration will take place. The rate of evapotranspiration will be a function of the type of vegetation, climate (temperature, relative humidity, net radiation and wind velocity) and moisture content within the soil cover.

The rate of infiltration will mainly be a function of the hydrological characteristics of the soil cover material. In the case of low permeability material (barrier layer), limited infiltration will take place and additional precipitation will result in run-off. However, cyclic wetting and drying will result in the degradation of the liner leading to a subsequent increase in permeability. It is therefore necessary to assess the long-term stability of a barrier layer. In the case of a less-stable barrier layer, the layer should be protected against desiccation by a protective layer.

The assessment of impacts from coal discard on groundwater resources is further complicated by the fact that the deterioration of groundwater quality is generally a delayed process. Acid Mine Drainage could occur years after rehabilitation has been completed. It could be argued that soil covers should therefore perform optimally only after this delayed period. However, since soil covers also prevent oxygen diffusion into the discard material, acid formation and pH buffering processes could be limited or, at least, significantly delayed if effective soil covers are put in place after backfill/stockpiling.

5. EXPERIMENTAL SETUP & PROCEDURE (METHODS AND MATERIALS)

The experiment site is located at the Ngagane station close to Newcastle in KwaZulu-Natal and was constructed during 1993. By constructing the experimental site close to the Natal coalfields, representative climatic conditions could be obtained which could assist in the calibration of data. The climate is characterised by a seasonal rain, mostly falling in short, high intensity thundershowers. (Refer to Section 6.1: Climatic Results)

Ten experimental cells were constructed by the Department of Water Affairs. Each cell was constructed to have a 10×10m surface area with an overall depth of 3m. All cells have been constructed to drain freely and have a collection system so that the quantity of outflow (recharge) can be measured. The cells are divided by one-meter thick low-permeability clay barriers. In addition, the cells were constructed to have different slopes to test the effect of run-off. Some cells were constructed without any soil covers while others have been constructed with either thin (300mm) single soil covers or thick (1 000mm) complex soil covers. The cell configurations of the ten experimental cells are summarised in Table 2.

Table 2: Configuration of the ten experimental cells

Cell	Layers (top to bottom)	Slope
1	No cover, unvegetated, uncompacted spoils	1%
2	No cover, unvegetated, compacted spoils	1%
3	No cover, vegetated, uncompacted spoils treated with lime	1%
4	300mm Avalon soil, uncompacted, vegetated; uncompacted spoils	1%
5	500mm Avalon soil, compacted, vegetated; uncompacted spoils	1%
6	300mm Avalon soil, uncompacted, vegetated; 700mm Estcourt soil, compacted; uncompacted spoils	1%
7	300mm Avalon soil, uncompacted, vegetated; 700mm Avalon soil, compacted; uncompacted spoils	1%
8	700mm Avalon soil, uncompacted, vegetated; 300mm Estcourt soil, compacted; uncompacted spoils	1%
9	300mm Avalon soil, uncompacted, vegetated; 700mm Estcourt soil, compacted; uncompacted spoils	5%
10	300mm Avalon soil, uncompacted, vegetated; 700mm Estcourt soil, compacted; uncompacted spoils	10%

The purpose of vegetation of rehabilitated discard dumps is firstly to protect against erosion and secondly to increase evapotranspiration, thereby preventing soil moisture movement into the discard material. The vegetation of the experimental cells comprised of a combination of stolan type and tufted type grasses. The former is preferred to protect against erosion while the latter are conducive to high evapotranspiration rates.

It can be argued that the lysimeters are not representative of a coal discard dump since the coal discard dumps are typically more than 30m in thickness, thereby creating substantially more storage for moisture. However, the main objective of this experiment was to compare different soil cover configurations in terms of their ability to limit water entering the coal discard and to compare actual with hypothesised outflow results. In this regard, the experimental set-up was deemed representative. In addition, the coarse grained coal discard material acts as a capillary barrier, which limits upward movement of water into the soil cover. Once the storage has been filled, most water in the discard material will probably manifest as recharge (or outflow).

5.1 Monitoring program

The monitoring program was started after the experimental cells were constructed in 1993 and continued up to 1996. Recent funding by the mining industry enabled continued monitoring up to the end of 1999. Weekly monitoring is conducted by the Department of Water Affairs and Forestry and the data is send to the authors for analyses. Long term monitoring is crucial in the calibration and verification of unsaturated flow models. In addition, although not being the main focus of the project, long-term bi-weekly measurements of leachate quality could prove to be highly valuable in understanding and ultimately predicting leachate quality under different soil cover scenarios. The monitoring programme comprises of weekly measurements of weather data, oxygen and carbon-dioxide readings, measurement of outflow (recharge) through the experimental cells and water quality analyses of leachate.

Continual, hourly climatic data, that includes rainfall, temperature, relative humidity and net radiation, have been measured by means of a self logging weather station at the site. The weather data was then transferred and analysed. In addition, measurements are made from a backup rainfall meter and A-pan evaporation pan.

Neutron access holes were installed in each cell. Neutron probe measurements were made periodically to measure in situ soil moisture content.

The leachate collection system comprises of a combination of continual tipping buckets and flow meters installed at each experimental cell. Tipping buckets are equipped with mechanical counters to enable weekly measurements of the leachate quantity.

Water quality analyses are conducted by the Institute for Water Quality Studies at Roodeplaat dam. The leachate samples are analysed for pH, electrical conductivity, total dissolved solids, total alkalinity, total hardness, sodium, potassium, calcium, magnesium, sulphate, chloride, nitrate, ammonia, fluoride, phosphate, silica, manganese and iron.

Weekly measurements of oxygen and carbon dioxide were made at all experimental cells. Perforated stainless steel canisters were installed in fixed locations with tubes protruding at surface to connect the gas monitor for measurement.

5.2 Soil property measurements

Two soil types, Avalon and Estcourt, were used in the experiments for cover material. These soils occur naturally in the area and are commonly used for rehabilitating purposes. The physical, hydraulic and chemical properties of the soils were tested. Both laboratory and in situ techniques were used to determine the hydraulic properties of the soil. Similar tests were conducted on the coal material.

The coal material comprises of more than 70 % gravel and sand size particles, and less than 10% clay. Avalon soils were selected to act as a drainage layer and a layer to sustain vegetation while the Estcourt soils were selected to act as a clay barrier and prevent infiltration. However, the foundation indicator test results indicated that both soils comprised of a sandy clay to clayey sand and both soils were markedly similar insofar particle-size distribution and Atterberg limits are concerned. The soils can be described as a clay (CL) by the USCS system and a sandy clay loam to clay loam by the soil texture classification chart.

The chemical and mineralogical analysis of the cover material revealed that Avalon and Estcourt soils are similar both chemically and mineralogically. However, the Estcourt soils do contain more smectite type clay minerals, which could cause instability of the soil structure due to cracking. In addition, and perhaps more significantly, the Estcourt soils do not contain geothite, an iron hydroxide that contributes to the stabilisation of the soils. These aspects could explain the different physical behaviour of the soils. It could also indicate that Estcourt soils are more prone to structural damage, for instance, desiccation cracking.

The coal material was characterised by a low bulk density and low particle densities indicating carbon in the material. The compacted Estcourt soils exhibited lower porosity and higher bulk densities than compacted Avalon soils.

Both in situ and laboratory techniques were applied to obtain the saturated hydraulic conductivity. The tests include laboratory permeability tests on remoulded compacted samples, laboratory controlled outflow tests (Lorentz et al., 1995) to determine both saturated and unsaturated hydraulic conductivity, both large diameter (1 000mm) and small diameter (300 mm) double ring infiltrometer tests and Guelph permeameter tests. The in situ tests provided representative saturated hydraulic conductivity values. Higher in situ conductivity values, relative to the laboratory tests, could indicate the presence of preferential pathways. The results of the saturated hy-

draulic conductivity tests are summarised in Table 3.

The results indicate that in situ values are up to a factor of 18 000 higher compared to laboratory permeability tests. The large differences could be ascribed to preferential pathways along microfractures and possible degradation of the soil covers. Results of the outflow tests (conducted on undisturbed soil samples) and in situ tests are markedly similar for the respective soil layers, the exception being compacted Estcourt soils exhibiting higher hydraulic conductivity for the large diameter infiltrometer tests compared to in situ tests conducted on smaller soil volumes. This could indicate that the large diameter infiltrometer test intercepted widely spaced preferential pathways which were not intercepted during the other tests resulting in higher hydraulic conductivities and therefore more representative values. However, experimental error can not be excluded.

In general, the uncompacted coal exhibits the lowest saturated hydraulic conductivity, though similar to the compacted coal. Uncompacted Avalon soils exhibit saturated hydraulic conductivities similar to the compacted coal. Compacted Avalon soils are characterised by saturated hydraulic conductivities lower by a factor of approximately 10 compared to uncompacted Avalon soils. The saturated hydraulic conductivity of compacted Estcourt soils in turn is approximately 10 times lower compared to the compacted Avalon, though results of the large diameter infiltrometer tests indicate similar in situ saturated hydraulic conductivities.

Although uncompacted Avalon and compacted coal discards exhibit similar saturated hydraulic characteristics, the water retention and unsaturated hydraulic conductivity characteristics differ significantly. The soils have a much higher water-retention capability than the coal discard. This implies that moisture moving from the soil layers into the coal discard will quickly drain, and will therefore not be removed by means of evapotranspiration. Compacted Estcourt soils exhibit the highest water retention characteristics. This implies that water retained in Estcourt soils could be removed by evapotranspiration before infiltrating the coal discard.

The tension infiltrometer and outflow results indicate that the coal discard is much less conductive at low soil suctions. This indicates that water has been drained and that the retained water is strongly adsorbed to the coal particles. Although uncompacted Avalon soils have saturated hydraulic conductivities higher than compacted Avalon soils, where as both soils have similar unsaturated hydraulic conductivities at low suctions, probably indicating preferential flow along large pores in the Avalon soils. The Estcourt soils have higher unsaturated hydraulic conductivities than Avalon soils at corresponding suctions. This is expected, since

Type of test	Remoulded sample		Undisturbed sample		
	Falling head tests	Outflow tests	Guelph permeameter	300 mm Double Ring	1 000 mm Double Ring
Area of sample tested	39.6 cm^2	23.8 cm^2	78.5 cm^2	706.9 cm^2	7 854.0 cm^2
Uncompacted coal	3.20×10^{-5}	3.16×10^{-3}	7.30×10^{-3}	3.16×10^{-3}	1.00×10^{-2}
Compacted coal	4.70×10^{-5}	5.64×10^{-3}	5.00×10^{-3}	5.64×10^{-3}	2.78×10^{-3}
Treated coal	N.A.	2.01×10^{-3}	4.00×10^{-3}	2.01×10^{-3}	1.00×10^{-2}
Uncompacted Avalon	N.A.	3.20×10^{-3}	2.28×10^{-3}	3.20×10^{-3}	3.33×10^{-3}
Compacted Avalon	2.35×10^{-8}	3.07×10^{-4}	N.A	3.10×10^{-4}	4.15×10^{-4}
Compacted Estcourt	6.15×10^{-8}	1.46×10^{-5}	2.00×10^{-5}	1.50×10^{-5}	3.33×10^{-4}

higher water retention capabilities have been measured at corresponding soil suctions.

6. EXPERIMENTAL RESULTS

6.1 Climatic results

The type of rainfall associated with the Northern Natal areas is highly seasonal in nature. A Mean Annual Precipitation (MAP) of 1010mm has been recorded over the four-year period varying between 801 to 1311mm. The rainfall is highly seasonal in nature and most of the rainfall occurs during the October to April months (89% MAP). Rainfall intensities are high and 38.8% of all rainfall days, which contributes 84,2% of total rainfall, recorded more than 20mm/day while seven days, which contributes 25.2% of total rainfall, recorded more than 100mm/day. These rainfall events occur mainly as thundershowers implying that most of the high intensity rainfall occur within a few hours. The 1995/1996 rainfall season was particular wet with 1311mm recorded and most of this rainfall occurred during November to February. Although high rainfall has also been recorded for the 1996/1997 rainfall season, the rainfall was more equally spread along the rainfall months and the year was also characterised by a high winter rainfall. These events had significant impacts on the outflow quantity recorded during these months.

Average temperatures varied little and ranged between 13 and 20°C for winter and summer months respectively. However, large temperature fluctuations (up to 30°C) were recorded for some 24-hour periods. Minimum temperatures of –7°C and maximum temperatures of 30°C were recorded.

6.2 Outflow results

The experimental cells were constructed to differentiate between the different physical aspects of the covers that affect the infiltration and outflow volume. Actual outflow results, expressed as an percentage of rainfall, are summarised in Table 4. Based on these aspects and the theory of unsaturated flow, a hypothesised ranking regarding the order of effectiveness of the experimental cells could be established.

The outflow results confirms the hypothesis that covered cells are effective in reducing outflow through the discard material. The uncovered cells (Cells 1, 2 and 3) remain the least effective cells with an average outflow of 26% of MAP. The compacted cell performed slightly better than uncompacted cells.

The outflow results have shown that outflow can be reduced by approximately 39% by covering coal discard with at least 300mm of vegetated, low permeable material and more by covering coal discard with 500mm to 1 000mm cover material. Little difference in outflow has been noted between the thin, single-layered soil covers and thick double-layered soil. Outflow ranged between 9.5 and 19.5% with the cell 7 (300mm uncompacted and 700mm compacted Avalon soil) being, in contrast to the original hypothesis, the most effective cell.

The outflow results further indicate that a definite change in the effectiveness of some soil covers occurred with time. Cells 9 and 10 outperformed all cells up to the 1995/1996 rainfall season after which the effectiveness of these soil covers decreased significantly. Similar trends have been identified for cells 6, 7 and 8. Initially, cell 6 outperformed cells 7 and 8. Since the 1995/1996 rainfall season, cell 7 outperformed both cells 6 and 8.

It does not appear that vegetation had a significant effect on the outflow volume from the cells. Although vegetation generally has a positive effect (due to increased evapotranspiration), this could have been off-set by increased infiltration by means of preferred infiltration along plant root holes.

Table 4: Outflow results (expressed as an percentage of rainfall) of the respective experimental cells

	Oct 93 Sep 94	Oct 94 Sep 95	Oct 95 Sep 96	Oct 96 Sep 97	Total	Hypothesised
Cell 1	16.81%	11.91%	24.78%	34.52%	25.82%	22%
Cell 2	13.12%	15.02%	21.40%	35.30%	21.34%	21%
Cell 3	12.96%	17.99%	19.30%	55.80%	31.89%	20%
Cell 4	8.51%	2.26%	17.76%	29.16%	19.05%	16%
Cell 5	4.87%	0.64%	11.53%	23.59%	13.57%	12%
Cell 6	6.77%	2.96%	15.86%	8.82%	14.65%	7%
Cell 7	9.47%	3.45%	14.97%	1.32%	9.46%	11%
Cell 8	7.15%	2.88%	16.69%	28.64%	19.53%	10%
Cell 9	0.05%	0.00%	17.57%	22.76%	13.51%	4%
Cell 10	0.77%	0.00%	15.29%	33.75%	16.80%	3%

The reasons for the decrease in soil cover effectiveness are as yet not fully understood. It is highly likely that the very dry winter preceding the 1995/1996 resulted in desiccation cracks appearring especially in the Estcourt clayey soils. With the high intensity 1995/1996 rainfall season, water flowed preferentially along the desiccation cracks before the cracks could close. This could explain the effectiveness of both cells 5 and 7 (comprising of Avalon soils) during this season. In addition, the effect of burrowing animals (burrowing into the moist, clayey material) could have contributed to the flow through these cells. Animal holes, approximately 50 – 90mm diameter have been observed on the surfaces of some cells during field visits. In addition, other preferential mechanisms, such as fingering and funneling processes could have contributed to the increased outflow through cells 9 and 10.

6.3 Water Quality Results

The leachate qualities recorded at each cell in this investigation have been influenced by the physical, chemical and biological processes mentioned earlier. It can therefore be assumed that the recorded leachate qualities are indicative of the ability of the different dump covers to inhibit or accelerate the various geochemical processes active in each case. In order to evaluate the performance of the different dump covers with respect to leachate quality, a statistical and comparative analysis were conducted on the recorded leachate quality.

The leachate is in general characterised by high salinity levels and neutral pH levels (except Cell 1), and can be classified as Mg-Ca-ALK-(SO_4) type water. Major ions analysed for in solution are Ca, Mg, Na, SO_4, and Cl. The following key trends were identified:

1. ABA analyses of the discard material indicated that the discard material has greater acid generating potential, than neutralising potential suggesting that the discard material will produce acid leachate given suitable circumstances and time. An acid break-through curve, substantiating the ABA analyses, has been recorded at Cell 1. The pH level at Cell 1 dropped over a period of only two weeks from an average 7.4 to an average 3. This trend is indicative of how quickly the base assimilative capacity is used up in the discard material as well as that the pH, after it was buffered at pH 7.4, is now buffered at pH 3.

2. Increasing trends of salinity have been recorded at Cells 1-5 indicating acid generation and pH buffering processes. At these cells levels of EC, TDS, Mg and SO_4 concentration increased over time.

3. Trends of decreasing salinity were recorded at cells 6-10 indicating a significant reduction in the rate of pyrite oxidation and pH buffering processes. The recorded EC, TDS, Mg and SO_4 concentration levels at these cells decreased over time.

4. There is a statistical significant correlation between Mg and SO_4, and Mg and EC. This correlation reflects the correlation between the production of acidity and the buffering of pH by Ca and Mg carbonate minerals. The reason why there is no significant correlation between Ca and EC and or SO_4 could be related to the precipitation of gypsum controlling the Ca concentration in solution.

From these results it can be concluded that the single layered dump covers (Cells 4 and 5) are less effective in preventing the production of acidity and salinity from the discard material. However, these single layered covers inhibit some of the acid generating processes by inhibiting the ingress of water and oxygen into the discard material. This means that an acid breakthrough in leachate quality will occur but will only take place in a number of years. Multi layered covers on the other hand have the capacity to prevent the production of acidity by establishing permanent anaerobic conditions in the discard material. This is confirmed by the oxygen monitoring results showing that zero oxygen was recorded at cells 6-10 for most of the monitoring period. In restricting the ingress of water to 7%MAP and completely preventing the migration of oxygen into the discard material, the multi layered covers succeed in preventing acid generating processes and subsequent generation of salinity through pH buffering processes.

7 PREDICTION OF LEACHATE QUANTITY: UNSATURATED FLOW MODELLING

Since flow in the vadose zone is dynamic and transient in nature, numerical modelling by means of computer models offers the best tool to understand, simulate and ultimately predict flow of water and solutes through the vadose zone. Since flow in the vadose zone is predominantly downward (or upward), flow through the vadose zone can be presented by a one-dimensional model. Transient flow in the vadose zone can be described by the Richards equation, which is based on Darcy's law (modified from Feddes et al., 1988):

$$\frac{\partial \varphi}{\partial t} = \frac{1}{\grave{e}'(\varphi)} \frac{\partial}{\partial z} \left[K(\varphi) \left(\frac{\partial \varphi}{\partial z} + 1 \right) \right] - \frac{S}{\grave{e}'(\varphi)} \qquad [1\text{-}1]$$

$S =$ a sink term and represents the volume of water taken up by plant roots (cubic volume water per cubic volume soil per unit time), ψ = soil suction (kPa), θ = volumetric water content in the soil (m^3/m^3), z = soil depth, (m)

A boundary condition is defined at the top of the system where water infiltrates the system by means of precipitation or irrigation and where water leaves the system by means of evapotranspiration. Evapotranspiration can be directly measured or can be calculated by means of a number of equations including the Penman, Priestly and Taylor, Monteith-Rijtema and the Makink equations. Rainfall is generally directly measured. Another boundary condition is defined at the bottom of the system and this will depend on the particular situation. A free-drainage lower boundary has been assumed for the experimental cells.

A number of numerical solution schemes can be used to simulate flow of water in the vadose zone. The finite-difference technique is the most frequently used technique to solve extended Richards' equation. Other techniques, including finite element, are also used to simulate flow of water through the vadose zone. Forty-five models have been identified that are able to handle unsaturated flow conditions. The most popular models include HELP, FEMWATER, LANDFIL, SEEP/W and ACRU. However, many of these models are analytical models and are not able to simulate dynamic conditions. It was decided to use the model, SWACROP, to simulate flow of water through the soil covers and coal discard material.

The numerical modelling results comprise of estimated daily outflow results in response to rainfall and evapotranspiration processes. These results were compared with actual weekly outflow measurements made at the experimental site.

In general, the numerical model has over-estimated outflow through the experimental cells.

However, general trends in outflow quantities have in general been well predicted by the model. The over-estimation of the unsaturated model is expected, since the model has of yet not been calibrated with actual data. A new Water Research Commission project has been launched in 1999 with the aim of calibrating the current unsaturated flow model.

Possible reasons for the inaccurate predictions of outflow could be as follows:

1. Rainfall at the experimental site occurs mainly as high intensity, short-duration thundershowers. The model however only accepts daily rainfall data with the result that rainfall is averaged by 24 hours. This results in the under estimation of surface runoff and over estimation of infiltration;

1. Possible increased run-off at sloped experimental cells is not adequately addressed by the model;

2. Hydraulic properties of the soil cover exhibit clear bi-modal hydraulic characteristics. This implies that saturated hydraulic conductivities are much higher than unsaturated hydraulic conductivities near saturation. These bi-modal characteristics are not accurately described by traditional equations describing unsaturated conductivity as a function of moisture content/hydraulic head;

3. The model does not take preferential flow into account. The model assumes constant permeabilities and therefore does not take soil cover degradation into account.

Although the numerical model does not at present predict outflow accurately, the results do hold great promise. General outflow trends have been accurately predicted by the model. It is of the authors opinion that due to the many factors influencing unsaturated flow, numerical modelling is the only way to assess and predict soil cover performance.

8 CONCLUSIONS, COST IMPLICATIONS AND FURTHER RESEARCH

The research has shown that unsaturated flow processes are highly complex. Both climatic aspects and the hydraulic properties of the soil cover and coal discard have to be considered in the assessment of outflow (or recharge) originating from rehabilitated opencast mines and discard dumps. Rainfall is highly erratic with approximately 90% of rainfall occurring in summer and mostly precipitate in the form of high intensity thundershowers. Huge environmental stresses are exerted on the soil covers and these complicate the assessment of outflow (or recharge) considerably. The research has shown that high intensity rainfall seasons, such as the 1995/1996 rainfall season do have a considerable impact on recharge and leachate generation with all experimental cells recording more than 10% recharge for the year. In contrast, less intense rainfall patterns, such as the 1996/1997 rainfall season, have a smaller impact on recharge and leachate generation. Since the rainfall patterns generally differ considerably from European and American climate, models developed in those countries might not be applicable in South Africa.

Although most unsaturated flow models assume constant hydraulic properties, the research has shown that soil cover systems are indeed highly dynamic. Desiccation cracks develop in response to cyclic wetting/drying of the soil cover and bioactivities (root penetration and burrowing animals) could contribute significantly in soil cover degradation, especially in dry seasons (such as the 1995 season) where plant roots and burrowing animals penetrate the barrier layer in search of moisture. Since rainfall generally occurs as high intensity thundershowers, it is likely that flow occur predominantly along preferential pathways with unsaturated flow contributing little to outflow (or recharge).

The outflow results should be interpreted with caution. Although the results indicate little difference in outflow between single-layer and complex-layer soil covers, it is important to note that complex soil layers, with the exception of experimental cell 8, began to outperform single layers during the 1996/1997 rainfall season. This can be attributed to the fact that the low permeable barrier layer is protected against desiccation and degradation. The poor performance of the steep-sloped experimental cells 9 and 10 are as yet not well understood, but it is obvious that preferential flow processes are at work.

Caution should be exercised when extrapolating these results to actual coal discard dumps. The fact that actual discard dumps are typically ten times the thickness than the experimental cells implies that much more storage will be available and that recharge rates will possibly be lower. However since coal discard acts as an capillary break, little water will be removed from the discard by means of evapotranspiration and outflow quantities might be similar to that of experimental cells once the initial storage capacity has been surpassed.

8.1 Recommendations for the placement of covers to limit infiltration

Based on the results of the research, the following general recommendations regarding the placement of soil covers to limit infiltration into the discard material can be made. Recommendations regarding technical and construction details of the soil cover are not made since these fall outside the scope of the project and are widely available.

1. The soil cover system should be designed to perform in harmony with the climate of the area. While single-layer compacted clay liners could be considered in humid conditions, they are not re-

commended for arid conditions or climates characterised by cyclic rainfall patterns;

2. Due to the many factors influencing soil cover design, assessment of soil cover effectiveness should be made by means of numerical modelling techniques. Numerical models based on the finite-difference approach are preferred. The model program HELP could be used for screening assessment of the proposed soil cover but finite-difference based models should be used for the long-term assessment of outflow (or recharge).

3. Rehabilitated opencast mines and discard dumps should be compacted if possible. This could reduce the permeability and reduce outflow

4. According to Daniel and Koerner (1993) single-layer compacted clay liners are not barrier systems of choice. It is recommended that a geomembrane or geomembrane overlying geosynthetic clay liner be used as cover system where economically feasible;

5. Where clay materials are to be used as barrier materials, the barrier layer should be protected against desiccation, plant root penetration and burrowing animals.

6. A protective layer of at least 600mm of compacted soil should be used to protect the barrier against desiccation cracking.

7. Rehabilitated opencast mines and discard dumps should be shaped to assure that a maximum slope is attained. This will increase runoff, thereby reducing infiltration into the discard material. However, slopes should also be designed to prevent erosion and this implies an optimum slope of 1:3 - 1:5:

8. The research indicates that vegetation does not significantly reduce outflow rates. However, vegetation should be established as part of soil cover design to act as erosion protection.

9. Long term degradation of soil covers should be expected. Funds should be available to repair soil covers damaged by for instance erosion and settlement of the discard material.

10. Chemical and mineralogical studies on the proposed soil cover material should form part of the geotechnical investigation. Soil that is less prone to structural degradation should be used for soil cover material;

11. The barrier layer should contain few smectite type clay minerals and significant soil binding properties such as iron oxides. This will add to the structural integrity of the soil cover material;

8.2 Future research

1. Continued monitoring at the experimental site is crucial in order to study long-term acid formation and pH buffering processes for the different soil cover scenarios;

2. Continued monitoring is also crucial to study the effects of long-term soil cover degradation;

3. Existing unsaturated flow models should be calibrated to suit South African climatic situations;

4. The calibration of unsaturated flow models should include the continues measurement of soil moisture with depth in response to infiltration and evapotranspiration processes;

5. The experimental site should make provision for the measurement of surface run-off;

6. A number of rehabilitated opencast mines and discard dumps should be included in future research in order to assess actual cover performance and to broaden the present database.

8.3 Acknowledgements

A number of institutions were involved either directly or indirectly through associated research projects. The research programme were conducted by Wates Meiring & Barnard and funded by the Water Research Commission. A new research project, funded by the WRC, has been launched in 1999 with the main objective to calibrate existing unsaturated flow models. The Department of Water Affairs and Forestry was involved in the construction of the experimental cells and for the weekly monitoring at the site. The research made use of the results of another WRC project by the University of Stellenbosch, Department of Microbiology, conducted concurrently with this project in which the occurrence of iron oxidising bacteria under different soil cover scenarios was investigated. The unsaturated hydraulic properties of the soil covers and coal discard were assessed and tested by the University of Natal, Department of Agriculture Engineering. A number of institutions including the Institute for Soil, Climate and Water (ISCW), the Chamber of Mines Vegetation Unit and the University of Witwatersrand, Department of Civil Engineering were involved in various stages during the project.

9 REFERENCES

Anglo American Corporation 1986. *Amcoal surface rehabilitation. Soil depth, soil compaction, and ripping studies: Land rehabilitation, 1984/1985 Season.* Report no. Gen 41/86

Bredenkamp, D.B., Botha, L.J., Van Tonder, G.J. & Van Rensburg, H.J. 1995. *Manual on quantitative estimation of groundwater recharge and aquifer storativity.* TT73/95, Water Research Commission, Pretoria

Brink, A.B.A 1983. *Engineering Geology of Southern Africa; Volume 3: The Karoo Sequence.* Silverton: Building Publications.

Daniel, E.D. & Koerner, R.M. 1993. Cover systems In Daniel (Ed) *Geotechnical Practice for Waste Disposal:* London: Chapman & Hall.

Feddes, R.A., Kabat, P., Van Batiel, P.J.T., Bronswitji, J.J.B. & Halbertsma, J. 1988. Modelling soil water dynamics in the unsaturated zone – State of the art. *J. Hydrol.* 100:69-111

Fredlund, D.G., Xing, A. & Huang, S. 1994. Predicting the permeability function for unsaturated soils using the soil-water characteristic curve. *Can. Geotech. J.* 31:533-546

Helling, C.S. & Gish, T.J. 1991. Physical and chemical processes affecting preferential flow in Preferential Flow In Gish, T.J. and Shirmohammadi, A. (Eds) *Proceedings National Symposium; American Society of Agricultural Engineers, St. Joseph:* MI, pp 77

Lorentz, S.A., Ballim, F. & Musto, J.W. 1995. *Hydraulic and physical properties of soil covers for mine and waste dump rehabilitation: Report to Wates, Meiring & Barnard.* Department of Agricultural Engineering, University of Natal, Pietermaritzburg.

Miyazaki, T. 1993. *Water flow in soils.* New York: Marcel Dekker Inc.

Rykaart, E.M. & Wates, J.A. 1999. *The performance of soil covers for opencast mine and waste dumps in South Africa: Draft Final Report* Project no K5/575, Water Research Commission, Pretoria

Steffen Robertson & Kirsten (B.C.) Inc. 1989. *Draft Acid Rock Drainage Technical Guide. Volume 1 – Technical Guide. Prepared for the British Columbia AMD Task Force.* Report No. 66 002/1

Stone, A.W. 1991. Aquifers and landfills - a formula for peaceful co-existence? *Conf. Proc. Groundwater and Pollution,* Midrand, August.

Travis, C.C. & Doty, C.B. 1990. Can contaminated aquifers at Superfund sites be remediated? *Environ. Sci. Technol.* 24(10):1464-1466

US Environmental Protection Agency (1989). *Technical Guidance Document, Final covers on Hazardous Waste Landfills and Surface Impoundments,* EPA/530-SW-89-047, Washington DC

Van Genuchten, M.Th. 1980. A closed-form equation for predicting the hydraulic conductivity of unsaturated soils. *Soil Sci. Soc. Am. J.* 44:892-898

Van Schalkwyk, A. & Vermaak, J.J.G. 1999. *The relationship between the geotechnical and hydrogeological properties of residual soils and rocks in the vadose zone. Draft final report* Project No. K5/701, Water Research Commission, Pretoria

Vermaak, J.J.G. 1997. The application of geotechnical data and techniques to estimate important hydrogeological properties of the vadose zone In *Geology for engineering, urban planning and the environment proceedings* Organised by SAIEG, Midrand, South Africa.

2 Foundation engineering and lateral support
Conception des fondations et soutènements

Geotechnics for Developing Africa, Wardle, Blight & Fourie (eds) © 1999 Balkema, Rotterdam, ISBN 90 5809 082 5

Interface stiffness effects on settlement of column-reinforced composite ground

M. Alamgir
Department of Civil Engineering, Bangladesh Institute of Technology (BIT), Khulna, Bangladesh

N. Miura
Department of Civil Engineering and Institute of Lowland Technology, Saga University, Japan

ABSTRACT: To consider the effects of stiffness of the interface layer, between the superstructure and the composite ground, formed by columnar inclusions, on settlement, and to widen the scope of use of columnar inclusions under varied situations, a simple theoretical model has been developed and discussed in this paper. The model incorporates the compressibility of granular fill, nonlinearity of the material properties, interaction as well as the stress transfer between the column and the surrounding soil along the depth and the possible slip along column-soil interface. Evaluations are to illustrate the influence of various parameters on the predicted settlements. This analysis reveals that the compressibility of the granular fill has an appreciable influence on the settlement response of the column-reinforced ground as long as the modulus of the granular fill is less than the optimum value. For the validation of the proposed model, predicted results are compared with those obtained from finite element analysis, with the approaches in vogue and with laboratory and field test results.

1 INTRODUCTION

Composite ground formed by the installation of columnar inclusions, such as sand compaction piles or stone columns, by in-situ deep mixing of lime or cement, may be designed to possess enhanced load carrying capacity, greater resistance to settlement, improved slope stability and resistance to liquefaction. So far, the emphasis has mainly to consider the geometrical parameters of the composite ground, relative stiffness of the columnar inclusion and of the soft ground around, with the tacit assumption that the settlement of the composite ground is uniform (Priebe 1976, Aboshi et al. 1979, Balaam and Booker 1981 and others). This equal strain consideration is valid if the superstructure load is transferred through a rigid base at the interface. Even as late as 1994, in the Foundation Engineering book by Hansbo (1994) analysis by equal strain theory, due to the conventional provision of a rigid base, is further endorsed.

In the transfer of load of flexible air field and highway pavements and embankments on composite ground, the rigid base considerations in the analysis are not tenable. In order to widen the scope of use of columnar inclusions under varied situations, simple methods of analysis with different interface media is imperative. It is the cardinal aim of this paper to consider the effects of stiffness of the interface layer between the superstructure and the composite ground on settlement, culminating in a simple theoretical method of analysis.

The deformation response of composite ground for the free strain condition has been recently accomplished (Poorooshasb, Alamgir and Miura 1996a). In this approach to the deformation response of composite ground the analysis is performed considering the deformation properties of the column material and the surrounding soft clay. The interaction shear stresses between the column material and the surrounding soil are considered to account for the stress transfer between the column and the soil. The solutions have been obtained by imposing compatibility between the displacements of the column and the soil for each element of the column-soil system.

Not withstanding the practical need to incorporate the effects of the rigidity of the interface to assess settlement of the composite ground, Madhav and Van Impe (1994) recognized the need to have a gravel bed during the construction stage itself, and hence the need for rigorous analysis. The analysis provided by Madhav and Van Impe (1994) has many deficiencies such as (i) granular fill is considered as incompressible so that it can deform only by shear, (ii) there is no slip at the column-soil interface, and

(iii) the stress transfer along the column-soil interface is not considered.

In this paper, a model, simple in concept, is proposed to predict the effect of granular fill on the settlement response of the column-reinforced ground. The model incorporates the compressibility of the granular material and the possible slip that may occur along the column-soil interface. The model also takes into account the nonlinearity of the material properties, the interaction as well as stress transfer between the column and the surrounding soil along the depth. The governing equations are solved numerically by the finite difference method. The results predicted using the proposed method are compared with those obtained from a finite element analysis and some existing approaches for representative cases. The predicted results are also compared with experimental results obtained from both the laboratory and field tests. A parametric analysis is also conducted to examine the influence of various parameters on the predicted behaviour of column-reinforced ground.

2 PROBLEM AND THE PROPOSED SOLUTION

Consider a general case in which the columns extend up to the bedrock in a homogeneous soil media (vide Fig.1). The composite ground is covered by a layer of granular fill and uniformly loaded over the entire area. The granular fill, the column and the surrounding soil are characterized by the deformation moduli E_f, E_c and E_s that may vary with depth linearly or in any other form and the constant

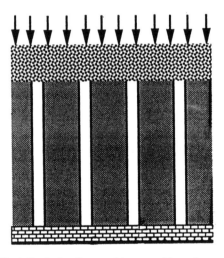

Fig.1 Typical soft ground improved by columnar inclusions and covered by a layer of granular fill.

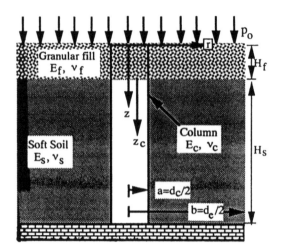

Fig.2 Idealization of the foundation system.

Poisson's ratios v_f, v_c and v_s, respectively. These properties are not influenced by columns and remain constant throughout the loading process.

2.1 Idealization of Foundation System

For simplification of the problem, the "unit cell" consideration is employed here (vide Fig.2). Here, p_o is the pressure applied over the reinforced ground, H_f is the thickness of the granular fill, L_c and $d_c(=2a)$ are the length and diameter of the cylindrical column, respectively and H_s is the total thickness of soil media. In order to reduce the complexity each influence zone is approximated by a circle of effective diameter, $d_e(=2b)$. Balaam and Booker (1981) relate d_e, to the spacing of the columns, s_c, as

$$d_e = c_g s_c \qquad (2.1)$$

where c_g is the geometry dependent constant equal to 1.05, 1.13 and 1.29 for triangular, square and hexagonal arrangements, respectively.

Further idealization is that in the top layer, the column material and the surrounding material have the same properties to the extent of the thickness of the granular layer, whereas in the second layer the column material and the surrounding soil are assigned different properties is as in composite ground created (vide Fig.2).

2.2 Analysis

Due to radial symmetry, only two displacement vectors, radial and axial, need be considered. For convenience, it is assumed that the radial displacement component is so small that it can be neglected. As the consolidation or the displacement of the system progress upon the load application, it is expected that the surrounding soil layer compresses vertically without any lateral deformation. Since the proposed foundation model is based on the radial symmetry, the following integro-differential equation recently proposed and adopted by Poorooshasb, Alamgir and Miura (1996a) is employed here for the analysis of the foundation system considered in this investigation.

$$\frac{\partial w(r,z)}{\partial z} = -\frac{1}{E(r,z)}\int_0^z \left[\begin{array}{c}\frac{\partial}{\partial r}\left\{G(r,z)\frac{\partial w(r,z)}{\partial r}\right\}+ \\ \frac{1}{r}\left\{G(r,z)\frac{\partial w(r,z)}{\partial r}\right\}\end{array}\right]dz+\frac{p_0}{E(z)}$$

(2.2)

For time dependent problems (for example those that include radial drainage), the parameters E and G change with time and are also functions of the r coordinate. In problems where time and the stress level do not play a role it may be assumed that E and G are not functions of r. Under these conditions Eq.(2.2) reduces to;

$$\frac{\partial w(r,z)}{\partial z}+\frac{1}{E(z)}\int_0^z G(z)\left\{\frac{\partial^2 w(r,z)}{\partial r^2}+\frac{1}{r}\frac{\partial w(r,z)}{\partial r}\right\}dz=\frac{p_0}{E(z)}$$

(2.3)

which is the governing equation of the problem. The integro-differential equation is developed based on the fundamental equation of equilibrium which can be written as (Fig.3):

$$\frac{\partial \sigma_{zz}}{\partial z}+\frac{\partial \sigma_{rz}}{\partial r}+\frac{\sigma_{rz}}{r}=0$$

(2.4)

2.3 The boundary conditions

The following boundary conditions of the problem, provide the additional equations needed for solution.

1. Boundary condition at the top of granular fill: This boundary condition has already been incorporated in deriving Eq.(2.3) in terms of applied stress (p_0) and cannot be any further aid in solution.

2. Boundary condition at the base: The base is considered as an undeformable bearing strata and such at this level all the displacements are zero i.e. $w(r,z)=0$ at $z=H_s$.

3. Boundary condition at the outside boundary of unit cell: Due to the symmetry of load and geometry, the shear stress at the outside boundary of the unit cell is zero i.e. $[\partial w(r,z)/\partial r]=0$ at $r=b$.

4. Boundary condition at the column-soil interface: The shear traction developed at the interface gives rise to a force of magnitude F_d as

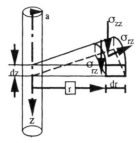

Fig.3 Equilibrium of a soil element around column.

$$F_d = 2\pi a\int_0^z G(z)\frac{\partial w(a,z)}{\partial r}dz$$

(2.5)

where, a is the radius of column and $w(a,z)$ is the vertical displacement component at a radial distance $r=a$ and at a depth z. The force F_d causes an additional axial strain in the column of magnitude $F_d/[A_c E_c(z)]$, where, A_c is the cross sectional area of column and $E_c(z)$ is its deformation modulus which may vary with depth linearly or by any other function. The axial strain in the column at a depth z is equal to $[\partial w(a,z)/\partial r]$ assuming that no slip takes place along the column-soil interface and one obtains the boundary condition as at the column-soil interface in the following form in case of no slip;

$$\frac{\partial w(a,z)}{\partial z}=\frac{p_0}{E_c(z)}+\frac{2\pi a}{A_c E_c(z)}\int_0^z G(z)\frac{\partial w(a,z)}{\partial r}dz$$

(2.6)

Since the natural soils have a finite and measurable shear strength and the column-soil interface has a finite adhesive strength, slip or local yield would invariably occur when the shear stress reaches the adhesive (or yield) strength. To account for this possible slip, a failure criterion needs to be introduced. One such criterion, which is very commonly used in geotechnical engineering problems, is that the shear stress that is mobilised along the column-soil interface cannot exceed the limit of $K_0(\gamma z+p_0)\tan\delta$, where K_0 is the coefficient of earth pressure at rest, δ is the angle of friction between the column and the soil. Thus the boundary condition at column-soil interface may be stated as;

$$\frac{\partial w(a,z)}{\partial z}=\frac{p_0}{E_c(z)}+\frac{2\pi a}{A_c E_c(z)}\int K_0(\gamma' z+p_0)\tan\delta dz$$
$$0\leq z\leq z'$$

(2.7)

$$\frac{\partial w(a,z)}{\partial z}=\frac{p_0}{E_c(z)}+\frac{2\pi a}{A_c E_c(z)}$$
$$\left[\int_{z'}^z G(z)\frac{\partial w(a,z)}{\partial r}dz+K_0(\frac{1}{2}\gamma' z'^2+p_0 z')\tan\delta\right]$$
$$z'\leq z$$

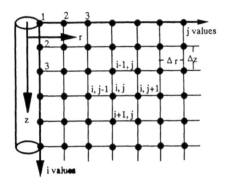

Fig.4 Finite difference mesh used in the analysis.

where, z' is the depth above which slip will take place. The development of boundary conditions at the column-soil interface for no slip and slip and the method to determine the depth z', are detailed in the papers, Poorooshasb, Alamgir and Miura (1996a, b).

2.4 The Numerical Scheme

The finite difference numerical scheme used earlier by the Author is simple and demonstrated briefly here. For a detailed account reference may be made to the papers by Poorooshasb, Alamgir and Miura (1996a, b). The finite difference network shown in Fig.4 is assumed to have $jmax$ columns i.e. $j=1, 2, 3, \ldots\ldots jmax$, spaced at Δr intervals, and $imax$ rows i.e. $i=1,2,3,\ldots imax$, spaced at Δz intervals. The origin of the r axis is the axis of the column and that of z at the top of the granular fill. Consider the general nodal point (i,j), the finite difference form of Eq.(3) may be written as

$$w(i+1,j) + \sum_{m=1}^{j}\alpha(m,j)\{w(m,j-1) - 2w(m,j) + w(m,j+1)\}$$

$$- w(i,j) + \sum_{m=1}^{j}\beta(m,j)\{w(m,j+1) - w(m,j-1)\} = \frac{p_0}{E(i)}\Delta z$$

where, $\qquad\qquad\qquad\qquad\qquad$ (2.8)

$$\alpha(m,j) = \frac{G(m-1)(\Delta z)^2}{E(i)(\Delta r)^2} \text{ and } \beta(m,j) = \frac{G(m-1)(\Delta z)^2}{E(i)(2r\Delta r)}$$

To express Eqs.(2.6) and (2.7) in their finite difference form a path similar may be followed.

3 RESULTS AND DISCUSSIONS

Analysis has been done to illustrate the influence of deformation modulus and thickness of granular fill and the spacing of columns on the settlement

response of composite ground covered by a layer of granular fill. Evaluations are made for the no slip case only although the proposed model can be used for the case considering possible slip along the column-soil interface. The values of the parameters considered are $H_f/H_s=0.00$ to 0.333, $L_c/d_c=20$, $d_e/d_c=2.5$ to 10, $p_0/E_s=0.10$, $E_f/E_{so}=1$ to 100, $E_c/E_{so}=50$, $m_s/E_{so}=0.10$, $v_f=0.30$ and $v_s=0.40$. Here, E_{so} is the deformation modulus of soil at the surface which varies linearly with depth at the rate of ms per meter. Some of the recent data available regarding the thickness and the type of granular layer suggested and adopted are shown in Table 1.

Table 1 Suggested thickness of granular fill.

Granular fill materials	Thickness H_f(m)	References
Sand and thin layer of gravel	2.00	Brons & De Kruijff (1985)
Gravel	0.30-1.0	Mitchell & Huber(1983)
Sand and gravel	0.70	Mitchell & Huber(1983)
Sand	2.00	Venmas (1990)
Sand, gravel or crushed stone	0.30-1.0	Barksdale & Bachus (1983)
Sand- medium to fine	0.90	Bachus & Barksdale (1984)
Coarse sand & crushed stone	1.00	Bhandari (1983)

3.1 Settlement of Reinforced Ground

The influence of thickness and deformation modulus of granular fill on the response of treated ground is presented in Figs.5 and 6. The changes in settlement profiles with relative thickness of granular fill i.e. H_f/H_s are shown in Figs.5 for the relative moduli of granular fill, $E_f/E_{so}=5$. Settlement profile of the reinforced ground changes significantly with the placement of granular fill over it. In the case of

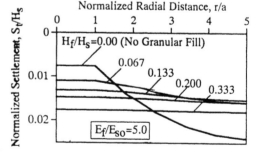

Fig.5 Influence granular fill thickness.

126

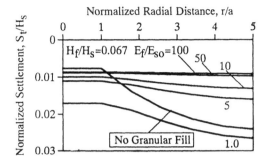

Fig.6 Influence of granular fill stiffness.

E_f/E_{so}=5.0, the value of S_t/H_s increases in the close region of the column, r/a=0.0 to 1.5, beyond which it decreases up to the boundary of the zone of influence, r/a=n, as shown in Fig.5. The value of S_t/H_s changes from 0.0076 to 0.0177 at r/a=0.0 and 0.024 to 0.0181 at r/a=5.0 for the value of H_f/H_s=0.000 to 0.333. The overall settlement of the reinforced ground tends to become uniform for higher value of H_f/H_s. The presence of a layer of granular fill significantly affects the magnitude of settlement. These results also reveal that the used of thick granular fill is not a necessity to obtain uniform settlement if the granular fill is placed is of high deformation modulus.

The influence of relative moduli of granular fill over the settlement response of the reinforced ground is presented in Fig.6 for a relative thickness of granular fill, H_f/H_s=0.067. The settlement profile of column reinforced ground for the no granular fill situation is also plotted. The relative moduli of the granular fill E_f/E_{so}, is varied from 1 to 100. In all the cases, the S_t/H_s decreases with the increase of E_f/E_{so}. From Fig.6, it can be seen that the settlement profile of the reinforced ground remains nonuniform as long as the value of E_f/E_{so} is less than 100. The value of S_t/H_s varies from 0.017 to 0.0087 at r/a=0.0 and 0.026 to 0.0091 at r/a=5.0 for increase in the value of E_f/E_{so} from 1 to 100. The result presented in Fig.6 reveals that the overall settlement on the reinforced ground practically uniform at the values of E_f/E_{so}=100. The lower value of E_f/E_{so} produces higher overall settlement of the improved ground which is not desirable for the satisfactory performance. Both figures show that granular fill with E_f/E_{so}=1.0, produces higher settlement of reinforced ground than that of no granular fill case. The findings reveal that the compressibility of the granular fill can not be ignored for the rational evaluation of its role over the settlement response of reinforced ground. Shukla (1995) also observed that

the compressibility of the granular fill has an appreciable influence on the settlement response of the geosynthetic-reinforced granular fill-soft soil system as long as the stiffness of the granular fill is less than fifty times that of the soil.

3.2 Stress Concentration in the Column

The effect of granular fill placed over the reinforced ground on the concentration of applied stress i.e. the stress taken by the relatively stiffer and stronger column p_c, is depicted in Figs.7 and 8. Figure 7 shows the variation of the magnitude of the normalized stress p_c/p_o, at the top of column with the relative moduli of granular fill E_f/E_{so}=1 to 100 for H_f/H_s=0.20 and column spacings, n=2.5, 5.0 and 10. These results indicate that the load taken by the column increases nonlinearly with the increase of E_f/E_{so} and attains almost a constant value beyond E_f/E_{so}=50. The rate of increase of p_c/p_o is considerably higher at the range of E_f/E_{so}=1 to 20 than that of at E_f/E_{so}=20 to 100. The value of p_c/p_o increases from 1.0 at E_f/E_{so}=0, to 2.62, 4.19 and 6.39 at E_f/E_{so}=100, for the values of n=2.5, 5.0 and 10, respectively.

Figure 8 shows the variation of p_c/p_o along the depth of the column for the value of E_f/E_{so}=0, 5, 10 and 100, H_f/H_s=0.20 and n=5.0. The value of p_c/p_o increases with depth and the rate of increase decreases with increasing value of E_f/E_{so}. The value of p_c/p_o is almost constant along the whole length of the column for E_f/E_{so}=100. A high constant value at this level indicates that no stress gets transferred along the depth for the higher value of E_f/E_{so}, which, of course, can be expected because the surface settlement is practically uniform for the value of E_f/E_{so}=100. From Fig.8, it is found that p_c/p_o changes from 1.0 to 4.17 for z_c/L_c=0.0 and 1.0 for no granular fill condition but these values are 4.185 and 4.19 for E_f/E_{so}=100. Since the success of this

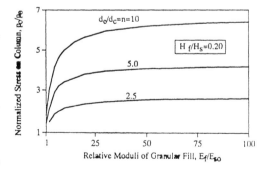

Fig.7 Effect of granular fill on the load sharing.

Normalized Stress in column, p_C/p_0

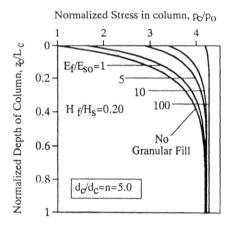

Fig.8 Load sharing by column along the depth.

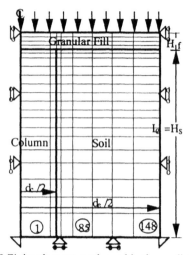

Fig.9 Finite element mesh used in the predictions.

ground improvement technique predominantly depends on the concentration of stress in the columns, these findings reveal an encouraging response of column-reinforced soft ground when an engineered layer granular fill forms the interface.

Settlement response of the composite ground ranges from flexible to rigid loading conditions depending on the stiffness of overlaying granular fill. It is revealed that a layer of granular fill having high stiffness results in uniform settlement even for low thickness. A layer of granular fill having lower modulus can also result in uniform settlement provided the appropriate thickness of layer is placed although the overall settlement is higher than that of without granular fill. Observations suggest that the compressibility of the granular fill can not be ignored. The analysis stated above points out that the design based on either "equal strain" (granular fill is perfectly rigid) or "free strain" (granular fill is perfectly flexible) theory without considering the actual situation may lead to considerable errors.

4 COMPARISONS

The theoretical verification is performed by comparing the results obtained from finite element analysis with the existing approaches developed in the past. The laboratory and field test results reported in the literature are also compared.

4.1 Verification by Finite Element Analysis

The program CRISP (Britto & Gunn1987) is used for the verification of the suggested model by finite element analysis. A typical finite element mesh

covering the solution region is shown in Fig.9. The applied load is of uniform pressure at the top. The boundary conditions for the mesh are the same as those considered in the proposed model. The media is assumed to be linear elastic with no slip between the column and the surrounding soil. To model the flexible loading condition, no granular fill over the reinforced ground is considered while for the rigid loading condition a granular fill having high stiffness i.e. $E_f/E_s=100$, $H_f/H_s=0.333$ and $v_f=0.30$, over the entire reinforced ground is considered.

Predictions are made for the value of parameters, $p_0/E_s=0.10$, $d_e/d_c=4.0$, $L_c/d_c=10$, $E_c/E_s=25$, $v_c=0.20$ and $v_s=0.40$. For no granular fill condition, the proposed model underpredicts the settlement compared to the finite element method from the center of column up to the radial distance. Beyond

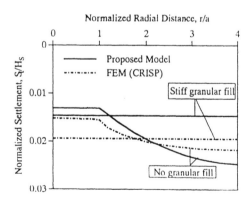

Fig.10 Comparison of predicted results with FEM.

128

that it overpredicts up to the outside boundary of the zone of influence i.e. at $r/a=4.0$, shown in Fig.10. The differences of settlements are 16.31% over the column and 11.85% at the boundary of the zone of influence. The model underpredicts the settlement by about 24.50% in the case of highly stiff granular fill. The results reveal that the predicted deformation patterns of the reinforced ground and distribution of shear stress obtained by the proposed model and finite element analysis are quite similar. Comparison of results with FEM for the distribution of interface shear stress and load sharing for both end bearing floating columns are given in Alamgir (1996).

4.2 Comparison with Existing Approaches

For a particular example, the predicted settlement of the treated ground as obtained by Madhav and Van Impe (1994) is compared with that obtained by this proposed model, Fig.11. In comparison, two extreme cases, namely, no granular fill and stiff granular fill, are considered for an end bearing column. The values of the parameters are $d_e/d_c=2.5$, $E_c/E_s=5.0$, $v_s=0.35$. The predicted values differ significantly from each other specially for the case of no granular fill. For stiff granular fill they are very close to each other, however, the proposed model overpredicts by 10.12%. For no granular fill, the proposed model underpredicts, in the column region, by 68.25% at $r/a=0.0$. But it overpredicts in the soil region by 39.55% at $r/a=2.5$. The proposed model considers the compatibility of the displacements at every nodal point and stress transfer between the column and the soil along the depth, while in Madhav and Van Impe (1994), the compatibility of the displacement is satisfied only at the top and the stress transfer along the column-soil interface is not being considered.

4.3 Comparison with Experimental Results

Laboratory test conducted by Leung and Tan (1993) on sand column is considered here for comparison. The design parameters are estimated based on the given information as $p_0/E_s=0.0375$, $d_e/d_c=4.0$, $E_c/E_s=12$, $v_s=0.30$, $\phi=14°$ and $\delta=\phi$. The variation of settlement, S_f/H_s, radial distance, r/a, is presented in Fig.12. There exists a reasonable agreement between the measurements and the predictions except near the boundary of the mold. The investigators in their discussion pointed out that the effect of friction along the inner surface of the mold might reduce the magnitude of settlement near it. It is observed that the predictions considering slip at the interface are close to the measured values than with no slip.

A series of field test results collected and collated

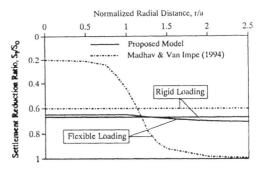

Fig.11 Comparison of results predicted by proposed model with those of Madhav and Van Impe (1994).

from published literature are compared with the analysis by the proposed model. As all the relevant data are not well documented, the average values of the parameters are being considered for all of them for predictions. For the predictions the values of the parameters are $p_0/E_s=0.10$, $E_c/E_s=5$ to 30 and $v_s=0.30$. It is considered that the reinforced ground is covered by a layer of well compacted granular fill having a reasonable high value of stiffness. These considerations ensure uniform settlement over the entire composite ground. This basis is employed in the analysis for validation of the test results in which only one settlement value is reported over the entire reinforced area. In Fig.13, the data points are plotted for the measured values while the predictions by the proposed model are indicated as full lines for the different E_c/E_s, ranging from 5 to 30. The variation of S_f/S_o with d_e/d_c for the test data is considered. These variations may be due to the differences in site conditions and methods of granular column installation. These analysis made with available limited data can be regarded primarily only to illustrate the general validity of the proposed model.

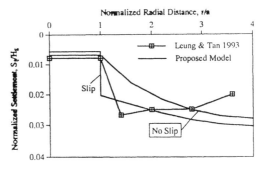

Fig.12 Prediction versus measurement of settlement profile of reinforced ground in the laboratory.

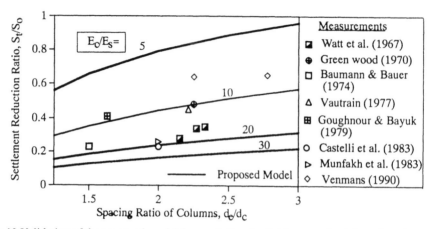

Fig.13 Validation of the proposed model for predicting the field test results (after Alamgir 1996).

5 CONCLUSIONS

The model emerging out in this in depth analysis to evaluate the settlement response of a foundation system incorporating the compressibility of the granular fill, interaction as well as the stress transfer between the column and the soil along the depth and the possible slip, is simple for adoption in practice. There exists good agreement between the predicted and observed responses. Further parametric study shows that the proposed approach can be used successfully to demonstrate the effect of overlaying granular fill on the settlement response of column-reinforced ground. Granular fill placed over the composite ground is very effective in reducing both the total and differential settlements and hence its role needs to be accounted for the design.

REFERENCES

Aboshi, H., E.Ichimoto, M.Enoki & K.Harada 1979. The compozer – A method to improve characteristics of soft clays by inclusion of large diameter sand columns. *Proc. Intl. Conf. on Soil Reinforcement*, Paris, 1:211-216.

Alamgir,M. 1996. *Analysis of soft ground reinforced by columnar inclusions*. Ph.D. Thesis, Dept. of Civil Engg., Saga university, Saga, Japan.

Bachus, R.C. & R.D.Barksdale 1984. Vertical and lateral deformation of model stone columns. Proc. *Intl. Conf. on In-situ Soil and Rock Reinforcement,* Paris, 99-110.

Balaam, N.P. & J.R.Booker 1981. Analysis of rigid raft supported by granular piles. *Intl. J. for Numer. and Anal. Meth. In Geomech.* 5:379-403.

Barksdale, R.D. & R.C.Bachus 1983. *Design and construction of stone columns, Vol.1.* Report No. FHWA/RD-83/026, NTIS, Virginia, USA.

Bhandari, R.K.M. 1983. Behaviour of a tank founded on a soil reinforced with stone columns. *Proc. 7th ECSMFE,* Helsinki, 1:65-68.

Britto, A.M. & M.J. Gunn 1987. Critical state soil mechanics via finite elements. *Ellis Horwood Ltd.*

Brons, K.F. & H.De Kruijff 1985. The performance of sand compaction piles. *Proc. XI ISSMFE,* Sanfransisco, 3:1683-1686.

Hansbo, S. 1994. *Foundation Engineering.* Elsevier Science B.V., Amesterdam.

Leung, C.F. & T.S.Tan 1993. Load distribution in soft clay reinforced by sand columns. *Proc. Intl. Conf. soft soil Engg.* Guangzhou, China, 779-784.

Madhav, M.R. & W.F.Van Impe 1994. Load transfer through a gravel bed on stone column reinforced soil. *Geotechnical Engineering,* 24(2):47-62.

Mitchell, J.K. & T.R.Huber 1983. Stone column foundations for a waste treatment plant-A case history. *Geotechnical Engineering,*14(2):165-185.

Poorooshasb, H.B., M.Alamgir & N.Miura 1996a. Negative skin friction on rigid and deformable piles. *Computers and Geotechnics.*18(2):109-126.

Poorooshasb, H.B., M.Alamgir & N.Miura 1996b. Application of an integro-differential equation to the analysis of geotechnical problems. *Structural Engineering and Mechnics,* 4(3):227-242.

Priebe, H. 1976. Estimating settlement in a gravel column consolidated soil. *Die Bautechnik,* German, 53:160-162.

Shukla, S.K. 1995. *Foundation model for reinforced granular fill-soft soil system and its settlement response.* Ph.D. Thesis, Department of Civil Engineering, IIT, Kanpur, India.

Geotechnics for Developing Africa, Wardle, Blight & Fourie (eds) © 1999 Balkema, Rotterdam, ISBN 90 5809 082 5

Recent advances in design of shallow and deep foundations

O. Maréchal, D. Remaud, J. Garnier & S. Amar
Laboratoire Central des Ponts et Chaussées, Nantes, France

ABSTRACT: Centrifuge modelling may be an efficient tool to provide experimental data that are needed in the validation processes of design codes. The paper presents the results of two centrifuge research programmes. The first one deals with bearing capacity of shallow footings subjected to inclined and eccentric loads and sometimes installed in or near slopes. The reduction coefficients are i_δ for load inclination, i_e for load eccentricity and i_β for proximity of a slope. In most codes of practice, it is proposed to combine these three coefficients in the following formula : $i_{e\delta\beta} = i_e \times i_\delta \times i_\beta$. The validity of this principle has been investigated by centrifuge model tests. The second programme focuses on the group effects in horizontally loaded piles. Different techniques are used throughout the world to determine the P-y curves for a single pile from laboratory or in situ tests results. The paper gives the main results obtained on group effects on load-deflection curves, bending moment and P-y curves.

1 INTRODUCTION

Geotechnical centrifuge model testing constitutes a powerful tool for the study of many geotechnical problems. It may be useful for developing an understanding of the basic of soil structure interactions. It is indeed very well suitable when a large number of tests are needed (Corté et Garnier, 1986) as demonstrated in the two following parametric studies.

2 BEARING CAPACITY OF SHALLOW FOUNDATIONS

This work comes within the framework of studies conducted by the LCPC in the field of surface foundations (Amar et al. 1995, Bakir 1995). Its main object is the study, on centrifuged models, of the bearing capacity of shallow foundations placed near a slope and subjected to both eccentric and inclined load.

The bearing capacity is obtained by applying a reduction coefficient, $i_{e\delta\beta}$, to a reference bearing capacity q_{ref} (same foundation on horizontal soil, submitted to a vertical central load) :

$$q_r(e, \delta, \beta) = i_{e\delta\beta} \times q_{ref} \qquad (1)$$

The expression for this coefficient reduces to :

$$i_{e\delta\beta} = i_e \times i_\delta \times i_\beta \qquad (2)$$

where i_e, i_δ, i_β, are the reduction coefficients for each basic loading cases (eccentricity, inclination, slope).

The validity of these methods has been investigated through some usual loading cases.

2.1 Experimental program

Eight series of tests were performed using rectangular containers (1200mm × 800mm).

Each container contained 6 foundations of width B, divided up in two lanes delimited by three vertical thick glass plates as shown on Figure 1.

These glass plates were used in order to simulate two-dimensional conditions.

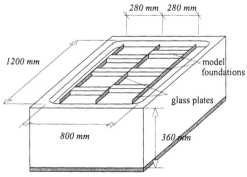

Figure 1. View of a container equipped with glass plates.

The composition of a basic container is detailed below :
- a reference test : footing on horizontal soil supporting a central, vertical load (C.V.Reference)
- an eccentric vertical loading (E.V)
- an inclined central loading (C.I)
- a vertical central loading with the footing placed near a slope (C.V.Slope)
- a combined loading : footing near a slope supporting an eccentric inclined load.

The notations used are presented on Figure 2.

The sand used in all the tests is white Fontainebleau sand, which main characteristics are presented in table 1.

The average density obtained by the raining technique for the eight samples is : γ_d = 16.03 kN/m^3 (I_D = 0,7).

All tests were performed in the geotechnical centrifuge at the LCPC Nantes. Its main characteristics are (Corté et Garnier, 1986):
- maximum model mass : 2 tons at 100 g, maximum acceleration : 200 g.
- radius (distance axis/ basket platform) : 5.50 m.

The centrifuge tests were performed at 50g on strip footings 40 mm wide (2 m in prototype scale) and L/B = 7.

2.2 *Experimental results*

Loads are applied step by step until failure. Table 2 sums up the parameters investigated in the eight containers (total of 46 tests). All results presented below are given in prototype conditions (Marechal et al., 1998).

The prototype load/settlement curves obtained from the tests performed in container n°4 have been gathered on Figure 3.

According to the curves and results obtained, for the sand used in this study and for 2 meters wide foundations, it can be stressed that :
- all the observed failures are sudden with a clearly marked pressure peak,
- the repetitiveness is fairly good. As an example, the maximum difference to the average peak pressure, for all the reference tests on shallow foundations, is less than 13% (Table 3).

Table 3 sums up the bearing capacity of the reference test and the bearing capacity reduction coefficients obtained in each container. Combined loading case coefficient, $i_{e\delta\beta}$, is compared with the product of the elementary coefficients (2).

On an other hand, results show that the global combined coefficients $i_{e\delta\beta}$ are close to the product of the basic coefficients : $i_e \times i_\delta \times i_\beta$.

Table 2. Experimental conditions

Container	tg β	D/B	d/B	e/B	δ
1 - 3	1/2	0	1	+1/8	15°
4 - 5	1/2	0	1	+1/8	20°
6	1/2	3/8	1	-1/8	15°
7	1/2	0	2	-1/8	15°
8	1/2	1/2	2	-1/8	15°

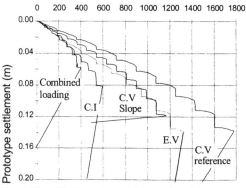

Figure 3. Load/Settlement curves for container n°4.

Table 3. Bearing capacity reduction coefficients.

Container n°	q_{ref} (kPa)	i_e	i_δ	i_β	$i_{e\delta\beta}$	$i_e \times i_\delta \times i_\beta$
1	2016	0,85	0,47	0,67	0,27	0,27
2	1776	⌈0,83	0,52	0,64	⌈0,31	⌈0,28
		⌊0,82			⌊0,28	⌊0,27
3	1768	0,82	0,50	0,68	0,28	0,28
4	1779	0,75	0,34	0,66	0,23	0,17
5	1941	0,75	0,28	0,64	⌈0,18	0,13
					⌊0,19	
6*	2246	0,82	0,54	⌈0,55	0,20	⌈0,24
				⌊0,59		⌊0,26
7	1573	0,84	0,56	0,84	0,34	0,40
8*	2596	⌈0,79	0,57	0,60	0,31	⌈0,27
		⌊0,93				⌊0,32

* Embedded foundations (see Table 2).

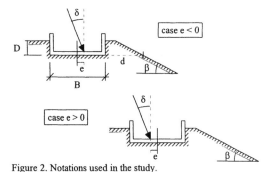

Figure 2. Notations used in the study.

Table 1. Fontainebleau sand characteristics.

γs (kN/m^3)	γdmin (kN/m^3)	γdmax (kN/m^3)	c (kPa)	φ (degrees)
26	13.95	17.06	0 to 6 *	39° to 40° *

* for γ_d = 16 kN/m^3 and normal stress ranging from 50 to 300 kPa.

3 GROUP EFFECTS IN LATERALLY LOADED PILES

In the present study, the piles are installed at 1g. Sand is rained around the piles suspended in the container. This procedure is an approximate simulation of an in situ installation without soil displacement. Thus, the study is limited to bored piles.

3.1 Test procedures

In the following sections, all results are presented at prototype scale, except where noted.

The model piles are designed for a centrifugal acceleration of 40g. They have an outside diameter B of 18 mm and a total length of 380 mm. The penetration depth in the soil is $D = 300$ mm.

Therefore they simulate a prototype pile of 720 mm in diameter with an embedded length of 12 m and classified as a flexible pile ($EI = 473$ MN.m²).

The piles are equipped with 20 pairs of gauges giving the bending moment profiles during lateral loading. The study is based on the analysis of these experimental data.

For a given lateral load, the measured distribution of bending moment M, with depth z, is used to derive the reaction profile of the soil P, by double derivation, and the displacement profile y, by double integration, according to :

$$P = \frac{d^2 M}{dz^2}$$ (3)

and

$$y = \iint \frac{M}{EI} dz$$ (4)

The soil used is a fine white Fontainebleau sand. The sand is placed in the containers (1200 mm × 800 mm) by raining, using an automatic hopper. The density achieved is 16.3 kN/m³ ($I_D = 0.89$) for the present tests.

Several tests were performed to investigate different experimental configurations : single pile or pair of piles with 2B, 4B and 6B spacing "s". Pile head is free.

The lateral force is applied by a hydraulic servo-actuator with a metallic cable connected to the pile at the height of 40 mm above the soil surface (1.6 m in prototype scale).

The load is applied in about 20 steps of 80 N each (128 kN m in prototype scale).

3.2 Experimental results and analysis

The experimental P-y curves derived by double derivation and double integration of the bending moment profiles are first validated.

For that purpose, they are input in the software PILATE-LCPC (Frank et al., 1990) to predict the behaviour of a single pile under lateral loads. A good agreement is observed between the experimental data and the back calculated ones (deflections, bending moment profiles as seen in Figure 4).

- Group effect on load-deflection curves

Figure 5 shows the lateral load applied to each pile in the group and the load supported by a single pile for the same deflection.

The group efficiency is a good criterion to qualify the group effect. It is calculated in the different configurations for the same displacement at the load level as follows :

$$e = \frac{\text{total load on the group for a given displacement}}{\text{n} \times \text{load on single pile for the same displacement}}$$ (5)

We get 0.81, 0.87 and 0.95 respectively for a pile spacing of 2B, 4B and 6B.

The load distribution in the group is determined from the bending moment measured at the soil surface. The leading pile takes 59% of the lateral load applied at 2B, 56% at 4B and 53% at 6B spacing.

The pile-soil-pile interactions are small as soon as the pile spacing is greater than 6B.

- Group effect on bending moment profiles

The bending moments in the leading pile are very similar to the ones of the single pile with a scatter of ±5% (Remaud et al., 1998). In the trailing pile, with pile spacing rising from 2B to 6B, they tend to be equal to those of the single pile.

The group effect is characterised by higher maximum bending moments in the leading pile than in the trailing pile (at 2B spacing maximum bending moments on leading piles are 20% higher).

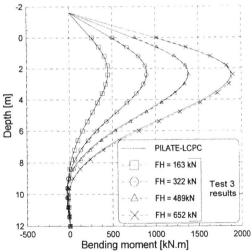

Figure 4 - Validation of the P-y curves - Bending moments.

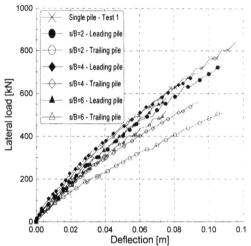

Figure 5 - Loading curves at the load level.

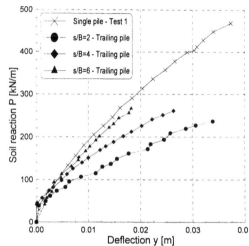

Figure 6 - *P-y* reaction curves of the trailing pile at z=2.4 m.

These results also confirm the conclusion of the previous section : the group effect is small for a pile spacing equal to 6B considering the experimental scatter.

- Group effect on P-y curves

The *P-y* curves are calculated with the procedure previously developed.

Results show that the leading *P-y* curves are not different from the single pile one (Remaud et al., 1998). No significant group effect is observed on leading piles even at small spacing.

The *P-y* curves of the trailing piles are plotted in Figures 6. As seen before, the group effect is more significant on the trailing pile when compared to the single pile.

"P-multipliers" P_m are commonly used for group effects (Brown et al. 1990) :

$$P_{\text{group pile}} = P_m \times P_{\text{single pile}} \qquad (6)$$

Thus a first qualitative analysis was done on the trailing piles. The P_m coefficient was calculated at four different depths ranging from 0.6 to 2.4 m. The observed average values obtained at maximum displacements are then :

- s/B=2 : $P_m = 0.52$
- s/B=4 : $P_m = 0.70$
- s/B=6 : $P_m = 0.93$

4 CONCLUSIONS

These results lead to confirm that :

- For shallow foundations : the reduction coefficient applied to a foundation submitted to a combined loading can be calculated as the product of the basic coefficients corresponding to each loading

condition (proximity of slope, eccentricity and/or inclination of the load,...) : $i_{e\delta\beta} = i_e \times i_\delta \times i_\beta$.

- For laterally loaded piles a clear group effect is shown for pile spacing less or equal to 6B :
    - the total load for the group is reduced (by about 20% for 2B spacing) ;
    - the load and bending moments on the trailing pile are reduced. For 2B spacing the trailing pile supports 41% of the total load and the maximum moment is about 25% less than in the single pile ;
    - group effects also affect the P-y curves. For 2B spacing the mobilised soil reaction on the trailing pile is 50% that of a single pile.

REFERENCES

Amar, S., F. Baguelin, & Y. Canepa 1995. Experimental study of the behaviour of shallow foundations. Proc. 11th ARCSMFE. pp. 85-106. Cairo, Egypt.

Bakir, N.E., Garnier, J. & Y. Canepa 1995. Etudes sur modèles centrifugés de la capacité portante de fondations superficielles. *Rapp. Rech. LPC*, GT59, 188 p.

Brown, D. A. & C.-F. Shie 1990. Numerical Experiments into Group Effect on the Response of Piles to Lateral Loading, *Computers and Geotechnics*, N° 10, pp. 211-230.

Corté, J.F. & J. Garnier 1986. Une centrifugeuse pour la recherche en géotechnique. *Bull. Liaison Lab. P. et Ch.* Nov. pp.5-28.

Frank, R. & J.-C. Romagny 1990. PILATE-LCPC *Version 10.4 - Notice d'utilisation*, Ed. LCPC.

Marechal, O., J. Garnier, Y. Canepa, & A. Morbois 1998. Bearing capacity of shallow foundations near slopes under complex loads. *Centrifuge 98*, Ed. Kimura, Kusakabe & Kimura, A.A. Balkema Publishers, Rotterdam, Netherlands, pp 459-464.

Remaud, D., J. Garnier & R. Frank 1998. Laterally Loaded Piles in Dense Sand : Group Effect, *Centrifuge 98*, Ed. Kimura, Kusakabe & Kimura, A.A. Balkema Publishers, Rotterdam, Netherlands, pp 533-538.

Geotechnics for Developing Africa, Wardle, Blight & Fourie (eds) © 1999 Balkema, Rotterdam, ISBN 90 5809 082 5

Theoretical analysis of the behaviour of clays around a pressuremeter

R. Bahar
Institut de Génie Civil, Université Mouloud Mammeri, Tizi-Ouzou, Algerie

Y. Abed
Institut de Génie Civil, Université de Blida, Algerie

G. Olivari
Département de Mécanique des Solides, Ecole Centrale de Lyon, France

ABSTRACT : In this paper, the behaviour of clay around a pressuremeter is analysed by means of a theoretical study based on an elasto-plastic model in accordance with the theory of generalised standard materials. Menard pressuremeter tests, carried out on clays, are then analysed in order to determine the undrained cohesion. The obtained results are then compared with those determined with other methods.

1 INTRODUCTION

It is well known that there is a formal analogy between the behaviour of ductile metals, such as mild steel, and the behaviour of cohesive soils under undrained conditions; the straining happens without volume change. This framework is well suited to the use of elasto-plastic models such as the generalised Prager model which gives a realistic response on the unloading path. From an analytical representation of the total stress-strain curve obtained during an undrained triaxial test, it is shown how it is possible to identify all the model parameters. The result of this concept is that the proposed model depends finally on three parameters only, taking into account the assumptions of incompressibility and plane strain, and gives a realistic response on the unloading pressuremeter path without any new parameter. The model is then introduced in a computer code to identify the parameters of soil behaviour from a pressuremeter test. Finally, it leads to the determination of the undrained cohesion of normally consolidated isotropic clay, in good agreement with the values obtained by other methods.

2 THE GENERALISED PRAGER MODEL

In the incremental theory of plasticity, the Prager model describes the behaviour of a yielding material by a yield surface which gives the stress state under which yielding first occurs, a work hardening rule which specifies how the yield surface is changed during plastic flow, and a flow rule which relates the plastic strain increment to the state of stress and to the stress increment. Iwan (1967) has proposed a generalisation of the Prager model by considering a chain of elementary links associated in series. Each link represents a classic Prager model. Consequently, the constitutive equations are :

- equation of the yield surface f_k:

$$f_k \left(s_{ij} - X_{ij}^k \right) = S_k^2 \qquad (1)$$

where

k used as a superscript denotes the kth element of the chain,

s_{ij} denote the components of the deviatoric stress tensor,

S_k is the threshold stress of the kth yield surface,

X_{ij}^k is a tensor specifying the axis of this kth surface.

- evolution law of the hardening variables, X_{ij} :

$$dX_{ij}^k = C_k \, d\varepsilon_{ij}^p \qquad (2)$$

C_k is the modulus of the kth element of the chain,

$d\varepsilon_{ij}^p$ is the plastic strain increment tensor.

- plastic strain increment $d\varepsilon_{ij}^p$:

$$(d\varepsilon_{ij}^p)^k = \left(\frac{\partial f^k}{\partial \sigma_{ij}} \right) d\lambda_k \quad d\lambda_k > 0 \quad \text{if } d\sigma_{ij} . n_{ij}^k > 0 \quad (3a)$$

$$(d\varepsilon_{ij}^p)^k = 0 \quad \text{if} \quad d\sigma_{ij} n_{ij}^k \leq 0 \quad (3b)$$

where n_{ij}^k is the normal external vector to the kth yield surface and $d\sigma_{ij}$ is the stress increment tensor. $d\lambda_k$ is a factor given by:

$$d\lambda_k = \frac{1}{C_k} \frac{\dfrac{\partial f^k}{\partial \sigma_{ij}} d\sigma_{ij}}{\left(\dfrac{\partial f^k}{\partial \sigma_{lm}} \right)\left(\dfrac{\partial f^k}{\partial \sigma_{lm}} \right)} > 0 \quad (4)$$

If the plastic strain of an element is different from zero, this element is called an active element. The number of activated links, n^*, is such that : $n^* \leq n$. The model is thus defined by the compliance of n elastic elements and their associated n yield surfaces. Eventually, to complete the model, a single linear elastic element of compliance J_o, and a single yield surface of threshold stress S_∞. can be introduced and may be connected in series to the chain. Thus, the model is defined by $2n+2$ parameters which can be represented by a discreet spectrum of compliance as shown in Figure 1. Moreover; The components of the tensors X_{ij} which specify the state of the model must be given. In the virgin state, all the residual stresses are equal to zero, hence for each link, the hardening variables are also equal to zero, $X_{ij}(k)=0$. Since each of the n Prager elements will individually obey a linear work-hardening law, their combined action leads to a piecewise linear behaviour with kinematics hardening for the material as a whole.

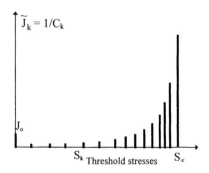

Figure 1. Compliance spectrum of the model.

3 MODELLING OF THE UNDRAINED BEHAVIOUR OF CLAY

The framework of the short-term behaviour of an incompressible medium is adopted. The considered stresses are thus total stresses. The material is supposed to be isotropic and normally consolidated or with a low value of the overconsolidation ratio. The criterion used obeys the condition of plastic incompressibility. Referring to the results of Habib (1953), obtained on Provins clay, it is admitted that the rupture happens for the same value of the deviator in compression and in tension. Therefore the undrained cohesion is sufficient to characterise the plastic flow criterion. Consequently, the Von Mises criterion is used :

$$\frac{2}{3} (s_{ij} - X_{ij})(s_{ij} - X_{ij}) - K^2 \leq 0 \quad \text{with } K=2C_u \quad (5)$$

where s_{ij} is the deviatoric stress tensor and C_u is the undrained cohesion.

3.1 Triaxial curve and generalised Prager model

The response of this model on a triaxial path is a polygonal line which can be considered as a discretisation of the experimental curve. Then, if an analytical representation for this curve is chosen, the parameters of the Prager model can be identified easily. The following relationship, proposed by Olivari and Bahar (1995), is used :

$$\varepsilon_d^p = - A \left(\ln(1 - R) + (1 - 2R) \frac{R}{1 - R} \right) \quad (6a)$$

$$R = \frac{\sigma_1 - \sigma_3}{(\sigma_1 - \sigma_3)_f} \quad (6b)$$

where σ_1 and σ_3 are the principal total stresses, $(\sigma_1 - \sigma_3)_f$ is the asymptotic value of stress difference which is related closely to the strength of the soil and A is a parameter defining the curvature of the curve as shown in Figure 2 (A > 0).

As a consequence to this expression, the non dimensional yields of the different links are fractions of the units of the chosen discretisation. Then, the compliance $\tilde{J}_k$ can be easily obtained from the successive secant moduli of the experimental curve. The compliance intervening in the behaviour law are written as :

$$J_o = \frac{1}{E} \qquad J_k = \sqrt{\frac{2}{3}} \frac{\tilde{J}_k}{2 \rho_k C_u} \qquad (7a)$$

$$0 < \rho_k < 1 \qquad (7b)$$

Consequently, the model is totally defined with three parameters only, the Young's modulous E, the parameter A, and the undrained cohesion C_u, the Poisson ratio being equal to 0.5.

4 MODEL RESPONSE ON A PRESSUREMETER PATH

The expansion of a pressuremeter probe in a homogeneous soil is considered. The hypothesis of soil incompressibility and plane strain in the axial direction are assumed. The axisymmetry imposes that the stress increments in the directions r, θ and z are principal (Fig. 3). The boundary conditions can be specified either in displacements or stresses.

$$dU_r = dU_0 \quad \text{or} \quad d\sigma_r = \Delta p_0 \quad \text{for} \quad r = r_0 \qquad (8a)$$

$$dU_r = 0 \quad \text{or} \quad d\sigma_r = 0 \quad \text{for} \quad r = r_\infty \qquad (8b)$$

where r_0 is the initial borehole radius, U_0 is the radial displacement at the cavity wall and p_0 is the applied pressure.

There are two concentric annular zones around the probe : - the first one is bounded by a circle of radius r_e in which the material is subject to elasto-plastic straining, - the second one is located beyond r_e in which the material behaves elastically.

4.1 Solution of the elastic problem

The solution, which is developed in the work of Boubanga (1990), is adapted to the case of the existence of two zones. The obtained results are as follows (ν=0.5) :

$$d\sigma_r = \frac{2}{3} E \frac{r_e}{r^2} dU_e \qquad d\sigma_\theta = -\frac{2}{3} E \frac{r_e}{r^2} dU_e$$

$$d\sigma_z = 0 \qquad (9)$$

The straining happens without volume change and the mean stress remains constant.

4.2 Solution of the elastoplastic problem

The plastic incompressibility implies that the yield surfaces are represented by a cylinders in the principal stress space (Von Mises criterion). Thus, the incremental strain vector $d\varepsilon_{ij}^p$ is normal to the sections of the yield surfaces characterised by the circles of a radius S_k and a centre X_{ij}^k. Furthermore, the condition of plane strain imposes that the strain path is normal to the ε_z axis. The two conditions lead to the following relationships between the deviatoric increments :

$$d\varepsilon_r = C(r) ds_r \quad d\varepsilon_\theta = -d\varepsilon_r \quad d\varepsilon_z = 0 \qquad (10a)$$

$$ds_r = \frac{d\varepsilon_r}{C(r)} \quad ds_\theta = -ds_r \quad ds_z = 0 \qquad (10b)$$

$$C(r) = \frac{3}{2E} + 2 \sum_{k=1}^{n^*} \frac{J_k}{S_k^2} (s_r - X_r^k)(s_r - X_r^k) \qquad (10c)$$

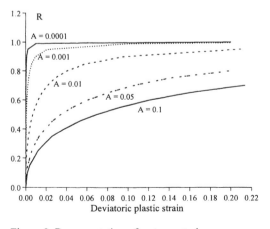

Figure 2. Representation of a stress-strain curve.

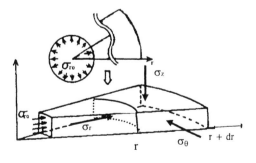

Figure 3. Equilibrium of a soil element.

S_k being a constant proportional to the threshold stress of the kth yield surface, which is a function of the undrained cohesion C_u. The coefficient $C(r)$ is a function of the distance r from the studied point of the soil mass to the axis of the borehole, because the number, n^*, of activated links of the model depends upon this distance. Further, to solve the problem completely, an unknown isotropic pressure $p(r)$ may be introduced :

$$d\sigma_r = dp(r) + ds_r \qquad d\sigma_\theta = dp(r) + ds_\theta$$
$$d\sigma_z = dp(r) \qquad\qquad\qquad (11)$$

This pressure is determined from the equilibrium equations and boundary conditions. Therefore, the following expressions are deduced :

$$dp(r) = r_0\, dU_0 \left[\left(\frac{1}{r^2\, C(r)} \right)_{r_e}^{r} + 2 \int_r^{r_e} \frac{dr}{r^3\, C(r)} \right] \qquad (12)$$

$$dU_0 = \frac{\Delta p_0}{\dfrac{r_0}{r_e^2\, C(r_e)} + 2\, r_0 \int_r^{r_e} \frac{dr}{r^3\, C(r)}} \qquad (13)$$

4.3 Numerical implementation

From the analysis of the pressuremeter test results, a computer code is developed in order to automatically identify the three parameters of the proposed model, particularly, the undrained cohesion C_u. This code can be used in two different ways :
- given the parameters of the model, the corresponding pressuremeter curve can be deduced
- given the pressuremeter curve, parameters of the model can be identified by means of an optimisation method.

4.3.1 Evaluation of the proposed model from pressuremeter test

The simplex algorithm is used to optimise the parameters A, E and C_u to produce the best fitting curve to the data (Nelder et al., 1965). At the beginning of the optimisation procedure, it is necessary to define initial values of E, C_u and A. For this purpose, analytical formulations taking into account analytical simplifying assumptions are

considered. Elastic formulation allows to the initial value of E to be determined from the measured pressuremeter modulus, elastic plastic formulation allows the initial value of C_u to be determined from the measured limit pressure p_l. Using the Duncan and Chang (1970) representation of the stress-strain curve for the undrained response of the soil, and after some algebraic manipulations, the following relationship can be obtained for the initial value of the parameter A

$$A = \sqrt{6}\, \frac{R_f C_u R^2}{E_i (1 - R_f R)\left(\ln(1 - R) + (1 - 2R)\dfrac{R}{1-R} \right)}$$

where C_u and E_i are the initial values of Young's modulus and undrained cohesion defined above, R_f is a reduction factor normally less than one ($R_f = 0.7$). R is defined from pressuremeter test results. A numerical study shows some evidence to suggest that there is a relationship berween R and the parameter β used to classify the soils proposed by Baguelin et al. (1978).

Figure 4 compares for a typical optimisation case the responses associated with the initial and optimal values of the three parameters with the experimental data. It can seen that the optimal theoretical response is very close to the experimental data, indicating that the constitutive model is able to describe the stress-strain behaviour of the soil around pressuremeter adequately. Figure 4 also compares the theoretical and finite element simulations using the optimal parameters. The finite element program Press'Ident developed at the Ecole Centrale de Lyon, taking into account the proposed model is used. In this case, because exact incompressibility cannot be enforced in a finite element program using the displacement method, the Poisson's ratio for the soil was set at 0.49 rather than 0.5. It can be noted that the numerical and theoretical curves are very close.

4.3.2 Distribution of total stresses at various levels of cavity expansion and stress path

Experimental data simulating the pressuremeter test using a large hollow cylindrical specimen indicate that the total vertical stress does change during expansion cavity (Thevanayagam et al., 1994). The results of other analytical studies simulating a plane strain cylindrical cavity expansion in normally consolidated clays also indicate changes in total vertical stresses in all elements surrounding the cavity cavity. One set of results for the total stresses σ_r, σ_θ and σ_z, using the proposed model is shown on Figure 5. Figure 5c indicates that the total vertical stress σ_z initially decreases at very low strain levels and subsequently increases. The predicted stress near the borehole is shown on Figure 6, plotted on (p,q) space, where p and q are mean and deviator stress respectively.

4.3.3 Response of the model on the unloading model path

The unloading response of the proposed model on the pressuremeter path is very easy to obtain. Indeed, the Prager model complies with the masing rule, so that the unloading curve has the same shape as the first loading curve, except that the scale is increased by a factor of two. This result, which cannot be obtained using the Duncan model, seems to be realistic for isotropic clays, as shown on Figure 7.

5 APPLICATION EXAMPLES

Bordj-Menail clay, Fucino clay

The proposed method has been applied to Menard pressuremeter results performed in uniform clay deposit. 44 Menard pressuremeter test performed in Bordj-Menail clay in four boreholes at the same site were kindly supplied to us by the National Laboratory of Housing and Constructions, LNHC, of Rouiba. Figure 8a shows an identification example concerned with the test realised to a depth of six meters in the borehole number one. The undrained cohesions obtained using the proposed method are presented in Figure 8. The undrained cohesion values for four interpretation methods, Press'Ident with Duncan model (Boubanga, 1990), Amar & Jézéquel (1972), Prevost and Hoeg (1975) and subtangent method (Baguelin et al., 1978), are shown plotted against depth in Figure 9. It can be noted that; above 6 m depth, the proposed method gives a values relatively similar to those deduced by the others method. Below 6 m, the values differ.

The proposed approach is used to interpret the fucino V2P14 test (SBP) (Ferreira et al 1992). The final curve matching is presented in Figure 10. The Ferreira et al (1992) method gives results very close to the proposed method (C_u = 116 kPa). The Jefferies method gives C_u = 100 kPa.

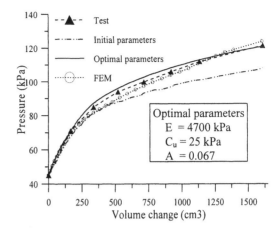

Figure 4. Pressure expansion curves (experimental, theoretical and finite element computation).

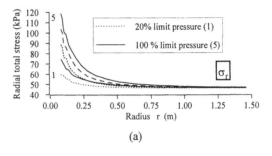

(a)

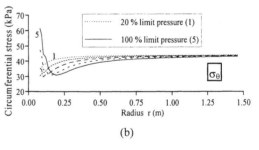

(b)

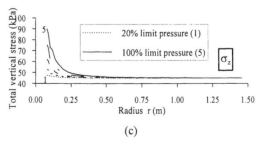

(c)

Figure 5. Distribution of total stresses at various levels of cavity expansion.

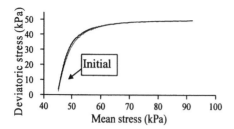

Figure 6. Stress path near the boring wall.

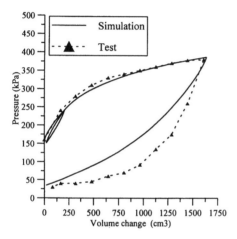

Figure 7. Pressuremeter test with unloading.

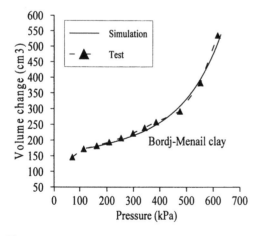

Figure 8. Identification of clay Parameters : Bordj-Menail clay (depth, 6m).

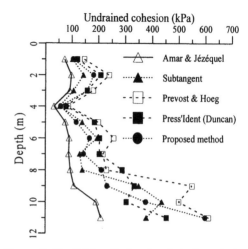

Figure 9. Bordj Menail clay borehole 1, undrained cohesion versus depth.

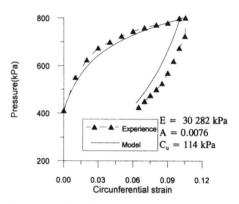

Figure 10. Identification of clay Parameters : Fucino V2P14 test.

6 CONCLUSION

A method to interpret pressuremeter results in clay has been presented. The proposed model is specifically developed to describe the undrained behaviour of clays. Introducing an analytical formulation of the total stress-strain curve obtained during an undrained triaxial test, the model depends finally on three parameters only. The response of the model on the pressuremeter path is in good agreement with experimental data. An application example of the proposed method has shown realistic values compared to those obtained by other means. However, Further research is needed to verify it for various clay types with both field and laboratory results.

REFERENCES

Amar, S. & J.F. Jézéquel 1972. Essais en place et en laboratoires sur sols cohérents : Comapaison des résultats. *Bulletin de Liaison des Ponts et Chaussées* 58 : 97-108.

Baguelin, F. J.F. Jezequel & D.H. Shields 1978. *The pressuremeter and foundation engineering*, Trans Tech Publications, Switzerland

Boubanga, A. 1990. Identification de paramètres de comportement des sols à partir de l'essai pressiométrique. Thèse de doctorat. E. C. de Lyon.

Duncan, J.M. & C.Y.Chang 1970. " Non linear analysis of stress and strain in soils" *J. Geotech. Engg Div.* SM5 : 1629-1653.

Habib, P. 1953. La résistance au cisaillement des sols. Thèse Paris, Publiée par la Doc. technique du bâtiment et des travaux publics.

Ferreira, R. & P.K. Robertson 1992. Interpretation of undrained self-boring pressuremeter test results incorporationg unloading, *Can. Geotech. J.* 29 : 918-928.

Iwan, W.D. 1967. On a class of models for the yielding behaviour of continuous and composite systems. *J. of applied Mechanics*. 612-617.

Nelder, J..A. & R. Mead 1965. A simplex method for function minimisation. *The computer journal*, 7 : 308-313.

Olivari, G. & R. Bahar 1995. Response of generalized Prager's model on pressuremeter path. *Proc. 4th Int. Sym. on pressuremeter*. A.A Balkema, 207-213, 17-19 mai 1995, Sherbrooke, Canada.

Prevost, J.H. & K. Hoeg 1975. Analysis of pressuremeter in strain-softening soil. *J. Geotech. Engng. Div.* vol 101, GT8 : 717-731.

Thevanayagam, S. J.L. Chameau & A.G. Altschaeffl 1994. Some aspects of pressuremeter test interpretation in clays, *Geotechnique* 44, 2 : 319-334.

Geotechnics for Developing Africa, Wardle, Blight & Fourie (eds) © 1999 Balkema, Rotterdam, ISBN 90 5809 082 5

The design and performance of driven-cast-in-situ piles in Dar es Salaam, Tanzania

E.J. Braithwaite
Franki Africa (Pty) Limited, Johannesburg, South Africa

E. Fenn
Franki Africa (Pty) Limited, Dar es Salaam, Tanzania

ABSTRACT: Several large projects incorporating piled foundations have recently been completed in Dar es Salaam, Tanzania. Most of these have utilized driven-cast-in-situ piles with bulbous bases. Full scale static load tests have been conducted on most of these projects. This paper reports on the results of these tests and gives recommendations for the design of this pile type in local conditions.

1.0 INTRODUCTION

The city of Dar es Salaam, Tanzania, is underlain by deep clayey and silty sands interbedded with gravelly coral horizons. The sub-soils are typically firm or medium dense although consistency does vary both vertically and horizontally. The soils are typically normally consolidated and quite highly compressible. Coupled with a high water table, these poor soil conditions have presented significant challenges to engineers involved in the design of heavy foundations. In the past, due to a lack of suitable alternatives, raft foundations have been almost mandatory for all buildings in excess of about 3 stories.

With the recent introduction of piling and other ground improvement techniques to the Tanzanian market, Engineers are now able to compare the costs and benefits of several foundation options to the advantage of their clients.

2.0 GEOLOGICAL SETTING

According to the 1:125000 geological map of Dar es Salaam, QDS No. 186, the city is underlain mainly by recent (holocene) deposits of silty and clayey sands. These are in turn underlain by similar coastal deposits originating in the Mio-Pliocene period. In general these deposits are characterized by deep silty and clayey sands with interbedded gravelly coral limestone deposits. Coral Limestone bedrock may be encountered at relatively shallow depth near to the coast but is erratically developed.

A typical soil profile along with illustrative soil parameters is contained in table 1.

The existence of several perched water tables is relatively common and is related to the local presence of aquifers and or confining layers. Piezometric measurements usually indicate a phreatic surface about 4.0 to 5.0m below surface. This implies that lower confined aquifers are under artesian pressure.

Table 1 : Typical soil profile and illustrative soil properties

Depth	Description	SPT N	Young's Modulus (kPa)
0.0 – 0.5	Fill, topsoil, hillwash etc		
0.5 – 4.5	Loose silty clayey SAND.	4 – 20 (Ave 10)	3500 – 10000
4.5 – 7.0	Loose / medium dense clayey gravelly (coral) SAND	10 – 30 (Ave 20)	5500 – 15000
7.0 – 12.0	Medium dense / firm to stiff clayey SAND / sandy CLAY	10 – 30 (Ave 15)	4000 – 12000
12.0 -	Medium dense to dense gravelly clayey SAND	20 – 40 (Ave 30)	10000 – 30000

The topography within the city center is relatively flat and is within about 8m of mean sea level.

3.0 DESCRIPTION OF PILES

Driven cast-in-situ piles with bulbous bases have recently been preferred to other possible pile types such as precast driven or continuous flight auger piles. The reasons for this are partly economic and partly the reliability of the bulbous base pile.

The main feature of this pile type is the enlarged base formed at the toe of the pile. In forming the enlarged base, the end bearing area is increased considerably. In addition, the displacement achieved when expelling the plug and forming the enlarged base compacts and pre-conditions the soil surrounding the base resulting in greatly increased end-bearing. These factors combine to enable the pile to develop its design working load at relatively low deflections.

The pile itself is constructed by driving an open ended tube into the ground by means of a heavy cylindrical drop hammer. The hammer is repeatedly raised and dropped onto a dense gravel plug at the toe of the tube. The plug draws the tube into the ground and prohibits the entry of water or foreign matter. On reaching a suitable founding horizon, the tube is withdrawn slightly and the plug is expelled enough to allow a bulbous base to be formed below the tube by expelling dry mix concrete with direct blows of the hammer until the consolidation of the base, or the size of the base meets the pre-determined criteria related to the working load on the pile.

After introducing the reinforcing cage, the shaft is formed by filling the tube with high slump concrete. The tube itself is then withdrawn.

The installation process is illustrated schematically in figure 1.

A total of six tests performed on piles constructed in this manner, have been referenced for this paper. Five of the piles were on different sites, while two piles from the same site were tested. All the piles had a nominal shaft diameter of 520mm and an assumed bulbous base diameter of 760mm. Pile lengths varied from 4.75m to 11.0m. The sites are all within a 7km radius of the city center.

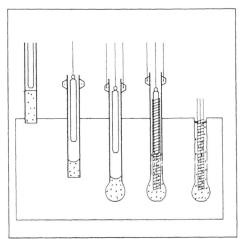

Figure 1 : Sequence of installation for driven cast-in-situ pile.

4.0 PILE TEST PROCEDURE

The piles were tested by means of static compression tests in which the load is gradually applied to the head of the pile while the deflection of the pile head is monitored.

In the first cycle the load was increased in 25% increments up to the design working load, held for 12 hours and then unloaded in 25% increments back to zero. The

intermediate load increments were maintained until two successive readings 30minutes apart indicated that the head deflection had not changed by more than 0.1mm. The load was then kept at zero for a period of one hour, after which the deflection was checked and the second load cycle begun. The second cycle was similar to the first but the maximum load was 1.5 times the design working load. After unloading from the second cycle, the residual deflection was monitored over a 12 hour period.

The configuration of the test apparatus is shown schematically in figure 2.

In all cases the reaction to the test load was provided by four tension piles. In two cases, test piles 1 and 4, the full test load of 1.5 times design working load could not be attained due to tension failures in the couplings of the reaction piles.

5.0 RESULTS OF LOAD TESTS

The pertinent details of each pile test are summarized in Table 2, while the results of

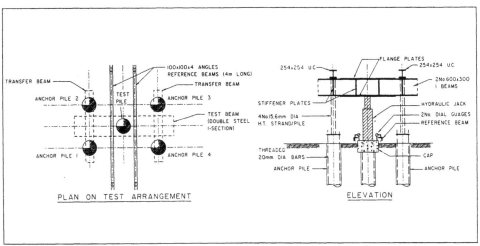

Figure 2 : Test pile arrangement.

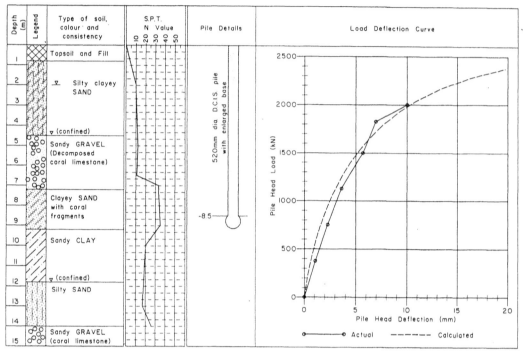

Figure 3 : Test Pile no. 1

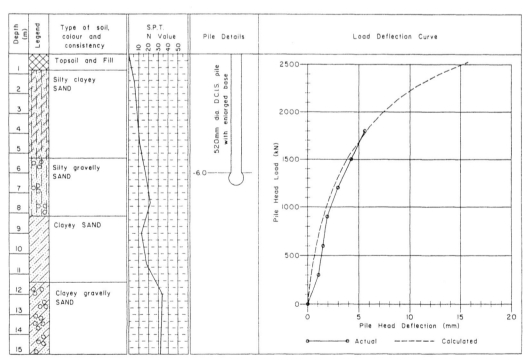

Figure 4 : Test Pile no. 2

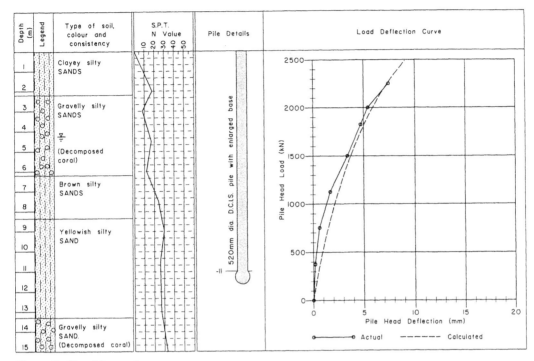

Figure 5 : Test Pile no. 3

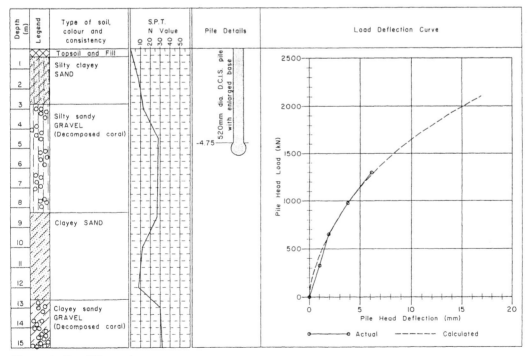

Figure 6 : Test Pile no. 4

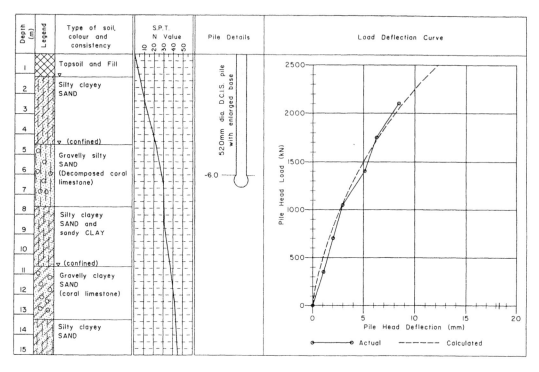

Figure 7 : Test Pile no. 5

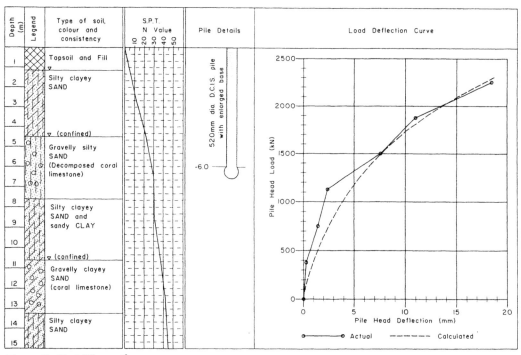

Figure 8 : Test Pile no. 6

Table 2 : Details of Test Piles

	Pile no. 1	Pile no. 2	Pile no. 3	Pile no. 4	Pile no. 5	Pile no. 6
Shaft length (m)	8	6	11	4.75	6	6
Shaft diameter (mm)	520	520	520	520	520	520
Base diam (mm) (assumed)	760	760	760	760	760	760
Working load (kN)	1400	1200	1500	1300	1500	1400
Settlement at WL (mm)	4.0	3.6	3.4	6.1	7.6	4.9
Settlement at 1.5 x WL (mm)		5.6	7.43		17.45	8.42

the test programs are summarized in figures 3 to 8.

6.0   BACK ANALYSIS OF RESULTS

The results of the pile tests were modeled using the *Loaddefl* computer program (Everett 1991). Using this program the various parameters can be altered until the calculated load/deflection curve approximates the actual. The calculated curves are superimposed on the actual test results in figures 3 to 8.

Figures 9 to 12 provide useful summaries of design parameters suggested by the back analyses.

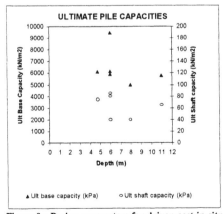

Figure 9 : Design parameters for driven cast-in-situ piles with bulbous base in typical conditions in Dar es Salaam.

Figure 9 indicates that, aside from one outlier, an ultimate base capacity of between 5000 and 6000 kPa may be assumed for bases formed in sandy material with an SPT blow count of about 25 to 30. Assuming that the average soil profiles reflected on the test sites are reasonably representative of Dar es Salaam, this is probably a good enough estimate for preliminary design purposes.

Average ultimate shaft friction values of between 40 and 85 kPa were achieved. Again values in the range 60 to 70 kPa appear to be appropriate for preliminary designs in a typical profile.

"Ultimate" pile capacity is defined as the load at which the pile head settles by 10% of the shaft diameter.

Also of interest is figure 10 which may be used to provide a rough guide to the shaft length (in a typical profile) likely to be required to limit settlements at working load to any given levels.

For example a pile with a design working load of 1200 kN founded (in typical soil conditions) at 7m below surface may be expected to settle of the order of 1200/300 = 4mm at working load.

It is emphasized that these recommendations are based on typical soil conditions and must of course be referenced to site specific conditions.

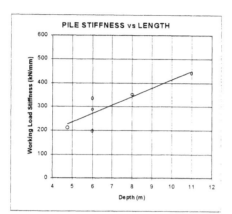

Figure 10 : Pile stiffness at working load related to shaft length.

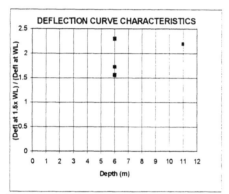

Figure 11 : Ratio of settlement at 1.5 x WL to settlmemt at WL.

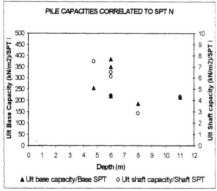

Figure 12 : Ultimate base and shaft capacities related to SPT N (Uncorrected).

Figure 11 confirms that the pile head settlements at 150% working load are likely to be of the order of twice the working load settlement. This is a useful rule of thumb when assessing static load test results or in estimating the effects of additional load to be carried on existing piles.

Figure 12 correlates the calculated ultimate shaft and base resistances with the average SPT data available for each site. It is found that the ratio of ultimate base capacity to SPT N (averaged in the vicinity of the base) is between 200 and 400. The corresponding ratio for ultimate skin friction was found to be in the range 3-8. This is in good agreement with Meyerhof (1956) and compares well with values suggested by Byrne et al (1995).

7.0 CONCLUSIONS

The series of pile tests reported confirm the applicability of the driven cast in situ pile with expanded base for the geotechnical conditions prevalent in Dar es Salaam. The load / settlement performance of these piles can be predicted with a relatively high level of confidence using the *Loaddefl* program.

8.0 REFERENCES

Byrne, G.P., Everett, J.P., and Schwartz, K. (1995), A guide to Practical Geotechnical Engineering in Southern Africa. 3rd Ed. Frankipile, Johannesburg.

Everett, J.P. (1991), Load Transfer Functions and Pile Performance Modelling, Geotechnics in the African Environment, Blight et al. (eds), Balkema, Rotterdam.

Meyerhof, G.G. (1956), Penetration Tests and Bearing Capacity of Cohesionless Soils, Journal of the Soil Mechanics and Foundation Division, ASCE, Vol. 82, SM1:1-19.

Geotechnics for Developing Africa, Wardle, Blight & Fourie (eds) © 1999 Balkema, Rotterdam, ISBN 90 5809 082 5

Un cas d'instabilité de sol de fondation: Exemple des marnes gonflantes de Rufisque, Sénégal

I.K.Cisse
Laboratoire de Mécanique des Sols et des Matériaux, Ecole Supérieure Polytechnique, Thiès, Sénégal

M.Fall
Institut des Sciences de la Terre, Université Cheikh Anta Diop, Dakar, Sénégal

A.Rahal
Géomécanique, Thermique, Matériaux (G.T.Ma), I.N.S.A., Génie Civil, Complexe Scientifique de Rennes-Beaulieu, France

RESUME : Les désordres causés sur les bâtiments légers par le gonflement des marnes sont observés depuis plusieurs années à Rufisque.
Il a fallu attendre les années 80, suite à une expertise des bâtiments dégradés, pour qu'on se rende compte qu'il s'agissait de sols gonflants, dont le comportement suivant les saisons , était assimilé à une « respiration ».
L'analyse du cas du Centre de Santé de Gouye Mouride (Rufisque) a montré que dans le cas des sols gonflants, même de faibles pressions de gonflement sont à redouter. Une méthode de conception des fondations simple suffit cependant pour mettre à l'abri de désordres dont la remise en état conduit à des dépenses plus élevées qu'une réalisation de l'ouvrage suivant les règles de l'art prenant en compte le caractère expansif du sol-support.
En fait, il n'y a aucune relation entre la résistance au cisaillement et la pression de gonflement ; une portance suffisante ne met pas à l'abri de désordres éventuels.

INTRODUCTION

Les désordres causés sur les bâtiments légers par le gonflement des marnes sont observés depuis plusieurs années à Rufisque. Cependant, jamais l'expansion de ces sols n'a été mise en cause par les victimes qui attribuaient les fissures observées aux vibrations consécutives au dynamitage pour extraction des matières premières nécessaires à la fabrication du ciment, par l'usine productrice qui se trouve à proximité. Pour d'autres, loin du site de l'usine, c'est l'effet de la « respiration » du sol qui est à l'origine des dégradations saisonnières (fissurations de murs et de planchers).
Il a fallu attendre les années 80, suite à une expertise des bâtiments dégradés, pour qu'on se rende compte qu'il s'agissait de sols gonflants, dont le comportement suivant les saisons, était assimilé à une « respiration ».
Depuis 1983 et jusqu'en 1985, ces marnes ont fait l'objet d'études géotechniques qui ont confirmé leur caractère expansif auquel on pouvait apporter des remèdes préventifs [1].
L'étude que nous présentons ici a porté sur un Centre de Santé à construire dans un quartier de Rufisque où la presque totalité des constructions en maçonnerie et béton subissent des fissurations à ouvertures périodiques. Après une identification et une mise en évidence du caractère gonflant du sol de fondation, nous proposons une méthode de construction garantissant à la structure toute la stabilité vis à vis de l'expansion.

1. IDENTIFICATION GEOTECHNIQUE

1.1 Détermination des caractéristiques d'état, pondérales et de gonflement

Deux sondages (figure 1) ont été effectués avec 4 prélèvements qui ont révélé l'homogénéité du sol puisque les prélèvements 1 et 3 d'une part , 2 et 4 d'autre part sont identiques au vu des résultats d'identification (tableau 1).
Nous nous sommes rassurés de la continuité de la deuxième couche par un trou au fond des sondages (figure 1) et par enquête auprès des habitants.
La mesure de la pression de gonflement s'est faite par la méthode de prégonflement bien connue [1] et certains paramètres (w, gh, gd) ont été mesurés avant et après essai. L'examen des résultats du tableau 1 révèle un degré de saturation des sols supérieur à 90 % due à une inondation du site antérieurement à notre intervention.

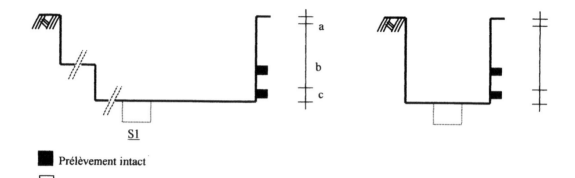

<u>S1</u>

■ Prélèvement intact

☐ Trou en fond sondage

Figure 1: Coupe des sondages effectués : S1 S2

Couche	Epaisseur (m)	Nature sol
a	0,40	terre végétale
b	1,10	argile noirâtre
c	---	argile grisâtre à beige

Tableau 1 : récapitulatif des essais d'identification

Numéro prélèvement		1	2	3	4
Description visuelle		marne noirâtre	marne grisâtre à beige	marne noirâtre	marne grisâtre à beige
W (%)	initiale	25,20	28,30	26,11	28,50
	finale	27,07	29,41	29,24	29,50
γ_h (KN/m³)	initiale	19,20	19,00	18,80	19,50
	finale	20,50	20,20	19,90	19,70
γ_d (KN/m³)	initiale	15,30	14,80	14,90	15,20
	finale	16,10	15,60	15,40	15,20
γ_{sat} (KN/m³)		19,15	18,98	18,80	19,71
γ_s (KN/m³)		25,70	26,00	26,00	26,10
W_{sat} (%)		26,45	29,10	28,65	29,71
S_r (%)		95,27	97,25	91,13	95,93
n		0,405	0,431	0,427	0,425
e		0,680	0,757	0,745	0,74
W_l (%)		82,14	130,22	73,60	124,17
W_p (%)		39,70	65,34	37,20	59,28
I_p (%)		42,44	64,88	36,40	64,89
P_g (bars)		1,20	0,72	1,21	0,69
Ig (%)		50	50	50	50

Les valeurs d'indice de plasticité (Ip) élevées classent nos sols dans le domaine des limons très plastiques. Le coefficient d'activité Ac qui, est égale à 1,75 indique que notre argile est bien active d'une part et sa position dans la classification de Skempton (1953) et Mitchell (1976) confirme (ou montre) que nous avons affaire à une montmorillonite calcique d'autre part ; la pression de gonflement varie quant à elle entre 1,20 bars et 0,70 bars lorsque la profondeur passe de 1 à 1,20 m. Cependant, dans le cas des constructions légères, même une pression de gonflement de 1 bar est à redouter du fait de la distribution particulière des contraintes en profondeur [2].

Dans le cas des sols à pression de gonflement supérieure à 1 bar, c'est en général ce paramètre qui est prépondérant par rapport à la portance ; mais il vaut mieux s'en assurer en déterminant les variables de résistance (cohésion, angle de frottement interne) dans des conditions correspondant aux différentes situations possibles.

1.2 Détermination des paramètres de dimensionnement des fondations

Compte tenu de l'homogénéité du sol, nous avons choisi d'effectuer les essais sur les prélèvements 2 et 3

1.2.1 Essai de cisaillement à la boîte de Casagrande
Trois situations possibles ont été simulées

1er cas : Echantillon à l'état naturel, consolidé puis cisaillé

Cette situation correspond à la construction du bâtiment sur le site dans son état actuel. La démarche consiste à consolider à des pressions choisies à priori (0,5 bar, 1bar , 2 bars et 3 bars) et à cisailler à chaque fin de consolidation avec une vitesse aussi lente que possible. On a ainsi un essai de type consolidé-drainé avec mesure du tassement.

2ème cas : Echantillon saturé puis consolidé puis cisaillé

Cette situation correspond au cas où, pour une raison ou pour une autre, le sol d'assise est saturé au moment du début des travaux, comme cela pourrait l'être après une forte pluie ou une inondation. On réalise encore un essai de type consolidé-drainé.

3ème cas : Echantillon saturé et consolidé en même temps, puis cisaillé

Cette situation correspond au cas où l'on construit pendant qu'il pleut par exemple (en hivernage) ou au cas où le site est inondé pendant la construction. L'essai consiste à inonder en même temps qu'on met la charge de consolidation. Suivant l'importance de la charge, on notera un début de gonflement puis de tassement ou pas de gonflement du tout.
Dans les trois cas, les tassements sont négligeables

et les paramètres de cisaillement mesurés sont récapitulés au tableau 2 .

Tableau 2 : Résultats des essais de cisaillement

Moyen de traitement avant cisaillement	$N°$	z (m)	C' (bars)	ϕ' (°)
1er cas : consolidé	3	1,00	0,30	27°
à l'état naturel	2	1,50	0,60	22°
2ème cas : saturé	3	1,00	0,18	6°
puis consolidé	2	1,50	0,40	5°
3ème cas : saturé	3	1,00	0,20	15°
et consolidé	2	1,50	0,55	10°

1.2.2. Essais de compressibilité à l'œdomètre

L'objectif de ces essais est de déterminer l'état de consolidation (σ_c) du sol et son coefficient de compression (C_c) dont la connaissance est nécessaire pour prévoir le comportement sous charges ultérieures.
Les résultats sont récapitulés au tableau 3

Tableau 3 : Résultats des essais de compressibilité

$N°$ prélèvement	3	2
Profondeur z (m)	1,00	1,50
σ'_c (bars)	1,20	1,90
C_c	0,188	0,294
P_g (bars)	1,20	0,72

1.2.3. Analyse des résultats

S'agissant des essais de cisaillement, les meilleures caractéristiques (c', ϕ') sont obtenues dans le premier cas, ensuite vient le cas où l'échantillon est consolidé en même temps qu'il est saturé (3ème cas). Le cas où l'échantillon est d'abord saturé puis consolidé donne les moins bons résultats.
Cette dégradation de la qualité du sol est liée à l'augmentation de la teneur en eau ; autrement dit, même après consolidation, il n'est pas possible d'évacuer l'eau déjà introduite dans l'échantillon. Il faut signaler que cela est une caractéristique des argiles surconsolidées qu'on ne peut presque jamais remettre dans leur état initial de matériau intact par compactage ou consolidation. En effet, une fois remaniée, on ne peut reconstituer l'état initial du matériau surconsolidé, état lié à son histoire géologique, cela a été confirmé par ailleurs [1].
Pour ce qui est des essais de compressibilité, l'examen du tableau 3 montre que les contraintes de consolidation augmentent avec la profondeur, de

même que les coefficients de compression. Cela est normal car plus le sol est sous-jacent, plus les

pressions naturelles des terres appliquées sont importantes et plus la compacité est élevée et l'énergie interne de gonflement importante.

Les valeurs des pressions de gonflement trouvées sont tout à fait comparables à celles obtenues par la méthode de prégonflement : 1,15 bars au lieu de 1,21 bars à 1,00 m et 0,69 bars au lieu de 0,72 bars à 1,50 m.

Il reste maintenant à examiner l'état de consolidation des sols et les paramètres de dimensionnement des fondations.

1.2.3.1 Etat de consolidation des sols

Il faut signaler que contrairement aux sols non expansifs, pour les sols expansifs - surconsolidés en général -, et pour des fondations superficielles, la portance calculée à partir des caractéristiques à court terme est souvent supérieure à la portance calculée avec les caractéristiques à long terme ; cette remarque est très importante pour la vérification de la stabilité des fondations. Auparavant, vérifions cependant l'état de consolidation.

Les prélèvements étant faits à 1,00 m et 1,50 m, nous comparons les pressions des terres à ces niveaux avec les pressions de consolidation.

- à 1,00 m de profondeur :
$\sigma'_0 = \gamma'_1.z_1 = (18,8 - 10) \times 1 = 8,8$ KN/m^2 = 0,088 bars
- à 1,50 m de profondeur
$\sigma'_0 = \gamma'_2.z_2 = (18,98-10) \times 1,5 = 13,47$ KN/m^2 = 0,13 bars

Or, à 1 m, on a $\sigma'_c = 1,20$ bars et à 1,50 m, $\sigma'_c = 1,90$ bars

Donc, dans tous les cas, le sol est surconsolidé puisque $\sigma'_c > \sigma'_0$.

1.2.3.2 Répartition des contraintes en profondeur σ'_M sous la semelle S_1

Cette semelle est la plus chargée d'après le plan béton armé fait sur la base d'un *taux de travail* de 1,2 bars (120 KPa) et fourni par le bureau d'études.

a) Descente de charges (sur S_1)

Charges apportées par le poteau :
$1,10 \times 1,065 \times 2,69 = 3,151$ t
Poids propre de la semelle :
$2,50 \times (0,70)^2 \times 0,15 = 0,184$ t
soit Q = 3,335 t que nous prendrons égale à 4 tonnes (40 KN

b) Vérification des dimensions de la semelle

La section doit être telle que :
$$S \geq \frac{Q}{s_s} = \frac{40}{120} = 0,33 m^2$$

Le poteau reposant sur la semelle étant carré, cette dernière aura la même géométrie ; on a donc :
$L^2 = 0,33$ m^2, soit L = 58 cm = 0,58 m
Si on choisit une section de 0,70 m, on est largement en sécurité.

c) Calcul des contraintes sous la semelle

Nous supposons le fond de fouille à -1 m :
A -1m, on a : $\sigma'_M = \sigma'_0 + \Delta\sigma$ avec $\sigma'_0 = 0$
avec
$$\Delta\sigma = q = \frac{Q}{(0,70)^2} = \frac{4}{(0,70)^2} = 8,16 \,{}^t\!/_{m^2} \approx 0,82 bar$$

A-1,50m, on a : $\sigma'_M = \sigma'_0 + \Delta\sigma$
avec $\Delta\sigma = 4.q.I$ (à l'aplomb du centre de la semelle)
Pour une semelle carrée de 0,70 m et pour un point situé à -0,50 m de la base de la semelle, on a :

$$\left.\begin{array}{l} a = \dfrac{0,70}{2} = 0,35 \\ b = a \\ z = 0,50 \end{array}\right\} \Rightarrow I = 0,124$$

Par ailleurs, $\sigma'_0 = \gamma'.z_1 = (18,8 - 10) \times 0,5 = 4,40$ KN/m^2 = 0,044 bar
donc, $\sigma'_M = \sigma'_0 + 4.q.I = 0,44 + 4 \times 0,82 \times 0,124$ # 0,45 bar
A-2 m, on a : $\sigma'_0 = 8,98 \times 1 = 8,98$ KN/m^2 = 0,089 bar
$\Delta\sigma = 4.q.I = 4 \times 0,82 \times 0,048 = 0,156$ bar
soit $\sigma'_M = 0,089 + 0,156$ # 0,24 bar
A-2,50 m,
$$\begin{cases} I = 0,023 \\ z = 1,50 \end{cases}$$
et $\sigma'_0 = 1,5 \times 8,98 = 13,7$ KN/m^2 = 0,14 bar
$\Delta\sigma = 4.q.I = 4 \times 0,082 \times 0,023 = 0,075$ bar
d'où $\sigma'_M = 0,14 + 0,075$ # 0,21 bar
Le tableau 4 fait le résumé de la répartition des contraintes.

Tableau 4 : Répartition des contraintes sous la semelle

z (m)	σ'_M (bars)
1,00	0,82
1,50	0,45
2,00	0,24
2,50	0,21

2. Comportement du sol d'assise sous l'effet des charges et du gonflement

2.1 Vérification du tassement

Pour cela, nous comparons le supplément de contrainte apporté par la charge q avec la contrainte de consolidation (tableau 5).

Tableau 5 : Comparaison entre contraintes appliquées et de consolidation

z (m)	$\Delta \sigma$ (bars)	σ'_c (bars)
1,0	0,82	1,2
1,50	0,40	1,9

On constate que dans les deux cas, $\sigma'_c > \Delta \sigma$.
Donc, les risques de tassement sont négligeables. En effet, les fondations surchargeant un sol surconsolidé sans que les contraintes supplémentaires ($\Delta \sigma$) apportées au poids des terres ne dépassent σ'_c, les tassements sont très faibles, voire négligeables. Ceci a déjà été vérifié pendant les essais de cisaillement.

2.2 Vérification des contraintes admissibles

Nous avons calculé la contrainte admissible dans les trois cas de cisaillement, avec un coefficient de sécurité de 3. Les résultats sont indiqués au tableau 6. On constate que dans le 2ème cas et lorsqu'on est fondé à 1,00 m, la contrainte admissible (0,62 bars est inférieure à la contrainte appliquée (0,82 bars) , donc, il y a problème de portance. On peut faire remarquer que le choix à priori du « taux de travail » de 1,2 bars n'est donc pas justifié.
Mais , il n'y a pas lieu de s'inquiéter avant d'avoir procédé à la vérification du risque de soulèvement.

Tableau 6 : Comparaison entre contraintes appliquées et admissibles

Cas	z (m)	q_{ad} (bars)	$q_{app} = \Delta \sigma$ (bars)
1er	1,00	3,68	0,82
	1,50	4,89	0,044
2ème	1,00	**0,62**	**0,82**
	1,50	1,28	0,044
3ème	1,00	1,15	0,82
	1,50	2,22	0,044

2.3 Vérification du risque de soulèvement

Il s'agit de comparer à plusieurs niveaux, la pression de gonflement à la contrainte due au poids propre des terres et à la charge sur la semelle ($\sigma'_M = \sigma'_0 + \Delta \sigma$).
Le tableau 7 montre que quelque soit la profondeur, la pression de gonflement reste supérieure à la contrainte due au poids des terres et à la surcharge q.

Tableau 7 : Comparaison entre pressions de gonflement et pressions appliquées

z (m)	Pg (bars)	σ'_M (bars)
1,00	1,20	0,82
1,50	0,72	0,45
2,00	< 0,72	0,24

La figure 2a montre l'évolution des pressions sous la semelle, évolution conforme à la théorie (figure2b). Ainsi, même dans le cas où on avait un défaut de portance, ce n'est pas celle-ci qui est prépondérante puisqu'il y a risque de soulèvement (Pg > σ'_M). Dans les autres cas où on n'a pas de défaut de portance, on a toujours
Pg > σ'_M. En d'autres termes, c'est la pression de gonflement qui est le paramètre prépondérant par rapport auquel il faut dimensionner les semelles et non pas la contrainte admissible.

2. Solution proposée

La solution que nous proposons vise à amoindrir, voire annuler les mouvements verticaux de la semelle qui pourraient être préjudiciables au bâtiment projeté. Nous voulons éviter tout contact direct entre la base de la semelle ou le plancher bas et le sol gonflant. Ensuite, on accompagnera la solution préconisée de certaines mesures visant à maintenir l'équilibre de l'humidité sous le bâtiment.
Nous donnons la description de la solution proposée et schématisée à la figure 3.
1°) Après excavation de la fouille destinée à recevoir la semelle (à -1,25 m), on met en place une couche de remblai de gravier classique (calcaire ou grès résistant) ou de latérite granulaire sur une épaisseur de 25 cm. La semelle viendra reposer sur ce remblai qui repose lui-même sur le fond de fouille compacté. La couche de matériau granulaire absorbera les mouvements éventuels et les vibrations consécutives au dynamitage dû à l'extraction des matières premières pour la fabrication du ciment à la SO.CO.CIM. En effet, elle a une porosité ouverte relativement importante par rapport à un matériau fin.
Le remblayage du reste de la fouille se fera avec le sol excavé initialement et bien compacté. En effet, les mouvements verticaux étant éliminés et le gonflement latéral empêché - c'est comme si le sol

était fretté -, il n'y a pas de risque de soulèvement du sol au dessus de la semelle.

2°) Pour éviter le soulèvement de la dalle, on la fera porter sur un vide sanitaire de 20 à 30 cm ; en effet, cela suffit pour absorber les poussées verticales du sol en dessous, cette dalle sera dimensionnée et bien armée comme un plancher appuyé sur 4 côtés.

3°) Aux mesures liées au non soulèvement de la dalle, on ajoutera les suivantes pour le maintien de l'équilibre de l'humidité, à savoir :

■ ceinturage du bâtiment à l'aide d'un trottoir périphérique (voir figure 3) ;

■ drainage éventuel si le sol est en pente vers le bâtiment ;

■ éloignement des arbres : éviter de les planter à moins de 1,5H (H = hauteur de l'arbre arrivé à maturité) ;

■ utilisation de canalisations flexibles et non

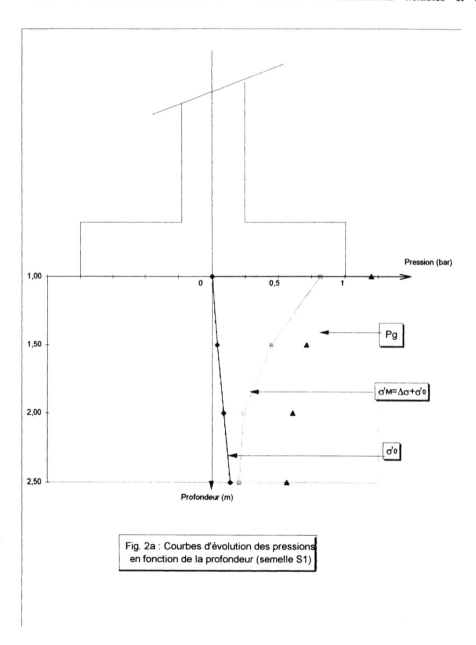

Fig. 2a : Courbes d'évolution des pressions en fonction de la profondeur (semelle S1)

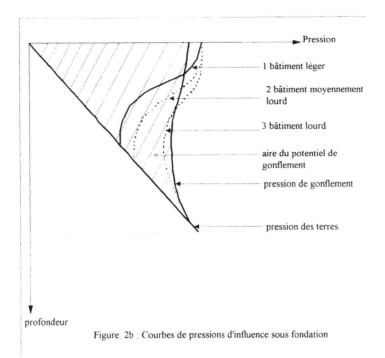

Figure. 2b : Courbes de pressions d'influence sous fondation

encastrées au bâtiment, avec des joints étanches et souples puis posés sur un lit épais de matériaux inertes (grave par exemple).

Cette solution n'a pas fait l'objet d'une évaluation économique mais son coût est certainement moins élevé que les frais d'une reprise en sous-œuvre qu entraînerait son non respect.

4. Remarques

Pour des raisons de recherche d'économie, le maître d'ouvrage avait proposé de se passer de la couche de latérite et de se fonder à 1,50 m en pensant que les pressions de gonflement qui diminuent avec la profondeur d'encastrement seraient inférieures aux pressions appliquées. De nouveaux essais dont les résultats sont récapitulés au tableau 8 ont révélé que la pression de contact était effectivement plus élevée que celle de gonflement à ce niveau mais qu 'au delà, c'est - à dire à 2,00 et 2,50 m, la distribution de la

Tableau 8 : Evolution des contraintes
à partir de 1,50 m.

z (m)	Pg (bars)	σ'_M (bars)
1,50	0,72	0,82
2,00	0,61	0,45
2,50	0,57	0,24

contrainte appliquée est telle que les pressions d'expansion dépassent celles dues à la surcharge appliquée.

Cela confirme l'idée selon laquelle la prise en compte d'une pression de contact supérieure à la pression de gonflement au même niveau ne suffit pas pour mettre à l'abri de futurs déboires.

Il ne serait plus intéressant d'augmenter davantage l'encastrement au delà de 2,50 m puisqu'on serait dans le domaine des fondations profondes qui coûtent plus chers.

CONCLUSION

L'étude du cas du Centre de Santé de Gouye Mouride (Rufisque) a montré que dans le cas des sols gonflants, même de faibles pressions de gonflement sont à redouter. Une méthode de conception des fondations simple suffit cependant pour mettre à l'abri de désordres dont la remise en état conduit à des dépenses plus élevées qu'une réalisation de l'ouvrage suivant les règles de l'art prenant en compte le caractère expansif du sol-support.

C'est le lieu de signaler qu'il n'y a pas de relation entre la résistance au cisaillement et la pression de gonflement ; une portance suffisante ne met pas à l'abri de désordres éventuels.

Il faut rappeler également que les contraintes sous

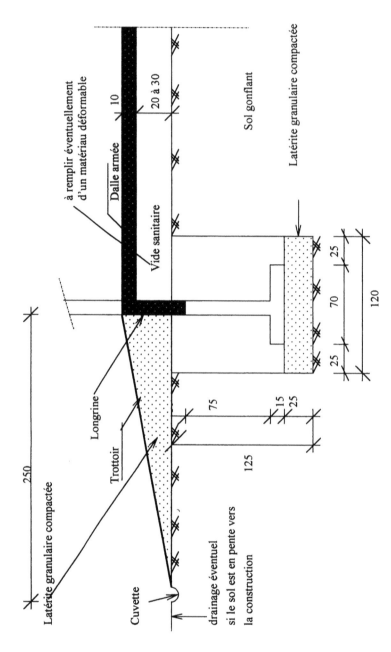

Figure 3 : Technique de fondation proposée à Gouye Mouride : semelle S1
(toutes les dimensions sont en centimètres)

une fondation dues à une charge extérieure diminuent en profondeur et qu'il ne suffit pas de s'opposer au gonflement en surface . Même si la pression de gonflement est inférieure en surface, elle peut dépasser celle due à la charge appliquée en profondeur, d'où les risques dangereux de soulèvement.

Il faut signaler que le bâtiment construit depuis 1993 sur la base de la méthode proposée ici n'a jusqu'à présent pas subi de dégradations visibles inhérentes aux dispositions constructives proposées.

Vu le succès enregistré, nous espérons que les futurs maîtres d'ouvrages prendront conscience de l'importance du phénomène de gonflement en adoptant les multiples solutions proposées par ailleurs suivant le cas [3].

REFERENCES

[1]- Cissé I. 1985. *Caractérisation et méthodes de construction sur sols gonflants : Application aux marnes de Rufisque (Sénégal)*, Thèse de doctorat de géologie de l'ingénieur, INPL de Nancy

[2]- Mariotti M.1981. *Le gonflement des sols argileux, Influence de l'expansibilité ou de l'instabilité sur les structures et recommandations pour délimiter les effets*, C.E.B.T.P.- Information- Documentation B.C.E.O.M. N° 18 :19-33

[3]- Mouroux P., Margron P., Pinte J.C.1976. *La construction économique sur sols gonflants* Editions manuels et méthodes.

Possibilité d'utilisation de la boue rouge de Friguia dans la construction

M.S. Diané

Centre de Recherche Scientifique, Conakry-Rogbanè, Guinée

RESUME: Cet article traite de l'efficacité de l'utilisation de pouzzolane d'un type de résidus de bauxite comme remplaçant partiel du ciment dans le mortier ou le béton pour la réalisation de constructions économiques dans les pays en développement. Il s'agit du produit du traitement chimique de bauxite concassée à l'Usine d'Alumine de Friguia (Guinée). L'étude a porté sur l'hydratation du ciment portland avec 20 et 30 % de remplacement de cette pouzzolane. Celle-ci est obtenue par déshydroxylation du résidu sus-cité à la température de 700 - 800°. La mise en oeuvre de techniques expérimentales physico-chimiques et minéralogiques (Analyse thermique différentielle, analyse chimique, sédimentométrie) et l'étude des résistances mécaniques sur éprouvettes cubiques ont permis:
- d'apprécier le comportement de la pouzzolane de boue rouge;
- d'établir la réactivité chimique de la boue rouge;
- de tenter d'optimiser les quantités de l'activateur chimique (hydroxyde de calcium) permettant d'obtenir les meilleures résistances.

ABSTRACT: This paper deals with the effectiveness of the use of pozzolana from bauxite residue as a partial replacement for cement in mortar or concrete for low-cost constructions in developing countries. This material was produced by the chemical treatment of crushed bauxite at the Usine d'Alumine in Friguia (Guinea). The study relates to the hydratation of Portland cement 20 or 30 % of which has been replaced by pozzolana. This is obtained by dehydroxylation of the above residue at 700 - 800 °C. The implementation of experimental physicochemical and mineralogical techniques (differential thermal analysis, chemical analysis, sedimentology) and investigation of the mechanical strength of cubic samples have:
- demonstrated the effectiveness of this pozzolana for construction purpose;
- determined their chemical reactivity;
- attempted to optimise the dosage of chemical activator (calcium hydroxide) in order to achieve the highest strengths.

INTRODUCTION

De nos jours encore, l'économie de la Guinée dépend essentiellement de l'exploitation des ressources minières et en particulier des gisements de bauxite dont les réserves sont estimées à 2/3 des réserves mondiales.

Ainsi le pays compte actuellement trois usines d'exploitation qui sont: la Compagnie des Bauxites de Guinée (CBG), la Société des Bauxites de Kindia (SBK) et l'usine d'alumine de Friguia-Kimbo qui est la première usine d'aluminium en terre africaine.

Si les deux premières unités exportent le minerai brut, la troisième, quant à elle, exporte l'alumine, procédant ainsi à un traitement chimique préalable du minerai sur place par le procédé Bayer avant son exportation.

La boue rouge, produit de la décantation de la solution obtenue après le concassage, le broyage et l'attaque du minerai à la soude caustique concentrée, est composée d'eau (65-70 %), de particules minérales (25 %) et d'une quantité relativement faible de substances solubles. A Friguia, la boue rouge, après avoir causé d'importants dégâts dans le fleuve Konkouré où elle s'écoulait directement, est aujourd'hui recueillie dans un lac artificiel de volume supérieur à 5 millions de mètres cubes autour duquel la végétation est détruite dans un rayon de 50 à 100 mètres. Après la précipitation de la masse essentielle des particules, la phase liquide se jette dans le fleuve Konkouré à travers des ouvertures pratiquées dans le barrage. Dans le plan d'eau de l'océan adjacent à l'estuaire de ce fleuve on observe une teneur en alumine très élevée (Diané, 1989). On estime à près de 700.000 t/an la production de boue rouge à l'usine de Friguia.

Or on sait que (Hammond. 1974) quand des argiles, des schistes ou d'autres matériaux terreux sont soumis à la cuisson, à une certaine température optimale ils acquièrent des propriétés pouzzolaniques. Finement broyés et en présence d'humidité, ces propriétés rendent lesdits matériaux aptes à réagir avec l'hydroxyde de calcium pour former un composé cimentier même si eux-mêmes ne possèdent pas de propriétés cimentières.

La bauxite fait partie de la gamme des matériaux qui exhibent des propriétés pouzzolaniques après un traitement thermique. Ainsi en 1909 Le Chatelier rapporte avoir identifié la bauxite dans la composition d'un mortier chaux/pouzzolane prove- nant de ruines anciennes à Baux dans la vallée du Rhône en France. Lea rapporte également que Ferrarri a déclaré avoir constaté que la bauxite calcinée constitue une excellente pouzzolane.

Ainsi une pouzzolane est un matériau siliceux, alumineux ou ferrugineux qui, bien que ne possédant pas de propriétés cimentières en lui-même, peut, dans certaines conditions de cristallinité et de structure, réagir avec la chaux en présence d'humidité dans les conditions normales de température et de pression pour donner des produits cimentiers (Hammond, 1976). La bauxite étant une matière première abondante en Guinée une solution présentée dans cette étude est l'élaboration de pouzzolanes artificielles par calcination des boues rouges à 700-800° C.

Par ce procédé le "Building and Road Research Institute de Kumasi (Ghana) a mis au point une pouzzolane performante par recyclage des déchets de bauxite obtenus par lavage à l'eau de la bauxite concassée (Hammond, 1974, 1926, 1976, 1985,1974).

Pendant que tout le ciment utilisé en Guinée est importé, la boue rouge pourrait être utilisé dans le béton et le mortier comme pouzzolane, c'est-à-dire un remplacement partiel du ciment. Ce qui contribuerait, du coup, à protéger l'environnement, à développer une industrie de matériaux de construction basée sur l'utilisation maximum des ressources locales et à réaliser des économies, à résistance égale et un effet bénéfique sur la chaleur d'hydratation et sur les réactions alcali-granulats (Ambroise, 1984).

PROGRAMME EXPERIMENTAL

L'expérimentation a porté sur les résidus de la transformation de la bauxite en alumine à l'usine de Friguia. Ces résidus de bauxite ont été récupérés et séchés au soleil avant la détermination de leur densité et de leur composition granulométrique (tableau I). Une portion a été ensuite soumise à la cuisson dans un four à des températures de 700 et 800°C et maintenue à ces températures pendant 8 heures. Les produits de ce traitement thermique rejoignent la température ambiante à la vitesse propre de refroidissement du four en l'absence de tout apport extérieur. ILs sont ensuite broyés à la finesse de mouture du ciment dans un broyeur à boulets pour devenir de la pouzzolane de boue rouge. L'activité pouzzolanique du matériau traité thermiquement avait été évaluée par l'exécution d'essai de compression sur des éprouvettes cubiques de mortier de ciment dans lequel on avait procédé au remplacement de proportions définies de ciment par de la pouzzolane de boue rouge. L'analyse sédimentométrique (tableau I), l'analyse chimique (tableau II) et l'analyse thermique différentielle (fig.1 et 2) avaient été effectuées sur la pouzzolane, ainsi que sur la boue rouge à l'état naturel.

RESULTATS ET PRODUITS

Identification géotechnique

Les propriétés géotechniques de la boue rouge sont données dans le tableau I. Dans ce tableau, les résultats de l'analyse granulométrique montrent que le pourcentage de fines dans la boue rouge, est de 60% dont 12% d'argile. Ce pourcentage de fines contribue, entre autres, à l'élévation de la limite de liquidité (42%) et de l'indice de plasticité (14%).

Analyse chimique

Les résultats de l'analyse chimique de la boue rouge à l'état naturel, effectuée au Geological survey Department d'Accra (Ghana), sont donnés dans le tableau II. Dans ce tableau on voit que les composantes essentielles de la pouzzolane de boue rouge sont la silice (SiO_2), 7,06; l'alumine (Al_2O_3), 14,68 et l'oxyde de fer (Fe_2O_3), 53,89%. Comparées à celles du ciment dont la teneur en silice, alumine et oxyde de fer sont, respectivement de 20,80 ; 6,14 et 2,52%. il apparaît évident que, contrairement au ciment. la pouzzolane est pauvre en silice et riche en alumine.

De l'analyse chimique il ressort que la boue rouge de Friguia est un sol latéritique contenant de la sidérite et de l'oxyde de fer, avec des traces de silice, de calcium, de magnésium et de titane.

Cependant La bauxite est une roche caractérisée par la présence d'une grande quantité d'alumine libre sous la forme d'un ou de plusieurs oxydes hydratés, en particulier la gibbsite, la bohémite et le diaspore, mélangés avec des impuretés généralement constituées de fer, d'alumine et de silice. La quantité d'alumine, qui est la composante commerciale la plus importante, est inversement proportionnelle à celle de ces impuretés. Ainsi la composition chimique de la bauxite varie dans des proportions très faibles.

Tableau I: Propriétés géotechniques de la boue rouge de Friguia

Couleur	Rougeâtre
Densité	2,71 g/cm^3
Granulométrie	
Sable	40 %
Limon	48 %
Argile	12 %
Limites d'Atterberg	
Limite de plasticité	42 %
Limite de liquidité	28 %
Indice de plasticité	14 %

Une comparaison de la composition de la pouzzolane de bauxite avec celles d'autres pouzzolanes artificielles (Hammond, 1974) montre qu'à l'exclusion de toutes les autres pouzzolanes artificielles qui contiennent une grande proportion de silice (plus de 50% environ), la proportion de silice dans la pouzzolane de bauxite est de 14% environ. L'alumine forme environ la moitié et l'oxyde de fer le tiers de sa composition. Un autre trait caractéristique du tableau est l'absence de l'oxyde de titane de la composition de toutes les pouzzolanes sauf la pouzzolane de bauxite.

Tableau II: Composition chimique de la boue rouge de Friguia

Composantes	Teneur %
SiO2	7,06
Al2O3	14,68
Fe2O3	53,89
CaO	2,1
MgO	0,25
TiO2	3,12
P2O5	1,21
CO2	14,6

Analyse thermique différentielle

Cette analyse apporte des informations sur l'état de cristallisation du matériau, mais surtout permet de contrôler le degré de cuisson et d'identifier les hydrates formés lors de la réaction d'hydratation de celui-ci après cuisson. Le principe consiste à enregistrer les transformations endothermiques (déshydratation) et exothermiques (recristallisation) d'un matériau au cours d'une montée linéaire de température. Les résultats de l'analyse thermique différentielle de la boue rouge à l'état naturel et après cuisson à 800°C, selon les données de l'Industrial Research Institute d'Accra, sont donnés sur les figures 1 et 2 respectivement.
Les pics endothermiques (Fig. 1) à 1O6; 274,

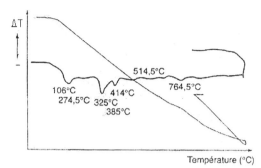

Figure 1 : Courbe ATD de la boue rouge

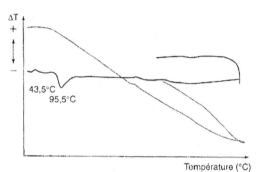

Figure 2 : Courbe ATD de la pouzzolane de boue rouge

3; 325; 414; 514 et 761,5°C correspondent, respectivement, à l'expulsion de l'eau adsorbée, la conversion de la gibbsite et de la bohémite en alumine, celle de la gibbsite en bohémite et du diaspore en alumine, et l'expulsion du cristal d'eau adsorbé.

Le pic endothermique à 95°C (Fig.2) correspond à l'expulsion de l'eau adsorbée du matériau. De ces deux graphiques il ressort que l'échantillon de boue rouge à l'état naturel se présente comme de la bauxite exhibant toutes les transformations minéralogiques relatives aux trois principaux minéraux qui la composent. Il en est de même pour celui soumis à la cuisson à 800°C mais qui a atteint une phase de transformation irréversible et, de ce fait, n'exhibe plus aucun autre pic à part celui indiquant l'expulsion de l'eau adsorbée à la température de 95°C

Essai de compression simple

Dans le but d'étudier l'interaction entre le ciment Portland et la pouzzolane de boue rouge hydratée, il a été préparé du mortier de ciment et de sable standard dans la proportion en poids de C:S =3 et avec un rapport E/C = 0,4. Dans ce mortier on a procédé au remplacement de 20 et 30 % en poids du ciment par de la pouzzolane de boue rouge. Pour l'évaluation du développement de résistance du maté- riau obtenu, des éprouvettes cubiques de dimensions 6,45 X 6,45 X 6,45 cm avaient été confectionnées et soumises à l'essai de compression simple. Au cours de l'hydratation chaque éprouvette avait été soigneuse- sement compactée sur une table vibrante en vue d'évi- ter la ségrégation des produits d'hydratation. Des périodes d'hydratation de 28 et 60 jours avaient été choisies pour l'évaluation de la résistance à la compression. Les essais ont été réalisés conformé-ment à la norme britannique BS 12 1958.

Avant d'examiner l'influence de la boue rouge sur la résistance à la compression du mortier, il convient de décrire brièvement le processus d'hydratation du ciment, dont Hammond (1987) donne le schéma suivant:

Le ciment portland ordinaire est composé des minéraux suivants: C_3S^*, C_2S, C_3A et C_4AF auxquels on ajoute 4 - 5 % de gypse ($CaSO_4 .2H_2O$).

$$\begin{array}{ll} C_3S & C\text{-}S\text{-}H \\ \quad + H_2O \longrightarrow \\ C_2S & Ca(OH)_2 \end{array}$$

$$\begin{array}{ll} C_3A & C_4AS_3H_{32} \text{ (Etringite)} \\ \quad + H_2O \longrightarrow \\ C_4AF & C_4ASH_{12} \text{ (Monosulpho-} \\ \text{aluminate hydraté)} \end{array}$$

$$\begin{array}{ll} & C_2AH_8.C_4AH_{13} \\ CaSO_4.2H_2O \longrightarrow \\ & C_3AH_6 \end{array}$$

* - La notation simplifiée utilisée dans le texte est celle de la chimie des ciments, à savoir:
$C = CaO$; $S = SiO_2$; $A = Al_2O_3$; $H = H_2O$.
$S = SO_3$; $F = Fe_2O_3$

C_3S constitue la composante principale, tandis que les phases C-S-H et $Ca(OH)_2$ sont les plus importantes du point de vue des propriétés de structure et de résistance de la pâte de ciment. Les phases aluminates se combinent avec le gypse, contribuant ainsi à l'hydratation accélérée et à l'affaissement du ciment.

Le trisulphoaluminate tétracalcique hydraté (étringite) est le premier formé de tous les produits à partir de C_3A et il exige plus d'eau. Le C_4AF réagit directement pour donner le monosulphoaluminate hydraté. Cette hydratation est plus lente que celle de C_3A, mais plus rapide que celle de C_3S.

Les aluminates hydratés C_2AH_8 et C_4AH_{13} peuvent exister séparément ou en solution solide avec le sulphoaluminate hydraté. Thermodynami-quement instables, ces produits peuvent se transformer en C_3AH_6 qui est stable. Cependant,

celui-ci ne se forme généralement pas pendant l'hydratation du ciment.

Par définition la pouzzolane réagit chimiquement avec l'hydroxyde de calcium pour former un composé cimentier stable. La chaux est mélangée avec la pouzzolane pour former du ciment chaux-pouzzolane. Pendant l'hydratation du ciment Portland, l'hydroxyde de calcium est libéré, si une pouzzolane est ajoutée à du ciment Portland ordinaire, cet hydroxyde de calcium libéré réagit avec la pouzzolane pour former un produit liant. On profite de cette réaction chimique pour la production de ciment de pouzzolane par le remplacement de proportions définies de clinker par celui-ci. Ainsi, l'activité pouzzolanique est déterminée par la quantité de $Ca(OH)_2$ que la pouzzolane est capable de libérer.

Dans l'évaluation de la pouzzolane les deux propriétés de résistance et de fixation de la chaux du matériau doivent être considérées séparément. Cette exigence équivaut à la détermination, dans des conditions standards, de la résistance mécanique du mortier de béton et à l'estimation du degré de réduction de l'hydroxyde de calcium libre. Ainsi le processus de l'hydratation du ciment avec de la pouzzolane de boue rouge a été suivi de l'essai à la compression simple. Les résultats des essais sont donnés dans le tableau III et sur la figure 3.

De ces derniers il ressort que, pour les mélanges considérés, toutes les valeurs de la résistance à la compression sont inférieures à celles du mortier ordinaire. Cependant A.A.Haammond (1974) a montré que dans le cas du mélange chaux: eau:pouzzolane = 1:2:4, la résistance à la compres-sion est supérieure à celle du béton ordinaire pour les déchets de bauxite.

Par ailleurs la résistance à la compression des différents mélanges augmente avec l'augmentation de l'âge des éprouvettes alors qu'elle diminue avec l'augmentation du pourcentage de pouzzolane. Elle atteint son maximum pour la température optimale de 700°C.

Comme on le voit la résistance du mortier de ciment avec 30% de remplacement du ciment par de la pouzzolane de bauxite est assez intéressante. L'importance de ce fait est qu'une économie peut être réalisée par le remplacement d'un pourcentage raisonnable de ciment par de la pouzzolane.

Tableau III: Résistance à la compression du mortier de ciment contenant de la pouzzolane de boue rouge à 700°C et 800°C (E/C=0,4)

Age	Résistance à la compression, MPa				
(jours)	100% de ciment	80% de ciment 20% de pouzzolane		70% de ciment 30% de pouzzolane	
		700°C	800°C	700°C	800°C
28	24,04	10,47	8,50	6,57	6,6
60	23,87	16,5	14,23	16,23	7,77

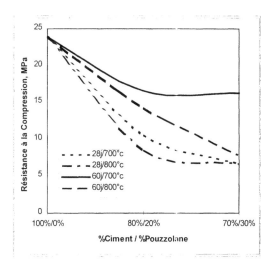

Figure 3 : Effet du remplacement du ciment par la pouzzolane de boue rouge sur la résistance du mortier de composition C : S = 1:3 ; E/C = 0,4

Dans le cas de la pouzzolane de déchets de bauxite, A.A. Hammond (1974) avait établi qu'une économie d'environ 30% peut être obtenue par le remplacement d'environ 20-30% de ciment par de la pouzzolane.

D'autre part un certain nombre de qualités du béton, telles qu'une faible chaleur d'hydratation, une bonne maniabilité, une diminution de la porosité et une bonne résistance aux attaques chimiques, peuvent être obtenues.

CONCLUSION.

Par un traitement thermique approprié et un broyage conséquent il est possible de produire de la pouzzolane assez réactive à partir de la boue rouge de Friguia. Les ciments de boue rouge hydratés ont des niveaux de résistance acceptables pour leur utilisation dans l'habitat. En conséquence la boue rouge de Friguia offre des perspectives de recyclage comme ciment pouzzolanique dans la mesure où elle contient une fraction argileuse suffisante. La réalisation de cet objectif contribuerait, non seulement à la protection de l'environnement dans les zones minières, mais aussi à la promotion de la construction d'habitat à coût modéré.

REMERCIEMENTS:

Ce travail a été effectué au Building and Road Reasearch Institute (BRRI) de Kumasi (Ghana) grâce à l'assistance que Dr. Gidigasu, Directeur de l'Institut, et Dr. Hammond, chef de la Division des Matériaux de Construction ont bien voulu m'apporter, aussi bien pour la mise en route du projet que pour l'exécution du programme. Qu'ils trouvent ici l'expression de ma profonde gratitude.

Le Réseau Africain des Institutions Scientifiques et Technologiques (RAIST) de Nairobi (Kenya) a bien voulu soutenir généreusement ce travail. Je lui suis très profondément reconnaissant.

Je remercie très sincèrement l'Académie des Sciences du Tiers-Monde (TWAS) pour son aimable assistance dans la mise en oeuvre du projet et l'organisation du voyage.

Je remercie enfin la Compagnie Britannique d'Aluminium d'Awaso (Ghana) pour avoir autorisé la visite de l'usine et le prélèvement d'échantillons de déchets de bauxite pour les travaux de recherche.

REFERENCES:

AMBROISE J. (1984), Elaboration de liants pouzzolaniques à moyenne température et étude de leurs propriétés physico-chimiques et mécaniques, Thèse de doctorat es-sciences, INSA de Lyon

DIANE M.S. (1989), Etude comparée de la boue rouge et des déchets de bauxite en vue de leur utilisation dans la construction, CERESCOR, Conakry, 15 p.

DIANE M.S., ROSSIKHINE M.M. (1987), De l'utilisation de la boue rouge dans la construction, *Bulletin du centre de Rogbané* N°7, Conakry, Juillet, pp.93-95.

FERRARRI F. (1931), in *Le Industrie del Cements*, Vol 28, n° 11, p. 88

HAMMOND A.A. (1976), Bauxite-waste in Building, *Building Research and Practice*, March/April, pp. 80-83

HAMMOND A.A. (1981), Durability of bauxite-waste Pozzolana cement concrete, American Society for Testing and Materials 1926, Raca Street, Philadelphie, Pa. 29103,

HAMMOND A.A. (1987), Hydratation products of bauxite-waste pozzolana cement, *The International Journal of Cement Composites and Lightweight Concrete*, Vol. 9, No. 1. February HAMMOND A.A. (1985), Mineralogical changes occurring in bauxite-waste conversion to pozzolana, Proceedings of the conference on development of Low-Cost and Low-Energy Cons-truction Materials, Vol.1, Rio de Janeiro, Brazil, 9-12 July, pp.401-15.

HAMMOND A.A. (1974), Pozzolana from bauxite-waste, *Achievement of cost reduction in public construction*, Paper No.4/5b, Building and Road Research Institute Kumasi, June.

HAMMOND A.A. (1974), Pozzolana from bauxite-waste; Conférence on cost reduction in public construction Sector, Accra, June 25-28, Building and Road Research Institute, Kumasi, pp.1-18.

LEA F.M. (1970), in The Chemistry of cement and Concrete, Edward Arnold, London, pp. 414-453.

Le Chatelier H. (1909), 7th International Congress of applied Chemistry, London, Sec.2, p.19.

.

Geotechnics for Developing Africa, Wardle, Blight & Fourie (eds) © 1999 Balkema, Rotterdam, ISBN 90 5809 082 5

Reinforcement of subsoil with sand and geotextile

A. Edinçliler
International Union of Local Authorities, Section for the Eastern Mediterranean and Middle East (IULA-EMME), Istanbul, Turkey

E. Güler
Department of Civil Engineering, Boğaziçi University, Istanbul, Turkey

ABSTRACT: In this study, the effects of the reinforcement of the subsoil with sand and geotextile on the deformation and settlements of soft soil have been investigated. A sand layer with its thickness proportional to the base width of the foundation has been placed on the subsoil and the most effective placement of the geotextile has been investigated. The analysis has been carried out using a finite element program based on Critical State Theory, called CRISP. The results have shown that a reinforced sand fill over the soft soil and a geotextile reinforcement placed on this subbase increases the load carrying capacity of the subsoil. Failures and excessive settlements can be considerably avoided and safety of the building can be improved with the proper choice of the placement and properties of the sand fill and the reinforcing geotextile.

1 INTRODUCTION

There are several improvement techniques available to increase the bearing capacity of soft soils. It is common practice to use piles to support foundation loads by transferring the load of the structure onto stronger and incompressible soils or alternatively a coarse soil layer is placed over soft soil in order to spread the loads over a larger area. However, penetration of the coarse materials into soft soil layer can cause some limitations in the use of this technique.

Using a geotextile layer as separator between soft soil and sand fill can be a solution to these type problems. The inclusion of the geotextile also has a reinforcing effect by preventing lateral movement of the granular layer (Bauer and Preissner 1990). Test results conducted by several researchers about this subject showed that the bearing capacity of soil is a function of the ratio of sand layer thickness to foundation width. The same test results indicate that the type of geotextile is the most important parameters affecting the bearing capacity of subsoil (Dembicki and Jermolowicz 1988).

The purpose of this study is to find out the optimum solution in the case of improvement of foundation soil using a sand fill and geotextile reinforcement. The effect of the depth of sand layer and the place and specifications of geotextile used as a reinforcement on the deformations and displacements taking place on the soil have been investigated. To find out the best solution, several analysis have been considered using the finite element program based on Critical State Theory, called CRISP (Britto 1990). With the help of finite element models, displacements and deformations at each step of loading and consolidation stages have been obtained.

2 FINITE ELEMENT MODELS REPRESENTING FOUNDATION CONSTRUCTION

The simulation of soft soil and foundation constructed on soft soil was modeled using 268 elements (Fig.1). Because of symmetry about the vertical axis only one half needs to be considered in the models. The model of bulk soil is 6 m. high and 10 m. wide. The foundation modeled has 1 m. width as B/2 and 0.5 m. thickness.

It has been assumed that the foundation applies a pressure of 250 kPa. The appropriate fixity codes are specified for all the element sides along the boundaries. The lateral and vertical displacements restricted along the bottom boundary. On the sides horizontal displacements are restricted.

y
1.0 m.
Foundation

0.5 m.

6.0 m.

10 m.

x

Figure 1. Finite Element Model Representing the Foundation Design

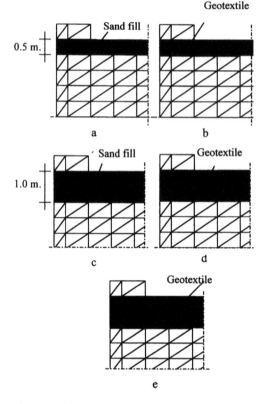

Figure 2. Placement of sand fill and Geotextile a) Model 1, b) Model 2, c) Model 3, d) Model 4, e) Model 5.

Five different finite element analyses have been used (Fig.2). In the first analysis (Model 1), a sand layer has been used which has a thickness of B/2 (Fig. 2a). In the second analysis (Model 2), placing a 6 m-long-geotextile between sand layer and the soft soil has been considered (Fig. 2b). In the third analysis (Model 3), the placing of 1 m. deep sand layer on the soft soil has been modeled (Fig.2c). In the fourth analysis Model 4), a 6 m-long-geotextile has been placed in the middle of a sand layer of 1 m. thickness (Fig. 2d). In the last Analysis (Model 5), the case of placing geotextile between a 1m. thick sand layer and soft soil has been investigated (Fig.2e).

Soft soil was a fully saturated soil. The soil has been modeled by Modify Cam Clay Soil Model which simulates elasto-plastic soil behavior. The following parameters have been used to represent the soil: $\kappa = 0.05$, $M = 1.00$, $\gamma = 20$ kN/m³, $\lambda = 0.30$, $\nu = 0.30$, $k_x = 1.9$ E-9 m/s, $e_{cs} = 2.95$, $\gamma_w = 10$ kN/m³, $k_{xy} = 0.1$ E-9 m/s.

Sand fill has been modeled as an elastic material which has the following parameters: $E_x = 1.E4$ kPa, $\nu_x = 0.25$, $G_{hv} = 4.E3$ kPa, $E_y = 1.E4$ kPa, $\nu_y = 0.25$, and $\gamma = 18$ kN/m³. A low quality sand fill was used during the Analysis. It represents a sand which can be found in the very near vicinity of the construction site.

The properties of the geotextile reinforcement used in the models are: $E = 3.E6$ kPa, $\nu = 0.31$, and $A = 0.001$ m². The properties of elastic materials representing the foundation are: $E_x = 2.E7$ kPa, $\nu_x = 0.16$, $G_{hv} = 8.6.E7$ kPa, $E_y = 2.E7$ kPa, and $\gamma = 20$ kN/m³.

The in-situ pressures are given in Table 1. The groundwater level is at 0.5 m. depth.

Table 1. In-situ stresses along the depth.

Effective Stresses (kPa)	σ_x'	σ_y'	σ_z'	P_c
Depth (m)				
0.0	0	0	0	0.0
0.5	5.5	9	5.5	11.4
6.0	39.0	64.0	39.0	79.8

3. FINITE ELEMENT ANALYSIS

To observe the soil behavior under the foundation, three increment blocks have been used. The first increment block representing foundation construction contains five increments, lasting for two days. The second increment block simulating the loading consists of ten increments, lasting for three months and comprises of the application of the

Table 2. Horizontal and vertical deformations at critical nodes.

Nodes	3		25		26		40		147	
Analysis/ Deformations	x	y	x	y	x	y	x	y	x	y
				Deformations after loading (mm)						
Analysis 1	0.00	-49.70	9.40	-52.60	15.70	-38.90	10.50	14.40	0.20	-56.90
Analysis 2	0.00	-42.70	6.00	-45.80	12.60	-36.00	8.70	10.60	0.20	-49.00
Analysis 3	0.00	-29.00	2.50	-27.00	6.70	-25.00	3.80	2.40	0.20	-31.00
Analysis 4	0.00	-28.00	2.00	-27.00	6.30	-25.00	3.30	2.30	0.20	-31.00
Analysis 5	0.00	-25.60	1.60	-25.00	4.60	-23.00	2.50	1.20	0.10	-28.00
				Deformations after consolidation (mm)						
Analysis 1	0.00	-126.00	13.60	-156.00	22.80	-117.00	17.50	-23.30	0.20	-160.00
Analysis 2	0.00	-98.60	7.90	-126.00	15.00	-86.30	12.80	-17.40	0.20	-130.00
Analysis 3	0.00	-74.00	2.70	-72.00	9.10	-70.00	3.00	-19.00	0.10	-76.00
Analysis 4	0.00	-74.80	2.10	-71.70	8.60	-69.00	2.60	-19.00	0.10	-75.00
Analysis 5	0.00	-67.00	1.40	-65.60	6.10	-63.00	9.00	-18.00	0.10	-69.00

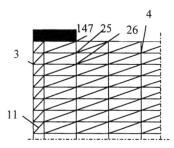

Figure 3. Critical points subjected to excessive deformations.

foundation load in increments of 25 kPa. The third increment block represents the consolidation phase and takes 15 years with the same increment number as the second increment block. In this manner, consolidation time and each calculation step can be evaluated to determine the failure loads.

4. NUMERICAL RESULTS

Some critical points were chosen where the results are investigated further (Fig.3). Horizontal and vertical deformation values at the critical points are shown in Table 2. When the values given in Table 2 are observed, it is seen that maximum

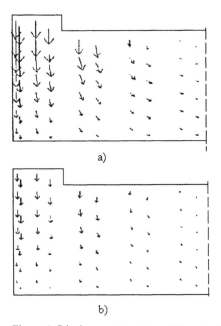

Figure 4. Displacement vectors at the end of loading a) Model 1, b) Model 5.

deformations in both directions occurred in the case of placing a sand fill of B/4 on soft soil.

On the other hand, the smallest deformations have been observed for Model 5 (Fig. 4).

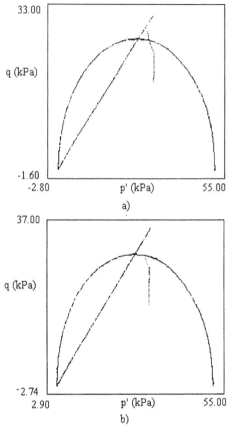

Figure 5. Stress paths at Element 11 a) Analysis 1, b) Analysis 5.

It can be stated that geotextile reinforcement reduces the amount of settlement during the loading and also the consolidation phases (Table 2).

The stress paths at the Element 11 subjected to maximum and minimum deformations in Analysis 1 and 5 are given in Fig. 5. From the Figure, it is observed that using a sand fill of B/4 thickness causes yielding of soil. While the soil approaches yielding in the case of using a B/4 sand fill, the soil can carry the loads safely when a geotextile is placed between the sand fill and soft soil.

5. RESULTS

The aim of this study is to investigate the effects of reinforcement on deformations and settlements of soft soil. The analyses revealed the following results:

1. If the sand layer thickness is chosen as B/2 of the foundation width, deformations and settlements taking place on the soft soil will decrease significantly.

2. Geotextile reinforcement further decreases the settlement and increases the bearing capacity.

3. Placing the geotextile at the base of the sand fill instead at the centre of it proves to be more effective in terms of reducing settlements.

4. When placing the geotextile reinforcement between the sand fill and the soft soil, besides the reinforcements function of geotextile, it can be used as a separation material to prevent the mixture of two different material. As a result, safety of the structure can be improved.

REFERENCES

Bauer, A. & Preissner, H. 1986. Studies on the Stability of a Geotextile Reinforced Two-Layer System. *Third International Conference on Geotextile*. Austria:1079-1083.

Britto, A. M. & Gunn, M.J. 1987. *Critical State Soil Mechanics via Finite Elements*, Ellis Horwood, Chichester, England.

Dembicki, E. & Jermolowicz, P. 1988. Model Tests of Bearing Capacity of a Weak Subsoil Reinforced by Geotextiles. *First Indian Geotextile Conference on Reinforced Soil and Geotextiles:* 67-72.

Geotechnics for Developing Africa, Wardle, Blight & Fourie (eds)© 1999 Balkema, Rotterdam, ISBN 90 5809 082 5

A geotechnical study of 'Hippo mud' in the Durban area, South Africa

C.I.Ware & C.A.Jermy
School of Geological and Computer Sciences, University of Natal, Durban, South Africa

ABSTRACT: A geotechnical study of organic Holocene estuarine silty clays (locally known as Hippo Mud) in the greater Durban area, South Africa, is presented. The study area occurs in the lower reaches of the Mhlangane River and comprises a soil profile, up to 15 m, of highly compressible Hippo Mud overlain by more recent alluvial sediments and underlain by weathered Permian shale bedrock. Various laboratory tests were undertaken to predict the soil's physical behaviour and facilitate comparisons with tests from other areas underlain by similar material so that prediction could be made about its properties. Geochemical tests show the samples to contain smectite and illite clay minerals, plagioclase and quartz, with the provenance terrain being the lower and middle Ecca Group, which dominate the drainage basin. Other areas where these soils are likely to be found are in incised coastal valleys north of Durban where the Permian Ecca Group shales form the drainage basin.

1 INTRODUCTION

The physical and chemical properties of a particular type of compressible estuarine clayey material, known locally as "Hippo Mud", have been quantified through various laboratory tests. The results are then compared to similar material in a hope that the soil's behaviour could be more easily predicted, if encountered elsewhere.

Hippo Muds are estuarine Holocene very dark grey silty clays with intercalated sands and clayey silts that occur within the greater Durban area. This material is generally associated with drowned coastal river valleys which have lower and middle Ecca Group rocks comprising their drainage basin.

The study area of a typical Hippo Mud deposit is found within the lower Mhlangane River valley at Sea Cow Lake, Durban, South Africa (Figure 1). The soil succession comprises approximately 10 m of this compressible material on shale bedrock and its residuum and is overlain by more recent alluvial clays. A portion of the N2 Outer Ring Road crosses a section underlain by this Hippo Mud in the study area and considerable settlement occurred.

2 SITE DESCRIPTION AND GEOLOGY

The study area (Figure 1) is located in Sea Cow Lake which is a low lying dammed off tributary valley (Mhlangane River valley) of the lower Mgeni River. The Mhlangane River runs almost parallel with the coast and converges with the Mgeni River, in the vicinity of the Springfield Flats, and flows eastwards to the Indian Ocean. The Mhlangane and Mgeni valleys have been deeply incised into bedrock by river action due to glacio-eustatic sea level fluctuations during the Quaternary. With the rise in sea levels to its present level since the last glacial event (maximum 18 000 years ago), this valley system first became a flooded lagoonal estuary under tidal influence, being partially filled with estuarine sediments (from about 6 000 to 4 000 years ago), and after a hiatus of about 2 000 years, the closure of the Mhlangane River enabled the so called Sea Cow Lake to develop. Lacustrine/alluvial sediments primarily being deposited in this lake, filling it, and forming a flood plain. This is confirmed by radiocarbon dating of organic matter at various depths in the sediment profile (R. Maud, pers. comm.)

The Mhlangane River drainage basin contains rocks of the Ecca Group which has been intruded by Post Karoo dolerite. The Ecca Group includes the lower (Pietermaritzburg) and middle (Vryheid) formations which were deposited under deltaic conditions during the Permian. The Pietermaritzburg Formation (PF) comprises closely jointed dark grey to black fissile shale and siltstone, and outcrops most extensively to the north of the Mgeni River. Mineralogically the shale, in its unweathered state, mainly consists of five minerals; quartz, illite,

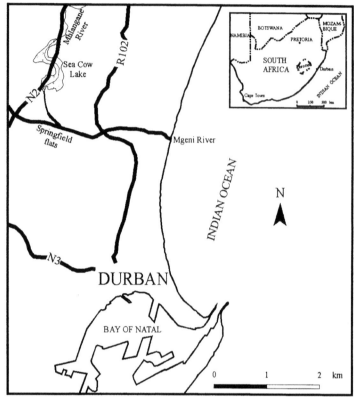

Figure 1. Locality map of study area, Durban, South Africa.

chamosite, chlorite and chalybite (Maud, 1983). In the Durban area, it is approximately 300 - 400 m thick (King & Maud, 1964), and generally thickens in a southerly direction. Further northwards, it thins out rapidly, until it wedges out completely along the northern rim of the Karoo basin (Visser, 1989).

The Vryheid Formation follows conformably, by way of transition, on the Pietermaritzburg shales, and outcrops exclusively to the north of the Mgeni River. It comprises thick beds of closely jointed yellowish to white, cross-bedded arenite and grit, which alternate with beds of soft, dark-grey sandy shale and a few coal seams (Visser, 1989). According to Maud (1983) the arenites comprise well rounded quartz grains, feldspar and muscovite, whereas the shales resemble those of the PF, but with additional feldspar and muscovite. It obtains thicknesses of approximately 300 m in the Durban area, but thickens northwards (Maud, 1983).

Within the study area, the PF is overlain by a moderate to deep soil profile (Figure 2) of Holocene silty sands and clays totalling 18 m in places. The estuarine sediments (Hippo Mud) were deposited in a lagoonal environment and consists mainly of very soft to firm becoming stiff with depth, compressible sandy clayey silts and silty clays which generally

occur from 2 to 5 m below the existing ground level. Occasionally sandy lenses occur within the more clayey horizons. The distinction between the overlying lacustrine/alluvium sediments and the estuarine sediments are the highly compressible nature and the lower strength characteristics of the estuarine sediments. This is the result of the generally higher clay content and lower average grain size resulting in a generally higher natural moisture content in comparison with the overlying lacustrine/alluvial sediments.

The lacustrine/alluvium sediments overly the estuarine sediments in low lying areas comprising the Mhlangane flood plain. These sediments comprise compressible clayey silts and silty clays and extend to depths of greater than 5m below the existing ground level. In addition some sand horizons occur within the clay horizons, and recent alluvial sediments occur adjacent to the Mhlangane River.

The provenance rocks of Hippo Mud in Sea Cow Lake have been attributed to the outcropping lower and middle Ecca Group rocks of the drainage basin, which contribute colluvial material of silt and clay to the streams (Francis, 1983). In similar marginal lake situations on the major rivers north of Durban,

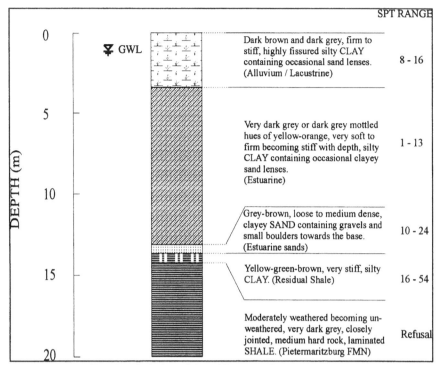

Dark brown and dark grey, firm to stiff, highly fissured silty CLAY containing occasional sand lenses. (Alluvium / Lacustrine) — SPT RANGE 8 - 16

Very dark grey or dark grey mottled hues of yellow-orange, very soft to firm becoming stiff with depth, silty CLAY containing occasional clayey sand lenses. (Estuarine) — 1 - 13

Grey-brown, loose to medium dense, clayey SAND containing gravels and small boulders towards the base. (Estuarine sands) — 10 - 24

Yellow-green-brown, very stiff, silty CLAY. (Residual Shale) — 16 - 54

Moderately weathered becoming un-weathered, very dark grey, closely jointed, medium hard rock, laminated SHALE. (Pietermaritzburg FMN) — Refusal

Figure 2. Generalized borehole log of the Mhlangane River basin, Sea Cow Lake.

however, the silt and clay can originate also from other rock types such as granite-gneiss and basalt.

In the lower Mgeni River valley, thick beds of Hippo Mud are intercalated with sands which were derived from Archaean granitic rocks outcropping in the Valley of a Thousand Hills some 15 to 30 km upstream (Francis, 1983). The bedding units vary from 2 m thick sheets to 15 m thick bodies (Brink, 1985) and overly a boulder bed which rests on the Pietermaritzburg shales and obtain thicknesses of up to 40 m. The bedding units thin out eastwards and occur as isolated 1-2 m thick lenses. Southwards, towards the city centre, they become subordinate to the grey, brown and red clayey and silty sands of the Harbour Beds (Brink, 1985).

3 METHODS OF INVESTIGATION

This was conducted in two stages, namely an academic approach conducted in 1997, and, later during the second half of 1998, a more detailed site investigation was conducted for the proposed development of the low lying Sea Cow Lake area. The academic approach was chiefly concerned with compiling known data of the estuarine material and conducting various tests to identify the physical characteristics and strength related properties, as well as to assess its chemical makeup. The more

detailed site investigation was carried out by Drennan, Maud and Partners for Moreland Developments (Pty) Ltd and included 19 boreholes, 189 Dynamic Cone Penetrometer tests (DCP) and the excavation of 95 inspection pits using a Hella HE 220 LC mechanical backactor, down to depths of 6 m. Various laboratory tests on samples obtained from the boreholes and inspection pits were analysed to assess strength related properties, settlement and particle size distribution. Laboratory tests were conducted by the authors (1997) at the University of Natal, Durban and by Drennan, Maud and Partners' Soil Laboratory during the later half of 1998. X - ray diffraction (XRD) samples were analysed by the Council for Geoscience.

4 LABORATORY RESULTS & DISCUSSION

The laboratory data together with known data of Hippo Mud is summarized in Table 1. A typical Hippo Mud can be described as a compressible very dark grey, very soft to firm, silty organic clay of estuarine origin. It has a high plasticity and a clay content generally greater than 60%. In general the Linear Shrinkage is about 10 % and it has very low strength characteristics, obtaining soaked CBR values of less than 1.5 % at 100 % Mod AASHTO with a swell of some 5 %. The value of the

Table 1. Summarized laboratory test data for Hippo Mud in the greater Durban area.

TESTS	Units	n	This study Range	Francis (1983) Range	Jones et al (1975), Jones (1975a), Brink (1985) Range
Clay fraction (< 2µm)	%	6	45 - 71	35 - 55	45 - 67
Plasticity Index (*PI*)	-	28	18 - 37	15 - 30	21 - 43
Liquid Limit (W_L)	-	28	36 - 64	40 - 70	56 - 70
Natural moisture content (*w*)	%	12	30 - 60	45 - 80	50 - 93
Linear Shrinkage (*LS*)	%	9	6 - 13	9 - 13	-
pH (electrometric method)	-	4	6.7 (Ave)	-	-
Organic carbon (H_2O_2 method)	%	3	2.5 (Ave)	-	-
Specific gravity (Density bottle method)	-	3	2.72 (Ave)	-	-
Degree of Saturation (*S*)	%	3	90 - 100	92 - 100	-
Quick Undrained Shear Strength (C_u)	kPa	3	23 - 30	20 - 40	16 - 27
Angle of internal friction (ϕ')	°	5	7 - 30	-	20 - 33
Apparent cohesion (c')	kPa	-	0 - 5	-	0 - 1
Shear vane test (C_{ui})	kPa	5	18 - 25	-	10 - 25
Shear vane test (C_{ur})	kPa	5	4 - 7	-	-
Coefficient of volume compressibility (m_v)	m²/MN	6	0.06 - 0.42*	0.10 - 0.97	0.15 - 1.76*
Back analysed (Actual value) m_v	m²/MN	-	-	0.10 - 0.60	-
Coefficient of consolidation (c_v)	m²/year	6	0.19 - 1.36**	0.1 - 5***	0.23 - 1.29**
Back analysed (Actual value) c_v	m²/year	-	-	5 - 20	1 - 5
Over-consolidation ratio	-	-	-	< 1 - > 2	1 - 2
Predicted time for 90 % ground settlement for 14m fill height, mentioned in text	years	-	± 110	1 - 69	50 - 150
Actual time for 90 % ground settlement for 14m fill height, mentioned in text	years	-	-	0.3 - 0.8	1.4
Unsoaked CBR at 100 % Mod AASHTO	%	1	77.4	-	-
Soaked CBR at 100 % Mod AASHTO	%	1	1.3	-	-
Standard Penetrometer Test (SPT)	Blows	30	1 - 13	-	-
Dynamic Cone Penetrometer (DCP)	Blows	189	0 - 15	-	-

* For 100 - 200 kPa pressure range ** For 200 kPa pressure increment *** For 80 - 300 kPa pressure range

maximum dry density is 1716 kgm^{-3} at an optimum moisture content of 21.2 %.

Consolidated drained triaxial tests reveal a normally consolidated to slightly over-consolidated material with a cohesion value (c') of 0 - 5 kPa and an angle of internal friction (ϕ') of 7 - 30°, however higher values of c' and ϕ' were obtained, namely 12 kPa and 36° respectively. The higher values can be attributed to the relatively low natural moisture content (w) of the sample tested, approximately 26 - 27 %.

Undrained unconsolidated triaxial tests were carried out to obtain the undrained shear strength (C_u). The values obtained varied from 23 to 30 kPa, however again lower natural moisture content samples (31 - 39 %) revealed higher values of between 62 and 96 kPa. Shear vane tests reveal values of C_{ui} of between 18 and 25 kPa and values of C_{ur} of between 4 - 7 kPa. It was noted that material with lower natural moisture contents reveal higher values of C_{ui} up to 35 kPa and C_{ur} up to 9 kPa.

SPT results range from 1 - 13 blows and show a very soft material becoming firm with depth. DCP results confirm this and range from 0 - 15, with values 0 - 5 mainly occurring at depths of about 1.2 - 1.8 m, which corresponds to the water table.

Geochemically the soil comprises quartz, plagioclase, illite and smectitic clay minerals using XRD methods and confirmed using XRF analyses. The authors have not included the geochemical make up of the material as this falls out side the scope of this paper. However the geochemical results do correlate with materials derived from the lower and middle Ecca Group shales.

Using the conventional one dimensional consolidation theory and the laboratory consolidometer coefficients the estimated time to achieve 90 % consolidation for a fill height of 14 m, and having the strength parameters ($\phi = 25°$, c = 10 kPa, γ = 19.5 kgm^{-3}) with the upper alluvial/ lacustrine sediments in situ strength parameters (ϕ = 15°, c = 12 kPa, γ = 19.5 kgm^{-3}) and lower estuarine sediments (Hippo Mud) in situ strength parameters ($\phi = 0°$, c = 20 kPa, γ = 18 kgm^{-3}), is in the order of 110 years for the lower estuarine sediments. This however will vary on the position of the fills and the actual subsoil conditions below the fills. It is interesting to note that the upper alluvial / lacustrine clayey material will achieve 90 % consolidation within about 7.5 years. In addition, excess pore water pressure must be known and monitored and dissipated as this will adversely affect fill stability, and fills over 5.75 m must be specially engineered as they pose stability problems.

Settlement analyses carried out for the in situ estuarine / lacustrine sediments using both SPT tests and laboratory tests inferred settlements of 101 - 375 mm for fill heights of 3.0 - 5.75 m over a period of 110 years or so. According to Jones *et al* (1975) settlement of a test section of embankment for the N2 road was predicted using one dimensional consolidation theory, with friction ratios taken into account, to be 1.1 ± 0.3 m, over a period of 50 - 150 years. When actual settlements were measured,

values of 1.3 m were obtained but settlement took place within only 500 days. Similarly, the final embankment was predicted to settle 1.4 ±0.3 m and was observed to have settled 0.9 m, although it was noted to still be settling when this figure went to press (Jones, 1975a).

The Revised US Classification of this material is a A-7-6 indicating the material unsuitable for use in construction. In terms of TRH14 of 1985, it fails on all criteria to obtain a classification and thus can be considered a >G10 material.

5 CONCLUSIONS

The study area compared favourably with other results quoted in the text of similar material found within the greater Durban area and coastal KwaZulu-Natal, and can thus be considered a representative example of a Hippo Mud. The latter is the local name given to compressible very dark grey, very soft to firm, silty organic clay of estuarine origin and being derived from Permian Ecca shales. It comprises quartz, plagioclase, illite and smectitic clay minerals, which were deposited due to marine transgression during the Holocene under a lagoonal environment. It has very low strength characteristics and high clay content (generally >60%) with an over all smaller average particle size which makes it distinguishable from the more recent overlying alluvial / lacustrine clayey materials in the study area. Due to its poor geotechnical properties, it is considered unsuitable for construction purposes and requires specialised foundations and construction techniques, especially in those areas underlain by soil profiles of appreciable depth. In the past, very considerable settlement problems have been experienced in the construction of road fill embankments in the regions where these have been underlain by Hippo Mud. To deal with the long settlement periods that would otherwise be involved in these situations resort has been had to sand and hand drain instillations in the Hippo Mud and surcharge loading thereof.

6 ACKNOWLEDGEMENTS

The authors would sincerely like to thank the following companies and persons for their assistance, without which this paper could not have been written.

(1) Drennan, Maud & Partners for sharing their knowledge and information of local geological and engineering conditions.
(2) Moreland Developments (Pty) Ltd for making available their reports on the geotechnical

investigation of the Sea Cow Lake area, dated October 1998.

(3) Director, Council for Geoscience, for granting permission to perform the XRD analyses, and Ms M.T. Atanasova for performing the analyses.

7 REFERENCES

Brink, A.B.A. (1985). *Engineering Geology of Southern Africa.* Vol. 4. Building Publications, Pretoria. 108-153.

Francis, T.E. (1983). *The engineering geology of the lower Umgeni valley, Durban.* MSc Thesis, University of Natal, Durban.

Jones, G.A. (1975a). *Deep sounding - its value as a general investigation technique with particular reference to friction ratios and their accurate determination.* Proceedings of the Sixth Regional Conference for Africa on Soil Mechanics & Foundation Engineering, Vol. 1, Durban, South Africa. 167-175.

Jones, G.A. (1975b). *Discussion on paper - Embankments on soft alluvium - settlement and stability study at Durban, by G.A. Jones, D.F. Le Voy and A.L. Mc Queen.* Proceedings of the Sixth Regional Conference for Africa on Soil Mechanics & Foundation Engineering, Vol. 2, Durban, South Africa. 134-139.

Jones, G.A., Le Voy, D.F., and Mc⁻ Queen, A.L. (1975). *Embankments on soft alluvium - settlement and stability study at Durban.* Proceedings of the Sixth Regional Conference for Africa on Soil Mechanics & Foundation Engineering, Vol. 1, Durban, South Africa. 243-250.

King, L.C. & Maud, R.R. (1964). *The Geology of Durban and Environs. Geological Survey Bulletin Number 42.* Government Printer, Pretoria, South Africa. 54 pp.

Maud, R.R. (1983). *The geology of the Ecca Group in the Durban Area.* Symposium on Geotechnical Engineering in areas in Natal underlain by Ecca shale. 2-10.

TRH14 (1985). *Technical Recommendations for Highways 14, Guidelines for Road Construction Materials.* Department of Transport, Pretoria. 51-55.

Visser, D.J.L. (1989). *Explanation of the 1:1000000 Geological Map, Fourth Edition, 1984: The Geology of the Republics of South Africa, Transkei, Bophuthatswana, Venda and Ciskei and the Kingdoms of Lesotho and Swaziland.* Government Printer, Pretoria. 491 pp.

Geotechnics for Developing Africa, Wardle, Blight & Fourie (eds) © 1999 Balkema, Rotterdam, ISBN 90 5809 082 5

Experimental and numerical investigation of soil-nailed wall structures

J. Lai Sang & F. Scheele
Department of Civil Engineering, University of Cape Town, South Africa

ABSTRACT: The design of soil-nailed structures is based on limit equilibrium considerations. As in all limit state approaches wall deflections and the kinematics in terms of the nail/soil interaction and ground movements are not accounted for. This paper presents the main findings of a comprehensive investigation of full-scale laboratory experiments on soil-nailed walls of different heights in Cape Flats sand. The measurements of the forces along the nails and wall deflections during construction of the wall and subsequent loading tests to failure allow insight into the performance and thus overall stability of the structure. The experienced failure patterns of the various structures validate the log-spiral mechanism for the selected nail length to wall height ratios and load configuration. To support these observations numerical studies (3-D finite element analysis) were undertaken, which contribute to an enhanced understanding of the mechanics of a soil-nailed structure.

1 INTRODUCTION

Soil nailing is a relatively new method of in-situ earth reinforcement and has undergone a remarkable development over the past decades. It has been gaining popularity worldwide as a result of its economic competitiveness compared to other earth support systems. In South Africa the first soil nailing application was the support of a 6m deep excavation in Krugersdorp (Schwartz & Friedlaender 1989).

Essentially, the reinforcement is provided by the introduction of a pattern of closely spaced grouted steel bars, called 'nails', into the ground as the soil-nailed structure is constructed downward in a sequence of excavation and nailing phases. During and after the installation of the nails, the ground deforms, and the soil and nails interact. The nails are stressed in tension thereby reducing the loads in the soil mass which cause the soil to fail in shear or by excessive deformation.

The design of a soil-nailed structure is based on the limit equilibrium method, which is used conveniently in practice. The philosophy of the method is to ensure stability of the supporting structure and retained soil material in terms of activating and resisting forces and moments. Generally, different assumptions with respect to the failure mechanisms and slip surfaces are made and the least safe 'scenario' is anticipated. Therefore, in order to obtain accurate results it is important that the failure surface chosen represents a close approximation of the actual failure situation of the envisaged structure.

In the search and development of realistic failure concepts, the instrumentation of full-scale experiments on soil-nailed structures is imperative. This approach allows not only the measurements of forces but also of deformations at any stage of construction, an entity not taken cognisance of in limit analysis approaches.

Four full-scale laboratory experiments on soil-nailed walls of different heights in Cape Flats sand were undertaken and are briefly presented. The measurements of wall deflections and forces along the nails during construction and subsequent loading tests to failure provide an insight into the performance of the structure, the soil nail interaction and, thus, the mechanism of the structures. The observations were compared with design calculations based on limit equilibrium concepts. The soil-nailed walls were also modelled using ABAQUS, a multi-purpose finite element package, to verify the experimental results.

2 LIMIT EQUILIBRIUM DESIGN METHODS

The application of the limit equilibrium stability analysis in the design of soil-nailed structures was first proposed by Stocker & Gässler (1979). The method is based on the consideration of one or more rigid bodies in translation or rotation in a two-dimensional space (plane strain conditions). Today, a wide range of design theories (Gässler 1988, Juran et al. 1990), design codes (South African Code of

Practice 1989, HA 58/94 1994, BS 8006 1994) and recommendation (Project Clouterre 1993) are available for the design of soil-nailed structures. The rigid body mechanism is common to all these methods.

Several forms of trial surfaces are generally examined to determine the critical failure surface. The assumed surface may be a straight line (single or double wedge), a circular arc or a logarithmic spiral. The single wedge analysis is the most basic and can be readily applied. The double wedge mechanism is a widely used approach taking into account the balance of forces. Herewith, a wide range of potential failure surfaces can be covered. The analysis is relatively simple either by hand or using a computer. At present, only those two translational (wedge) analyses are recommended in the South African Code of Practice (1989).

The rotational (circular arc and logarithmic spiral) failure analyses are based on the equilibrium of moments. The circular slip, in analogy to Taylor's method (Lambe & Whitman 1969), is relatively rarely applied whilst the use of a log-spiral failure surface is considered as an even more sophisticated analysis.

3 EXPERIMENTAL RESEARCH APPROACH AND FINDINGS

The soil-nailed wall structures were accommodated in a large test pit with the overall inside dimensions of 6.0m in length, 1.5m in width and 3.0m in depth. The longitudinal view and cross-section of the test pit with the loading device are illustrated in Figure 1. The sides of the test pit were lined with low friction plastic sheeting to approximate plane strain conditions. A uniform and homogeneous, dense soil mass, Cape Flats sand of a unit weight of $16.5\pm0.2kN/m^3$ was achieved by compaction. The respective shear strength parameters were cohesion, c', of $3kN/m^2$ and angle of internal friction, ϕ', of $41°$.

The soil-nailed structures were constructed in vertical excavation steps of 500mm. 40mm diameter

boreholes, inclined at $15°$ with a horizontal spacing of 0.45m, were augered (without casing) to a depth of 1m. 6.2mm reinforcing steel bars were inserted and gravity grouted. A face plate was fixed to the extruding end of each reinforcing bars. The excavation faces were stabilised with a gunnite layer of about 5mm in thickness. These construction sequences were then repeated to achieve the desired height of the soil-nailed wall structure.

Three different wall heights were adopted for the experimental tests, namely 1.25m, 1.75m and 2.25m. Additionally, the maximum free standing height of an unreinforced excavation as well as the ultimate loading of an unsupported excavation of 1.25m were established in separate tests.

Each soil-nailed wall consisted of three columns of nails with the middle column being instrumented. Strain gauges were placed on the nails to measure the force distribution along the nails and linear potentiometers were located at the face of the wall to measure horizontal displacements. The instrumentation was linked to a data capturing system, which allowed continuous monitoring during the construction of the wall and the subsequent loading tests to failure.

During the excavation stages of the soil-nailed wall, the soil behind the excavation face relaxes from its initial in-situ state and thus expands laterally with maximum displacements at the top of the wall. The displacement increased rapidly with the increasing excavation depth as shown in Figure 2. At the end of the 2.25m excavation, the maximum face displacement was about 0.1% of the total wall height.

The nails, interacting with the sand, absorb the stresses and strains as the wall deforms. The interaction results in the distribution of forces along the nails as the excavation proceed as shown in Figure 2. Again, as the displacements increase with each excavation step, the tensile nail forces also increase. A

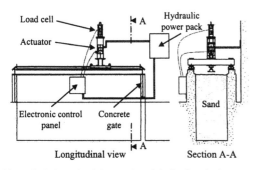

Figure 1. Schematic of the test pit and the loading device

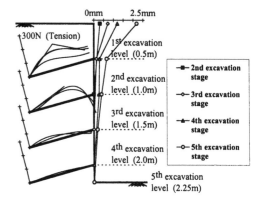

Figure 2. Nail force distribution and displacement after every successive excavation

178

maximum tensile force of about 200N was recorded at mid-length of the upper nail while the lower nail was barely stressed.

Following the construction of the soil-nailed wall, a rigid loading plate of 0.5m wide was positioned on the ground surface at a distance of 0.5m from the excavated face. The wall was loaded to failure at a rate of 20kN per minute. Failure was initiated by a localised collapse of the wall face between the last two rows of nails and below as shown in Figure 3. The observation of a local failure was common to all soil-nailed wall experiments.

The development of the wall face displacement and force distributions along the nails during the loading test is presented for the 1.75m soil-nailed wall in Figure 4. Measurements from the excavation stages are excluded for clarity. The wall sustained an ultimate load of 87kN with the evolution of a vertical crack at the far end of the loading plate. At this stage, the wall facing displaced by a further 0.6% of the wall height in addition to the displacement measured during construction.

Generally, the location of the peak tensile force along the nails coincides with the failure surface through the reinforced soil mass (Project Clouterre 1993). Therefore, based on the limit equilibrium design methods, three different failure surfaces were drawn, as shown in Figure 4. The double wedge was not analysed as it does not apply to this particular loading configuration. With the indicators of the location of the failure surface, namely the vertical crack, local collapse and peak nail forces, it appears that the logarithmic spiral is the most appropriate mechanism. This was substantiated by observations made after the walls collapsed.

The results of the various walls under investigation are shown in Figure 5. The improvement in the load carrying capacity of soil-nailed walls compared to their respective unreinforced walls of similar

height was remarkable. The unreinforced soil mass collapsed after a 1.75m excavation whilst the respective reinforced wall could sustain a load of up to 87kN before failure. This confirms that soil nailing can effectively be used to strengthen and improve the performance of weak soils.

The failure load of the soil-nailed walls decreases with increasing wall height, as shown in Figure 5. This indicates that the effectiveness of the reinforcement decreases with decreasing ratio of nail length to depth of excavation (nail length being constant in all walls). The test data was extrapolated using a best fitting logarithmic function (a correlation coefficient close to unity was achieved). If the height of the wall is reduced to zero, the situation is representative of a bearing capacity consideration. An ultimate bearing load of about 325kN was obtained using the general bearing capacity equation (Brinch Hansen 1970). Plate load tests carried out by Schwaeble (1991) on similar sand bedding were in close agreement with the above predicted load. According to the logarithmic extrapolation, a maximum wall height of 3.25m can be constructed using a similar pattern of nails. The investigation of the sliding stability of a soil-nailed block of similar shape resulted in a maximum height of 3.4m for the state of limit equilibrium.

A computer program was developed for the design of soil-nailed structures based on limit states principles. The calculations require the assumption of a constant skin friction value for the soil/nail interface. In pull-out tests on independent nails, the ultimate load per metre length of nail was established as 5.5kN, which is in terms of skin friction in the order of 40kPa (on average).

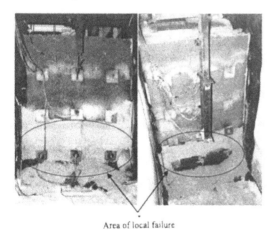

Figure 3. View of failed soil-nailed walls of 1.75m and 2.25m in height

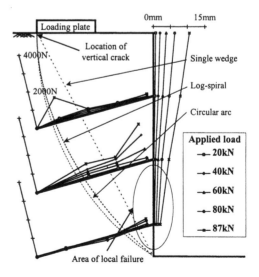

Figure 4. Test results for a 1.75m high soil-nailed wall

179

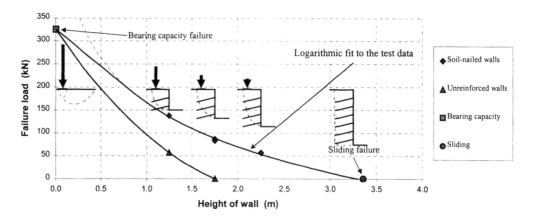

Figure 5. Failure loads and wall heights of the test series

In Figure 6, the predicted failure loads of the wall structures based on the assumption of three different slip surfaces (single wedge, circular arc and logarithmic spiral shaped) are presented. Also shown are the load carrying capacities of the test series with the respectively fitted (solid) lines.

For nail length to wall height ratios, L/H, of 0.5 to 1.2, the generally recommended ratios for soil-nailed structures (Project Clouterre 1993), the predictions of the failure load based on the assumption of a log-spiral failure surface are in relatively good correlation with the experienced values in the test pit. These ratios translate to wall heights from 2.0m to 0.8m in the test environment of this study. Even better matching results can be achieved if the effect of the surcharge on the nail carrying capacity of the nails is taken into account. An adjusted skin friction value due to the increase in the vertical stresses

experienced at the soil/nail interface is then applied in the analysis.

The other two analyses using a single wedge or circular arc failure surface provide almost identical, very conservative results for ratios L/H ≤ 0.8 (wall height, H ≥ 1.25m). It should be noted that the predicted failure loads assuming the single wedge failure mechanism rapidly change over from 'safe' to critical values for ratios of L/H > 0.8. If the ratio is greater than 1 the analysis even overestimates the strength of the wall structure. The same applies for high walls with short nails when the ratio L/H < 0.4. For those nail/height wall configurations none of the investigated failure surfaces appear suitable since the reinforcing action of the nails has no bearing on the considered failure mechanisms, and the walls fail as solid soil-nailed blocks.

4 FINITE ELEMENT ANALYSIS

The finite element method is not a readily available tool for design purposes. However, it can be used to support experimental observations. A three-dimensional model was discretised as shown in Figure 7.

The sand was modelled using a Drucker-Prager/Cap constitutive law with the specific material parameters, which allows volumetric plastic strain to occur and controls the volume dilatancy when the material yields in shear. Interface elements were placed between the soil and the nails and soil/nail interaction was simulated using the classical Coulomb friction model. The construction sequence of the soil-nailed structure was followed as in situ; after each excavation step (i.e. removal of soil elements) the nails were installed by assigning the respective properties to the particular nails and interfaces.

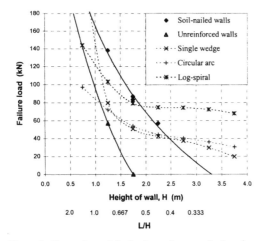

Figure 6. Comparison of the load carrying capacity based on the limit equilibrium method with the experimental results

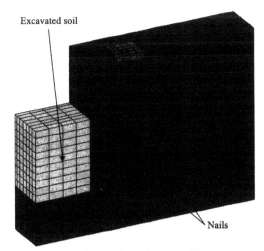

Figure 7. 3-D mesh of the finite element model

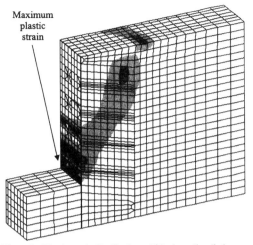

Figure 9. Plastic strain distribution within the soil-nailed mass

The contour plot of the stresses along the nails (see Fig. 8) shows that there is a concentration of tensile stresses near the uppermost nail ends below the loading plate. The magnitude of the stresses decreases and the peak shifts towards the facing, the further down the nails are located. This compares well with the observations made along the instrumented nails.

The contour plot of the plastic strain within the soil-nailed structure (see Fig. 9) indicates a localised region of instability between the last two rows of nails and the toe of the wall. This corresponds to the localised failure observed in the collapsed soil-nailed walls. The progression of the plastic strain within the soil-nailed mass to the surface with decreasing magnitude indicates that the failure surface might start at the toe of the wall and propagates to the ground surface, behind the loading plate.

Therefore, it might be sensible to increase the nail density at the toe of a soil-nailed structure by reducing the horizontal spacing of the adjacent nails. However, this should be confirmed in further experiments.

5 CONCLUSIONS AND RECOMMENDATIONS

The experimental investigation validates the log-spiral mechanism for the selected nail length to wall height ratios and load configuration.

The test results indicate that failure mechanisms based on circular arc shaped and particularly the linear (single wedge) failure surfaces have no substance for design considerations in similar soil conditions, geometry and surface loading.

The finite element model can be used to substantiate the experimental observations and contribute to a further insight into the mechanics of a soil-nailed structure.

The finite element model suggests that the failure surface initiates from the toe of the structure and propagates to the top surface.

Increasing the nail density near the base of the structure might alleviate the localised failure at the toe of a soil-nailed structure.

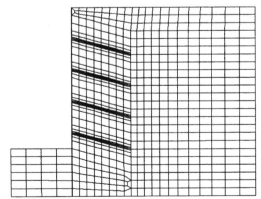

Figure 8. Stress distribution along the nails

REFERENCES

Brinch Hansen, J. 1970. A Revised and Extended Formula for Bearing Capacity, *Danish Geotechnical Institute Bul.*, No. 28, Copenhagen.

BS 8006 1994. *Code of Practice for Strengthened/Reinforced Soils and other Fills*, British Standards Institution, Document No. 94-104-108.

Gässler, G. 1988. Soil-nailing - Theoretical basis and practical design. In T. Yamanouchi, N. Miura and H. Ochiai (ed.), *Theory and Practice of Earth Reinforcement*: 283-288. Rotterdam: Balkema.

HA 68/94 1994. Design methods for the reinforcement of highway slopes by reinforced soil and soil nailing techniques, *Design Manual for Roads and Bridges*, Vol. 4, Section 1, Part 4, HMSO.

Juran, I., G. Baudrand, K. Farrag, & V. Elias 1990. Kinematical limit analysis for design soil-nailed structures, *Journal of Geotechnical Engineering*, Vol. 116, No. 1: 54-72.

Lai Sang, J. 1999, *Experimental and numerical investigation of soil-nailed wall structures*, M.Sc. Thesis (in prep.), Department of Civil Engineering: University of Cape Town.

Lambe, T.W. & R.V.Whitman 1969. *Soil Mechanics*. New York: John Wiley and Sons: 357-362.

Project Clouterre 1993, *Soil nailing recommendations-1991*, For Designing, Calculating, Construction and Inspecting Earth Support Systems Using Soil Nailing, French National Research Project Clouterre, Presses de l'Ecole Nationale des Ponts et Chaussées, Federal Highway Administration: Washington.

Schwaeble, R. 1991, *Determination of ultimate bearing capacity beneath various foundation specimens*, B.Sc. Thesis, Department of Civil Engineering: University of Cape Town.

Schwartz, K. & E.A. Friedlaender 1989. Soil nailing in South Africa, GeoDrilling/Technics, *Journal of the South African Drilling Association*, No. 17: 4-9.

South African Code of Practice 1989. *Lateral Support in Surface Excavations*, The South African Institution of Civil Engineers Geotechnical Division, 2nd Edition: 60-64.

Stocker, M.F. & G. Gässler 1979. Ergebnisse von Großversuchen über eine neuartige Baugrubenwand-Vernagelung, *Tiefbau, Ingenieurbau, Straßenbau*, No. 9: 677-682.

Geotechnics for Developing Africa, Wardle, Blight & Fourie (eds) © 1999 Balkema, Rotterdam, ISBN 90 5809 082 5

Behaviour of double footings resting on cemented sand underlain by clay

N.E. Marei
Construction Research Institute, Delta-Barrage, Kanater, Egypt

ABSTRACT: A deposit of very soft clay is one of the problematic soils because of its low bearing capacity and high settlement. Very soft clay is found in many regions in Egypt especially in the North coast of Egypt. In this study, the enhancing the performance of double strip footings resting on very soft clay using cemented sand cushion is investigated. The cemented sand cushion consists of a mixture of 90% sand, 8% cement, and 2% rice husk ash by weight. An experimental modeling program was conducted. The main parameters affecting the bearing capacity and settlement of the cemented sand cushion underlain by very soft clay were studied. The studied parameters are thickness of cemented sand cushion, distance between the centerline of the double footings, and the outside length of the cemented sand cushion measured from the centerline of the footing.

1 INTRODUCTION

Several soil improvement techniques are used to improve the soil bearing capacity and settlement conditions of shallow foundation supported by weak layers. The improvement techniques include chemical stabilization, compaction, replacement of weak soil, and reinforcement of soil beneath foundation. During the last decade, several laboratory model tests were carried out to investigate the increasing of the bearing capacity of single strip footing resting on compacted sand fill material underlain by soft clay with different types of geosynthetics inclusions at the interface or with more than one layer within the sand fill (Das (1988), Sridharam et al (1991), Tan (1993), Shuklaad Chandra (1994), Abdel-Hamid (1995), and Saleh (1996)). Many types of materials may be utilized as reinforced material as long as they have the principal properties needed to improve the performance of the composite system. Natural pozzolanic materials (Fly ash, silica fume, and rice husk ash) are used alone or with other cementing materials to improve the geotechnical properties of the soil. In this research mix of cement and rice husk ash (RHA) is used to reinforce the sand cushion under the double strip footings underlain by very soft clay. Rice husk is a residual produced in significant quantities in the world. Bureau of Agricultural Economics (1983) estimated that the world produces 406.6 million tons of rice or 16.3 million tons of ash. RHA is consists essentially of non-crystalline matter, which under ambient conditions hydrates slowly even in the presence of the alkaline solution, which is formed by the dissolution of cement minerals. Rice husk ash is used alone or with other cementing materials to improve the geotechnical properties of the soil. Rahman (1988) stated that adding RHA to latertici soils improve its geotechnical properties. El-Kasaby (1992b) found that the stabilization of Egyptian fine-grained soil with cement-RHA improves the geotechnical properties of the soil. Marei (1998), found that reinforced the sand soil with cement-RHA mix increases the shear strength of the soil. Marei used this reinforced sand in an experimental model as a cushion over very soft clay to enhance the performance of single strip footing. Marei found that a mixture of 90% sand, 8% cement and 2% rice husk ash by weight is the optimum mix to enhance the engineering properties of the cemented sand. This optimum mix is taken to make the cemented sand cushion in the present study. Rice husk ash is abundant and cheap material in Egypt and using it as replacement part of the cement content to cement the sand cushion leads to low cost of cementation, higher strength, and more durability of cemented sand cushion. Laboratory model was constructed to study the main parameters affecting the bearing capacity and settlement of double strip footings resting on the cemented sand cushion underlain by very soft clay.

2. PROPERTIES OF THE USED MATERIALS

In this research the very soft clay in Sahel El-Tina area is used. Sahel El-Tina is located between longitudes 32°20ʺ and 32°35ʺ East and latitudes 31°00ʺ and 32°14ʺ North. Soft clay disturbed samples were taken from depth 1.0 m to 1.5 m below the ground surface. The samples were taken using the open excavation method. The water table is about 1.0 m from the ground surface. The undrained shear strength of the very soft clay in the field, which was measured by the vane shear device, is about 2.0 kN/m^2. According to the laboratory investigations the total soluble salts in the deposit is (23170 p.p.m). The cocentration of sodium chloride in the soil is 11500 p.p.m. Clay minerals are Motmoillonite (49%), Kaolinite (27%), and Illite (24%). The clay contains 38%silt and 62% clay. The natural water content, liquid limit , and plastic limit were found to be 76%, 63%, and 24% respectively. It can be noted that, the natural water content of the very soft clay is more than the liquid limit. This can be explaining by the fact that this clay was formed during the recent periods in the form of marina deposits and the main clay mineral is montmorillonit. This clay mineral has a flocculate structure, which allows water storage into it. The initial void ratio of the clay is 2.4. The values of the compression index and heaving index were 0.64 and 0.05 respectively. The deposits were found to be normally consolidated.

The sand used in this present research is brought from Naser city and was air-dried. The portion of the sand passing from a sieve with an opening of 0.8 mm and retained on a sieve with an opening 0f 0.06 mm were used in the experimental work. The sand is poorly graded sand (SP). The uniformity coefficient (Cu), the coefficient of curvature (Cc), and the specific gravity of the sand are 2.68, 1.14, and 2.64 respectively. The D_{50} is 0.38 mm. The angle of internal friction (ϕ) of the sand is 32°.

Rice husk ash can be obtained in two ways: burning the rice husk in furnaces under controlled temperature and retention time or burning the rice husk in an open furnace. In this research the second method was adapted to obtain rice husk ash because this method is very economic and can be used in the field. The physical properties and chemical composition of the used rice husk ash and Ordinary Portland Helwan cement (OPC) are summarized in table (1).

Table 1. Physical properties and chemical compositions of RHA and OPC.

Parameter	RHA	OPC
Specific gravity	2.1	2.84
Specific surface	8790 cm^2/g	3320 cm^2/g
%passing opening sieve 0.063 mm	100 %	100 %
Color	White gray	Dark gray
Silicon dioxide	78.025 %	22.025 %
Aluminum oxide & iron oxide	10.500 %	8.100 %
Calcium oxide	0.086 %	63.8 %
Sulphate oxide	0.375 %	2.1 %
Loss on ignition	6.201 %	3.3 %
Other oxides and impurities	4.813 %	0.7 %
Electric conductivity	0.75 mS/cm	

3 TESTING PROCEDURES AND PROGRAMS

The experimental model consisted of testing prespex tank, footing, and loading system. A prespex tank stiffened with a steel frame with inner dimensions of 2.97 meters long, 0.7 meters wide and 0.92 meters deep was constructed. Two footing models are made of steel. Each of them is with dimensions of 5 cm thick, 10 cm wide, and 69 cm long. The loading system was designed to be rigid and capable of sustaining the loading stresses without suffering from deflection. The load is applied at the centerline of the footing. Eleven experimental model testing were carried out. The procedures of each test were as follows: The clay was placed in a tank in layers of 10 cm thickness and was pressed by the metallic plate to prevent the formation of air voids and to obtain the required density (γ=1.79 t/m^3). Then by using a wooden rod with a diameter of 4 cm the surface of each layer was scratched by making a series of grooves so that each clay layer interlock with next one and lipping at the layer surface is prevented. After reaching the required height, which is marked on the side of the tank, the surface was scraped by a wooden plate (5 × 10× 69 cm) in order to provide a horizontal surface. Since the studied clay is very weak, compaction of the fill cushion in the field can not be done using heavy compactors. Therefore hydraulic compaction technique should be utilized in this case. In this technique sand, cement, and RHA were mixed first in dry state (2% RHA +8% cement + 90% sand). This mix was placed in the tank and then water was added until all the mix is submerged in the water. Then the water was sucked from the fill through a vertical drain system using a suction pump. This process causes the cemented sand to be compacted under hydraulic pressure. After 7 days from the placement time of the cemented sand cushion, the two footings were placed in the requirement position. The loads on the footings were applied in small loading increments equal 5.20κN/m². Each load increment continued for 24 hours. The test terminated when the failure occurs. The ultimate bearing capacity (UBC) in any test was defined as the stress causes cracks in the soil layer (cemented cushion or clay soil). To determine the increase in the ultimate bearing capacity and the decrease in the settlement in case of cemented sand cushion related to that clay, two dimensionless parameters were employed: bearing capacity ratio (BCR) and settlement ratio (SR). BCR and SR were defined according to eaquation (1) and (2) respectively.

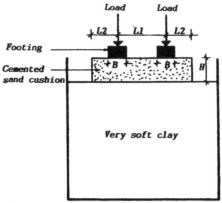

Figure (1) Outline of the study parameters

$$BCR = qu/qc \qquad (1)$$

$$SR = S/Sc \qquad (2)$$

Where:
qu = UBC for different study cases.
qc = UBC for the case of clay.
S = Failure settlement for different study cases.
Sc = Failure settlement for the case of clay.

4 EXPERIMENTAL RESULTS AND ANALYSIS

The ultimate bearing capacity values of the footing resting directly on the very soft clay is 9.42 KN/m². The main parameters affecting the bearing capacity and settlement of the very soft clay were investigated. These parameters are: the outside length of the cemented sand cushion measured from the centerline of the footing (L2), the distance between the centerline of the double footings (L1), and the thickness of cemented sand cushion (H). The outline of the study parameters is shown in figure (1).

4.1 Outside length of the cemented sand cushion

Loading tests were conducted to investigate the effect of changing the length of the cemented sand cushion measured from the centerline of the footing (L2), expressed in a dimensionless form L2/B, on the bearing capacity and settlement values. B is the width of the footing. In the performed tests the H/B and L1/B ratios were constant and equal 0.5 and 3.5 respectively. The tests were carried out at L2/B

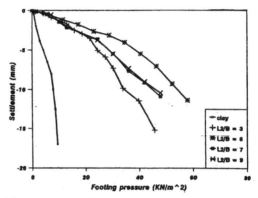

Figure (2) Load-settlement curves for different L2/B ratios (H/B=0.5, L1/B=3.5)

equal to 3, 5, 7, and 9. Figure (2) shows the load-settlement relationship for the footing at different L2/B ratios. It is noticed from this figure that the improvement of the load-settlement relationship due to increasing L2/B ratio is more pronounced at high settlement values. The load settlement curve in case of L2/B =5 is the flattest curve and the behavior of the load settlement curves in both cases of L2/B =7 and L2/B =9 is the same. The optimum outside length of the cemented sand cushion (L2) required to mobilize the maximum improvement in the bearing capacity of the loaded double strip footings and minimize the settlement is 5 times the footing width (B) as shown in figure (3). In figure (4), the effect of changing L2/B ratio on BCR and SR is illustrated. The BCR have the values of 4.85, 6.15, 5.08, and 5.08 for the cases of L2/B ratio equal 3, 5, 7, and 9 respectively. SR values were 0.89, 0.66, 0.64, and 0.61 for the same cases

185

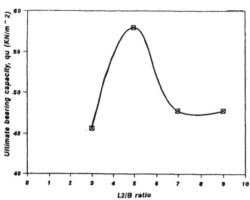

Figure (3) Effect of L2/B on the ultimate bearing ratio (H/B=0.5, L1=3.5)

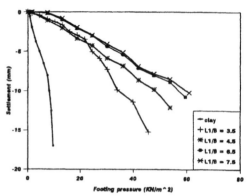

Figure(5) Load-settlement curves for different L1/B ratios (H/B=0.5, L2/B=3)

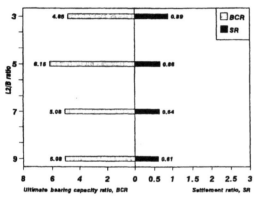

Figure (4) Effect of L2/B ratio on the BCR and SR (H/B=0.5, L1/B=3.5)

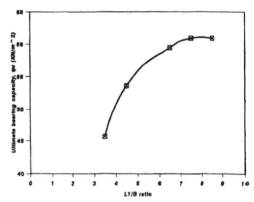

Figure (6) Effect of L1/B ratio on the ultimate bearing capacity (H/B=0.5, L2/B=3)

respectively. From the above values, it can be seen that the cemented sand cushion has a great effect in increasing the BCR and decreasing the SR values. In the case of H/B=0.5 and L1/B=3.5, the optimum outside length of the cemented sand cushion (L2) is about five times the footing width (B).

4.2 Distance between the double footings

To investigate the effect of changing the distance between the centerline of the double footings (L1) on both bearing capacity and settlement under the footings, some experimental tests were carried out. The ratios H/B=0.5 and L2/B=3 were kept constant in these tests. The ratio L1/B was taken 3.5, 4.5, 6.5, 7.5, and 8.5. The comparison between load-settlement curves for different L1/B ratios is shown in figure (5). The load settlement curves became more flat when the L1/B ratios increased. There is

no significant change in the load-settlement relationship at L1/B=6.5 and L1/B= 7.5. This can Explained as above L1/B=6.5 each footing behave as single footing. The bearing capacity increases with increasing L1/B ratios as shown in figure (6). The effect of L1/B ratios on the BCR and SR is illustrated in figure (7). It can be easily seen from this figure that the increase in BCR when increasing L1/B ratio from 3.4 to 4.5 was more than that when increasing L1/B ratio from 6.5 to 8.5. It is also noted that the settlement ratio is nearly constant for cases of L1/B equal 6.5, 7.5, and 8.5.

4.3 Cemented sand cushion thickness

To study the effect of changing cmented sand cushion thickness on both bearing capacity and settlement under double strip footings, some loading tests were carried out. In the tests performed the L1/B=3.5 and L2/B=3.0 were kept

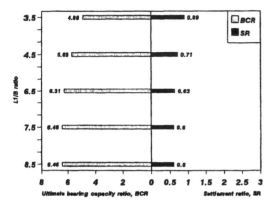

Figure (7) Effect of L1/B ratio on the BCR and SR (H/B=0.5, L2/B=3)

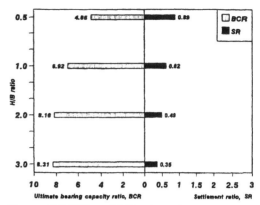

Figure (10) Effect of H/B ratio on the BCR and SR (L1/B=3.5, L2/B=3)

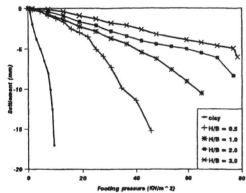

Figure(8) Load-settlement curves for different H/B ratios (L1/B=3.5, L2/B=3.0)

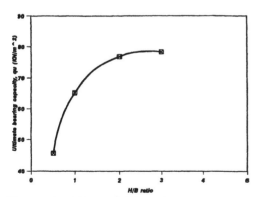

Figure (9) Effect of H/B ratio on the ultimate bearing capacity (L1/B=3.5, L2/B=3.0)

constant and the cemented sand cushion thickness was presented in dimensionless ratio H/B. The tests Were carried out at H/B equal to 0.5, 1.0, 2.0, and

3.0. Figure (8) shows the load-settlement relationship for double strip footing resting on cemented sand cushion underlain by very soft clay. It is noticed from this figure that when increasing the thickness of the footing the load settlement curves became more flat and the ultimate bearing capacity increases. The bearing capacity increases as the ratio H/B increases as shown in figure (9). This is because at low ratio, most of the failure wedge (the zone of active and radial shear deformation) passes through the clay deposits which highly affects the performed of the foundation system. When H/B ratio is large, the failure wedge beneath the footing does not extend beyond the thickness of the cushion and the shear resistance along the failure plane is developed the strong cemented sand zone. The change in BCR and SR with H/B is illustrated in figure (10). It can be seen from this figure that there was no significant increase in the bearing capacity when increasing the H/B ratio from 2 to 3. But there is margin gain in reducing the settlement as shown in figure (10).

CONCLUSIONS

The outside length main conclusions that can be derived from the results obtained in this research are as follows:
-There is an optimum of the cemented sand cushion (L2) measured from the centerline of the footing required to achieve the maximum load sustained by double strip footings and minimum settlement.
- Increasing the distance between the centerline of the double footing (L1) increases the load sustained by the footings. In the case of H/B=0.5 and L2/B=3.5 each footing of the double footings

behaved as single footing when the distance between the centerline of the footing is about 6.5 times the width of footing.

-The load sustained by the double strip footings resting on cemented sand cushion underlain by very soft clay increases significantly with increasing the ratio of the thickness of the cemented sand cushion (H) to the width of the footing (B) till it reaches 2.0. No significant gain in the load is achieved when increasing the value of H/B ratio from 2 to 3.

ACKNOWLEDGEMENT

The authors would like to express their deep gratitude to the director and staff of the Construction Research Institute for their valuable assistance in performing the laboratory tests.

REFERENCES

Abdel-Hamid, A.E. 1995, Improvement of very soft clay using sand cushion reinforced with geogrids, M.Sc. Thesis, Cairo University, Egypt.

Bureau of Agricultural Economics Situation and Outlook.1983, RICE, Australia Government Publishing Service, Canberra, pp. 1-10.

Das,B.M.1988, Shallow foundation on sand underlain by soft clay with geotextile interface, Geosynthetics for soil improvement, ASCE, National Convention, Nashville, Tennessee, USA, pp. 112-126.

El-Kasaby, E.A. & AbdAlla, E.A.1992b, Stabilization of fine-grained soil with cement-rice husk ash mixture, 1st.International Conference of the Transportation Depte faculty of Eng., Alexandria University, Alexandria.

Marei, N.1998, Enhancing the performance of foundations resting on very soft clay using cemented and composite sand cushions, PhD thesis, Cairo University, Egypt.

Rahman, M.N.1988, Effect of cement-rice husk ash mixtures on geotechnical properties of Lateritic soil, Japanese society of soil and Foundation Engineering, Vol.27, No. 2,pp.61-65.

Saleh, N.M.1996, Study of strip footing on geosynthetic-Reinforced sand underlain by weak clay, PhD thesis, Cairo University, Egypt.

Shukla, S.K & Chandra, S. 1994, Study of settlement response of a geosynthetic-reiforced compressible granular fill-soft soil system, Geotextile and Geomembranes, Vol.13, No. 9, pp.627-639.

Sridharam, A.et al.1991, Technique for using fine-grained soil reinforced earth, Journal of geotechnical engineering division, ASCE, Vol.117,No.GT 8,pp.1174-1190.

Tan,S.K.et al. 1993, Measurment of interface friction between a jute geotextile and a clay slurry, geotextiles and geomembrances, Vol. 12, No. 4, pp.363-376.

Geotechnics for Developing Africa, Wardle, Blight & Fourie (eds) © 1999 Balkema, Rotterdam, ISBN 90 5809 082 5

Experimental study of a strip footing on collapsible soil

M. Mashhour
Faculty of Engineering, Zagazig University, Egypt

M. Saad & A. Aly
Military Technical College, Cairo, Egypt

E. El-Bahhi
Ministry of Housing and Reconstruction, Alexandria, Egypt

ABSTRACT : Eighteen plane strain model tests were carried out on a rigid footing resting on an extended uniform collapsible soil formed of well graded sand mixed with 20% Kaolinite by weight . The tests were performed in a special rig allowing for increments of vertical loads to be applied to the footing . A uniform flow of water across the whole width of the model could also be achieved in both downward and upward directions. soaking pressures (which is defined as the pressure at inundation) of 50,100,200 and 400 kN/m^2 were applied to the footing and the corresponding collapse settlement of the footing as well as surface movements were measured for both cases of downward and upward flow . Two cases of soil treatment were also studied under the above conditions, namely soil replacement with and without geogrid reinforcement. It was found that collapse settlement increases with the increase in soaking pressure .The collapse settlement increases as the depth of wetting increases in case of downward flow but the strain is less at lower levels Generally , the maximum collapse strain in case of downward flow was greater than that of upward flow.Soil treatment using sand replacement with and without reinforcement resulted in a significant reduction in collapse settlement.

1. INTRODUCTION

The accurate prediction of collapse settlement depends strongly on the understanding of the behaviour of foundations resting on collapsible soils upon wetting. For a given soil type and initial conditions two main factors were found to greatly affect the mechanism and value of collapse settlement (Aly, 1998). These are soaking pressure and flow characteristics . The effect of the applied pressure prior to wetting was studied by many investigators (e. g. Booth, 1975 and Adnan et al, 1992) .

The effect of flow characteristics on collapse and the estimate of the depth of wetted zone were studied by several investigators (e.g. El-Ehwany et al 1990 , El-Saadany, 1986, Bond and Gollis - George, 1981 and Stauffer, 1981) .

In the present paper the effect of both soaking pressure and direction of flow on the collapse settlement are studied for the case of a plane strain footing model resting on collapsible soil .Two methods of soil treatment were also considered, namely soil replacement and soil replacement with geogrid reinforcement .

2. TESTING EQUIPMENT AND MATERIALS

2-1 *Experimental model*

The plane strain model consisted of a plex - glass sided box with dimensions of 800 mm width 700 mm height and 100 mm breadth . A steel tank 800 mm x 50 mm x 100 mm was fixed beneath the box to act as a water reservoir . A porous plate 2.0 mm thick was fitted between the soil sample inside the plex-glass sided box and the tank in order to distribute the water uniformly .In the case of upward flow the water was supplied from a constant head tank 1100mm above the soil surface. In the case of downward flow water was allowed to flow through the top surface through a porous pad .

The footing was 100 mm wide , It was placed in the centre of the box on top of the soil sample , and loaded with concentric normal loads using a special set up with a lever mechanism . Average footing displacement was measured using two dial gauges. Two additional gauges were used to measure soil surface deformations at two points away from the centre. Friction between the two box sides and the soil was reduced using a film of lubricant agent .

2.2 *The Tested soil*

The soil used was a well graded sand (Cu = 6.11 , Cc =1.94). It was mixed with 20% Kaolinite by weight . This particular ratio was found to give a significant measurable collapse (collapse potential Cp (Jennings and Knight , 1975) as measured in the oedometer = 15.5%) .The properties of the soil mixture was as follows : natural water content = 0.7% , dry unit weight = 17 kN/m^3 ,max dry density = 21.59

kN/m³ , min dry Density = 14.78 kN/m³ , specific gravity = 2.71, cohesion for dry case = 83 kN/m² , cohesion for 10% W.C. = 26 kN/m² ,angle of friction for dry case = 46, ⁰ angle of friction for 10% W.C. = 33.5 ⁰ .

2.3 Properties of sand replacement :

The sand replacement was a well graded sand with dry unit weight of 18.7 kN/m³

2.4 Properties of Reinforcing geogrid

Polymeric geogrid Netlon CE121 was used . Its properties are : mesh thickness = 3.3mm , mesh aperture size = 8 x 6 mm , weight = 730 gm/m² , tensile strength = 7.68 kN/m , extension at max. load = 20.2% , elongation at 1/2 peak strength = 3.2% .

3 TESTING PROCEDURE AND VARIABLES

3.1 Testing procedure

The footing was loaded by incrementally increasing normal loads while the soil was still in dry condition For each load increment the dial readings of deformation were recorded . This process was continued until the required pressure for soaking was reached , when flow of water was allowed through the soil, and the vertical deformation was then recorded with time .

3.2 Test Variables

The following table shows the tested variables.

Table 1 . Tested variables .

SERIAL NUMBER	UPWARD FLOW	DOWNWARD FLOW	SOAKING PRESSURE (kN/m2)				SAND REPLACEMENT	REINFOR CEMENT
			50	100	200	400		
1	*		*					
2	*			*				
3	*				*			
4	*					*		
5		*	*					
6		*		*				
7		*			*			
8		*						
9	*	*	*			*	*	*
10	*			*			*	*
11	*				*		*	*
12	*					*	*	*
13		*	*				*	*
14		*		*			*	*
15		*			*		*	*
16		*				*	*	*
17	*				*		*	
18		*			*		*	

4. RESULTS AND DISCUSSION

4.1 Effect of Soaking Pressure on Collapse Settlement under Downward Flow Condition

Collapse settlement is the additional settlement due to soaking .Water was allowed to infiltrate downwards through the soil by gravity under a small pres-

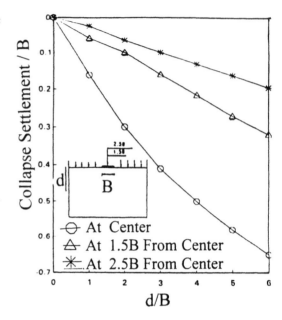

Fig . (1) Collapse settlement versus depth of wetting - downward flow - untreated soil - soaking pressure = 200 kN/ m² .

sure head of 50mm , and collapse settlement started immediately as the water front advanced downwards a decreasing rate from the soil surface . Fig (1) shows a sample of the results of collapse settlement versus the depth of wetting .

The maximum collapse settlement (at d/b=6) increases with the increase of soaking pressure (Fig 2) Since increasing imposed pressure accelerates the deformation of cemented particles. This reduces the void ratio and results in higher collapse .

4.2 Effect of Direction of Flow

In the case of upward flow , water was allowed to flow through the soil against gravity under a head of 1100 mm . Collapse settlement started when the water front was at depth= 1 B , 1.5B , 2.25B and 3B under the footing for soaking pressures of 50,100, 200 and 400 kN/m² respectively .

Fig.(3) shows collapse settlement versus height of wetting for upward flow , soaking pressure = 200 kN/m²

The soil below the upward wetting front line became partially saturated and hence soft compared to the dry hard cemented sand part above it . This dry part behaved like a beam on elastic foundation which suffered bending cracks. The depth of the beam (i.e. depth of wetting front) at which the cracks occurred is related to the magnitude of the soaking pressure.

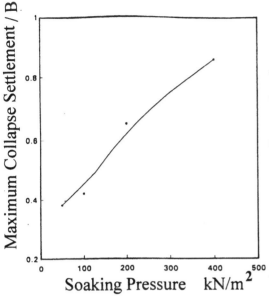

Fig. (2) Maximum collapse settlement versus soaking pressure downward flow .

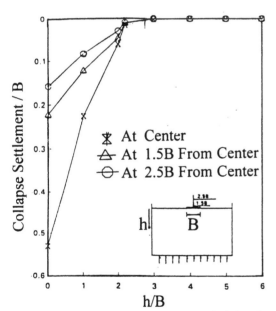

Fig . (3) Collapse settlement versus height of wetting - upward flow - untreated soil - soaking pressure = 200 kN/ m^2 .

Fig.(4) shows the maximum collapse settlement against soaking pressure for upward flow case .

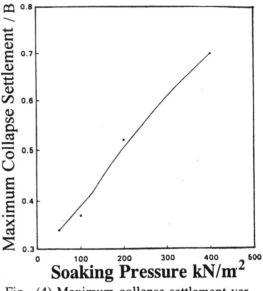

Fig . (4) Maximum collapse settlement versus soaking pressure - upward flow .

4.3 *Effect of soil Treatment*

Two methods of treatment were used . Replacement of the top collapsible soil layer depth = 1.0B, with clean sand, and sand replacement with one layer of geogrid reinforcement at the interface between replaced sand and original collapsible soil. Collapse settlement of the footing was significantly reduced as shown in the following table.

Table 2. Reduction (%) in collapse Settlement due to soil Treatment .

Method	Downward Flow				Upward Flow			
	Soaking Pressure (kN/ m^2)				Soaking Pressure (kN/ m^2)			
	50	100	200	400	50	100	200	400
Sand Repl. + Reft.	46.2	34.1	30.5	20	64.6	36.5	29.2	18.5
Sand Repl. only			20.4				25.2	

4.4 *Progress of Water Front*

The water movement is governed by partially saturated flow. The gradient of the suction is initially high and decreases with time. This decrease explains the decrease in infiltration rate with time where soil properties remain relatively constant Fig(5). In the case of downward flow the water moves due only to gravity (gravitation potential). As in most natural

191

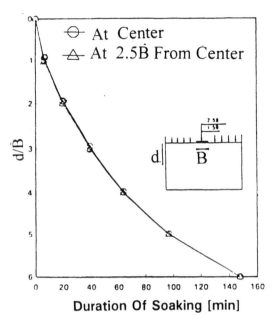

Fig. (5) Progress of wetting front- time relationship - downward flow - untreated soil - soaking pressure = 200 kN/m^2.

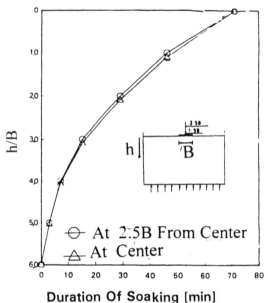

Fig. (6) Progress of wetting front - time relationship - upward flow - untreated soil - soaking pressure = 200 kN/m^2.

soils the surface layer is more permeable than the subsoil. Entrapped air will decrease permeability by decreasing effective porosity. This agreed with the findings by Parra and Bertand, (1960).

The infiltration process is characterized by relatively sharp fronts of water movement.

The main governing factor to the progress of upward water flow in the soil sample was the pressure head due to the elevated water tank.

Fig (6) shows the level of wetting front against time for untreated collapsible soils in the experimental model.

It can be observed that the rate of water front propagation decreases with time. This may be due to the decrease of the difference in head between the level of water tank and that of water front in soil sample (i.e. decrease in pressure head).

Part of the water movement in soil sample in the liquid phase may be due to gradients of capillary potential,which arises from differences in water content.

There are no significant difference between the rate of water front propagation in case of soaking pressure = 200 kN/m^2 and that of the others soaking pressure cases.

5- CONCLUSIONS

1 - As the soaking pressure value was increased, the collapse strain increased. However the maximum collapse strain in case of downward flow exceeds that in case of upward flow by about 18%.

2 - In case of downward flow, as the wetting front advances into the soil, the collapse strain occurs in the wetted soil layer and the collapse strain is less for the lower layer.

3 - In case of upward flow, The soil part below the wetting front line became partially saturated sand while the soil part above this line was still dry cemented sand. This upper dry layer of cemented sand behaves as a beam on elastic foundation. It develops a bending crack due to imposed footing pressure, then the settlement occurs as the crack propagates.

4. In general, the sand replacement technique, with and without reinforcement, results in a significant reduction in collapse settlement This reduction decreases with the increase of soaking pressure.

5. In case of upward flow, The rate of settlement is higher for the case of treated soil than for natural untreated soil due to decreasing the depth of dry collapsible cemented soil by using sand replacement.

REFERENCES

1 - Adnan, A.B., and Erdil , R.T. (1992) . "Evaluation and Control of collapsible Soils". Journal of Geotechnical Engineering, Vol. 118 , No . 10, pp. 1421-1504.

2 - Aly , A.I. (1998) . " Foundations" on collapsible soil" Ph.D. thesis Military Technical college Cairo , Egypt.

3- Bond , W.J., and Collis-George, N. (1981) ." Pounded Infiltration into Simple Soils Systems : I. The Saturation and Transition Zone in the Moisture Content Profiles ". Soil Science, 131 (4) , 202 - 209 .

4- Booth . A.R. (1975) . " The Factors Influencing Collapse Settlement in Compacted Soils" . Six Regional Conference for Africa on Soil Mechanics and Foundations Engineering , Durban . South Africa, pp.5763 - 5769 .

5- El-Ehwany, M. and Houston , S. (1990) . " Settlement and Moisture Movement in Collapsing Soils". Journal of the Soil Mechanics and Foundation Division, ASCE, Vol. 116, No. 10, October 1990 , pp. 1521-1535.

6- El- Saadany, M.M. (1986) ." Some Factors Affecting Engineering Behaviour of Light Weight Dry Soil". M.Sc. thesis, Faculty of Engineering, Al-Azhar University.

7- Jennings, J.E., and Knight , K. (1975) . " A Guide to Construction on or with Materials Exihibiting Additional Settlement Due to Collapse of Grain Structure ". Sixth regional Conference of Africa on Soil Mechanics Foundation Engineering, Durban, South Africa , September .

8- Stauffer, F. (1981) " Infiltration into Layered Soils : Experiments and Numerical Simulation " Proc. of Euromech., A.A. Balkema , Delft, The Netherlands .

Geotechnics for Developing Africa, Wardle, Blight & Fourie (eds) © 1999 Balkema, Rotterdam, ISBN 90 5809 082 5

Model footing tests on unsaturated soils

S.Y.Oloo
CSIR-Transportek, Durban, South Africa

ABSTRACT: Matric suction increases the shear strength of soils by increasing the apparent cohesion. Foundations placed on unsaturated soils would therefore be expected to have higher bearing capacities than those placed on the same soils in the saturated state. In order to quantify the increase in bearing capacity arising from matric suction, laboratory tests involving the determination of the failure loads of model footings placed on homogeneous soils subjected to different levels of matric suction were carried out. The results indicated a significant increase in bearing capacity with increasing matric suction and compared reasonably well with theoretical predictions based on independently measured values of ϕ^b and a modified Mohr-Coulomb failure criterion.

1 BACKGROUND

1.1 *Bearing capacity of saturated soils*

Terzaghi (1943) determined the bearing capacity of shallow footings resting on the surface of dry soil by considering the equilibrium, under an assumed general shear failure mechanism, of soil under the footing. The bearing capacity was given by:

$$q_f = cN_c + 0.5B\gamma N_\gamma \tag{1}$$

where q_f = the bearing capacity; c = cohesion of the soil; γ = unit weight of the soil; B = width of the footing; and N_c, N_γ are bearing capacity factors dependent on the soil friction angle, ϕ.

If the soil is saturated, with a static water table above the foundation level or at a depth less than the width of the footing, semi-empirical solutions based on the principle of effective stress exist for the determination of bearing capacity (Terzaghi 1925, Meyerhof 1955).

1.2 *Bearing capacity of unsaturated soils*

Most foundations are placed above the water table where soils have negative pore water pressures. Soils with negative pore water pressures are unsaturated and their state of stress cannot be expressed in terms of effective stresses. The state of stress in unsaturated soils is usually expressed in terms of the net normal stress, $(\sigma - u_a)$, and the matric suction $(u_a - u_w)$.

The bearing capacity of unsaturated soils can also be determined by considering the equilibrium of the soil under the footing (Oloo 1994, Oloo et al. 1997). In this case the stresses acting on the soil are expressed in terms of $(\sigma - u_a)$ and $(u_a - u_w)$. Since matric suction does not have a gravitational component, the influence of matric suction on the stability of the soil will arise only from its contribution to the shear strength.

Fredlund et al. (1978) proposed the following equation for the shear strength of an unsaturated soil:

$$\tau = c' + (\sigma - u_a)\tan\phi' + (u_a - u_w)\tan\phi^b \tag{2}$$

where ϕ^b is the friction angle associated with the matric suction of the soil.

The effect of matric suction is to pull particles of the soil together. The contribution of matric suction to shear strength is due to the forces between the soil grains arising from this pull. Matric suction therefore contributes to the cohesion of the soil. The components of the equation for the shear strength of the unsaturated soil can then be rearranged as follows:

$$\tau = [c' + (u_a - u_w)\tan\phi^b_t] + (\sigma - u_a)\tan\phi' \tag{3}$$

The value in parenthesis is the total cohesion of the soil.

The relationship between ϕ^b and matric suction is nonlinear (Gan et al. 1988). If matric suction is uniform within the soil layer, the bearing capacity of unsaturated soils can be obtained by substituting the total cohesion for cohesion in Equation 1 as follows:

$$q_f = [c' + (u_a - u_w)\tan\phi^b]N_c + 0.5B\gamma N_\gamma \qquad (4)$$

2 MODEL FOOTING TESTS

2.1 Testing program

A laboratory testing program was set up to provide experimental verification of Equation 4. It involved the determination of the bearing capacity of model footings placed on the surface of soils subjected to different levels of matric suction. The results of the testing program are summarised below.

2.2 Properties of test soils

The quantification of the influence of matric suction on the bearing capacity of model footings requires soils that desaturate over a wide range of matric suction. Fine grained soils have a wide desaturation range over which significant, and therefore, measurable changes in shear strength occur. Such soils would be ideal for model footing tests. On the other hand, model footing tests require relatively large soil samples that have to be brought to matric suction equilibrium within reasonable periods of time. Due to their low coefficients of permeability, fine grained soils would require relatively long periods of time to equilibrate. The selection of suitable soils for model footing tests must compromise between the two apparently conflicting criteria.

A Botkin Pit silt and an Indian Head till were used in the testing program. Soil-water characteristics determined using the axis translation technique for the two soils are shown in Figure 1. The Botkin Pit silt has a smaller desaturation range than the till.

It was important that model footing tests be carried out on soil samples in which ranges of constant matric suction were maintained. Constant matric suction was generated in the silt samples using both the axis translation technique and compaction at different water contents. Due to the slow desaturation rate in the till samples, constant matric suction was generated using only compaction over a range of water contents. The relationship between compaction water content and matric suction was determined for both soils to facilitate the assignment of matric suction values to each compaction water content. The compaction water content versus matric suction relationships for the two soils are shown in Figure 2.

The shear strength parameters, c', ϕ', and ϕ^b are required for the prediction of bearing capacity using Equation 4. The saturated shear strength parameters, c' and ϕ', were determined on compacted specimens

using consolidated-drained triaxial tests. Average values of effective cohesion and angle of friction determined were 2.5 kPa and 28.1° for the silt and 16.9 kPa and 21.9° for the till, respectively.

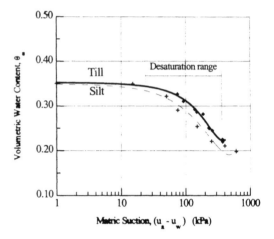

Figure 1. Soil-water characteristics.

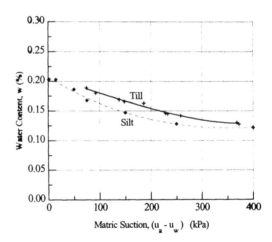

Figure 2. Compaction water content versus matric suction curves.

The unsaturated shear strength parameter, ϕ^b, was determined using two different methods depending on whether constant matric suction was generated using compaction over a range of water contents or using the axis translation technique. For samples compacted over a range of water contents, ϕ^b was determined in consolidated-undrained triaxial tests. Specimens compacted at a given water content were

196

placed in a shear box which was subsequently sealed with silicon grease to reduce water loss through evaporation. The specimens were sheared at a fast rate to ensure minimal changes in water content during the shearing process. Matric suction values were assigned to each compaction water content level using the compaction water content versus matric suction curves and the results analysed using the method proposed by Oloo et al. (1996). Typical results are presented in Figure 3 and 4.

The unsaturated shear strength parameter, ϕ^b, for samples in which constant matric suction was generated using the axis translation technique was determined in the modified direct shear apparatus (Gan et al. 1988). The results of the modified direct shear tests are presented in Figure 5 and 6.

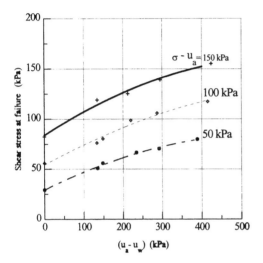

Figure 3. Variation of shear stress at failure with matric suction for compacted silt samples.

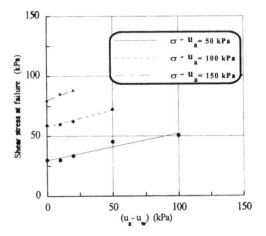

Figure 5. Variation of shear stress at failure with matric suction for silt samples as determined in the modified direct shear test.

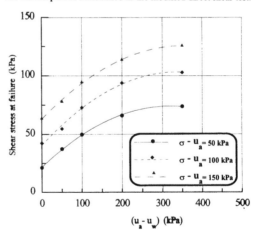

Figure 6. Variation of shear stress at failure with matric suction for till samples as determined in the modified direct shear test.

2.3 Sample Preparation

Special sample preparation techniques were used to ensure uniform compaction and distribution of matric suction in the specimens for model footing tests. Samples of silt and till in which constant matric suction was to be generated by compaction and different water contents were air dried and thoroughly mixed with the required amount of distilled water to achieve a pre-determined water content. The speci-

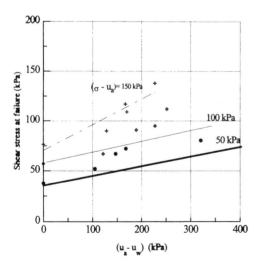

Figure 4. Variation of shear stress at failure with matric suction for compacted till samples.

mens were sealed and allowed to cure for a week in a temperature and moisture controlled room. After curing, the silt specimens were statically compacted in two layers to a dry unit weight of 17.5 kN/m³ into steel moulds, 150 mm diameter by 150 mm high. The till samples were compacted in 3 layers in moulds, 300 mm diameter by 150 mm high, to the same dry unit weight. The moulds were then sealed against evaporation and the specimens allowed to cure in a temperature and humidity controlled room for 2 weeks. At the end of this period it was assumed that uniform matric suction conditions prevailed in the compacted soils in the moulds.

Specimens of silt in which matric suction was to be generated by equilibration under a constant air pressure using the axis-translation technique were also statically compacted into moulds. A compaction water content of 18 percent was used. Compaction was carried out in two equal layers to attain a dry unit weight of 17.5 kN/m³. The specimens were then kept flooded with water in a temperature and humidity controlled room until such a time that saturation occurred, usually within 10 days.

Upon saturation, the top caps of the moulds were fitted and the required level of air pressure applied. The specimens were allowed to come to equilibrium at matric suction levels of 25, 50, 75 and 100 kPa. During the equilibration process, water flowing out from the mould was continuously monitored. Equilibration was assumed to be complete when the flow stopped. The air pressure was sustained until the footing loading tests commenced.

2.4 Model Footing Tests

The specimens were transferred from the temperature and humidity controlled environment to the triaxial loading machine. Square (30 mm width) and circular (30 mm diameter) brass footings were placed on the specimens and loaded at a strain rate of 0.18 mm/min. The shortest possible time was taken between the transfer of the specimens from the temperature and humidity controlled room and the start of the loading. This, together with the relatively fast shearing rate of 0.18 mm/min., ensured that the matric suction attained by the specimens after equilibration remained the same during the model footing tests.

The model footings were loaded at a constant rate in a triaxial loading frame. The vertical displacements of the model footings were measured using linearly variable displacement transducers (LVDT) while the loads were measured using a load cell. Both measurements were recorded using a data acquisition system. The tests were terminated when the settlements reached 20 percent of the footing width.

3 TEST RESULTS

3.1 Definition of failure load

Typical applied pressure versus settlement ratio curves for model footings on till specimens compacted to different water contents are shown in Figure 7. The general shape of these curves was typical of all model footing tests. The most significant feature of the curves was a lack of a clearly defined failure load since failure occurred by punching rather than general shear. At low water contents and high matric suction, failure was accompanied by slight tilting of the footings and cracking at the soil surface.

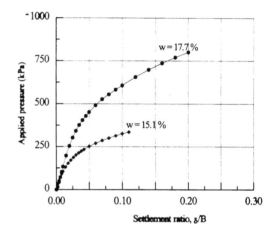

Figure 7. Typical load-settlement curves during model footing tests.

In the absence of clearly defined failure loads a number of proposals have been put forward for the ultimate load criterion (De Beer 1970, Vesic 1963, 1973). The definition of the ultimate load based on the minimum slope of the load versus settlement curve (Vesic 1963) was adopted for the analysis of the model footing tests.

3.2 Bearing capacity results

The variations of bearing capacity with matric suction for specimens of silt and till are shown in Figures 8 and 9, respectively. There is a general increase in bearing capacity with increasing matric suction for both soil types. At high levels of matric suction, the relationship between bearing capacity and matric suction tends to be nonlinear. It is also apparent from the figures that the differences in bearing capacity of circular and square footings are not significant.

4 THEORETICAL ANALYSIS

The determination of bearing capacity using Equation 4 is based on the assumption of a strip footing failing in a general shear failure mechanism. Square and circular footings were used in the footing tests and the predominant failure mechanism was punching as opposed to general shear failure. The use of Equation 4 for the analysis of the model footings requires the adjustment of the equation to give due consideration to the differences in failure mode and footing shape.

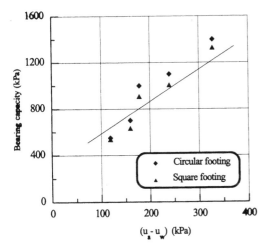

Figure 8. Bearing capacity versus matric suction for silt samples.

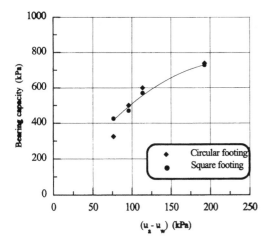

Figure 9. Bearing capacity versus matric suction for till samples.

There are no rational methods for analysing foundations that fail in modes other than general shear. Vesic (1973) proposed correction factors to account for soil compressibility and scale effects that give rise to other failure mechanisms. The proposed equations give no more than a qualitative estimation of the reduction in bearing capacity caused by compressibility effects. Terzaghi (1943) suggested the reduction of the shear strength parameters by a constant factor of one third to account for punching and local shear failure. This simple proposal by Terzaghi was used in the analysis of the model footing tests.

The derivation of bearing capacity equations for footing shapes other than strips is fraught with mathematical difficulty. Some solutions have been proposed (Cox 1962, Cox et al. 1961) for axisymmetric loading on purely cohesive soils. No rational mathematical solutions exist for other shapes or for soil with both friction and cohesion. It is common practice to use empirical correction factors to account for shape in such soils. A shape factor with respect to cohesion of $(1 + N_q/N_c)$ and with respect to self-weight of 0.6, as proposed by De Beer (1970), were used in the analysis.

A constant value of ϕ^b of 7.8° and 13.2° for silt and till, respectively, were used to simplify the analysis for the determination of bearing capacity. The results are presented in Figures 10 and 11 as comparisons between experimental and theoretical variations of bearing capacity with matric suction.

Experimental and theoretical variations of bearing capacity with matric suction exhibit similar trends. The predicted variation of bearing capacity with matric suction is linear because a constant value of ϕ^b was used in the analysis for each soil. The variation of the experimental bearing capacities with matric suction tends to be nonlinear at high values of matric suction.

5 COMPARISON WITH MEASUREMENTS

The experimental bearing capacities are lower than predictions for both silt and till samples (Figs 10, 11). In both cases, however, the theoretical and experimental results are reasonably close.

The lack of coincidence between the predicted and experimental results can be attributed, in part, to the lack of accurate methods to account for differences in foundation shape and mode of failure. The correction factors used were proposed for saturated soils. In applying them to the present analysis, it was assumed that they held for unsaturated soils as well as they do for saturated soils. The empiricism inherent in the determination of these correction factors for saturated soils, and their arbitrary application to unsaturated soils, precludes the possibility of a match between predicted and experimental values.

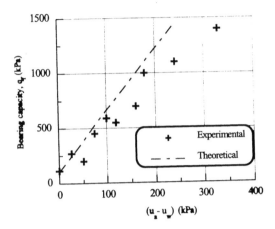

Figure 10. Theoretical versus measured bearing capacities for silt samples.

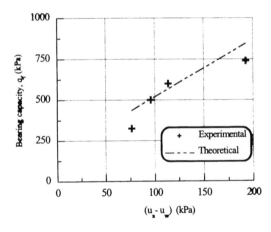

Figure 11. Theoretical versus measured bearing capacities for samples of till.

6 CONCLUSIONS

1. Matric suction increases the bearing capacity of unsaturated soils.
2. The relationship between bearing capacity and matric suction is nonlinear at high levels of matric suction.
3. The bearing capacity of unsaturated soils can be analysed in terms of the stress state variables, (σ-

u_a) and (u_a-u_w) by assuming that matric suction increases the cohesion of the soil by an amount equal to (u_a-u_w)$\tan\phi^b$.
4. Predictions of bearing capacity based on the above assumption compare reasonably well with the results of the model footing tests.

REFERENCES

Cox, A.D. 1962. Axially-symmetric plastic deformation in soils-II. Indentation of ponderable soils. *International Journal of Mechanical Sciences* 4: 371-380.

Cox, A.D., Eason, G. & Hopkins, H.G. 1961. Axially symmetric plastic deformations in soils. *Philosophical Transactions of the Royal Society of London,* Series A *(25):* 1-45.

De Beer, E.E. 1970. Proefondervindelijke bijdrage tot de studie van het gransdraagvernogen van zand onder funderringen op staal. English Abbreviation. *Geotechnique* 20(4): 387-411.

Fredlund, D.G., Morgenstern, N.R. & Widger, R.A. 1978. The shear strength of unsaturated soils. *Canadian Geotechnical Journal* 15(3): 316-321.

Gan, J.K-M., Fredlund, D.G. & Rahardjo, H. 1988. Determination of shear strength parameters of an unsaturated soil using the direct shear test. *Canadian Geotechnical Journal* 25: 500-510.

Meyerhof, G.G. 1955. The influence of roughness of base and ground-water conditions on the ultimate bearing capacity of foundations. *Geotechnique* 5(3): 227-242.

Oloo, S.Y. 1994. A bearing capacity approach to the design of low-volume traffic roads. *Ph.D. Thesis*, University of Saskatchewan, Saskatoon.

Oloo, S.Y. & Fredlund, D.G. 1996. A method for determination of ϕ^b for statically compacted soils. *Canadian Geotechnical Journal* 33:272 – 280.

Oloo, S.Y., Fredlund, D.G. & Gan, J.K-M. 1997. Bearing capacity of unpaved roads. *Canadian Geotechnical Journal* 34: 398-407.

Terzaghi, K. 1925. *Erdbaumechanik auf bodenphysikalischer grundlage (soil mechanics on soil physics basis).* F. Deuticke, Vienna.

Terzaghi, K. 1943. *Theoretical Soil Mechanics.* Wiley, New York.

Vesic, A.S. 1963. Bearing capacity of deep foundations in sand. *Highway Research Record* 39: 112-153.

Vesic, A.S. 1973. Analysis of ultimate loads of shallow foundations. *ASCE Journal of Soil Mechanics and Foundation Engineering Division* 99(SM1): 45-73

Geotechnics for Developing Africa, Wardle, Blight & Fourie (eds) © 1999 Balkema, Rotterdam, ISBN 90 5809 082 5

Friction characteristics of sand/geotextile interfaces for three selected sand materials

D. Kalumba
Makerere University, Kampala, Uganda

F. Scheele
University of Cape Town, South Africa

ABSTRACT: Extensive direct shear, direct interface and pull-out tests of a non-woven, needle punched geotextile in Cape Flats, Klipheuwel and Munich sand were undertaken to study the interface behaviour as a function of confining pressure. The respective responses are summarized in terms of shear stress/horizontal displacement and pull-out resistance/front displacement relationships showing the frictional performance of the geotextile in these sands of various grain characteristics. Specific emphasis and detailed analyses went into pull-out experiments in which local displacements of the specimens were measured, allowing the derivation of force/displacements and thus skin friction distributions along the geotextile.

1 INTRODUCTION

Design of geotextile reinforced soil structures requires the knowledge of both the mechanical behaviour of the composite material (comprising of soil and geotextile reinforcement) and the behaviour at the soil/geotextile interface. The soil/geotextile interface behaviour depends on the geofabric's material composition, thickness and surface texture, as well as the soil's particle size and shape, structure, fabric, grading and condition (Makiuchi & Miyamori, 1988; BS8006:1994). The characteristics of sands with emphasis on grading and grain size, interacting with a particular geotextile were investigated in this study.

The interface friction parameters were established by a series of interface tests carried out using two types of test methods, the direct shear test and a specifically developed pull-out test.

Conclusions are drawn regarding the interface friction properties for the type of sand in which the geotextile is embedded and some consequences for the design of reinforced soils pointed out.

2 MATERIAL DESCRIPTION

The geotextile used in the test series was a 100% polyester, continuous-filament, needle punched, non-woven geofabric manufactured locally and therefore readily available. Some physical properties, provided by the manufacturer (Kaytech Geotechnical & Industrial Fabrics, 1995) are listed in Table 1.

Table 1. Properties of geotextile.

Property	Unit	Grade U34
Mass per unit surface	g/m^2	270
Thickness under normal stress of 0.5kPa	mm	2.30
Thickness under normal stress of 200kPa	mm	1.05
One-directional tension (50 x 200 mm)	kN/m	14
One-directional tension (100 x 200 mm)	kN/m	18

Table 2. Mechanical properties of sands.

Property	Cape Flats	Klip-heuwel	Munich
Specific gravity, G_s	2.67	2.70	2.76
Optimum moisture content (%)	7.0	10.8	8.7
Maximum dry density (kg/m^3)	1660	1915	2040
Coefficient of uniformity, C_u	2.26	8.24	7.80
Coefficient of grading, C_g	0.62	1.27	1.12
Internal angle of friction (°)	41	54	48.5
Residual angle of friction (°)	30	35	35

Three different types of sand were selected: Cape Flats, Klipheuwel and Munich sand. Cape Flats sand is a uniform, medium quartz sand, with the larger grains subrounded, while the medium and small sized particles are subangular. Klipheuwel is a well graded coarse-grained quartz sand, with a high percentage of bulky, angular particles. Munich sand is a fairly well graded, generally with rounded and subrounded as well as angular particles of relatively big diameters.

The mechanical properties of these sands were determined by conducting standard tests in the Geo-

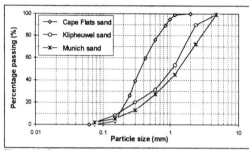

Figure 1. Grain size distribution of sands.

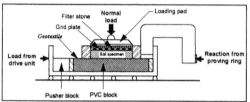

Figure 2. Cross-section of the modified direct shear apparatus.

technical Laboratory at the University of Cape Town. The summary of the results is given in Table 2. The respective grain size distributions are shown in Figure 1.

3 APPARATUS AND TEST PROCEDURE

3.1 Direct shear testing

The direct shear tests on sand were conducted using a 100mm square *Wykeham Farrance SB1* constant rate of strain shear apparatus, according the procedures laid down in BS 1377: Part 7: 1990.

The sand/geotextile investigation utilised the same shear apparatus, however modified to allow the inclusion of the geotextile specimen on which the top rigid metal frame (100mm square), containing the re-

spective sand materials, was sheared as shown in Figure 2.

Initial interface test conditions were achieved by compacting the sand at the optimum moisture content to approximately its maximum dry density in the top frame *'sitting'* on the geotextile which was tightly stretched onto a sandpaper which, in turn, had been glued to the surface of a PVC block.

In total, 19 direct shear tests were conducted at three normal pressures, namely 25kPa, 50kPa, and 100kPa at a displacement rate of 1.2mm/min. Nine of those involved sand only, while another nine were sand/geotextile interface tests. One *'zero'* shear test, in which the *'empty'* top frame of the shearbox was sheared on the geotextile sheet, was undertaken to be able to eliminate the shear stress contribution of the steel frame in the evaluation of the shear strength parameters.

The shear responses are presented in terms of shear stress versus horizontal displacement in Figure 3.

3.2 Pull-out testing

The tests were performed in a specially manufactured test apparatus where a 150mm by 200mm geotextile specimen, sandwiched in a sand bed, was pulled out in a highly controlled manner. For each particular sand material under investigation, tests were conducted at the same bulk density and moisture content as previously undertaken in the direct and direct interface shear tests.

The test apparatus consisted of an aluminium frame with inner dimensions of 288mm x 300mm x 50mm (length x width x depth) with fitted pressure pads on either side of the frame. The pressure pads consisted of parallel, flexible tubes which could be pressurised for the purpose of applying a uniformly distributed pressure to the sand, thus, exerting a uniform normal pressure to the geotextile. Figure 4 shows the cross-section of the apparatus.

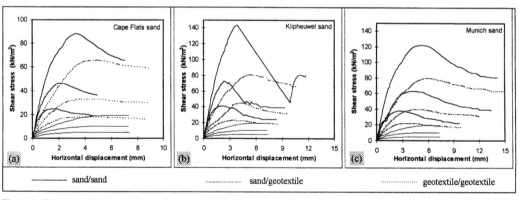

Figure 3. Responses of the various materials in direct shear at normal pressures of 25kPa, 50kPa and 100kPa.

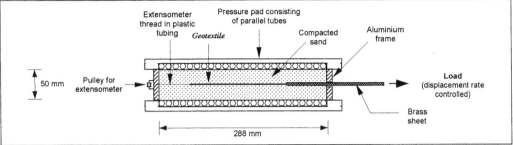

Figure 4. Schematic of the pull-out apparatus.

The test device was mounted into a computer controlled universal material testing machine and the pull-out load applied to the geotextile specimen embedded in the different sands, in a displacement controlled manner. The responses in terms of pull-out resistance versus front displacement of the geotextile in the various sands confined at the indicated pressures are shown in Figure 5.

Special emphasis was placed on the local displacement measurements of the geotextile specimen during load application. Using three extensometers, attached to the geotextile at 25mm, 75mm and 150 mm from the front (pull-out) end, the displacements of the geofabric were monitored as the load was applied which allowed the calculation of stretch (elongation) of the specimens. The stretching behaviour of the geotextiles in different sands is also shown in Figure 5 as dashed lines.

For each interface configuration, experiments were conducted at applied normal pressures between 0 and 140kPa, all at a displacement rate of 10 mm/min.

4 RESULTS

4.1 *Direct shear testing*

4.1.1 *Shear stress-horizontal displacement (Fig. 3)*

Sand/sand:

All three types of sand showed a classical relationship: the higher the normal pressure, the more intense the interlock of particles and the higher the frictional resistance for sliding to occur.

At the three normal pressures, Klipheuwel sand developed the highest shear stresses, followed by Munich sand, while the responses of Cape Flats sand were significantly lower.

Cape Flats and Munich sands exhibited smooth shear - displacement relationships while Klipheuwel sand developed a pronounced softening behaviour for normal pressures above 50kPa. In the residual stages '*saw-toothed*' responses were observed arising from a repeated process of build-up of resistance, local yielding at the points of contact and collapse of the structure consisting of coarse and angular shaped sand particles.

The plots demonstrate that, in general, the displacement required to fully mobilise the shear stresses increases with an increase in the confining pressure. At any particular normal pressure, this displacement was highest in Munich sand. The shear displacements to peak were about the same in Klipheuwel sand and Cape Flats sand.

Sand/geotextile:

The shear-displacement relationship of the interface tests are less '*dramatic*', the initial steepness and the peak shear values at all confinements are significantly lower than the respective responses in the direct shear tests. A smooth drop off, if at all, to the residual stress is reached at relative large displacements which appears to be a function of the confinement and the grain size distribution.

It is of interest to note that the residual strength values are about the same for the sand/sand and the sand/geotextile interface i.e. the sand material in the shear and interface shear zone is at this stage of shearing at its critical volume.

The highest shear strengths were observed in the Klipheuwel/geotextile tests and were slightly higher than those in the Munich/geotextile tests. Both well graded sand materials interact intimately with the fabric, the angular broken grains interlock with the fibres resulting in an increased shear resistance.

4.1.2 *Shear stress - normal stress relationship*

The limiting conditions in terms of strength envelopes were assessed for both direct shear and direct interface shear tests according to standard procedures. The resulting strength parameters are summarised in Table 3.

Table 3. Summary of shear strength parameters determined in direct shear and pull-out tests.

Sand	Sand/sand direct shear				Sand/geotextile direct shear		Pull-out test	
	Peak parameters		Residual parameters		Peak parameters		Peak parameters	
	Cohesion, c_s' (kN/m²)	Friction, ϕ' (°)	Cohesion, c_{sr}' (kN/m²)	Friction, ϕ_r' (°)	Adhesion, c_g (kN/m²)	Friction, δ_d (°)	Adhesion, c_p (kN/m²)	Friction, δ_p (°)
Cape Flats	2.8	41.0	2.5	30.0	1.6	32.5	9.0	27.5
Klipheuwel	5.3	54.0	5.4	35.0	3.6	37.5	41.0	37.0
Munich	8.2	48.5	3.5	35.0	2.5	37.5	13.6	29.0

Sand/sand:

As expected, the Klipheuwel sand, due to its grain characteristics and the high coefficient of uniformity, showed the highest angle of internal friction, ϕ', followed by Munich and Cape Flats sands. It is apparent that a more balanced distribution of particle sizes results in a higher degree of interlocking. Thus, Cape Flats sand being a uniformly graded material produced the lowest shearing resistance.

Klipheuwel sand and Munich sand exhibited the same residual angle of internal friction, ϕ_r', of 35°. The minor differences in the residual values within the different categories of sands, suggest that the critical void ratio of the sand materials in the shear zone is of the same order to which the void ratio decreases as the shear strain increases.

Sand/geotextile:

Addressing the sand/geotextile interface tests, Klipheuwel and Munich sands showed the same peak interface friction value, δ_d, of 37.5° which was somewhat higher than the interface friction angle experienced in the Cape Flats sand/geotextile tests.

For any sand/geotextile test arrangement, the interface friction angle, δ_d, was significantly lower than the respective angles of internal friction, ϕ', with friction angle ratios, $\tan\delta_d / \tan\phi'$, of 0.6 to 0.75. This can be explained by the fact that, although in the direct shear test failure is forced to take place along a predetermined shear plane, the process of shear and yielding involves a relatively big shear volume within the sand test specimen and, therefore, a higher shear strength mobilisation is developed. The shear activities in the interface test, however, are in fact confined to a very thin shear zone along the surface of the geotextile specimen.

A comparison between the sand/geotextile interface friction angle, δ_d, of each test material and the corresponding residual angle of internal friction, ϕ_r', showed an average deviation of only 2.5°. Thus, based on these experiments, it can be concluded that the sand/geotextile interface tests become obsolete for the purposes of design once direct shear tests are carried out to very large strains.

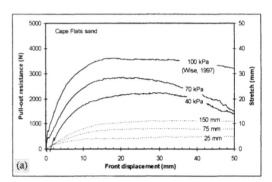

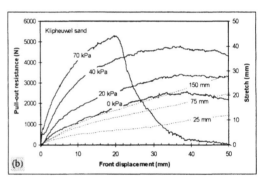

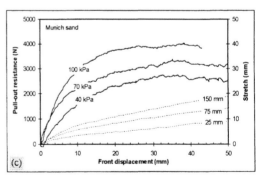

Figure 5. Pull-out resistance and stretch of geotextile (at σ = 70kPa in Cape Flats and Munich sand and σ =40kPa in Klipheuwel sand) versus front displacement.

4.2 Pull-out testing

4.2.1 Pull-out resistance-front displacement(Fig. 5)

All tests showed that at any particular confinement, there was an increase in the pull-out resistance with increasing displacement of the geotextile specimen till a maximum point was reached beyond which, the pull-out resistance either decreased to, or remained constant at a residual value.

The responses after the peak point are characteristic for the type of failure of the geotextile in sand. In a slippage failure, the pull-out resistance stayed either constant or gently decreased with increasing displacements indicating the gradual expulsion of the geotextile specimen. In Klipheuwel sand, at a confinement of 70kPa, a rupture failure was experienced. After reaching a peak resistance, a rapid drop off was observed, the geotextile tore apart, and only the front-section was pulled out of the test apparatus.

None of the other interface pull-out tests experienced rupture failure at that confinement. This implies that the '*critical*' confinement above which rupture failure is experienced varies from sand to sand depending on the respective soil mechanical properties, above all, grain size, grain size distribution and particle shape.

It was noted that for any particular normal pressure, the pull-out resistance of the geotextile specimen was highest in the two mentioned sands of significant coarse components; Klipheuwel and Munich sand. Well graded, coarse sands show higher resistance than the uniformly graded sands.

4.2.2 Skin friction distribution

Local displacements of the geotextile specimens were measured as the pull-out tests progressed. The measurements enabled the study of the stretching characteristics of the geofabric in the different sands. Applying an extrapolation procedure to approximate the constantly changing deformation characteristics of the geotextile as it stretched in the respective sands, allowed the back-prediction of the pull-out force/displacement relationship, and thus enabled the study of skin friction distribution along the geotextile specimen during pull-out (Kalumba, 1998). Figure 6 shows the distribution of interface skin friction along the 150mm long geotextile specimen confined at 70kPa in the different sand materials (confined at 40kPa in Klipheuwel sand) at peak pull-out resistance (failure).

The experimental observations together with the back analysis of these pull-out tests show that at failure, the maximum skin friction along the geotextile specimen occurs at the pull-out end in all the differ-

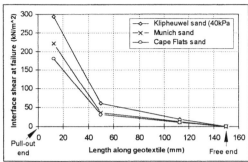

Figure 6. Sand/geotextile interface shear distribution along the test specimen length.

ent sand materials. The skin friction is by no means constantly distributed along the specimen as suggested in the design code of reinforced soils (BS8006:1994). The maximum was always close to the pull-out end, and steeply decreased to zero towards the free end.

The highest skin friction values were observed in Klipheuwel sand (although at a confinement of 40 kPa in the order of 300kPa), followed by Munich sand with about 220kPa. The lowest skin frictions were observed in Cape Flats sand. It can be concluded that the soils with the highest angles of internal friction generate the highest skin friction distributions along the test specimen.

4.2.3 Shear stress - normal stress relationship

The maximum pull-out resistances at a particular confinement were read off the observed relationships in Figure 4, and the maximum average (constant) interface shear stresses were calculated. The computed stress values were then plotted against the corresponding applied normal pressures in the standard Mohr-Coulomb representation. The graph (not included here) showed again that the interface shear stress at failure for the sand/geotextile test configurations increased linearly with increasing applied normal pressure. The interface shear strength parameters of the pull-out tests were quantitatively summarised in Table 3.

The interface shear strength in Klipheuwel sand, with a bulk density of 2094 kg/m^3, was highest with a δ_p of 37.0° followed by Munich sand with 29.0° at a bulk density of 2185 kg/m^3. Cape Flats sand had the lowest value of interface friction angle of 27.5°. At any particular confinement, Klipheuwel had the highest interface shear stress resistance to pull-out, followed by Munich sand. The high shear strength in Klipheuwel and Munich sands was attributed to their grading, grain size and shapes. The grading and angularity of these sands enabled the grains to penetrate and interlock with the geofabric with high intensity increasing the resistance to pull-out during testing.

The sand/geotextile interface friction angles, δ_p, obtained in pull-out tests are significantly lower than the internal friction angles, ϕ', of the respective sand materials. In general, it is safe to say that based on these results, the interface friction angle ratio, defined as $tan\delta_p$ upon $tan\phi'$, is in the order of 0.5.

5 CONCLUSIONS

1. The direct shear interface friction angles, δ_d, of the investigated sand and geotextile materials are generally lower than the angles of internal friction, ϕ', of the respective sands. The friction ratios were in the order of 0.60 and 0.75 depending on the soil material. The results of the studies show that interface shear angles obtained in the direct shear test investigations were about the same as the residual angles of friction of the respective sands.

2. Interface friction angles, δ_p, obtained from pull-out tests are lower than the ones obtained in the direct shear interface tests, δ_d. The interface friction angle ratios were in the order of 0.5 to 0.6.

3. It was observed, in both the direct shear and the pull-out laboratory tests that the interface friction angles are a function of sand grading, grain sizes and particle shape. The coarser and better graded a particular soil is, the higher the respective angles of internal friction, interface friction in direct shear as well as in pull-out.

REFERENCES

BS 1377 : Part 7: 1990. *British Standard Methods of Test for Soils for Civil Engineering Purposes, Part 7. Shear strength tests (total stress).* British Standards Institution, 1990.
BS 8006: 1994. *Code of Practice for Strengthened/Reinforced Soils and Other Fills.* Document No. 94/104108, British Standards Institution, June 1994.
Kalumba, D. 1999. *Effect of grading and grain size on the friction characteristics of a sand/geotextile interface.* MSc (Eng.) thesis, University of Cape Town.
Kaytech Geotechnical & Industrial Fabrics, 1995. *Kaymat Geotextiles; Manufactured in South Africa.* General Civil Engineering Applications, No. 5.4/93, Proprint.
Makiuchi, E. & T. Miyamori 1988. *Mobilization of Soil-Geofabric Interface Friction.* Proceedings International Geotechnical Symposium on Theory and Practice of Earth Reinforcement, Fukuoka, Japan, 129-134.

Reinforced Earth[(R)] structures for headwalls to crushing plants in Africa

A.C.S. Smith
Reinforced Earth (Pty) Limited, Johannesburg, South Africa

ABSTRACT: The use of Mechanically Stabilized Embankments (MSE) for the construction of Head Walls and/or wingwalls to mine crushing plants in Africa is increasing. The advantages of MSE construction is that all material, except the locally available earth backfill, can be easily and efficiently shipped to remote locations in containers or break bulk on trucks and in aeroplanes. The structures are not sensitive to poor foundation conditions and consequently in most cases no special foundation preparation is required. No difficulty has been encountered in finding backfill conforming to standard MSE specifications. This paper describes the use of MSE for this application. It briefly describes various projects in several countries using precast concrete, steel and weldmesh cladding types.

1 INTRODUCTION

MSE structures were introduced into Southern Africa in 1975. The material rapidly found application for headwalls to crushing plants and numerous structures were constructed in South Africa, Namibia, Botswana, Lesotho and Malawi. After the end of South Africa's political isolation the use of MSE for headwalls spread to other more remote African countries such as Angola, Mali, Guinea and the Democratic Republic of Congo. A list of projects in Africa using the Reinforced Earth® system of MSE is given in Table 1.

2 APPLICATION

Crushing, screening and processing plants are gravity plants in which the ore is fed into the primary crusher at a height which could range between a few metres and 40 m from the bottom level of the plant. An earth ramp is required for the trucks to offload their ore. In order to prevent the earth ramp from enveloping the plant a headwall with associated wingwalls is required.

The headwall and the wingwalls can be constructed in MSE or alternatively, the headwall can be integrated into the crusher building and only the wingwalls are of MSE construction.

When Reinforced Earth® systems are used then a choice is made between 3 types of cladding viz.
- precast concrete cladding elements
- steel cladding elements with elliptical cross-section
- weldmesh cladding backed by rock or geo-fabric

The highest known straight up, non-tiered cladding on any structure to date is
- 38 m for steel cladding in Indonesia
- 27 m for precast concrete cladding in South Africa
- 18 m for weldmesh cladding in Australia

Higher walls are possible but, beyond a height of 15m, say, the cladding should be stepped, even if only a distance of 300 mm, in order to relieve stress in the cladding due to internal settlement of the backfill both during and after construction.

These systems are illustrated in the examples given in Figs 2, 3, & 4.

A slab, generally of reinforced concrete, with an upstand at the front is constructed on top of the MSE structure. The purpose of this slab is to
- provide an all-weather riding surface
- distribute the wheel loads of what could be very large vehicles
- provide a kicker against which trucks can reverse in order to tip.

Table 1. Summary of MSE Tipwalls – Reinforced Earth® systems only

Date	Name	Location	Description	Max Height m	Type of cladding	Client
1976	CDM - 50G	Namibia	Headwall	20	Concrete	De Beers
1976	Grootegeluk	Ellisras NP	Wingwalls	20	Concrete	Iscor
1976	CDM No 3	Namibia	Headwall	13	Concrete	De Beers
1976	Silicon Smelters	Pietersburg NP	2 Tier Headwall	16,5	Concrete	Silicon Smelters
1977	Beestekraal I	NW Province	Headwall	8,3	Concrete	PPC
1977	Koingnaas	NW Cape	8 Tiered Wingwalls		Concrete	De Beers
1978	Kimberley I	Kimberley NW	Wingwalls	6	Steel	De Beers
1978	Jwaneng I	Botswana	Headwall	6,7	Steel	De Beers
1979	Tweepad	NW Cape	Wingwalls	2Tier 41	Concrete	De Beers
1979	Machadadorp	E Transvaal	Headwall	6	Concrete	S A R
1979	Beestekraal II	BeestekraalMine	Raise Headwall	3	Concrete	PPC
1979	Jwaneng II	Botswana	Headwall	8	Steel	De Beers
1979	Kimberley II	NW Cape	Headwall	7	Steel	De Beers
1979	Marquard	Free State	Headwall	6	Steel	RG Nuttall
1980	Aggenuys	NW Cape	Headwall	13	Concrete	Goldfields
1982	Rooikraal	Gauteng	Headwall	13	Steel	D& H
1982	Silicon Smelt II	N Province	Headwall	6	Concrete	Silicon Smelters
1983	Fairview	Mpumalanga	Headwall	6	Concrete	Gencor
1985	NCD	Kwa-Zulu Natal	Wingwall	6	Steel	N C D
1986	Wastetech	Gauteng	2 Tier Headwall	8	Steel	Wastetech
1986	Tweepad II	NW Cape	Wingwalls	2 Tier 41	Concrete	De Beers
1987	Wolwekrans	Mpumalanga	Headwall	7,5	Concrete	Rand Mines
1987	Koffiefontein	NW Cape	Headwall	7	Steel	De Beers
1987	Grootegeluk II	N Province	Wingwalls	2 Tier 25	Concrete	Iscor
1989	Auchas	Namibia	Headwall	6	Concrete	De Beers
1989	Samadi Mine	NW Province	Headwall	13,5	Concrete	Samadi Mine
1989	Premier Mine	Gauteng	Headwall	12	Concrete	De Beers
1990	SavmoreColliery	Kwa-Zulu Natal	Headwall	8,3	Concrete	Genmin
1990	Weltevreden	NW Province	Headwall	10,5	Concrete	Genmin
1991	Dull Coal	N Province	Wingwalls	24	Concrete	Iscor
1991	Lone Rock	Mpumalanga	Headwall	6	Weldmesh	Lone Rock
1991	Auchas II	Namibia	Headwall	14	Concrete	De Beers
1991	Sasol 1 Tip	Free State	Headwall	5	Concrete	Sasol
1991	Kromdraai	Mpumalanga	Wingwalls	15 of 25	Concrete	AAC
1992	Luzamba	Angola	Headwall	11,4	Weldmesh	Endiama
1993	Sigma	Free State	Headwall	6,8	Concrete	Sasol
1994	Duvha	Mpumalanga	Wingwalls	13	Concrete	Eskom
1995	Sadiola Mine	Mali	Wingwalls	16	Weldmesh	AAC
1996	Catoca	Angola	Headwall	10,5	Concrete	Endiama
1997	Siguiri Mine I	Guinea	Headwall	13	Concrete	Ashanti
1998	Orapa	Botswana	Wingwalls	16; 10 Sur	Concrete	De Beers
1998	Namakwa Sands	SW Cape	Headwall	10	Concrete	Namakwa Sands
1998	Lethlekhane	Botswana	Wingwalls	2 Tier 20	Concrete	De Beers
1998	Congo	DR Congo	Headwall	12	Steel	Miba Mine
1998	Siguiri Mine II	Guinea	Headwall	14	Concrete	Ashanti
1998	Kroondal	NW Province	Headwall	13,5	Concrete	Kroondal
1998	Bulyanhulu	Tanzania	Headwall	11,7	Weldmesh	Placer Dome
1998	Samancor	Gauteng	Headwall	9	Concrete	Samancor

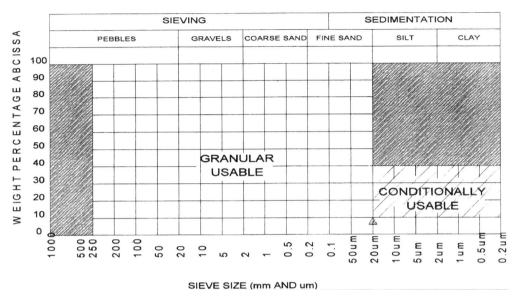

Chemical and Electro-chemical Properties		
	Out of water	In Water
Resistivity (ohm mm)	≥ 10 000	≥ 30 000
pH	5 to 10	5 to 10
Chlorides (ppm)	< 200	< 100
Sulphates (ppm)	< 1 000	< 500
Sulphides (ppm)	< 300	< 100

Figure 1.   Material suitable for use in MSE structures

3 BACKFILL

The backfill required for the MSE structure should meet the mechanical and electro-chemical requirements shown in Fig 1.   The African continent presents the full range of potential backfill materials.   Since   the service   life of structures is generally only 30 years or so, backfills with higher salt contents than would be used for more permanent structures may be considered.   Backfill may be categorized into 4 groups.

3.1 The Best
• Free draining sands or gravels ie < 5% by weight passing the 75 micron sieve.

• Mine waste rock or tailings, provided it has not become polluted in the process.

3.2 Good
• Granular material according to Fig 1.   either from transported material or weathered residual rock.

3.3 Satisfactory

Intermediate materials according to Fig 1.   These materials are water sensitive and local deformation and bulging of the face due to post construction settlement and excess pore water pressure should be anticipated.
• Lateritic gravels, pebble markers.

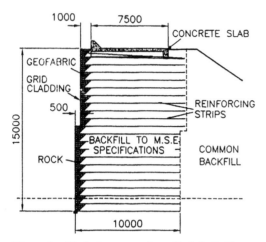

Figure 2a. Typical cross section - Sadiola Gold Mine, Mali

3.4 Unacceptable

• Material with grading falling outside that shown in Fig 1. such as silts and clays.
• Organic material

Care should be taken when using :

shales - they may deteriorate when exposed to the atmosphere

rocks - which, oxidize when in contact with air or water, producing aggressive groundwaters.

The experience to date is that backfill materials are readily found near the Site. It is advisable to use good quality backfill for structures higher than say 12 m.

4 ADVANTAGES OF MSE FOR THIS APPLICATION

• An MSE structure is a flexible one and can absorb some differential settlement. This means that an MSE structure is not particularly sensitive to the nature of the foundation and can be founded on relatively poor foundations without the need for any foundation improvement. Should foundation improvement be required, however, then undercutting and replacement with good quality compacted backfill to the underside of the MSE structure should suffice. On the other hand piling may be required for a more rigid structure such as a reinforced concrete one.
• Local backfill can generally be found to meet the specifications.
• All materials except the earth can be supplied in containers to far-away and remote locations. The cladding and strips are not bulky and are sized to fit into containers. Should 20 ft containers be used then the longer length reinforcing strips can be cut and punched at

Figure 2b. Weldmesh cladding backed with rock – Sadiola Gold Mine, Mali

Figure 3a.   Elliptical steel cladding - Miba Mine, Democratic Republic of Congo

Figure 3b.   An example of a steel cladding wall as used at Miba Mine above.

both ends and spliced together into the longer lengths on site with fish plates and bolts.

Should precast concrete cladding walls be required, then the moulds, inserts, reinforcing strips and jointing material can be supplied in containers and the panels precast on site.

- No special skills are required for the construction and local labour can readily be used.

- The construction of the approach ramp can be planned to be built at the same time as the headwall structure. This results in time savings and allows the earthworks contractor to finish off the earthworks in one period of time. The bulk earthworks for headwalls is often undertaken by the Mine itself with massive earthmoving equipment and little compaction control. The interface between the Mine bulk

Figure 4. Precast concrete cladding -
Siguiri Mine, Guinea

earthworks and the formal layers and densities
required for the MSE structure need to be
carefully managed. If sufficient attention is
given to this then considerable time and cost
savings can be realized.
- An advantage, on occasion, is that no
aggregate is on site at the time that the
headwall is constructed. It may be the
headwall for the aggregate plant. In this case
steel or weldmesh cladding is an ideal system
since only local labour and earthmoving
equipment is required.

5 CONCLUSIONS

- MSE walls have proven themselves to be able
to meet the technical requirements for
headwalls and to be cost effective. They utilize
local earth, labour and earthmoving plant to
construct headwalls quickly, easily and
efficiently.
- Careful control is however required to
coordinate the Mine and the Contractor and to
ensure that the structures are built according to
the drawings and specifications.
- Africa is a mining continent and a considerable
number of headwalls and/or wingwalls are
required. MSE walls should be considered for
these applications.

Geotechnics for Developing Africa, Wardle, Blight & Fourie (eds) © 1999 Balkema, Rotterdam, ISBN 90 5809 082 5

Foundations engineering with the method of soil thermal stabilization by microwave energy

A. Z. Zhusupbekov & S. F. Shevzov
Karaganda Metallurgical Institute, Temirtau, Kazakhstan

J. I. Spektor & O. L. Denisov
Ufa Oil University, Russia

ABSTRACT: One of the ways of foundations engineering perfection and prognosis of existing buildings and structures bases and foundations refusals is a quite new way of soil thermo-stabilization with microwave energy when heat is transfered into the soil mass by means of radiation. This way of soil stabilization differs from already known thermal ones when thermal energy is transfered into the soil around the heat-carrier by means of convection through the subsoil macropores.

1 INTRODUCTION

Superhigh frequency radiation pemeates into the soil up to a significant depth that can be varied, it is accompanied with the intensive absorption of the electromagnetic energy and is followed with the elementary minerals particles vibration against crystal lattice nodes. As a result, the volume heating of soil mass at relatively small time period takes place.

Small heat conductivity and dielectric soil properties are positive factors for soil heating in superhigh frequency field (Spektor & Denisov 1998).

2 LABORATORY TESTS

As a result of the laboratory tests on clay soil samples heat treatment with the help of the installation КИЭ - 5 of power 5 kWt and radiation frequency 2450 MHz, the dependences of compressive strength R on temperature T have been obtained (Fig. 1.). Figure 1 shows that after superhigh frequency heating, soil samples become water-resistant at T over 300^0 C. In temperature interval of 400 ...900 ^{0} C the samples strength sharply increases from 3,6 up to 11,5 MPa. At $T= 900$ ...1000 ^{0} C the formation of new minerals and their sintering occurs. The strength increases up to 18,6 MPa (1,6 times). In bend joint at $T = 1000^0$ C clay fraction melts and the number of phase contacts increases. In temperature interval of 1000 ...

1200 ^{0} C samples strength sharply decreases, this is connected with weakened ties formation, cracks and macropores occurance. Clay samples become frost-resistant at $T = 600^0$ C after 25 freeze-thaw cycles.

At frost-resistance coefficient $K_м = 0,75$ the samples are considered to be frost resistant.

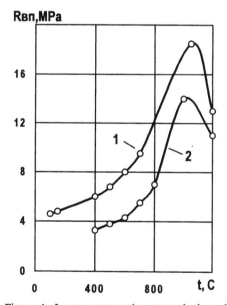

Figure 1. Loam compressive strength dependence on temperature of superhigh frequency heating
1 - in air - dried state; 2 - in water - bearing state

The tests showed that after heating at T = 800 ... 1000 ^{0} C the soil is transformed into a very strong material, near to brick or concrete of mark $M50$ ... $M100$ or class $B3,5$ ... $B7,5$. It is stated that the most suitable for heat treatment with superhigh frequency field are clay soils (loam and clay). Their compressive strength after calcination at T more than 800 ^{0} C becomes 22,6 MPa and bending sterngth becomes 2,1 MPa. Temperature field in soil mass are calculated separately for two processes: drying and calcination, as drying is accompanied with the effects of heat - and moisture transfer and calcination is accompanied only with heat exchange. Method of finite differences was used for joint solution of differential equations of heat - and moisture transfer as well as eqation of Fourier heat conduction for the process of soil mass calcination under the influence of superhigh frequency radiation and equations of electromagnetic field propagation in dielectrics (Zhusupbekov et al. 1995).

Figure 2 shows temperature distribution in soil mass treated with superhigh frequency field through the wall of the borehole for the generator of power 50 kWt and different radiation frequency "f". As one can see in Fig. 2, the effective depth of heat treatment at T = 360 ^{0} C is 0,2 - 0,8 m depending on electromagnetic field radiation frequency. So, at f =433 MHz we may obtain piles from thermo-stabilized soil with diameter up to 1...1,8 m. The pecularities of such piles construction technology are two step heat influence on leader borehole walls with the help of radiator of superhigh frequency field at given time interval, radiator extraction from the borehole and filling it with the inert material. At the first step the soil is heated at T = 150 ^{0} C with the least radiator power, at the second step the soil is heated at the greatest power and T = 1000 ^{0}C.

3 GEOTECHNOLOGY

Figure 3 shows the technological scheme of positioning piles from thermo-stabilized soil. At site a leader borehole is bored, or the borehole is made means of driving an element that is left in the borehole (step 1). Then a waveguide with the radiator is lowered into the borehole. The borehole mouth is closed with the protective casing with a plate and socket pipe for the evaporated moisture diversion with the vacuum pump. Then the soil is heat treated by heating it up to T = 1000 ^{0} C at radiator power 50 kWt (step III). At step IV the borehole is filled with sand, soil gravel or concrete and the head is formed. Figure 3 (step V) shows the ready pile from the thermal stabilized soil.

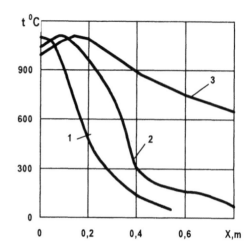

Figure 2. Dependence diagrams of temperature distribution in soil on the distance from heatig surface:
1- f = 2450 MHz; 2 - f=915 MHz; 3 - f=433 MHz

In order to engineer new foundations from such piles or to avoid the geostability loss of existing structures bases and foundations while their strengthening or reconstruction, a special mobile superhigh frequency installation must be used, that combines operations on earth excavation, soil heat treatment and fillng the cavity with the inert material.

The installation consists of energetic unit, displacement mechsnism and base vehicle. The unit is an industrial installation КИЭ -2 with waveguide cable, slot radiator and control box. The displacement mechanizm provides the energetic unit movement in different planes. Base vehicle is equipped with the metal body for radiation protection and control desk.

For preliminary evaluation of bearing capacity of piles from the thermo-stabilized soil we shall use the soil static sounding results and data of calculated load transmitted onto the pile, N. Having taken the pile embedding depth "h" and that the tip of such pile acts as a shallow foundation, and thermal-stabilized soil acts as a solid stiff body, where specific cohesion along the lateral surface doesn't depend on the vertical pressure, we have the following allowance for clay soils:

$$R = 0,1\, q_s ; \qquad f = f_s, \qquad (1)$$

where R and f are soil resistances under the pile bottom end and along the pile lateral surface, accordingly;

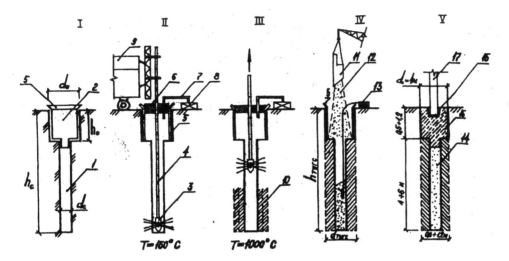

Figure 3. The technological diagram of foundations construction from piles of thermo-stabilized soil with the energy of superhigh frequency field usage:
1- borehole; 2- borehole widened part; 3- radiator; 4- waveguide; 5- casing; 6- plate; 7- socket pipe; 8- vacum-pump; 9- mobile superhigh frequency installation; 10- thermo strengthened soil mass; 11-bucket; 12-inert material; 13-vibrator with flexible; 14-fill; 15-excavation for column; 16-cap; 17-column;

q_s and f_s are soil resistances under the probe tip and along the prob lateral surface, accordingly. The from the equation

$$\frac{0,1\pi D^2}{4}\, q_s + \pi Dh f_s = 1,25\,N \qquad (2)$$

determine the required pile diameter D

$$D = \frac{6,4}{q_s}\sqrt{9,9\,h^2 f_s^2 + 0,4\,q_s\,N} - \frac{20\,hf_s}{q_s} \qquad (3)$$

According to pile diameter let's fix the technological parameters of soil heating with superhigh frequency installation, i.e. power, radiation frequency, heat treatment time and regime and other parameters of borehole walls irradiation. Let's take the control temperature at this diameter to be 150-200 ^{0} C, when soil sample superimposed load doesn't influence the shear force. At last, control static pile test is carried out.

4 CONCLUSIONS

The above method of soil stabilization with the energy of superhigh frequency field for bases strengthening and foundations engineering provides high strength, water - and frost resistance, durability of soil obtained, heat treament process controllability, ecological purity, transport expenses and main construction materials (concrete and steel) wastes decrease up to minimum, compared with already known ways of thermal treatment.

REFERENCES

Spektor, J.I. & O.L. Denisov 1998. *Mathematic simulation of the process of thermal stabilization of ground massives by the field of super higher frequency.* Russian J. Soil Mech. and Found. Eng. Moscow. 1: 8-11.

Zhusupbekov, A.Z., Babin, L.A., O.L. Denisov & J.I. Spektor 1995. *Foundation Engineering with the method of soil thermal stabilization.* Proc. X-th Asian Reg. Conf. on Soil Mech. Found. Eng., Aug. 29 - Sept. 2, Beijing, China, vol. 2: 210-2

3 Transport geotechnics

La géotechnique dans les moyens de transport

Geotechnics for Developing Africa, Wardle, Blight & Fourie (eds) © 1999 Balkema, Rotterdam, ISBN 90 5809 082 5

Long-term behaviour of soils stabilised with lime and with cement

H. Brandl
Technical University, Vienna, Austria

ABSTRACT: Soil stabilisation with lime and with cement has proved suitable for roads, highways, railways, airfields, and for several special measures (e.g. embankments, backfill of retaining structures), and it is increasingly used for contaminated land and slope stabilisation. The paper focuses on comprehensive test series with cohesive soils which have been running since the year 1963. Laboratory tests have involved the repeated determination of the following parameters: density, Atterberg limits, Proctor values, compressive and tensile strength, permeability, shear strength, and freezing-thawing behaviour. Comparative tests have been performed with lime, lime + cement, and with cement to demonstrate the different behaviour of soil stabilisations. Short and long-term reactions between soil and additives are also reported, including the results from site observations, and finally, practical recommendations are given.

1 INTRODUCTION

Soil stabilisation with lime and/or cement causes an initial and a long-term change of soil parameters. The modification occurs rapidly after addition and mixing of the additive and usually is finished within 24 hours, although it sometimes takes up to 72 hours depending on the clay minerals. The long-term stabilisation/cementation may last over many years depending on the following factors:

- Mechanical, mineralogical, and physico-chemical properties of the untreated soil;
- Quantity and properties of the additive(s);
- Elapsed time between mixing and compaction;
- Degree of compaction and water content during compaction;
- Post treatment (protection against drying or wetting);
- Elapsed time since compaction.

Strictly speaking three degrees of intensity of soil "stabilisation" should be distinguished, involving an increasing amount of additives and an increasing quality of site equipment (e.g. precise spreaders with a proportioning device, efficient mixers with direct injection of the necessary quantity of water in the mixing chamber of the rotovator/pulverizer) or mixing-in-plant:

- Treatment = modification to improve workability and compactibility;
- Stabilisation = to gain sufficient resistance against water and weather (drying–shrinking, freezing-thawing);
- Cementation = to improve essentially stress-strain behaviour and resistance against short-term impacts and long-term fatigue.

Stabilisation against chemicals is especially important in environmental geotechnics, e.g. for immobilising contaminants in the soil. Chemical swelling of lime soil mixtures can occur due to the presence of sulphates in the soil or the groundwater. In practice "stabilisation" is still generally used as umbrella term for all degrees of soil treatment

Cohesive soils, especially with a high content of multi-layered clay minerals undergo longer and more intensive changes in behaviour than non-cohesive granular mixtures with non-active fines. Therefore, three silty-clayey soils have been selected for this paper. They were used for road and highway construction in the years 1963-1965, and the structures are still under heavy traffic. About six thousand samples were produced then from materials which were intensively homogenised and stored air-tight at constant water content before being mixed with lime or cement. Some index data of three typical cohesive soils are given in Table 1. The additives were slaked lime and portland cement 275(H). Curing of the stabilised specimens has been performed at + 20°C for the standard case.

2 ALTERATION OF SOIL PROPERTIES

2.1 *Compactibility*

The samples for determining stress-strain behaviour and the resistance toward swelling-shrinking, soaking, freezing-thawing, and erosion-leaching were compacted in a special Proctor cylinder with a slenderness of h : d = 2 : 1. The results from soils 1 to 3 and from many others may be summarised as follows:

Figure 1. Cross section through soil samples (soil 1, see Table 1) stabilised with 7.5 % cement.
Left: compaction immediately after mixing
Right: compaction 24 hours after mixing.

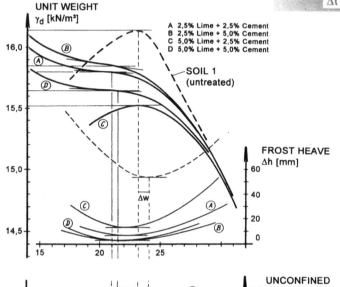

Figure 2. Proctor curve, frost heave, and compressive strength after frost tests on soil 1, stabilised with lime + cement. Curing time after compaction: 28 days. Abnormal Proctor curves occur frequently with such "combined" stabilisation.

Commonly, the optimum water content increases with lime, but decreases with cement. If the soil-lime mixture remains uncompacted over a longer period (>2 to 4 days, depending on clay content and temperature) and carbonates, the optimum water content drops because the agglomerating particles behave like "sandy" grains. The Proctor density mostly decreases when adding lime, but may increase after stabilising with cement, if the soil-cement mixture is compacted immediately. The compactibility of cement stabilisations decreases significantly with elapsed time between mixing and compaction, and this simultaneously worsens practically all soil characteristics. Fig. 1 illustrates the different structures of the stabilised soil.

Cohesive soils with a high plasticity index and water content are not suitable for a conventional stabilisation with cement. But the addition of lime plus cement can combine beneficial properties derived from both materials (Brandl, 1967). Lime prepares soil for cement and should therefore be added at least some hours before cement. In exceptional cases, a simultaneous mixing of both additives is also possible. The overall effect of this combined stabilisation is to produce more workable material by breaking down large clods of clay and

Table 1. Soil properties.

SOIL PROPERTIES	Soil 1	Soil 2	Soil 3
SOIL CLASSIFICATION	clay A-7-6	silt A-4	silt A4 (clay A-6)
MAIN MINERAL COMPONENTS	muscovite, quartz, illite, feldspar, montmorillonite	much dolomite, calcite; quartz, feldspar, muscovite	quartz, muscovite, montmorillonite, illite, feldspar
S-value (mval / 100g) (exchange with NH_4Cl)	7.4	6.7	11.2
GRAIN SIZE (%)			
sand 2 - 0.063 mm	21	15	30
silt 0.063 - 0.002 mm	54	73	48
clay < 0.002 mm	25	12	22
ATTERBERG LIMITS (%)			
liquid limit w_L	57.5	26.6	31.5
plastic limit w_P	25.2	21.4	16.0
plasticity index I_P	32.3	5.2	15.5

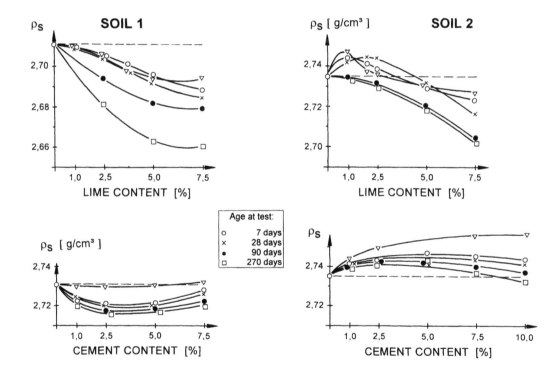

Figure 3. Influence of additive and time of curing on the density, ρ_s, of solid particles. Soil 1 and 2.

reduce plasticity with lime, and increase strength and freezing-thawing resistance with cement. Consequently, the combination of a low lime content plus a higher cement content provides better results than a high lime content plus a low cement content. Usually, about 2 to 3 % lime is sufficient; a smaller addition is difficult to distribute and mix homogeneously.

Many soils, especially clays, exhibit fairly unusual Proctor curves if stabilised with lime plus cement. A plausible optimum water content for compaction can be deduced then from compressive strength – water content curves before and/or after frost tests as indicated in the example of Figure 2.

The increasing number of heavy trucks, increasing axle loading and tire pressure, and

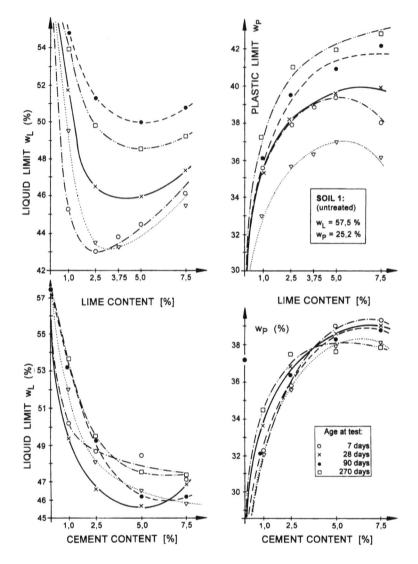

Figure 4. Influence of additive and time of curing on Atterberg limits. Soil 1.

changes in axle configuration contribute significantly to the increase in pavement loading. Furthermore, weather influence has required regeneration of many roads with pavements containing stabilised soil. This old material has proved very suitable now for recycling and on-site stabilisation with cement. Due to the former long-term reactions between soil and first additive, such material now requires only a small content of secondary binding agents. The Proctor curves frequently exhibit unusual shape without a clear optimum, e.g. similar to Figure 2.

2.2 Density of solid particles

The density of solid particles is an indicator of soil-additive reactions and is required for assessing the void ratio of the compacted material. It may increase or decrease after lime or cement is added, and it changes with time (Fig. 3) due to the physico-chemical reactions in the compacted mixtures. Mostly it decreases with time, and the relevant alteration process is finished within one to three years.

2.3 Atterberg limits

Flocculation of the clay particles, a substantial reduction in the thickness of the adsorbed water layer, and cation exchange are the main initial reactions when adding lime to a soil. The clay exhibits reduced plasticity, as evidenced by a significant reduction of I_p that is usually caused by a considerable increase of the plastic limit w_P. The

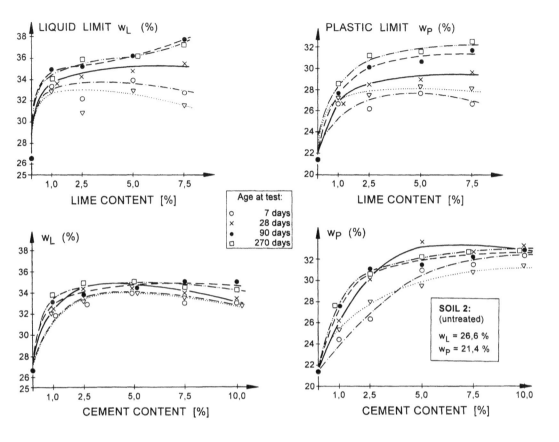

Figure 5. Influence of additive and time of curing on Atterberg limits. Soil 2.

liquid limit of clays with high physico-chemical activity (e.g. sodium montmorillonite) decreases, whereas silts rich in natural calcium exhibit an increase of w_L. Most of the time a threshold value exists: increase or decrease change after this point is reached. In exceptional cases, these reactions also occur when adding cement, but to a lesser degree (Figs. 4,5).

Sometimes the plasticity index of lime stabilised soils may be higher than that of untreated soil. This was observed in carbonate soils and when adding a (too) high quantity of lime.

The Atterberg limits hardly change in the long-term. Tests performed within a period of 35 years showed more or less a scatter within those values which were obtained already during the first year after mixing. The greatest changes could be observed within one to two days, depending on the mineralogical composition of the clay particles. This proves the long-term stability of secondary agglomerates which is an important advantage with regard to a possible re-use of such materials during road regeneration.

2.4 Hydraulic conductivity

The hydraulic conductivity of silt and clay stabilised with lime or cement decreases significantly with time. Contrary to non-cohesive soils, an initial increase occurs at first, even after adding cement, but still more after adding lime (Figs. 6,7). This is caused by the flocculation of the clay particles and the textural change from plastic clay to a friable material which is granular in nature. In the long-term, silts of low plasticity gain relatively quickly lower permeability than the untreated material if stabilised with cement. Due to the physico-chemical long-term reactions in fine-grained soil, the hydraulic conductivity changes over a long period, especially in case of clays with a high percentage of multi-layered minerals and a high quantity of added lime. Small changes could be observed even after 30 years (Fig. 6), which can be deduced mainly from the long-term development of secondary minerals and changes in the micropore structure. An intricate network of secondary minerals coating and bridging aggregates

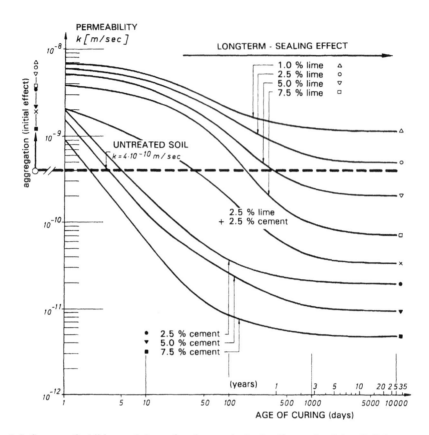

Figure 6. Influence of additive and time of curing on the hydraulic conductivity. Soil 1. Compaction at 100 % Standard Proctor density at w_{opt}. Test series since 1963.

closes progressively the interaggregate pores and therefore reduces gradually the hydraulically available void ratio.

When soil-cement mixtures were stored loosely for 24 hours before compaction, secondary agglomerations caused a large void ratio even if higher compaction energy than usual was applied (Fig.1). Furthermore, long-term crystallisation and hydraulically relevant void ratio reduction occurred to a lesser degree. Therefore, the difference in hydraulic conductivity of immediately and delayed compacted soil-cement mixtures increased with time. Δk was 0.5 to 1.5 powers of minus ten during the first days after compaction and 1.5 to 2.5 powers of minus ten in the long-term (1 to 35 years).

2.5 Compressive and tensile strength

Compressive and tensile strength of lime stabilised soils increase with lime content up to a peak (threshold) value. When exceeding this optimum, strength decreases again, because the particles swim in a matrix without forming a binding texture. With increasing curing time, the threshold value increases and occurs at a higher lime content (Figs. 8,9). This behaviour depends on the long-term reactions and can be observed for compressive and tensile strength likewise, also after exposure of the samples to water (soaking or underwater) or freezing-thawing.

Soil stabilisation with cement does not exhibit a threshold value, at least not within economically relevant quantities of the binding agent. It gains higher strength than lime stabilisation and can be designed to be frost resistant already one to four weeks after compaction which is not possible for lime stabilisation.

The time dependant increase of strength is significantly reduced if the stabilised soil is water-saturated shortly after compaction, or if cement stabilised soil is compacted some hours/days after mixing. Usually the long-term strength increase of in-time and properly compacted mixtures occurs within the following border lines during the first to second year:

224

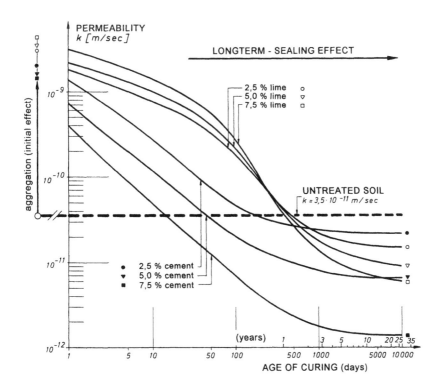

Figure 7. Influence of additive and time of curing on the hydraulic conductivity. Soil 2
Compaction at 100 % Standard Proctor density at w_{opt}. Test series since 1963.

$$q_{u,t} = \tan\alpha \, \log (t/7) + q_{u,7} \qquad (1)$$
$$q_{u,t} = q_{u,7} \, (t/7)^{\tan\alpha} \qquad (2)$$

where $q_{u,t}$ = compressive strength at time t
  $q_{u,7}$ = compressive strength at 7 days
  t = curing time (reaction in days)
  $\tan\alpha$ = gradient of the straight line of
    cementation (="cementation factor").

Lime stabilisation tends towards equation (1), unless the soil contains a significant portion of pozzuolanic components. Cement stabilisation behaves more in accordance with equation (2), at least during the first 6 months after compaction. The gradient of cementation decreases after 1 to 2 years, especially for stabilisation with cement, which reaches its final value within 10 years at the latest. Lime stabilisation exhibits longer lasting strength increases, mainly depending on the mineralogical composition of the clay and on the quantity and quality of added lime. This can be explained by the long-term development of secondary minerals, gradual carbonation, crystallisation, etc. Minimal increases were measured even after 30 years for clays with a high portion of multi-layered minerals.

Tensile strength behaves widely similar to compressive strength, and equations (1) and (2) are fully valid also for this parameter. But an essential difference lies in the behaviour of clays treated with too small lime content. A quantity of less than 1 to 3 % may cause a tensile strength which is lower than that of the untreated soil (Fig. 10). In this case the strength-reducing effect of flocculation and textural change is greater than the positive effect of cementation. This behaviour explains the frost susceptibility of soils treated with a small quantity of lime, which most of the time provides sufficient modification (workability) but not stabilisation. An increased hydraulic conductivity at small lime contents (Figs. 6,7) additionally favours this negative effect by facilitating water movement to the 0°C - isotherm and formation of ice lenses.

On the other hand, too high tensile strength of cement stabilised courses should be avoided because this may cause an unfavourable crack pattern due to shrinking and thermal contraction. The cracks are more closely spaced and therefore narrower, the smaller the tensile strength. Under traffic, pumping can occur more easily at wide

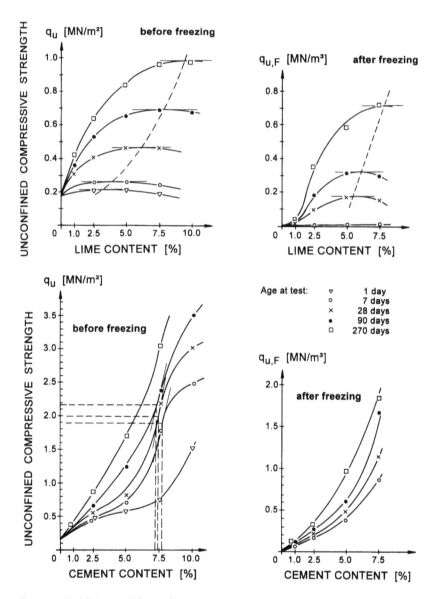

Figure 8. Influence of additive and time of curing on the unconfined compressive strength before (q_u) and after freezing ($q_{u,F}$). Soil 1.

cracks with the development of voids beneath the sub-base. Stabilised sub-bases can then cantilever and fail. Furthermore, at cracks, frost can increasingly damage the sub-base.

Strain at yielding and failure is reduced by soil stabilisation with lime and especially with cement. Brittleness increases with time which may cause fatigue cracking of cement stabilised sub-ballasts of railways or highly loaded sub-bases of expressways. Accordingly, long-term investigations have disclosed that pumping of foundations on

cohesive subgrade can be prevented better by placing non-woven geotextiles rather than by a cement stabilised layer.

Finally, it should be mentioned that the ratio of unconfined compressive strength, q_u, to tensile strength, σ_T, of cohesive soils increases significantly after adding lime or cement, especially if only 1 to 2.5 % are added. Values of $q_u : \sigma_T$ = about 10 to 20 were measured then, whereas values of 7 to 12 occurred for properly stabilised soil. Usually, this ratio increases with time.

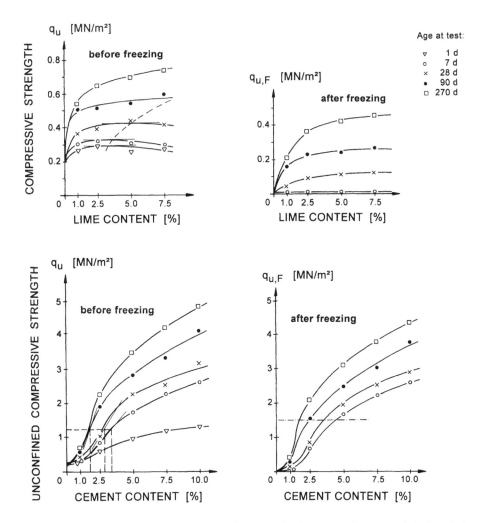

Figure 9. Influence of additive and time of curing on the unconfined compressive strength before (q_u) and after freezing ($q_{u,F}$). Soil 2.

2.6 *Shear strength*

The angle of internal friction, Φ, and especially the residual shear strength, Φ_r, of clays and silts increase significantly when adding lime. In case of cement stabilisation or non-cohesive soils, this effect is less, but nevertheless essential. If the untreated material exhibits a low residual strength, the improvement is especially effective regarding the post failure behaviour (Fig. 11). Moreover, the peak value increases and the shear deformation decreases. The increase of Φ and Φ_r occurs already during the initial state of reactions between soil and additive. Furthermore, a relatively small quantity of lime (about 2 %) or cement (2 - 4 %) is sufficient to achieve the maximum available values of Φ and Φ_r, because secondary aggregates are formed already at

a small lime content. Higher contents of additives hardly provide a relevant improvement of the frictional behaviour, but raise the cohesion (Fig. 12). Contrary to Φ and Φ_r, cohesion clearly increases with time – similar to compressive strength. Consequently, small long-term improvements of lime stabilised plastic clays could be observed even after 10 to 15 years.

Shear strength and stress paths of cement stabilised soils are strongly influenced by the time, Δt, elapsed between mixing and compaction. Figure 13 illustrates this for a clayey silt (soil 1) stabilised with 5 % cement. Cohesion decreases significantly with Δt, whereas the friction angles differ less. In the short-term, Φ and Φr may be even somewhat higher for too late compacted material due to the friction behaviour of the secondary aggregates.

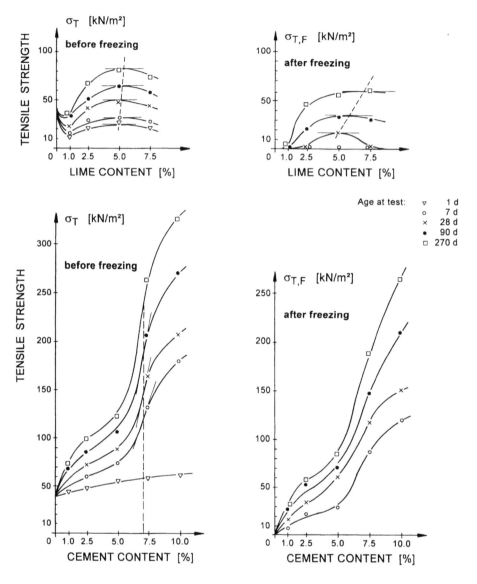

Figure 10. Influence of additive and time of curing on direct tensile strength before (σ_T) and after freezing ($\sigma_{T,F}$). Soil 1.

The stress paths of cement stabilised soils compacted several hours after mixing (Fig. 13b) become increasingly similar to those obtained from lime stabilisation (Fig. 14). Furthermore, they exhibit strong grain rearrangement and shear displacement after exceeding that shear stress which corresponds to the internal friction of the untreated material. On the other hand, cement stabilisations, which were compacted immediately after mixing exhibit a relatively constant void ratio during shearing (e.g. Fig. 13a).

2.7 Freezing-thawing resistance.

Freezing-thawing resistance of untreated and stabilised soils was investigated by freezing tests with an open system. The groundwater temperature was kept constant at + 4° C, and the air temperature varied between + 20° C and - 24° C. Standard tests were performed at -24° C over one week.

There is a fundamental difference between the frost susceptibility of soils stabilised with lime on the one hand and with cement on the other hand.

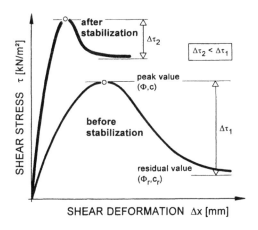

Figure 11. Influence of soil stabilisation with lime or cement on the shear stress-strain behaviour of clays with a low residual strength. Schematical.

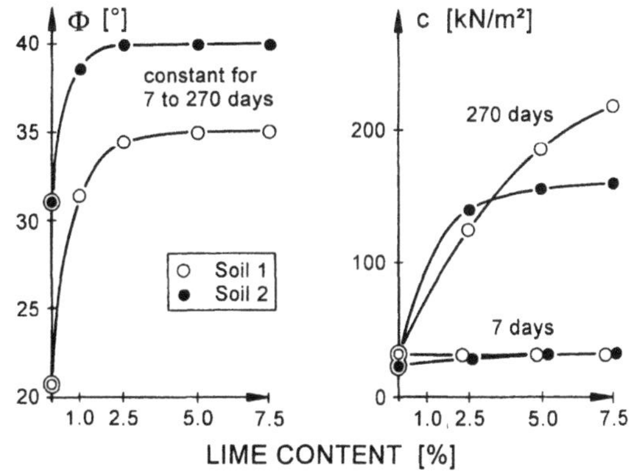

Figure 12. Influence of lime content and time of curing on the shear parameters. Soil 1 and 2.

Figure 13a.

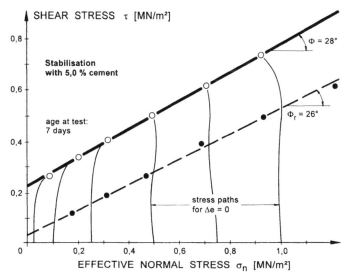

Figure 13. Results of direct shear tests on stabilised soil 1 (clayey silt) conducted under drained conditions with a constant void ratio e during first shearing to obtain the peak values Φ. Samples were compacted in the Proctor cylinder and then consolidated in the shear box before shearing which started 7 days after compaction. The peak value Φ was determined by slow shearing and the post-failure minimum Φ_r with quick shearing.

In the long-term, the friction angles Φ and Φ_r increase only by 1° to 2° (3°) at the maximum. Cohesion c increases similarly to compressive strength.

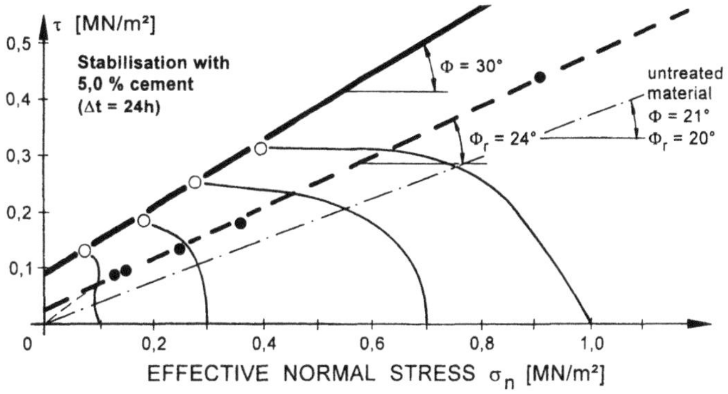

Figure 13b.

229

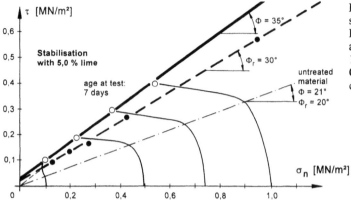

Figure 14. Similar to Fig. 13, but soil 1 stabilised with 5 % lime. In the long-term, the friction angles Φ and Φ_r increase only by 1° to 2° (3°) at the maximum. Cohesion c increases similarly to compressive strength.

Adding lime usually causes even greater frost heave and loss of bearing capacity than that of untreated soil, if frost penetrates within one month after compaction. Exceptions are soils with a high portion of natural pozzuolanic components and a high ion exchange capacity – for instance soil 3 in Figure 15.

Frost resistance increases rapidly with curing time if more than 1 to 2 % lime is added. However, if the lime content is too small, frost susceptibility may remain even in the long-term higher than that of the untreated soil (Fig. 15). This can be explained by the low tensile strength and the high permeability of poorly lime treated soils. Too much lime is also of disadvantage, although the threshold values of minimum frost heave and maximum compressive strength after freezing gradually move towards higher lime contents in the long-term.

Soil-lime mixtures should be compacted between 8 to 30 hours after mixing (depending on clay minerals and temperature). This favours initial reactions to improve workability and avoids too much carbonation of the loose material. If the mixture carbonates intensively before compaction, the layer exhibits high frost susceptibility. In such cases, no relevant long-term cementation can be expected, and frost heaving may remain even higher than that of the untreated soil.

Contrary to lime, cement improves the freezing-thawing resistance from the very beginning and even with small quantities. Moreover, a high cement content does not cause a decrease of this resistance by exceeding a peak value. But too much cement should be avoided because it favours reflective cracking of the compacted layer. The maximum obtainable freezing-thawing resistance decreases with increasing time between mixing the cement and compacting the mixture. Therefore, a quick compaction and proper post-treatment (no evaporation; no loading in the initial phase) are essential for a high quality.

The compressive and tensile strengths of soils after freezing-thawing are widely influenced by the damaging effect of ice lenses. This refers to untreated and stabilised soils likewise. Consequently, lime stabilisation, most of the time, loses all its bearing capacity if it is frozen within 28 days after compaction. But it gains sufficient strength in the long-term (Figures 8, 9). Cement stabilisation commonly exhibits high strength even after one week, and the bearing capacity increases with time - widely similar to equations (1), (2). Contrary to lime stabilisation, no threshold value exists, and a higher cement content increases also the frost resistance.

3   LIME/CEMENT-SOIL REACTIONS

The reactions between lime or cement and cohesive soil can be divided into two distinct processes: initial reactions (modification of the soil) and long-term reactions (stabilisation, cementation of the soil). This simplified distinction involves numerous reactions:

- Exchange, adsorption, and chemosorption of cations and anions;
- Proton exchange (partly with subsequent bond of ions);
- Alteration of the ionic concentration in the pore water (alteration of pH value, hydrolysis);
- Development of dipole water molecules (on the surface of the silicate layers of clay minerals);
- Alteration of the electric surface forces of the clay minerals; alteration of the mutual attraction and repulsion between the particles;
- Potential alteration;
- Alteration of soil water (free, adsorbed, inner-crystalline, chemically bound);
- Alteration of viscosity and capillary forces;
- Neutralisation of acids;

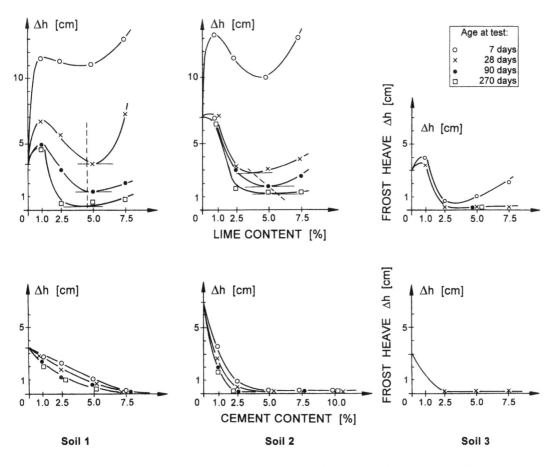

Figure 15. Influence of additive and time of curing on the frost heave, Δh. Soils 1,2,3 (see Table 1).

- Development of new chemical compounds, intricate network of secondary minerals;
- Development of gels and progressive crystallisation;
- Solidification, cementation, hardening and strengthening processes.

This list is neither in chronological sequence nor in order of intensity of the alteration of the mechanical, physical, and colloid-chemical properties. The reactions are partially reversible; many influence each other, overlap, or occur simultaneously.

Due to this complexity, the characteristics of soil-lime/cement mixtures exhibit wide scatter and are strongly time dependent. Therefore, a precise prediction of the alteration of soil parameters caused by lime, cement, or chemicals is practically not possible, although some rough, specific correlations exist. But such correlations are not generally valid: Neither the dielectric dispersion properties of clay-water-electrolyte systems nor

clay activity (I_A), specific surface, S-value (ion exchangeability), amount of Ca^{++}-ions and anions, pH-value, etc. of the untreated soil allow a reliable prognosis of lime reactivity of a soil or of the long-term reaction in a lime-soil mixture.

The portion of pouzzolanic components and the content of semi-movable silica (determined with concentrated hydrochloric acid) represent, relatively, the best suitable parameters to asses roughly lime-soil-reactions.

Similar difficulties exist regarding the prognosis of cement-soil reactions on the basis of untreated soil and cement parameters. Though some specific correlation exists, they are not sufficient to assess quantitatively the behaviour of stabilised mixtures. For instance, the plastic limit of soils may decrease or increase after adding lime or cement.

Consequently, only direct geotechnical tests provide reliable results. They comprise the determination of Atterberg limits, Proctor values, swelling and shrinkage, stress-strain behaviour,

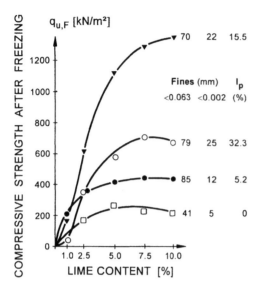

Figure 16. Unconfined compressive strength after freezing-thawing versus lime content of four silty soils. Percentage of fines and plasticity index as parameters. 270 days of curing before freezing-thawing.

freezing-thawing resistance, permeability, volume change upon soaking, erosion and leaching, long-term deterioration, etc. The test program has to be adapted to the specific requirements of design and should involve short-term as well as long-term tests.

Figure 15 illustrates this for the compressive strength after freezing: Four silty soils showing no correlation between plasticity index or content of the fines on the one hand and bearing capacity after freezing on the other hand.

4   CONCLUSION

Long-term research has disclosed that properly designed, constructed, and post-treated layers of soils, stabilised with lime or cement, exhibit appropriate behaviour even after 35 years. Lime has proved suitable especially for subgrades and sub-bases or sub-ballasts of roads and railways, whereas cement should be preferred for sub-bases and bases, especially if exposed to freezing-thawing cycles and high traffic loads. Sub-ballasts stabilised with cement have frequently failed in the long-term due to fatigue cracking. Pumping under dynamic railway loads has begun then again, finally requiring the removal of the rigid, cracked layer. Pumping can be better avoided by placing a non-woven geotextile than a layer of stabilised soil.

Reflective cracking of cement stabilisation can be minimised by several methods. An innovative technique is relaxation rolling with continuous compaction control (Brandl,1997,1999) which has proved suitable over about four years. This "micro-crack rolling" involves a slight over-compaction that creates a fine-meshed network of numerous surface-near thin cracks which do not propagate into the bituminous top layer. Furthermore, continuous compaction control which uses the roller as measuring tool makes an intensive and very homogenous compaction possible which is essential for a long life time of stabilised layers, especially under high traffic loads.

Experience has shown that the minimum amount of lime or cement added on the site should be at least 3 %, even if smaller contents were sufficient from laboratory tests. Efficient mixing to produce a uniform lime or cement addition is necessary to minimise the quantity of additive(s) to achieve the improvements sought. For the mixed-in-plant technique a quantity of 2.5 % could be tolerated, but less would involve a high risk of locally insufficient quality.

The mineralogical composition of the fines plays an essential role, especially in case of soil stabilisation with lime. The higher the portion of mulit-layered clay minerals (especially sodium montmorillonite) and the quantity of additives are, the longer will the long-term reactions last, which can sometimes be observed even after 20 to 30 years (e.g. changes in hydraulic conductivity). Due to the complex reactions between soil and additives, reliable prognoses of the behaviour of the mixtures and the assessment of threshold values are only possible on the basis of direct geotechnical tests and should not be deduced on the basis of a more or less questionable correlation between mineralogy, chemistry, soil physics, and soil mechanics.

REFERENCES

Brandl, H. 1967. The influence of freezing on the characteristics of soils stabilised with lime or/and cement. *Report No. 8, Institute for Soil Mech, & Ground.Eng.*, Technical University Vienna. (In German).

Brandl, H., & D. Adam. 1997. Sophisticated continuous compaction control of soils and granular materials. *Proceedings XIVth ICSMFE, Hamburg*. Rotterdam: Balkema.

Brandl, H. 1999. Mixed-in-place stabilisation of pavement structures with cement and additives. *Proceedings XIIth European Conference of ISSMGE, Amsterdam*. Rotterdam: Balkema.

Geotechnics for Developing Africa, Wardle, Blight & Fourie (eds) © 1999 Balkema, Rotterdam, ISBN 90 5809 082 5

Propriétés physico-méchaniques des matériaux locaux utilisés dans les corps de chaussée en Côte d'Ivoire

A.Gueï
Institut National Polytechnique (INP-HB), Yamoussoukro, Côte d'Ivoire

RESUME : Les matériaux locaux utilisés en corps de chaussée (Base et fondation) en Côte d'Ivoire, sont des graveleux latéritiques sur le socle et des sables argileux dans la region du bassin sédimentaire.
Ces sols sont plus ou moins connus du point de vue propriétés physiques. Cela a permis d'adopter une classification. On distingue en général trois classes de sables argileux (Sc1, Sc2 et Sc3) et trois classes de graveleux latéritiques (G1, G2 et G3).Concernant les propriétés mécaniques, elles sont mal connues. L'unique propriété mécanique utilisée dans le dimensionnement des chaussées est l'indice portant CBR.
Dans le present communiqué, après un bref aperçu des propriétés physiques de ces sols, un accent particulier est mis sur le comportement mécanique (petites et grandes déformations).

ABSTRACT :Local materials used in the body of the pavement (basic and grade) in Côte d'Ivoire, are the lateric gravel on the shield and sand-clay in the sedimentary basin region.
These soils are more and less known as far as their physical characteristics are concerned. This has led to one classification. There exist three classes of sand-clay (Sc1, Sc2 and Sc3) and three classes of lateric gravels (G1, G2, G3). Regarding the failure parameter for determining the size of the pavement, the Californian Bearing Ratio (CBR) is solely used.
This study deals with the physical and the mechanical properties of these soils. It focuses on their behavior in the small strains (compressibity) and the big strains (shearing) domains for the same energy of compaction.

1-INTRODUCTION

Les matériaux locaux utilisés en corps de chaussée (fondation et base) en Côte d'Ivoire, sont les sables argileux dans la région du bassin sédimentaire (région du sud) et les graveleux latéritiques dans les régions du socle (région du Centre, du Nord, de l'Est et de l'Ouest).Les propriétés physiques de ces sols ont fait l'objet de nombreuses études réalisées par le Laboratoire du Bâtiment et des Travaux Publics (LBTP 1974-1983). Ces études ont permis d'adopter une classification. On distingue en général trois classes de sables argileux (Sc1, Sc2 , Sc3) et trois classes de graveleux latéritiques (G1, G2, G3) Les propriétés mécaniques, hormis la portance (CBR) sont mal connues.Il n'existe pas de relations par classes entre les propriétés physiques et les propriétés mécaniques.

Les propriétés mécaniques (comportement en petites et grandes déformations), faisant l'objet de cette communication sont les résultats des essais de compressibilité à l'oedomètre et des essais de cisaillement CU avec mesure des pressions interstitielles sur des échantillons compactés à différentes énergies de compactage pour les graveleux latéritiques.

2- LES SABLES ARGILEUX

Ce sont les principales formations du bassin sédimentaire situé au sud de la Côte d'Ivoire (voir figure1). Ce sont des sables argileux latéritiques ocres ou rouges de 30 à 60 m de puissance, reposant sur une alternance d'argiles bariolées souvent violettes et des sables grossiers.

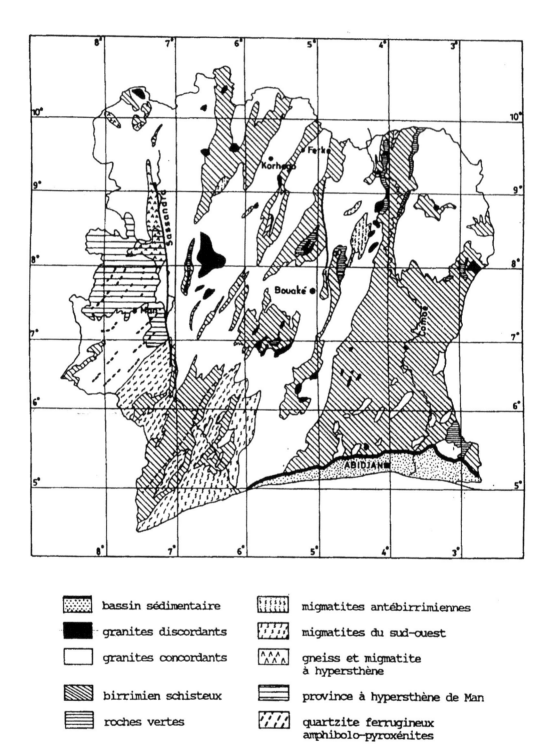

Legend:

- bassin sédimentaire
- granites discordants
- granites concordants
- birrimien schisteux
- roches vertes
- migmatites antébirrimiennes
- migmatites du sud-ouest
- gneiss et migmatite à hypersthène
- province à hypersthène de Man
- quartzite ferrugineux amphibolo-pyroxénites

Figure1 : esquisse géologique de la Côte d'Ivoire d'après la carte géologique à 1/1 000 000 de la SODEMI(1965) (BRGM, 1986)

Ces sables argileux ferruginisés peuvent être rapprochés de la « terre de barre » décrite au Togo-Benin par BOLGARSKY et TASTET. L'examen des propriétés physiques et mécaniques permettra de compléter la classification existante (Gueî & Ahoua Don Mello1997)

2-1 *Propriétés physiques et indice portant CBR*

La figure 2 (LBTP, 1978. Rapport de recherche N° RR 8- septembre 1978) ci-dessous représentent les fuseaux granulométriques des différents types de sables argileux, montre que ces sols qui ont une granulométrie assez étalée peuvent se prêter mieux au compactage.

Les propriétés physiques déterminées par les essais classiques de laboratoire sont resumées sur le tableau N°1 ci-dessous. Sur ce tableau, les indices portant CBR sont déterminés sur des éprouvettes compactées à l'optimum proctor modifié et conservées sous l'eau pendant 4 jours.

Tableau N°1 Classification des sables argileux
(Guei & Ahoua Don Mello, 1987)

Propriétés	classes		
	Sc1	Sc2	Sc3
%F	10-20	20-30	30-40
Ip	5-15	15-20	20-30
CBR	30-60	20-50	10-30
Moyenne	50	35	17

Les caractéristiques de portance varient en fonction de la teneur en eau de compactage. Les figures ci-dessous montrent la variation du CBR en fonction de la teneur en eau pour différentes énergies de compactage pour un sable argileux de type Sc1.

Les valeurs des CBR des sables argileux quel que soit le type, sont assez faibles pour être utilisés comme couche de base pour des trafics intenses. A titre d'exemple, les récommandations ivoiriennes exigent un CBR supérieur à 60 pour la couche de base. Ces sables argileux peuvent être utilisés en couche de base après stabilisation par un liant hydraulique (à 4% de ciment CPA- 350 par exemple).

2-2 *Propriétés mécaniques*

Concernant les propriétés mécaniques des sables argileux, les résultats dont nous disposons sont relatifs aux échantillons partiellement remaniés

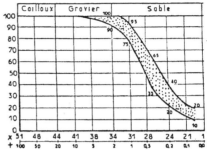

Sables Sc1 (sables argileux grisâtres)

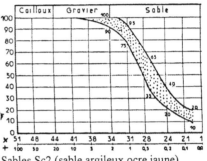

Sables Sc2 (sable argileux ocre jaune)

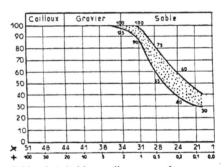

Sables Sc3 (sables argileux rougeâtres

X Module AFNOR
+ Dimensions en mm

Figure 2 Fuseaux granulométriques des sables argileux

prélévés à l'emplacement des fondations de certains ouvrages dans la ville d'Abidjan.

2- 2-1 Compressibilité

Du point de vue compressibilité, ce sont des matériaux moyennement compressibles. L'indice de compression Cc est inférieur à 0.10. On a trouvé une valeur moyenne de l'ordre de 0.06 sur un ensemble

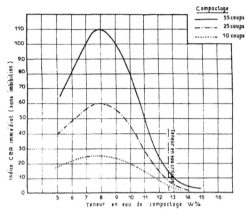

CBR à sec

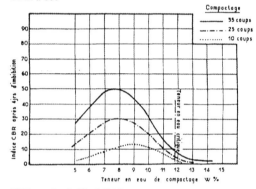

CBR après 4j d'imbibition

Figure 3 . Variation du CBR en fonction de la teneur en eau de compactage (Sable argileux Sc1)

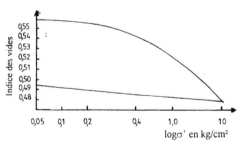

Figure 4. Courbe de compressibilité type d'un sable argileux

de 20 échantillons. La figure 4 ci-dessous représente la courbe de compressibilité typique d'un sable argileux de type Sc1 (moins plastique).

L'indice de compression moyen de ce sable argileux est de l'ordre de 0.07. Le coefficient de perméabilité verticale pour la contrainte en place est de l'ordre de 10^{-5}cm/s.

2-2-2 Comportement en grandes déformations

Les sables argileux offrent une bonne résistance au cisaillement. Sur un ensemble de 15 échantillons prélévés à l'emplacement de certains ouvrages à Abidjan, on a trouvé une valeur moyenne de $\phi'=32°$ et $C'=0.04$ MPA. Pour les sables argileux de type Sc1, on a trouvé les valeurs moyennes suivantes : $\phi'=34°$ et $C'=0.03$ MPA sur un ensemble de 6 échantillons. Quant aux sables argileux de type Sc2 on a trouvé une valeur moyenne de $\phi'=31°$ et $C'=0.06$ MPA. Pour les sables argileux de type Sc3, nous ne disposons pas de valeurs fiables.

Ces résultats montrent que les sables argileux du type Sc1 (peu plastiques) ont un angle de frottement interne effectif assez élevé et une cohésion faible. La cohésion effective semble croître avec la plasticité par contre l'angle de frottement interne décroît.

Le tableau 2 ci-dessous donne un aperçu de l'ensemble des propriétés physiques et mécaniques des sables argileux.

Tableau 2 :Propriétés physiques et mécaniques des sables argileux.

Propriétés	Type de sables argileux		
	Sc1	Sc2	Sc3
%F	10-20	20-30	30-40
Ip	5-15	15-20	20-30
CBR	30-60	20-50	10-20
Cc	0.07	< 0.15	-
ϕ'	34°	30°	
C'(MPA)	0.03	0.06	-
Ig	0	<3	>3
Classification HRB-AASHO	A_{2-4}	A_{2-6} àA$_{2-7}$	A_{7-6}
USCS	CL	CL	CH

Dans ce tableau, le paramètre Ig represente l'indice de groupes.

3- LES GRAVELEUX LATERITIQUES

Les graveleux latéritiques sont des sols que l'on trouve très souvent dans la partie supérieure de l'horizon pédologique B. Ils sont beaucoup plus fréquents dans les régions tropicales où les conditions suivantes sont reunies. Ces conditions sont : Une alternance de saisons pluvieuses et sèches, une forte pluviométrie, une température moyenne annuelle assez élevée, une bonne couverture végétale etc.. .

Ces sols sont en général constitués de rognons de quartz ou de pizolithes, de concrétionslatéritiques plus ou moins friables et en proportions variables pris dans une matrice d'argile rougeâtre à marron ou grise (Menin Messou, 1980). Ce mélange de matériaux, véritable tout venant à granulométrie très étalée est utilisé comme matériau de fondation ou de base (traité au ciment pour certains trafics) dans presque toutes les structures de chaussées souples en Côte d'Ivoire.

La figure 5 ci-dessous représente le fuseau granulométrique type d'un graveleux de forêt.

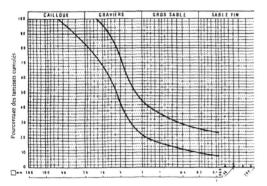

Figure 5 : Fuseau granulométrique d'un graveleux de forêt

3-1 *Classification des graveleux latéritiques selon les propriétés physiquues*

On distingue en général en fonction des propriétés physiques trois types de graveleux : G1 (moins plastique) G2 et G3. Le tableau 3 ci-dessous résume les principales propriétés physiques (Menin Messou. 1980 et 1995)

Tableau N°3 : Propriétés physiques des graveleux latéritiques

Propriétés	Type de graveleux		
	G1	G2	G3
Indice de plasticité	5-15	15-25	25-35
Pourcentage des fines	5-15	15-25	25-35
CBR à 95% OPM après 4j d'imbibition	30-80	20-50	15-40
Densité sèche max à l'OPM	2.10-2.30	2-2.25	1.90-2.20
Teneur en eau	5-8%	9-10%	8-12%

3-2 *Propriétés mécaniques des graveleux compactés*

Les propriétés mécaniques d'un sol étant fonction de la nature et de la compacité, nous avons trouvé opportun d'étudier l'influence de l'énergie de compactage sur les différentes propriétés mécaniques.

3-2-1 Influence de l'énergie de compactage

Pour étudier l'influence de l'énergie de compactage, des éprouvettes d'un graveleux de type G2 (les plus rependus) ont été moulées à différentes énergies de compactage : 95% OPM, 97% et 100%. Les essais réalisés ont donné les resultats suivants :

a) Compressibilité : Les courbes de compressibilité présentées sur la figure 6 ci-dessous, montrent que l'indice de compression Cc décroît avec l'énergie de compactage alors que la contrainte de préconsolidation croît avec l'énergie. Les échantillons étant moulés à différentes énergies de compactage dans un moule CBR.

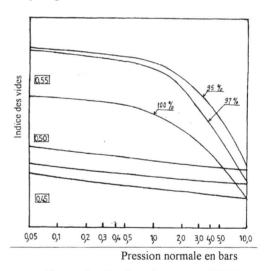

Pression normale en bars

Figure 6 : Courbes de compressibilté des graveleux latéritiques

b) La résistance au cisaillement : Les paramètres de cisaillement C et ϕ étant fonction du chemin des contraintes suivi lors de leur détermination, les paramètres qui nous paraissent représentatifs sont les paramètres effectifs (C' et ϕ') . Les figures 7 et 8 ci-dessous montrent que l'angle de frottement ϕ' décroît avec l'énergie de compactage alors que la cohésion effective croît avec l'énergie. Le coefficient des pressions

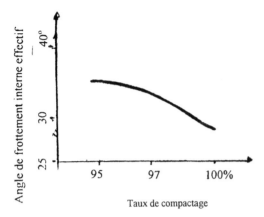

Figure 7 : variation de l'angle
de frottement interne en fonction du
taux de compactage

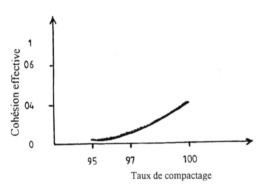

Figure 8 variation de la cohésion effective
en fonction du taux de compactage

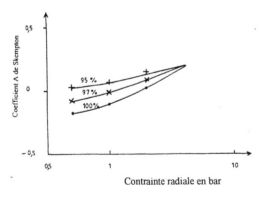

Figure 9 : Variation du coefficient A de Skempton
en fonction du taux de compactage

interstitielles dû à SKEMPTON, le coefficient A
semble décroître avec l'énergie de compactage pour

une contrainte latérale donnée. Ceci lorsque les
contraintes latérales sont faibles. Cela est exprimé
sur la figure 9. Ces trois resultats montrent
nettement l'influence de la surconsolidation. Par
suite du compactage, le sol a subi une
surconsolidation dont l'influence n'est pas
négligeable. La courbe intrinsèque d'un sol remanié
et recompacté représentée ci-dessous montre bien
que dans le domaine surconsolidé, les pentes des
tangentes à la courbe intrinsèque sont faibles.

L'étude de l'influence de l'énergie de compactage
montre que les propriétés mécaniques d'un sol
remanié et recompacté sont fonction du degré de
surconsolidation. La figure 10 ci-dessous est une
représentation typique de la courbe intrinsèque d'un
sol remanié et recompacté.

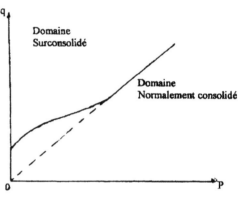

Figure 10. Courbe intrinsèque d'un sol remanié
et recompacté

3-3 *Classification des graveleux latéritiques en fonction des propriétés mécaniques (C' et ϕ')*

Des essais du type CU avec mesure des
pressions interstitielles et de type CD réalisés sur
des échantillons de graveleux compactés à 95%

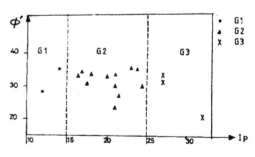

Fig. 11 : Variation de ϕ' en fonction de Ip pour
différents types de graveleux compactés à 95%
OPM.

OPM, ont donné les resultats qui sont rassemblés sur les figures 10 et 11.

A partir de ces resultats on peut affirmer que pour une énergie de compactage donnée, l'angle de frottement effectif décroît avec la plasticité alors que la cohésion effective semble croître. C'est à dire à même énergie de compactage, les graveleux G3, plus plastiques ont un angle de frottement effectif faible et une cohésion effective plus elévée que les graveleux G2 et G1 moins plastiques.

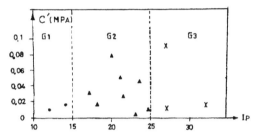

Figure 12 : Variation de la cohésion C' en fonction de lp pour différents types de graveleux compactés à 95% OPM.

Les figures 11 et 12 ci-dessus ont permis de dresser le tableau ci-dessous pour des graveleux compactés à 95% OPM. Bien sûr moyennant certaines réserves car la dispersion est assez importante.

Propriétés	Type de graveleux		
	G1	G2	G3
φ'	25-35°	20-35°	20-30°
C' (Mpa)	<0.02	0.02-0.08	0.02-0.1

4- CONCLUSION

Cette étude qui s'interesse aux propriétés fondamentales des sables argileux et des graveleux latéritiques, matériaux locaux utilisés en corps de chaussée en Côte d'Ivoire, est une approche de l'étude des propriétés mécaniques des principales familles de sols ivoiriens.Elle a permis de mettre en évidence que les propriétés mécaniques (c et φ) , fonction de l'énergie de compactage, croissent lorsque la plasticité augmente pour la cohésion effective c' et décroissent lorsque la plasticité augmente pour l'angle de frottement interne effectif φ'.Ces resultats méritent d'être confirmés par l'exploitation d'un nombre assez important d'essais triaxiaux sur différents types de sables argileux et de graveleux. Des essais triaxiaux cycliques sont aussi nécessaires pour mieux représenter le comportement de ces sols sous charges roulantes occasionnant des effets de fatigue. L'objectif final sera de vérifier si la classification adoptée à partir des propriétés physiques est aussi applicable aux propriétés mécaniques . Dans le cas contraire , redéfinir une autre classification qui puisse à la fois tenir compte des propriétés physiques et mécaniques.

Références bibliographiques

Tastet J.P, 1986. Environnements sédimentaires et structuraux quaternaires du littoral du golfe de Guinée (Côte d'Ivoire, Togo, Benin). Thèse de doctorat ès sciences. Université de Bordeaux 1.

Menin Messou, 1995, Notes sur les chaussées
        Cours de routes ENSTP.

Menin Messou, 1980 ; Comportement mécanique d'une couche de base en graveleux latéritiques améliorés au ciment. Cas des routes de Côte d'Ivoire. Thèse de Docteur-Ingénieur. ENPC 19 décembre 1980.

Guei A. & Ahoua Don Mello, 1987. Etude en laboratoire de quelques propriétés géotechniques des sables argileux du Continental Terminal Ivoirien. 9th Regional Conference for Africa on soil Mechanics and Foundation Engineering. Lagos/September/1987. Ed. Balkema.

Thomas J. J. 1978. Illustration graphique des caractéristiques et propriétés des principales familles de sols ivoiriens. Abidjan : LBTP éd. Rapport de recherche RR 8 .

J. Couland. Caractéristiques mécaniques de quelques graveleux latéritiques compactés.Abidjan : LBTP éd. Rapport de recherche RR 51. juin 1983

BRGM, 1986 . (d'après SODEMI 1965) Catalogue de la cartographie géologique mondiale. Géocarte-Information, N°9 (Côte d'Ivoire, Guinée), P. 61.

LBTP (dossiers N°: 84/B/6224, 83/F/5624, 87/B/7015, 86/B/6958, 83/F/6622, 87/B/7159, 86/B/6778, 85/B/6374, 86/B/6936).

LBTP : Laboratoire du Bâtiment et des Travaux Publics, 04 BP 03 Abidjan 04 Côte d'Ivoire.

Geotechnics for Developing Africa, Wardle, Blight & Fourie (eds) © 1999 Balkema, Rotterdam, ISBN 90 5809 082 5

Contribution à l'étude des chaussées sur sols gonflants – Etude de cas

S. Haddadi, N. Laradi & F. Kaoua
Institut de Génie Civil, Université des Sciences et de la Technologie Houari Boumediène, Alger, Algérie

RESUME : Notre étude est orientée vers une approche technico-économique en vue de résoudre certains problèmes rencontrés lors de la mise en œuvre de sols comme assise (sols supports, remblais, couches de forme, etc.). Une étude de cas est traitée ici en utilisant un enrobé bitumineux (0/15), avec un bitume de classe 80/100 et des granulats calcaires provenant de la carrière de Keddara (sud d'Alger) et des granulats basaltiques provenant de la carrière de Cap Djinet (est d'Alger) . Nous donnons également, les différentes recommandations pour la réalisation des chaussées sur sols gonflants.

1 INTRODUCTION

Dans les études géotechniques relatives à un projet de construction routière, le gonflement d'un sol a un caractère aussi important que son tassement. Les variations dimensionnelles, qui résultent de ce phénomène constitue un souci permanent pour les ingénieurs routiers et les géotechniciens

L'amélioration des sols argileux passe le plus souvent par l'intermédiaire d'un traitement. Parmi les traitements possibles on notera :
- la préhumidification,
- la substitution,
- le traitement à la chaux et au ciment,
- la stabilisation par ajout de sable.
- etc.

2 ETUDE EXPERIMENTALE

2.1 *Etude expérimentale 1 : Stabilisation par ajout de sable au sol support..*

Nous présentons les résultats d'une application de la stabilisation par ajout de sable à un sol gonflant.

a- Influence de l'ajout de sable sur la limite de liquidité : La variation de la limite de liquidité en fonction du pourcentage du sable ajouté est illustrée dans la figure 1. Nous remarquons une diminution de Wl en fonction du pourcentage de sable ajouté. Cependant cette diminution est plus rapide pour les

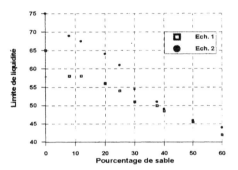

Figure 1- Influence de l'ajout de sable sur Wl.

fortes concentrations en sable et ralentit au fur et à mesure que la concentration en sable augmente.

b- Influence de l'ajout de sable sur l'indice de plasticité : L'influence de l'ajout de sable sur l'indice de plasticité est représenté dans la figure 2. Comme en (a) on observe une diminution de Ip en fonction du pourcentage de sable ajouté. De plus, cette diminution est pratiquement linéaire.

c- Influence de l'ajout de sable sur la pression de gonflement : Les résultats obtenus sont très encourageants. Mais il est encore plus intéressant de voir l'influence de cet ajout de sable sur la pression de gonflement du sol. Aussi nous avons mesuré expérimentalement les pressions de gonflement de l'échantillon 1 correspondant à quelques pourcentages de sable ajouté (20, 30, 35, 40 et 50%).

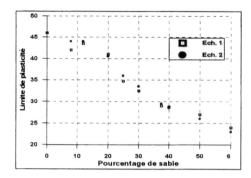

Figure 2- Influence de l'ajout de sable sur Ip.

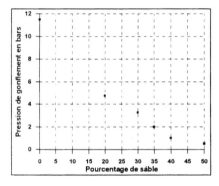

Figure 3- Influence de l'ajout de sable sur la
pression de gonflement.

La figure 3 montre la variation de la pression de
gonflement en fonction du pourcentage de sable
ajouté. Cette figure fait ressortir des résultats très
intéressants puisque un ajout de sable de 20% réduit
la pression de gonflement de plus de 60% (de 11,5 à
4,5 bars). De plus, pour un même rang de teneur en
sable de 20 à 40%, la pression de gonflement passe
de 4,5 à 0,86 bar soit 20% environ. Donc la
stabilisation est un phénomène à rendement
décroissant.

Mais néanmoins, l'utilisation des sols argileux
pose de nombreux problèmes tant en remblais que
sol support des chaussées dont un certain nombre de
facteurs influent sur le traitement, parmi ces
facteurs, ceux qui sont liés au sol ou au couple sol-
produit de traitement, sont les plus importants. En
effet, quelque soit le soin apporté à la mise en
œuvre, le résultat dans certains cas risque d'être
médiocre si les réactions argile-liant ne se produisent
pas ou se réalisent mal.

Pour éviter ce risque d'échec toujours onéreux,
deux solutions s'imposent :

    - une étude paramétrique.

    - une étude minéralogique.

Cette étude doit être poursuivie afin de voir
l'influence de la granulométrie du sable sur
l'efficacité du traitement et également la façon de
mettre en œuvre cette technique sur le terrain.

L'étude a justifié expérimentalement la relation
entre, d'une part, la limite de liquidité et la pression
de gonflement et, d'autre part, l'indice de plasticité et
la pression de gonflement à savoir qu'ils varient dans
le même sens.

Elle a confirmé aussi, l'intérêt du sable en tant
que matériau stabilisant les sols gonflants.

2.2 *Commentaires* : Les résultats obtenus dans les
différents travaux réalisés, Kaoua et al, 1994, Kaoua
et al, 1992, Kaoua et al, 1995, montrent la non
maîtrise de la construction en génie civil et la non
compréhension du phénomène, ainsi que sa
négligence, sa répercussion sur les catastrophes qu'il
a engendrées tant sur le plan humain que sur le plan
matériel.

2.3 *Etude expérimentale 2: Influence des paramètres*
de composition sur le comportement des enrobés.

2.3.1 *Le matériau* : Le matériau utilisé est un enrobé
bitumineux (0/15) de pourcentage, en poids de
bitume par rapport au granulat, de 6%.

Le bitume est un bitume pur de classe 80/100, sa
densité est de 1,008 à 20 °C .

Les agrégats sont des granulats calcaires
provenant de la carrière de Keddara, et des granulats
basaltiques provenant de la carrière de Cap Djinet.
Nous avons reconstitué les courbes
granulométriques à partir des classes granulaires 0/3,
3/8 et 8/15.

Les caractéristiques des granulats sont regroupées
dans le tableau 1.

Nous constatons, que les matériaux des carrières
choisies présentent, des insuffisances en qualités
intrinsèques (la forme, la dureté, l'usure par
frottement). La propreté n'est pas, également,
respectée.

Le vieillissement se manifeste, soit au cours de la
fabrication des enrobés, soit pendant leur durée de
vie. On a coutume de considérer l'évolution des
bitumes sous deux formes : durcissement physique
ou durcissement chimique.

Dans notre étude, il s'agit d'analyser les
caractéristiques de deux bitumes de même classe
mais d'âges différents. Ces deux bitumes sont, d'une
part, un bitume frais 80/100 et d'autre part, un
bitume âgé d'une année de classe initiale 80/100.

Nous avons effectué des essais de pénétrabilité et
de ramollissement qui permettent la classification
des deux bitumes.

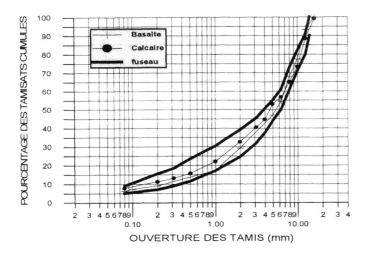

Figure 4 - Courbes granulométriques

Tableau 1- Caractéristiques des granulats.

	Cap Djinet (Basalte)			Keddara (Calcaire)		
ESSAIS	Classe granulaire			Classe granulaire		
	0/3	3/8	8/15	0/3	3/8	8/15
Propreté (%)		2.33	1.57		2.07	2.70
Equivalent de Sable (%)	61.73			39.92		
Aplatissement (%)		38.51	20.51		18.69	6.19
Los Angeles (%)		16	14		20	16
Micro-Deval (%)		29	16		21	18
Densité spécifique	2.61	2.63	2.84	2.72	2.77	2.76

243

Tableau 2.

Bitume	Bitume frais	Bitume âgé
Pénétration (1/10mm)	98	51
TBA °C	43	47
Classe du bitume	80/100	40/50

Tableau 3.

CLASSES	POURCENTAGE DE MATERIAUX	
	Cap Djinet	Keddara
8/15	40 %	35 %
3/8	30 %	35 %
0/3	30 %	30 %
Bitume	6 %	6 %

Les résultats obtenus sont représentés dans le tableau 2.

Dans le tableau 2, il est montré clairement que l'âge du bitume influe considérablement sur la classe de ce dernier. En effet, le bitume, initialement de classe 80/100, devient de classe 40/50 (Haddadi,1995).

Ceci s'explique par le fait qu'il l'a eu évaporation des éléments volatiles, ce qui a conduit au durcissement du bitume.

Nous avons confectionné des mélanges hydrocarbonés de formules (0/14), (tableau3).

Le pourcentage de bitume est le même que le bitume soit frais ou âgé.

Notons que ces pourcentages représentent la classe 0/14 de l'enrobé qui rentre dans un fuseau granulométrique et qui répond aux normes routières internationales.

Nous constatons, que les matériaux des carrières choisies présentent, des insuffisances en qualités intrinsèques, ce qui confirme les résultats présentés par ailleurs (Boularak et al, 1993 et Chaib et al, 1993. Ces insuffisances sont dues, surtout, à la très forte demande qui dépasse de loin l'offre. Ce qui ne permet pas aux carrières d'effectuer les contrôles nécessaires de qualité).

Nous étudions, de façon globale, l'influence de l'âge du bitume, de la nature des granulats et des paramètres de composition, sur les caractéristiques thermomécaniques des enrobés bitumineux. A cet effet, nous avons procédé à des essais de stabilité Marshall sur des éprouvettes de mélanges hydrocarbonés.

Nous avons confectionné 15 éprouvettes semblables dans chaque série.

Au vu des résultats obtenus, il ressort que:

1°) quel que soit l'état et la nature des granulats, l'association du bitume âgé dans l'enrobé bitumineux provoque une diminution de la stabilité, de la compacité, et une augmentation du fluage.

2°) la nature des granulats influe sur les caractéristiques mécaniques de l'enrobé bitumineux. Les matériaux de nature calcairo-dolomitique, mélangés au bitume, présentent comparativement aux granulats basaltiques de meilleures stabilité, de compacité et des fluages plus faibles.

3°) le lavage et le calibrage n'influent pas notablement sur les caractéristiques mécaniques des mélanges hydrocarbonés.

4°) la teneur en filler, associée à l'énergie de compactage, influe sur la stabilité, la compacité et le fluage des mélanges hydrocarbonés.

5°) la chaux introduite comme filler d'apport présente des résultats acceptables pour les caractéristiques mécaniques des enrobés bitumineux.

2.3.2 *La* sollicitation : Pour l'étude de l'influence des granulats sur le comportement des enrobés bitumineux, on soumet les éprouvettes confectionnées, conservées à température ambiante à la sollicitation choisie et représentée sur la figure 4, 24 heures après leur fabrication.

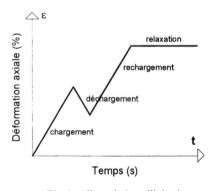

Fig.4- Allure de la sollicitation

On charge à vitesse constante à raison de 1 % /min. Arrivé à σ/2, on décharge à vitesse constante de 1 % /min. jusqu'à σ = 0, puis on recharge, également à la même vitesse, jusqu'à la rupture. Dès que la contrainte chute sensiblement, on maintient l'éprouvette sous la déformation axiale constante pendant un certain temps jusqu'à la stabilisation de la contrainte (fig.3).

2.3.3 Résultats : Les résultats des essais sur des éprouvettes d'enrobé bitumineux, de granulats de Cap Djinet ou de Keddara, mélangés au bitume sont présentés sur les figures 5, 6 et 7..

Les résultats obtenus nous permettent de remarquer :

- que les enrobés bitumineux obtenus avec des granulats calcairo-dolomitiques, présentent de meilleures résistances (fig.6 et 7).

- de plus petites déformations sont enregistrées par les mélanges hydrocarbonés, à base de granulats de la carrière de Keddara (fig.8 et 9).

Nous pouvons conclure que les granulats calcairo - dolomitiques de Keddara, mélangés au bitume, présentent de meilleures performances (résistance et déformation) que ceux basaltiques mélangés au même bitume (même classe).

Les résultats des essais sur des éprouvettes d'enrobé bitumineux nous a permis de mettre en évidence les observations suivantes :

- la nature calcairo - dolomitique des granulats, lorsque ceux-ci sont mélangés au bitume, assure un meilleur comportement que les granulats de nature basaltique ;
- quelle que soit la sollicitation appliquée, au cours de la relaxation, la contrainte chute jusqu'à se stabiliser au voisinage de 0.5 MPa. Ce résultat est en concordance avec ceux déjà trouvés par les autres chercheurs (Laradi et Haddadi, 1997 et Haddadi, 1995), sur le comportement des matériaux granulaires ou présentant un squelette granulaire.

3 RECOMMANDATIONS POUR LA REALISATION DES CHAUSSEES SUR SOLS GONFLANTS

La réalisation des chaussées sur sols expansifs présente des difficultés particulières nécessitant des dispositions de construction spéciales, nous donnons en fin de ce rapport les différentes recommandations pour la construction des chaussées sur sols gonflants:

- les chaussées construites sur ce genre de sol doivent être simples et leur environnement doit être

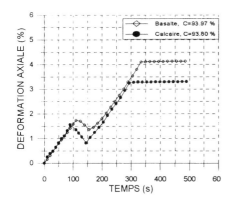

Fig.5 : exemple de sortie graphique de la sollicitation.

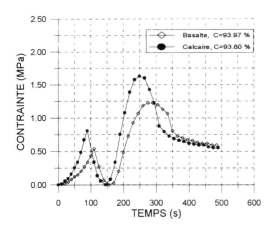

Fig.6 : Résultats, courbe σ = f(t).

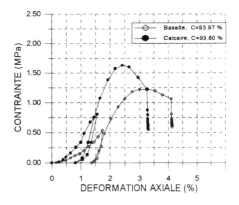

Fig.7 : Résultats, courbe σ = f(ε).

245

conçu de telle sorte que les variations de la teneur en eau du sol soient réduites au minimum.

- La plate-forme doit être terrassée avec une pente transversale d'au moins 5 % et si possible traitée à la chaux avant compactage, qui doit rester modéré et réalisé à une teneur en eau supérieure à l'Optimum Proctor Modifié (OPM).

Une imprégnation et un sablage protégeant la plate-forme de la dessiccation, les membranes imperméables sont mises en place, dans le même but. Si l'on veut qu'une meilleure butée reste disponible, les remblais doivent être évités ; la dessiccation du sol sous la chaussée progresse plus difficilement si la route est en déblai.

- L'exécution des accotements fera l'objet de précautions spéciales, car c'est généralement par la que débute la dégradation d'une route sur sols gonflants. Ils seront constitués du même matériau que le corps de chaussée et une pente transversale d'au moins 6% et imprégnés sur toute leur largeur ou de préférence revêtus d'un bi-couche. La largeur de l'accotement sera d'eau moins 2,5m.
- Les fossés seront surdimensionnés et imperméabilisés, leur fil d'eau sera à 20m ou 30m sous le niveau de la plate-forme.
- Les arbres doivent être déracinés jusqu'au moins 5m du fossé car ils contribuent à la dissociation du sol en période sèche.
- Le corps de chaussée aura une épaisseur d'au moins 60 cm, la surcharge qu'il représente s'oppose au gonflement du sol.
- Les assises traités doivent être évités et on doit utiliser des concassés en couche de base et des graves naturelles ou concassées en couche de fondation. Ces matériaux ont malheureusement une perméabilité souvent assez élevée qui est néfaste. Ils devront être protégés des infiltrations par des revêtements très étanches mais restant souples.
- Les enduits superficiels sont les revêtements les mieux adaptés que l'on peut réutiliser périodiquement en même temps que l'on recharge, si besoin est la couche de base.
- Ne sera mis en place, un tapis d'enrobés, que si le trafic l'exige et après stabilisation du gonflement.
- Il est recommandé de parfaire l'imperméabilisation du revêtement par un coulis bitumineux, éventuellement clouté pour améliorer le gonflement.

4 CONCLUSIONS

A l'époque où le réseau routier Algérien est en plein essor, où il est en train de s'étendre et de se renforcer à une cadence jamais atteinte, où les méthodes d'auscultation constituent une éclatante confirmation de la valeur d'idée théoriques, construire une route sur sols gonflants reste un domaine non encore exploité et il est impératif de s'y pencher de prés afin de palier aux problèmes des dégradations qui peuvent surgir et avoir des conséquence graves sur les usagers et l'économie algérienne.

La mécanique des sols seule ne peut pas traiter tous les problèmes posés par les sols argileux, rencontrés en géotechnique routière car cette dernière utilise des mélanges complexes, tant par leur nombre que par la nature des constituants.

Les phénomènes de retrait et de gonflement de certains sols tiennent à la nature minéralogique des argiles dont ils sont constituées. Ces matériaux, appartenant généralement au groupe de la montmorillonite, ont une structure à feuillet qui peuvent reversiblement absorber et rejeter de l'eau selon les conditions hydriques du milieu. Les variations de volume et les pressions qui en résultent peuvent être énormes.

Une bonne corrélation entre la composition minéralogique et les caractéristiques géotechniques peut nous renseigner sur le caractère du minéral-argileux.

Il est donc possible de lutter contre les effets du gonflement en permettant certains mouvements à la structure dans la limite où ces dernières sont admissibles. En fait la connaissance de la pression de gonflement n'est plus la priorité, si l'on peut adapter des ouvrages aux mouvements résultant de la variation de la teneur en eau des sols expansifs, ou par un traitement aux liants hydrauliques, telle que la chaux qui est très répandue et efficace.

Nous pouvons dire que nous pouvons construire sur sols gonflants. Cette construction exige au préalable une classification des sols juste et une analyse minéralogique complète. Mais l'Ingénieur doit toujours penser au coût de réalisation et les études doivent être économiques et proportionnelles à l'importance du chantier.

Pour l'Algérie, en ce qui concerne la construction routière, il y a beaucoup de choses à revoir.

REFERENCES

Boularak M. , Benachour Y. 1993. *Rôle des laboratoires dans la chaîne de qualité* - 1ères Rencontres Nationales sur les Granulats, Décembre 1993, Alger, pp. 82-108.

Chaib K., Louaheb A. 1993. *Ressources en granulats*- 1ères rencontres nationales sur les granulats. Décembre 1993 à Alger, pp. 6-26.

Haddadi S. (1995) . *Contribution à l'étude du comportement des matériaux traités aux liants hydrocarbonés : influence des paramètres de composition* - Thèse de Magister, USTHB, Février 1995. 119 p.

Kaoua F. , Laradi N. 1994. Contribution à l'étude de la stabilisation des sols gonflants par ajout de sable. *Algérie Equipement, Revue technique de l'Ecole nationale de Travaux Publics de Kouba N° 15*, pp.12-15

Kaoua F.& Laradi N. 1992, Comportement des sols gonflants. *Rapport de recherche*, Projet de recherche, I.G.C. U.S.T.H.B. 25 p.

Kaoua F., Laradi N. & Behar R. 1995. Etude des chemins de contrainte et de déformation parcourus dans un massif de fondation. *XI African Regional. Conference on soils mechanic and fondations Engineering*. Cairo, Egypt. 11-15 Décembre, pp.

Laradi N. & Haddadi S. 1 997. Influence des propriétés rhéologiques sur le comportement mécanique d'un tuf - sable - bitume. 1er Congrès Arabe de Mécanique 97. Damas, Syrie, pp..

Geotechnics for Developing Africa, Wardle, Blight & Fourie (eds) © 1999 Balkema, Rotterdam, ISBN 90 5809 082 5

A brief review of chemical soil stabilisation by labour intensive methods using sulphonated petroleum products

W.P.C.van Steenderen
Department of Civil Engineering, University of the Witwatersrand, Johannesburg, South Africa

ABSTRACT: Sulphonated petroleum product (SPP) soil stabilisers act as compaction aids and as stabilisers. Successful stabilisation depends on there being sufficient active clay in the soil to permit ion exchange reactions to take place. Several tests useful in identifying soils that will react with the chemical stabilisers, are discussed. These tests do not give estimates of strength, so conventional California Bearing Ratio tests have been used to measure changes in strength due to treatment of the soil with the stabiliser. The dosage should be optimised to give maximum strength gain at the expected field density.

Due to their good solubility in water, chemical stabilisers are easily applied via the water needed for compaction. The solution then migrates through the soil without needing much mechanical mixing, which makes chemical stabilisers suitable for use on labour intensive construction sites. If the chosen stabiliser improves the soil strength sufficiently, in situ or local soils can be used instead of transported gravels.

1 INTRODUCTION

Road engineers face questions when considering stabilisation of poor soils. The commonly known stabilisers, cement and lime, are costly and can only be considered if the soil meets certain conditions. If the soil does not meet these conditions, must it then necessarily be removed? In certain cases ion exchange stabilisers (eg sulphonated petroleum products) can be used. These are used in small quantity and their cost is modest. However, the ion exchange stabiliser may not stabilise that soil but simply make matters worse. This is a very real dilemma, as sulphonated petroleum based ion-exchange stabilisers work well only with certain soils and may lower the strength of other soils. Simple testing of the California Bearing Ratio (CBR) of the stabilised soil(s) can measure the effect of the stabiliser and provide the engineer with the needed confidence in the product.

There are at present in South Africa, a number of sulphonated petroleum based ion-exchange stabilisers (SPPs) available. They come in widely differing concentrations and chemical compositions and hence need to be tested individually. Certain products work better on some soils than do other products. If a particular soil reacts poorly or not at all with one product, it may be worthwhile to try another.

Due to their good solubility in water, SPP stabilisers are easily applied via the water needed for compaction. Further, the soapiness of the solution helps it to migrate through the soil and is therefore not entirely dependent upon good mechanical mixing for its dispersion. This makes SPP stabilisers very suitable for use on labour intensive construction sites. If the chosen stabiliser improves the soil strength sufficiently, then in situ or local soils can be used instead of transported gravels. The effect will be twofold: lower road construction costs and a higher proportion of total costs expended on labour (McCutcheon, 1993) thus directly benefitting the local community.

2 STABILISERS

There are many chemical soil stabilisers on the South African market today. Several of these work on an ion exchange principle. This paper covers only sulphonated petroleum products (SPPs). This group of stabilisers generally acts in two ways on a soil: as a stabiliser and as a compaction aid.

The ion exchange stabilisation process is similar to the action of lime on the clay fraction of a soil. Lime exchanges stable calcium ions for more reactive sodium and hydrogen ions on the clay surface. The ion exchange reactions result in a modified clay that is less susceptible to the influence of water. In the case of lime, cementation follows the ion exchange if sufficient free lime is available, by the formation of calcium alumina-silica hydrates [Savage, 1986]. The stabilising action of sulphonated petroleum based chemical stabilisers similarly results in a modified clay, less susceptible to the influence of water. However, the exact nature of the chemical and physical processes is unknown but none the less real, as large strength gains will testify.

In addition to the ion exchange reactions with the soil, SPP stabilisers also act as compaction aids on most soils. During manufacture the sulphonation process produces wetting agents or detergents that reduce the surface tension of the water in the soil and lubricate the soil particles, allowing them to slide past each other during compaction. This results in lower energy being needed to achieve a specified density or a higher density for the same energy input. However, there are disadvantages if the dosage of stabiliser is high and the soil is "over-lubricated", as the shear strength is then reduced. The amount of stabiliser "used up" in the ion exchange reaction depends on the clay content and on the activity of the clay. The "unused" portion of the stabiliser seems then to be responsible for excessive lubrication and subsequent reduction in strength.

From the paragraphs above it may be concluded that the dosage of the stabiliser is critical. Too little stabiliser may not take advantage of the soil's potential strength and yet too much stabiliser will result in reduced strength. Optimisation of the dosage is clearly required.

manufacturers recommend limits to the values as a starting point for determining soils which can be successfully stabilised.

Paige-Green [Paige-Green and Bennett, 1993] measured the cation exchange capacity (CEC) of the soil in order to identify the potential of the soil to react with the stabilisers . He shows that 2:1 clay minerals are the only ones with useful cation exchange capacities, but adds that capacity alone is not enough: the cations must be capable of being exchanged. The cation exchange capacity is difficult to measure as the test is susceptible to changes in the pH of the soil and to the organic content.

X-ray diffraction (XRD) analysis has been used to identify clay minerals present in the soil which could then be compared to a known database. From this comparison an estimate could be made of the required dosage of the SPP stabiliser.

However, neither the CEC nor XRD test results can be related to the strength of the soil. This can only be measured by strength tests like the California Bearing Ratio (CBR). The CBR test is time consuming and costly and requires large quantities of soil, but does give reasonably realistic values of the strengths that can be achieved in the field.

The time taken by the chemical reaction of the SPP on the soil is the subject of some controversy. Theoretically ion exchange reactions are very rapid [Paige-Green and Bennett, 1993], but some manufacturers imply that the reactions may take up to 60 days to reach completion. Limited testing has shown differences in reaction time of different SPPs. Until the rates of the reactions are properly explored under controlled conditions, it is recommended that some damp curing be applied in the field and to laboratory specimens prior to CBR testing. The period of cure should emulate expected field conditions. A period of damp cure of three to seven days is recommended until better knowledge of the reaction process is available.

3 TESTS

The standard "indicator" tests [TMH 1, 1986, test methods A1 to A6] which measure the grading and Atterberg limits, are useful as the proportion of clay in the soil sample can be measured. The plasticity index gives an estimate of the clay activity. SPP stabiliser

4 RECOMMENDED CBR-BASED TESTS

Due to the differences between stabilisers, it may be necessary to try several stabilisers before deciding which gives the best result with the particular soil in question.

Precautions must be taken when handling the

stabiliser concentrates as some stabilisers are highly acidic. However, once diluted, the stabilisers are less aggressive and relatively safe to handle. The manufacturers generally recommend dilution by a factor of 100 to 500.

It is recommended that standard Mod.AASHTO moisture/density tests [TMH 1, 1986: test method A7] are performed and followed by standard soaked or unsoaked California Bearing Ratio (CBR) tests [TMH 1, 1986: test method A8]. The use of the soaked or unsoaked CBR test requires a knowledge of the field conditions. If the road layer is likely to become saturated in service, then the soaked CBR test must be used to simulate the field conditions and measure the lower CBR value. If the road will perform in relatively arid conditions, then the unsoaked CBR is relevant and measures the higher strength which may realistically be expected. The unsoaked CBR test is normally performed directly after compaction and thus measures the strength at the moulding moisture content.

The results of these tests will give a bench mark value for the maximum density, optimum moisture content (OMC) and CBR. These tests should then be repeated using soil treated with the chosen stabilisers at various dosage rates. The optimum dosage is that amount of SPP which results in the highest CBR strength at the expected field density. Note that density has some influence upon the optimum dose: lower densities seem to require marginally higher dosages.

It has long been good practice to prepare specimens for compaction for subsequent CBR testing, the day before compaction is to take place. This allows the compaction water to permeate into and through the soil. The moisture content of the soil is then much more uniform and more uniform compaction can be achieved, which probably results in lower test variations. The same approach is recommended when testing the action of SPP stabilisers on soils.

Typical SPP stabilisers are administered in small quantity: often between 50 and 200 ml of stabiliser per (compacted) cubic metre of soil. Due to the relatively small specimens prepared in a laboratory, the quantity of stabiliser needed for each specimen is small. Typical amounts of stabiliser concentrate range from 0.60 ml for 200 ml/cu.m dosage to 0.15 ml for 50 ml/cu.m dosage and even less for other products. These minute quantities of concentrate are best administered in highly diluted form, so that they can be accurately measured. The stabilisers dissolve readily in water, so dilute solutions can be easily prepared.

After compaction the specimens should be damp cured by sealing the specimens in plastic to prevent moisture loss, for the recommended period. CBR penetration can then be done at the moulding moisture content or the specimens can be soaked in a water bath for the standard four days before penetration.

5 LABOUR INTENSIVE APPLICATION

Water soluble soil stabilisers were found to be particularly suited to labour intensive road construction as the stabilisers were applied via the water needed for compaction. Several methods of application of diluted stabilisers were suggested (during discussions with a variety of people), many of which were tried on the Mamelodi Roads Pilot Project during 1993 [Van Steenderen and McCutcheon, 1994]. The successful methods were:

(i) Watering cans: 12 litre watering cans were used and were found highly successful for small awkward areas where other methods could not reach. However, control of the application rate of the solution of stabiliser needed attention. It was found best to clearly demarcate areas small enough for one watering can full (at the calculated dosage) and then to ensure that only one watering can was applied there and only there. For occasional awkward areas the additional supervision needed could be given, but the method did not lend itself to large areas. The application was slow and had to be supervised at both the filling point and the application point.

(ii) Hand-drawn bowser: a 210 litre drum was mounted on a trolley and fitted with a spray bar [ILO: 1981, pages 9.27 to 9.29]. Stabiliser concentrate was measured off and poured in and the drum was filled with water and shaken to mix the solution. A team of five men was needed to handle the full drum on its trolley across the loose stony soil. The flow was controlled with a valve on the spray bar. The trolley was run up and down the roadway at a slow pace until empty. It was found advisable to apply at least four full drums across a section as variations in the spread rate tended then to cancel out. This method was used at length on the Mamelodi Roads Pilot Project.

(iii) Conventional (machine based) construction contracts use a normal water bowser to dispense the stabiliser. This method or variations on the theme can be used on labour-intensive projects. Towed water bowsers or tanks on trucks fitted with spray bars are suitable alternatives.

After the solution had been applied, the area was left for a while to allow the water to soak in and was then mixed with spades or forks. The soil was generally turned over twice to achieve a fairly uniform mixture. The soil was then left overnight if possible, to be remixed once, raked to shape and compacted.

6 COST COMPARISONS

As each road project has unique aspects, a plea is made that the true cost of every operation and its alternatives should be determined before financially based decisions are made. SPP stabilisers can then be compared to cement or lime stabilisation and be compared to the rates that would normally be paid for the substitution of the stabilised layer with gravel transported from a borrow pit.

7 CONCLUSIONS

Standard laboratory California Bearing Ratio tests can be used to measure the influence of chemical stabilisers on soils. The test should be performed on the natural soil and then repeated on the soil treated with the chosen chemical at various dosage rates to optimise the application.

Stabilisers that can be dissolved in water can be readily and cheaply applied labour-intensively by water bowser or watering can. Reasonable control over the application rates can be exercised.

REFERENCES

ILO, 1981. Guide to Tools and Equipment for Labour-based Road Construction. International Labour Office. Geneva.

McCutcheon, R T, 1993. Employment creation in South Africa: The potential and the problems. *Annual Transportation Convention: Volume 3B, Labour Based Construction, Paper 1*. Pretoria.

Paige-Green, P & Bennett, H, 1993. The use of sulphonated petroleum products in roads: state of the art. *Annual Transportation Convention, Volume 4C, Pavement Engineering Paper 5* Pretoria.

Savage, J M L. 1986. Stabilisation experience and practice in Natal. Notes for TRH 13 Symposium, Pretoria.

TMH 1 1986. Standard methods of testing road construction materials. CSIR National Institute for Transport and Road Research, Second edition. Pretoria

Van Steenderen, W P C & McCutcheon, R T, 1994. The application of chemical stabilisers to labour intensive road construction. *Research Report RR 93/339*, Department of Transport. Pretoria.

4 Monitoring, laboratory and field testing

Contrôle et le suivi, essais de laboratoire et in-situ

Geotechnics for Developing Africa, Wardle, Blight & Fourie (eds) © 1999 Balkema, Rotterdam, ISBN 90 5809 082 5

Quelques corrélations entre essais in situ et essais de laboratoire pour certaines argiles Algériennes

R. Bahar
Institut de Génie Civil, Université Mouloud Mammeri, Tizi-Ouzou, Algérie

F. Kaoua
Institut de Génie Civil, Université des Sciences et de la Technologie Houari-Boumediène, Alger, Algérie

T. Aissaoui
Laboratoire National de l'Habitat et de la Construction, Rouiba, Algérie

RESUME : Cette communication présente les résultats d'une série d'essais pressiométriques, d'essais au péné-tromètre statique et d'essais en laboratoire réalisés sur deux sites de la région d'Alger, constitués d'argiles molles à raides. Une étude des corrélations entre paramètres pris deux à deux, à l'échelle d'un sondage ou du site, est présentée.

ABSTRACT: The proposed paper presents the results of Menard pressuremeter (MPT), static penetrometer (CPT) and laboratory tests performed at two sites consisting of soft to stiff clays. The sites are located in Algiers region. An analysis of the correlations between parameters, taken two by two, along a vertical or on the whole site, is presented.

1 INTRODUCTION

Les essais in situ tels que le pressiomètre Ménard, le pénétromètre statique, sont fréquemment utilisés pour le dimensionnement des fondations de bâtiments et d'ouvrages d'art en Algérie. A ce jour, de nombreux essais in situ ont été réalisés sur des argiles molles à très raides et roches tendres pour caractériser le comportement mécanique des sols.

Dans les études de mécanique des sols, il est rare que l'on n'ait besoin que d'une seule propriété du sol. Parmi les paramètres physiques et mécaniques qui décrivent les propriétés des sols, certains sont reliés par des expressions mathématiques. Pour d'autres paramètres il n'existe pas de telles relations fonctionnelles. Pourtant, on sent bien que les valeurs des paramètres d'un sol ne sont pas totalement indépendantes : plus un sable est dense, et plus sa résistance à la pénétration est grande, c'est-à-dire que son angle de frottement interne, sa pression limite et son module pressiométrique sont élevés.

Dans la littérature, Il existe beaucoup de travaux qui traitent des corrélations entre les paramètres des sols (Bahar 1998, Nuyens 1995, Abdul Baki et al. 1993, Briaud 1991, Cassan 1988, Holtz & Krizek 1972).

Cette communication présente les résultats d'une série d'essais pressiométriques, d'essais au pénétromètre statique et d'essais en laboratoire réalisés sur deux sites de la région d'Alger. Une étude des corrélations entre paramètres pris deux à deux, à l'échelle d'un sondage ou du site, est présentée.

Cette étude était compliquée par le fait que la reconnaissance géotechnique n'avait pas été prévue pour l'analyse des corrélations et que les sondages utilisés pour analyser celles ci n'étaient pas toujours très proches.

2 DESCRIPTION DES SITES

2.1 *Site de Dar-El-Beida*

Le site de Dar El Beida, destiné à la construction de la future gare centrale d'Alger, se situe à faible distance d'Alger (20 km environ); le relief est plat et peu incliné nord-sud.

L'ensemble des sondages permettent de distinguer quatre types de dépôts :
- un remblai sur toute la surface du site, atteignant une épaisseur de 4 à 5 m par endroit;
- des passages argileux grisâtres carbonatés et argilo-limoneux par endroit de moyenne à faible consistance, atteignant 20 m d'épaisseur dans la partie sud du site et 9 m dans la partie centre, avec des nuances beige et marron grisâtre;
- des alluvions grossières (cailloux, galets) mélangés avec des argiles par endroit;
- des lentilles de sable fin légèrement argileux gréseux en profondeur, d'une épaisseur importante par endroit.

La Figure 1 montre la situation des sondages effectués sur le site de Dar-El-Beida. Parmi les essais mécaniques en place, on compte 14 sondages pressiométriques et 41 sondages pénétrométriques. De même, il a été réalisé 18 sondages carottés. L'ensemble des échantillons soumis aux essais de laboratoire sont prélevés sur des argiles carbonatées grisâtres molles et les argiles carbonatées grisâtres à graves calcaires.

Les essais pénétrométriques ont été réalisés au moyen du pénétromètre statique lourd Borro de 20 t. La pointe du cône a un angle de 60° et une surface à la base de 10 cm². La surface latérale du manchon est de 150 cm². Le fonçage a été réalisé à la vitesse de 2 cm/seconde. Les mesures se font tous les 20 centimètres. Les essais pressiométriques ont été réalisés à l'aide d'un pressiomètre Ménard en utilisant une sonde de volume initial de 535 cm³. Les forages pressiométriques ont été exécutés à l'aide d'une tarière mécanique.

Pour l'étude présentée dans cet article, sur les échantillons prélevés, les résultats des essais de laboratoire suivants, 74 granulométries, 74 densités, 61 essais oedométriques, 74 essais de cisaillement rectiligne à la boîte de Casagrande, ainsi que les résultats de 8 sondages pressiométriques (PR-01 à PR-08) et 8 sondages pénétrométiques statiques ont été exploités.

Les échantillons étudiés présentent les caractéristiques suivantes :

- Le pourcentage des particules inférieures à 2 microns varie entre 25 et 76%. L'analyse des différentes limites d'Atterberg, montre que le sol est de plasticité moyenne à élevée avec un indice de plasticité variant entre 16 et 46.6%. Les argiles étudiées sont inactives à actives par endroit.

- Partie sud : la teneur en eau varie de 15.3 à 32%, le degré de saturation est supérieur à 93% et le poids volumique du sol sec compris entre 14.1 et 17.6 kN/m³. L'indice de consistance est compris entre 0.53 et 1.14. Le coefficient de compressibilité est compris entre 0.11 et 0.32. Le sol est surconsolidé sur les six premiers mètres. On note un sol sous-consolidé à normalement consolidé à partir du 6 m. L'indice de gonflement est compris entre 0.011 et 0.106. Les échantillons étudiés présentent un angle de frottement variant de 12 à 29° et une cohésion C comprise entre 25 et 130 kPa.

- Partie nord : la teneur en eau varie de 14 à 23.7%. Le degré de saturation est supérieur à 95%. Le poids volumique du sol sec compris entre 16.1 et 18.9 kN/m³. L'indice de consistance est compris entre 0.95 et 1.13. Le coefficient de compressibilité est compris entre 0.09 et 0.97 à l'exception des résultat du sondage 10. L'indice de gonflement est compris entre 0.021 et 0.77. Le sol est en général gonflant. Les échantillons étudiés présentent un angle de frottement variant entre 12 et 29 ° et une cohésion C comprise entre 20 et 80 kPa. Le sol est surconsolidé sur les six premiers mètres. A partir de 6 m de profondeur, on note un sol normalement consolidé à surconsolidé.

Les Figures 2 (a), (b) (c) présentent les profils de pression limite conventionnelle de Ménard, p $_L$, du module pressiométrique, E_p, et de la résistance de pointe, q_c, en fonction de la profondeur. Dans la partie sud, le rapport E_p/p_L est compris entre 5.8 et 16 dans les six premiers mètres. Il est en général inférieur à 8 à partir de 6 m de profondeur. Dans la partie nord, ce rapport varie entre 7.3 et 11.6 dans les six premiers mètres et entre 5 et 25 à partir de 6 m de profondeur.

2.2 Site de Bordj-Ménail

Le site de Bordj-Ménail se situe à 70 km à l'Est d'Alger. Cinq (05) forages pressiométriques et dix (10) sondages au pénétromètre statique lourd ont été réalisés sur ce site. La Figure 3 illustre le plan de situation de ces forages.

L'identification visuelle des échantillons prélevés révèle un sol de nature argileuse entre la surface du sol et 12 m de profondeur, légèrement sableux ou limoneux par endroit.

Les Figures 4 (a), (b) (c) présentent les profils de pression limite conventionnelle de Ménard, du module pressiométrique et de la résistances de pointe en fonction de la profondeur. Le rapport E_p/p_L est en général supérieur à 10. Il est compris entre 6 et 40.

3 CORRELATIONS ENTRE PARAMETRES

La régression linéaire simple est souvent utilisée en mécanique des sols, dans le but d'analyser les relations des différents paramètres physiques et mécaniques des sols en expliquant un paramètre au moyen d'un ou plusieurs paramètres.

, Compte tenu de leur importance dans la connaissance des différents caractéristiques des sols, nous avons analysé ces corrélations entre paramètres géotechniques sur les deux sites décrits ci-dessus.

La relation linéaire cherchée a pour forme générale Y=AX+B. Le calcul de la corrélation entre les paramètres des sols deux à deux est effectué dans les cas suivants :

- corrélation entre les propriétés deux à deux par sondage;

- corrélation entre les propriétés des sondages voisins;

- corrélation entre les propriétés deux à deux sur tout le site, en prenant en compte les couples de valeurs des sondages voisins qui existent dans la zone étudiée.

Parmi les nombreux couples de paramètres dont les relations ont été analysées de façon systématique, certains sont de même nature et ont des corrélations générales qui ne sont pas attachés au site ou au sondage (on peut donner comme exemple les couples de paramètres de nature minéralogique : w_L, w_p, et I_p). D'autres paramètres sont liés à l'état actuel du sol et

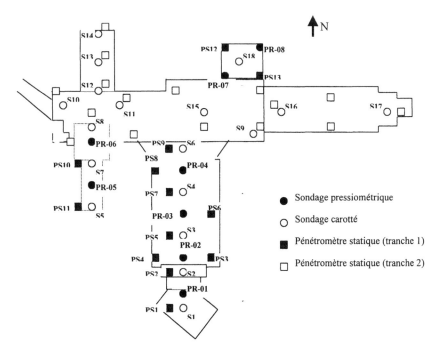

Figure 1. Implantation des sondages.

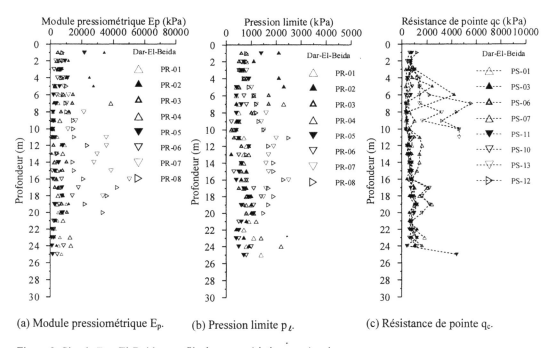

(a) Module pressiométrique E_p. (b) Pression limite p_ℓ. (c) Résistance de pointe q_c.

Figure 2. Site de Dar-El-Beida : profils des caractéristiques mécaniques.

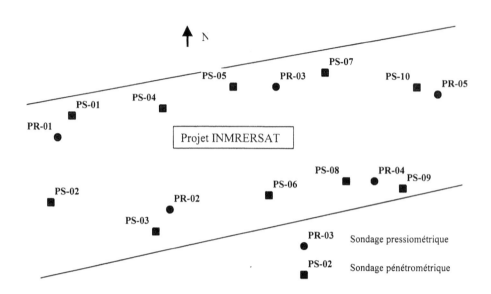

Figure 3. Site de Bordj-Ménail : implantation des sondages.

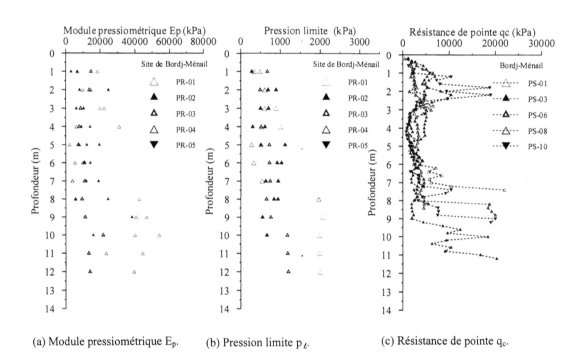

(a) Module pressiométrique E_p. (b) Pression limite p_ℓ. (c) Résistance de pointe q_c.

Figure 4. Site de Bordj-Ménail : profils des caractéristiques mécaniques.

pas seulement à sa nature : teneur en eau, résistance au cisaillement, etc.

3.1 Corrélations entre paramètres physiques

Les Figures 5 (a) à 5 (i) et le tableau 1 présentent les valeurs des paramètres dont on a étudié les corrélations et la droite de régression calculée. L'ensemble des résultats obtenus dans cette étude permettent de formuler les commentaires suivants :

pour ce qui concerne les corrélations deux à deux dans le même sondage, on retrouve :
- une forte corrélation entre teneur en eau naturelle et poids volumique du sol sec représentée par une droite de régression décroissante (le coefficient de corrélation R > 0.95 pour tous les sondages) ;
- une forte corrélation entre le poids volumique du sol humide et le poids volumique du sol sec avec une régression croissante (R varie de 0.95 à 0.99 pour tous les sondages);
- une forte corrélation entre la limite de liquidité et l'indice de plasticité (R > 0.95 pour tous les sondages) ;
- les coefficients de corrélations obtenus entre l'indice de plasticité, la limite de liquidité et le pourcentage des particules de diamètre inférieur à 2 microns sont faibles en général.

Pour ce qui concerne les corrélations entre deux paramètres sous l'ensemble du site, on notera que les niveaux de signification obtenus sont assez élevés en général à l'exception des régressions entre (I_p, w_L) et le pourcentage des particules de diamètre inférieur à 2 microns (Tableau 1)

Tableau 1. Paramètres des régressions linéaires entre les propriétés deux à deux pour tout le site.

Site de Dar-El-Beida				
Y	X	Corrélation	R	Nb. points
w	w_L	$w = 2.16\ w_L$	0.97	54
w	γ_h	$w = -5.1\ \gamma_h + 126.5$	-0.92	74
w	γ_d	$w = -3.55\ \gamma_d + 82$	-0.97	74
w	I_p	$w = 0.73 I_p$	0.96	74
w_L	I_p	$w_L = 0.71\ I_p - 6.95$	0.97	54
γ_h	γ_d	$\gamma_h = 0.64\ \gamma_d + 9.60$	0.98	74
w_L	I_c	$w_L = 56.3\ I_c$	0.985	54
w_L	<2%	$w_L = 0.15 <2\mu + 48.5$	0.20	37
I_p	<2%	$I_p = 0.002 <2\ \mu + 20.2$	0.29	49

3.2 Corrélation entre pressiomètre - pénétromètre statique

Site de Dar-El-Beida

Dans cette analyse, nous avons pris en compte que les résultats d'essais réalisés dans les couches d'argiles du site de Dar-El-Beida. Sur les Figures 6

(a), (b) et (c), nous avons reporté les résultats de 135 essais réalisés dans les argiles saturées plastiques à très plastiques de résistance variée. Le nuage de points obtenus, accuse une certaine dispersion, mais suggère une relation linéaire.

Le tableau 2 présente les corrélations entre la pression limite nette $p*_l$ et la résistance de pointe nette $q*_c$ provenant de deux sondages voisins. Les corrélations obtenues entre $p*_l$, $q*_c$ et le module pressiométrique E_p en considérant les parties sud et nord du site, sont données par les équations suivantes :

- partie sud

$$q*_c = 1.1\ p*_l \qquad (R=0.72) \qquad (1)$$

$$E_p = 4.9\ q*_c \qquad (R=0.66) \qquad (2)$$

$$E_p = 8.8\ p*_l \qquad (R=0.91) \qquad (3)$$

- partie nord

$$q*_c = 2.3\ p*_l \qquad (R=0.91) \qquad (4)$$

$$E_p = 4.17\ q*_c \qquad (R=0.92) \qquad (5)$$

$$E_p = 13.6\ p*_l \qquad (R=0.91) \qquad (6)$$

On note un niveau de signification élevé de la corrélation entre $p*_l$ et $q*_c$: on obtient 8 coefficients supérieurs à 0.6 sur les huit (8) couples de sondages voisins. Pour ce qui concerne les corrélations entre deux paramètres sous l'ensemble du site, on notera que les niveaux de signification obtenus sont en général assez élevés. Le rapport $q*_c/p*_l$ trouvé est différent de celui donné par d'autres auteurs pour les argiles. Les résultats de $E_p/q*_c$ obtenus sont à rapprocher de ceux de Van Vambecke (1975) qui donne une valeur de 4.5 dans les argiles normalement consolidées et une valeur entre 5 et 7 dans les argiles surconsolidées.

Tableau 2. Paramètres des régressions linéaires (Y=AX) entre $q*_c$ et $p*_l$ par sondages voisins.

Sondage	Nb.Points	$q*_c/p*_l$	R
PR-01-PS01	22	1.05	0.92
PR-02 -PS-03	17	0.96	0.94
PR-03 - PS-06	17	0.89	0.64
PR-04-PS-07	15	1.34	0.83
PR-05-PS-11	23	1.17	0.70
PR-06-PS-10	22	1.26	0.87
PR-07-PS-13	10	1.96	0.94
PR-08-PS-13	10	2.83	0.92

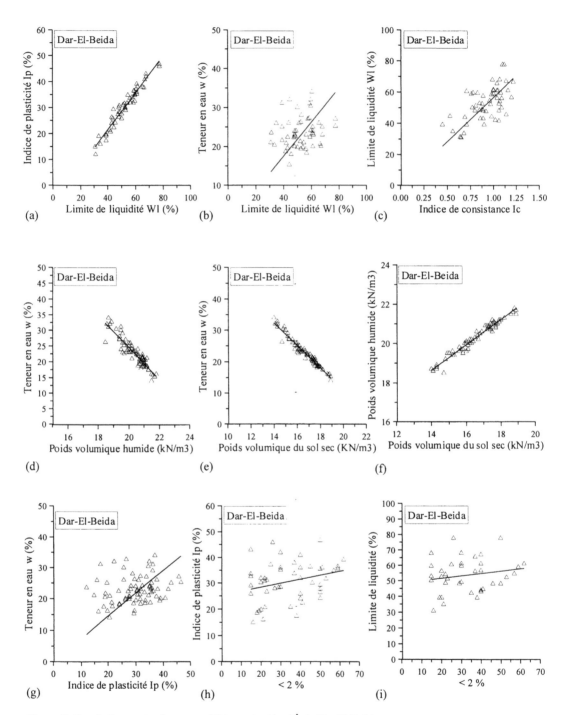

Figure 5. Corrélations entre deux propriétés sur tout le site de Dar El Beida.

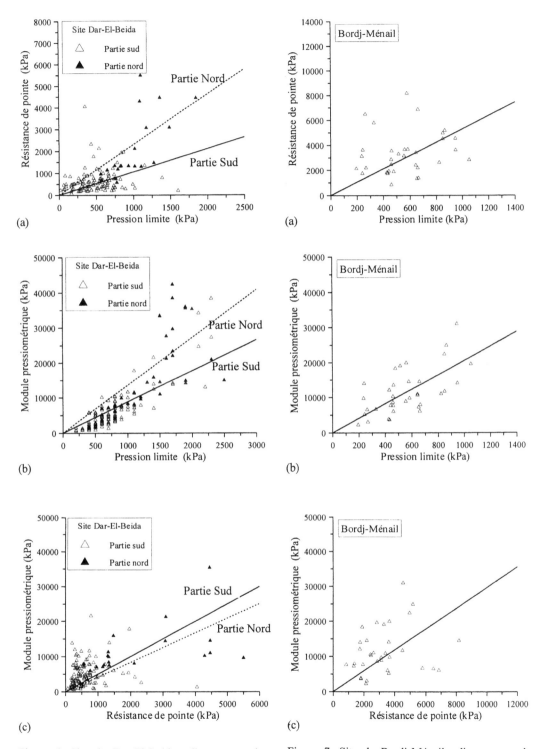

Figure 6. Site de Dar-El-Beida : diagrammes des corrélations entre p*$_l$, q$_c$* et E$_p$.

Figure 7. Site de Bordj-Ménail : diagrammes des corrélations entre p*$_l$, q$_c$* et E$_p$.

Site de Bordj-Ménail

Sur le site de Bordj-Ménail, on distingue deux couches d'argiles. La première est comprise entre le terrain naturel et 7.5 m de profondeur et la deuxième couche est comprise entre 7.5 m et 12 m de profondeur. Le tableau 3 présente les corrélations entre la pression limite nette p^*_l et la résistance de pointe nette q^*_c provenant de deux sondages voisins pour la première couche. On note un coefficient de corrélation élevé. Pour les cinq (05) couples de sondages voisins, on obtient des coefficients supérieurs à 0.8.

Sur les Figures 7 (a), (b) et (c), nous avons reporté les résultats d'essais réalisés dans la première couche. Les droites de régression obtenues entre la pression limite nette, la résistance de pointe nette et le module pressiométrique pour les deux couches sont données par les équations suivantes:

- première couche (0 – 7.50 m):

$$q^*_c = 5.29 \ p^*_l \qquad (R=0.86) \qquad (7)$$

$$E_p = 2.96 \ q^*_c \qquad (R=0.83) \qquad (8)$$

$$E_p = 20.5 \ p^*_l \qquad (R=0.92) \qquad (9)$$

- deuxième couche (7.50 – 12 m):

$$q^*_c = 3.1 \ p^*_l \qquad (R=0.86) \qquad (10)$$

$$E_p = 4.15 \ q^*_c \qquad (R=0.83) \qquad (11)$$

$$E_p = 21.3 \ p^*_l \qquad (R=0.92) \qquad (12)$$

Tableau 3. Paramètres des régressions linéaires (Y=AX) entre p^*_l et q^*_c par sondages voisins.

Sondage	Nb.Points	q^*_c/p^*_l	R
PR-01-PS01	7	6.32	0.95
PR-02 -PS-03	6	7.00	0.87
PR-03 - PS-06	7	5.60	0.83
PR-04-PS-07	7	4.52	0.88
PR-05-PS-11	6	3.68	0.93

4 CONCLUSION

L'analyse des corrélations entre les paramètres physiques deux à deux par sondage et sur tout le site montre une bonne linéarité entre les paramètres étudiés.

Pour les argiles de Dar-El-Beida, la corrélation obtenue entre la résistance en pointe et la pression limite conventionnelle de Ménard est différente de celle donnée par d'autres auteurs. Les résultats de E_p/q^*_c obtenus sont à rapprocher de ceux de Van Vambecke (1975).

Pour les argiles de Bordj-Ménail, les corrélations obtenues sont assez proches de celles données par d'autres auteurs pour les argiles normalement consolidées à surconsolidées (Briaud, 1992).

La présente étude a été menée avec l'objectif de présenter quelques résultats d'étude d'argiles algériennes. Elle est donc qu'une ébauche qui pourrait être étendue ultérieurement à des centaines de résultats non encore exploités.

REMERCIEMENTS

Les résultats d'essais présentés dans cette communication nous ont été aimablement fournis par le Laboratoire National de l'Habitat et de la Construction de Rouiba, Algérie.

REFERENCES

Abdul Baki, A., J.P. Magnan & P. Pouget 1993. Analyse probabiliste de la stabilité de deux remblais sur versant instable. *Etudes et recherches des laboratoires des Ponts et Chaussées.* GT56. Série géotechnique.

Bahar, R. 1998. Properties of clays from Menard pressuremeter test results, *Proc. of Int. Symposium on Geotechnical Site Characrerization. ISC'98.* Vol. 2 : 735-740. Atlanta : USA.

Briaud, J.L. 1992. The pressuremeter. Rotterdam: Balkema.

Cassan, M. 1988. Les essais in situ en mécanique des sols. Tome1. Editions Eyrolles.

Holtz, R.D. & R.J. Krizek 1972. Statistical evaluation of soil test data. *Proc. 1st ICASP.* 229-266. Hong Kong.

Nuyens, J. 1995. Essais en place dans l'argile des Flandres. *Revue Française de géotechnique*, n° 70. 37-54.

Van Vambecke, A. 1975. Les corrélations entre caractéristiques géotechniques. *Annales des Travaux publics de Belgique*, n° 4.

Geotechnics for Developing Africa, Wardle, Blight & Fourie (eds) © 1999 Balkema, Rotterdam, ISBN 90 5809 082 5

Industrial liquid wastes and the integrity of calcareous sand from Ain Helwan, Cairo

A. M. Radwan & F. A. Baligh
Civil Engineering Department, Helwan University, El-Mataria, Cairo, Egypt

A. S. El-Sayed
Drainage Authority, Ministry of Public Works and Water Resources, Egypt

ABSTRACT: A laboratory study was conducted to investigate the effect of the percolation of water contaminated with chemicals from industrial activities on the integrity of calcareous sand particles. Distilled water and six prepared chemical solutions, which represent the most common waste effluents in the studied region, were used in the leaching of natural soil samples under a certain effective stress. A special laboratory setup was developed to facilitate the control of changes in axial strain and permeability of soil samples during the leaching process. Test results show that the integrity of calcareous sand particles is influenced by being exposed to some industrial liquid wastes, depending on the ability of chemicals in liquid wastes to increase the dissolution of carbonate salts from calcareous soils. This causes possible erosion to soil grains and changes in relationships established between soil particles leading to their compression.

1 INTRODUCTION

Recently, many factories have been constructed in the Ain Helwan region, Cairo. Most of these factories do not have their own waste treatment facilities or sewerage networks. Large amounts of industrial liquid wastes are discharged directly to the soil in the surrounding environment. Due to the calcareous nature of soils in the study region, a possible reduction in soil strength may be expected. The objective of the present investigation, is to develop a better understanding of the interaction between calcareous sand soil and some chemicals in the industrial liquid wastes. An experimental study was conducted to investigate the axial strain response during the percolation of different chemical solutions. The dissolution of carbonate salts from soils causes a mass-loss in the soil matrix leading to ground collapse.

2 BACKGROUND

2.1 Calcareous soils in Ain Helwan region

The term calcareous identifies soils having a percentage of eight or more by weight, as a divalent carbonate, usually estimated in a calcium carbonate form. Calcareous soils in Ain Helwan region are pebbly sands and silts cemented mainly with carbonate salts resulting from the precipitation of calcium carbonate found in ground water when it emerges into the atmosphere and loses CO_2. Many layers of cemented soils and gravel are precipitated from evaporating water at the surface, which are called caliche deposits.

During construction of Helwan University from 1994 to 1996, many caves, caverns and cavities were observed in the excavated sections. These voids derived from the dissolution of calcium carbonate salts from soil to the water flowing above water table. Many case histories show clearly that the dissolution of sulfate, chloride, and carbonate salts results in settlement of structures (Salih & Trenter, 1989).

2.2 Chemical interaction between water and carbonate sediments

The solubility of $MgCO_3$ and $CaCO_3$ depend on the temperature, the pH, the concentration of other salts in the soil solution and on the concentration of carbon dioxide in soil air. Water is an agent of solution by virtue of its acidity. Water becomes a dilute carbonic acid (H_2CO_3) by dissolving carbon dioxide, it normally acquires CO_2 as it falls through the atmosphere and acquires more as it seeps through soils. Carbonic acid attacks calcium carbonate in soil by stripping off the Ca^{++} and Mg^{++}, which are then carried off in solution. The solvent then picks up bicarbonate (HCO_3^-). Water that drains humus-rich wastes also picks up organic acids, and falling through industrial or urban air pollution turns it into sulfuric acid. Finally, liquid wastes seeping out of industrial complex or urban wastes seems to contain various organic and inorganic salts and can be highly acidic. The following chemical equations are given to represent the reaction between calcium ca bonate salts and some salts or acids in water:

$$CaCO_3 + H_2O \longrightarrow Ca^{++} + HCO_3 + OH^-$$

$$CaCO_3 + 2CH_3COONa \longrightarrow (CH_3COO)_2Ca + Na_2CO_3$$

$$CaCO_3 + (NH_4)_2SO_4 \longrightarrow CaSO_4 + (NH_4)_2CO_3$$

$$(NH_4)_2CO_3 \rightleftharpoons NH_3\uparrow + CO_2\uparrow + H_2O$$

$$CaCO_3 + 2HCl \longrightarrow CaCl_2 + H_2O + CO_2$$

$$CaCO_3 + 2CH_3COOH \longrightarrow (CH_3COO)_2Ca + H_2CO_3$$

$$H_2CO_3 + CaCO_3 \xrightarrow{Wet} Ca(HCO_3)_2$$

$$Ca(HCO_3)_2 \xrightarrow{dry} CaCO_3 + CO_2 + H_2O$$

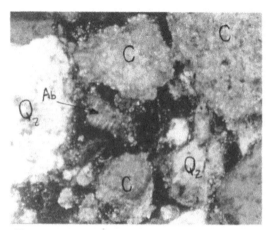

Photo 1. Calcareous sand structure (after Hussein 1989)

3 EXPERIMENTAL INVESTIGATION

3.1 Soil Samples:

The natural soil used in this study is a non-plastic, calcareous, angular, and well graded sand. This soil is classified as (SW-SM) according to the Unified soil classification system. The gradation curve and index properties for this soil are shown in Figure 1 and Table 1, respectively. The maximum and minimum densities of the soil established in accordance with (ASTM-SPT 523) procedure are 1.86 and 1.55 gm/cm^3, respectively. Chemical analysis of the soil in this region indicates that the major salts presented are $CaCO_3$ and $MgCO_3$. Quartz, calcite, and dolomite are the most insoluble materials. Figure 2 shows the percentage of calcium carbonate in each sand fraction. It is worth noting from the results that the percentage of calcium carbonate present in the coarse fraction is more than it's content in the fine fraction. This indicates the presence of carbonate salts as a coating for coarse sand particles, in addition to it's presence in silt bonds between grains as shown in Photo 1.

3.2 Chemical Solutions:

The following chemicals were used in leaching solutions to represent the most severe liquid wastes which discharged from various industrial activities.
1. Hydrochloric acid: was chosen to represent inorganic acids and used at 6% HCl concentration.
2. Acetic acid: chosen to represent organic acids and used at a 1%, 3%, 6% CH_3COOH concentrations.
3. Ammonium Sulfate: represented inorganic

salts. It was received in crystalline form and mixed with water as a 10% NH_4SO_4 solution.
4. Sodium Acetate: represented organic salts. It was received in crystalline form and was mixed with water. It was used in a solution of 1 normality of CH_3COONa.

3.3 Chemical analysis:

The chemical analysis of leaching and leached-out solutions were carried out according to the Egyptian Standard Methods, (1995). The determination of the individual cations of the soluble salts in water included Na^+, K^+, Ca^{++}, Mg^{++}. The electrical conductivity (EC) is measured based on the principle that the amount of electrical current transmitted by a salt solution under standardized conditions (25^0C) increases as the salt concentration in the solution is increased. The EC values can be converted to equivalent (TDS) (total dissolved solids) through the approximate relationship:

$$\mathbf{TDS}(mg/l) = 640 \times \mathbf{EC}(mmhos/cm).$$

4 APPARATUS

A special apparatus was developed to simulate the case of a long-term leaching process for soil under foundations. The apparatus is shown schematically in Figure 3. It consisted mainly of :
1. Soil cylinder with an inside diameter of a - proximately 106mm, and 250mm long. The cylinder was connected with a copper tube at 15mm from the top edge of the cylinder. This tube was connected to a suitable drain hose. The steel cylinder was welded (at the bottom) to a circular perforated steel disk 10mm thick and 150mm diameter.

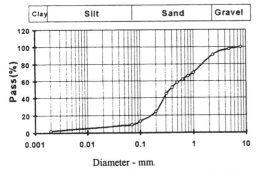

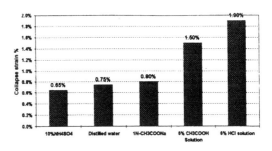

Fig. 1. Particle size distribution for natural soil from Ain Helwan

Fig. 4 Comparison between collapse strain due to inundation of natural soil by different liquids at normal stress of 88 kN/m²

Table 1. Specific gravity and index properties of fine graind soils

Fine part of Natural soil from	Description	Gs	% Passing From # 200	% Silt Content	% Clay Content	L.L %	PI %	Free Swell %
Ain Helwan region	Gray to brown (calcareous)	2.67 gm/cm3	9	7	2	19	N.A*	3

* N.A = not applicable for measuring

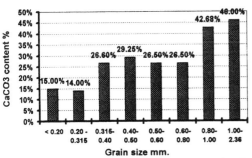

Fig. 2 Percentage of Caco₃ content in each fraction from calcareous sand

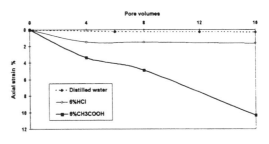

Fig. 5 Axial strain due to leaching with distilled water & different chemical solutions

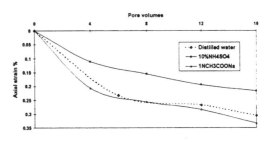

Fig. 6 Axial strain due to leaching with distilled water & different chemical solutions

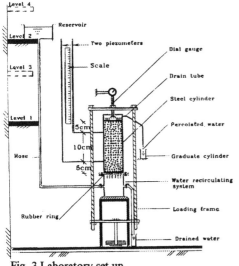

Fig. 3 Laboratory set up.

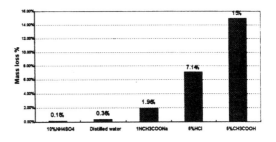

Fig. 7 Percentage of mass-loss after leaching with different chemical solutions

2. Water recirculation System consisted of a base steel cylinder with two valves (inlet and outlet). The inlet valve was connected to a storage reservoir, which was placed at a constant level according to the test procedure.

3. Loading system consisted of a loading frame, which is used to pressurize a circular perforated top cap. Two layers of filter were used at the top and the bottom of the soil sample. One was a synthetic filter "polypropylene 360" with 106mm diameter placed in direct contact with soil and followed by a wire screen No. 100. These were used to prevent fines-escape during the seepage of liquids through soil sample.

5 TEST PROCEDURE

Five natural soil samples were prepared by raining a known amount of soil into five leaching cells and vibrating the soil until the required density was attained. All samples were tested at an initial relative density of 60%. Then, a stress of 88 kN/m^2 was applied on the soil samples through the loading frame. After 24 hours, the reservoirs were filled with leaching solutions (6% HCl, 6% CH_3COOH, 10% NH_4SO_4, 1N CH_3COONa, and Distilled water) at level 1 to saturate all specimens for 48 hours. This would ensure good contact of the specimens with the side of the steel cylinder before leaching. After saturation of all samples, the inlet valve was closed and the inlet water level was then raised to a constant elevation to produce an initial hydraulic gradient of 5 in each leaching cell. When the set-up was ready for testing, the inlet valve was opened to start seepage of the liquid through the soil sample. The volume of leachate was collected in a bottle incrementally. A leached-out volume of 3000 cm^3 (about 6 pore volumes) was considered as a datum for axial strain measurements. After passing 3000 cm^3 of leaching fluid through the specimen, the inlet valve was closed and the outlet valve was opened. The specimen was left for 24 hours without leaching. The dial gauge reading was recorded at the end of this increment. After that, leaching was started again at the same inlet level with the same leaching solution for another increment. During each test increment the time required for seeping of each 1000 cm^3 (about 2 pore volumes) was recorded for permeability calculations. The chemical analysis was carried out for each leachate volume. After the passage of the required leachate volume, the soil samples were extruded from the cells and dried to measure the loss of weight due to the leaching process as follows:

Mass-loss (%) = (Wd1-Wd2) x 100 / (Wd1)

Where :

Wd1 = dry weight of the sample before testing
Wd2 = dry weight of the sample after leaching.

The sieve analysis were carried out on the extruded soil samples.

6 TEST RESULTS AND DISCUSSIONS

The initial collapse strain for the five tested samples due to the initial saturation is shown in Figure 4. It is clear from this figure that the presence of acids in water causes an increase in collapsibility of calcareous soil at inundation. Figures 5 and 6 demonstrate an increase in axial strains when the quantity of leachate volume increases. The samples leached with acidic solutions show a larger increase in axial strain during leaching, than the samples leached with distilled water, or salt solutions.

Figure 5 shows an increase in the axial strain of the sample leached with 6% acetic acid solution to be greater by 30 times than that for the sample leached with distilled water, and about 6 times more than that for the sample leached with 6% HCl solution. This can be due to the increase in the dissolving rate of organic carbonate salts found in the calcareous soil sample due to the percolated organic acid solution.

Figure 6 shows the axial strain curves for soil samples leached with distilled water, 1N_CH_3COONa and 10% NH_4SO_4 solutions. The sample leached with 1N_CH_3COONa solution is compressed more than the two other samples. This indicates that the increase of leaching efficiency with the presence of organic salts in water is greater than in the presence of mineral salts.

The previous observations are confirmed by measuring the percentage of mass loss accompaning the axial strain after leaching. The mass loss resulting from the leaching process in the five tested samples are shown in Figure 7. The increase in the mass loss of calcareous soil samples is due to the increase of dissolving the carbonate salts in calcareous soil during leaching. According to the chemical rule which says "like dissolve like" the organic carbonate in calcareous soils dissolved in organic solutions is more than in other mineral solutions.

The permeability measurements of collected leachate volumes in the five tested samples are shown in Figure 8. The permeability values decreased with the increase of pore volumes collected. The permeability of soil samples leached with acidic solutions (between $3x10^{-6}$ and $1x10^{-7}$) is less than

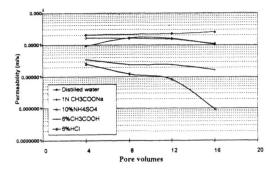

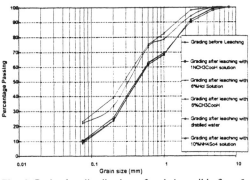

Fig. 8 Permeability variation during leaching with different chemical solutions

Fig. 9 Grain size distribution of real site soil before & after leaching with different chemical solutions

Table 2 Composition and contents of imposed chemical solutions

Leaching Solutions	pH Number	EC mmhos/cm	COD	Soluble Cations & Chloride (mg/l)				
				Na^+	K^-	Mg^{--}	Ca^{--}	Cl^-
6%-HCl	0.22	663.46	13702	72	1.64	0	0	57333
1%-CH₃COOH	3.3	0.9	0	2.53	50.7	2.28	87.4	17.8
3%-CH₃COOH	3.3	0.9	0	2.53	50.7	2.28	87.4	17.8
6%-CH₃COOH	2.95	2.75	5946	69	1.64	0	0	90.5
1N-CH₃COONa	8	72.37	5850	24840	1.64	1.08	17.4	2716
10%-NH₄SO₄	5.25	109.14	590	575	0.55	0	0	74.6
Distilled H₂O	6.94	0.001	0	0	0	0	0	0

Table 3 Pore fluid chemistry of the samples before & after leaching with different chemical solutions

Leaching Solutions	pH Number	EC mmhos/cm	COD mg/l	Soluble Cations & Chloride (mg/l)								Percentage of CaCO₃ content
				Na^+	K^-	Mg^{++}	Ca^{++}	Cl^-	SO_4^-	CO_3^-	HCO_3^-	
Natural soil	8	19.85	1380	4002	27.7	393	801	5740	3879	0	275	25.0%
Sample leached with 6%-HCl	8.1	2.44	2070	138	13.7	158	515	212	1699	0	232	28.7%
Sample leached with 6%-CH₃COOH	7.5	15.39	1426	1829	9.4	936	5746	151	10464	0	13542	29.0%
Sample leached with 1N-CH₃COONa	8.1	17.69	874	8050	22.2	31.2	221	254	14160	0	3782	23.4%
Sample leached with 10%-NH₄SO₄	6.9	51	1035	513	25.4	1192	800	4260	1680	0	420.9	23.0%
Sample leached with Distilled H₂O	7.9	16	1100	1840	27.3	300	700	1775	3830	0	500	24.0%

that in other samples (between 1 and 2×10^{-5}). The noticeable decrease in permeability may be due to (i) an increase in carbon dioxide within the specimen generated by the reaction of the acid solutions with $CaCO_3$ in soil, and/or (ii) some plugging up of voids by amorphous materials formed, if some fines are dissolved by acidic solutions.

The leaching process was stopped after the infiltration of 8 liters of liquid through a soil sample (about 16 p.v.). Then, soil samples were sieved to compare between the grain size distribution of soil samples before and after leaching Figure 9. The comparison shows an increase of fines content in the soil samples leached with acidic solutions from 9% (by weight) before leaching to 22% after leaching. The results indicate that erosion of calcareous sand particles during leaching by acidic solutions is more than its erosion when leached with other solutions.

7 CALCAREOUS SOIL-CHEMICAL INTERACTIONS

The compositional contents of the leaching chemical solutions are shown in Table 2. The analysis of pore fluid chemistry before and after leaching the tested samples is shown in Table 3. In order to investigate the leaching effects on soil, the electrical conductivity (EC) of the leachate resulting from samples were measured and compared with the electrical condu - tivity of the leachate resulting from the reference sample (leached with distilled water) and its original value in the leaching solution as shown in the graphs of Figure 10.

Breakthrough curves for different ions determine the relative concentration in the leachate comming out from the soil sample. If the contaminant conce - tration in the leaching solution is termed "Ci" (which is kept constant during a test), and that in the leachate collected from cells is "Co", then the rel - tive concentration is "Co/Ci". When the relative concentration is plotted versus the pore volume it represents the breakthrough curve for the contaminant.

The breakthrough curves of total dissolved solids (TDS) are shown in Figure 11, and those for calcium and magnesium cations are shown in Figures 12 and 13. The breakthrough value of total dissolved solids was found to be (Co/Ci=10) at pore volume No. 4. While the breakthrough values at the same pore volume do not exceed 1.0 for the other tested samples. It is interesting to note that the breakthrough value of the leachate resulting from the soil sample leached with 6% HCl solution is 0.20. This decrease in the relative concentration can be due to the neg -

tive adsorption of chloride, calcium and/or magnesium on soil particles under high acidity conditions. Calcium and magnesium ions in the leachate collected from this soil sample do not exist. The breakthrough curves in Figures 12 and 13 show that the increase in the breakthrough value of calcium and magnesium ions in the leachate resulting from the soil sample leached with acetic acid solution is higher than others. At leachate volume equivalent to (4×500 cm^3), the relative concentration of calcium ions is equal to 7000, and equals 800 for magnesium ions. These values represent high removal of calcium and magnesium ions from the calcareous soil sample during the percolation of acetic acid solution.

8 INVESTIGATION OF LEACHING CHARACTERISTICS USING DIFFERENT CONCENTRATIONS OF ACETIC ACID SOLUTIONS

The previous results show the effect of leaching with a 6% acetic acid solution on the increase of leaching efficiency and compressibility of tested soils during leaching. Another test series was done to investigate the effect of leaching by different concentrations. Four samples (with the same initial relative density of 60%) were leached with distilled water (under 88 kN/m^2 effective stress) for approximately six pore volumes (3000 cm^3 distilled water). This resulted in a similar initial state condition for all cells before leaching with an organic acid solution.

The axial strain due to leaching of the four ident - cal samples by distilled water and 1%, 3% and 6% acetic acid solutions are plotted versus the collected leachate volume in Figure 14. The plotted curves clarify the increase in axial strain of soil sample with the increase in concentration of acetic acid in the leaching solutions. Figure 15 represents EC variation in the leachate collected during the test compared to EC value for the different leaching solutions. The compositional analysis for 1%, 3% and 6% acetic acid solutions are given in Table 2. The gradual increase in the electrical conductivity of leachate with the gradual increase in the concentration of acetic acid in leaching solution is observed from the results in Figure 15.

The permeability measurements during leaching the samples with 1%, 3% and 6% acetic acid solutions are presented in Fig. 16. The average permeability of the soil sample leached with 6% acetic acid solution is equivalent to 2.5×10^{-6} m/s which is lower than the permeability of the soil samples leached with 3% and 1% acetic acid solutions which are equivalent to 7×10^{-6} and 1×10^{-5} m/s respectively. This reduction in permeability with the increase in

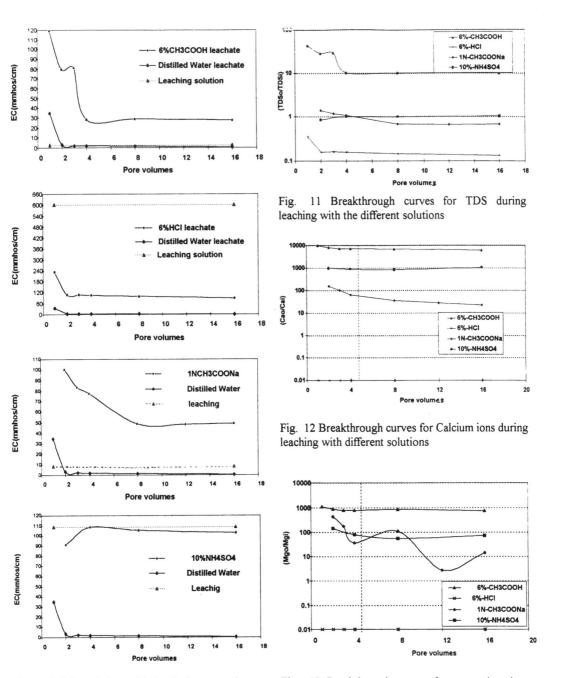

Fig. 11 Breakthrough curves for TDS during leaching with the different solutions

Fig. 12 Breakthrough curves for Calcium ions during leaching with different solutions

Fig. 13 Breakthrough curves for magnesium ions during leaching with different solutions

Fig. 10 EC variation with leached pore volumes resulting from tested samples

concentration of acetic acid in leaching solution is due to the strength of the chemical reaction between calcareous sand particles and acetic acid.

In order to ensure the logical interpretation for the increase of axial strain and leaching efficiency with the increase in acetic acid concentration in leaching solution, The percentage of mass-loss was estimated after leaching as shown in Figure17. The percentages of mass-loss are 5.5, 10, and 17% in soil samples leached with 1, 3, and 6% CH3COOH solutions, respectively, and 0.36% for soil sample leached with distilled water. The relation between compression ratio and mass loss after leaching of the tested samples is shown in Figure 18.

9 CONCLUDING REMARKS

The investigated parameter in this experimental study was the effect of the quality of infiltrating liquid on leaching efficiency. Based on test results, the following conclusions can be drawn:

1. Leaching of calcareous sand soil can lead to its compression.
2. Axial strain response of calcareous sand soils is affected by the presence of acids in liquids more than other chemicals.
3. The stability of a calcareous sand unit during the percolation of different liquids through soil, is conditioned by chemical interaction between calcareous soil particles and the percolated liquids.
4. Organic acid solutions (represented by acetic acid) cause a major leaching effect on calcareous soils.
5. Leaching of organic carbonate salts from calcareous sand soils increases with the increase of organic acid concentration in leaching solution.
6. The compression of soil samples due to leaching is directly proportional to the mass loss occuring during leaching.

REFERENCES

Airey, D. W. 1991. Proceeding of the 9th Asian Regional Conference on Soil Mechanics and Foundation Engineering, Vol. 2, No. 9, pp. 297. Bangkok, Thailand, December.

Al-Sanad, H., & B. Al-Bader 1990. Laboratory study on leaching of calcareous soil from Kuwait. ASCE, Journal of Geotechnical Engineering, Vol. 116, No. 12., December.

Beckwith, G. H. & L. A. Hansen 1982. Calcareous soils of southwestern United States. Geotechnical properties, behavior, and performance of calcareous soils, ASTM STP 777, K. R. Demars & R. C. Chaney, Eds., American Society for Testing and Materials, pp. 16-35.

El-sayed, S. A.1999. A study of some environmental impacts on soil properties in Ain Helwan region. M.Sc. Thesis, Civil Engineering Dept., Helwan University, El-Mataria, Cairo Egypt.

Hussein, M. H. 1989. Effect of grading and relative density of calcareous sand on its geotechnical design parameters. M.Sc. Thesis, Civil Engineering Dept., Helwan University, El-Mataria, Cairo Egypt.

Salih, A. 1989. Geotechnical problems associated with the rising groundwater in Arriyadh. Proc, 2nd Symp. on Geotech. Problems in Saudi Arabia, 1, 235-285.

Trenter, N. A. 1989. Some geotechnical problems in the Middle East. Keynote Address, 1st Reg. Conf. Civ. Engrg., Univ. of Bahrain, 1, 1-22.

Yong, R. N., D. S. Elmonayeri & T. S. Chong 1985. The effect of leaching on the integrity of a natural clay. Engineering Geology, 21, pp. 279-299.

Yong, R. N., A. M. O. Mohamed & B. P. Warkentin 1992. Principles of contaminant transport in soils. Elsevier scientific publishing co., Amsterdam.

Geotechnics for Developing Africa, Wardle, Blight & Fourie (eds) © 1999 Balkema, Rotterdam, ISBN 90 5809 082 5

Calibration pressure cells to measure pressures in structures retaining granular solids

G. E. Blight & A. R. Halmagiu
University of the Witwatersrand, Johannesburg, South Africa

ABSTRACT: The investigation of horizontal pressures in full size retaining structures by means of built-in pressure cells requires the development of methods for calibrating the pressure cells. Three different calibration methods are described and their drawbacks and advantages analysed. In the first method, the pressure cell, covered by the calibration material, is loaded by means of a rigid platen pressing onto the calibration material. In the second method, the pressure cell, covered by the calibration material, is loaded by means of an air bag, and in the third method the pressure cell is loaded directly by the air bag, without any covering material. The calibrations obtained for a single pressure cell in conjunction with the three methods of calibration, are compared and the most important factors affecting the calibrations are analyzed. A comparison is made with an ealier theory that indicates that the registration of a pressure cell depends on the ratio of the compressibilities of the calibration material and the cell itself. While the theory is valid, it is shown here that the calibration depends primarily on the compressibility of the pressure cell.

1 THE THREE CALIBRATION METHODS

Three calibration methods were used with three different pressure cells in over fifty calibration tests. The origins of the three pressure cells were as follows:

Cell 1: Tokyo Sokki Kenkyuio - Japan (TML-KD2D). This is a mercury-filled pressure cell with a pressure range from 0 to 200 kPa and working temperature range from 0 to 60°C.

Cell 2: J Smid - Czech Republic. This pressure cell measures normal and shear stresses. The pressure cell is based on a strain-gauged proving ring with a pressure range from 0 to 100 kPa and a working temperature range from 10 to 40°C.

Cell 3: A heat resistant pressure cell built to the design of the authors and having a pressure range from 0 to 200 kPa and a working temperature range from 0 to 140°.

To calibrate the pressure cells a 350 mm square calibration box with rigid steel walls was used. The cells were mounted one at a time in the box by bedding them in with a non-shrink grout. The grout supporting the pressure cell was carefully screeded to form a surface as nearly coplanar with the pressure-sensitive face of the cell as possible.

For each of the three pressure cells three different calibration methods were used. For two of the methods, cement powder, cement clinker and sand were used as the calibration materials, ie these materials were in contact with the pressure cell and were used to transmit the calibrating pressure to the cells.

The three calibration methods were as follows:

Method 1. The pressure cell, covered by the calibration material, was loaded by means of a rigid loading platen. The platen was pressed against the fill via a loading arm on which dead weights were hung. A sketch showing methods 1 is given as in Figure 1.

Method 2. The Pressure cell, covered by the calibration material, was loaded by means of an air bag. The air bag was inserted between the loading platen and the calibration material covering the pressure cell. The platen was locked in position and air at controlled and measured pressure was applied via the air bag. Figure 2 shows the application of method 2. The air bag was made

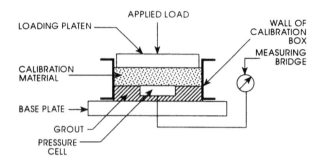

FIGURE 1: Sketch of the installation used in method 1.

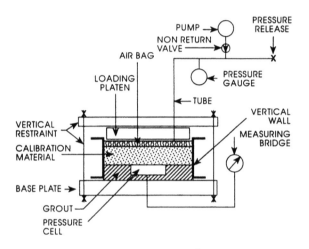

FIGURE 2: Sketch of the installation used in method 2.

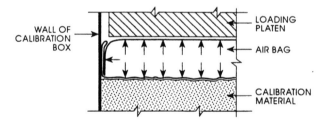

FIGURE 3: Detail of the folded edge of the air bag.

out of welded sheets of polyethylene. To make sure that a uniform pressure would be created over the whole surface of the material, the bag was made bigger than the calibration box and the edges were folded to obtain the required surface area. Figure 3 shows a detail near the box wall of a folded edge of the air bag under pressure.

Method 3. The pressure cell was loaded directly by means of the air bag. Method 3 is similar to method 2, but without having the calibration material between the air bag and the pressure cell. The pressure in the air bag was, in this method, the actual pressure applied to the sensitive face of the pressure cell. if R is the registration ratio of the

272

pressure cell (defined as $R = \sigma_c / \sigma_A$ where σ_c is the pressure recorded by the cell and σ_A is the applied pressure), then in method 3 the two pressures are equal and R is unity, ie the pressure cell registers perfectly.

For all three methods, three cycles of loading and unloading, using the same load increments were applied. It will be noted that the largest applied pressure did not exceed 60 kPa, which is the range of pressure in the silos for which the pressure cells were to be used. This is a much smaller pressure range than that used previously by Blight and Bentel (1988). (0 to 200 kPa.)

Peattie and Sparrow (1954) and Hvorslev (1976) developed a theory based on compatibility between the elastic compression of a layer of material in front of the pressure cell and the elastic compression of the pressure cell. This showed that the registration ratio of a pressure cell depends on the ratio of the compressibility of the material which is in contact with the pressure cell and the compressibility of the cell itself. According to the theory a cell will under-register in a silo during the filling process because the fill will expand towards the cell, as the cell compresses, thus locally reducing the stress in the fill in contact with the cell. A series of tests reported by Blight and Bentel (1988) lent support to Hvorslev theory.

2 RESULTS OF THE PRESENT TESTS

The three pressure cells all displayed similar behaviour in the tests made by the three calibration methods. The calibration curves obtained with the Czech pressure cell when tested by the three methods are shown in Figure 4 and are typical of those obtained for the other two cells.

For method 3 the response of all the pressure cells under the three cycles of loading and unloading was a straight line and no hysteresis was observed. Calibration by this method has a registration ratio R = 1 (perfect registration) and will be used as a basis for comparison. As should be expected, an identical response for all the pressure cells was recorded by method 3, no matter how many times the test was repeated.

The biggest hysteresis occurred with method 1. Repeated tests showed big differences between calibration coefficients and from perfect registration, even using the same material to transfer the pressure to the cell. The data summarized in Table 1 shows differences between repeat tests by method 1 that approach 65%. An explanation for these inconsistent results could be a non-uniform pressure applied by the loading platen to the calibration material. Because it is difficult to smooth the surface of the material perfectly, a non-uniform stress distribution that differs from test to test can be created between the calibration material and the pressure cell. In fact, this is almost impossible to avoid. The differences in response for the heat resistant pressure cell are closer together than for the other two cells. The surface area of the heat resistant pressure cell was more than five times the area of the other two cells and thus could probably better average the non-uniform stress distribution.

To obtain a uniform pressure over the calibration material the air bag was introduced and method 2 developed. Using method 2 the behaviour of the material in the shear box approaches much more closely the conditions under which a pressure cell would operate in a silo. Hysteresis still occurs with method 2 and hence the almost straight line for the first loading was taken as the calibration line.

The variation of the calibration coefficient from the perfect registration was small for repeated tests or for tests made with different calibration materials. Estimating the influence of the friction occurring between the wall of the calibration box and the calibration materials showed that a calculated response of the pressure cell which allowed for the effects of wall friction came very close to accounting for the hysteresis that we observed during the tests. The differences between the tests were much smaller (not exceeding 10%) compared with the 65% maximum difference when method 1 was used. The calibration coefficients were also much closer to the perfect registration, but there was no constant factor between perfect registration and the other tests. This was especially so for the Japanese pressure cell for which the perfect registration value should be the biggest.

3 DISCUSSION AND CONCLUSIONS

It appears that calibration factors for pressure cells used to measure pressure in granular or powdery materials and at small pressures (in the range of the horizontal pressures which usually occur in retaining structures) are influenced in only a minor

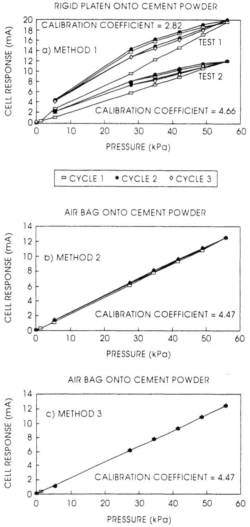

FIGURE 4: Responses for a pressure cell tested by three methods.

Table 1 Results for tests made by calibration method 1.

Type of pressure Cell	Calibration material	Cement powder	Sand	Cement Clinker	Air bag
		Calibration Factor			
		0.078	0.069		
Japanese			0.086		0.106
			0.095		
			0.076		
Czech		2.82	3.14		4.47
		4.66	2.73		
Heat resistant		0.099	0.085	0.128	0.092
		0.082	0.089		

way by the properties of the calibration material.

A comparison of two different calibrations made by method 1 under the same conditions but using pressure cells with different stiffnesses illustrates the importance of cell stiffness (Figure 5). The two calibrations were made with the heat resistant pressure cell, for which the stiffness can be adjusted. if it is assumed that for both tests the same amount of hysteresis was produced by friction of the material on the calibration box walls, the difference in hysteresis must be produced by the difference in stiffness of the pressure cells. As can be seen in Figure 5, the bigger the displacement of the sensitive face of the

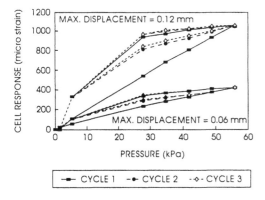

FIGURE 5: The influence of pressure cell displacement
on calibration

pressure cell (ie the lower the cell stiffness) the bigger the hysteresis. The most important factor affecting pressure cell accuracy thus appears to be the compressibility of the pressure cell, the calibration material having a lesser influence.

Theoretically, a perfectly stiff pressure cell will have a perfect registration in any condition, and this was demonstrated by the prototype "zero strain" pressure cell developed and tested by Blight and Bentel (1988). So far, however, this cell has not been developed beyond the laboratory stage.

REFERENCES

Blight, G E and Bentel, G M 1988 Measurements on Full Size Silos. Part 2: Pressures. *Bulk Solids Handling*, 8, 3: 343-346.

Peattie, K R and Sparrow, R R 1954 The Fundamental Action of Earth Pressure Cells. *Journal of the Mechanics and Physics of Solids*, 141-155.

Hvorslev, M J 1976 *The Changeable Interaction Between Soils and Pressure Cells; Tests and Reviews at the Waterways Experiment Station*. Technical Report S - 76 - U S Army Engineer Waterways Experiment Station, Vicksburg, USA.

Geotechnics for Developing Africa, Wardle, Blight & Fourie (eds) © 1999 Balkema, Rotterdam, ISBN 90 5809 082 5

Apport de la modélisation en centrifugeuse à l'analyse des pieux chargés latéralement

A. Bouafia

Maître de Conférences, Université de Blida, Alger, Algérie

RESUME: L'article présente quelques résultats d'une importante campagne expérimentale menée sur la centrifugeuse du centre de Nantes du LCPC (France) pour la modélisation du comportement des fondations profondes sollicitées horizontalement. Les résultats de l'étude paramétrique montrent que les pieux flexibles présentent un mécanisme de comportement différent de celui des pieux rigides. L'interprétation des courbes de moment fléchissant le long du pieu ont permis de construire les courbes de réaction P-Y et d'analyser le module de réaction latérale. Enfin, certains aspects particuliers du problème de flexion des pieux tels que l'effet de la proximité du talus et le comportement en grands déplacements dans le sable sont présentés.

ABSTRACT: This paper presents some results of an intensive experimental programme undertaken on the LCPC centrifuge, centre of Nantes (France) for the physical modeling of deep foundations behaviour under horizontal loading. The parametric study shows that flexible piles follow a behaviour mechanism different from this of rigid piles. The interpretation of bending moment curves along the pile has allowed to obtain the P-Y curves and analyze the subgrade reaction modulus. Thereafter, some particular aspects of laterally loaded piles like the proximity effect of a sloppy soil and the large displacements of piles in sand are studied.

1 INTRODUCTION

Actuellement, la modélisation géotechnique en centrifugeuse connaît un essor mondial considérable. En mécanique des sols, du fait de la forte influence du niveau des contraintes sur la réponse du sol et de la méconnaissance de sa loi de comportement, il est indispensable que les contraintes en modèle et en prototype soient identiques. Pour que la réduction des dimensions de l'ouvrage à l'échelle 1/N ne modifie pas son comportement, les lois de similitude stipulent qu'il faut augmenter N fois les forces de masse du prototype, ce qui peut s'obtenir par centrifugation du modèle. Depuis 1986, la modélisation en centrifugeuse est mise en oeuvre par le LCPC (Laboratoire Central des Ponts & Chaussées), centre de Nantes, dans le cadre de ses recherches sur le comportement des pieux selon un vaste programme expérimental.

Le comportement en flexion des fondations profondes est un problème d'interaction sol-pieu assez complexe vu son caractère tridimensionnel et la multitude des paramètres physiques mis en jeu. La littérature traitant de ce problème est assez riche. Cependant, les approches théoriques de calcul sont insuffisantes pour décrire la réalité. L'interaction sol/pieu y est modélisée par des schémas simplistes qui ne tiennent pas compte de certains facteurs pouvant influencer le comportement du pieu, comme par exemple, le mode d'installation du pieu dans le sol et la rugosité de l'interface sol/pieu.

Le présent article se propose de répondre à un certain nombre de questions à partir de l'interprétation des essais de chargement horizontal des modèles centrifugés de pieux implantés dans le sable du Rheu (France).

2 DESCRIPTION DES MOYENS D'ESSAIS

2.1 *Présentation de la centrifugeuse*

La centrifugeuse du LCPC a été décrite, ailleurs en détails (Garnier, 1990, Corté et Garnier 1986). En service depuis 1986, la centrifugeuse est conçue spécialement pour les recherches et essais en géotechnique. Ses performances peuvent être caractérisées par un rayon de 5.50 m, une accélération centrifuge maximale de 200 fois celle de la gravité terrestre (G) et une masse maximale embarquable du modèle de 2000 kg à 100xG et de 500 kg à 200xG.

2.2. Modèles de pieux

Trois modèles de pieux tubulaires métalliques ont été conçus pour l'expérimentation en centrifugeuse. Les caractéristiques des prototypes simulés sont résumées au tableau 1.

Chaque modèle a été instrumenté par 12 paires de jauges montées en 1/2 pont, à 3 fils. Le collage des jauges est fait à l'extérieur du pieu, sur deux génératrices diamétralement opposées. Les jauges ont été couvertes par une couche protectrice épaisse de 1 à 2 mm contre l'humidité et le frottement du sable. Les jauges de chaque modèle ont été étalonnées selon le schéma d'une console. Un écart de 5% au plus a été constaté entre les réponses mesurée et théorique des jauges. La surface de contact des modèles de pieux P_1 et P_2 a été rendue parfaitement rugueuse par collage du sable d'essai.

2.3 Construction du massif de sable

Le sable utilisé provient du site expérimental du Rheu (Rennes, France). Selon la classification USCS, il s'agit d'un sable propre mal gradué SP. Ce sable a été écrêté avec un tamis de 2.5 mm afin d'éliminer les gros grains et d'obtenir un massif à granulométrie plus homogène. Les essais de cisaillement à la boite à différentes densités ont permis d'obtenir l'angle de frottement interne de ce sable.

La caractérisation du massif de sable en centrifugeuse s'effectue par un contrôle de la densité, et par des essais de pénétration statique à l'aide d'un pénétromètre miniature embarqué en centrifugeuse.

Tableau 1. Caractéristiques des pieux prototypes

Pieu	P_1	P_2	P_3
Echelle 1/N	1/17.8	1/20	1/25
Fiche D (m)	5	5	3.1-5
Diamètre B (m)	0.50	0.90	0.50
Elancement D/B	10	5.5	6.25-10
Rigidité $E_p.I_p$ (MN.m^2)	56.6	741	628
Cote de l'effort e (m)	1.0	1.0	2.25

Outre la mesure de la résistance en pointe, l'essai permet à partir du diagramme pénétrométrique d'apprécier l'homogénéité du massif. Pour garantir l'homogénéité des massifs de sable, le LCPC a développé un appareil constitué d'une trémie mobile utilisant le principe de pluviation dans l'air a hauteur de chute et débit fixés.

2.4 Système de chargement latéral

Le chargement du pieu est exercé par le dispositif de chargement des fondations développé au LCPC. Comme l'illustre la figure 1, l'effort latéral est fourni par un câble métallique souple passant sur une poulie, et mis en tension par déplacement d'une masse mobile. Le moteur assurant le déplacement de la masse est commandé depuis la salle de contrôle en agissant sur son alimentation électrique, à l'aide d'une source de tension continue. Le principal avantage de ce dispositif outre sa simplicité de fonctionnement réside dans la possibilité d'appliquer des paliers d'effort quels que soient les déplacements du pieu.

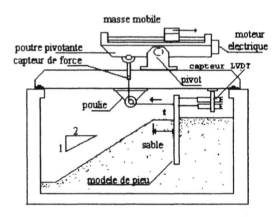

Figure 1. Schéma du modèle de pieu (cas de l'étude de l'effet du talus).

3 ETUDE PARAMETRIQUE

3.1 Programme d'essais sur modèles

Une étude paramétrique a été entreprise sur les modèles de pieux P_1 et P_2 afin de mettre en évidence l'effet de certains paramètres sur le comportement du pieu, comme :
1- la densité du massif sableux.
2- la rugosité de surface du pieu.
3- le mode d'installation du pieu dans le massif.
4- la teneur en eau ω du sable.

Le tableau 2 récapitule les données des différents essais menés sur les modèles centrifugés. Dans ce qui suit, les résultats présentés sont relatifs aux grandeurs prototypes.

3.2 Analyse des résultats d'essais

3.2.1 Déplacements en tête

La figure 2 récapitule les différents effets étudiés sur les déplacements en tête Y_H du pieu P_2. Ce dernier, considéré comme court et rigide, présente une sensibilité à la densité du sable.

Ainsi, les déplacements diminuent considérablement avec la densité du massif en un rapport de 0.27.

Tableau 2. Paramètres étudiés

Essai	I_d %	ω %	Installation	rugosité
1	57	0	pluviation autour du modèle	surf. lisse
2	63	0	idem	idem
3	93	0	idem	idem
4	95	0	idem	idem
5	93	0	fonçage	idem
6	94	0	fonçage	idem
7A	91	0	fonçage	sable
7B	91	7	forage en sable humide	collé en surface
8A	92	0	battage	idem
8B	92	7-9.5	battage en sable sec	idem

La rugosité de surface du pieu foncé fait diminuer les déplacements d'un rapport moyen de 0.80. Elle peut s'expliquer par une meilleure mobilisation des contraintes de cisaillement du sol avec les surfaces rugueuses.

Notons que la teneur en eau est mesurée juste après l'essai, au voisinage des pieux. La comparaison entre les déplacements des pieux battus dans le sable montre que les déplacements augmentent avec la teneur en eau, contrairement à ce qui était attendu sur l'effet de la cohésion capillaire apportée par l'eau. En passant d'un massif sableux sec à un massif humide de teneur en eau moyenne de 7 %, les déplacements augmentent selon un rapport moyen de 1.75.

Scott et Ting ont aussi constaté séparément, au cours des essais en centrifugeuse que les déplacements augmentent avec la teneur en eau (Bouafia,1990). De telles constatations expérimentales rejoignent les recommandations de Terzaghi de considérer des valeurs de module de réaction d'un sable saturé plus faibles que celles d'un sable sec, quelle que soit la densité du sable (Bouafia et Garnier, 1991).

La même figure montre aussi que le pieu P_2 est pratiquement insensible au mode de mise en place dans le sable.

Quant au pieu P_1, la figure 3 montre que l'effet de la densité est moins marquant. Le rapport des déplacements atteint 0.60 à 200 kN.

Le rapport des déplacements du pieu rugueux et ceux du même pieu lisse est de 0.90 à 200 kN. L'effet de la rugosité paraît moins important que dans le cas du pieu P_2. La même constatation a été faite par Lyndon et Pearson, au cours de leurs essais en centrifugeuse de l'Université de Liverpool, en 1988 (Bouafia, 1990). Les déplacements en tête augmentent avec la teneur en eau. Au cours des essais 8A et 8B (w =9.5 %) le rapport moyen des déplacements est de 1.50.

Quant à l'effet de l'installation du pieu, on a constaté que les déplacements en tête du pieu foncé (7A) sont plus faibles que ceux du pieu battu (8A) avec un rapport moyen de 0.75. Les déplacements du pieu foncé lisse (5) sont plus faibles que ceux avec sable en pluviation (4), avec un rapport variant de 0.5 à 0.75.

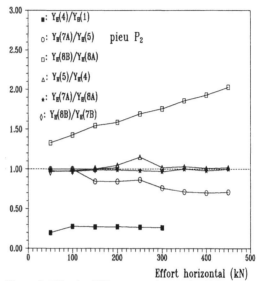

Figure 2. Effet des différents paramètres sur les déplacements du pieu P_2.

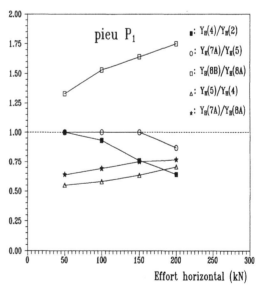

Figure 3. Effet des différents paramètres sur les déplacements du pieu P_1.

Il se dégage de ces constatations que le comportement du pieu P_1, est sensible au mode d'installation, contrairement au pieu P_2. Ainsi, le pieu P_1 paraît affecté par tous les paramètres précédents. Par contre, le pieu P_2 n'est sensible qu'à la densité, la teneur en eau du sable et la rugosité de surface.

Une étude paramétrique numérique par le biais du logiciel PILATE (PIeu sous efforts LATEraux) basé sur les courbes P-Y, a montré que le comportement du pieu P_2, considéré comme court et rigide, dépend essentiellement des conditions en pointe. Au contraire, le pieu P_1 considéré comme long et flexible, dépend plutôt des conditions latérales le long du pieu.

3.2.2 Moments de flexion

Il a été constaté que le moment fléchissant le long des deux pieux étudiés dépend peu des propriétés du sol, du mode d'installation du pieu ou de la rugosité d'interface sol/pieu.
En littérature, la notion du "point fixe" fut introduite pour définir le point de cote d tel que :

$$M_{max} = H.(e+d) \qquad (1)$$

Sa connaissance permet d'évaluer directement le moment maximum. La valeur moyenne de d est de 0.77 m pour P_1 et de 0.90 m pour P_2, quel que soit l'effort latéral, avec un coefficient de variation de 4%.

4 CONSTRUCTION DES COURBES P-Y DE REACTION LATERALE

4.1 Méthode de construction et résultats

A l'heure actuelle, la notion du module de réaction connaît une large application dans le calcul des fondations et des soutènements. La théorie du module de réaction se base sur la courbe P-Y, qui décrit la loi d'interaction sol/pieu. Géométriquement parlant, la réaction latérale P du sol est au signe près, la courbure du moment fléchissant. Une légère incertitude sur la courbe des moments va engendrer par conséquent, des variations importantes de la réaction latérale le long du pieu. Outre l'incertitude expérimentale sur les moments mesurés, les valeurs des réactions obtenues dépendent fortement de la courbe retenue pour le lissage des points expérimentaux. Par contre, le déplacement latéral résultant d'une double intégration des moments en est beaucoup moins sensible.
La méthode de lissage retenue est celle des fonctions spline quintiques paramétrées par un coefficient d'ajustement définissant le degré de fidélité du lissage aux points expérimentaux. Au cours de cette étude, ce coefficient a été choisi

tel que l'équilibre du pieu soit satisfait. Le logiciel SLIVALIC-5 du LCPC a été utilisé à cette fin.
La figure 4 illustre une courbe P-Y typique pour le pieu P_1. Le centre de rotation est à 3.5 m, soit 7 diamètres. Il est remarquable de noter qu'il coïncide exactement avec le zéro de la pression latérale, confirmant ainsi l'hypothèse de Winkler. On constate en outre le caractère non linéaire prononcé de ces courbes, même pour des faibles niveaux de déplacements.
Un calcul a rebours a été mené sur PILATE en introduisant les courbes P-Y ainsi obtenues. Une très bonne concordance a été constatée entre les réponses calculée et mesurée, ce qui présente une validation de la méthode de construction des courbes P-Y adoptée dans cette étude.

4.2 Etude du module de réaction latérale

Le module tangent E_{ti} du pieu P_1 illustré à la figure 5, augmente à peu près linéairement avec la profondeur, ce qui est conforme à certaines constatations expérimentales et recommandations théoriques dans la littérature des massifs pulvérulents.
Les valeurs du module tangent initial du pieu P_1 sont comparées au module de réaction de Ménard. Les caractéristiques pressiométriques du sable ont été déduites par corrélation avec celles de l'essai de pénétration statique réalisé avec le mini-pénétromètre embarqué, au cours des essais sur modèles. Cassan (1978), Van Wambecke et Al

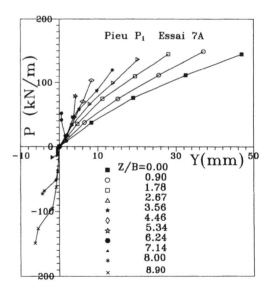

Figure 4. Courbes P-Y typiques du pieu P_1.

(1982) ont recommandé la corrélation suivante entre qc et le module pressiométrique E_m :

$$\frac{E_m}{q_c} = 1.5 \qquad (2)$$

Selon la figure 5, une très bonne concordance est à remarquer entre les modules de réaction.

L'institut japonais des recherches portuaires PHRI recommande une formule simple décrivant la courbe P-Y comme suit :

$$P = K_s.Z.Y^{1/2} \qquad (3)$$

La formule est dûe à Kubo (1965) qui a suggéré de déduire le coefficient K_s à partir de l'essai SPT.

Suite à des essais en centrifugeuse sur du sable sec et dense de Toyoura (Japon), Terashi et al (1989) ont proposé une loi simple qui permet de déterminer K_s à partir du diamètre du pieu.

Selon la figure 5, les valeurs des modules tangents déterminés à partir d'essais sur modèles, sont en très bonne concordance entre elles. Dans les deux expériences, les massifs de sable ont été construits par pluviation dans l'air.

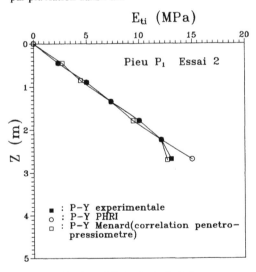

Figure 5. Profils du module de réaction latérale.

5 EFFET DE LA PROXIMITE D'UN TALUS SUR LA RESISTANCE LATERALE DU PIEU

5.1 Programme d'essais sur modèles

La présence des pieux implantés près d'un talus est assez fréquente dans la pratique (autoponts, quais, ...). La littérature traitant de ce problème complexe est assez rare. Trois autres paramètres géométriques interviennent : la hauteur du talus, son angle et la distance du pieu par rapport la tête du talus. Dans ce qui suit, on présente les résultats d'une étude en centrifugeuse du comportement du pieu P_2 implanté à différentes distances de la tête d'un talus sableux très dense (I_D= 95 %) ayant une hauteur de 3.30 m et une pente 2 :1, tel que montré en figure 1.

Le massif de sable est construit par pluviation autour des modèles dans le conteneur. Le terrassement du talus se fait après le remplissage du conteneur par du sable, par arasage progressif, à l'aide d'une règle guidée, jusqu'à l'obtention de la pente fixée. Avant la réalisation du talus, un modèle de pieu a été chargé dans le sol horizontal pour servir de référence.

5.2 Analyse des résultats d'essais

La figure 6 illustre l'effet de la distance t sur les déplacements en tête Y_H du pieu. Le coefficient I_{YH} est défini comme suit :

$$I_{YH} = \frac{Y_H(t)}{Y_H(\infty)} \qquad (4)$$

En tête du talus, les déplacements sont pratiquement triplés quel que soit le niveau d'effort. Au-delà d'une distance de 6.5 diamètres de la tête du talus, les déplacements n'en sont plus affectés et se stabilisent. Le comportement du pieu devient celui du cas du sol horizontal (Bouafia et Bouguerra, 1996).

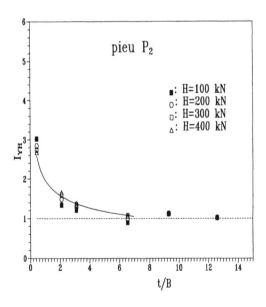

Figure 6. Effet de la proximité du talus sur les déplacements.

La proximité du talus a un faible effet sur les moments de flexion. Le moment maximum du pieu en tête du talus (t=0) augmente seulement de

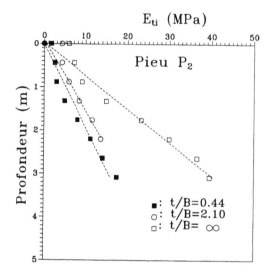

Figure 7. Effet de la proximité du talus sur le module de réaction latérale.

10% par rapport au cas du sol sans pente.

L'interprétation des courbes de moment selon la méthode décrite auparavant a permis de construire les courbes P-Y et d'en déduire le module tangent E_{ti} de réaction latérale. Selon la figure 7, la proximité du talus cause une réduction du module de réaction du pieu en tête du talus d'environ la moitié par rapport à celui du sol horizontal. Dans tous les cas, l'allure pratiquement linéaire du profil est conservée.

6 ETUDE DES GRANDS DEPLACEMENTS DU PIEU

6.1 *Programme d'essais sur modèles*

L'objectif de cette étude expérimentale est l'analyse de l'effet de l'élancement du pieu P_3 sur son comportement en grands déplacements, pour un diamètre et une cote de l'effort latéral donnés.

Le modèle est battu dans un massif de sable dense à l'aide d'un batteur comportant un mouton de 1.03 kg pouvant coulisser sans frottement sur la tige du batteur. La hauteur de chute a été fixée à 1.02 m. Le modèle est battu à différentes fiches dans le sable sec et dense consolidé au préalable par centrifugation.

Les essais de chargement sont effectués pour 4 valeurs d'élancement D/B, soit 6.25, 7.5, 8.75 et 10. Le chargement est assuré par un vérin hydraulique

ayant un déplacement fixé à une vitesse minimum de 1 mm/s, afin d'assurer un chargement statique continu. Chaque modèle est chargé horizontalement jusqu'à des déplacements de l'ordre de 4 diamètres.

6.2 *Analyse des résultats d'essais*

Les courbes de chargement obtenues pour les différentes fiches sont regroupées dans la figure 8. Elles ne font pas apparaître le palier d'effort

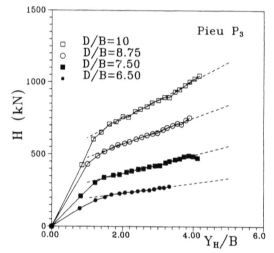

Figure 8. Courbes de chargement du pieu P_3.

horizontal correspondant à la rupture, comme c'est traditionnellement admis dans le cadre de la théorie des milieux rigides plastiques.

En outre, une certaine bilinéarité est à constater sur les courbes, correspondant à un écrouissage positif. Il est à rappeler qu'un tel schéma de réponse est usuellement rencontré dans les fondations superficielles ancrées dans les massifs sableux lâches en rupture par poinçonnement.

L'effort latéral ultime H_u est défini conventionnellement par le seuil de la variation linéaire sur la courbe de chargement.

Diverses approches de calcul de l'effort latéral ultime existent en littérature. Le tableau 3 en résume quelques unes avec leurs hypothèses. Une analyse détaillée est donnée par Bouafia (1990).

Comme le montre la figure 9, l'effort ultime est en bonne concordance avec ceux des méthodes de Ménard (pressiomètre) et de Dembicki.

Pour le reste des approches, le passage de la butée à la contre-butée se fait brusquement, avec des pressions très élevées, ce qui ne correspond pas au niveau de déplacements faibles dans cette zone. Le schéma de réaction latérale de Ménard paraît à ce titre, plus réaliste que celui des autres méthodes. Cependant, il est insuffisant puisqu'il néglige les contraintes de cisaillement à l'interface sol/pieu.

Tableau 3. Résumé de quelques approches

Auteur	Hypothèses principales
Blum (1932)	rupture par coin
Hansen (1961)	rupture par coin en surface et Ecoulement horizontal en profondeur.
Reese (1974)	idem
Menard (1962)	analogie pieu/pressiometre
Broms (1964)	pression des terres amplifiée
Dembicki (1977)	formule semi-empirique de pression des terres
Pender (1988)	idem
Petrasovits (1972)	idem

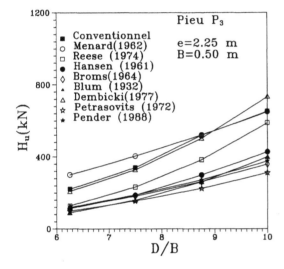

Figure 9. Comparaison des prédictions des méthodes.

7 CONCLUSIONS

La modélisation en centrifugeuse a été appliquée à l'analyse du comportement des pieux sous un effort latéral dans le sable.

L'étude parametrique a montré qu'un pieu flexible présente une sensibilité à certains paramètres physiques différente de celle d'un pieu court et r - gide.

Le lissage des courbes de moment par des splines quintiques, les dérivations et les intégrations successives de ces courbes ont permis de construire les courbes P-Y.

La modélisation en centrifugeuse a enfin été appliquée à deux problèmes particuliers de sollicitation latérale. L'étude de l'effet de la proximité d'un talus sur les déplacements et les moments a montré qu'il existe pour le pieu rigide étudié une distance limite de 6.5 diamètres par rapport à la tête du talus, au-delà duquel l'effet du talus disparaît.

La dernière partie concerne l'étude des grands déplacements latéraux du pieu dans le sable. Un modèle de pieu instrumenté par des jauges de déformation et battu à différentes fiches, a été sollicité jusqu'à des déplacements de l'ordre de 4 diamètres. Les efforts latéraux ultimes ont été déterminés conventionnellement. La comparaison avec les méthodes théoriques d'équilibre limite montre que les schémas classiques de profils de pression ultime sont très éloignés de la réalité. Ils conduisent le plus souvent une sous-estimation des efforts latéraux d'environ 50 %. L'étude d'autres configurations sol/pieu en centrifugeuse permettra certainement d'approfondir les constatations énoncées dans cette étude.

8 REFERENCES

Bouafia, A. 1990. *Modélisation des pieux chargés latéralement en centrifugeuse.* Thèse de Doctorat en génie civil de l'Université de Nantes, 267 pages.

Bouafia, A. & Bouguerra, A. 1996. Effet de la proximité d'un talus sur un pieu court et rigide chargé horizontalement. *Revue Française de géotechnique* No.75, juin 1996, PP 47-56.

Bouafia, A.& Garnier, J.1991. Experimental study of P-Y curves in sand, Comptes rendus du congrès international Centrifuge'91, Boulder, Colorado,13-14 Juin 1991, PP 261-268.

Corté, J.F .& Garnier, J.1986. Une centrifugeuse pour la recherche en géotechnique. *Bulletin de liaison des Ponts et Chaussées,*N0.146,Nov-Déc 1986.

Garnier,J . 1990. La centrifugeuse du LCPC : un nouveau moyen d'essai en macro-gravité, Colloque International ASTELAB' 90,Paris,11-15 Juin 1990 PP 417-422.

Terashi, M. et Al. 1989. Centrifuge modelling of a laterally loaded pile. Comptes rendus du 12ᵉ Congrès international de mécanique des sols, Rio De Janeiro, Vol 1, PP 991-994.

Van Wambecke et Al. 1982. Correlations between the results of static or dynamic probing and pressuremeter tests. Comptes rendus du 2ᵉ Symposium Européen sur les essais de pénétration, Amsterdam.

Geotechnics for Developing Africa, Wardle, Blight & Fourie (eds) © 1999 Balkema, Rotterdam, ISBN 90 5809 082 5

On the determination of vane shear strength of soft soils

M. Bouassida & S. Boussetta
National School of Engineers of Tunis, ENIT, Tunisia

ABSTRACT: The present norms for determination of the undrained cohesion employing vane tests lead often to a classical value which overestimates the soil strength mobilized at the time of yielding. Based on the definition of the mobilized strength, it has been verified that the classical value is a lower bound of the undrained cohesion. In order to make a realistic determination of this characteristic using vane tests, we propose to limit the rotation of the vane at a given maximum value for which the torque value at yielding is retained. Applying this approach for "torque-amount of rotation" curves, from laboratory vane tests carried out for a Tunisian reconstituted clay, a lower value of undrained cohesion is deduced regarding the classical value corresponding to the maximum torque used by the present norm. Based on this recommendation, it has been verified that the classical value of undrained cohesion is overestimated up to 40%.

1 INTRODUCTION

To carry out laboratory tests (direct shear, classical triaxial shear), cored samples are usually needed for determination of undrained shear strength of soft soils. Such sampling method generates disturbance which modifies the soil mechanical characteristics corresponding to the undisturbed state (Shogaki et al 1994). The field vane test has been devised to allow a direct determination of undrained cohesion of soft soils to avoid disturbance occurring both at the time of sampling process and manipulation of specimens to carry out laboratory tests. The laboratory vane test has been also introduced to allow direct determination of the undrained cohesion on the soil existing in the case of the sample. The vane test is performed in two steps. At first, the vane is introduced in the soil at a fixed depth. Secondly, a rotation at a constant rate is applied to the vane (Fig. 1). This loading induces the reaction of the soil which is transmitted from the rod to the couplemeter by a spring. From recorded angular deflection red on the graduated scale and knowing the spring stiffness, the value of torque is determined. The recorded data are plotted for giving the variation of the torque M (in N.m) as a function of the amount of rotation α (in degrees) of the vane (Fig. 2). The undrained cohesion C_u corresponding to the undisturbed soil strength is determined from the peak value of torque, without consideration of the corresponding amount of rotation. Nevertheless, it has been indicated by Bjerrum (1973) that the

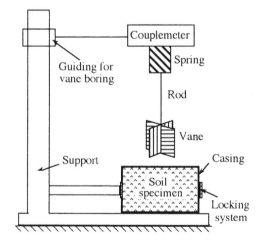

Figure 1. Scheme of a laboratory vane apparatus

classical method of determination of the undrained cohesion leads to an overestimated value from vane test data.

Some correlation have been established particularly by Bjerrum (1973) and Mesri (1975) to avoid overestimation of the undrained cohesion value by using reduced factors depending on soil parameters as the plasticity index, or the overconsolidation ratio. The aim of these corrections is to provide a reliable undrained cohesion value corresponding to the

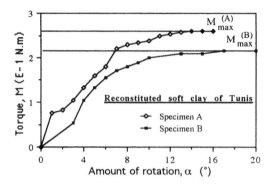

Figure 2. Evolution of torque vs amount of rotation from laboratory vane tests

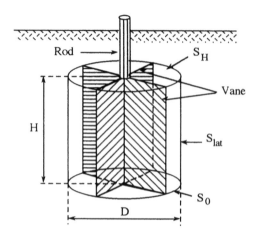

Figure 3. The assumed "yielding cylinder" model

mobilized strength during soil yielding. But these correlation are not usually valid (Tanaka 1994).

We are interested in studying this problem because of usefulness of the vane test which is sometimes the unique tool for determining undrained cohesion of soft soils. Using concepts of continuum mechanics, it has been verified that classical value of C_u constitutes rather a lower bound of this characteristic (Ben Achour 1998). Thus, the classical value of the undrained cohesion C_u^0, derived from the vane test is, on one hand, overestimated from the practical viewpoint, and, on the other hand, it could be theoretically higher.

The aim of the present article is to clarify this paradoxical situation and to undertake a process for making a useful proposal ensuring a reliable determination of undrained cohesion value.

In this work, the problematic of C_u determination, in the framework of vane tests, is explained. An approach is proposed to give a solution for the problem in question. Based on results of two experimental laboratory investigations made on samples of a reconstituted soft clay of Tunis, it has been pointed out to examine well the evolution of "torque-amount of rotation" curves issued from vane tests in order to determine correctly the undrained cohesion value for soft soils.

2 PROBLEMATIC OF C_u DETERMINATION FROM THE VANE TEST

In terms of the vane test conditions, the undrained cohesion, assumed as an isotropic characteristic, is determined classically after the moment equilibrium equation (in cylindrical co-ordinates) of a soil cylinder with D = diameter and H = height (Fig. 3).

By denoting $\tau(r)$ the soil shear strength developed on the cylinder boundary which is constituted by the identical surfaces S_0 and S_H and the surface S_{lat}

($0 \leq r \leq D/2$), one obtains:

$$M = \int_{S_{lat}} \tau(r) \, R^2 \, dz \, d\theta \; + 2 \int_{S_0} \tau(r) \, r^2 \, dr \, d\theta \qquad (1)$$

where M = resisting torque resulting from the rotation assigned to the rod.

The torque due to soil shearing on the rod is not taken into account, its contribution remains insignificant regarding terms intervening in eq. (1), (Ben Achour 1998). Assuming that the maximum soil shear strength is mobilized uniformly on the cylinder boundary, from which one obtains the classical expression of the undrained cohesion by setting $\tau(r) = C_u$ in eq. (1):

$$C_u^0 = \frac{2 \, M_{max}}{\pi \, D^3 (\frac{H}{D} + \frac{1}{3})} \qquad (2)$$

where M_{max} = maximum torque corresponding to soil yielding, it is determined from the peak value of the "torque-amount of rotation" curve (Fig. 2).
C_u^0 = mean undrained cohesion of the soft soil.

As a matter of fact, the undrained cohesion of clayely soils is an anisotropic characteristic. During soil yielding in vane tests, C_u^v = the lateral shear strength is different from C_u^h = the horizontal shear strength. From these two characteristics, the soil strength anisotropy is defined by the ratio:

$$k = \frac{C_u^h}{C_u^v} \qquad (3)$$

After Tristan-Lopez (1981), values of k ratio rang in the interval [0,5 ÷ 2,5]. Taking into account the soil strength anisotropy, one obtains the expression of the vertical undrained cohesion:

$$C_u^v = \frac{2\,M_{max}}{\pi\,D^3(\frac{H}{D} + \frac{k}{3})} \qquad (4)$$

For the "yielding cylinder" model related to vane tests, in the case (H/D) ≥ 1, shear strength of the soil is mostly due to the contribution of vertical undrained cohesion. In addition, from expression (4), it can be noted in the case k < 1 (i.e. $C_u^v > C_u^h$), the corresponding value of vertical undrained cohesion becomes greater, and as a consequence the mean undrained cohesion has a higher value. Hence, by taking into account strength anisotropy, it has been verified, by the use of the "yielding cylinder" model, that the mean undrained cohesion could be overestimated. In the case of a reconstituted soft clay of Tunis, the overestimation of C_u has been verified (Ben Achour 1998).

In the following, undrained cohesion of the soil is assumed to be isotropic. By reconsidering C_u determination, in eq. (1) it should be written that soil shear strength on the cylinder boundary is lower than or equal to C_u, hence:

$$\tau(r) \le C_u \qquad (5)$$

In fact, the maximum shear strength C_u is not surely mobilized everywhere on the cylinder boundary. This implies that C_u^0 value calculated from expression (2) is a lower bound of C_u, i.e., from eqs (1) and (5), we have:

$$C_u \ge \frac{2\,M_{max}}{\pi\,D^3(\frac{H}{D} + \frac{1}{3})} \qquad (6)$$

This approach is equivalent to that used by yield design theory utilizing the approach from outside by stresses (Salençon, 1990).

By comparing expressions (2) and (6), it appears that theoretical value of the mean undrained cohesion could be more overestimated. Hence, it arises a paradox inciting the re-examination of the classical determination method of C_u.

Obviously, if C_u value is really overestimated, it is owed to the torque maximum value deduced from recorded measurements. So, in order to make an accurate determination of C_u it is advised to look over again the exploitation of the "torque-amount of rotation" curve plotted from vane test measurements. This approach is detailed in the following paragraph.

3 A PROPOSAL FOR AN ADEQUATE DETERMINATION OF C_u VALUE

Based on hypotheses used in the classical method of determination of C_u, a lower value of this

characteristic could be advanced if a torque value lower than M_{max} is considered in eq. (2).

Following recommendations of the French norms NF P 94-072 and NF P 94-112 (currently used), the value of M_{max} should correspond to the maximum recorded torque, without any restriction about the corresponding amount of rotation of the vane, (Afnor 1995a & Afnor 1995b). However, for direct shear test, in which yielding surface is prescribed by NF P 94-071-2 norm (hence similar to the vane test where cylindrical yield surface is assumed), it is postulated in the case of levelled "shearing force-horizontal displacement" curve that soil shear strength is fixed at most for a maximum horizontal displacement δ_{max} = 5 mm. This case corresponds for a circular shear box with diameter d = 6 mm and to a ratio (δ_{max}/d) equal approximately to 8,33%. Using the same reasoning in the case of the laboratory vane test, a limitation on the amount of rotation of the vane should be done for a given value α_{max} for which the torque value corresponding to the yielding phase is fixed (Fig. 4).

Such approach, needing restriction on soil strains, has the advantage to be valid for all kinds of soft soils. Hence, in order to attribute a realistic value for C_u, there is no need to use correlation which are valid only for a given class of soft soils.

For example, in the case of Japanese maritime soft clay, it has been verified (Tanaka 1994) that the undrained cohesion does not depend on the plasticity index; then Bjerrum's correction factor could not be used to propose a reliable undrained cohesion value from vane tests data.

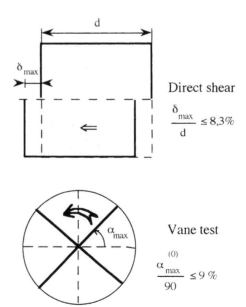

Figure 4. Analogy between direct shear and vane test

The proposed approach is applied for the analysis of the results obtained from an experimental investigation done on a Tunisian reconstituted soft clay, at the soil mechanics laboratory of National School of Engineers of Tunis.

4 ANALYSIS OF RESULTS OBTAINED FROM EXPERIMENTS ON A TUNISIAN RECONSTITUTED SOFT CLAY

The clay tested in this study is initially sampled at the disturbed state from the south lake of Tunis at 5 m depth. Reconstitution of this soft clay has been made in two steps. At first, starting from a mixture of clay and water, the clay used is obtained from the passing through 100 μm sieve. After desiccation, the resulting material is powdered and mixed again with water. A new soil paste is then obtained, at water content by 120%, and mixed thoroughly by an electric mixer under vacuum condition. Secondly, the new mixture is placed in cylindrical plexiglas cells and it is consolidated under oedometric path by applying gradual loading increments to a final stress of 50 to 60 kPa (for soil consolidation, one loading increment takes nine days). At the end of the loading, cells are unloaded gradually. Then, steel tubes (with 7 cm in diameter and 15 cm in height) are introduced in these cells of reconstituted soft clay of Tunis. Finally, the steel tubes are extracted as to obtain cored samples. Following this procedure, homogeneous soft soil specimens have been made with a water content of 51 %, (Bouassida 1996). A typical gradation curve of the reconstituted soft clay is shown in Fig. 5. Physical and mechanical parameters of this soil are given in table 1.

The values of undrained cohesion and frictional angle have been determined from classical undrained triaxial tests. It should be noted that physical and

mechanical characteristics determined from the two experimental investigations are quasi-identical.

In the following, this soil is assumed as purely cohesive with undrained cohesion $C_u = 8$ kPa and with zero undrained frictional angle. On the basis of these referenced values, validation of C_u determination from vane tests will be done.

Laboratory vane tests have been made using a model apparatus "Wykeham Farrance Engineering Limited serial N° 152".

For the two investigations, sizes of the vane are respectively H = D = 12 mm and H = 2 D = 24 mm. Tests have been conducted at the same constant rate of rotation, approximately 10° per minute.

The recorded results from laboratory vane tests are plotted in Figs 6, 7, 8 and 9 giving the evolution of the torque as a function of the amount of rotation of the vane.

Undrained cohesion values calculated from expression (2) are given in tables 2 and 3.

For each test, two values of the undrained cohesion are proposed. The first one corresponds to the maximum recorded torque value, and the second one corresponds to a maximum fixed value of the amount of rotation, denoted α_{max}, of the vane. For the first investigation (H = D) and the second one (H = 2 D) we have chosen respectively $\alpha_{max} = 10°$ and $\alpha_{max} = 6°$.

4.1 Comments

From vane test results, by assuming that soil yielding occurs for the recorded maximum torque value, the undrained cohesion of the tested soft clay ranges from 8.74 kPa to 12.8 kPa.

This margin of C_u values, although being representative for soft soils, shows up the risk in overestimating C_u value from vane tests with regard to the value $C_u = 8$ kPa determined from undrained triaxial tests.

From Fig. 6 (case of specimen 1), it should be noted that for the amount of rotation $\alpha = 10°$ the value

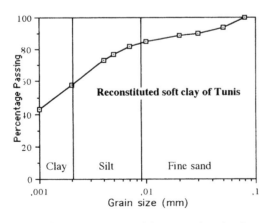

Figure 5. Gradation curve of the reconstituted soft clay of Tunis.

Table 1. Characteristics of the reconstituted soft clay of Tunis.

Experimental investigation	First (*)	Second (**)
Total unit weight (kN/m³)	16.6	16.6
Specific gravity	2.64	2.64
Water content (%)	51	59
Liquid limit (%)	73	74
Plasticity index (%)	47	46
Undrained cohesion (kPa)	8	8
Undrained frictional angle (°)	3	1

(*) Boussetta et al (1994) ;   (**) Jouini et al (1998)

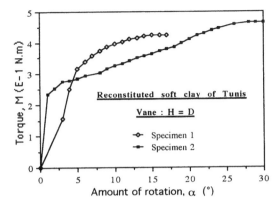

Figure 6. Evolution of torque vs amount of rotation (First investigation).

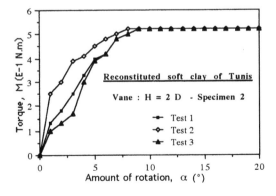

Figure 7. Evolution of torque vs amount of rotation (Second investigation).

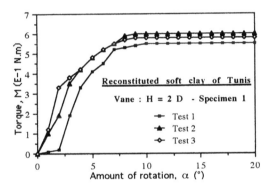

Figure 8. Evolution of torque vs amount of rotation (Second investigation).

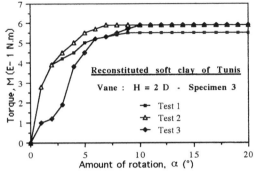

Figure 9. Evolution of torque vs amount of rotation (Second investigation).

Table 2. Values of C_u recorded for the reconstituted soft clay (First investigation: H = D = 12 mm).

Specimen	C_u (kPa) " "	- $\alpha (M_{max})(°)$	α_{max} (°) -	- ΔC_u (%)
1	9.0	-	10	-
	12.8	30	-	42.2
2	10.9	-	10	-
	11.7	≥ 15	-	7.3

C_u = 9 kPa represents well the value of undrained cohesion determined from triaxial tests. Nevertheless, the use of NF P 94-072 norm (Afnor 1995a) leads to C_u = 12.8 kPa. This value corresponding to a high amount of rotation ($\alpha = 30°$ as shown in table 2) overestimates the value of undrained cohesion by 42.2%. From vane test results it follows that, by limiting the amount of rotation to a given maximum value α_{max}, the corresponding torque which characterizes the soil yielding, leads to a representative undrained cohesion value, and avoids overestimation of C_u.

From tables 2 and 3, it should be noted that maximum torque value is obtained for a widely margin of values of rotation, i.e. :

$14° \leq \alpha_{max} \leq 30°$ for   H = D,

$7° \leq \alpha_{max} \leq 10°$ for   H = 2 D.

This remark remains valid also for field vane tests (with vane dimensions H = 2 D = 12.3 cm), done in a relatively thick layer of the soft clay of Tunis (GETU 1998). As indicated in table 4, the recorded values of rotation related to soil yielding are too different:
$5,4° \leq \alpha_{max} \leq 14°$.

Table 3. Values of C_u recorded for the reconstituted soft clay (Second investigation: H = 2 D = 24 mm).

Specimen	Tests	C_u (kPa) "	α_{max} (°) $\alpha(M_{max})$(°)	ΔC_u (%)
	1	7.56 9.24	6 ≥ 10	- 22.2
1	2	8.74 10.08	6 ≥ 9	- 15.3
	3	8.74 9.74	6 ≥ 9	- 11.4
	1	7.06 8.74	6 ≥ 9	- 23.8
2	2	8.06 8.74	6 ≥ 8	- 8.4
	3	7.06 8.74	6 ≥ 9	- 23.8
	1	8.74 9.24	6 ≥ 9	- 5.7
3	2	9.54 9.91	6 ≥ 7	- 3.6
	3	8.74 9.91	6 ≥ 10	- 13.4

ΔC_u = relative difference giving overestimation of C_u between the two determination methods.

Table 4. Variation of C_u with depth from field vane tests carried out for a soft clay layer of Tunis.

Layers	Depth (m)	$\alpha(M_{max})$ (°)	C_u (kPa)
Filling in	1	--	--
0 to 1.6 m	2	9	10
	3	10.4	9
	4	11	9
	5	13	11
	6	10	9
	7	12.4	11.7
	8	12	14.6
	9	9	12.8
	10	8	10
Greyish	11	6	12.4
clay with	12	10	17.9
shells	13	12	11.3
	14	5.4	33
	15	8	29
1.6 m to	16	10	40
19.5 m	17	11.4	37
	18	13	37
	19	14	40
Bed rock	20	-	-

will be indicated to fix the torque value corresponding to the soil yielding.

4.2 Recommendations

From the experimental investigations here described, the following recommendations are announced:

1. As specified by the present norms, in order to determine the undrained cohesion from laboratory vane tests, it is advised to carry out tests on soft clays with a vane characterized by H = 2 D. Therefore, a lower mean value of C_u is obtained with respect to the case when vane sizes H = D.

2. By carrying out laboratory vane tests on soft soils, the determination of a reliable undrained cohesion is done by considering a torque value corresponding to a limited amount of rotation. For vane sizes H = D, the recommended maximum amount of rotation corresponds to $(\alpha_{max}/90) \leq 9$ %. While for vane sizes H = 2 D the recommended maximum amount of rotation corresponds to a ratio $(\alpha_{max}/90) \leq 6$ %.

The approach here proposed concerns results recorded from laboratory vane tests. It is useful to extend it for results issued from field vane tests for which the appropriate maximum amount of rotation

5 CONCLUSION

In order to derive a non overestimated value of undrained cohesion, it has been indicated that the torque at soil yielding does not correspond always to the maximum recorded value. This result is verified on laboratory vane tests done on a reconstituted soft clay of Tunis. Indeed, by limiting the amount of the rotation of the vane related to the state of small strains (as proposed $\alpha_{max} = 10°$ for H = D, and $\alpha_{max} = 6°$ for H = 2 D), a lower value, than that obtained by the maximum torque value, of the undrained cohesion is determined.

Such approach, being valid for all kinds of soft soils, avoids the need to use correlation established to correct the classical undrained cohesion value. In fact, these correlation are usually established for a given category of soft soils.

The procedure here proposed needs to be validated by performing further investigations with laboratory and field vane tests on the reconstituted soft clay of Tunis, and on other soft soils as well. Such procedure will contribute to a reliable determination of the undrained cohesion for soft soils.

REFERENCES

AFNOR 1995a. Recueil de Normes Françaises. Essais de reconnaissance des sols, t. 1. France.

AFNOR 1995b. Recueil de Normes Françaises Essais sur ouvrages géotechniques. Dimensionnement. Exécution, t. 2. France.

Ben Achour, W. 1998. Sur la détermination de la cohésion non drainée des sols mous à partir de l'essai scissométrique. Mémoire de DEA, LMCS, ENIT.

Bjerrum, L. 1973. Problems of soil mechanics in unstable soils. Proc. of 8th ICSMFE. 3: 111-159.

Bouassida, M. 1996. Etude expérimentale du renforcement de la vase de Tunis par colonnes de sable - Application pour la validation de la résistance en compression théorique d'une cellule composite confinée. Rev. Franç. de Géotech. 75: 3-12.

Boussetta, S. & Makhlouf, S. 1994. Etude expérimentale du renforcement de la vase de Tunis par colonnes de sable. Projet de Fin d'Etudes; département de Génie Civil, ENIT.

GETU, 1998. Rapport géotechnique du projet de construction de la banque de l'Habitat. Société Géotechnique Tunisie.

Jouini, Z. & Baya R. 1998. Etude expérimentale d'une vase de Tunis. Application au calcul des ouvrages. Projet de Fin d'Etudes; département de Génie Civil, ENIT.

Mesri, G. 1975. New design procedure for stabilty of soft clays. Discussion. ASCE, GT4. 101: 409-412.

Salençon, J. 1990. An introduction to the yield theory and its applications to soil mechanics. Europ. J. of Mechanics, A/ Solids. 9; 5: 477-500.

Shogaki, T. & Kaneko, M. 1994. Effects of sample disturbance on strength and consolidation parameters of soft clay. Soils and Foundations J. 34, 3: 1-10.

Tanaka, H. 1994. Vane shear strength of a Japanese marine clay and applicability of Bjerrum's correction factor. Soils and Foundations J. 34; 3: 39-48.

Tristan Lopez, A. 1981. Stabilité d'ouvrages en sols cohérents anisotropes. Thèse Dr. Ing., Ecole Nationale des Ponts et Chaussés, France.

Geotechnics for Developing Africa, Wardle, Blight & Fourie (eds) © 1999 Balkema, Rotterdam, ISBN 90 5809 082 5

Electronic geotechnical data – Formatting for the future

R.J.Chandler
Key Systems Geotechnical (UK), Association of Geotechnical and Geoenvironmental Specialists

R.Hutchison
Environmental Services Group Limited (UK), Association of Geotechnical and Geoenvironmental Specialists

ABSTRACT: Ready access to geotechnical site investigation data in electronic format leads to significant advantages in our ability to visualise and understand the data. In the context of design and analysis we now have the ability to produce data plots, logs, 2D sections, 3D geological models, Rapid Retrieval Archives and GIS within the budget and time scales of small projects. All this has been made possible by the AGS data transfer format. Although the AGS data format has become widely used as an electronic version of the data section of a site investigation report, many engineers do not fully understand the powers of analysis that are available by obtaining data in this format. This paper gives a brief history and overview of the AGS data format and describes uses for the file format from simple design charts to national databases within Geographical Information Systems (GIS).

1 INTRODUCTION

The development since the mid 1980's of the personal computer (PC) has led to a vastly increased capacity to store and manipulate data. This in turn has given rise to enormous growth in the availability of affordable engineering software. This technological revolution has had a major impact on the Site Investigation (SI) industry, which can now deploy tools to use data more effectively in design and construction. An important consequence is that electronic data is more compact and more versatile than paper records, holding out the promise of substantial efficiency gains in the transfer and archiving of information. However, a major bottleneck in the flow of such information is at the human interface and unfortunately this is not just at the point of data collection. The wide range of applications for the handling of site investigation data makes compatibility between systems difficult to achieve, often resulting in data having to be regenerated.

This paper describes how the Association of Geotechnical and Geoenvironmental Specialists (AGS) resolved this problem by devising a common data transfer format, now generically called "The AGS Format".

2 STIMULUS

In the early 1990's as the industry began to use PC's in earnest, various interested parties sought to define the way in which SI data was formatted. Not only did all the producers have their own formats - based on what their systems could deliver, but the receivers of the data all wanted it in a format which could be immediately imported into their software. Matters were further complicated by the diversity of software species being used to store and manipulate data, which included databases, spreadsheets and even word processors.

The chaos that was rapidly developing, with multiple formats unique between some parties and alien to others, is illustrated in Figure 1.

Amid growing realisation that some form of standardisation was needed, the AGS set up a working party to try to address the issue on behalf of its members. The objective was to agree a format that was sufficiently versatile to meet a broad range of needs, but with rules that were rigorous enough to satisfy software developers. The comparative simplicity of the concept is illustrated in Figure 2.

3 THE "PERFECT JOURNEY"

3.1 *What is "The Perfect Journey"?*

To date the AGS Format has been extremely successful in addressing the "chaotic" problem. It has become the UK industry standard, and increasingly, an international standard. What was not envisaged at the time of its creation was the extent of the other benefits that would spin off. These we have termed "The Perfect Journey"

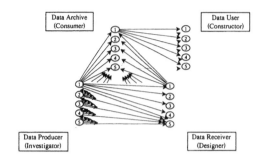

Figure 1 - Before AGS

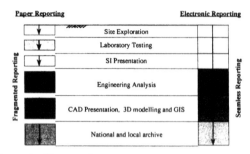

Figure 3 - The data Journey

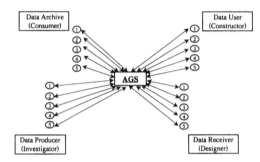

Figure 2 - After AGS

3.2 *"In a Perfect World"*

The most efficient data system has no learning curve and only requires each item of data to be entered once. The data can then be manipulated by the system to carry out all the required tasks or calculations. The system can also import and export any type of file format to ensure its compatibility with all the other systems.

Unfortunately such a system does not exist and anyway would not be very practical. In practice, combinations of commercial programs are used to carry out different aspects of geotechnical interpretation and design. Most of these processes require the ground investigation information to be entered as design parameters and so it is important to ensure that the systems used are "compatible.

Incompatibility of computer programs, or non-availability of suitable electronic data, can often result in some, or all, of the information being re-entered into a second system. The paper-reporting journey illustrated in Figure 3 shows that data can be entered up to 6 times within a single project. In order to reduce the number of times the data "dies" and has to be regenerated, it will be useful to review an example of "perfect" electronic reporting.

The perfect journey is the seamless transmission of data from the first point of generation to the end user. It starts with electronic on site logging of all the site and hole parameters. This information is transmitted

back to the office and can be reported to the client as and when required. The sample records are then sent electronically to each laboratory along with the samples.

When the samples are received at the laboratory the data records are transferred to the laboratory management system. Once the required tests have been completed the results are added to the records, which are then returned to the originator with the paper report.

The ground investigation specialist imports the laboratory test results into his ground investigation database and plots the logs, sections and location plans for the client's report as appropriate. He also produces an electronic version of this report for transmission to the client.

The availability of electronic records enables the client to use the data in a far more efficient and productive manner. Once the electronic data have been received, the client can import them into his data visualisation and interpretation programs, design packages, spreadsheets. The information can then be transferred into CAD and GIS systems and passed onto the company or government archive.

4 AIMING FOR PERFECTION

With a common data standard this perfect journey is now achievable for everyday geotechnical projects. The format has permitted substantial improvements in:-

1 The reliability and speed of data production
2 The flexibility to present data in a wide range of formats and styles
3 The rapid access to data records

4.1 *Reliability and speed of data production*

Site work is one of the major data sources of any ground investigation. All aspects of the day to day work will need to be recorded and presented to the client. Details of hole progress, water strikes, sample references, depths and stratigraphy, to name but a few, are usually entered into the engineer's log-book before

294

being entered onto a computer system.

Although this is usually the first instance of double entry (once into the log-book and then again into a computer) it is often one of the last instances to be rectified. On site electronic logging can remove this immediate data death but still fills most companies with the fear of losing data or operating mud covered laptops in the rain. Robust computer technologies with built-in mobile data transmitters are now becoming available to the geotechnical engineering community. This permits the site data to be entered directly and backed up over a phone line to an office network.

Alternatively, the written site data could be tabulated in a format that can be easily entered into a spreadsheet or other software capable of producing AGS files. The re-entry process is then made as simple as possible and can be carried out by any member of staff with minimal training.

Passing an AGS file containing the sample reference data to the laboratory ensures that the sample referencing system used is consistent with that generated on site. This data can then be imported into the laboratory management system saving time and eliminating re-entry errors. This system can be of even greater assistance if specialised testing is required, e.g. chemical contamination, as the data can be passed onto the specialised laboratory, eliminating yet another instance of double entry of data.

Access to the large quantity of data produced by the laboratory is crucial for the design of any geotechnical project. It is of great benefit to the data users if the parameters generated by the laboratory can be transferred via an AGS data file and imported directly into their design and interpretation software. This saves time and errors, but relies on a standard file format to be truly successful.

4.2 Presentation of Data

Computer software now generates almost all borehole logs contained in a ground investigation report, whether by a simple spreadsheet program or a specialised data handling system. Whichever method is used the majority of time is spent entering the data into the system.

By using an AGS file for site and laboratory records all the data can be imported directly into the system and any style of log can be selected and printed. With the efficiency of electronic systems it is often possible to include laboratory data or graphs that were once felt to be too time consuming to include within a report.

Similarly, computer systems enable the same data to be presented in a number of different formats without further data entry. For example, borehole logs can be produced with just the field data on, or with the engineer's interpretation and the laboratory tests. The data can also be incorporated with ease into site plans or section drawings.

The main attraction for electronic data is efficiency. An engineer logging electronically on site, can electronically transfer the information to the office to have logs produced and the information emailed to the data user within hours of the hole being completed. If the data is emailed to the user in AGS file format, then they can effectively have an interactive ground investigation report simply by importing all the data into an appropriate ground investigation software package.

It is at this point in the data journey that the efficiency of electronic reporting, with its inherent reliability, far exceeds that of the traditional paper route. This is before we even consider its main benefit - the accessibility of the data.

4.3 Access to the Data

The traditional paper-reporting route can produce reams upon reams of paper, which the user then needs to assimilate in order to interpret and visualise the geotechnical environment.

When an investigation report (Figure 4) contains several hundred borehole logs, laboratory summary tables or other graphical representations, this can be a very time consuming task. Companies who receive "dead" paper data will actually go through the report by hand collating the data they require into a spreadsheet program for further analysis - yet another example of double entry! But, critically, time constraints will often require a substantial cull of the data actually used.

The same quantity of data stored in AGS format on a single CD-ROM would enable software to import the electronic data for all the logs, laboratory, chemical and in situ tests and allow the user to query, present and collate the data in any report style with ease. Reliable importation is only possible however if a standard file format is used.

With some programs it is possible to import an AGS data file and print location plans, logs, and reports as shown in Figures 5,6 and 7 within 5 minutes of receiving the data. If the data have been imported into a sophisticated ground investigation package it is also

Figure 4 - Volumes of Paper Reports.

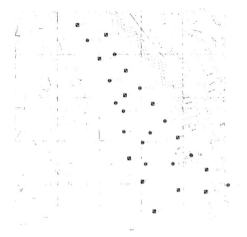

Figure 5 - Site And Borehole Plans

possible to produce contaminant analysis of the data (Figure 8) and logs with multiple borehole data represented for comparison. (Figure 9)

4.4 *Have We Achieved the Perfect Journey?*

The perfect journey has been achieved numerous times in the UK and other countries by using the AGS data format. Data producers are now increasingly using computer software to collect and store the data and the data users have access to a rapidly developing range of analytical software based on the format. The AGS Format has provided the catalyst, allowing all parties and software packages to communicate reliably and effectively.

Sampling

Depth	Type	Casing Depth	Date/ Water	SPT N (Cu)
0.00-0.40	B		10/02 0097 DRY	
0.40-1.20	B			
1.20-1.65	C	1.20	DRY	10
1.20-1.65	B			
1.65-2.20	B			
2.20-2.65	SD	2.20	DRY	6
2.65-3.30	B			
		2.20	DRY	
2.65			11/02	
		2.20	DRY	
2.85	W			
3.30-3.75	C	3.10	2.85	3
3.30-3.75	B			
3.75-4.40	B			
4.40-4.85	SD	4.30	3.20	2
4.85-5.50	B			
5.50-5.95	C	5.50	WET	33
5.50-5.95	B			
5.95-6.50	B			
6.50-6.95	SD	6.35	5.10	50
6.95-7.60	B			
7.60-7.90	SD	7.20	5.50	54/ 150
7.90-8.60	B			
8.60-8.90	SD	7.20	3.20	60/ 150
8.90-9.70	B			
9.70-10.00	SD	7.20		65/ 150
10.00	D			
10.00		7.20		

Strata

Description	Depth (Thickness)	Level	Legend
	G.L.	47.04	
MADE GROUND: Soft to firm brown sandy (fine and medium) clay with some subangular to subrounded fine to coarse gravel of mudstone, brick, china, concrete, glass and quartzitic gravel with abundant fibrous rootlets.	(0.40) 0.40	46.64	
MADE GROUND: Firm dark grey brown slightly sandy (fine and medium) ashy clay with some to much subangular fine to coarse gravel of brick, wood, ash, coal and quartzitic gravel. (Organic Odour).	(1.50) 1.90	45.14	
MADE GROUND: Soft red brown and grey brown slightly sandy (fine) clay with some subangular fine to coarse gravel of brick, coal, glass, slag and quartzitic gravel with pockets of dark grey ashy clay.	(0.40) 2.30	44.74	
MADE GROUND: Soft brown slightly sandy (fine to coarse) clay with some subangular fine to coarse gravel of glass and brick with occasional decayed vegetation. (Odourous)	(1.00) 3.30	43.74	
MADE GROUND: Very soft dark grey silty clay (Sludge Cake) with a little subrounded fine to coarse gravel of slag and abundant decayed vegetation. (Organic Odour) At 3.75m: with some subangular fine to coarse gravel of glass At 4.85m: with abundant decayed wood	(2.20) 5.50	41.54	
Stiff to very stiff brown mottled red brown slightly sandy CLAY with some subrounded fine to coarse quartzitic gravel.(Possible Mercia Mudstone)	(0.90) 6.40	40.64	
Stiff to very stiff red brown mottled grey green silty CLAY. (Mercia Mudstone Grade IV)	(1.20) 7.60	39.44	
Very stiff red brown occasionally mottled grey green slightly silty CLAY with some subangular fine to coarse mudstone lithorelicts with occasional dark grey staining. (Mercia Mudstone Grade III)	(2.40) 10.00	37.04	
End of Borehole.			

Equipment: Cable Percussion

Borehole Dia (mm): 150 to 10.00m — Casing Dia (mm): 150 to 7.20m

Groundwater

No.	Struck	Behaviour	Sealed
1	3.00	Rose to 2.85m in 20 mins slow	NS

Ground Level 47.04 m OD

Drilled by BC
Logged by TA
Checked by CV

Remarks
Service inspection pit hand dug to 1.20m.
Standing for 15mins while engineer checked position clear of services with STW personnel.
Chiselling from 9.45 to 10.00m (60mins).
Permeability test carried out at 5.65-6.30m.
Piezometer installed with sand surround from 5.90 to 4.50m and tip at 5.50m.

See key sheet and appendices for explanations

Form 1.0

Borehole Record — Exploration Associates — Project — Contract — Borehole BH7(1 of 1)

Figure 6 - Typical Borehole Log

Key Systems Geotechnical
Systems House
Burnt Meadow Road
Redditch
B98 9PA
Tel 44 1527 68888
Fax 44 1527 62880
HoleBASE@key-systems.com
www.keygeotech.co.uk

Quinley Gasworks Solid Contaminants Report

Solid Contaminant	No of Smps	Max Concen.	Min Concen.	Mean Concen.	Standard Deviation	Dutch IV	Smpls Exc'g	ICRCL Trig 1	Smpls Exc'g	ICRCL Trig 2	Smpls Exc'g	ICRCL Trig 3	Smpls Exc'g
pH Units	431	8.5	2.4	4.6	1.0	NA	NA	NA	NA	NA	NA	NA	NA
Total Sulphate	509	3.0700	0.0300	0.0429	0.0082	NA	NA	NA	NA	NA	NA	NA	NA
Arsenic, Total	38	350.0000	0.0000	101.4211	117.1570	0.0000	38	200.0000	10	0.0000	38	0.0000	38
Barium, Total	32	3030.000	0.0000	990.6250	718.2267	0.0000	32	0.0000	32	0.0000	32	0.0000	32
pH	5	5.0000	2.0000	4.0000	1.4142	0.0000	5	0.0000	5	0.0000	5	0.0000	5

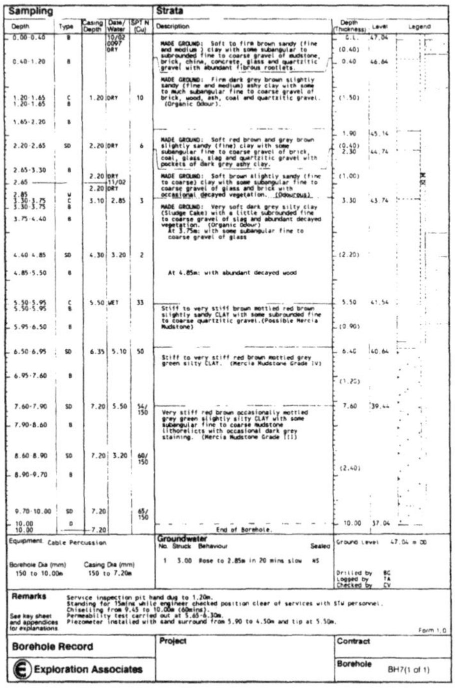

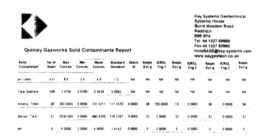

Figure 7 - Sort And Reporting Data Via A Query Report.

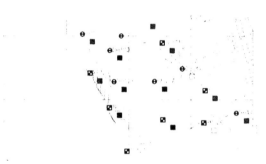

Figure 8 - Contamination hot spots and contours

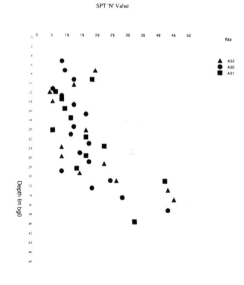

Figure 9 - Multiple BH plots Enable Fast *Collation Of Data Across The Site*

Early compatibility testing is the key to successful electronic data transfer. All too often the compatibility issue is not raised until the end of the project and then time constraints prevent successful data transfer. Despite the evident benefits of using a common data transfer format, some organisations are still specifying their own format, while others may receive their data in standard format but then proceed to retype it. Worst of all, is where an intermediate party specifies data in AGS Format but then declines to pass it on to subsequent potential users such as constructors. Fortunately, as the AGS Format becomes increasingly recognised by users and software writers alike, these problems are rapidly diminishing.

5 THE AGS FILE FORMAT - SIMPLICITY AND EFFECTIVENESS

5.1 Introduction

As we have seen, the key to the "Perfect Journey" is a recognised data standard. Within the UK, and increasingly in other countries, this has been achieved by the use of the AGS Format. Since its introduction in 1992, the twin virtues of simplicity and flexibility have been seen as pivotal to its success.

5.2 The Underlying Philosophy

The purpose of the AGS Format is to provide a means of transferring geotechnical and geoenvironmental data between parties. From the outset the fundamental consideration has been that potential users of the Format should be able to use standard software tools to produce the data file. These tools may range from simple text editors and word processors, through spreadsheet packages to sophisticated database systems. In order to ensure the widest possible level of acceptance it was also agreed that the Format should use the American Standard Code for Information Interchange (ASCII).

To minimise file size and eliminate potential conflicts it was also considered that the data file should only contain fundamental data such as exploratory hole and test data required to be reported by the relevant British (or other National) Standard or similar recognised documents. Interpreted or derived data should not be transmitted – any further processing being at the discretion, and under the control of the receiving user.

To provide maximum flexibility and to enable hardcopy printout for checking or other purposes, a "data dictionary" approach as first proposed by Greenwood (1988) has been utilised as the foundation for the Format. This concept is both compatible with the needs of software developers and simple to implement.

5.3 File Format

The AGS Format file is a text file, containing tables of information on all aspects of the geotechnical site investigation. Each table comprises a *Group* name, column *Headings* and data *Variables* as illustrated in Figure 10.

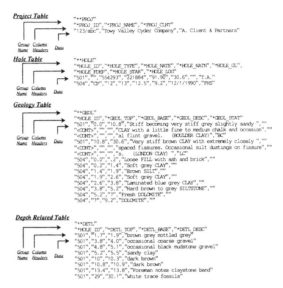

Figure 10 - Example AGS file

Group Names

The AGS data book, AGS(1992), AGS(1994), contains all the tables required for a site investigation. The group name is shortened to a four-letter code, often the first letters of the table description. Group names must be on a line by themselves and surrounded by quotes. A list of AGS(1994) Group names is included in Appendix A.

Column Headings

The column headings (or field names) must be listed below the group name line as shown in Figure 10. Every column must be surrounded by quotes and separated by a comma thus allowing programs to determine where the start and end of each data item is. Each header is abbreviated to a code listed in the AGS data book.

The column header codes allowed for the hole table are shown in Figure 11, only the codes marked with a "*" are mandatory within a table. The hole group in Figure 10 contains 8 column headers. Additional column headers from figure 11 could be added at the user's discretion.

Data Variables

The third section of the hole table actually contains

297

Exploratory Hole Data

Group Name : HOLE	Hole information			
Status	Heading	Unit	Description	Example
*	HOLE_ID		Exploratory hole name / number	327/18A
	HOLE_TYPE		Type of exploratory hole	CP ***
	HOLE_NATE		O.S. National Grid six figure Easting	523145
	HOLE_NATN		O.S. National Grid six figure Northing	178456
	HOLE_GL	m	Ground level relative to Ordnance Datum	16.23
	HOLE_FDEP	m	Final depth of hole	32.60
	HOLE_STAR	dd/mm/yyyy	Date of start of excavation	18/03/1991
	HOLE_LOG		The definitive person responsible for logging the hole	DPG
	HOLE_REM		General remarks on hole	Chiselled 1 - 1.5m
	HOLE_LETT		Ordnance Survey letter grid reference	TQ 123 456
	HOLE_LOCX	m	Local grid x co-ordinate	565
	HOLE_LOCY	m	Local grid y co-ordinate	421
	HOLE_LOCZ	m	Level to local datum	-106.6
	HOLE_DIAM		Deleted	
	HOLE_CASG		Deleted	
	HOLE_ENDD	dd/mm/yyyy	Hole end date	22/03/1991
	HOLE_BACD	dd/mm/yyyy	Hole backfill date	22/03/1991
	HOLE_CREW		Name of driller	A.B. Driller
	HOLE_ORNT	deg	Orientation of hole (degrees from north)	50
	HOLE_INCL	deg	Inclination of hole (measured positively down from horizontal)	90
	HOLE_EXC		Plant used	JCB 3CX or Hand
	HOLE_SHOR		Shoring/support used	None, Speedshore
	HOLE_STAB		Stability	Stable during excavation
	HOLE_DIMW	m	Trial pit width	0.9
	HOLE_DIML	m	Trial pit length	2.5

Figure 11 - Hole Table Definition

Field Name	Description	Value
HOLE_ID	Exploratory hole name/ number	504
HOLE_TYPE	Type of exploratory hole	CP
HOLE_NATE	National Grid six figure Easting	12
HOLE_NATN	National Grid six figure Northing	13
HOLE_GL	Ground level relative to National Datum	2.5
HOLE_FDEP	Final depth of hole	9.2
HOLE_STAR	Date of start of excavation	12/1/1990
HOLE_LOG	The definitive person responsible for logging the hole	FHS

Figure 12 - Summary Of BH504 Hole Information

the data for all the holes in this project. Each column must again be surrounded by quotes and separated by a comma. The information for the second hole in the example file is summarised in Figure 12.

A full list of the data file rules are contained in the AGS data book and can be downloaded from their web site.

5.4 *Flexibility*

The file structure described above, with its use of the data dictionary concept and subdivision into small groups, allows a high degree of flexibility. New groups can readily be added without disturbing existing groups, which is particularly useful in the international context, allowing local requirements to be incorporated. Similarly, fields can be added to existing groups to take account of particular needs as appropriate.

6 WWW.AGS.ORG.UK AND ITS RELATIONSHIP TO THE FORMAT

To meet the rapidly changing needs of its users the Format must continue to develop. Promulgation of changes has been achieved by the publication of a First Edition in 1992, AGS(1992), and a Second Edition in 1994, AGS(1994), both in hardcopy form. However, the recent expansion and broadening of the user base has made more frequent updates necessary. Since this has made hardcopy updates no longer practical it was decided to make use of the AGS web site to publish the Third Edition in 1999. While placing the Format in open access on the web site permits more frequent updates, all changes are subject to rigorous control and notification procedures.

The flexibility introduced by the use of the web site has also allowed the possibility of supplementary versions tailored to particular national requirements. This eliminates the dependency on UK conditions and requirements and allows true international usage.

7 CASE HISTORIES.

The Format has enabled many organisations, both in the UK and overseas, to attain efficient data management and data access that have allowing their projects to flourish. Below are two case studies that illustrate the wide range of possibilities the Format offers.

7.1 *Estimating the Clean Up.*

The AGS Data Format allows contamination test levels to be stored and transferred in the same way as laboratory test data. Thus large quantities of contamination and gas data can be viewed in the same context as the soil data.

One of the first contracts Geotechnical Developments (UK) Ltd's fully applied the AGS Data Format involved the investigation of gas generation/migration on a large, proposed housing development. Although earlier investigation had been carried out and general assessments made by others, a full review of the ground and gas data was required. This enabled the SI to establish source, pathways, target and to carry out the necessary risk assessment.

With over 12 months of gas data (from in excess of 50 monitoring points) using previous investigations together with recent installation monitoring it was essential to be able to import the data with the minimum of data entry. The AGS files imported straight into the AGS compatible system and they were able to integrate all the old data with the new. Immediately the ground, groundwater and gas data could be modelled providing contour, isopachyte, flow path, and concentration/plume plots, as seen in Figure 13. With time related data, this could also be further modelled

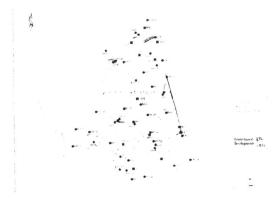

Figure 13 - Countours Of Contamination

to show gas concentrations changing through time. This was then presented in the form of a computerised video sequence.

Input of soils data via an AGS Format File, in particular the Made Ground, allowed relationships between the in situ organic soils (potential sources of gas) and actual gas concentrations at various levels within the ground to be viewed.

Rapid access to the data allowed areas of the site having no measurable elevated concentrations of gas (methane or carbon dioxide) during the monitoring phases to be quickly identified. Combined with soil data this information provided early confidence in possible areas of low risk for development.

This work utilised the Key Systems HoleBASE package where available data can be linked together to enable the ground to be better interpreted and more quickly assessed. End product drawings, plans, logs, graphs and images can be produced accurately and to high quality for inclusion in in-house assessments and external reports. The improved access to data provides added value to the client and a greater confidence in data interpretation.

7.2 Submarine cables and the AGS format.

In recent years the huge growth in electronic transfer of data and the internet has generated a boom in the installation of transoceanic submarine telecommunications cables. Fibre optic cables offer the capability to transmit vast quantities of digital data around the globe literally at the speed of light.

SAGE Engineering Ltd was involved one such global cable project called Sea Me We 3, which encompassed the USA, Europe, the Middle East and South East Asia. Projects of this scale involve national governments, local telecommunications providers, and global institutions such as the World Bank and installation contractors. The common factor linking all these organisations is the requirement for fast track information to enable rapid assessment of cost, choice of route and design.

Figure 14 - MCPT

In February 1998, a Miniature Cone Penetration Test (MCPT), see Figure 14, survey for an American client was undertaken as part of the Sea Me We 3 Project. The site was in the Indian Ocean and involved some 140 MCPT's at 1-kilometre intervals to assess trenchability.

The reporting requirement for the testing meant that data had to be processed on board the vessel and transmitted to the office for interpretation and final presentation. The major problem to overcome was the quantity of data that could be transmitted daily using the ship's satellite communications system. However, satellite communication is very slow and expensive.

Previous experience had shown that data files generated by the Excel based postprocessor were up to 4 Megabytes for each test and data was usually written to CD-ROM or ZIP disc and taken off the vessel at monthly re-supply port calls.

SAGE had been using Key System's HoleBASE package for final presentation of data in the UK Bath office for some time although the potential to create and export AGS format data had not been effectively utilised. The solution to the problem proved to be very simple. The vessel was equipped with a copy of the software and the final data was converted into the AGS data format on completion of each test. This effectively reduced the quantity of data from 4,000,000 Kilobytes to around 25 Kilobytes, a quantity easily handled by the ship's satellite system.

With the AGS format enabling easy transmission of data by email, the tests were sent back to the UK twice daily. The AGS data was imported directly into the office AGS compatible software without further proc-

299

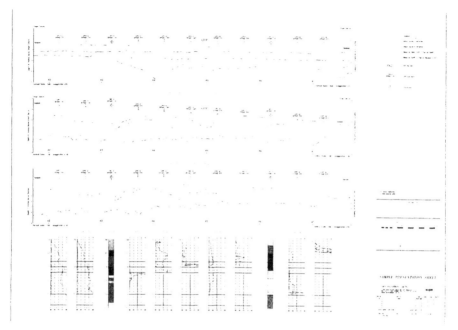

Figure 15 - CPT data superimposed along pipeline route

essing. After checking and interpretation, the final output was then emailed to the client as a Windows metafile and AGS file for immediate import into a CAD survey drawing as illustrated in Figure 15.

This approach led to the client fulfilling the fast track reporting requirement of the contract. It also made field operations more cost effective by allowing a large portion of the final reporting to be completed by field staff.

8 INTERNATIONAL ADOPTION

While the Format was originally conceived to address a problem developing in the UK, similar problems have occurred elsewhere, notably in Hong Kong and Singapore. In the Mid 1990's The Hong Kong authorities adopted the AGS Format with a few minor additions to reflect local geology and practices. Since this time all site investigation data must be reported to the authorities in AGS Format, thus enabling a central archive to be generated and maintained.

At the time of writing the Singapore government was also starting to implement the AGS system on their government contracts.

9 CONCLUSIONS

With the advent of the AGS Format the geotechnical industry has, for the first time, the ability to maximise its use of the data obtained from ground investiga-

tions. It is no longer necessary to regenerate or re-enter data at each stage of the design process. Data now needs only to be entered once, at source. Not only does this eliminate transcription errors but it also allows immediate access to the data for use within tight construction time-scales. Coupled with the ready availability of powerful software for sorting and presenting data, the AGS Format gives users access to reliable data and the ability to fully exploit it in the design process.

10 ACKNOWLEDGEMENTS

The authors acknowledge the work of the AGS Working Party, which was responsible for the development of the Format and its implementation. Without their hard work and enthusiasm the project would never have achieved so much.

The authors would also like to thank Neil Cash from SAGE Engineering UK and Chris Eaton from Geotechnical Developments UK for their valuable additions to this paper.

Thanks must also be given to Mike Rothery of Key Systems Geotechnical for general guidance and production of Figures 5 – 9.

11 REFERENCES

Association of Geotechnical and Geoenvironmental Specialists, 1992 "Electronic Transfer of Geotechnical Data From Ground Investigations"

Association of Geotechnical and Geoenvironmental Specialists Second Edition, 1994 "Electronic Transfer of Geotechnical Data From Ground Investigations", ,

Greenwood J.R. (1988) "Developments in computerised ground investigation data", Ground Engineering,

Greenwood J. R., Rothery M.G.W. (1992) "Collection, Transfer, Storage and Presentation of Ground Investigation Data" Geotechnique et Informatique,

APPENDIX A – AGS GROUP NAMES (1994)

Group Name	Description	Group Name	Description
CBRG	CBR Test - General	IRDX	In Situ Redox Test
CBRT	CBR Test	IRES	In Situ Resistivity Test
CHEM	Chemical Tests	ISPT	SPT Results
CHLK	Chalk Tests	IVAN	In Situ Vane Test
CLSS	Classification Tests	MCVG	MCV Test - General
CMPG	Compaction Tests - Gen.	MCVT	MCV Test
CMPT	Compaction Tests	POBS	Piezometer Readings
CNMT	Contaminant Testing	PREF	Piezo Details
CONG	Consolidation Test - Gen.	PROB	Profiling Instr. Readngs
CONS	Consolidation Test	PROF	Profiling Instruments
CORE	Rotary Core Information	PROJ	Project Information
DETL	Stratum Detail Descriptions	PRTD	Pxmeter Test Data
DPRB	Dynamic Probe Test	PRTG	Pxmeter Test - General
DREM	Depth Related Remarks	PRTL	Pxmeter Test - Ind. Loops
FRAC	Fracture Spacing	PTIM	Hole Progress by Time
FRST	Frost Susceptibility	PTST	Lab Permeability Tests
GAST	Gas Constituents	PUMP	Pumping Test
GEOL	Stratum Descriptions	RELD	Relative Density Test
GRAD	PSD Analysis Data	ROCK	Rock Testing
HDIA	Hole Diameter by Depth	SAMP	Sample Ref Information
HOLE	Hole Information	SHBG	Shear Box - General
HPGI	HP Gauge Inst Details	SHBT	Shear Box Testing
HPGO	HP Gauge Observations	STCN	CPT Run
ICBR	In Situ CBR Test	SUCT	Suction Tests
IDEN	In Situ Density Test	TNPC	Ten Percent Fines
INST	SP Instrument Installation	TRIG	Triaxial Test - Gen.
IOBS	SP Instrument Readings	TRIX	Triaxial Test
IPRM	In Situ Permeability Test	WSTK	Water Strike Details

Geotechnics for Developing Africa, Wardle, Blight & Fourie (eds)© 1999 Balkema, Rotterdam, ISBN 90 5809 082 5

Assessing the stiffness of soils and weak rocks

C.R.I.Clayton
Department of Civil and Environmental Engineering, University of Southampton, UK

ABSTRACT: In the light of recent developments in our understanding of their stress-strain behaviour, this paper reviews requirements for determining the mass stiffness of soils and weak rocks, both from laboratory and field measurements. The effects of bonding, fissuring and jointing, non-homogeneity, anisotropy, non-linear stress-strain behaviour and creep are noted. Examples are given of their importance when planning an investigation to determine stiffness. Techniques that have recently been used for high-quality stiffness determination during UK site investigations are described. Examples of specialist equipment and the data produced are given, with particular emphasis on the need to test a sufficiently large volume of soil, at relevant stress or strain levels, whilst eliminating the effects of bedding.

1 INTRODUCTION

In the early days of geotechnical engineering, in the first half of this century, it was the failure of engineering works that pre-occupied most engineers. There was an urgent need to understand and limit the problems that were occurring in harbours, canals and tunnels. At the present time our knowledge has increased to the point where most failures occur primarily as a result of poor design and construction management, rather than because of shortcomings in the engineering sciences of soil and rock mechanics. Attention has therefore turned to the prediction of displacements, since these will often be an important factor in dictating the cost of a particular piece of construction. Good calculations of displacement require accurate assessments of the stiffness of the ground. This paper therefore reviews recent developments in stiffness determination, both using laboratory and field techniques, in the light of the needs of practising engineers.

2 SOIL AND WEAK ROCK BEHAVIOUR

Over the past 30 years or so our conceptual models of the behaviour of soils and weak rocks have been considerably advanced. Soils and weak rocks have complex behaviour, typically displaying the effects of

- bonding
- fissuring or jointing
- non-homogeneity
- anisotropy
- non-linear stress-strain behaviour
- creep.

Additionally they may

- yield and undergo strain localisation as stresses increase
- be partially saturated.

The first six of these factors have a significant effect on stiffness, and therefore on sampling and testing strategy, as discussed below.

The challenge for the practising engineer is to recognise the complexity of soil and weak rock behaviour, adopting a computational and design strategy which, whilst delivering a satisfactory outcome, uses the simplest possible methods to characterise the ground and calculate the displacements caused by construction. The key is therefore to understand which elements of a design are most critical to the safety and economy of a project, and then to identify if and how modern geotechnical knowledge and techniques can enhance its success. There is nothing wrong in using traditional, empirical techniques provided their deficiencies are understood, and they are judged to be appropriate to the overall design strategy. But given a good knowledge of the techniques available for investigation and analysis, appropriate geotechnical engineering can often help speed up construction, reduce cost and enhance the end value of a project.

2.1 Bonding and structure in soils and weak rocks

Traditionally, soils have been regarded as materials whose behaviour is dominated by effective stress, whilst rocks have been thought of as cemented or bonded. Recent improvements in sampling and testing practice have shown us, however, that virtually all geotechnical materials, including fills, display the characteristics of bonding (Leroueil and Vaughan, 1990, Clayton and Serratrice, 1997). A wide range of diagenetic processes are known to operate within the depositional environment (for example, see Clayton, 1983), and many of these are at work very early in the life of a material, whether it be placed in a marine or terrestrial environment, naturally or as man-made fill. Thus bonding (also sometimes referred to as "structure") has engineering significance for a number of reasons:

- stiffness and strength are greater than for an un-bonded material, permitting greater loading or less support during construction
- the early diagenesis of materials (before significant overburden pressure is applied) can mean that high void ratios are "locked in", creating "structure-permitted space" (Burland, 1990) which can produce very large strains or runaway failure if the material becomes over-loaded
- the stress-strain behaviour of the bonded material will be considerably different from what is experienced once the material becomes de-structured, whether by yielding under load or by remoulding during excavation or sampling. Before destructuring the material may be

expected to behave in an approximately elastic manner.

Extreme examples of the effects of de-structuring bonded high-porosity materials can easily be found in the literature; the sensitivity of Canadian and Norwegian quick clays is a classic example, but the effect of remoulding the English Chalk can be, if anything, more dramatic. This rock will often become a slurry when worked by earthworks plant, and in the past this has led to enormous disruption of contracts for highway construction (Clayton, 1977). The warning is clear; care must be taken when characterising soils and rocks. Inappropriate sampling and testing practices may give completely misleading values of stiffness.

2.2 Fissuring and jointing

Joints and fissures introduce significant and unwelcome complexities into the behaviour of soils and weak rocks. They are generally both weaker and more compressible than the intact material around them, and normally occur in a limited number of orientations ("sets"). The orientations and spacings of discontinuities may make the properties of the soil or rock in the mass anisotropic, and it is then important to know not only their orientations in relation to each other, but also in relation to the problem that is to be solved. In terms of strength, the classic example is, perhaps, the effect of discontinuity orientation on mine excavation stability (for example see Markland's work, in Hoek and Bray, 1981). Numerical modelling of a fractured material is more complex and demanding in terms of computer resource than modelling a continuum.

Discontinuities can also have a large effect on mass compressibility. But whereas fissures will significantly reduce the strength of a material as weak as a firm to stiff clay (Table 1), it appears that they do not have a major effect on the compressibility of such materials. There has been considerable success in estimating movements around excavations in London clay from the results

Table 1   Estimates of the effects of fissuring on the undrained strength of stiff London clay (data from Clayton, Matthews and Simons (1995)

Method of assessment of undrained shear strength	Strength relative to that back-figured from a short-term slope failure
Back-analysis of slope failure	1.0
Horizontal 600mm x 600mm shear box test	1.6
305mm diameter triaxial tests	1.6 - 1.7
152mm diameter triaxial tests	1.6 - 1.9
102mm diameter triaxial tests	1.6 - 2.0
38mm diameter triaxial tests	2.4 - 3.6
intact clay c. 15mm dia. Tests	4.8 - 12.0

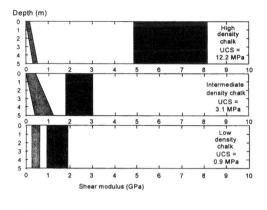

Depth (m)

High density chalk
UCS = 12.2 MPa

Intermediate density chalk
UCS = 3.1 MPa

Low density chalk
UCS = 0.9 MPa

Shear modulus (GPa)

■ Stiffness derived from field surface wave velocity measurements
■ Stiffness derived from laboratory tests on intact specimens of chalk
UCS = unconfined compressive strength

Figure 1. Effects of joints on the mass stiffness of fractured weak rock with varying intact strength (Clayton, Gordon and Matthews, 1994)

of special triaxial tests (Jardine et al, 1991). However materials with only slightly higher strength, such as very high porosity chalk (unconfined compressive strength = 0.9 MPa) show very significant decreases in compressibility as a result of the presence of discontinuities (Figure 1).

Whereas there has been reasonable success in developing laboratory tests to measure the strengths along pre-existing failure planes and discontinuities, measuring the change in stiffness caused by joints has proved to be more problematic. Representative stiffnesses can only be obtained by testing a sufficiently large in-situ mass of the material. As a rough guide, results of plate tests on weak rock (Lake and Simons (1975)) have suggested that it is necessary to test a volume which has a minimum dimension of at least 6 times the joint or fissure spacing. Furthermore, it is normally necessary to test the material *in situ* since any movement between joint faces, during sampling or test preparation, will remove the effects of previous loading, making the material seem much more compressible that it is *in situ*.

2.3 *Non-homogeneity*

Engineers tend to prefer to think of deposition of sediments as occurring in wide uniform basins, in an environment dominated by mechanical processes such as consolidation. But as any basic text on physical geology shows, depositional environments are more likely to be complex, both because of physical variations on a small and a large scale, and because a wide range of biologically- and chemically-driven processes will occur. This is

particularly true in the early life of a deposit and during weathering / alteration.

One therefore needs to be wary of extrapolating the results of a few small tests when characterising the material around large-scale construction. Fortunately, in many materials the effects of consolidation (gravitational compaction) and late-stage diagenesis are sufficient to overcome many of the characteristics introduce in the early life of the deposit, thus making the ground reasonably uniform laterally. Even so, completely uniform deposits of soil or weak rock are extremely unusual. At the very least, where the depositional environment was large and sufficiently unvarying, it is likely that the changes of effective stress and of weathering will produce strengths and stiffnesses which vary significantly with depth. In addition, experience suggests that early diagenesis, very coarse particles (gravels, cobbles and boulders), varves, silt-covered fissures, laminations, thinly inter-bedded soils and rocks, and uneven effects of weathering and alteration can produce significant variations in properties at all scales. When deciding on a strategy for determining stiffness it is necessary to bear these factors in mind.

2.4 *Anisotropy*

In soils the effective horizontal and vertical stresses are rarely the same, and therefore it can be expected that the stiffness of an un-bonded soil deposit will be different in the horizontal and vertical directions. Thus in the near-surface over-consolidated London clay the horizontal effective Young's modulus has been estimated to be about 1.6-2.0 times its vertical value, as a result of a K_o values of about 2-3 (Atkinson, 1973, Wroth, 1971). This has important implications for the prediction of movements around excavations at critical inner-city sites (Cole and Burland, 1972).

Some researchers have suggested that the horizontal total and effective stresses will vary with bearing, as a result of regional tectonism (Dalton and Hawkins, 1989). However, modelling of full, three-dimensional anisotropy is not practical, since it requires the determination of some 21 parameters, even for straight-forward linear elasticity. The assumption that stiffness will be the same in all horizontal directions, on the basis that there is unlikely to be major variation laterally during deposition, is obviously an over-simplification. But even for this 'cross-anisotropy' it is normally difficult to determine more than 2 of the 5 elastic parameters required for analysis.

Nonetheless, an assumption of isotropic stiffness behaviour is unlikely to be correct. Anisotropy can be introduced not only by variations in effective stress but also by depositional effects (for example, flat particles deposited face down, and inter-layered deposits of clays and limestone will have a higher horizontal than vertical stiffness, whilst for uniform normally-consolidated clays the reverse will be true). Fractures introduce anisotropy into weak rocks, since sub-vertical and sub-horizontal sets rarely have the same frequency, aperture or infill.

2.5 *Non-linear stress-strain behaviour*

The acceptance that soil behaves in a non-linear way, even at very small strain levels, arose as a result of finite element back-analyses of field observations (Simpson O'Riordan and Croft, 1979) and because of the development of instrumentation to measure small strains locally (i.e. without bedding effects) on triaxial specimens (Brown and Snaith, 1974, Burland and Symes, 1982, Clayton and Khatrush, 1986, Daramola, 1978). The realisation that the relatively low stiffnesses being obtained from laboratory tests might not be simply due to the impossibility of obtaining undisturbed samples has led to a revolution in sampling, testing and instrumentation practices during the past decade. Recently (Clayton and Heymann, in press) sufficiently high-accuracy triaxial instrumentation has been developed which has allowed us to see that the linear elastic range of soils and intact weak rocks typically only extends to an axial strain level of about 0.002%, in triaxial compression. Traditional stiffnesses measured using external deformation gauges (e.g. dial gauges or LVDTs) in the triaxial

apparatus are now known to be significantly affected by bedding and system compliance (Figure 2), up to strain levels as high as 0.5-1.0%.

When considering how to measure stiffness, important factors to consider are that:

- the strains developed around typical construction (tunnels, retaining structures) where there is an adequate margin of safety against failure are small, and often of the order of 0.1 - 0.01% (Jardine, Potts, Fourie and Burland, 1986, Mair, 1993).
- the secant modulus of high quality samples of soil will fall by about one half in this strain interval
- the stiffness measured at 0.01% axial strain may be about 10 times greater than that determined on the same specimen using external strain measurement, as a result of bedding, system compliance, and material non-linearity (see for example

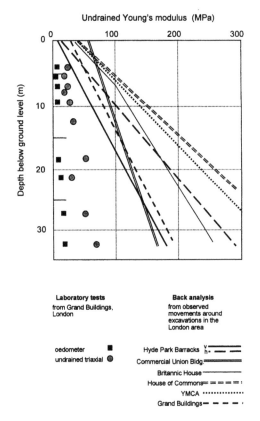

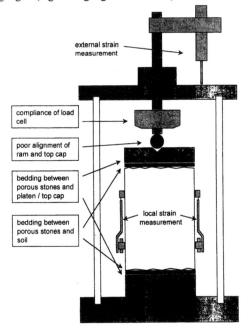

Figure 2. Sources of bedding and compliance in the triaxial apparatus

Figure 3. Comparison of stiffnesses back-analysed from measured movements around excavations in London clay with those determined from conventional sampling and laboratory testing (Clayton et al., 1991)

306

Figure 3 which compares stiffnesses back analysed from observations of displacements around excavations in London with those obtained from triaxial and oedometer testing).

Therefore when making measurements of stiffness, it is essential to ensure both that bedding is eliminated and that stiffness is measured at a strain level relevant to that expected around the proposed construction. It is normally considered prudent to measure the degradation of stiffness under increasing strain, so that this can be modelled in any analysis if required.

2.6 *Creep*

Despite the fact that creep behaviour has been recognised for many years (e.g. Buisman, 1936), geotechnical models generally ignore time-dependent deformation, except when produced by changes in effective stress, during consolidation or swelling. Unfortunately many soils and weak rocks exhibit significant creep under constant load. Well-known examples include organic materials (peat, etc.), sands and silts, and fractured weak rocks. The common link between these materials is their high permeability, which allows excess pore pressures to be rapidly dissipated, thus showing the true time-dependent behaviour (independent of effective stress) of the material itself.

Materials which creep significantly present a huge challenge. As yet we have no fundamental theory to allow us to understand and predict the magnitude of this behaviour. We have no reasonable scaling laws to allow us to extrapolate from the results of short-term loading tests to long-term construction behaviour. All that can be done is to identify those materials that are known to undergo significant time-dependent behaviour under constant effective stress conditions, and make allowances on the basis of our limited experience. In fractured weak rocks, for example, it seems likely that long-term settlements of 2-3 times the immediate settlement should be allowed for under foundations (Burland and Lord, 1970).

3 STIFFNESS MEASUREMENT

A good approach to stiffness measurement should consider both field and laboratory techniques, and

Table 2. Pros and cons of in-situ and laboratory testing

In-situ testing	Laboratory testing
Advantages	
• tests can be carried out on materials which cannot be sampled	• tests are carried out in a well-regulated environment
• results can be obtained quickly, during the course of fieldwork	• stress and strain levels are controlled, as are drainage boundaries and strain rates
• appropriate methods can test large volumes of ground, including the effects of large particles and of joints	• effective stress testing is straightforward
• estimates of horizontal in-situ stress can be made	• effects of stress path and strain level can be examined
	• drained bulk modulus can be measured
Disadvantages	
• drainage boundaries are not controlled, sometimes making it difficult to know whether tests are drained or undrained	• testing cannot be carried out if undisturbed samples are unobtainable
• stress paths and strain levels are often poorly controlled	• test results are typically only available some time after the completion of fieldwork
• the long test period of consolidation or drained tests in clays makes them inconvenient and expensive	
• pore pressures are not measured, so the effective stress is unknown	

the possibility of using more than one method of measurement. In-situ and laboratory techniques have complementary attributes, as Table 2 shows.

In selecting a technique to be used for determining as accurately as possible the stiffness of the ground, the following factors need to be considered:

1. the material tested should be in as undisturbed state as possible. In principal this will mean that
   a)   it should not have been significantly strained during test preparation
   b)   changes in effective stress should have been minimised
   c)   joints and fissures should not have been allowed to open
   d)   moisture content should not have changed

2. the volume of material to be tested should be large enough to include the effects of grain size, joints and fissures. As a general rule, the minimum dimension of the loaded soil or rock (for example the diameter of the triaxial specimen , the height of an oedometer specimen, the diameter of a pressuremeter, the diameter of a plate) should be at least 6 times the maximum particle size or 6 times the fissure or joint spacing

3. the number of tests conducted, and their locations should take into account the expected non-homogeneity of the ground, relative to the size of the test

4. bedding and compliance effects should be avoidable, either by measuring away from the loading position (e.g. local strain or sub-plate measurement), and/or by placing plaster or some similar bedding material at the contact between the soil/rock and the loading platen (Figures 2 and 4).

5. the amount of loading, whether judged in terms of stress or strain, and its direction should be similar to that expected from construction effects. This is particularly important because non-linearity, plastic straining and anisotropy (whether induced by stress, fracturing patterns, or layering of soils and rocks) are always likely to be present.

Conventional techniques of stiffness measurement often cannot deliver accurate estimates of the mass in-situ stiffness of the ground because they fail to address most of these problems. For example, oedometer tests are carried out on specimens which are disturbed by sampling, the specimen size is often too small to account for even small-scale fabric, and bedding is not avoided. In response to this, best UK practice has introduced new drilling and sampling

techniques, and has adopted a new range of methods of measuring stiffness, both in the field and in-situ. The advantages and disadvantages of some of these techniques are described below.

3.1 Small-strain triaxial testing

Local strain instrumentation has now become widely used in the UK, particularly where stiffness values are required for numerical modelling of foundations, tunnels and excavations in stiff clays. The general configuration of the local strain instrumentation on a triaxial specimen is shown in Figure 2. Points to note are that:

- the local strain gauges are best set on the mid-third of the specimen, if platen friction effects are to be avoided. If one is simply concerned with bedding and compliance then it is acceptable for the pads to be closer to the platens
- in clays it is common to use local strain instrumentation in conjunction with mid-

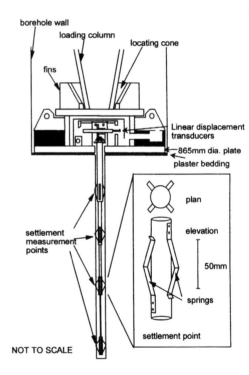

Figure 4. Methods of avoiding bedding effects in plate testing, in a down-hole test (after Marsland and Eason, 1973

plane pore pressure measurement, since this ensures that effective stresses are correctly measured. For stiff clays a flushable high-air-entry probe is necessary, because of the high level of suction in the clay during test set-up.

- local strain instruments should have an overall accuracy of better than about 3 micron, if stiffnesses in the strain range of 0.01% to 0.1% are to be measured on 200mm high specimens, using a 70mm gauge length
- two common forms of local strain instrument are the Hall effect device (Clayton and Khatrush, 1986) and the miniature submersible LVDT (for example, Cuccovillo and Coop, 1997). Hall effect gauges are favoured because they are small, and are easier to mount on the specimen. Given excellent calibration facilities (Heymann, Clayton and Reed, 1997) LVDTs can provide far better accuracy, however. Recently Heymann (1998) has achieved accuracies of about 0.03 micron (better than 0.001%.axial strain), and this has permitted the linear-elastic stiffness plateau to be measured (Figure 5, from Clayton and Heymann, in press).

A range of stress paths has been used for local strain tests, as described by Jardine et al (1991). Of these, the most common in practice is the relatively simple unconsolidated undrained test. This involves an initial determination of the effective stress in the specimen, obtained by elevating the cell pressure until mid-plane pore pressure measurement indicates a B value of about unity, followed by shearing at a fixed rate of strain (typically 5%/day). In carrying out such tests it is vital that the specimen is allowed to settle fully under the cell pressure, before loading commences, so that creep from previous loading is reduced to an acceptably small level.

Whilst local strain triaxial testing can provide good estimates of the undrained Young's modulus or the shear modulus for vertical loading, this technique cannot be applied to determine horizontal stiffness, if stiffness anisotropy is thought to be large. Furthermore, the success of the testing is entirely dependent upon a source of high-quality "undisturbed" samples. When advanced techniques such as these are to be used in the UK, samples are generally obtained either from pushed thin-walled tubes, or from wireline rotary coring. In southern Africa block sampling is widely used (see Heymann and Clayton, in this conference, for example) so that very high quality samples should be fairly readily available. Further limitations are that

- element tests of this size cannot reasonable be used on hard fissured soils or fractured weak rocks, because the specimen size is too small to represent the mass
- also because of specimen size, the technique is only satisfactory for relatively fine-grained soils

Nonetheless, good predictions of displacements around excavations have been made using data collected in this way (Jardine, et al., 1991) for stiff and very stiff clays.

3.2 *Geophysical testing*

Seismic geophysical testing is rapidly gaining popularity in the UK, the US and the Far East (Hong

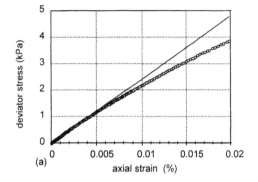

(a)

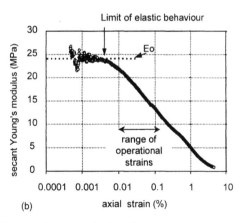

(b)

Figure 5. Stress-strain data from a triaxial test on a high-quality sample of Bothkennar clay (after Clayton and Heymann, in press)

a) deviator stress v. local axial strain

b) variation of secant Young's modulus with strain for the same data

Kong, Korea and Japan) as a means of field stiffness measurement. The most frequent techniques currently in use are the parallel cross-hole method, the down-hole method, and the surface-wave method. The principles behind these three techniques are summarised in Figure 6 (from Clayton, Matthew and Simons, 1995).

In relatively soft and saturated ground (soils, rather than rocks) the velocity of a seismic compressional or primary (i.e. first-arriving, P) wave will be a function of the stiffness of the pore fluid, since this is much greater than that of the skeleton (i.e. B=1). The velocity of a compressional wave in water is about 1500m/s, and P-wave velocities measured in soils are normally not much greater than this. To determine the stiffness of soils, therefore, it is necessary to measure either the shear-wave (second-arriving, S) velocity, or the Rayleigh surface-wave velocity, which is very close to the S-wave velocity. In contrast to the P-wave, the Rayleigh and S-wave velocities impose shear distortion without volumetric strain, and their speed is therefore a function of the shear modulus (G) of the soil (as opposed to its undrained bulk modulus).

Seismic techniques used in geotechnical engineering therefore must be able to generate S-waves or Rayleigh waves, and detect their arrival at various distances from the source. Simple methods that can be used to generate P and S waves are shown in Figure 7.

The relationship between shear wave velocity and shear modulus is a simple one for an isotropic material:

$$G = V_s^2 \rho$$

where G is the shear modulus, E = G(1+2ν), V_s is the shear wave modulus, and ρ is the bulk density of the material. The strain level imposed by seismic methods is generally considered to be very small. Auld (1977) has estimated their shear strain magnitude as between 10^{-6} and 10^{-4} %, and as can be seen from Figure 5, this means that shear and Rayleigh waves will travel at speed which is a function of the very small strain stiffness, G_o. For this reason, and the fact that these are dynamic techniques, the stiffness obtained from geophysics has traditionally been thought by geotechnical engineers to be a gross over-estimate of that required for analysis of civil engineering problems. However, there is increasing evidence (for example Clayton, Gordon and Matthews, 1994, Heymann, 1998, Matthews, Clayton and Own, in press, Mukabe, Tatsuoka and Hirose, 1991,) that because of the bonded nature of soil, the stiffness of soils and weak rocks at 0.01% strain (which is required for design) will typically be only 20% to 50% below E_o (see Figure 5(b) for example). The apparently high stiffness obtained from geophysics, when compared

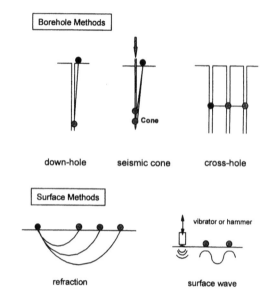

Figure 6. Seismic methods used for stiffness profiling

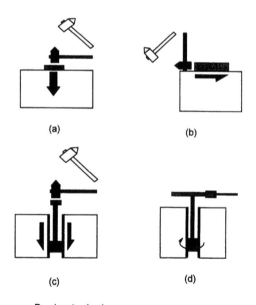

Dominant seismic energy

(a) compressional 'P' waves
(b) horizontally polarized shear (S) waves
(c) vertically polarized shear waves
(d) horizontally polarized shear waves

Figure 7. Simple methods of producing compressional- and shear-wave energy for shallow seismic surveys (after Clayton et al., 1995)

310

with values from traditional testing, turn out to be largely a function of the inappropriately high strain levels and the effects of bedding in the latter.

As with any other technique, seismic methods of determining stiffness have their strengths and weaknesses. Advantages are that

- the soil or rock is tested at in-situ effective stress levels, and (because of the large volume tested compared with the size of any holes or probes) in an undisturbed state
- the volume tested is sufficiently large to provide mass stiffnesses which take into account joints. For example, excellent predictions have been made of the settlements beneath 1.8m diameter loaded plates on fractured rock, using shear modulus values obtained from surface wave tests (Matthews, 1993). Coarse-grained soils, or indeed rock fills, can also be tested.
- seismic methods can also be used to obtain estimates of stiffness anisotropy when this is induced by layering (Pinches and Thompson, 1990) although because the calculated stiffness is a function of the square of the seismic velocity, only high levels of anisotropy can be detected (Hope, Clayton and Butcher, 1999)

The major disadvantages are that the techniques and equipment are unfamiliar to most engineers, and that background noise (for example from roads, railways and industrial plant) may mean that testing has to be conducted at times of least activity, such as at night, and at the weekend. However, even under unfavourable conditions it is usually possible to obtain reasonable stiffness profiles, as the data in Figure 8, for a site underlain by mudstones and sandstones with unconfined compressive strengths of the order of 2MPa and a fracture spacing typically around 200mm, shows.

Given the relatively stiff bonded residual soils that cover much of Africa, it seems likely that cheap surface methods of geophysical stiffness determination would perform particularly well, giving relatively deep measurements for quite modest energy inputs.

3.3 Self-boring pressuremeter testing

In high-quality investigations of stiff clays in the UK it has become commonplace to combine self-boring pressuremeter testing with laboratory small-strain stiffness measurement, in order to obtain parameters for design. The pressuremeter is used not only to obtain the horizontal modulus, but also most importantly (since this is required for elasto-plastic finite element analyses) the in-situ horizontal total stress, from which K_o is derived. The undrained shear strength is also obtained as a "by-product" of the analysis. After many years of experience it seems clear that expert operators can routinely obtain good results from stiff clays with this tool. Unfortunately, results are often produced which show a great deal of scatter (for example, Figure 9), such that neither a stiffness profile nor the K_o values can be defined with sufficient certainty for the designer. This may occur for a number of reasons:

- the equipment used is not of sufficiently good quality
- the amount of disturbance caused when inserting the pressuremeter is large, or varied. It is rare, for example, for site specific investigations of the effect of cutter position to be carried out
- unload-re-load loops are not carried out at similar strain and pressure levels, so that the moduli derived from them are not comparable.

Self-boring pressuremeter and dilatometer tests are increasingly being used in weak fractured rocks. Here the problems of interpretation caused by the presence of fractures daylighting along the test pocket are very large (Habberfield and Johnson, 1993), and may prevent any meaningful interpretation of the data in terms of shear modulus unless joint spacing is less than about 20mm. At the same time, bedding and disturbance will reduce the measured stiffness, but this may be compensated to some extent by the fact that the loaded area is small

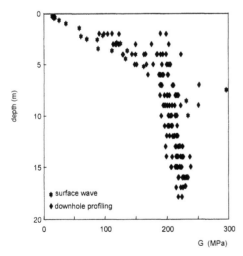

Figure 8. Stiffness / depth profile determined using Rayleigh wave and down-hole seismic methods (Hope et al., 1998)

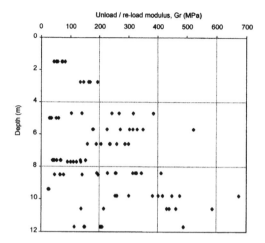

Figure 9. Pressuremeter unload - reload modulus v. depth, for a weak rock (Hayward, in preparation)

(typically less than 100mm in diameter), so that it is the intact stiffness rather than the bulk modulus which, in the absence of joints daylighting in the test pocket, is being measured.

Figure 10 shows the pressuremeter data as in Figure 9, but normalised with respect to both the cavity strain during the unload-reload loop, and the radial pressure at the point of unloading. The scatter in Figure 9 is removed, indicating the importance of

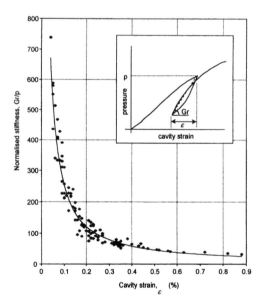

Figure 10. Normalised stiffness as a function of cavity strain (Hayward, in preparation)

strain level and pressure in controlling the measured stiffness. But problems remain in using the test results. These data are from the same site as the geophysical data shown in Figure 8. It is clear (because of the bonding in the weak rock, and its fracture spacing) that the pressuremeter should be measuring the intact stiffness of the rock, and that this should be relatively insensitive to changes in effective stress level. Based upon loading tests in other bonded materials, it seems likely that this material should fail at strains of around 0.1 to 0.5%. It can therefore be concluded that the strain levels used during the tests are generally too high to produce relevant stiffnesses for design. The strong dependency of stiffness shown by the tests is a function either of bedding, or of the fact that the pressures used during the test (which sometimes exceeded 7 MPa) were sufficiently large to have destructured the intact material immediately around the test pocket.

If the self-boring pressuremeter has the disadvantage of requiring sophisticated use and interpretation, it also has many advantages. Tests can be carried out at a variety of depths and locations. In theory (using elasto-plasticity) the results of the test can be interpreted to give fundamental parameters (Mair and Wood, 1987) and to determine the influence of strain level on stiffness. Because it is an in-situ test, results are available relatively quickly.

3.4 Plate testing

Plate testing is quite widely used in many parts of the world, but is often considered excessively expensive unless carried out at small diameter. The technique of plate testing is well known in the geotechnical community, but despite this there has been a fair amount of mis-use over the years. When carrying out such tests it is important to remember that

- excavation to the level at which the plate is to be located must not produce significant disturbance. The area on which the plate is to be founded must be carefully cleaned, preferably using hand tools, to remove excavation debris
- the plate must be properly bedded on the soil or weak rock, normally using a fluid plaster, placed in the thinnest possible layer
- the size of the plate must be sufficient to cope with the grain size of the soil, and the fissure/joint spacing for hard soils and weak rocks
- any datum used for settlement measurement must not be significantly

affected by daily temperature changes, since tests are likely to last for some time

- tests should be carried on for sufficient time to allow the effects of creep to be determined

Failure to attend to any of these matters will almost certainly produce incorrect results.

Despite their apparent simplicity, there a major difficulties in interpreting plate test results. Large variations in contact pressure, between the plate and the soil or rock, occur because of the stiffness and method of loading of the plate. Markedly different stress paths and strain levels are applied at different positions under the plate. It is therefore impossible to interpret the load-settlement data on a sound theoretical basis in terms of modulus in a material such as soil or fractured weak rock, where behaviour is stress path and strain level dependent. Accepting these problems, and interpreting plate results in terms of a rigid punch on an isotropic linear elastic material, there still remains the need to extrapolate the results in order to predict the settlement of much larger foundations. Probably the most sensible method is to regard plate loading as an element test, and to carry out a number of tests at different locations and depths below ground level, in order to obtain a stiffness/depth profile. Unfortunately it is rarely possible to justify expenditure on down-hole plate tests, and so it is necessary to resort to extrapolation (e.g. Lake and Simons, 1970, Terzaghi and Peck, 1948).

Nonetheless, the plate test continues to make an important contribution, if for no other reason that it models a footing, and can be carried out on materials which are impossible to sample, such as granular soils, fractured weak rocks, rockfill, and waste. When using the results, however, one should be cautious, bearing in mind that any stiffness derived from the test is more of a "figure of merit" (Ward et al, 1969) than a true measure of stiffness.

4 CONCLUSIONS

Soils and weak rocks have complex stiffness behaviour. Measurements can be affected by bonding, fissuring and jointing, non-homogeneity, anisotropy, non-linear stress-strain behaviour, bedding and creep. If values of stiffness appropriate to the geotechnical design of foundations, tunnels and excavations are to be obtained, then sampling, laboratory and in-situ testing strategies need to take these factors into account.

With the increasing need for accurate measurements of stiffness, new techniques of measurement have been developed, and traditional ones re-evaluated. Each has advantages and disadvantages, and these, coupled with some basic requirements for the determination of realistic stiffness values, must be borne in mind during the development of any testing strategy. An ability to obtain high quality samples of many cohesive or bonded soils, coupled with the use of local small-strain measurement in the triaxial test, has brought about a revival of laboratory stiffness measurement. At the same time the potential value of seismic geophysical techniques has become appreciated, and their use has grown, and our ability to interpret the self-boring pressuremeter has improved.

It is suggested that a good test programme will use more than one type of test, and where samples can be obtained will often involve combinations of field and laboratory testing. This paper has illustrated some of the equipment and techniques now available, and how the data they produce can be interpreted within the framework of modern geomechanics knowledge.

ACKNOWLEDGEMENTS

The author gratefully acknowledges the advice and help given by Prof. G. Heymann of University of Pretoria, Dr. M.C. Matthews of University of Surrey, and Toby Hayward of University of Southampton, during the preparation of this paper.

REFERENCES

Atkinson, J.H. (1973) 'The deformation of undisturbed London clay' PhD Thesis, University of London.

Auld, B. (1977). 'Cross-hole and down-hole VS by mechanical impulse' Proc. A.S.C.E., Journal of the Geotechnical Engineering Division, vol. 103, GT12, pp. 1381-1398.

Brown, S.F. and Snaith, M.S. (1974) 'The measurement of recoverable and irrecoverable deformation in the repeated load triaxial test' Geotechnique, vol. 24, no. 2, pp. 255-259.

Buisman, A.S.K. (1936) 'Results of long duration settlement tests' Proc. 1st Int. Conf. Soil Mech. and Found. Engng, Harvard, vol. 1, pp. 103-105.

Burland, J.B. (1990) 'On the compressibility and shear strength of natural clays' Geotechnique, vol. 40, no. 3, pp. 329-378.

Burland, J.B. and Lord, J.A. (1970) 'The load-deformation behaviour of Middle Chalk at Mundford, Norfolk: a comparison between full-scale performance and in-situ and laboratory measurements' Proc. Conf. on In Situ Investigations in Soils and Rocks, British Geotechnical Society, London, pp. 3-15.

Burland, J.B. and Symes, M.A. (1982) 'A simple axial displacement gauge for use in the triaxial apparatus' Geotechnique, vol. 26, pp. 371-375.

Clayton, C.R.I. (1977) 'Some properties of remoulded Chalk' Proc. 9th Int. Conf. Soil Mech. and Found. Engng, vol. 1, pp. 65-68.

Clayton, C.R.I. (1983) 'The influence of diagenesis on some index properties of the Chalk of England' Geotechnique, vol. 33, no. 3, pp. 225-241.

Clayton, C.R.I. and Heymann, G. (in press) 'The stiffness of geomaterials at very small strains' Submitted to Geotechnique.

Clayton, C.R.I. and Khatrush, S.A. (1986) 'A new device for measuring local strains on triaxial specimens' Geotechnique, vol. 36, no. 4, pp. 593-597.

Clayton, C.R.I. and Serratrice, J.F. (1997) 'The mechanical properties and behaviour of hard soils and soft rocks' Proc. Int. Symp. on "Geotechnical Properties of Hard Soils - Soft Rocks" (1994), (ed. Anagnostopoulos, A., Schlosser, F., Kalteziotis, N. and Frank, R.). General Report - Session 2., vol. 3, pp. 1839-1877. Balkema, Rotterdam.

Clayton, C.R.I., Gordon, M.A. and Matthews, M.C. (1994) 'Measurements of stiffness of soils and weak rocks using small strain laboratory tests and field geophysics' Proc. Int. Symp. on Pre-failure Deformation Characteristics of Geomaterials, Sapporo, Japan, vol. 1, pp. 229-234. Balkema, Rotterdam.

Clayton, C.R.I., Matthews, M.C., and Simons, N.E. (1995) 'Site Investigation' Blackwell Science, Oxford, 584 pp.

Clayton, C.R.I., Edwards, A. and Webb, D.L. (1991) 'Displacements in London clay during construction' Proc. 10th Eur. Conf. Soil. Mech. Found. Engng, Florence. vol. 2, pp.791-796.

Cole, K.W. and Burland, J.B. (1972) 'Observations of retaining wall movements associated with a large excavation' Proc. 5th European Conf. Soil Mech. and Found. Engng, Madrid, vol. 1, pp.445-453.

Cuccovillo, T. and Coop, M.R. (1997) 'The measurement of local axial strains in triaxial tests using LVDTs' Geotechnique, vol. 47, no. 1, pp. 167-171.

Dalton, J.C.P. and Hawkins, P.G. (1989) Fields of stress - some measurements of in-situ stress in a meadow in the Cambridgeshire countryside' Ground Engineering, vol. 15, no. 4, pp. 15-23.

Daramola, O. (1978) 'The influence of stress history on the deformation of sand' PhD thesis, University of London.

Haberfield, C.M. and Johnson, I.W. (1993) 'Factors influencing the interpretation of pressuremeter tests in soft rock' Proc. Symp. on Geotechnical Engineering of Hard Soils - Soft Rocks, Athens, vol. 1, pp. 525-532. Balkema, Rotterdam.

Heymann, G. (1998). 'The stiffness of soils and weak rocks at very small strains' PhD Thesis, University of Surrey, UK.

Heymann, G. and Clayton, C.R.I. (1999) 'Block sampling of soils: some practical considerations' 12th African Regional Conference of the ISSMGE. Durban.

Heymann, G., Clayton, C.R.I. and Reed, G.T. (1997) 'Laser interferometry to evaluate the performance of local displacement transducers' Geotechnique, vol. 47, no. 3, pp. 399-407

Hayward, T. (in preparation) Ph.D. Thesis, Dept. of Civil and Environmental Engineering, University of Southampton

Hoek, E. and Bray, J.W. (1981) 'Rock slope engineering' 3rd Edition. Instution of Mining and Metallurgy, London.

Hope, V.S., Clayton, C.R.I. and Butcher, A.P. (1999) 'In situ determination of Ghh at Bothkennar using a novel seismic method' Quarterly Journal of Engineering Geology, vol. 32, no. 2, pp. 97-106.

Hope, V.S., Clayton, C.R.I. and Sutton, J.A. (1998) 'The use of seismic geophysics in the characterisation of a weak rock site' Proc. 1st Int. Conf. on Geotechnical Site Characterisation (ISC'98), Atlanta (ed. Robertson, P.K. and Mayne, P.W.), vol. 1, pp. 479-484.

Jardine, R. J., Potts, D. M., St. John, H. D. and Hight, D. W. (1991). Some practical applications of a non-linear ground model. Proc. 10th Eur. Conf. Soil Mech. Found. Engng, Florence. vol. 1, pp.223-228.

Jardine, R. J., Potts, D. M., Fourie, A. B. and Burland, J. B. (1986). Studies of the influence of non-linear characteristics in soil-structure interaction. Géotechnique, vol. 36, no. 3, pp. 377-396.

Lake, L.M. and Simons, N.E. (1970) 'Investigations into the engineering properties of chalk at Welford Theale, Berkshire' Proc. Conf. on In Situ Investigations in Soils and Rocks, British Geotechnical Society, London, pp. 23-29.

Lake and Simons (1975) 'Some observations of the settlement of a four storey building founded on chalk at Basingstoke, Hampshire' Proc. BGS Conf. on Settlement of Structures, Cambridge, pp. 283-291.

Leroueil, S., and Vaughan, P.R. (1990) 'The general and congruent effects of structure in natural soils and weak rocks' Geotechnique, vol. 40, no. 3, pp. 467-488.

Mair, R. J. (1993). 'Developments in geotechnical engineering research: application to tunnels and deep excavation' Unwin Memorial Lecture. Proceedings of the Institution of Civil Engineers, Civil Engineering, vol.93, pp.27-41.

Mair, R.J. and Wood, D.M. (1987) 'In-situ pressuremeter testing: Methods Testing and Interpretation' CIRIA Ground Engineering Report. Butterworths, London.

314

Marsland, A. and Eason, B.J. (1973) 'Measurements of the displacements in the ground below loaded plates in deep boreholes' British Geotechnical Society Symp. on Field Instrumentation in Geotech. Engng, vol. 1, pp. 304-317.

Matthews, M.C. (1993) 'The mass compressibility of fractured chalk' PhD Thesis, University of Surrey.

Matthews, M.C., Clayton, C.R.I., and Own, Y. (in press) 'The use of field geophysical techniques to determine geotechnical stiffness parameters' Submitted to Geotechnical Engineering, Proc. I.C.E.

Mukabe, J.N., Tatsuoka, F., and Hirose, K. (1991) 'Effect of strain rate on small strain stiffness of kaolin' Proc. 26th Japanese National Conf. on Soil Mech. and Found. Engng, Nagano, pp. 659-662.

Pinches, G.M. and Thompson, R.P. (1990) 'Crosshole and downhole seismic surveys in the UK Trias and Lias' Proc. Conf. on Field Testing in Engineering Geology. Eng. Geology Special Pub. No. 6, pp. 299-308. Geological Society, London.

Simpson, B., O'Riordan, N.J., and Croft, D.D. (1979) 'A computer model for the analysis of ground movements in London clay' Geotechnique, vol. 29, no. 2, pp. 149-175.

Terzaghi, K., and Peck, R.B. (1948) 'Soil Mechanics in Engineering Practice' 1st Edition. Wiley, New York.

Ward, W.H., Burland, J.B. and Gallois, R.W. (1968) 'Geotechnical assessment of a site at Mundford, Norfolk, for a large proton accelerator' Geotechnique, vol. 18, pp. 399-431.

Wroth, C.P. (1971) 'Some aspects of the elastic behaviour of overconsolidated clay' Proc. Roscoe Mem. Symp. on Stress-Strain Behaviour of Soils, Cambridge, pp. 347-354.

Geotechnics for Developing Africa, Wardle, Blight & Fourie (eds) © 1999 Balkema, Rotterdam, ISBN 90 5809 082 5

Geotechnical properties of Qoz soils

Ahmed M. El Sharief, Yahia E-A. Mohamedzein & Yassir A. Hussien
Building and Road Research Institute, University of Khartoum, Sudan

ABSTRACT: Qoz Soils are wind blown deposits of fine silty sands which cover large areas in Mid-Western Sudan. The soils vary mainly in their silt content (7 to 25%). Several important infrastructure projects, such as the Continental Western Salvation Road, are under active development in Mid -Western Sudan. These important Structures will be supported on the relatively unfamiliar Qoz soils. This paper summarizes the basic properties of Qoz soils such as compaction properties, strength, compressibility and potential for collapse. Qoz soils attain relatively high dry densities when compacted. The maximum dry density increases linearly with silt content . The California Bearing Ratio CBR vary from 10 to 28% depending on the percentage of fines. The initial moisture content and silt content significantly affect the compressibility and collapse potential of these soils. The collapse potential may be rated as low to moderate.

1 INTRODUCTION

Sudan is the largest country in Africa with an area of about 2.50 million square kilometers. The climate in Sudan varies from desert climate in the North to tropical climate in the South. In between, regions of semi-desert, poor Savannah and rich Savannah exist. Rainfall is clearly seasonal with rainfall intensity averaging less than 10mm in the North to over 1500mm in the far South. The common soil types include (Figure 1): desert soils; alluvial and lacustrine soils; Qoz soils, semi-arid and alkaline soils and residual ferruginous deposits (Whiteman 1971).

Qoz is a local name given to the stabilized fine sands located west of the White Nile and North of the alluvial black cotton soils, Figure 1. In this respect Qoz soils are stabilized fine silty sand deposits which are extension of the non-stabilized desert sands and sand dunes in the North. The stabilizing agents are partly flora, partly thin surface crust of iron oxide and partly fine silt or clay (Whiteman 1971). These formations were formed during the early dry phases in the Pleistocene, but the main period of formation was during the Sebilian (Middle Palaeolithic) dry phase (i.e. 20000 - to 25000 years ago). Another phase of deposition was during late Holocene (3000 B.P.). Periods of very wet climate usually followed the dry climate and promoted stabilization. The formations are mostly of Nubian origin (Edmonds 1942).

The area covered by Qoz soils is considered to be one of the most potential for development in Sudan. Livestock breeding and agriculture are practiced by the local inhabitants. The existing settlements in the region are growing and many development projects are proposed and currently in the design or construction phases (i.e. roads, buildings and airports etc.). The Western Salvation Road (WSR) is the most important infrastructure project in the region. This continental road will connect eastern and western regions of Sudan with the Western African Countries. Geotechnical investigations have been made for some of these projects (specially the WSR) and a great deal of data have been collected. This paper focuses on the geotechnical properties of these soils which will affect their potential use as foundation and/or road building materials.

2 DATA SOURCES

The data analyzed covers a large area in the Qoz region. If we consider Ennahoud town as a center the data covers a circle of about 200km in diameter. The data were obtained from material reports of roads in the Qoz region prepared for preliminary and/or detailed phases of design and from geotechnical reports for structures built in the region. Laboratory tests were also carried by the authors at the Building and Roads Research Institute in Khartoum - Sudan.

SCALE :- 1:15,000,000

Figure 1. Soil deposits in Sudan (Whiteman 1971)

The sources include but are not limited to : material reports for El Obeid-Ennahoud; Ennahoud - AlFula - Al Muglad; Ennahoud - El daien roads; and geotechnical investigation reports for El Obeid Refinery and an Islamic Center at Almanara near El Rahad town (see Figure 1).

3 CLASSIFICATION OF QOZ SOILS

Qoz soils are nonplastic reddish brown sands. The color changes to yellowish brown, then light yellowish brown with depth. According to the Unified Soil Classification System they are classified as poorly graded silty sand (SM) and A-2-4 according to AASHTO classification. The soils pass the 1.00mm standard sieve and the percentage of fines (%<75 microns) mostly range between 7 to 25%. The major difference between these soils being the amount of fines or silt. Table (1) shows results of statistical analysis of the data of the fines content obtained from four projects. The analysis gives the mean, standard deviation and 95% confidence interval for the mean. The analysis is made using the statistical program SPSS. The mean value for fines content ranges between 15% to 17% for three projects and is 25.8% for El Obeid-Ennahoud road projects. The standard deviation is less than 5% for the three projects and 10.8% for El Obeid Ennahoud which shows a large scatter. No significant variation is found between the mean value of fines content except for El Obeid Ennahoud road project.

Table (1) Statistical Analysis of Fine Content

Group	Count	Mean	Standard Deviation	95% Conf. Int. for mean
Ennaoud-Daein	181	15.54	3.74	14.99 to16.09
Ennaoud-Almujlad	14	16.77	4.69	14.06 to19.50
Elobeid-Ennahoud	82	25.8	10.8	23.36 to28.1
Around Elobeid	78	15.5	4.69	14.45 to16.56
Total	355	17.93	7.6	17.14 to8.73

4 DENSITY AND STRENGTH

The insitu density and strength were measured or assessed at various locations within the Qoz region using data of Standard Penetration Test (CPT), Cone Penetration Test (CPT), Dynamic Cone Penetration Test (DCP) and in-situ density measurements by the sand cone method. The soils are loose in the upper 2.0 meters and change to medium dense and dense with depth. Manual excavation was easy in the upper layers. There is a general trend of increase in density with depth (Figure 2). At wadis (water courses) loose deposits usually extend to greater depths (up to 4.0m).

Minor spatial variability in density and strength were noticed within a specific site. Figure 2 shows typical CPT results (soundings, less than 30m apart at Almanara Islamic Center Site). The same observations were noticed in the investigation carried at El Obeid refinery site.

The DCP results carried during the preliminary investigations of Ennahoud-AlFula Road revealed a very loose to loose soil layers in the upper 1.5 to 2.0 meters. The density then improved with depth. Insitu density tests were also carried to 1.50 to 2.0 meters. The results showed in-situ bulk density values between 1.25 to 1.65 kg/cm^3.

The maximum dry density measured using British Heavy Compaction Test ranges between 19.5 to 21KN/m^3 for soils having variable fines content (8 to 25%). The maximum dry density relates well, i.e., increases, with fines content (Figure 3). The optimum moisture, however, was not sensitive to fines content.

The California Bearing Ratio (CBR) test results (Figure 4) showed the soaked CBR values to range between 10 to 28, depending on the fines content. The

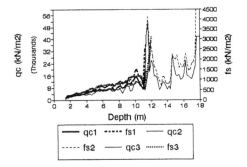

Figure 2. CPT results at Almanara site

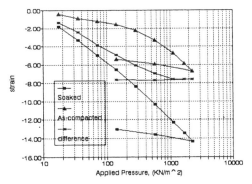

Figure 5-a. Measured compressibility for a Qoz sample (fines content 25, m.c. 4)

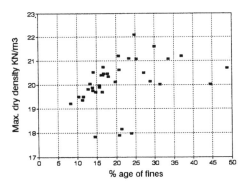

Figure 3. Maximum dry density versus fines content for Qoz soils

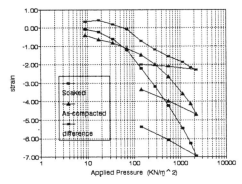

Figure 5-b. Measured compressibility for a Qoz sample (fines content 25, m.c. optimum)

soaked CBR generally decreases with increase in fines content. This is contrary to the trend observed for maximum dry density and indicates an effect of wetting on the strength of the Qoz soils as the silt or fines content increases.

5 COMPRESSIBILITY

Compressibility of a compacted Qoz soil was assessed using the double oedometer technique (Jenning and

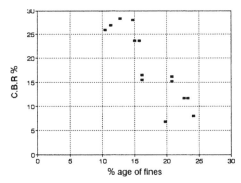

Figure 4. CBR versus fines content for Qoz soils

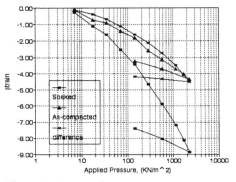

Figure 5-c. Measured compressibility for a Qoz sample(fines content 25,m.c. 7.5)

Knight 1957). A soil with a high silt content was selected (percentage fines equals to 25%). The soil was tested at the optimum moisture content, dry of optimum and wet of optimum. The technique involved preparing two nominally identical samples at the same initial moisture and density and incrementally loading one sample in the oedometer at its as compacted condition. The other sample, under a small seating load, was inun-

319

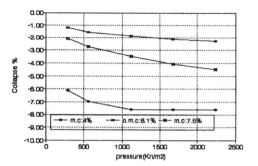

Figure 5-d. Measured compressibility for a Qoz sample (fines content 25, different m.c)

dated with water, allowed to come to moisture equilibrium and then incrementally loaded as for the first one. The loading sequence was identical for the two tests.

The test results are given in terms of strain versus applied pressure (Figures 5a,b and c). The difference in strain for each loading is also given and the results are summarized in Figure 5d. The difference in strain is a measure of the collapse potential of the soil (Lawton et al,1992). The results from these tests, therefore, show that Qoz soils are potentially collapsible. The highest collapse was measured for the driest soil and the least for the soil at its optimum moisture content. The maximum collapse was measured for the highest load. The amounts of collapse for the given loading, measured for the sample, dry of optimum , may be considered as medium to high.

6 CONCLUSIONS

Qoz soils are stabilized nonplastic wind-blown fine silty sands which cover large areas in mid-western Sudan. However the double oedometer technique was reported to slightly overpredict collapse potential (Booth 1977).

They mainly vary in their silt content (7 to 25%). They are loose to medium dense in the upper zones and change to dense and then very dense with depth. The maximum dry density using the British heavy compaction test increases with the increase in fines content while the soaked CBR shows a relative decrease with increase in fines. Qoz soils when properly compacted may be considered as good subgrade materials for roads.

Double oedometer tests on a Qoz sample with high silt content showed the soil to be potentially collapsible. The collapse potential increases with decrease in initial moisture and with increase of loading. Minimam collapse potential was measured at the optimum moisture content. Care should be excersized when placing foundation on loose dry Qoz soils.

7 REFERENCES

Booth A.R. (1977); Collapse Settlement in Compacted Soil, CSIR Res Report 324; Council for Scientific and Industrial Research , Pretoria, South Africa (Quoted from Lawton et al 1992).

Edmonds J.M. (1942) The distribution of Kordofan, Anglo-Eqyptian Sudan. Geo,Mag. 79 18-30 (Quoted from Whiteman 1971).

Jennings, J.E. & K. Knight, (1957), The prediction of total heave from the double oedometer test Transactions Symposium on Expansive Clays, South African Inst. of Civil Engineering pp-13- 19.

Lawton,E.C.; R.J. Fragaszy & M.D. Hetherington Review of Wetting. Induced Collapse in Compacted Soil JGED, ASCE, Vol. 118 No. 9pp 1376 - 1394.

Whiteman A.J. (1971) The Geology of Sudan Republic Carendon Press, Oxford.

Geotechnics for Developing Africa, Wardle, Blight & Fourie (eds) © 1999 Balkema, Rotterdam, ISBN 90 5809 082 5

Déformations permanentes sous chargements répétés de trois graveleux latéritiques du Sénégal Occidental

Meïssa Fall
Institut des Sciences de la Terre, Université Cheikh Anta Diop, Dakar, Sénégal

ABSTRACT: Several studies have demonstrated the susceptibility of laterite soils to degradation under load. Results of CIU triaxial tests on reconstituted samples of western Senegal laterites are presented during cyclic loading in the triaxial apparatus. The response of three lateritic soils to undrained cyclic triaxial loading is described. The cyclic stress level has been varied, while the frequency remained constant for all tests. In order to model the response of the soils under the traffic load the results are considered only in terms of the evolution of the permanent strain and also of the variation of the resilient modulus during loading. The three soils all acts alike.

RESUME: Le comportement sous trafic demeure un aspect relativement peu traité en ce qui concerne les sols latéritiques. Dans une chaussée, sous l'effet des chargements routiers, les graveleux latéritiques subissent des déformations permanentes, responsable de phénomènes d'orniérage et surtout une ruine prématurée. Cet article présente des essais triaxiaux cycliques effectués sur trois graveleux latéritiques compactés à l'Optimum Proctor Modifié (OPM). Les graveleux latéritiques sont d'abord soumis à des sollicitations monotones à l'appareil triaxial, dans le but de déterminer les paramètres de rupture sous chargement monotone. La démarche consiste à évaluer l'évolution des déformations en fonction du nombre de cycles ; mais surtout de chercher des approximations correctes pouvant relier la déformation axiale permanente avec le nombre de cycles. Les résultats obtenus ont permis de trouver des relations adéquates entre la déformation axiale permanente et le nombre de cycles

1 INTRODUCTION

La résistance au cisaillement des sols latéritiques compactés a fait l'objet de nombreuses études. Pour la plupart de ces travaux, le comportement des sols latéritiques est seulement testé sous sollicitations monotones au triaxial ou à la boîte de cisaillement. Et rarement des études ont été menées en sollicitations cycliques. L'étude a été réalisée sur trois graveleux latéritiques différents (tableau 1). Ces essais préliminaires ont permis de définir certains critères par la suite utilisables lors des chargements répètes, notamment la charge de rupture en sollicitations statiques.

2 ETUDE DE LA DEFORMATION AXIALE PERMANENTE

L'analyse des déformations permanentes demeure un des aspects les plus complexes de l'étude des sols sous sollicitations cycliques et il reste aussi à définir des relations adéquates entre la déformation permanente et le nombre de cycles ou le niveau des contraintes. Les figures 1 à 5 donnent l'allure de l'évolution générale de la déformation permanente en fonction du nombre de cycles d'application de la charge pour les trois matériaux. Cependant on peut remarquer que l'allure des courbes est partout identique, la déformation permanente est d'autant plus importante que le rapport de charge est élevé. Ceci est valable dans le cas de l'échantillon de Sébikotane. Pour les autres sols, l'amplitude (Am) de la charge intervient. Par exemple, on s'aperçoit que est valable dans le cas de l'échantillon de Sébikotane.

Tableau 1. Caractéristiques des trois graveleux latéritiques.

	Ndienné	Sébikotane	Yenne/mer
γ_{dmax} (kN/m^3)	20,9	20,75	17,7
WOPM (%)	8,6	8,9	8,5
Ip (%)	8	20	20,5
> 2 mm (%)	26	30	15
> 0,1 mm (%)	6	6	7
CBR (%)	113	82	86

Pour les autres sols, l'amplitude (Am) de la charge intervient. Par exemple, on s'aperçoit que pour l'échantillon de Yenne/mer à 50 kPa de confinement, la variation de cette amplitude (de 20 à 30 % Qr pour une même charge moyenne (Q_{moy})) entraîne des évolutions relatives très différentes. Mais la tendance globale est à l'augmentation graduelle de la déformation permanente quand le niveau de charge croît car en faisant varier l'amplitude de la charge, on fait varier de la même manière le niveau de charge. Dans le cas de Yenne/mer pour une étreinte latérale de 100 kPa, on observe le phénomène contraire qui semble être relié à la variation de l'amplitude. On observe que pour Ndienné, indifféremment de la contrainte latérale, l'amplitude et le rapport de charge déterminent la déformation axiale permanente qui diminue alors

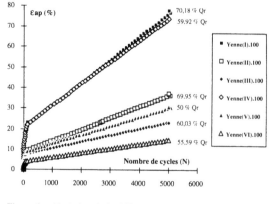

Figure 3 – Evolution de la déformation axiale permanente en fonction du nombre de cycle.

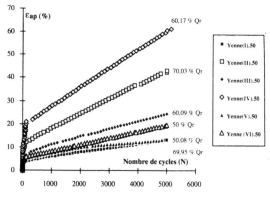

Figure 1 – Evolution de la déformation axiale permanente en fonction du nombre de cycle.

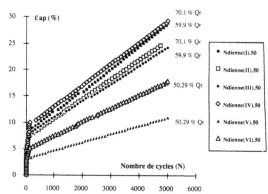

Figure 4 – Evolution de la déformation axiale permanente en fonction du nombre de cycle.

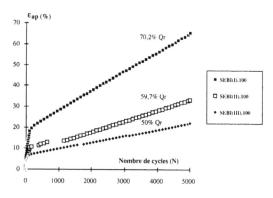

Figure 2 – Evolution de la déformation axiale permanente en fonction du nombre de cycle.

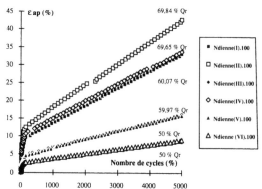

Figure 5 – Evolution de la déformation axiale permanente en fonction du nombre de cycle.

D'une manière générale la déformation permanente se traduit par :

- l'accroissement très important de la déformation permanente pour les premiers cycles ; la déformation axiale permanente est semblable pour les tous premiers cycles, un point d'inflexion est observé vers la centaine de cycles ;

- les déformations permanentes continuent de se développer avec le nombre de cycles tout à fait linéairement et dans aucun des cas nous n'observons après cette phase un quelconque point d'inflexion, donc il n'y a ni accélération ni rupture brutale.

Vu le nombre relativement important de cycles, on peut penser qu'à plus ou moins long terme ce comportement continuera de se manifester. Martinez (1990) observe ce genre de comportement sur des graves et l'attribue au « rochet » dans lequel cas la déformation permanente et l'énergie dissipée croissent progressivement à chaque cycle ; cette situation peut conduire à la ruine progressive de l'échantillon. Ces phénomènes se distinguent de l'accommodation où la déformation permanente atteint une borne et il y a dissipation de l'énergie ou encore de l'adaptation où ne se manifestent que des réponses élastiques et pour laquelle aussi bien l'énergie que la déformation permanente ont atteint une borne. L'observation de la variation de volume en cyclique peut aussi nous apporter des renseignements sur ces comportements. On sait que $\varepsilon_v = \varepsilon_1 + 2\varepsilon_3$. Aux figures 6 et 7 on observe que le comportement de la déformation permanente volumique est lié apparemment à l'amplitude et aussi à la contrainte latérale, ainsi pour tous les échantillons avec des amplitudes de 30% de Qr et à σ_3 de 100 kPa, le comportement est dilatant, et à 20% d'amplitude à σ_3 de 50 kPa les sols sont contractants.

Enfin, on notera que pour un rapport de charge élevée correspond de fortes déformations axiales permanentes et un comportement dilatant.

3 APPROXIMATION DE LA DEFORMATION AXIALE PERMANENTE

Pour nos échantillons, l'approche par une loi de type puissance comme :

$$\varepsilon_a^p = aN^b$$

donne des résultats satisfaisants (du moins concernant les coefficients de corrélation, qui sont toujours très élevés, généralement supérieurs à 90%) (figure 8). Cette approximation est le critère limite usuel limitant l'orniérage dû aux déformations permanentes du sol. Selon Martinez (1990) l'influence des contraintes se fait sentir surtout sur le paramètre **a**, qui est théoriquement égal à la déformation au premier chargement et qui selon lui ne représente pas son évolution. Mais on peut aussi remarquer que cette approximation sous estime ou surestime les déformations plus le nombre de cycles est élevé.

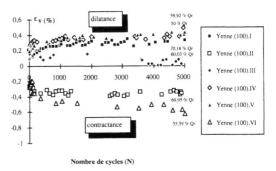

Figure 6 – Evolution de la déformation volumique permanente en fonction du nombre de cycle.

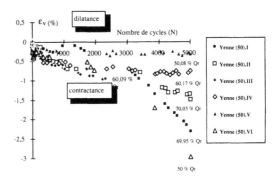

Figure 7 – Evolution de la déformation volumique permanente en fonction du nombre de cycle.

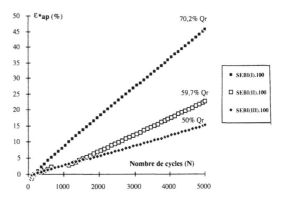

Figure 8 – Approximation de la déformation axiale permanente par une loi linéaire.

4 CONCLUSION

Les principaux résultats à retenir de cette étude sont les suivants :

• Pour les chargements étudiés, les trois graveleux latéritiques présentent des déformations permanentes très importantes qui se caractérisent par :

- une variation très rapide durant les premiers cycles (jusqu'au 144 ème cycle généralement) ;

- une absence de stabilisation après un grand nombre de cycle, ce phénomène étant attribué au rochet tel défini par Martinez (1990) sur des graves ;

• L'étude de la déformation permanente par la recherche d'une approximation meilleure en fonction du nombre de cycles a montré que la loi puissance constitue une bonne approche. Mais l'évolution quasi linéaire des déformations après les 100 premiers cycles, font penser qu'un ajustement linéaire peut être utilisé même si dans ce cas on considère que les tous premiers cycles ont une importance secondaire par rapport aux cycles suivants.

BIBIOGRAPHIE

Fall, M. 1993. Identification et caractérisation mécanique de graveleux latéritiques du Sénégal : Application au domaine routier. Thèse Doct. INPL en Génie Civil et Minier INPL (Nancy), 174 p.

Martinez, J. 1990. Contribution au dimensionnement rationnel des structures de chaussées souples et inverses. Comportement des graves non traités et des sols support. Thèse Doct. Etat. Univ. Montpellier.

Geotechnics for Developing Africa, Wardle, Blight & Fourie (eds) © 1999 Balkema, Rotterdam, ISBN 90 5809 082 5

The determination of soil-water characteristic curves from indicator tests

B.A. Harrison
P.D. Naidoo and Associates, Johannesburg, South Africa

G.E. Blight
University of the Witwatersrand, Johannesburg, South Africa

ABSTRACT: Soil-water characteristic curves were determined for 21 undisturbed soil samples, 11 of which were of transported origin and the remainder residual. Indicator tests were performed on the samples and relationships between them and the corresponding characteristic curve investigated. It was found that for soils of transported origin, the characteristic curves are related to the indicator tests. Little, if any, relationship is evident between the indicators and characteristic curves for residual soils. From this information a preliminary estimate of moisture content changes associated with suction changes can be assessed, and likely changes of volume, shear strength, permeability and water storage in undisturbed unsaturated transported soils subsequently estimated.

1 INTRODUCTION

The engineering behavior of an unsaturated soil depends on its water content, and changes in water content are produced by suction changes. This constitutive relationship between water content and suction can be expressed as a plot, usually referred to as a soil-water characteristic curve by engineers, and a soil-water retention curve by soil scientists.

The soil-water characteristic curve provides a useful conceptual aid to the understanding of unsaturated soil behaviour. It has also been employed for unsaturated shear strength determinations by Vanapalli et al (1996), and has been used as a field capacity curve for water balance computations by Blight & Roussev (1995). The derivation of permeability functions from the characteristic curve has been outlined by Gardner (1956), and the magnitude of suction changes associated with water content variations can be used in estimating soil volume changes (Fredlund & Rahardjo, 1993).

This paper illustrates how soil-water characteristic curves were derived in the pressure plate apparatus for undisturbed transported and residual soils. Factors influencing the form and shape of the curve are discussed, and its properties in relation to simple indicator tests examined.

2 DESCRIPTION OF SOILS TESTED

Undisturbed soil specimens were prepared from block samples extracted from trial holes excavated at various localities in the Eastern Cape, Gauteng and KwaZulu-Natal provinces of South Africa. In total 21 samples were prepared, 11 of which were of transported origin and the remaining 10 residual. The transported soils were located above a clearly defined pebble marker and were of alluvial, gulleywash, aeolian and hillwash origin. The residual soils were derived from the *in-situ* decomposition of andesite, dolerite, syenite, granite, norite and basaltic lava rocks.

Routine foundation indicator tests including grading and hydrometer analyses, Atterberg limits and linear shrinkage tests were carried out on each of the soil samples.

3 DETERMINATION OF THE SOIL-WATER CHARACTERISTIC CURVE

A pressure plate apparatus was used to derive the soil-water characteristic curve for each soil specimen. The apparatus uses the axis translation technique to generate a matrix suction within the soil. In the test, the pore water pressure (u_w) was maintained at atmospheric pressure i.e. $u_w=0$, and the pore air pressure (u_a) was raised to achieve the desired matrix suction (u_a - u_w).

The soils were prepared by trimming the block samples into 20mm diameter and 15mm deep copper rings. These were subsequently immersed in a container of water and saturated by soaking for 2 days. After this, the soil filled rings were placed onto the saturated ceramic in the pressure chamber, and a predetermined air pressure (u_a), or suction, applied.

Each specimen was left to equilibrate for one week, after which time it was removed from the chamber and its mass accurately determined on an analytical balance. The sample and ring were returned to the chamber and the procedure repeated for the next increment in pressure. Air pressures were applied in 10 and 50kPa increments and typically ranged from 10 to 900kPa.

After equilibrating at the final pressure, and determining the wet soil mass, the specimen was oven dried and the soil water content at each pressure increment, or suction, determined. This enabled a soil-water characteristic curve to be derived. Six of these curves for samples of increasing fines content are presented in Figure 1. The form or shape of these curves is typical of the 21 specimens tested.

4 A CONCEPTUAL FRAMEWORK TO THE CHARACTERISTIC CURVE

It is evident from Figure 1 that a roughly bi-linear relationship exists between suction and water content. A flat gradient at low suctions, or relatively high water contents, is followed by a smooth transition to a steeper gradient at high suctions. The pressure at which the two linear portions of the curve

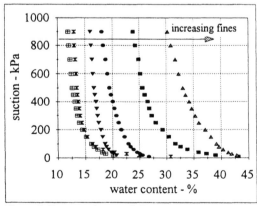

Fig.1: Soil-water characteristic curves

intersect is defined as the air entry suction. A somewhat simplistic description of the forces operative within the matrix of a homogeneous soil mass of reducing water content, as reflected by the characteristic curve, follows.

The flat portion of the curve reflects the outflow of water from the larger soil pores, which are unable to retain water against the applied suction. Capillary tension effects are thus the primary forces responsible for the release of water at low suctions. The water phase within the soil is continuous over this pressure range until the air entry suction is approached.

On approaching the air entry suction, at intermediate suctions, smaller pores are progressively emptied and, where present, the hydrated water layer enveloping secondary clay minerals reduces. The transition curve thus reflects matrix suction associated with capillary tension and adsorption. Soil water continuity progressively diminishes over this range as air filled channels begin to form within the soil voids.

At high suctions water retention is considered to be primarily due to adsorptive forces, with capillary tension operative in those voids that still retain capillary water. The soil matrix contains water in isolated pockets, which continuously reduce in size as the suction increases.

In view of the foregoing, certain soil properties play a significant role in determining the water content–suction relationship. These include homogeneity and distribution of void size, and the abundance and nature of the clay minerals present. With the exception of homogeneity, it can be argued that index tests are also related to these properties.

For example, the void size distribution is largely dictated by the particle size distribution, as determined in the sieve and hydrometer tests. The hydrometer test also quantifies the proportion of clay sizes ($<2\mu m$) present, whilst the Atterberg limits provide an indication as to their activity. In view of these observations, a relationship, albeit tenuous, should exist between the index tests and some property of the soil-moisture characteristic curve.

5 RELATIONSHIP BETWEEN INDICATOR TESTS AND THE CHARACTERISTIC CURVE

Burland (1990), Marinho & Chandler (1993) and Ridley & Peres-Romero (1998) have presented the characteristic curves of reconstituted compacted soils as a semi-logarithmic plot. They found a linear relationship exists between log suction and water content at suctions above about 100kPa. Similar relationships were established for the soils tested in this work, which are illustrated in Figure 2 for suctions ranging from 10 to 900kPa (1 to 90m of negative water head).

The gradient of the curve for suctions greater than 100kPa has been termed the Suction Capacity (C – in logkPa/%) by Marinho & Chandler (1993). Plots of C against liquid limit for reconstituted samples have been presented by the aforementioned authors, from which a substantially direct relationship was derived. A similar plot is presented in Figure 3, for all of the undisturbed soil samples tested in this work.

On initial inspection it appears that only a poor correlation exists. If, however, the residual soils are removed from the plot, with one or two exceptions, fairly strong interdependence for the transported soils is evident.

The suction capacity derived for each of the transported soils, above 100kPa suction, has been plotted against the corresponding plasticity index, linear shrinkage, clay fraction and grading modulus. The plots are presented in Figures 4 a) to d), from where it can be seen that satisfactory correlations exist. As shown in Figures 5 a) to d), the water content at a suction of 100kPa also correlates well with the corresponding plasticity index, linear shrinkage, clay fraction and grading modulus. Hence a value of any one of these familiar index parameters can be used, via Figures 4 and 5, to define the suction capacity and the water content at a suction of 100kPa, and thus the complete soil water characteristic curve for suctions from 100kPa upwards.

Attempts to correlate the air entry suction with index tests met with little success. As shown in Figure 2, the semi-logarithmic plots change slope markedly at suctions less than 100kPa. However, figures 6 a) to d) illustrate that the water content at 10kPa suction correlates very well with the grading modulus but less so with the other soil indicators. Thus, employing the information from Figures 5 and 6, or 6 d) in particular, the lower leg of the soil-water characteristic curve can be established and hence the entire curve for drying.

It should be emphasized that curves such as those shown in Figures 1 and 2 are drying curves. There is a considerable hysteresis between wetting and drying soil water characteristic curves and the above correlations should only be used for predictions of soil processes in which drying is expected to occur.

It is of interest to speculate how the index tests correlate with the water contents at suctions of 10 and 100kPa. In the latter case better correlation is achieved with the Atterberg limits and linear shrinkage than with the other indicator tests, whilst moisture contents at 10kPa suction correlate well with the grading modulus but less so with the Atterberg limits and linear shrinkage. This corroborates the conceptual model described earlier, where void size distribution, inferred from the grading modulus, substantially influences the development of capillary tension at low suctions whilst clay mineralogy, as reflected by the Atterberg limits and linear shrinkage, plays a more predominant role in the retention of water, through adsorption, at the higher suctions.

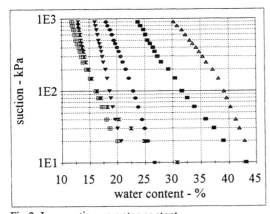

Fig.2: Log-suction vs water content

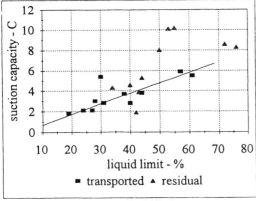

Fig.3: Suction capacity vs liquid limit

6 CONCLUSIONS

Undisturbed samples of homogeneous transported soils were found to exhibit reasonable correlation between indicator tests, suction capacity and water

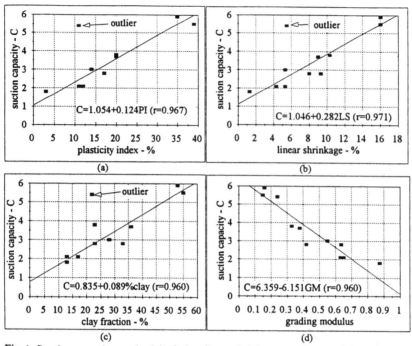

Fig.4: Suction capacity vs plasticity index, linear shrinkage, grading modulus and clay fraction.

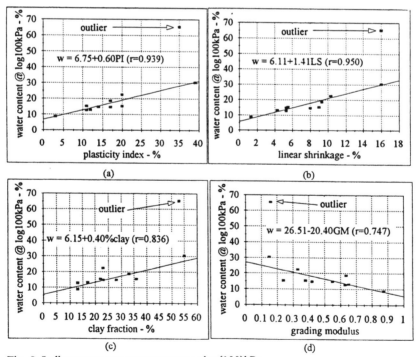

Fig. 5: Indicator tests vs water content at log(100)kPa

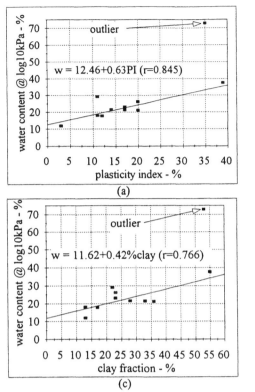

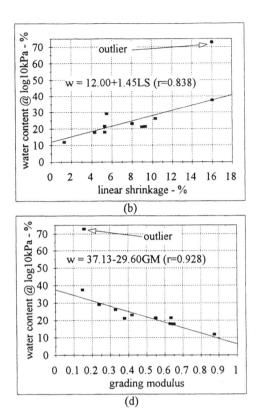

Fig. 6: Indicator tests vs water content at log(10)kPa

content at 10 and 100kPa suction. Similar relationships with residual soils, on the other hand, were not attainable due to their inherent saprolitic fabric and non-homogeneity, both natural features established during their formation.

Employing the relationships presented in this paper, a preliminary estimate of moisture content changes associated with suction changes can be established. This may provide useful information for determining the magnitude of volume change, shear strength, permeability and water storage in undisturbed unsaturated transported soils.

7 REFERENCES

Blight, G.E. & K. Roussev 1995. The water balance for an ash dump in a water-deficient climate. *1st Int. Conf. On Unsat. Soils*, Paris: Vol. 2, 833-840. Rotterdam: Balkema,

Burland, J.B. 1990. On the compressibility and shear strength of natural clays. *Geotechnique* 40(3): 329-378

Fredlund, D.G. & H. Rahardjo 1993, Soil mechanics for unsaturated soils. *John Wiley & Sons*, Toronto: 517

Gardner, W.R. 1956. Calculation of capillary conductivity from pressure plate outflow data. *Soil Sci. Soc. Am. Proc.* 20: 317-320.

Marinho, F.A.M. & R.J. Chandler, 1993. Aspects of the behavior of clays on drying. *ASCE*, Geotech. Spec. Pub. No. 39: 77-90.

Ridley, A.M. & J. Perez-Romero 1998. Suction-water content relationships for a range of compacted soils. *2nd Int. Conf. On Unsat. Soils*, Beijing: 114-118. Rotterdam: Balkema.

Vanapalli, S.K., D.G. Fredlund, D.E. Pufahl, & A.W. Clifton 1996. Model for prediction of shear strength with respect to matric suction. *Can. Geotech. J.* 33: 379-392.

Geotechnics for Developing Africa, Wardle, Blight & Fourie (eds) © 1999 Balkema, Rotterdam, ISBN 90 5809 082 5

Block sampling of soil: Some practical considerations

G. Heymann
University of Pretoria, South Africa

C. R. I. Clayton
University of Surrey, UK

ABSTRACT: Block sampling of soils can produce samples of the highest quality. However, during block sampling attention has to be given to a number of practical aspects to minimize disturbance to the material. Mechanisms by which disturbances can be introduced include strains imposed to samples during the act of sampling, swelling, stress relief and moisture content changes during storage.

This paper describes a number of techniques to obtain block samples. Techniques referred to include sampling of soil from test pits, auger holes and tunnels as well as down hole block sampling under high water table conditions. The mechanisms by which disturbance to samples can occur are discussed, and practical guidelines to minimize the level of disturbance are suggested.

1 INTRODUCTION

The quality of geotechnical design parameters obtained from laboratory testing is strongly dependant on the quality of the soil samples used during testing. So called "undisturbed" samples are generally taken by cutting blocks of soil or by pushing or driving tubes into the soil. However, even for such sampling methods, disturbance will occur. Some types of disturbance are unavoidable but other types can be minimized or even eliminated. The key to minimizing the level of disturbance during the sampling process is understanding the mechanisms by which sample disturbances occur. Some of the mechanisms which introduce sampling disturbance are discussed below.

2 SAMPLING DISTURBANCE

Disturbance may occur before, during and after the act of sampling. The significance of each mechanism that contributes towards disturbance depends on a number of factors including the sampling method, soil characteristics such as fabric, stiffness and strength as well as the geological features such as the presence of layering and the position of the water table.

2.1 Pre-sampling disturbance

The creation of a cavity such as a borehole or a pit in which to sample soil imposes strains on the soil in the vicinity of the cavity as a result of the reduction in total stress. When the relief in total stress is large relative to the undrained shear strength of the soil significant strains can be imposed. These may take the form of base heave where plastic flow of the soil at the bottom of the hole occurs (see for example Bjerrum and Eide (1956)). The shear strains imposed on the clay during base heave will result in a reduction in the mean effective stress of a soft clay. Lefebvre and Poulin (1979) calculated that for a soft clay with an undrained shear strength of 15kPa, a trench can only be excavated to 4m depth if a factor of safety of 2 is to be maintained against failure from base heave. Their results imply that the disturbance from base heave will increase with excavation depth. In a borehole the strains imposed from base heave may be reduced by keeping it filled with bentonite during the sampling process (Lefebvre and Poulin (1979)).

Swelling of a soil specimen and subsequent reduction in mean effective stress may occur prior to sampling. If the material is saturated, water will be drawn into the clay resulting in disturbance which is reflected in a reduction in mean effective stress. This type of disturbance was demonstrated by Ward et al. (1965). They showed that for block samples of a heavily overconsolidated clay taken during excavation of a deep shaft, the samples that were left in intimate contact with the base of the shaft when preparing the blocks by trimming, were wetter than the ones that were removed from the base prior to trimming. Hopper (1992) quantified the suctions which develop below the bottom of a borehole as a result of total stress relief. The largest suctions develop at a depth of approximately one borehole diameter. The amount of disturbance as a result of suction will depend on the suctions developed, the permeability of the material and the time allowed for swelling to occur.

2.2 Disturbance during sampling

Disturbance of the soil during the act of sampling takes place because of the changes in stress condition that occur. The change in stress has two constituents. Firstly, the total horizontal and vertical stresses are changed from the in-situ stress to zero, and secondly the deviatoric effective stress component is reduced to zero. Both these affects are unavoidable, but under certain conditions the imposed disturbance may be low.

After completion of sampling, the sample will experience zero total stress. If sufficiently high suctions can be sustained by the soil, the mean effective stress will remain similar to the in-situ mean effective stress. If the mean effective stress after sampling differs significantly from the in-situ mean effective stress, this indicates that the soil has undergone substantial shear strains during sampling. Therefore a change of the mean effective stress can be used as a guide to the level of disturbance introduced during sampling (Hight (1986), Chandler et al. (1992), Clayton et al. (1992)).

Most soils are subjected to anisotropic in-situ stress conditions. Consequently, the soil will experience shear deformation when the total stresses are reduced to zero, because the deviatoric effective stress component is changed. The level of disturbance is therefore a function of the anisotropy of the stress condition prior to sampling.

During tube sampling, the soil is displaced. The magnitude of the strains caused by this is strongly influenced by the geometry of the sampling tube.

Baligh (1985) calculated the strains imposed on a center line soil element for a simple sampler (see Figure 1). The soil experiences compressive strains before it enters the sampler and extension strains once inside. Around the periphery additional strains in the form of shear distortions occur. Clayton et al. (1995) illustrated that for stiff clay these distortions can lead to substantial damage to samples. Clayton and Siddique (1999) showed that in addition to the sampler geometry the cutting edge taper and inside clearance has a significant influence on the strains imposed during tube sampling. When viewed in the context that natural soils yield at axial strain levels of the order of 1% to 2% it is clear from Figure 1 that for most tube samplers, the soil will have undergone significant strains and yielding during sampling. It is therefore probable that significant disturbance of the soil in the form of destructuring will occur during tube sampling.

During block sampling, the soil surrounding the sample is cut away to isolate the sample from the rest of the in-situ material. When executed with care, this action does not impose significant strains on the soil to be sampled. The disturbance as a result of the act of sampling is therefore very much smaller for block sampling. This fact has been demonstrated by Hight (1993) and Hight and Jardine (1993). They found that the undrained shear strength of a fissured highly over consolidated plastic clay was significantly higher for samples taken by rotary coring when compared to that for samples taken by pushed thin-walled tube sampler. The effective cohesion intercept (c') was almost four times higher for the rotary cored samples compared to the tube samples. This clearly demonstrates the potential of block sampling to produce samples of low disturbance.

2.3 Disturbance after sampling

A number of factors may lead to disturbance of the soil after taking samples. Moisture loss, chemical changes and mechanical damage during transportation are the dominant mechanisms. With due recognition of these factors, and appropriate precautions, disturbance caused after sampling may be minimized.

Engineers often underestimate the significant impact of moisture content change on disturbance of sampled soil. Loss of moisture during storage increases the suction pressure of the pore fluid. Figure 2 illustrates the relationship between suction

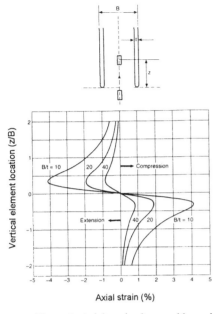

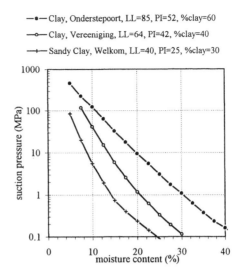

Clay, Onderstepoort, LL=85, PI=52, %clay=60
Clay, Vereeniging, LL=64, PI=42, %clay=40
Sandy Clay, Welkom, LL=40, PI=25, %clay=30

Figure 1. Axial strains imposed by a simple sampler. (After Baligh 1985).

Figure 2. Relationship between moisture content and pore fluid suction. (Williams and Pidgeon (1983) and Brackley (1975)).

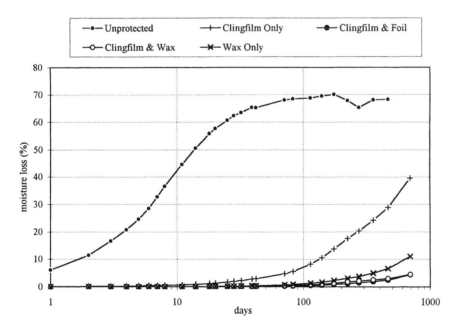

Figure 3. Moisture loss for various sealing methods during storage of London clay.
(Heymann (1998)).

pressures and moisture content for a number of soils. It is clear from Figure 2 that even a relatively small reduction in moisture content can result in a large increase in mean effective stress experienced by the soil. Such an increase in mean effective stress will lead to erroneously high strength and stiffness measurements. In extreme cases if the soil is allowed to reach its shrinkage limit, yielding may occur and cause irrecoverable damage.

Sample disturbance during storage can be minimized by protecting the soil against moisture loss by sealing it. A variety of materials have been used for this purpose. The use of wax as a means to seal samples has been reported by various authors (see for example Hvorslev (1949), Ward et al. (1965)). Hvorslev (1949) showed that wax provides a more efficient seal when brushed onto the soil in layers as opposed to pouring the liquid wax onto the soil. This has been explained on the basis that the numerous layers of brushed wax are applied at a lower temperature and are therefore less susceptible to shrinkage cracks than the poured wax (Clayton et al. 1995). Wax may also be used in conjunction with layers of cheese cloth or cling film. The cheese cloth or cling film layers act as reinforcement and therefore reduce the development of shrinkage cracks in the wax. The use of wax may be problematic in some cases. The molten wax has to be held at a temperature of between 60°C and 65°C. Temperatures higher than 70°C lead to evaporation of volatile hydocarbons which greatly reduces the sealing efficiency of the wax. A purpose-made stove, often powered by mains electricity is therefore required. This may be impractical on remote sites. In addition, wax is less efficient in warm countries. The wax will remain plastic at high ambient temperatures, and this will allow holes in the sealing layer to develop as it creeps.

Heymann (1998) investigated the use of aluminium foil combined with clingfilm as a means to seal samples of London clay against moisture loss. The rate of moisture loss was compared for a number of sealing methods including sealing by paraffin wax only, sealing by cling film only, sealing by layers of paraffin wax and cling film and sealing by layers of aluminium foil and clingfilm. The samples were kept in a soils laboratory where the temperature varied between 17°C and 24°C and the relative humidity between and 27% and 65% over a period of 2 years. The results are shown in Figure 3. An unprotected control sample was used against which the efficiency

of each sealing method could be judged. Moisture loss was rapid for the unprotected sample with a 6% loss after only one day and 50% moisture loss after 14 days. Similar results were found by Hvorslev (1949) on Boston blue clay. Figure 3 also shows how ineffective clingfilm is in protecting the clay, when used alone. This sample lost 7% of its moisture after 100 days. In contrast, the sample protected only by wax lost 1.5 % and the clingfilm - wax combination reduced the moisture loss to 0.6% over a period of 100 days. The clingfilm - wax combination produced better results than the wax in isolation because of the reinforcing effect of the clingfilm. However, of all the methods investigated, the aluminium foil - clingfilm combination proved to be the most effective protection against moisture loss and was marginally better than the clingfilm-wax combination. As noted above this has some practical implications, as the equipment required to use aluminium foil is less than that required for wax. In addition the characteristics of aluminium foil are not significantly influenced by high ambient temperatures.

One practical comment regarding the combined use of aluminium foil and clingfilm is warranted. Two to three layers of each is required and care should be taken to used clingfilm for the first layer. This ensures that the aluminium foil is not in direct contact with the soil and therefore the possibility of a reaction between the metal foil and the soil is avoided.

Figure 3 shows that regardless of the sealing method, sample moisture loss will occur if samples are held for very long periods before testing. After a period of six months, the sample which experienced the least moisture loss (aluminium foil - clingfilm) had lost approximate 1% moisture and after a period of 2 years the loss was approximately 5%. This indicates that if samples are to be stored for very long periods of time, additional measures will be required. One option is to store specimens at humidity levels for which the suction potential of air approximates the suction in the sample. La Rochelle et al. (1986) showed that when sealing soft clay samples with layers of clingflim and a wax - Vaseline compound and stored at a relative humidity of 90%, samples experienced no moisture loss over a period of 3 years. They argued that the relative humidity should not be more than 90% in order to avoid condensation and possible water ingress into the samples. Storing the sample at high humidity is only a feasable option for soft clays with relatively low suction levels. Hard clays with high suction may experience damage by swelling at high humidity conditions. Regardless of the method of storage, control tests should be conducted to monitor any moisture loss in the samples during long term storage (Schjetne (1971)).

3 BLOCK SAMPLING TECHNIQUES

Block sampling may be defined as a sampling method by which a volume of soil is isolated from the in-situ soil by progressively cutting away the soil surrounding the sample by a mechanical action. Block sampling therefore causes less disturbance during the act of sampling compared to tube sampling. Disturbances to block samples which occur before and after sampling therefore become more important.

Throughout much of the world block sampling has traditionally only been carried out in shallow test pits above the water table. However, recently a number of block sampling techniques have been developed to obtain samples at depth under various ground conditions. The following sections describe a number of these techniques.

3.1 Block sampling from test pits

Block sampling from test pits have been described by Hvorslev (1949). First a regular shaped block is trimmed to the desired size. An open box without a top or bottom is placed over the trimmed block and the void between the box and the sample is filled with wax or a wax - Vaseline compound. The top of the box is secured in place before cutting the bottom of the trimmed block from the in-situ material. The box is turned upside down and the sample bottom trimmed, filled with wax and the bottom plate is attached. USBR (1963) described a modified version of this method where the trimmed specimen is first sealed by layers of cheese cloth and wax before the box is paced over the sample.

3.2 Block sampling in auger holes

If the water table is deep, or if the soil is of low permeability, access to deep sampling positions may be gained by large diameter auger hole. Sampling from auger holes is a practice that is widely used in Southern Africa to sample residual soils. This stems from the fact that the material is often unsaturated and of sufficient strength to allow cutting and handling of intact blocks.

A technician is lowered down the auger hole in a safety cage and a specimen cut from the side of the hole using hand tools such as geological picks, spades etc. If a number of samples are to be taken from the same hole at various depths, the sampling disturbance can be minimized by drilling the hole to the required depth for each sampling position (or slightly deeper to allow the technician standing room) immediately prior to sampling. After the sample has been taken the hole is advanced to the next position. Samples are therefore taken in a downward sequence. This procedure will reduce the time that the soil to be sampled is in intimate contact with the surrounding soil after total stress relief occurred and high pore fluid suctions has been generated. Disturbance due to swelling will therefore be minimized. van der Berg (1998) suggested a method by which the hole is first drilled to the maximum depth and then progressively backfilled to the level of each sampling position. Samples are therefore taken in a upward sequence. This method minimizes plant movement and leads to an improved production rate. However, for this method the disturbance due to swelling will be more for the samples closer to the surface compared to the those for the downward sampling sequence. Depending on the time required to cut each sample and constraints on site, a balance should be found between production rate and sampling disturbance.

Space is confined down the auger hole and therefore the sample is normally brought to the surface before it is sealed and packed for transportation.

3.3 Block sampling in tunnels and shafts

Underground construction, where access to the excavation face is available, often presents the opportunity to obtain block samples of soil. As noted in the section above, consideration should be given to possible pre-sampling disturbance as a result of shear strains experienced by the soil due to the excavation process.
Sampling in tunnels present certain practical difficulties, particularly in the form of time and space constraints. Production rate in a tunnel is often of overriding importance as it may have a bearing on the tunnel stability. Access to the tunnel face may therefore only be available for a few minutes which makes the use of hand tools impractical. Heymann (1998) described a method by which blocks of London clay were cut from the station tunnels constructed as part of the express rail link between

Heathrow Airport and the city of London. The 7.9m diameter tunnel was constructed in a top-heading, bench and invert configuration. The samples were taken from the bench at a depth of 18m below ground level. The first step in the sampling process was to remove the 50mm temporary shotcrete lining from the tunnel face by excavator. The orientation of the in-situ clay was marked by painting horizontal lines on the exposed face. Two trenches were then excavated into the bench approximately one meter apart. This isolated a block of material approximately 1 meter wide, 1.5 meters high to a depth of 1 meter. Next the excavator bucket was used to apply a lateral force to the side of the exposed block. As a result of the blocky nature of the fissured clay, it tended to separate on pre-existing fissures, creating a number of smaller blocks. This part of the operation took between ten and fifteen minutes. Suitable samples were transported to the surface and covered with coatings of paraffin wax and clingfilm.

The availability of an excavator may significantly simplify the sampling process described in the previous paragraph. However, in smaller tunnels an excavator may not be available. Bracegirdle (1999) used a pneumatic chain saw to cut blocks from the face of a tunnel in over consolidated clay at a depth of 33m below ground level. The procedure is shown in Figure 5. Again, the time required for sampling was relatively short.

3.4 The Sherbrooke sampler

Block sampling from below the water table may be conducted by means of a Sherbrooke sampler. (Lefebvre and Poulin (1979)). The sampler consists of three cutting shoes at 120° intervals along its circumference (Figure 4). A mechanical or hand mechanism is used to rotate the sampler. It is advanced by rotation and simultaneous flushing of drilling fluid through the jets close to the cutting shoes. As a result an annular slot with an outside diameter of approximately 400mm diameter is cut. Once the sampler had been advanced by 350mm, a trigger is activated to release the three spring-loaded cutting blades at the bottom of the sampler. Rotation is continued and in the process the cutting blades separated the block sample from the bottom of the hole. Finally the block is lifted to the surface.

Lefebvre and Poulin (1979) suggest that either water or bentonite mud may be used as the borehole

fluid. Water, as opposed to bentonite, may lead to sample disturbance by means of base heave or swelling, as discussed above. The amount of disturbance will largely depend on the sampling depth, suctions developed, the permeability of the material and the time allowed for swelling to occur. Bentonite as a drilling fluid avoids these disturbance mechanisms, as the total stresses remain relatively constant during sampling.

Work carried out on a soft recent clay deposit in Scotland by Clayton et al. (1992) where various sampling techniques were compared, showed that the Sherbrooke sampler obtained the highest quality samples, outperforming even sophisticated thin-wall tube samples.

3.5 Rotary core sampling

The value of rotary coring techniques in providing samples of stiff and hard clay has long been known. For example, Seko and Tobe (1977) showed that the unconfined compressive strength of samples obtained using a modified double-tube core-barrel with mud flush were, on average, about twice those from conventional thin wall open-drive hammered samplers. As we noted above, Hight and Jardine (1993) found that the effective cohesion intercept, c', of London clay samples taken by wire-line rotary coring were almost 4 times those obtained from tube samples. Rotary coring has now become routinely used in the UK for high-quality sampling of London clay.

The quality of sample obtained using rotary coring is dependent upon a number of factors. Clearly the core-barrel must not be subjected to excessive thrust, because of the undesirable effects of displacement rather than coring. The driller will normally be aware if this is occurring, however, because blocking off will occur. It is important to use good quality mud (bentonite or polymer muds are preferred). A triple-tube barrel, incorporating an inner plastic liner is the norm, and inside clearance (which may often be excessive in rotary core-barrels) should be reduced to the levels normally specified for drive samplers. Finally, barrel vibration needs to be minimised, and the stability of the barrel should be maximised. It is perhaps for this reason that British practice has made so much use of wire-line techniques, which in normal practice have typically been used for deep mineral exploration. The success of wire-line sampling techniques also appears less sensitive to the skill of the driller.

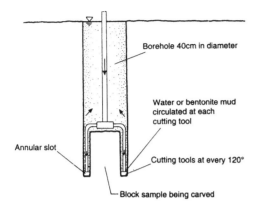

Figure 4. Sherbrooke down hole block sampler (Lefebvre and Poulin (1979)).

3.6 Block sampling of friable soil

Block sampling of friable soils often proves difficult when the soil can not sustain high enough suctions and exhibit a lack of cementation to keep samples intact after sampling. The test pit sampling method described above is well suited for sampling friable soils as the space between the box and the sample is filled with wax. This gives support to the sample and reduces the risk of crumbling.

Bica (1997) described a technique by which cylindrical block samples of residual granite were cut on site by using a sampling tube as a guide. The tube is placed with its cutting edge on a flat soil surface and a tool used to cut away the soil at the base of the tube to a depth of a few millimeters. This exposes soil in the shape of a cylindrical disk. The exposed soil is then advanced into the tube by applying a small downwards pressure and in the process the cutting edge of the sampling tube removes any excess material. Repeating the sequence creates a cylindrical soil sample inside the sampling tube. If sampling is carried out with due care the soil will fit snugly and thus be supported by the tube. In addition the tube will protect the sample against shocks during transportation. Care should be taken not to push the tube into the material before the disc has been trimmed, as this will impose strains in the soil similar to those produced by tube sampling. The sample should be protected against moisture loss by sealing the ends of the sampling tubes (see for example Clayton et al. (1995)).

Aoyama et al. (1983) (as reported by Mori (1985)) developed a method to reinforce friable soil before it is sampled. The method is shown schematically in Figure 6. A flat surface is prepared on which a bakelite square plate is placed. Nails are driven through closely spaced holes which are arranged in the form of a circle. Next the soil surrounding the sample is cut away by means of hand tools and the block sample is removed from the ground. Finally the sample is sealed against moisture loss. Aoyama

Figure 5. Pneumatic chain saw used to cut samples of London clay. (Bracegirdle (1999)).

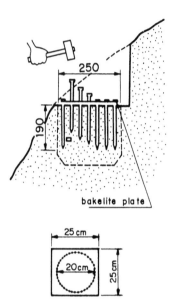

Figure 6. Reinforcement of friable soil prior to block sampling. (Aoyama et al. (1983)).

et al. (1983) used this method to sample residual granite. Comparison between the densities of the samples and the in-situ densities measured by radio isotopes showed good agreement. Care should be taken, however. Even without significant density changes, it would seem possible to damage the structure of soil as a result of the strains occurring when the nails are driven.

4 CONCLUSIONS

Block sampling techniques can now be carried out on a wide range of geomaterials and under various water table conditions, and can potentially produce samples of the highest quality. However, the mechanisms by which sampling disturbance can occur must be recognized by the practitioner in order that the necessary steps to minimize them can be taken. Disturbances to the soil may be introduced before, during or after the act of sampling. One of the main advantages of block sampling is the fact that the soil is isolated from the in-situ soil by a cutting action and therefore mechanical disturbance of the soil during sampling is minimized. This makes the mechanisms which induce disturbance before and after sampling of primary importance to block sampling. Such disturbances include total stress relief and swelling prior to sampling and moisture loss

during storage. Total stress relief is unavoidable during all types of sampling including block sampling. However, under certain condition, particularly when the soil can sustain high suctions, disturbances from these effect may be low. Disturbance from swelling may be kept within acceptable limits by good drilling or excavation practice and by minimizing the contact time between the sample and the surrounding soil. Moisture loss during storage may result in significant disturbance to the soil and may even cause the soil to yield. It has been demonstrated that moisture loss can be reduced by good sealing methods. A number of sealing methods were evaluated and evidence presented to show that of the sealing methods considered, aluminium foil when used in conjunction with clingfilm produces the most efficient seal against moisture loss.

5 REFERENCES

Aoyama, C, Nakayama, Y. and Nishida (1983). Sampling and triaxial testing undisturbed specimen of decomposed granite soil. 18th Annual Assembly of the Japanese Society of SMFE, pp.497-480 (in Japanese).

Baligh, M. M. (1985). Strain path method. Journal of Geotechnical Engineering, ASCE, Vol. 111, pp.1108-1136.

Bica, A. V. D. (1997). Personal communication.

Bjerrum, L. and Eide, O. (1956). Stability of strutted excavations in clay. Géotechnique, Vol. 6, No. 1, pp.32-47.

Bracegirdle, A. (1999). Personal communications.

Chandler, R. J., Harwood, A. H. and Skinner, P. J. (1992). Sample disturbance in London clay. Géotechnique, Vol. 42, No. 3, pp. 577-585.

Clayton, C R. I., Hight, D. W. & Hopper, R. J. (1992). Progressive destructuring of Bothkennar clay: implications for sampling and reconsolidation procedures. Géotechnique, Vol. 42, No. 2, pp. 219-240.

Clayton, C. R. I., Matthews, C. & Simons, N. E. (1995). Site Investigation, 2nd Edition. Blackwell Science, Oxford, 600pp.

Clayton, C.R.I., & Siddique, A. (1999). Tube sampling disturbance - forgotten truths and new perspectives. To be published in Proceedings of I.C.E. (Geotechnical Engineering).

Heymann, G. (1998). The stiffness of soils and weak rocks at very small strains. PhD Thesis, University of Surrey, UK.

Hight, D. W. (1986). Laboratory testing: Assessing BS 5930. Geological Society, Engineering Geology Special Publication No. 2, pp. 43-58.

Hight, D. W. (1993). A review of sampling effects in clays and sands. Proceedings on SUT Conference on Offshore Site Investigation and Foundation Behaviour, pp.115-146.

Hight, D. W. & Jardine, R. J. (1993). Small-strain stiffness and strength characteristics of hard London tertiary. Geotechnical Engineering of Hard Soils - Soft Rocks, Anagnostopoulos et al. (eds). Balkema, Rotterdam, p.533-552.

Hopper, R. J. (1992), The effects and implications of sampling clay soils. PhD thesis, University of Surrey.

Hvorslev, M. J. (1949). Subsurface exploration and sampling of soils for civil engineering purposes. Waterways Experimental Station, Vicksburg, USA.

La Rochelle, Leroueil, S & Tavenas, F. (1986). A technique for long-term storage of clay samples. Canadian Geotechnical Journal. Vol. 23, pp.602-605.

Lefebvre, G. & Poulin, C. (1979). A new method of sampling in sensitive clay. Canadian Geotechnical Journal. Vol. 16, pp.226-233.

Mair, R. J. & Taylor, R. N. (1993). Prediction of clay behaviour around tunnels using plasticity solutions. Predictive soil mechanics, Thomas Telford, London, pp. 449-463.

Mori, H. (1985). Sampling and testing of granitic residual soils in Japan. Brand E. W. and Phillipson, H. B. Eds. Balkema, Rotterdam, 194pp.

Schjetne, K. (1971). The measurement of pore pressure during sampling. Proc. Specialty Session on Quality in Soil Sampling, 4th Asian Regional Conference, Int. Soc. for Soil Mech. and Found. Engng, Bangkok, pp. 12-16

Seko, R. & Tobe, K. (1977) Research of stiff clay sampling. Proc. Specialty Session 2 on Soil Sampling, 9th Int. Conf. on Soil Mech. and Found. Engng, Tokyo, pp. 44-45.

US Bureau of Reclamation (1963). Earth Manual, US Department of the Interior, Denver, Colorado, USA.

van der Berg, J. P. (1998). Personal communications.

Ward, W. H., Marsland, A. & Samuels, S. G. (1965). Properties of the London clay at the Ashford Common Shaft: In-situ and undrained shear strength. Géotechnique, Vol. 15, No. 4, pp. 321-344.

Williams, A. A. B. & Pidgeon, J. T. (1983). Evapo-transpiration and heaving clays in South Africa. Géotechnique, Vol. 33, pp. 141-150.

Geotechnics for Developing Africa, Wardle, Blight & Fourie (eds) © 1999 Balkema, Rotterdam, ISBN 90 5809 082 5

Assessing the dispersivity of soils

C.A.Jermy
School of Geological and Computer Sciences, University of Natal, Durban, South Africa

D.J.H.Walker
Steffen Robertson and Kirsten, Johannesburg, South Africa

ABSTRACT: Dispersive soils are readily eroded because the clay particles deflocculate and detach from the clay surface. Clay dispersion is dependant on two main factors: the nature of the exchangeable cations on the clay surface and the presence of salts in the pore water. If the forces of repulsion due to the adjacent negatively charged clay surfaces overcome forces of attraction between the clay particles, the clay particles repel one another, deflocculate and go into suspension, even in still water. Dispersivity can not be identified using routine soil mechanics tests, but specialised physical and chemical tests have been devised to assess the potential dispersivity of a soil. However, since no single test can be used with complete certainty to identify dispersivity it is necessary to combine a number of the tests into a rating system, using a weighting system dependent upon the reliability of each test employed.

1 INTRODUCTION

Dispersion occurs in cohesive soils when the repulsive forces between the clay particles exceed the attractive forces, so that in the presence of water the particles repel each other, causing deflocculation. In nondispersive soils the clay particles adhere to one another and are only eroded by water flowing above a definite threshold velocity. The clay particles in dispersive soils readily deflocculate and go into suspension even in still water, and are therefore susceptible to erosion and piping. The dispersion phenomena is caused by a high percentage of exchangeable sodium on the surface of the clay minerals. Sodium is loosely held on the clay surface and imparts a negative charge to it. The adjacent negatively charged particles tend to repel one another, and if the forces of repulsion are great enough to overcome the forces of attraction, the clay particles deflocculate and detach from the clay mass and can be carried away. Dispersion is also affected by the presence of cations in the pore water; with a very high amount of pore water salts having a stabilising effect on clay dispersion.

Since they are easily eroded, special precautions need to be taken if dispersive soils are to be used for the construction of water retaining structures such as earth dams. In the event of a leak through an earth dam constructed with dispersive soils, the clay particles would deflocculate, be taken into suspension and be washed out, leading to the rapid enlargement of the leak to form a pipe, which could

result in the failure of the dam. The identification of dispersive soils is thus of great importance, so that adequate precautions may be taken prior to and during construction. In this investigation 94 samples were collected from colluvial soils in central and northern KwaZulu-Natal, where evidence of dispersion was seen in the development of gullies and dongas, and the presence of pipes and cavities in the soil.

2 TESTING OF DISPERSIVE SOILS

Dispersive soils cannot be identified by routine soil mechanics tests such as natural moisture content, particle size distribution and Atterberg limits, or by routine chemical tests such as soil pH or electrical conductivity. Thus, in order to assess the dispersivity of a soil, a number of specialised tests have been developed. These can be divided into physical and chemical tests. The physical tests show the effect of the dispersivity of the soil, that is the natural tendency of the soil to deflocculate in the presence of water. The chemical tests show the cause of the dispersivity, or the presence of sodium on the clay surface.

2.1 *Physical Tests*

Physical tests for dispersive soil identification include the pinhole test, crumb test, double hydrometer test, dispersion index, sticky point test, cylinder dispersion test and soil security test. However, only the first three are used as routine tests for dispersive soil identification. The physical tests show the interaction of the soil with water: the crumb and double hydrometer tests indicating the natural tendency of the soil to disperse, and the pinhole test giving an indication of the erodibility of the soil. Sherard and Decker (1977) and Elges (1985) assigned an order of preference to the results when they produced conflicting results. They gave the pinhole test the highest significance, followed by the crumb test and lastly the double hydrometer test.

2.1.1 *Crumb Test*

The crumb test was first described by Emerson (1964). It is a simple test requiring no special equipment and can be performed in the field as well as the laboratory. A small "crumb" of soil, 15 to 20 mm across, at its natural moisture content, or a remoulded crumb, is placed into a beaker containing distilled water. As the crumb begins to hydrate, the tendency for the colloidal sized particles to deflocculate and go into suspension is observed over a ten minute period. The reaction of the soil is then graded into four categories depending on the reaction of the soil in the water. It was found in this study that at least 40% of soils categorised as dispersive by the pinhole test gave a non-dispersive result in the crumb test.

2.1.2 *Double Hydrometer Test*

This test was developed by Volk (1937) to identify dispersive soils, by measuring the natural tendency of the clay to go into suspension. The percentage of particles in the soil less than 5 μm in size is determined using a standard hydrometer test, and a parallel test is run on an identical sample but with no chemical dispersant or mechanical agitation used. The amount of <5 μm soil particles that goes into suspension naturally in this test is expressed as a percentage of that measured in the standard test. The percent dispersion gives a quantitative assessment of the degree of dispersivity of the soil (Bell and Maud, 1994); with values in excess of 30% being regarded as dispersive and those with values between 15 - 30 % being of intermediate dispersivity. The degree of dispersivity of each soil determined by this test did not show good corellation with the other physical tests, or with the chemical tests in this study, and because of the uncertainty about the validity of the results, the double hydrometer test is not recommended for the identification of dispersive soils in KwaZulu-Natal.

2.1.3 *Pinhole Test*

The pinhole test was developed by Sherard *et al.*, (1976) and is widely considered to be the most reliable physical test for the investigation of dispersivity. A cylinder of soil is compacted at a moisture content equal to its plastic limit using a miniature compactor. A 1 mm pinhole is punched through the cylinder and water is allowed to flow through the pinhole under a constant head ranging from 50 mm to 1.0 m.

If the soil is nondispersive, the water flowing through the pinhole will appear clear, even when flowing at a high velocity under a high hydraulic head. The clay particles in dispersive soils will deflocculate and go into suspension, even in still water. The dispersed clay particles are washed out by the flowing water causing the pinhole to become enlarged and the water to become turbid. The soil dispersivity is classified according to the colour of the water flowing out of the pinhole and the size of the pinhole at the end of the test. Dark water and enlarged pinholes indicating dispersive soils, while clear water and uneroded pinholes indicated nondispersive soils.

The first parameter for classifying dispersivity was the size of the eroded pinhole at the end of the test. Highly dispersive (D1) soils were eroded to 2 - 3 mm in 5 minutes at a hydraulic head of 50 mm; moderately dispersive (D2) soils were eroded to 2 mm in 10 minutes and slightly dispersive (ND4) soils to 1.5 mm in 10 minutes. A nondispersive (ND1) soil showed no pinhole erosion after 5 minutes at a hydraulic head of 1.0 m.

2.1.4 *Spectrophotometer*

The second classifying parameter was the turbidity of the water (or effluent) flowing out of the pinhole. This was more difficult to quantify since the turbidity is normally assessed visually and classified according to its colour. It was found that it was very difficult to distinguish between the medium and dark effluents visually and only a qualitative assessment of the turbidity could be obtained. In order to provide a reliable and repeatable method of quantifying the turbidity of the effluent a spectrophotometer was employed. The spectrophotometer measures the percentage of light transmitted through or absorbed by the effluent, and since this is proportional to the amount of clay in suspension a direct measure of the turbidity can be obtained.

In this study measurements were taken at three wavelengths: 350 μm, 550 μm and 800 μm. For percentage light transmittance, the darker effluents were separated more at high wavelengths; the clearer ones at low wavelengths; the 550 μm wavelength light providing the most versatile measuring source. A greater range was obtained between the absorbance readings at the low wavelength, regardless of turbidity.

The use of the spectrophotometer was found to have distinct advantages over the visual method of assessing effluent turbidity. Firstly, it defined more accurately the boundary between the medium and dark effluents, which was very difficult to gauge visually. The division between the dark and light coloured effluent was very clearly defined using the spectrophotometer and was found to divide the dispersive and nondispersive soils. The second advantage of the spectrophotometer was in differentiating effluents with a bare hint of colour from slightly coloured effluents, which were almost impossible to classify visually.

2.2 Chemical Tests

The chemical tests show the cause of dispersion, or the relative amount of sodium to other cations, such as calcium, magnesium and potassium. Both exchangeable cations and saturation extract cations were determined. Exchangeable cations are those on the clay surface, whereas saturation extract cations are those present in the pore water. Exchangeable cation quantities were determined using the ammonium acetate extract method (Richards, 1954), and are expressed in milliequivalents per 100 gms of soil. Exchangeable sodium percentage (ESP) and exchangeable sodium ratio (ESR) were calculated from the exchangeable cations. The saturation extract cations were obtained by mixing the soil with distilled water and allowing the cation exchange complex to reach equilibrium. Saturation extract cations are expressed in milliequivalents per litre (meq/l). The cations present in the soil pore water are independent of the mineralogy of the soil and are the result of environmental conditions influencing the soils. Sodium adsorption ratio (SAR), total dissolved salts (TDS) and percent sodium (%Na) were calculated from the saturation extract cations. The cation exchange capacity (CEC) was also determined.

There is not a proportional relationship between exchangeable cations and saturation extract cations. On the whole, sodium is the least abundant exchangeable cation, and the most abundant saturation extract cation, possibly due to differing solubilities of the various cations. This relationship is reflected in the dispersivity tests: for exchangeable cations an exchangeable cation percentage greater than 5% indicates that the soil is dispersive, whereas for saturation extract cations the percentage sodium must be greater than 60% for the soil to be considered dispersive.

In the South African context, if the SAR is greater than 10 the soil is considered dispersive; if between 6 and 10, intermediate, and if less than 6, nondispersive (Harmse, 1980). However in Australia researchers consider all soils with an SAR greater than 2 to be dispersive. The results of this study indicate that the Australian criterion is more appropriate; dispersive soils generally had an SAR greater than 2, and a value less than 1.5 for nondispersive soils. The SAR values showed a good correlation with the pinhole test.

Sherard et al., (1976) proposed a chart with percent sodium (%Na) plotted against total dissolved salts (TDS) for assessing dispersivity. If the %Na is greater than 60% and the TDS greater than 1.0 meq/l the soil is regarded as dispersive. If the %Na is less than 40% and the TDS less than 0.2 meq/l, the soil is nondispersive. An intermediate dispersivity group is present between these two zones. In this study the TDS versus %Na chart identified the dispersive nature of the soils with a fair degree of accuracy. Very good agreement was found between soils that plotted within the dispersive zone and soils identified as dispersive or highly dispersive by other tests. However, soils plotted in the intermediate zone were found to be dispersive or nondispersive by other test methods.

The presence of exchangeable sodium is the main chemical factor contributing towards dispersive behaviour. There is some difference of opinion in the literature regarding the ESP value above which a soil should be regarded as dispersive. Sherrard (1973) considered soils with an ESP of between 7 and 10 to be moderately dispersive, whilst Harmse (1980) suggested that soils with and ESP greater than 5 are dispersive. Gerber (1983) devised a chart for assessing dispersivity using the ESP of a soil and its mineralogy, represented by the CEC of the clay fraction. He regarded soils with ESP values greater than 15% to be highly dispersive, whilst those with and ESP greater than 6% were dispersive. Values of ESP between 4 and 6% he regarded as marginal, whilst values below 4% indicate that the soil is nondispersive. Using the above criteria Gerber (1983) constructed the CEC vs ESP chart. The validity of the chart was verified empirically by plotting data from dispersive and nondispersive soils onto the chart and observing whether soils identified as dispersive or nondispersive by other tests plotted in the respective zones on the chart.

In this study, the CEC vs ESP chart was the most accurate of the chemical assessment methods (Figure 1). Since six categories of dispersivity are given, this chart should give an indication of the degree of dispersivity as well as simply wether the soil is dispersive or not. However, the highly dispersive and dispersive categories given by this chart did not correspond well with those given by the pinhole test; with many of the soils that plotted in the dispersive

343

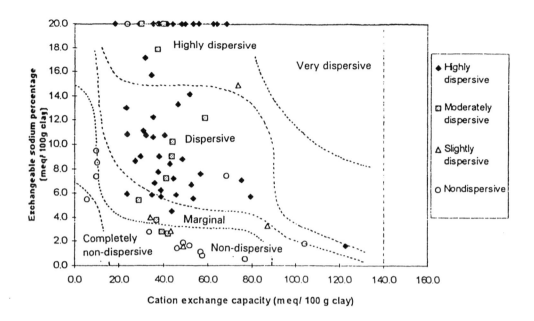

Figure 1 CEC vs. ESP chart to assess dispersivity

zone proving to be highly dispersive from the pinhole test. Even though this test was the most reliable of the chemical tests it would not be able to differentiate the degree of dispersivity without additional information from physical tests.

2.3 Discriminant Analysis

Craft and Acciardi (1984) questioned the reliability of the TDS vs %Na chart of Sherard *et al.*, (1976), which was then the currently accepted method for the identification of dispersive soils. They noted that variables other than pore water cations may be used to identify dispersive soils, and suggested the use of discriminant analysis which would allow the influence of a number of variables to be collectively assessed in determining soil dispersivity. Discriminant analysis requires a prior knowledge of the relationships between samples, and determines which parameters are best in separating the samples into the known groups. For the discriminant analysis of the dispersive soil testing data, the soils could readily be grouped according their dispersivity as determined by the pinhole test. Analysis could then be carried out using the results of the other physical and chemical tests to see which combination would cause the know groups to plot in distinct fields on the graphs.

The combinations used included those tests that generally showed good agreement with one another:

the pinhole test, crumb test, CEC vs ESP, TDS vs %Na and SAR value. Including the pH and electrical conductivity had no affect on the accuracy of the plots, but if the double hydrometer test results were included the overall accuracy of the plot was reduced. Discriminant analysis has an advantage that the data is represented graphically, making the classification easy to observe. It may be used to assess scatter within and between groups. For many of the plots the moderately dispersive soils plotted very close to the highly dispersive soils, showing that the two populations are closely related. The scatter of results for the slightly and moderately dispersive groups was greater than for the highly or nondispersive groups as there was greater variation between the results from the different test methods for these groups.

The discriminant analysis showed that using just the results of the physical tests or chemical tests alone did not produce a very good. A plot using all the test data gave a very good plot, but the most accurate plot was obtained using the results from the pinhole test, CEC vs ESP, TDS vs %Na and SAR value.

3 A RATING SYSTEM

Since no single test has been developed that can reliably identify all dispersive soils, it is widely accepted that a number of tests be conducted and

Table 1. New rating system for identification of dispersive soils

Pinhole Test	Class	Highly dispersive	Moderate	Slightly dispersive	Nondispersive
	Rating	5	3	1	0
CEC vs ESP	Class	Highly dispersive	Dispersive	Marginal	Nondispersive
	Rating	4	3	1	0
Crumb Test	Class	Strong reaction	Moderate	Slight reaction	No reaction
	Rating	3	2	1	0
SAR	Class	Over 2	1-5 - 2		<1.5
	Rating	2	1		0
TDS vs %Na	Class	Dispersive	Intermediate		Nondispersive
	Rating	2	1		0

their results combined to assess the dispersivity. Bell and Maud (1994) proposed a tentative rating system using the weighted results of a number of tests; the weighting depending upon the reliability of the test in identifying dispersivity. The present study has determined that pH and the double hydrometer test are not good indicators of dispersivity in KwaZulu-Natal, and have thus not been included in the new rating system. Only the tests found reliable in assessing dispersivity and useful in discriminant analysis have been included in the system. These are the pinhole test, CEC vs ESP, the crumb test, SAR value and TDS vs %Na.

The weightings assigned to each test correspond to the reliability of the tests. The pinhole test gave the best indication of the degree of dispersivity and was given the highest rating of 5 for highly dispersive (D1) soils. The CEC vs ESP chart gave the best indication of the cause of dispersivity, with soils plotting in the highly dispersive zone given a rating of 4. The crumb test was given a rating of 3 for a strong reaction. The two other chemical tests included, the SAR value and TDS vs %Na chart, were both assigned a maximum rating of 2 for dispersive soils. The total rating for the physical tests equals the total rating for the chemical tests, as they carry an equal importance in identifying dispersive soils.

If the total rating is equal to or greater than 12, the soil is classified as highly dispersive, and if the maximum of 16 is obtained the soil is termed extremely dispersive. If the rating is between 8 and 11 the soil is classified as moderately dispersive, and if between 5 and 7 the soil is classified as slightly

dispersive. A rating less than 4 indicates that the soil is nondispersive (Table 1).

4 CONCLUSIONS

Dispersive soils are readily eroded because the clay particles deflocculate and detach from the clay surface. This causes sodic soils to undergo piping and gully or donga erosion. Special treatment of the soils is necessary if they are to be used for construction purposes. Identification of dispersive soils is therefore of utmost importance so that chemical treatment of the soils and the appropriate construction techniques can be employed.

Dispersive soils tend to occur in those areas of KwaZulu-Natal with less than 850 mm precipitation per year, where they develop from a variety of rock types, especially mudrocks. Several physical and chemical tests have been developed to identify dispersive soils, however, no one test has proved successful in all circumstances. By combining together the results of a number of tests into a new weighted rated system it is now possible to identify dispersive soils with a greater degree of certainty than previously.

A new method of determining the turbidity of the effluent water from the pinhole test has been devised, which provides a quantitative measurement rather than the previous qualitative assessment based on a visual examination. The same technique could be used to quantify the results of the crumb test.

345

5 REFERENCES

Bell, F.G. & Maud, R.R. 1994. Dispersive soils: a review from a South African perspective. *Quarterly Journal of Engineering Geology.* 27. 195-210

Craft, D.C. & Acciardi, R.G. 1984. Failure of pore water analyses for dispersion. *Proceedings American Society Civil Engineers, Journal Geotechnical Engineering Division.* 110. 459-472

Elges, H.F.W.K. 1985. Dispersive soil. *The Civil Engineer in South Africa.* 27. 347-355

Emerson, W.W. 1964. The slaking of soil crumbs as influenced by clay mineral composition. *Australian Journal of Soil Research.* 2. 211-217

Gerber, F.A. 1983. *'n Evaluering van die fisies - chmiese eienskappe van dispersiewe grond en die methodes vir identifisering van dispersiewe grond.* Unpublished M.Sc. Thesis. University of Potchefstroom. 127pp

Harmse, H.J von M. 1980. Dispersiewe grond en hul ontstaan, identifikasie en stabilisasie. *Ground Engineering.* 22. 10-33

Richards, L.A. 1954. *Diagnosis and improvement of saline and alkali soils.* Agricultural Handbook. 60. US Department of Agriculture

Sherard, J.L. 1973. Embankment dam cracking. In. Hirschfield, R.C. & Paulos, S.J. (Eds) *Embankment Dam Engineering.* Wiley-Interscience Publications. New York. 339-343

Sherard, J.L., Dunnigan, L.P. & Decker, R.S. 1976. Identification and nature of dispersive soils. *Proceedings American Society Civil Engineers, Journal Geotechnical Engineering Division.* 102. 287-301

Sherard, J.L & Decker, R.S. 1977. Summary - Evaluation of symposium on dispersive soils. *Proceedings Symposium on Dispersive Soils, related piping and erosion in geotechnical projects.* ASTM Special Publication 623. 467-478

Volk, G.M. 1937. Methods of determination of degree of dispersion of the clay fraction of soils. *Proceedings Soil Science Society of America.* 2. 561-567

Walker, D.J.H. 1997. *Dispersive soils in KwaZulu-Natal.* Unpublished M.Sc. Thesis. Dept of Geology & Applied Geology, University of Natal, Durban. 120pp

Geotechnics for Developing Africa, Wardle, Blight & Fourie (eds) © 1999 Balkema, Rotterdam, ISBN 90 5809 082 5

Remedial measures for slope failure of canal system – A case study

K.Venkatachalam, R.K.Khitoliya & R.Chitra
Central Soil and Materials Research Station, New Delhi, India

ABSTRACT: Dispersive clays have a preponderance of sodium cation while ordinary clays have calcium and magnesium as the major cations in the pore water extracts. Potential problems with dispersive soils are of such magnitude that serious adverse situations may arise, if the materials are not properly identified and appropriate remedial measures not taken during design and construction stages. For identifying dispersive characteristics, four special tests viz. Sherard's Pin Hole, SCS Double Hydrometer, Analysis of Pore water extracts for Sodium, Potassium, Calcium and Magnesium cations and Crumb test are normally performed. The embankment of Sutlej Yamuna Link Canal system located in the State of Punjab, India experienced sloughing problem at various reaches. On investigation, it was found that the failure was due to the presence of erodible dispersive soil formations. Attempts were made to overcome this problem of dispersive soil by lime treatment.

1 INTRODUCTION

Certain natural clays are highly susceptible to piping by a process of colloidal erosion. These clays have predominance of dissolved sodium cations in the pore water, whereas ordinary, erosion resistant clays have calcium cations. Standard tests for classifying soils for engineering purposes do not identify the dispersivity nature of the clays. For identifying the dispersive characteristics, four special tests viz. Sherard's Pin hole, SCS Double hydrometer, Analysis of pore water extracts, Crumb test are normally conducted.

The Sutlej Yamuna Link (SYL) canal, was constructed to transfer the surplus water of Sutlej river to Yamuna river. The canal passes through two states viz. Punjab and Haryana. The embankment of the canal underwent sloughing at various reaches due to piping failure. The problem was more acute between RD 2 to 3 and RD's 67 to 83. The project authorities referred this problem to CSMRS for carrying out detailed geotechnical investigations for ascertaining the causes of failure and to suggest remedial measures to protect the canal embankment. This paper focuses on the field and laboratory investigations carried for dispersive soils from SYL canal system and treating it with lime for suggesting remedial measures.

2 FIELD INVESTIGATIONS

The field investigation works involved drilling and collection of undisturbed and representative soil samples from problematic reaches of canal embankment. In all 6 holes were drilled between RD 67.0 and 74.0 (Fig. 1 and 2), 61 undisturbed and 62 representative soil samples were collected. In addition, Standard Penetration tests and in-situ permeability tests were conducted wherever possible.

3 LABORATORY INVESTIGATIONS

The undisturbed soil samples (61) were subjected to extensive laboratory tests as detailed below:

* Mechanical Analysis
* Atterberg's Limits
* Insitu Dry Density and Moisture Content
* Specific gravity
* Triaxial shear (consolidated undrained drained with/without pore pressure measurements).
* One dimensional Consolidation

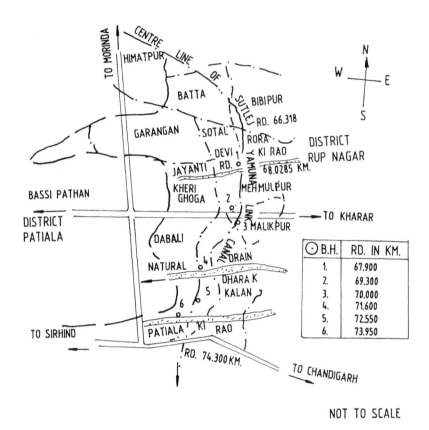

NOT TO SCALE

⊙ B.H.	RD. IN KM.
1.	67.900
2.	69.300
3.	70.000
4.	71.600
5.	72.550
6.	73.950

FIG. 1: INDEX PLAN OF SYL CANAL ALIGNMENT BETWEEN 67.000 TO 74.000 KM.

Based on the field identification tests/inspection it was expected that the soil formation mostly contains erodable dispersive soils and as such for taking up detailed investigation with particular reference to soil dispersivity identification tests and other basic tests as given below in the laboratory as many as 62 representative soil samples were supplied by the project authorities alongwith the values of in-situ dry density and moisture content pertaining to 8 selected cross section between RD 67 kms. and RD 74 kms.

*   Mechanical Analysis
*   Atterberg Limits
*   Sherard's Pin Hole
*   SCS Double Hydrometer
*   Chemical Analysis on pore water extracts
*   Crumb Test

Based on the test data generated, 4 potentially dispersive soil samples were selected and treated with percentages of hydrated lime viz. 0.5%, 1.0%, 1.5% and 2.0% by dry weight of the soil to arrive at suitable remedial measures.

4 TEST RESULTS - A REVIEW

4.1 Test on Dispersive Soils

Table 1 gives the range of physical and engineering properties of the embankment materials.The results of Mechanical Analysis and Atterberg limits indicate that the tested soil samples possess in general predominantly silt and clay with medium plastic characteristics. The values of insitu dry density/moisture content vary from 14.8 to 17.8 KN/m^3 and 11.4% to 26.8% respectively. The Specific gravity values of the tested soil samples vary from 2.62 to 2.79. The effective values of

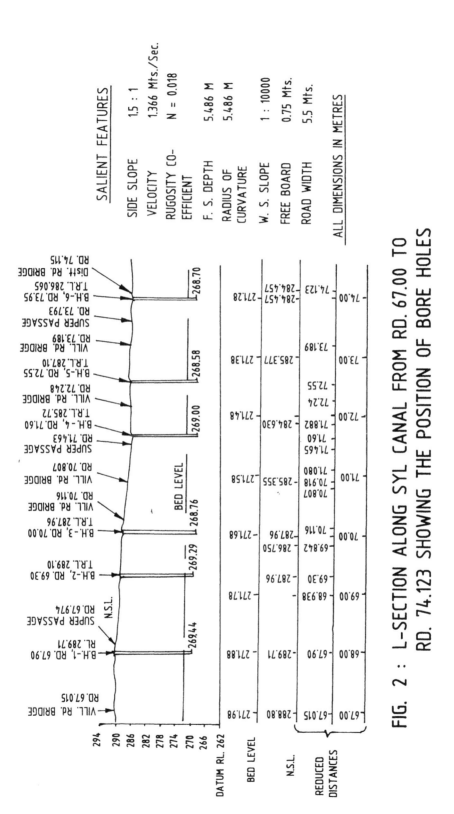

SALIENT FEATURES

SIDE SLOPE	1.5 : 1
VELOCITY	1.366 Mts./Sec.
RUGOSITY CO-EFFICIENT	N = 0.018
F. S. DEPTH	5.486 M
RADIUS OF CURVATURE	5.486 M
W. S. SLOPE	1 : 10000
FREE BOARD	0.75 Mts.
ROAD WIDTH	5.5 Mts.

ALL DIMENSIONS IN METRES

FIG. 2 : L-SECTION ALONG SYL CANAL FROM RD. 67.00 TO RD. 74.123 SHOWING THE POSITION OF BORE HOLES

Table - 1

Physical and Engineering properties of embankment materials

Properties	Range
Grain size	
. Clay size, %	2.0 - 46.4
. Silt size, %	6.6 - 77.2
. Sand size, %	0.1 - 70.0
. Gravel size, %	Nil
Index properties	
. Liquid limit, %	22.6 - 57.8
. Plastic limit, %	12.1 - 29.9
. Plasticity Index, %	6.9 - 36.1
Insitu dry density, KN/m^3	14.8 - 17.8
Natural Moisture Content, %	11.4 - 26.8
Specific gravity	2.62 - 2.79
Shear parameters	
. Effective cohesion, Kpa	16 - 38
. Effective angle of shearing resistance, degree	21.7 - 33.4

Table - 2

Consensus of Dispersivity Tests on soil samples

No.of Sample tested	SCS Dispersive Test	Crumb Test	Pin Hole Test	Chemical Analysis of Pore Water Extract	Consensus
62	⊘ 56	⊘ 14	⊘ 30	⊘ 11	⊘ 15
	◕ 6	◕ 19	◕ 11	◕ 19	◕ 43
	○ 0	○ 21	○ 21	○ 32	○ 4

Legend: ⊘ Dispersive
◕ Intermediate
○ Non-Dispersive

350

cohesion and angle of hearing resistance vary from 16.0 to 38.0 Kpa and 21.7° to 33.4° respectively. From the coefficient of compressibility values arrived based on the limited soil samples tested, it can be inferred that the soil strata is likely to exhibit low to medium compressibility characteristics. The consensus arrived at based on all the four special soil dispersivity identification tests determined by averaging the individual categories from all the four tests, with equal weight given to each test conducted on 62 soil samples as presented in Table-2 indicate that all the tested soil samples except 4 soil samples fall either in dispersive zone or intermediate zone. It can be therefore inferred from the above, that piping failure in the canal was due to dispersive phenomenon. Any use of such dispersive soils for canal reconstruction, would require pretreatment.

4.2 Mechanism of erosion

Erosion occurs when the external forces acting to abrade a soil exceed the internal forces holding soil together. The causative mechanisms can be due to either a large external force or to a degradation of the internal forces. Erosion is time dependent and usually continues until external and internal forces balance. It is necessary for a crack or flow channel to be present to allow a concentrated water flow to develop for internal erosion (piping) to occur in clays. This could occur because of differential settlements, fracturing or solutioning of chemicals. The dispersive property of the clay allows the finer particles to go into suspension on the introduction of water. Chemicals can be dissolved in water. The colloidal, suspended and dissolved matter is carried away in the water flow and the resultant soil structure is left with reduced resistance to shear.

4.3 Treatment of Dispersive Soil

As a trial experiment, few dispersive soils were taken up for pretreatment by adding lime in CSMRS laboratory. Four potentially dispersive soil samples were treated with hydrated lime equal to 0.5%, 1%, 1.5% and 2% by dry weight of soil and was thoroughly mixed, cured for few days. The lime treated soils indicated slight reduction in their plastic characteristics. Effect of treatment with lime is elucidated in Table-3. The findings based on the above test results indicate that these soil samples become non-dispersive with addition of 1% hydrated lime. However, a field application of 2% lime by dry weight of soil is suggested considering the potential variability of application and mixing.

Table - 3

Effect of Lime Treatment on Dispersive soil

Lab Sample No.89/166
Field No.73170-R-2
RL of Collection 278.30M

Sl. No	% of lime by wt.	Pin Hole test	SCS Dispersion	Chemical Analysis of Pore Water Extract	Crumb Test	Consensus
1	0.01	⊘	⊘	◐	⊘	⊘
2	0.5	○	◐	○	○	○
3	1.0	○	◐	○	○	○
4	1.5	○	◐	○	○	○
5	2.0	○	◐	○	○	○

Legend: ⊘ Dispersive
◐ Intermediate
○ Non-Dispersive

5 CONCLUSIONS

Based on the investigations carried out, it was found that in general the soil strata falls redominantly under CL and CI groups and as such the strata is impermeable. Even though the sand seams could be observed during deep drilling, it is presumed that these thin sand layers got mixed with thick clay layers during sampling and as such thin layers are not reflected in the logs of drill holes. Based on the limited soil samples tested for consolidation tests, it is inferred that the soil strata is likely to exhibit low to medium compressibility characteristics. On the basis of consensus drawn after conducting four dispersivity tests, it is inferred that embankment material is dispersive in nature. The removal of colloidal fines by seepage water followed by sloughing failure of the canal embankment is mainly due to inherent dispersive nature of the soil.

With addition of hydrated lime in dispersive soils, problems related to use of dispersive soils in hydraulic structures can be effectively tackled. Further, it was suggested to select a trial zone of about 100 mts. in the failure reaches and grout with 2% hydrated lime. While grouting the embankment, care has to be taken for avoiding hydraulic fracturing problem, if any, for which trial grouting has to be performed.

6 REFERENCES

Bureau of Reclamation, Dam Safety and Rehabilitation (1984), 4th Annual USCOLD Lecture, Phoenix, Arizona, Jan,24.

CSMRS Report "(1989) Geotechnical Investigations in the Failure Reaches (RD 67 to RD 74) of Sutlej Yamuna Link Canal Project, Punjab", Report No.S.I/89/9.

CSMRS Report "Soil Dispersivity Identification Tests on soils samples of SYL Canal Project, Punjab Report No.S.IV/89/7., Nov. 1989.

CSMRS Report (1990), Treatment of Sutlej Yamuna Link Canal's Embankment soil (Potentially Dispersive Soil) with hydrated lime, Report No. S-IV-89/7A, April.

Sherard, J.L. and Deckee, R.S., editors "Dispersive clays, Related Piping and Erosion in Geotechnical Projects", A symposium presented at the Seventy Ninth Annual Meeting of American Society for Testing and Materials, Chicago, iii, 27 June - July, 1976.

State of the Art Report on Dispersive Characteristics of Fine Grained Soils, (1986) Report No. RD-1- 86/SAR, 1986.

Venkatachalam, K, et.al, (1997) "Improvement of Dispersive Soils by mixing with hydrated lime, Second IISc - National Seminar on Geotechnical Engineering, Bangalore.

Venkatachalam, K., and Khitoliya R.K. (1997) Failure of Earthen Embankment due to dispersive soils; Seminar on Embankment, Dams and Slopes - Failure and Remedies (SEED-97), Kakinada, December 6-7.

Geotechnics for Developing Africa, Wardle, Blight & Fourie (eds) © 1999 Balkema, Rotterdam, ISBN 90 5809 082 5

The effect of type of reinforcement on the pullout force in reinforced soil

A.L. Kyulule
Department of Civil Engineering, University of Dar es Salaam, Tanzania

ABSTRACT: The paper examines the effect of type of reinforcement on the pullout force in reinforced soil. Three types of reinforcement materials were used in the pullout tests, namely, round steel bar mesh, ribbed steel bar and flat steel bar reinforcement. Five soil types were used in the investigation. The equipment for the pullout test consisted essentially of a metal box filled with soil. The test results of the investigations carried out in this work have indicated that round steel bar mesh reinforcement and ribbed steel bar and ribbed steel bar reinforcement give the highest values of pullout forces.

1 INTRODUCTION

The term 'reinforced soil' refers to a soil which is strengthened by a material able to resist tensile stresses and which interacts with the soil through friction and or adhesion (Jones 1985). Such reinforcement can also be achieved by relatively flexible, extendable, and sometimes compressible materials, such as non-woven fabrics or large quantities of individual fibres. The primary purpose of reinforcing a soil mass is to improve its stability, increase its bearing capacity, and reduce settlements and lateral deformations (Smith & Pole 1980).

Reinforced Soil is a well known technology which uses small quantities of construction materials compared with reinforced concrete. The technology also requires little equipment for construction. Long term experience in reinforced soil, however, is only available with non-cohesive coarse grained soils like sands and gravels (Vidal 1966).

A variety of materials can be used as reinforcing materials. Those that have been used successfully include steel, concrete, glass fibre, wood, rubber, aluminium and thermoplastics. Reinforcement may take the form of strips, grids (meshes), anchors and sheet material, chains, rope, planks and vegetation (Murray 1977).

This work investigates the effect of type of reinforcement on the pullout force in reinforced soil.

2 MATERIALS USED

2.1 *Soils*

Five soil types were used in the investigation. For easy reference and recognition, laterite soils from Kunduchi and Pugu Kajiungeni were designated as soil A and soil D respectively; tertiary soils were designated as soil B and soil E respectively; soil from weathered gneiss at Mufindi designated as soil C. Table 1 gives the summary of soil designation.

Table 1. Designation of soil samples

Name of soil	Designation
Laterite soil from Kunduchi	Soil A
Tertiary soil from Dar es Salaam University Campus	Soil B
Weathered gneiss rock from Mufindi	Soil C
Laterite soil from Pugu Kajiungeni	Soil D
Tertiary soil from Mbagala	Soil E

The particle size distribution of the various soils tested appears in Figure 1. According to the unified soil classification system (USCS) the five soils have been

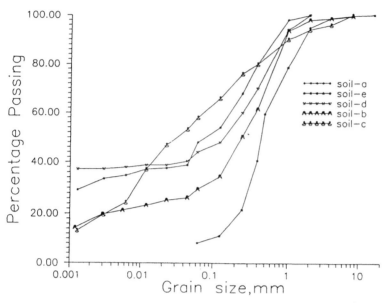

Figure 1. Particle size distribution curves

classified as SC, SC, SM, SC, and SW-SM respectively.

2.2 Reinforcements

Three types of reinforcements have been used in this investigation. These include: a mild steel flat bar, 40mm wide x 5 mm thick x 350 mm in length (designated as flat bar reinforcement); a mild steel round bar mesh reinforcement made of two 10 mm diameter longitudinal bars, placed 30 mm apart (centre to centre) and three transverse 10 mm diameter bars length 40 mm welded on the longitudinal bars, spaced 110 mm apart (designated as round bar mesh reinforcement); a ribbed mild steel flat bar, 40 mm wide x 5 mm thick x 350 mm in length with three 10 mm diameter bars length 40 mm welded on the flat bar (designated as ribbed steel bar reinforcement).

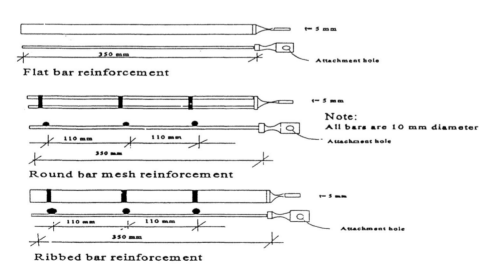

Figure 2. Types of reinforcement

Figure 2 shows the schematic drawing of the reinforcements.

3 EQUIPMENT AND TEST PROCEDURE

The equipment for the pullout tests consists of a box (made out of 5 mm thick mild steel) of size 492 mm x 130 mm x 149 mm with top plate of size 490 mm x 137 mm. The cover plate is provided with stiffening beams on top to ensure uniform distribution of normal load and to prevent bending its edges. The box has a side rectangular slot (on the side with dimensions 139 mm x 149 mm) through which the reinforcing materials can be inserted and pulled out during resting. This box is fastened to the hard floor using bolts and nuts. The box fastenings at the steel channel frame are designed to resist maximum pullout force caused by bond resistance between reinforcement and fill material during pullout tests, which would also be the pullout force in the proving ring. The required soil is first compacted in the bottom half of the box, then the reinforcement is placed carefully and finally soil is compacted in the top half of the box. Figure 3 shows the sequence of placing fill and reinforcement in the box.

After placing the cover plate the required normal load can be applied on the cover plate using a frame loaded through leverage load pivoted under the steel channel frame. The self weights of the loading frame, cover plate, and lever beam plus its attachments are included to constitute normal load together with added calculated weights to give desired normal stress. At higher normal stress the box is likely to bulge and hence reduction of the desired consolidation stress. This was taken care of by installation of a clamp to prevent longer sides (492 mm x 149 mm) of the box from bulging.

The reinforcing material in the horizontal position can be pulled out at the required speed and displacement can be measured by a dial gauge attached on the steel channel frame using magnetic steel block which is set to initiate readings on a transverse steel plate welded along the proving ring attachments. An electric powered motor drives, using a rubber belt, three sets of pulley systems connected to each other by rubber belts. The last set of pulleys in turn rotates two sets of gears. The last set of gears has been connected t a rotating screw shaft which effects pullout force to the reinforcing material through the proving ring and its attachments. The main function of the three sets of pulleys together with the two sets of gears is to reduce the electric powered motor speed to a desired speed. The proving ring is held in place using attachments fastened using bolts and nuts to the steel channel frame which also hold the rotating screw shaft in position. The attachment between the proving ring and screw shaft has a threaded hole through which the screw shaft rotates. Thus by the rotation of the screw shaft screwing and unscrewing effect results, depending on the rotational direction of the motor. Screwing effect results into tensioning the proving ring while the reverse effect results into compression of the same. The proving ring can be connected to the reinforcing element using another attachment by a pin of diameter 6 mm, length 50 mm placed in a hole at the end of the reinforcement and attachment. The proving ring measures the pullout force in tension using a gauge installed at its centre. The reinforcement is connected to a horizontal proving ring with a capacity of 1000 kg (10 kN). Dial gauge with sensitivity of 0.01 mm was used to measure horizontal displacement of the reinforcement.

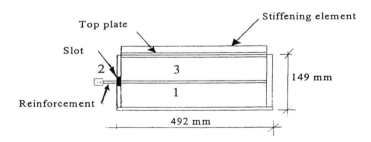

Note: The sequence of placing is indicated by numbers

Figure 3. Sequence of placing fill and reinforcement in the box

355

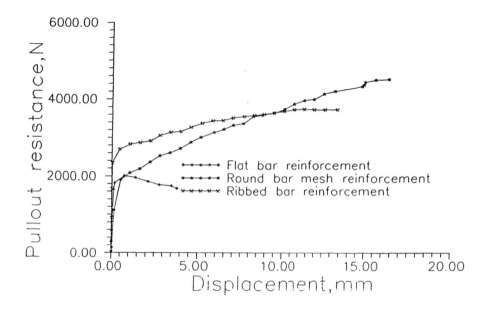

Figure 4. Pullout force versus displacement; Soil A.

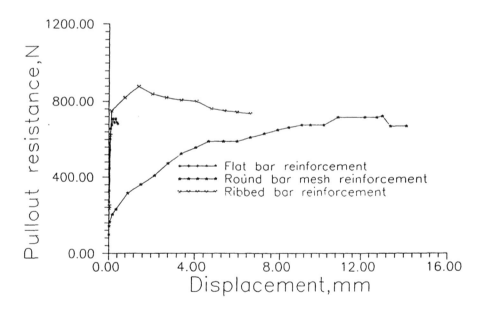

Figure 5. Pullout force versus displacement; Soil B.

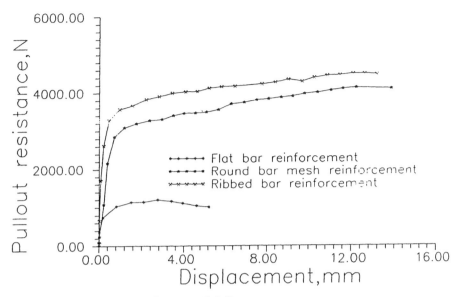

Figure 6.   Pullout force versus displacement; Soil C.

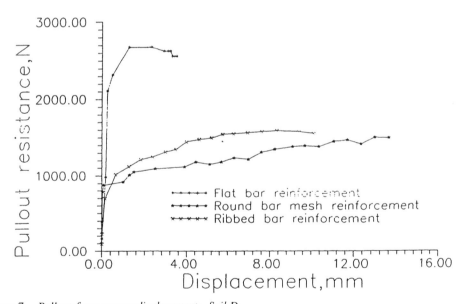

Figure 7.   Pullout force versus displacement: Soil D.

4 TEST RESULTS AND DISCUSSION

The test results are presented in the form of graphs that illustrate pullout forces versus displacements. Prior to performing pullout tests, each soil sample was thoroughly mixed at its optimum water content and compacted in the box to 95% of standard proctor compaction. All pullout tests were conducted under normal stress of 100 kN/m^2 and displacement rate of 0.11 mm/min. Figures 4 - 8 show the pullout force versus displacement results obtained using the three types of reinforcement for each soils type.

As can be seen in figures 4 - 8 different values of pullout forces and displacements were obtained for

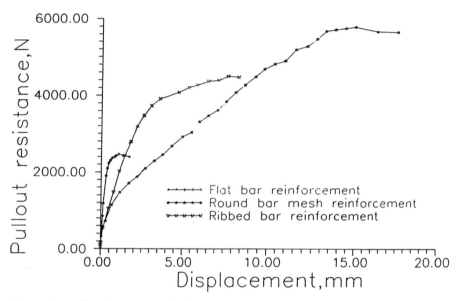

Figure 8. Pullout force versus displacement: Soil E

various types of reinforcement. Pullout forces of round bar mesh reinforcement and ribbed steel bar reinforcement are significantly higher than those of the flat bar reinforcement, indicating clearly the influence of the type of reinforcement on pullout force. The high pullout force is due to generation of local passive zones on transverse members of the reinforcements. It will be noted from figures 4 - 8 that a well pronounced peak in the pullout force versus displacement graphs can be observed only in the case of the flat bar reinforcement; for the remaining two types of reinforcement no clear peak is observed.

5 CONCLUSIONS

From the foregoing experimental work and discussion the following conclusions can be made:

1. Round steel bar mesh reinforcement and ribbed steel bar reinforcement give the highest values of pullout forces. This suggests round bar mesh and ribbed steel reinforcement to be the most effective type of reinforcement in reinforced soil technology.
2. A clear peak in the pullout force versus displacement relationship is observed only for flat steel bar reinforcement. For the ribbed steel bar and round steel bar mesh reinforcements, no clear peak is observed.
3. The displacement required to achieve maximum pullout force is highest for round steel bar

reinforcement; and it is least for flat steel bar reinforcement.

REFERENCES

Department of Transport 1978. Reinforced Earth Retaining Walls and Bridge abutment. *Technical Memorandum Bridges BE3/78.* London: Department of Transport.
Murray, R.T. 1977. Research at Transport and Road Research Laboratory to develop design for reinforced earth. *Joint Symposium for Reinforced Earth.* T.R.R.L. and Heriot Watt University.
Vidal, H. 1966. La terre armee. *Armaels de l' Institut Technique du Batimet et des Travaux Publics 19.*
Smith, G.N. and L.E. Pole 1980. Elements of Foundation design. *4th Edition, London, Granada Publishing Limited.*
Jones, C.J.F.P. 1985. Earth Reinforcement and Soil Structures. *Butterworth, London, England.*

Geotechnics for Developing Africa, Wardle, Blight & Fourie (eds) © 1999 Balkema, Rotterdam, ISBN 90 5809 082 5

Mobilised swelling pressure of an expansive soil

K. Mawire
Department of Civil Engineering, University of Zimbabwe

K. Senneset
Department of Geotechnical Engineering, Norwegian University of Science and Technology, Trondheim, Norway

ABSTRACT: This paper presents results of a constant-volume-swelling test (CVST) in which an unsaturated soil sample, confined to a fixed volume, was saturated. Components of the mobilised swelling pressure were measured in both the vertical and horizontal directions. A split-ring oedometer, designed to measure horizontal stress, was modified to accommodate swelling tests. These results, probably the first of a kind, are claimed to be fundamental in that they are a result of suction reduction only, without the interference of applied loads.

1. INTRODUCTION

Characterisation of the behaviour of expansive soils has been the focus of many researchers in the last thirty-five years. Laboratory experimental works mainly oedometer (with or without suction control) and, to a much lesser extent, triaxial or direct shear testing, have been reported in the literature. However, to the author's knowledge, there is no record of the direct measurement of the mobilised vertical and horizontal swelling pressures in expansive soils.

Vertical swelling pressure is commonly measured by subjecting a saturating soil sample to increasing external loads, to prevent volume change (Sridharan, Rao and Sivapullaiah, 1986). The vertical pressure due to the load is taken to equal the swelling pressure. This procedure introduces the mechanical effects of the external load onto the swelling process, a phenomenon that is essentially a result of physicochemical interaction at platelet level. This procedure best simulates the soil-structure interaction effects on the vertical swelling pressure, rather than measure the vertical swelling pressure.

In the case of horizontal swelling pressure, there is no record of direct measurement of horizontal swelling pressure. The few-recorded cases involved testing of a new apparatus or technique. In any case, the horizontal pressures measured were a combination of the effects of the swelling phenomenon and the applied vertical loads.

This paper presents result of a constant-volume-swelling tests (CVST) in which an unsaturated soil sample, confined to a fixed volume, was saturated. The components of the mobilised swelling pressure were measured in the vertical and horizontal directions. These results are part of a progressing laboratory and field test programme aimed at developing a rational unsaturated expansive soil model, and studying the expansive soil-structure interaction effects on raft foundations.

2. SPLIT-RING OEDOMETER

The split-ring oedometer is designed and developed at the Department of Geotechnical Engineering, Norwegian University of Science and Technology, Trondheim. Figure 1 shows a simplified cross-section of the oedometer. Figures 2 and 3 below show the photos of the oedometer with sample mounted.

The three parts of the split ring can be adjusted to clamp an undisturbed soil sample with a controlled contact pressure. In each of the ring segment, a sensitive LVDT indirectly records the horizontal stresses developed during the oedometer test (calibrated up to a horizontal stress of 1000kPa). An average lateral displacement of 21.7μmm for the three membranes was measured for an applied vertical pressure of 450kPa. This corresponded to a volume change of the soil in contact with the

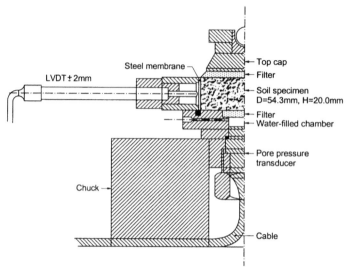

Figure 1. Simplified cross-section of the split-ring oedometer (after Senneset, 1989)

Fig. 2 Split ring oedometer with sample mounted

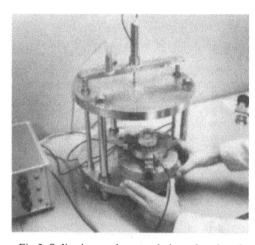

Fig.3 Split ring oedometer being placed under the loading ram

membranes of 0.18%, of the total sample volume. The measured eigendeformation of the three part steel ring was negligible. Hence, the test is described as a constant volume test, also with respect to horizontal displacements. Senneset (1982) gave the details of the split-ring oedometer, specifications and calibration.

3.   TEST PROCEDURE

An undisturbed soil sample, at specific water content and hence suction, was placed on a porous disc at the oedometer base, while the three ring parts were sufficiently separated to avoid sample disturbance. The ring segments were then adjusted towards the sample until they were just in contact with the sample. The slits between the three ring parts were sealed off with Loctite Silicone Sealant, to ensure a water- and pressure-tight chamber. The sample, with a porous disc and cap on top, was vertically locked in the ring with the loading ram, and fixed such that the sample could not swell in the vertical direction. Water was added through the base of sample, simulating a rise in water table. The mobilised swelling pressures were measured as the sample tried to expand within the confined chamber.

The test was run for 24 hours, after which there was no discernible increase in swelling pressure. A computer-controlled data logger captured and recorded the data from the LVDTs and the vertical load transducer.

4. SOIL DATA

Undisturbed samples were taken from a site within the University of Zimbabwe campus in Harare, and the soil data is tabulated in table 1 below.

Table 1: Soil data

Soil type	Black cotton soil
Sampling method	Block sampling
Sampling depth	1.0 m
In situ unit weight	16.48 kN/m^3
Lower Compactive Effort	1768 kg/m^3
Optimum water content	18 %
Grain density , G_d	2.58 kg/m^3
Liquid Limit, LL	65 %
Plasticity Index, PI	45 %
Linear shrinkage, LS	19 %
Silt fraction	28 %
Clay fraction	40 %
Water content at test	12 %
Saturation water content	38 %

5. TEST RESULTS

The mobilised swelling pressures from on specific test are given in fig.4. Figure 5 shows the change of the mobilised swelling pressure ratios, as the sample was saturated. The following definitions were adopted:

mean normal pressure, $P_m = (P_V + 2P_H)/3$

mobilised, $K_{swell} = P_H/P_V$,

where P_V is the mobilised vertical swelling pressure, and P_H is the mobilised horizontal swelling pressure, and K_{swell} is the swelling pressure ratio, defined similarly as the coefficient of earth pressure, K.

The following deductions can be made from this set of results.

1) The horizontal swelling pressure is consistently higher than the vertical swelling pressure. Maximum swelling pressures of 455kPa and 370kPa were recorded in the horizontal and vertical directions, respectively.
2) The initially very high K_{swell} value (>13) is partly a result of the "uncertainties" of contact pressure at the start of the test, and partly from the tendency of the horizontal pressure to be mobilised much faster than the vertical component.
3) K_{swell} value sharply drops from a high value (>13) to 2.6, and then gradually rises to a maximum value of 3 during saturation, before it drops to an equilibrium value of 1.2.
4) Swelling pressures tend to reach limiting value within 24 hours, for the soil type tested.

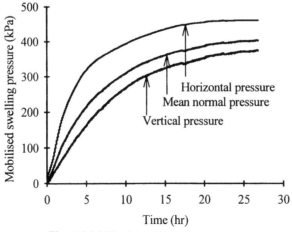

Fig. 4 Mobilised swelling pressures

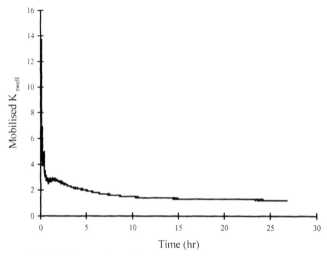

Fig.5. Change of swelling pressure ratio, K$_{swell}$ with time

6.   CONCLUSIONS

The horizontal and vertical components of swelling pressure on 12 samples of an undisturbed expansive soil were measured. The recorded pressures are in direct response to suction reduction only, without the interference of an external load. This response is a result of the fundamental behaviour of the soil.

The value of the horizontal swelling pressure is consistently higher than the vertical swelling pressure. The equilibrium values after saturation are such that the coefficient of swelling pressures, K$_{swell}$ is greater that one. The equilibrium swelling pressure obtained after 24 hours can be termed primary swelling pressure, since time effects are negligible.

When a sample is saturated from the bottom, simulating a steady rise in water table, the shape of the swelling pressure curve is hyperbolic.

The method of investigation and results presented in this paper serve to highlight a rational way of determining the swelling pressure of an expansive soil, at a given suction value.

7.   ACKNOWLEDGEMENT

The authors acknowledge the support of the Department of Geotechnical Engineering, Norwegian University of Science and Technology, Trondheim in this research work. The research is part of a collaborative programme between the University of Zimbabwe and Norwegian University of Science and Technology (NTNU). The research is funded by the Norwegian Council of Universities' Committee for Development Research and Education (NUFU).

8.   REFERENCES

Alonso, E.E., Gens, A., and Hight, D. W., 1987. Special Problem Soils. General Report, *Proceedings of the 9th European Conference on Soil Mechanics and Foundation Engineering*, Dublin: 1087-1146.

Brackley, I.J.A., 1973. Swelling pressure and free swell in compacted clay. *Proceedings of the 3rd International Conference on Expansive soils, Haifa.*

Brackley, I.J.A., 1975a. Swell under load. *Proceedings of the 6th Regional Conference for Africa on Soil Mechanics and Foundation Engineering*, Durban, vol.1: 65-70.

Brackley, I.J.A., 1975b. A model for unsaturated clay structure, and its application to swell behaviour. *Proceedings of the 6th Regional Conference for Africa on Soil Mechanics and Foundation Engineering*, Durban, vol. 1: 71-79.

Chen, F.H. and Ma, G.S., 1987. Swelling and shrinking behaviour of expansive clays, *Proceedings of the 6th International Conference on Expansive Soils*, New Dehli: 127-128.

Chen, F.H., 1988. Foundation on expansive soils. *Elsevier publishing company.*

Dakshanamurthy, V., 1979. A stress-controlled study of swelling characteristics of compacted expansive clays. *Geotechnical Testing Journal*, vol. 2 (1): 657-670.

Edil, T.B., and Alanazy, T.B.,1992. Lateral swelling pressures, *Proceedings of the 7th International Conference on Expansive Soils*, Dallas: 227-232.

Fourie, A.B., 1989. Laboratory evaluation of lateral swelling pressure. *Journal of Geotechni-*

cal Engineering, ASCE, 115(10): 1481-1486.

Holtz, W.G., and Gibbs, H.Jj., 1956. Engineering Properties of Expansive Clays, *Transactions of ASCE*, Vol. 121:641-663.

Habib, S.A., Kato, T. and Karube, D., 1992. One Dimensional swell behaviour of unsaturated soil, *Proceedings of the 7th International Conference on Expansive Soils*, Dallas: 222-226.

Khaddaj, S., Lancelot, L., and Shahrour, I., 1992. Experimental study of the swelling behaviour of heavily overconsolidated Flandres clays. *Proceedings of the 7th International Conference on Expansive Soils*, Dallas:239-244.

Komornik, A., and Zeitlen, J.G., 1965. An apparatus for measuring lateral swelling pressure in the laboratory. *Proceedings of the 6th ICSMFE*, 1:278-281.

Low, P.F., 1980. The swelling pressure of clay. II. Montmorillonites. *Soil Science Society of America Journal*, 44: 667-676.

Low, P.F., and Margheim, J.F., 1979. The swelling pressure of clay. I Basic concepts and empirical equations. *Soil Science Society of America Journal*, 43: 473-481.

Ofer, Z., 1981. Laboratory instrument for measuring lateral soil pressure and swelling pressure. *Geotechnical Testing Journal*, 4(4): 177-182

Ofer, Z. and Komornik, A., 1983. Lateral swelling pressure of compacted clay. *Proceedings of the 7th Asian Reg. Conf. SMFE, Haifa*, vol. 1: 56-63.

Scheiner, H. D., and Burland, J.B., 1987. Stress path during swelling of compacted soils under controlled suction. *Proceedings of the 6th International Conference on Expansive Soils*, New Dheli:155-159.

Sayed, A., Habib, S.A. and Karube, D., 1993. Swelling pressure behaviour under controlled suction. *Geotechnical Testing Journal*, Technical note.

Schreiner, H.D. & Burland, J.B., 1991. A comparison of the three swell-test procedures. *Geotechnics in the African Environment*.

Schreiner, H.D. & Burland, J.B., 1987. Stress path during swelling of compacted soils under controlled suction. *Proceedings, 6th International Conference on Expansive soils, New Dheli*.

Senneset, K., 1982. A new apparatus to study the influence of lateral deformation on constrained moduli of overconsolidated clays. *Geotechnical Division, The Norwegian Institute of Technology, Trondheim*.

Senneset, K., 1989. A new oedometer with splitted ring for the measurement of lateral stress. *12th International Conference on Soil Mechanics and Foundation Engineering*, Rio de Jeneiro, 1: 115-118.

Senneset, K. and Janbu, N., 1994. Lateral stress and preconsolidation pressure measured by laboratory tests. *13th International Conference on Soil Mechanics and Foundation Engineering*, New Dehli, 1: 309-312.

Snethen, D.R., 1980. Characterisation of expansive soils using soils suction data, *Proceedings of the 4th International Conference on Expansive Soils*, Denver: 54-75.

Snethen, D.R., 1984. Evaluation of expedient methods for identification and classification of potential expansive soils. *Proceedings of the 5th International Conference on Expansive Soils*, Adelaide: 22-26.

Sridharan, A., Rao, A.S., and Sivapullaiah, P.V., 1986. Swelling pressure of clays. *Geotechnical Testing Journal*, 9(4): 23-33.

Williams, A.A.B., Pidgeon, J.T., and Day, P., 1985. Expansive Soils. *The Civil Engineer in South Africa*, July 1985:367-377.

Xin, J.Z., and Ling, Q.X., 1992. A new method for calculating lateral swelling pressure in expansive soil. *Proceedings of the 7th International Conference on Expansive Soils*, Dallas: 233-238.

Yong, R.N., Sadana, M.L., and Gohl, W.B., 1984. A particle interaction model for assessment of swelling of an expansive soil. *Proceedings of the 5th International Conference on Expansive Soils*, Adelaide: 4-12.

Geotechnics for Developing Africa, Wardle, Blight & Fourie (eds) © 1999 Balkema, Rotterdam, ISBN 90 5809 082 5

Influence of reconsolidation stress history and strain rate on the behaviour of kaolin over a wide range of strain

J. N. Mukabi
Construction Projects Consultants Incorporated, Nairobi, Kenya

F. Tatsuoka
Department of Civil Engineering, University of Tokyo, Japan

ABSTRACT: In conventional laboratory testing , the reconsolidation procedures and shearing strain rate are known to influence the characteristics of clays and in particular their subsequent stress ~ strain behaviour. In this study the influence of reconsolidating Stress Ratio along five different stress paths (K_c=1.0,0.8,0.7,0.64 or 0.6), OCR and strain rate was evaluated over a wide range of strain from ε_a < 0.001% to post peak values of ε_a > 25% and the importance of so doing is reported. This study also shows that most existing theories characterizing the deformation and strength behaviour of clays which assume that failure for a normally consolidated clay occurs at the Critical State Line (CSL) irrespective of drain condition, strain rate and stress path traversed towards the CSL, are inadequate and may need modification.

1 INTRODUCTION

In most major civil engineering projects, structures that are constructed on or within the ground impose specific time dependent stresses and strains. Accurate prediction and simulation of the ground movement associated with the imposition of these stresses and strains is therefore of great importance.

1.1 *Effect of Consolidation Stress Ratio (CSR)*

Reconsolidation methods of clays have been known to affect their subsequent stress - strain relations under both drained and undrained shear conditions. Due to this, Tatsuoka and Kohata (1994) noted the importance of reconsolidating specimens to in-situ stresses through stress paths that retrace their recent stress - strain-time histories as precisely as possible. While investigating the effects of consolidation stress ratio and strain rate on the peak stress ratio of clay, Mukabi and Tatsuoka (1992) concluded that, possibly due to different structure formation, the stress ratio at q_{max} and the maximum stress ratio $(q/p')_{max}$ for Kaolin, increased as the consolidation stress ratio $K_c = \sigma_3' / \sigma_1'$ decreased. A study of Beaufort Sea clayey silt samples by Konrad (1985) indicated that the cyclic stress ratio at failure for tests with an initial stress ratio of 2 was 125% higher than for tests istropically consolidated, and further suggested that clayey silts subjected to anisotropic consolidation and cyclic loading may be sensitive to dy-

namic effect unlike sands. Mukabi and Tatsuoka (1992) also showed that, under montonic loading conditions in triaxial compression, the shearing stress ratio at failure K_f was a function of the initial consolidation stress ratio and that it decreased proportionally with decreasing K_c.

Rigorous research on the effect of various aspects of consolidation stress history over a wide range of strain i.e.; ε_a < 0.001% to ε_a > 20% is therefore considered extremely important.

1.2 *Effects of loading rate*

Various aspects of the effects of strain rate on the behaviour of soil over a range of strain levels have been investigated by a number of researchers, and their effects observed to affect certain engineering characteristics of the soils.

From cyclic simple shear tests on sands and clays, Tatsuoka (1978) and Hara and Kiyota (1979) investigated the effects of strain rate on the stress - strain relations at small strains and showed that the values of equivalent shear modulus (G_{eq}) and damping (h) were rather insensitive to loading frequencies between 0.1 and 10Hz. Mukabi (1990) and Tatsuoka and Shibuya (1992) also showed that the effects of loading rate on the G_{max} determined at strain levels in the region of ε_a < 0.001% for a range of loading rates of $0.002 \leq \varepsilon_a \leq 0.5$%/min. under monotonic loading conditions on Kaolin were virtually negligible. However, these effects became increasingly pronounced in the subsequent region of larger strains. Akai et al. (1975) showed that for axial strains larger than 0.05%, the effect of the strain rate on the stress - strain relation increased with increasing strain level and summarized this in the form:

$$q\{\varepsilon_1, d(\varepsilon_1)/dt\} = q[\varepsilon_1, \{d(\varepsilon_1)/dt\}_a] + \alpha.\log [\{d(\varepsilon_1) / dt\} / d(\varepsilon_1) / dt\}_a] \quad (1.1)$$

where $q\{\varepsilon_1, d(\varepsilon_1) / dt\}$: the deviator stress at a strain level of ε_1 and the strain rate $d(\varepsilon_1) / dt$, $q[\varepsilon_1, \{d(\varepsilon_1) / dt\}_a]$: q at the same strain level ε_1 but at a different strain rate $\{d(\varepsilon_1) / dt\}_a$ and α: coefficient representing the strain rate effect of a unit similar to that of q. Essentially, this would mean that when $\alpha=0$, there would be no apparent effect of axial strain rate. Given that α has the unit of stress and its appropriate application is limited to isotropically consolidated specimens, Tatsuoka and Shibuya (1992) introduced the parameter β in discussing the strain rate dependency of the stiffness of clays that can be applicable for a more general use. This is expressed as follows:

$$\Delta q(\varepsilon_1, d\varepsilon_1 / dt) = \Delta q\{\varepsilon_1, (d\varepsilon_1 / dt)_a\} + \beta. \Delta q_{max} \log\{d(\varepsilon_1) / dt\} / (d\varepsilon_1 / dt)_a\} \quad (1.2)$$

where $\Delta q(\varepsilon_1, d\varepsilon_1 / dt)$ and $\Delta q\{\varepsilon_1, (d\varepsilon_1 / dt)_a\}$ are the deviator stress differences $q-q_0$ ($=q$ at the initial value) at the same strain ε_1 but at different strain rates $d\varepsilon_1/dt$ and $(d\varepsilon_1/dt)_a$, and Δq_{max} is the peak value of Δq at a certain strain rate.

If not properly analyzed, the rate of loading can lead to gross misinterpretation of the behaviour and particularly strength and stiffness parameters of geomaterials leading to improper application of these parameters in design or modeling. As demonstrated in this study, it is vital to consider the effects of loading rate over a wide range of strain.

1.3 *Consideration of some crucial aspects of Critical State Soil Mechanics*

Critical State Soil Mechanics has been developed on the basis of Rendulics generalized principle of effective stress which states that for a soil in an initial state of stress and stress history there exists a unique relationship between voids ratio (e) and effective stress for changes in stress, ($\Delta\sigma_a'$ or $\Delta p'$). Most of the existing theories for deformation and strength characteristics of clays therefore assume this principle. Within this context, it is presumed that for a given normally consolidated clay, failure occurs at a unique line called the Critical State Line (CSL) defined by $q=Mp'$, without allowing the stress paths to locate above it at all stages irrespective of drain conditions, strain rate and stress path traversed towards the CSL. Most finite element based analytical tools

and simulation of such cases as multi-stage embankment design and construction widely employ Critical State models. While studying the influence of initial shear on undrained behaviour of normally consolidated kaolin, Ampadu (1988) concluded that the existing theory of Critical State Soil Mechanics alone cannot adequately explain the differences in behaviour between isotropically and anisotropically consolidated samples of the kaolin tested. It has also been reported by various researchers that the shapes and magnitudes of yield envelopes are influenced mainly by the composition, anisotropy and stress history of the clay features which have been inadequately modelled on the basis of the Critical State Soil Mechanics theory. Rigorous examination however, of the behaviour of clay that has been subjected through various stress ratios other than isotropic during consolidation and the corresponding relations is yet a subject to be exhausted. Part of this study therefore, attempts to further examine the effect of reconsolidation stress ratio on the behaviour of kaolin in relation to an isotropic state of stress. Modification of certain aspects of the existing theory of Critical State Soil Mechanics is also proposed.

2 TESTING PROCEDURE

A slurry made out of a mixture of a commercially sold kaolin powder and water at a $\omega_c = 160\%$, was thoroughly mixed under vacuum and consolidated one dimensionally in a large oedometer such as shown in Fig.1. The final index properties were LL=82.4%, PI = 43.6 and the specific gravity (Gs) = 2.65. Some results on a natural Tokyo Bay Clay (TBC) have also been included for comparison purposes. The TBC samples were obtained by using the triple tube sampling method from a clay layer of the Pleistocene Era, which exists at a depth of 36–38m below a 21.15 m deep seabed, and has a LL=50%, PI=20 and Gs=2.67. The Ko $(\sigma_3'/\sigma_1') = 0.54$. After saturation to Skempton B-values higher than 0.98

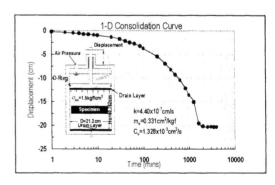

Fig. 1 One dimensional consolidation curve.

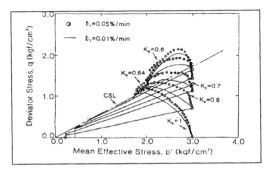

Fig. 2 Stress paths during consolidation for various consolidation stress ratios.

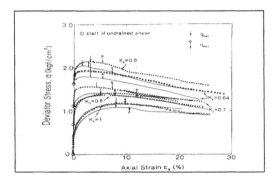

Fig. 3 Deviator stress ~ axial strain relations

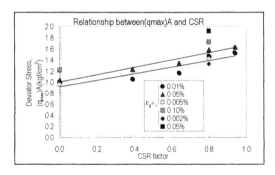

Fig. 4 Relationship between $(q_{max})_A$ and CSR factor

using the dry setting method, specimens (ϕ = 5cm, h = 12.5cm) were first normally consolidated in a triaxial cell along five different stress paths, namely Kc = σ_3' / σ_1' = 1.0, 0.8, 0.7, 0.64 (considered as Ko value) or 0.6 for the kaolin as shown in Fig. 2, and 1.0 and 0.54 for the TBC.

The overconsolidated specimens were initially reconsolidated along a stress path K = 0.64 and subsequently rebound through a stress path simulated by

$(K_o)_{rebound} = (K_o)_{NC} \times OCR^{\sin\phi}$ to OCR = 1.2 as shown in Fig. 8. The reconsolidation strain rate was at a constant ε_a = 0.001%/min to the same effective mean pressure p' = 3.0kgf/cm^2 for kaolin NC samples and either to the in-situ overburden pressure of p'= 1.525 kgf/cm^2 or NC$_{FIELD}$ conditions of p'= 3.0kgf/cm^2 for the TBC, prior to undrained shear. An automated stress path control triaxial testing equipment complemented with a system capable of applying very small load/unload/reload cycles within the region of very small strains, i.e. ε_a < 0.001% with virtually no backlashing, was applied throughout the testing programme. Basically two axial strain rates of ε_a = 0.01% or 0.05%/min. were adopted during shear.

3 TEST RESULTS AND DISCUSSIONS

Table 1 is summary of test results.

3.1 *Effect of CSR and strain rate on the stress-strain and stiffness characteristics*

The effective reconsolidation and undrained shear stress paths for kaolin are depicted on the p'~ q plane in Fig. 2. From the undrained shear stress paths, it can be seen that except for the very initial portions, the stress path formations diverge more from the isotropic one as consolidation becomes more anisotropic (K_c decreases) irrespective of the strain rate. The stress paths shown in Fig. 2 indicate that the stress ratio at large strains may be rather unique for the different consolidation stress ratios.

The entire stress - strain relations in Fig. 3 show that q_{max} increases both with increasing strain rate and decreasing (K_c). One of the possibilities that may explain the increase in q_{max} in respect to the CSR is the fact that, the undrained shear stress path for K_c <1 starts from a different initial state of stress which is located at a point S above the shear stress path for K_c = 1 as shown in Fig. 17. It is also quite likely that different structure formation occurred as a result of different consolidation stress histories. It can also be seen from Fig. 3 that, whereas the behaviour to failure is consistent both for K_c and strain rate, defining clearly increasing pre-failure resistance with increasing strain rate and decreasing K_c, the post-failure characteristics cannot be consistently defined in terms of η_{max} vs. strain level, unlike the stress ratio consistency in the p'~q plane in Fig. 2 earlier mentioned. This implies that the test data for samples under post-peak conditions may be unreliable when annalyzed in respect to stress–strain relations only.

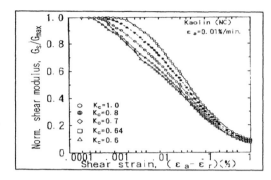

Fig. 5(a) $G_s/G_{max} \sim (\varepsilon_a - \varepsilon_r)$ relations for various stress ratios (ε_a = 0.01%/min)

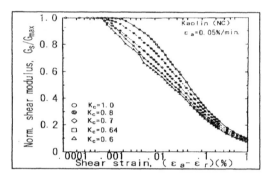

Fig. 5(b) $G_s/G_{max} \sim (\varepsilon_a - \varepsilon_r)$ relations for various stress ratios (ε_a = 0.05%/min)

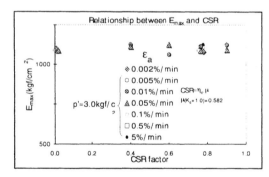

Fig. 6 Effect of CSR factor on E_{max} for various strain rates

In order to further analyze and verify the above stated facts, the relationship depicting the peak value $(q_{max})_A$ (ref. to Fig. 17) was estimated for evaluation in relation to undrained shearing conditions. Under this state of stress, which starts from the corresponding point A locating on the shear stress path for $K_c = 1.0$, the value $(q_{max})_A$ was estimated from the relation $(q_{max})_A = q_{max} \times (P'_A / P'_c)$. Values of $(q_{max})_A$

determined in this manner are plotted against the consolidation stress ratio (CSR) in Fig. 4.

The curves show that $(q_{max})_A$ increases almost linearly with the increase in the CSR factor (defined as CSR = η_c/μ (i.e. as K_c decreases). Under these circumstances, it is considered that the different structure formed during consolidation would be, as is evidenced from the axial strain-axial stress relations in Fig. 18(a), time - and stress - level dependent under continuos shearing. It was also observed in this study that both $\Delta p'$ and the radial strain at peak (ε_a = -2ε_r) decreased consistently with decreasing K_c. These results confirm the conformity of structures formed during consolidation to the pre-failure characteristics for clay sheared under undrained conditions. The above results also suggest that the possible effect of non-uniform deformation and associated non-uniform distribution of pore water on the measured stress values at the peak decreases as the consolidation stress ratio K_c traverses away from the isotropic conditions (i.e., $K_c = 1$). It can therefore be deduced that the reliability of the stress values measured at peak does not decrease as the consolidation stress ratio K_c decreases.

Figs. 5(a) and (b) depict normalized shear modulus (G_s/G_{max}) and shear strain ($\varepsilon_a - \varepsilon_r$) relations for various stress paths and two different axial strain rates, $\varepsilon_a = 0.001$ and 0.05%/min respectively. It can be seen from the decay curves in these figures that specimens consolidated at higher stress ratios (i.e. higher K_c or lower CSR), exhibit stiffer responses and have longer elastic ranges as can also be noted from Fig. 7 which shows the variation of elastic limit axial strain (ELS) with the CSR factor. Nevertheless, from Fig. 6, which compares the values of E_{max} defined from the initial linear portion of stress - strain relations at strain magnitudes less than about 0.001% for different consolidation stress ratios and strain rates, it may be seen that E_{max} is essentially constant and does not vary with the CSR.

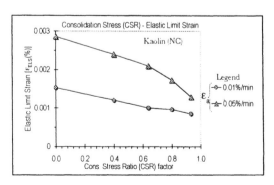

Fig. 7 Dependency of Elastic Limit axial strain (ε_a)$_{ELS}$ on CSR factor

	K_c (σ_3'/σ_1')	CSR η_c/μ	δ_{CSR} $(e^{CSR})^3$	$(q_{max})_{0.01}$ ($\sigma_1'-\sigma_3'$)	$(p'_m)_{0.01}$ $^1/_3(\sigma_1'+2\sigma_3')$	$(q_{max})_{0.05}$ ($\sigma_1'-\sigma_3'$)	$(p'_m)_{0.05}$ $^1/_3(\sigma_1'+2\sigma_3')$	$\phi'_{0.01}$ (°)	$\phi'_{0.05}$ (°)	$\mu_{0.01}$ $(q/p)_{max}$	$\mu_{0.05}$ $(q/p)_{max}$	$(\kappa)_{0.01}$ ϕ'/η_{qm}	$(\kappa)_{0.05}$ ϕ'/η_{qm}	λ'_{as}
KAOLIN	1.00	0	1	1.1	1.89	1.17	2.00	15.40	15.5	0.582	0.585	26.46	26.50	4.9
	0.80	0.40	3.31	1.18	2.04	1.38	2.18	15.60	16.6	0.578	0.633	26.99	26.22	6.8
	0.70	0.64	6.86	1.41	2.36	1.58	2.40	15.80	17.3	0.597	0.658	26.47	26.29	10.1
	0.64	0.81	11.39	1.80	2.67	1.92	2.58	17.60	19.3	0.674	0.744	26.11	25.94	11.1
	0.60	0.94	16.77	2.03	2.73	2.16	2.69	19.70	20.8	0.760	0.803	25.92	25.90	13.1
TOKYO BAY CLAY	1.00	0	1	2.96	1.95	-	-	37.3	-	1.518	-	24.57	-	1.15
	0.54	0.37	3.034	2.57	1.48	-	-	42.4	-	1.740	-	24.37	-	1.80
	0.54	0.42	3.525	3.17	2.08	-	-	37.5	-	1.524	-	24.61	-	1.92

Table 1 Summary of results

$\sigma_1', \sigma_3', q_{max}, p'_m$ measured in kgf/cm², $\phi' = Sin^{-1}\{(\sigma_1'-\sigma_3') / (\sigma_1'+\sigma_3')\}_{max}$

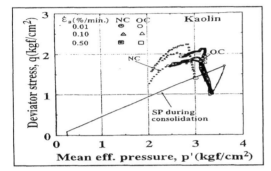

Fig. 8 Consolidation and shear stress paths for NC and OC specimens

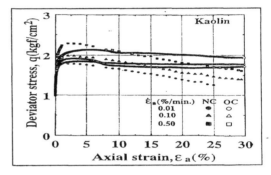

Fig. 9 Overall stress ~ strain relations

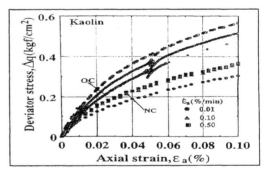

Fig. 10 Intermediate strain characteristics ($\varepsilon_a < 0.1\%$)

This characteristic can mainly be attributed to the changes in inter-particle contact forces, known to resist the initial particle yielding that develops within the entire assembly during consolidation through different stress histories.

3.2 Influence of OCR on stress-strain and stiffness behaviour

The effective reconsolidation and undrained shear stress paths for normally (NC) and lightly overconsolidated (OC) specimens sheared at three axial strain rates ($\varepsilon_a = 0.01$, 0.10 and 0.5%/min) are represented in Fig. 8. It can be seen that, whereas the NC specimens trace stress paths with constantly increasing maximum stress ratio, i.e. (q/p') as the strain rate increases, this trend is less obvious for OC specimens. Furthermore, after yield, the NC specimens show a decrease in q/p' while the OC shearing stress paths increase in stress ratio.

This difference may be associated with different structure-induced excess pore water pressure characteristics between NC and OC specimens. The stress - strain relations over the whole range of testing are shown in Fig. 9. It can be seen that although the NC specimens have higher maximum shear strengths (q_{max}) compared to the OC specimens, they exhibit a relatively less brittle response, larger post-peak strain softening and lower residual strength. Skempton (1984), for example, reported that the post-peak drop in strength of a normally consolidated clay is due only to particle re-orientation as is the case from these results. This again, is likely to be as a consequence of different structures formed during different consolidation

369

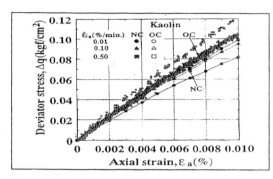

Fig. 11(a) Small strain characteristics (ε_a<0.01%)

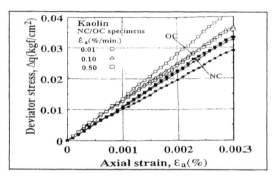

Fig. 11(b) Very small strain characteristics (ε_a<0.01%)

stress histories. On the other hand, in the post-peak region, the OC specimens exhibit a mild decrease in strength with relatively small displacements tending to the residual state which is distinctly characterized by large displacements at a virtually constant deviator stress; probably owing to the re-orientation of platy clay minerals parallel to the direction of shearing. From the foregoing argument and the stress - strain characteristics in Figs. 3 and 9, it can be deduced that the effects of different consolidation stress histories are eminent even in the post yield range and after large straining.

The corresponding stress - strain characteristics in the region of small (ε_a < 0.1%) to very small (ε_a< 0.001%) are shown in Figs. 10, 11 (a) and 11(b). The following observations can be made from these figures. (1) The OC specimens exhibit less non-linearity in comparison to the NC ones. (2) Longer linear elastic ranges can be observed for the OC than the NC specimens (ref. also to Fig. 15). (3) OC specimens tend to have slightly higher stiffness values from the very initial strain levels i.e., {$(G_{max})_{OC}$ = 444 and $(G_{max})_{NC}$ = 363 kgf/cm^2}over a wide pre-failure strain range from the start of shearing. This could be due to the difference in the initial mean effective stress (p_0') and final void ratio e_f at the end of consolidation. When corrected for void ratio dif-

ferences based on the empirical equation f(e) = (2.97 - e) ^{2}/ (1+e), then the difference becomes very small {$(G_{max})_{OC}$ = 433 compared to $(G_{max})_{NC}$ = 411}.

3.3 Strain rate effects on OC specimens

Figs. 9 ~11(b) show that, except in the linear elastic range of very small strains, i.e. ε_a<0.001%, the stress - strain behaviour of both NC and OC specimens is highly strain rate dependent. It can also be seen from these figures that for both specimens, strain rate tends to increase the brittle, but not necessarily the residual behaviour of kaolin. Another observation is that non-linearity tends to increase consistently with decreasing strain rate. However, no significant effect of strain rate on the post-peak strain softening i.e., the rate of decrease in q with straining, can be seen. This implies that unlike the stress history effect which is apparent over essentially the whole range of strain, the strain rate effects seem to be limited to the pre-failure region of strains. This may be interpreted as a mechanism characterizing the rate of displacement of the overall assembly to the intrinsic rate of particle movement termed "relative rate of particle displacement" in this study. Under this theory, it is considered that strain rate will only take effect when the inter-particle coefficient of friction, which is thought to increase with increasing stress level, reaches as limiting value just before the start of, and mainly within "zone 2" as described in the zoning mode proposed by Mukabi (1991). This may suggest that the fundamental component of strain rate effects is limited mainly to the pre-failure zones. Fig.12 shows that strain rate has a more pronounced effect on the maximum shear strength of NC than OC specimens but that this effect can be represented linearly per log cycle, for both consolidation histories. The influence of strain rate on the strain level dependency of stiffness is depicted by the decay curves plotted in Fig.13, while that on normalized modulus E/p_0' at different levels of strain and the Elastic Limit Strain (ELS) is shown in Figs. 14 and15 respectively. These results suggest that: (1) E_{max} is practically insensitive to the strain rate for both NC and OC specimens. (2) The quantitative effects of stain rate vary with the degree of the axial strain rate imposed. This can be deduced from the variation in the slopes in the log~log plot of $E_{sec}/p_0' \sim \varepsilon_a$ relation in Fig.14, where this effect is also seen to increases as the axial strain levels increase in the post-linear elastic zone. (3) Increasing the strain rate reduces the strain level dependency of stiffness (4) The elastic limit strain ($\varepsilon_a)_{ELS}$ increases with increasing strain rate but does so at a higher ELS level for the OC than NC specimen. In order to

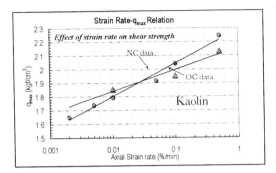

Fig. 12 Effect of strain rate on qmax of NC and OC

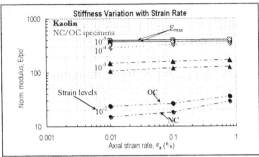

Fig. 14 Variation of stiffness with strain rate of Elastic Limit Strain.

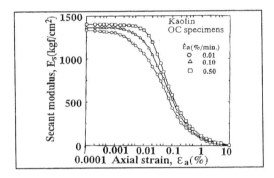

Fig. 13 Strain level dependency of stiffness (OC).

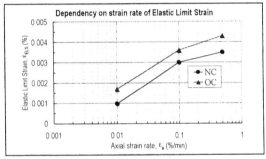

Fig. 15 Dependency on strain rate

effectively interpret the strain rate dependency of the stiffness of soils, Kim et-al. (1994) introduced parameter θ, a modified version of the β parameter proposed by Tatsuoka and Shibuya (1992). This parameter is represented as follows:

$$\Delta q(\varepsilon_1, d\varepsilon_1 / dt) = \Delta q\{\varepsilon_1(d\varepsilon_1 / dt)_a \}\cdot[1 + \theta.\log\{(d\varepsilon_1 / dt) / (d\varepsilon_1 / dt)_a \}] \quad (3.1)$$

Fig. 16 shows the values of θ, also known as the coefficient of strain rate effect, plotted against the axial strain obtained from various CU TC tests on clays as well as those of sedimentary soft rocks. The following observations can be made: (1) θ increases with strain level and the general trend of increase is fairly similar between clays and sedimentary soft rocks.(2) The parameter θ at ε_1 =0.001% is not perfectly zero in most cases. This means that the truly elastic strain limit is smaller than 0.001% for most soils. (3) Comparisons of the various soils at a similar strain level shows that θ decreases with increasing intrinsic stiffness of the material and with long term consolidation (4) For reconstituted kaolin, A specimens show less tendency of dependence on θ compared to the I ones within the region of axial strains (ε_a) less than 0.1%. This trend, however, drastically changes after this level of strains.

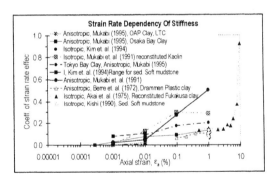

Fig 16. Strain rate dependendency of stiffness

3.4 Proposed modification of some aspects of Critical State Soil Mechanics

The crucial property of the Critical State Line (CSL) is that initially isotropically compressed samples will fail once the stress states reach the line, irrespective of the test path followed by the samples on their way to the Critical State Line; a theory which is widely applied particularly in the design and modelling of soft clays in relation to yielding. The projection of the CSL into the q: p′ plane is given by

371

q = Mp', where M is the gradient and $v = \Gamma - \lambda \ln p'$ in the $v:\ln p'$ space, where Γ is defined as the value of v corresponding to $p' = 1.0 KNm^{-2}$ on the CSL. Thus Γ locates the CSL in the $v:\ln p'$ plane in the same way that N, given by $N = v + \lambda \ln p'$, locates in the Normal Compression Line (NCL). This is what forms the basis of the original Roscoe and Hvorslev state boundary surfaces which characterize possible and impossible states of stresses to be achieved by clays under normally and lightly overconsolidated conditions respectively. This theory, however, does not adequately cater for samples that simulate an anisotropic state of consolidation history. This study attempts to introduce empirical methods of so doing by way of modifying the theory governing the CSL.

As was observed from the stress paths plotted in Fig. 2 on the $p'-q$ plane, as K_c decreases ($K_c < 1$), the stress path tends to intersect the CSL and to attain a higher stress ratio $\eta_{max} = (q/p')_{max}$. Mukabi and Tatsuoka (1992) introduced this phenomenon as "overshooting" of the CSL. In order to facilitate further discussion on this aspect, parameters shown in Fig. 17 are defined as $\eta_{max} = (q_m/p'_m) = (q/p')_{max}$, a value arbitrarily defined as the q/p' at the CSL for a sample sheared at $\varepsilon_a = 0.01\%/min$. when $K_c = 1$. $\mu = (q/p')_{cs}$ is the stress ratio at the CSL but not necessarily the ultimate conditions while the DOS factor, defined as the degree of "overshooting" the CSL, is calculated as DOS = $(\eta_{max} / \mu) - 1.0$. The consolidation stress ratio is given as $\eta_c = (q/p')_{cs}$ and facilitates in defining the distance to failure from the current consolidation state of stress which is $\mu - \eta_c$. $(q_{max})_A$ depicts the peak value of the shear strength had the undrained shear stress path started from point A, which is the intersection of the undrained shear stress path of $K_c = 1$ specimen and $K_c < 1$ consolidation path, instead of starting from point S. The Consolidation Stress Ratio (CSR) factor, described as the distance to failure from the current consolidation state of stress, is given by CSR = η_c/μ. Mukabi (1995) attributed the consequence of the "overshooting" phenomenon mainly to the formation of different time and stress-level related structures during consolidation along different K_c stress paths as is evidence from Fig. 18(a). Figs. 19 and 20 show the relations between η at q_{max} and the DOS factor with the change in the CSR factor respectively. The characteristics in these figures obviously defy those defined under the framework of Critical State Soil Mechanics (CSSM), which state that η at q_{max} is always constant and equal to M.

Three basic factors that may contribute to this behaviour are the clay structure in relation to the consolidation stress history, the initial state of stress in

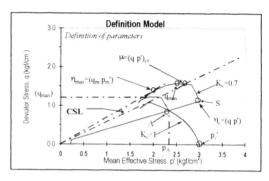

Fig. 17 Definition of stress parameters in p'~q plane

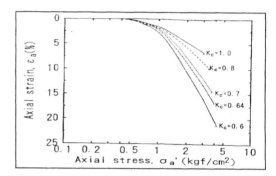

Figure 18(a) Consolidation axial stress-strain curves

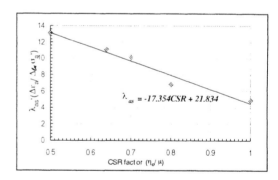

Fig. 18(b) Effect of CSR(η_c/μ) on $\lambda_{as}(\Delta\varepsilon_a/\Delta u\sigma_a')$

relation to the CSL and the fact that under a constant rate of strain during undrained shear, stress paths for higher K_c values traverse for longer periods to peak resulting in larger straining that contributes to larger structural changes. On the other hand, as K_c decreases from unity, the structure formed during consolidation is preserved to a larger degree, leading to the increase in peak strength and more brittle behaviour as earlier explained. Within the framework of this study, it was also considered important to in-

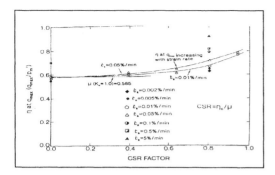

Fig. 19 η at q_{max} – CSR factor relations

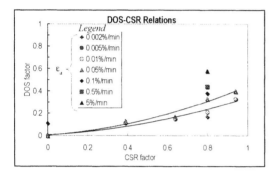

Fig. 20 Relationship between DOS and CSR

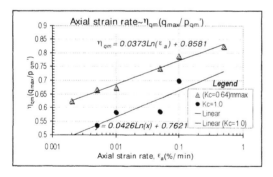

Fig. 21 Effect of axial strain rate on $\eta_{qm}(q_{max}/p_{qm}{}')$

vestigate the effects of loading rate on the peak stress ratio as well as the coupled effects or correlation of CSR and strain rate on the behaviour of the kaolin tested, in order to establish a clearer and better defined behaviour. As can be interpreted from previous Figs. 4, 19 and 20, which include other data of samples sheared at strain rates ranging between $0.002\%/\text{min} \leq \varepsilon_a \leq 5\%/\text{min}$, the tendency of increasing strain rate is to increase $(q_{max})_{A,}$ η at q_{max} as well as the DOS factor despite the CSR factor. With

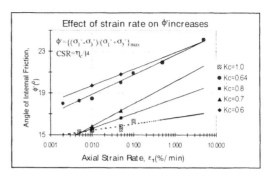

Fig. 22 Effect of axial strain rate and K_c on ϕ' at q_{max}

the increase in strain rate, these parameters seem to increase at a higher rate and larger CSR values. The effects of axial strain rate on two important failure parameters that delineate the CSL namely, μ_{max} (q_{max} / p'_{qm}) and the angle of internal friction (ϕ') are shown in Figs. 21 and 22 respectively.

The angle of internal friction, defined as ϕ' = arc sin $\{(\sigma_1' - \sigma_3')/(\sigma_1' + \sigma_3')\}_{max}$, representing shear resistance between the particles, is known to be the force that must be applied to cause a relative movement between the particles. This force is known to be frictional in nature and is the basis of the concept of failure of a soil in soil mechanics, represented in its most basic form as $\tau_{ff} = f(\sigma_{ff})$, where τ_{ff} and σ_{ff} are the shear and normal forces on the failure plane respectively. It has normally been reckoned that the value of ϕ' in triaxial compression tests is not greatly affected by the rate of loading. For sand, for example, Whitman and Healy (1963) compared tests conducted in 5 mins. and in 5 ms and found that tan ϕ' decreased at most by about 10% and probably only 1–2%. However, as can be seen from Fig. 22, as the axial strain rate increases, σ' increases with decreasing K_c until K_c attains a value $K_c = K_o$ (K_o = 0.64 in this case) whereby the tendency is towards a limiting value. In Fig. 23(a) it can be seen that a virtually linear relationship exists between ϕ' and the cube of the exponential of the CSR factor, denoted as $\delta_{CSR}[(e^{CSR})^3]$. Strain rate effects on this relationship are also apparent. Incorrect measurement of pore water pressure may be viewed as a possible reason for this difference. In other words, as the strain rate increases, gradients of pore pressure set up within the specimen being sheared undrained increase and hence the pore pressure measured at the end of the specimen may be different from the value within the central zone and may therefore be time dependent. However, Whitman (1963) reported results on tests that had been performed by applying a pore pressure measuring system with a rapid response time, which showed that pore pressure development was independent of strain rate. By plotting effective and total stress paths from AU and AD tests for five different strain rates. Mukabi (1991) showed that excess pore water pressure developed at a certain strain level was independent [A = $\Delta\mu$ /$\Delta\sigma_1$ = 0.6 (constant)] of strain rate and that this applied for both anisotropically (A) and isotropically (I) consolidated specimens. On the other hand, Casagrande and Wilson (1953) suggested that the mineral skeleton may, under rapidly applied loads, have a resistance to compression which approaches that of water resulting into structural viscosity. This would imply that the excess pore pressures set up during undrained shearing in soft, saturated soils would decrease as the strain rate increases. Possible explanations for the variation of ϕ' under varying strain rate in relation to I and A specimen behaviour would be: (1) The prevalence of large straining at the peak in relation to the value of σ_1' / σ_3' of the I specimens in comparison to the relatively smaller strains at failure of the A specimens indicates that the latters structure is less destroyed by the time the specimen attains peak conditions. This could mean that until the peak in σ_1' / σ_3' is reached, the overall yielding at interparticle points occurs to a much larger extent, compared to that of the anisotropically consolidated specimen whose structure may have been preserved to a larger extent. (2) The angle of internal friction ϕ' for "the structure" presumably depends on the material property which is a function of strain rate. (3) ϕ' is considered to be controlled by the bond strength of the structure formed during consolidation. Since the bond formation at the junctions of the soil particles is known to be time-dependent, ϕ' may be deemed to reduce as the time of loading increases. The deviation from an initial common stress for both I and A specimens and a tendency to move towards a common ultimate failure line as can be seen from Fig. 2 may further emphasize this point. In modelling the behaviour of clay within a modified framework of CSSM therefore, this study considered it important to establish a method which could cater for the effects of loading rate in relation to ϕ' as a function of CSR. The relationship of the CSR function (δ_{CSR}) and $\Delta_{\phi'}$, a function of the normalized angle of internal friction (ϕ'/q_{max}) in reference to that of K_c = 1 developed from Fig. 23(a), is represented by a linear equation in the form $\phi'/q_{max} = A_{\phi}\delta_{CSR} + B_{\phi'}$, where $A_{\phi'}$(=0.1628) and $B_{\phi'}$(= 7.9626 mean value in this case) are constants. In normalizing, the q_{max} was determined for the two (ε_a = 0.01 and 0.05%/min) respective strain rates from the linear regression of the NC line of the semi-log plot of ε_a ~ q_{max} shown in

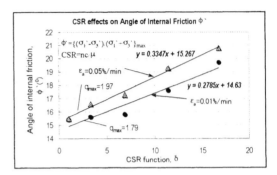

Fig. 23(a) CSR function (δ)-(ϕ')

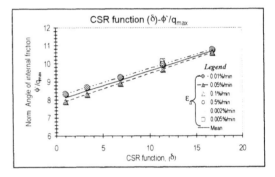

Fig. 23(b) CSR function (δ)-ϕ'/q_{max} relations.

Fig. 12. The linear relation in Fig. 23(b) is virtually similar and shows no dependency on the various strain rates (0.002 $\leq \varepsilon_a \leq$ 0.5%/min). This suggests that this equation uniquely relates CSR and ϕ' and may be applied in developing further mathematical relations that may aid in redefining the basic parameters related to the CSL without particular reference to strain rates effects.

Table 2 Summary of results calculated using proposed equations

ψ'	$(q_{max})_{cal}$	$(q_{max})_{cal}$	ϕ'	$(q_{max})\phi'$ calculated	$(q_{max})\phi'$ calculated	$(\lambda_a)_1$	λ_a calculated
0.582	1.10	1.16	1.00	1.10	1.17	1.00	4.9
0.622	1.27	1.36	1.24	1.375	1.45	1.39	8.39
0.648	1.53	1.56	1.456	1.60	1.70	2.06	10.48
0.669	1.79	1.73	1.658	1.82	1.94	2.27	11.96
0.685	1.88	1.93	1.852	2.04	2.17	2.67	13.1
1.518	2.96	-	1.00	2.96	-	1.00	1.15
1.615	2.39	-	0.822	2.43	-	1.57	1.91
1.627	3.38	-	1.253	3.71	-	1.67	2.01

q_{max}, p_m' in kgf/cm², K_1 = 1, μ_1 = 0.582, ψ' = $\mu/(K_1$-0.16CSR), ϕ' = $K_1/(K_1$-0.48CSR). $(q_{max})_{\phi'}$: value calculated from $(q_{max})_{K=1}$ x ϕ' (i.e., 1.1ϕ' for ε_a = 0.01%/min & 1.174ϕ' for ε_a = 0.05%/min). $(\lambda_a)_{cal} = (\lambda_a)_1$x(1+1.78CSR); where $(\lambda_a)_1$ is the value of λ_a at K = 1 [Isotropic(I) conditions].

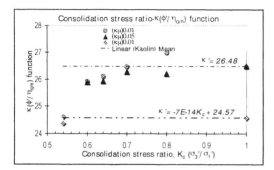

Fig. 24 Effects of CSR on κ function

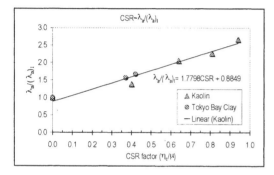

Fig. 25(a) CSR factor - $\lambda_a / (\lambda_a)_I$

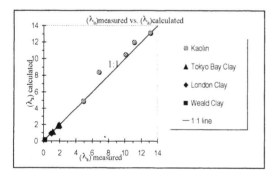

Fig. 25(b) λ_a measured-λ_a calculated relations

Now that the effects of loading rate have been characterized into a generalized state, a method of unifying the behaviour of anisotropically consolidated clay into a coherent form may be considered possible. In so doing, the following basic parameters and mathematical relations determined empirically from Figs. 24~25 as well as that from Fig. 23(b) are employed. The κ function which relates the angle of internal friction ϕ' determined from tests performed using different CSRs as shown in Fig. 24, is defined as $\kappa = \phi'/\eta_{qm}$. The plot shows that this parameter is virtually constant for all practical purposes for both

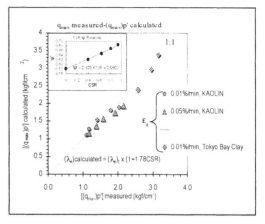

Fig. 26(a) $[(q_{max})_{p'}]$ measured vs. $[(q_{max})_{p'}]$ calculated.

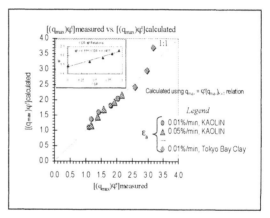

Fig. 26(b) $[(q_{max})_{q'}]$ measured vs. $[(q_{max})_{q'}]$ calculated.

kaolin and Tokyo Bay Clay and relates closely to a reference line considered to be analoguos to a modified common CSL (i.e., at ϕ' constant when $K_c = 1$). This implies that a proper approach may lead to a generalized state. From Figs. 17, 23(b) and 24, the ratio of deviation of the anisotropic $(CSL)_A$, as a function of the isotropic $(CSL)_I$ introduced as ψ' in this study, may therefore be expressed as:

$$\psi' = (\eta_{max})_I / \{K_I - (\Delta\phi'/\delta_{CSR}) \cdot CSR\} \qquad (3.2)$$

where the constants were determined in this study as $(\eta_{max})_I = 0.582$, $K_I = 1$, $\Delta\phi'/\delta_{CSR} = 0.16$. The expression $q = Mp'$ defining the conventional CSL can therefore be expressed in a modified form as:

$$q = \psi'p' \qquad (3.3)$$

since $M = \eta_{max}$. On the other hand, the modified λ constant which would project the modified CSL onto the v:lnp' space is determined from the relation in Fig. 25(a)

375

where $\lambda_a/(\lambda_a)_1 = A_\lambda CSR + B_\lambda (A_\lambda = 1.78, B_\lambda = 0.89)$.

$$\lambda_m = (\lambda_a)_1 \cdot (1 + 1.78CSR) \qquad (3.4)$$

$(\lambda_a)_1$ is the soil constant determined during conventional isotropic consolidation, i.e., the value of λ_a at $K_c = 1$. The modified Normal Consolidation and Critical State Lines in the v:lnp' space may therefore be defined as

$$v = N - \lambda_m \ln p' \qquad (3.5)$$

$$v = \Gamma - \lambda_m \ln p' \qquad (3.6)$$

respectively.

This implies that, with the determination of basic parameters from properly conducted IU tests, the conditions and parameters of failure for AU under TC may be recognized. An extrapolation of the consolidation constants to calculate the values of the stresses at failure in an undrained triaxial compression test ($v = v_1$) may then be given by equations (3.3) and (3.6) as

$$p'_f = \exp[(\Gamma - v_1)/\lambda_m] \qquad (3.7)$$

and

$$q'_f = \Psi' \exp[(\Gamma - v_2)/\lambda_m] \qquad (3.8)$$

Calculated values using this expression also showed a good agreement, i.e., (q_{max}) measured = 1.79 and $(q_{max})_{calc.} = 1.84$ kgf/cm^2 for $K_c = 0.64$. The calculated values plotted in Figs. 25(b), 26(a) and 27 were based on the modified equations proposed in this study. A satisfactorily good agreement may be seen from the comparison of measured and calculated values in all cases.

Another important expression proposed in this study is that developed from the theory that as the sample progresses to failure, the shearing stress path deviates from constant p_c' conditions at a virtually constant stress ratio. Incorporating this concept, a deviator stress factor φ' was developed empirically and is given as

$$\varphi' = K_1/(K_1 - R_d \cdot A_{\varphi'} CSR) \qquad (3.9)$$

where R_d is the shearing deviation stress constant ratio $(q'/p') \leq 3$ on average for this study for both A and I specimens. This parameter can be used to calculate q_f at failure of a specimen consolidated at $K_c < 1$ based on a value determined from $K_c = 1$ tests as $(q_{max})_{\varphi' \, calculated} = \varphi' \times (q_{max})_{K=1}$. An example of values determined by this method is shown in Fig. 26(b), which shows a good agreement between measured and calculated values. Fig. 27 shows a comparison of Critical State Lines based on calculated values compared to those that were measured. An appreciably close relation can be noted. However, these results show that a common CSL that would define failure conditions (i.e., $q = q_{max}$) irrespective of CSR

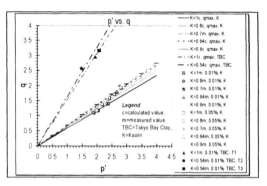

Fig. 27 Comparison of Critical State Lines based on calculated values compared to measured values.

on the p'–q plane would be a curvature and not a straight line.

4 CONCLUSION

The behaviour of kaolin was investigated from four basic aspects namely, CSR, OC, strain rate effects and their characterization within the Critical State Soil Mechanics. These results show that:-

(1) A "zone1" within the region of very small strains ($\varepsilon_a < 0.001\%$) exists even for soft clays. Within this region the behaviour is linear elastic and recoverable. It was shown that the behaviour in this zone, considered to be characterized by an inert state of particle contact, is insensitive to the effects of consolidation stress ratio, OCR and loading rate. It is considered that local interparticle yielding may start within the subsequent "Zone2".

(2) The behaviour of anisotropically consolidated clay cannot be adequately modelled within the framework of the conventional Critical State Soil Mechanics. Consequently, the existing theories based on these concepts need to incorporate some modification in modelling behaviour of clay that has been subjected to such consolidation stress histories.

(3) The modified approach of certain aspects of the CSSM proposed in this study yielded satisfactory results for the clays adopted.

ACKNOWLEDGEMENTS

The authors are highly indebted to Mr. Toda, Mr. Kimura, CPC staff, as well as Ruth Mukabi and Silvester Mukabi of Revtech Ltd.

REFERENCES

Akai, K., Adachi, T. and Ando, N. (1975): Existence of a unique stress-strain-time relation of clays, soils and Foundations, 15-1, pp. 1~16,

Ampadu, S.K (1988): The influence of initial shear on undrained behavior of normally consolidated Kaolin, Master Thesis, University of Tokyo.

Atkinson, J.H. and Bransby, P.L. (1978): The mechanics of soils-An introduction to critical state soil mechanics, Mcgraw Hill Publication Co-Ltd.

Burland, J.B (1990); On the compressibility and shear strength of natural clays, Geotechnique, 40-3 pp. 329~378

Casagrande, A. and Wilson, S.D. (1953) Effect of rate of loading on the strength of clays and shades at constant water content, Geotechnique, Vol. 2, pp. 251.

Graham,J., Crooks, J.H.A., and Lau, S.L.K. (1988)L: Yield envelopes: identification and geometric properties.

Hara, A. and Kiyota, Y. (1977): Dynamic shear tests of soils for seismic analysis, Proc. 9th Inter. Conf. On SMFE, Tokyo, Vol. 3, pp. 247-250

Kim, Y.S. Tatsuoka, F. and Ochi, K. (1994): Deformation characteristics at small strains of sedimentary soft rocks by triaxial compression tests, Geotechnique, 44-3, pp. 461-478

Konrad, J.M. (1985): Undrained cyclic behaviour of Beaufort sea silt, Proc. Conf. Arctic 85, ASCE, New York, N.Y., 830-837

Mukabi, J.N and Tatsuoka, F. (1994); Small strain behavior in triaxial compression of lightly over-consolidated Kaolin, Proc. 49th Annual Conf. Of JSCE, III, pp.296~297.

Mukabi, J.N. (1991): Behavior of clays for a wide range of strain in Triaxial compression, Msc. Thesis, University of Tokyo.

Mukabi, J.N. and Tatsuoka, F. (1992); Effects of consolidation stress ratio and strain rate on the peak stress ratio of Kaolin, the 27th Annual meeting of the JSSMFE, Kochi, pp. 655~648.

Mukabi,, J.N (1995): Deformation characteristics of small strains of clays in Triaxial, PhD Thesis, University of Tokyo

Schofield, A. and Wroth, p. (1968): Critical state soil mechanics, McGraw Hill publishing co. Ltd.

Skempton, A.W. (1985): Residual strength of clays in landslides, folded strata and the laboratory, Geotechnique 35, No. 1. 3-18

Tatsuoka, F. (1990): Some recent developments in triaxial testing systems for cohensionless soils, ASTM STP 977, p. 7-69.

Tatsuoka, F. and Kohata, Y. (1995): Stiffness of hard soils and soft rocks in engineering applications, Report of the IIS, University of Tokyo, Vol. 38, No.5.

Tatsuoka, F. and Shibuya, S. (1992): Deformation characteristics of soils and rocks from field and laboratory tests, Report of the IIS, University of Tokyo, Vol. 37, No.1

Whitman, Robert V. (1963): Some considerations and data regarding the shear strength of clays, ASTM STP No.361, Laboratory Shear Testing of Soil.

Geotechnics for Developing Africa, Wardle, Blight & Fourie (eds) © 1999 Balkema, Rotterdam, ISBN 90 5809 082 5

The bar linear shrinkage test – More useful than we think!

Phil Paige-Green & Dave Ventura
Division of Roads and Transport Technology, CSIR, Pretoria, South Africa

ABSTRACT: The bar linear shrinkage test (BLS) is a very simple, quick test to carry out and is a good indicator of the plasticity of a material. The rule of thumb that the BLS is approximately half of the plasticity index is not always true and this, together with the apparent poor reproducibility, resulted some years ago, in the BLS test being omitted from the manual of standard test methods (TMH1). Research since then has indicated that there is apparently a reason for the discrepancy in this rule of thumb and that the test is indicating more than just the plasticity. Refinements to the BLS test method, together with its incorporation into various specifications and prediction models, in preference to the plasticity index or liquid limit, have indicated that this, in fact, is a far more useful test than the other indicators of plasticity. A recently-developed, rapid method of carrying out the test using simple field equipment is introduced.

1 INTRODUCTION

The bar linear shrinkage (BLS) test (Method A4 in NITRR, 1979) is still carried out in conjunction with the Atterberg limits (liquid and plastic limit) as a matter of course in most soil engineering laboratories in South Africa, despite the test method having been removed from the current edition of TMH1 (NITRR, 1986).

The test is simple to carry out, requires minimal equipment and can give better precision (repeatability and reproducibility) than alternative tests. Research carried out on road materials over the last couple of decades has shown that the BLS is generally a better predictor of performance of road construction materials than the other traditional indicator tests such as liquid limit and plasticity index (PI). The PI, in particular, has a coefficient of variation of up to 75 per cent (Lay, 1981). A D2S% for the Liquid Limit of 25 per cent was determined by Weston (1978) and a value of 15 per cent by Sampson (1985). No D2S% for the PI was given by Weston but when the liquid limit and plastic limit were combined by Sampson (1985) to give the PI, a mean D2S% of 47 for 14 samples was obtained. In essence, the PI is determined by subtraction of two values which may differ in magnitude by as much as 400 per cent on a percentage basis (Heidema, 1957). Both of the input values for plasticity index could be in error due to many causes which cannot be continuously avoided and each one of the pair of these could be in error in opposite directions (Heidema, 1957). For this reason

the BLS test was mooted as long ago as 1959 by Wall (1959) as a control test for materials.

In this paper, the history of the BLS test in South Africa is summarised, and a case is made for the usefulness of the test. Improvements to the test method are proposed and a simplified method of carrying out the test in remote areas is suggested.

2 HISTORY

The earliest reference to the bar linear shrinkage test appears to be the discovery of the practical value of the results by the Texas Highway Department in about 1932 (Heidema, 1957).

The test reference to the BLS test in South Africa was in the *Standard Methods of Testing Materials and Specifications* published by the Department of Transport in 1958 (DOT 1958) which was an amendment of the 1948 version, a copy of which could not be located by the author. The reference to the test in this document was ASTM D427-39 and AASHO T92-42.

These tests both refer to the Shrinkage Factors from which the linear shrinkage is calculated, essentially from the cube root of the volumetric shrinkage.

The South African Department of Transport updated the Standard Methods of Testing Materials in 1971 (DOT, 1971) with no significant changes to the method but in addition to the ASTM and AASHO methods referenced in the earlier edition,

379

the California Standard Test Method 228-A was also referenced. The references to this test are the Soils Manual (USEO, 1942), Texas Highway Department (1953) and AASHTO T87 and T146 (California Department of Transportation, 1978). TMH 1 (NITRR, 1979) was issued by the Department of Transport (through the National Institute of Transport and Road Research) in 1979 with a reference only to the California Standard. Prior to revision of TMH 1 in 1986 (NITRR, 1986), discussion in the roads industry had indicated that the test was not very good as the results frequently differed from the "rule of thumb" that the BLS should be approximately half the PI (van Rooyen, 1967; Netterberg, 1969). Possibly as a result of this and more certainly as a result of the poor reproducibility of the results (Sampson, 1985), the test was omitted from the 1986 version of TMH 1 (NITRR, 1986).

The BLS as carried out in South Africa makes use of a standard shrinkage trough 150 mm long and 10 mm square cross-section. A sample of material passing the 40 mesh sieve (0.425 mm) is placed in the trough at a moisture content equivalent to the liquid limit and dried at 105 to 110°C until all shrinkage has stopped. The shrinkage of the material in the trough is then measured and expressed as a percentage of the trough length. No air-drying is allowed prior to testing.

Various other methods have been specified in other countries and make use of different troughs and techniques. The trough specified in the British Standard method (BSI, 1975) is 140 mm long and has a semi-circular cross-section with a radius of 12.5 mm. The drying process involves initial air-drying followed by drying at 60 to 65°C until shrinkage has "largely ceased" and then at 105 to 110°C until drying is complete.

The Texas Shrinkage Bar test (Texas DoT, 1993) and California method T228 (California DoT, 1978) utilize a 107 mm long trough with a 19 mm square cross-section and test materials "slightly more fluid" than their liquid limits (Heidema, 1957). A trough length of 127 mm has been described for this test by Senadheera et al, (1996). The material is air-dried slightly prior to testing to avoid bowing (Heidema, 1957; Senadheera et al, 1996).

As discussed above the linear shrinkage determined in the ASTM and AASHO (now AASHTO) methods is calculated from the volumetric shrinkage (V_s) as follows:

$$L_s = 100 \left[1 - \sqrt[3]{100/(V_s + 100)} \right]$$

No actual comparison of the results obtained by these two test methods has been carried out to the best of the authors knowledge. It has been shown, however, that the volumetric shrinkage which by definition occurs at a moisture content lower than the

plastic limit is frequently higher than the plastic limit, particularly with sandy and silty soils (Bauer, 1959; Netterberg, 1969; Bowles, 1988; Carter & Bentley, 1991; Sridharan &Prakash, 1998). Compounding these problems, various measurements of shrinkage are used. These are all determined during the shrinkage factors test (ASTM D427, 1982) and include the volumetric shrinkage, shrinkage limit, shrinkage index, shrinkage ratio, some of these parameters reflecting changes in volume whilst others represent moisture contents at which these changes occur. A full discussion on all of these parameters is beyond the scope of this paper.

Despite the exclusion of the test method from the current TMH 1 (NITRR, 1986), a survey (Sampson et al, 1992) indicated that all Road Authorities, major consulting firms and researchers include the test when Atterberg limits are determined. Two interviewees indicated that they did not use the test as a result of the poor reproducibility, but also to avoid the PI results being modified in the laboratory so that they are twice the BLS result.

It should be noted that irrespective of the consistency limit utilised, the test is carried out on a remoulded material (usually of limited grading) and the results are only directly applicable to similar materials i.e. borrow pit gravels, and not undisturbed in situ subgrade materials.

3 USEFULNESS

Work carried out on the use of calcretes in road construction in southern Africa (Netterberg, 1969) indicated that the bar linear shrinkage was the most reliable calcrete soil constant and this value weighted by the percentage passing the 0.425 mm sieve was used as one of the parameters required to specify calcrete base course materials. The weighted BLS was also used to specify gravel wearing course materials (Netterberg, 1978).

Paige-Green (1989) found that the BLS of gravel wearing course materials weighted by the percentage passing the 40 mesh sieve was the most significant indicator of plasticity/cohesion for all materials evaluated. This term was defined as the Shrinkage Product and limits were specified for gravel wearing course materials (CSRA, 1991).

Semmelink (1991), while developing models for the prediction of compactability, found that the BLS was one of the most significant material parameters. An evaluation of the importance of the BLS in the models to predict maximum dry density (MDD), optimum moisture content (OMC), zero air voids moisture content (ZAVMC) and critical moisture content (CMC) was carried out by Sampson et al (1992). The R-squared values of the prediction models decreased by between 0.2 and 0.3 by omitting the BLS parameter (also used as the weighted shrinkage (percentage passing 0.425 mm /100 x BLS)) from the models.

Analysis of the work carried out by Haupt (1980) and Emery (1985) during studies to determine subgrade moisture prediction models indicated that the inclusion of BLS produced as good, if not better, prediction models than the inclusion of any of the other Atterberg results. The initial correlation matrix which identified possible parameters to be included in the modelling process indicated that the BLS had a correlation coefficient of 0.755, the next best being PI at 0.753. Transformation of the BLS with the percentage fines (i.e. LS x P425$^{0.7}$) improved the correlation coefficient to 0.801. The best model to predict equilibrium moisture content included the transformed linear shrinkage parameter ($r^2 = 0.74$) whilst the use of BLS alone resulted in a model with an r^2 of 0.58, better than all other parameters except the OMC.

More recently the shrinkage product has been incorporated into the material selection process for emulsion treated bases (ETB) as an alternative to the PI (Verhaege et al, 1997).

In addition to this, current specifications and recommendations (CSRA, 1985; COLTO, 1996; COLTO, 1997; SATCC, 1997) all contain reference to and limits for the BLS.

The evidence thus suggests that the bar linear shrinkage test is a considerably more effective test to indicate material performance than the more traditional Atterberg limits.

It must be noted that no evaluation of the shrinkage limit is included in this report. The shrinkage limit is a more difficult test, not routinely specified in South Africa (it is included in BS 1377-Part 2 (BSI, 1985)) involving the drying of a standard soil pat and determination of its volume by mercury displacement (ASTM Method D427, 1982).

4 EVALUATION

As discussed earlier, the non-compliance with the rule of thumb that the BLS should be half the plasticity index was probably a contributory reason for the test being dropped from TMH 1 (NITRR, 1986). In order to investigate possible relationships between various parameters and the liquid limit a number of statistical analyses were undertaken.

A large number of test results on a wide range of materials was available for analysis. The relationships between the plasticity index, liquid limit, clay content and bar linear shrinkage of 940 results obtained from various laboratories were determined through correlation and regression analyses (Table 1).

Not surprisingly, reasonably good correlations between the Atterberg limits are obtained. However, significantly lower values are obtained from the correlations between the percentage of clay minerals

Table 1: Relationships between index properties

Properties					r
% < 2 μm and LL					0.652
% < 2 μm and PI					0.688
% < 2 μm and BLS					0.582
LL and PI					0.902
LL and BLS					0.815
PI and BLS					0.845
	Mean	Min	Max	σ	CoV
BLS/PI	0.449	0.13	1.67	0.165	0.368
BLS/LL	0.219	0	0.629	0.065	0.297

and the Atterberg limits. The average ratio between BLS and PI was found to be 0.449, close to the "expected" value of 0.5 but the standard deviation and range were large. The ratio of the BLS to LL was 0.22 with a much smaller standard deviation and Coefficient of Variation indicating that this ratio would probably make a more appropriate rule of thumb measurement.

A second database (where detailed clay mineralogy was available) was then evaluated in terms of the effect of the clay mineralogy on the PI and BLS relationships. The results of Atterberg limit testing, hydrometer analyses and detailed clay mineralogy were available for about 100 samples.

Difficulties were encountered in the analyses when non-plastic materials were involved. It was interesting to note that a number of materials which were non-plastic, i.e. either the liquid limit or the plastic limit could not be obtained by conventional methods, had measurable linear shrinkages. This was reported by Netterberg (1969) for calcretes where shrinkages up to 6 per cent were measured on materials which were classified as slightly plastic during laboratory testing.

5 DISCUSSION

An important characteristic of clay-bearing soils that will be used in the ensuing discussion is the concept of "activity", defined as the ratio of plasticity index to clay content i.e. that fraction of a soil material finer than 0.002 mm in diameter (Carter & Bentley, 1991). Kaolinite has an activity of 0.3 to 0.5; illite of about 0.9 and montmorillonite typically has a value in excess of 1.5. A high activity indicates the propensity of the clay mineral to adsorb relatively large amounts of water and is thus an indication of the swelling potential of the material (and consequently the shrinkage capacity on drying). The activity is strongly influenced by the exchangeable cation (Grim, 1962).

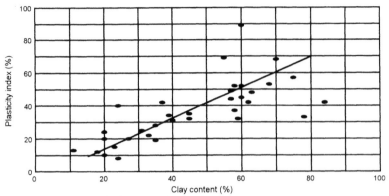

Figure 1: Relationship between PI and clay content (Croney, 1977)

Croney (1977) showed that some relationship existed between the plasticity index and the clay content of various materials (Figure 1) and suggested that the scatter was the result of the clay mineralogy of the material. Materials with active clay components such as the smectite minerals plot above the line and those with less active clays such as the kaolinite minerals plot below the line.

The relationships for a number of South African materials where clay mineralogy had been carefully quantified (mostly from Paige-Green, 1989) was evaluated. These results are shown in Figures 2 - 4 for the three major clay mineral groups commonly found in South Africa (the estimated proportions of the specific clay mineral are labelled at each point on the figure: 0 - absent; 1 = <5%, 2 = 5-15%; 3 = 15-30%; 4 = >30%)). Approximate activity lines (A) for these clay mineral groups are also plotted on the figures.

Although certain trends are evident, the confounding effect of the combination of different clays in the individual materials makes analysis difficult. Nonetheless, the materials with large kaolinite components generally plot below the A = 1 line, those with high smectite components, generally above the A = 1.5 line and illite-rich materials around the A = 1 line.

An analysis of the effect of the clay mineralogy on the ratio between the plasticity index and bar linear shrinkage has been carried out in the same way. The results of these are plotted in Figures 5 to 7.

The relationship is significantly better than that for the clay PI relationship although a significant spread of results is evident. Similar trends regarding the clay mineral type can be noted but the relationships are not as clear. It is thus concluded that the clay mineralogy is probably not the only cause of the scatter around the BLS:PI ratio of 0.5 (the reciprocal, 2, in the figures as plotted), but the arror associated with testing described in the next section also appears to be a contributory factor.

Netterberg (1969) pointed out that the nature of the exchangeable cation of clays, other than smectites, has only a negligible effect on the PI of the materials. Exchangeable sodium ions in the smectite mineral structure result in a significantly higher PI than calcium ions. On this basis, a similar conclusion can be drawn for the BLS of smectite rich clays.

It appears from the literature, however, that the gradation of the particles in the fine fraction of the material are probably the main cause of irregularities in the relationships. Grim (1962) shows a strong relationship between particle size (represented by specific surface area in m^2/g) and plasticity index. This has also been identified as having a significant impact on the shrinkage limit of soils (Sridharan & Prakash, 1998).

6 PROBLEMS WITH THE BLS TEST

6.1 Existing method
Various problems with the bar linear shrinkage test as described in TMH 1 were identified (Sampson et al, 1992) and a project to quantify these and propose suitable modifications for their improvement was carried out.

The use of a slot rather than a trough in order to eliminate (or at least reduce) bowing during the BLS test was first described in 1959 (Wall, 1959) and subsequently evaluated by Newill (1961) and Sampson et al (1992). This allows the material being tested to dry from two sides and makes the drying process (i.e. moisture loss) more even.

6.2 Repeatability and reproducibility
The linear shrinkage is repeatable to better than one percentage unit (Wall, 1959). Heidema (1957) indicates that the accuracy of the BLS closer than 2 per cent cannot be assured i.e a BLS of 20 may vary from 19.6 to 20.4 per cent. This is considerably better than the results possible from the PI test.

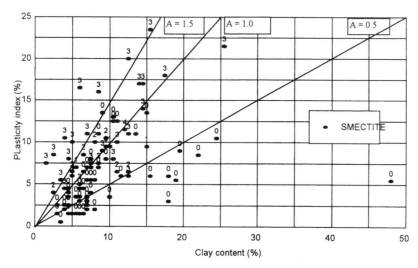

Figure 2: PI versus clay content by quantity of kaolinite

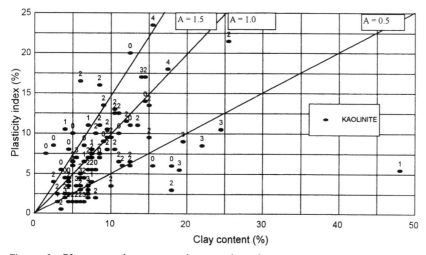

Figure 3: PI versus clay content by quantity of smectite

Work by Sampson (1985) indicated that the mean inter-laboratory precision of the BLS was 34.1 per cent, which although poor was considerably better than that of the plasticity index (46.8 per cent). Follow-up discussions with the participating laboratories indicated various problems leading to the poor reproducibility including (Sampson *et al*, 1992):
  cracking and bowing;
  variations in drying;
  variations as a result of the correction formula;
  lubrication of the troughs, and
  filling of the troughs.

The effect of high temperatures during drying (e.g. from heating on a sand bath to speed up the test) was not critical and it is not necessary to dry to constant mass as the shrinkage ceases at the shrinkage limit (Wall, 1959).

The effect of variations in the initial moisture content was found to be negligible provided that the material was approximately at the liquid limit (between 20 and 30 blows for closure in the Casagrande bowl) (Wall, 1959). Heidema (1957) in fact suggests that the material should be slightly wetter than the liquid limit at the start of the test.

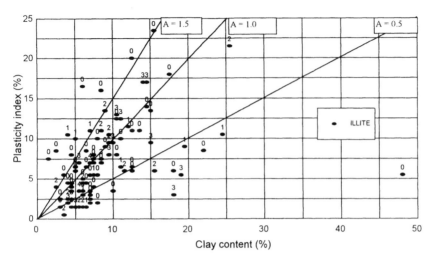

Figure 4: PI versus clay content by quantity of illite

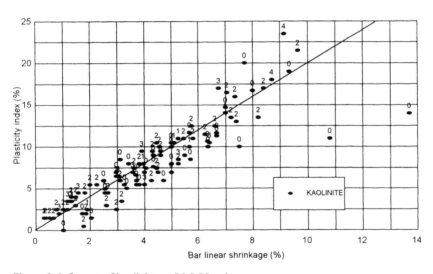

Figure 5: Influence of kaolinite on BLS:PI ratio

6.3 Recommended modifications to test method

Following the evaluation of the test method and the problems carried out by Sampson *et al* (1992), various recommendations regarding the test method were made.

The use of the slotted "trough" significantly reduces the bowing effect without affecting the result compared to the closed trough and was thus recommended by .

The drying method should not deviate from the standard method as air-drying prior to placing the trough in the oven reduces the BLS result significantly.

The correction formula given in TMH 1 (NITRR,

1979) should only be used for soils with a PI of less than 24 - a different factor is necessary for materials with PIs of between 24 and 30 and no correction is required for materials with a PI higher than 30.

Lubrication of the trough or slot is best carried out using a spayed silicone lubricating oil.

7 BLS TEST FOR LABOUR-BASED PROJECTS

A project to develop simple techniques for the control of materials in labour-based projects carried out for the International Labour Office (Paige-Green, 1998) resulted in a simplified test for the BLS,

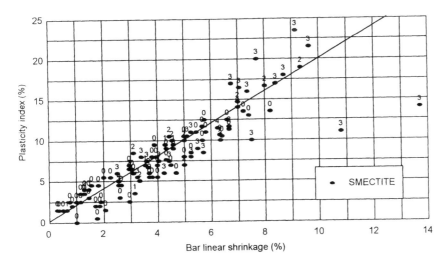

Figure 6: Influence of smectite on BLS:PI ratio

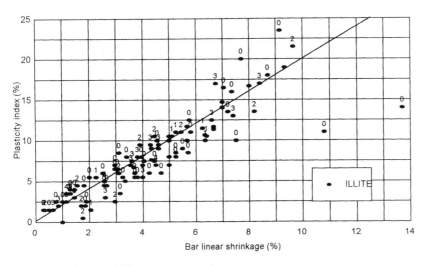

Figure 7: Influence of illite on BLS:PI ratio

among others. The test makes use of a simple drop-cone device modelled on the BS liquid limit method (BSI, 1975; Method 2a), to check that the material placed in the BLS shrinkage trough is at the correct moisture content, i.e. at the liquid limit, without actually determining the liquid limit. The material is placed in a slotted mould and dried in a solar oven. The apparatus necessary for this test, together with various other control tests is supplied as a kit ready for use where no electrical power is available.

A comparison of the results obtained from this method with the standard TMH 1 method (NITRR, 1979) on four materials typically used for unsealed roads produced the following results (Table 2).

These are considered to correlate adequately for this type of project.

Table 2: Comparison of results between TMH 1 and simplified BLS methods

Method	TMH 1	Simplified
Material	Bar linear shrinkage (%)	
Diabase	6.0	5.3
Andesite	4.2	3.6
Granite	5.0	4.0
Chert	5.2	4.7

8 CONCLUSIONS

Research and experience have shown the potential benefits of the bar linear shrinkage test for material classification and performance prediction in road design and construction.

The test as currently used in South Africa suffers from certain shortcomings, but modifications to the test method which will minimise or reduce these problems have been proposed and should be implemented as soon as possible.

There is no doubt that the test indicates more than just the shrinkage and any variation in results from the expected "half of the PI" more likely result from errors in the PI than the BLS. The effect of the clay mineralogy, nature of the exchangeable cation component of the clays and the particle size gradation of the fine proportion of the soil all influence the bar linear shrinkage.

Determination of the bar linear shrinkage, however, is quick, low-cost and more reproducible than the other Atterberg limits and it is suggested that far more use be made of the bar linear shrinkage. The test is less susceptible to operator error and requires significantly less training and experience than the traditional Atterberg limits.

9 REFERENCES

American Association of State Highway and Transportation Officials (AASHTO). 1982. *AASHTO Materials Part 2: Tests.* Washington, D.C: AASHTO, 13th Edition,

American Society for Testing Materials (ASTM). 1982. *Annual book of ASTM Standards,* Philadelphia: ASTM, Vol 11.

Bauer, E.E. 1959. *History and development of the Atterberg Limits tests.* Papers on soils - 1959 Meetings, ASTM STP 254, American Society for Testing Materials, pp 160-167.

Bowles, J.E. 1988. *Foundation analysis and design.* London: McGraw Hill.

British Standards Institution (BSI). 1975. *Methods of test for soils for engineering purposes.* London: BSI. BS 1377.

California Department of Transportation. 1978. *Standard test methods; Vol 1,* Sacramento: California DoT, 3rd Edition,.

Carter, M & Bentley, S.P. 1991. *Correlations of soil properties.* London: Pentech Press.

Committee of State Road Authorities (CSRA). 1991. *The structural design, construction and maintenance of unpaved roads.* Pretoria: CSRA, Technical Recommendations for Highways (Draft TRH 20).

Committee of Land Transport Officials (COLTO). 1996. *Structural design of flexible pavements for interurban and rural roads.* Pretoria: COLTO, Technical Recommendations for Highways (Draft TRH 4).

Committee of Land Transport Officials (COLTO). 1997. *Standard specifications for roads and bridges.* Pretoria: COLTO.

Croney, D. 1977. *The design and performance of road pavements.* London: HMSO.

Department of Transport (DOT). 1958. *Standard methods of testing materials and specifications.* Pretoria: Division of Civil Engineering, Department of Transport.

Department of Transport (DOT). 1971. *Standard methods of testing materials.* Pretoria: Division of National Roads, Department of Transport.

Emery, S.J. 1985. *Prediction of moisture content for use in pavement design.* PhD thesis, University of Witwatersrand, Johannesburg.

Grim, R.E. 1962. *Applied clay mineralogy.* New York: McGraw-Hill.

Haupt, F.J. 1980. *Moisture conditions associated with pavements in southern Africa.* MSc Thesis, University of Witwatersrand.

Heidema, P.B. 1957. The bar linear shrinkage test and the practical importance of bar linear shrinkage as an identifier of soils. *Proc 4th International Conference on Soil mechanics and Foundation Engineering,* London, Vol 1, pp 44-49.

Lay, M. 1981. *Source book for Australian Roads.* Melbourne: Australian Road Research Board.

National Institute for Transport and Road Research (NITRR). 1979. *Standard methods of testing road construction materials.* Pretoria: NITRR, CSIR.

National Institute for Transport and Road Research (NITRR). 1986. *Standard methods of testing road construction materials (2nd edition).* Pretoria: NITRR, CSIR.

Netterberg, F. 1969. *The geology and engineering properties of South African calcretes.* PhD thesis, University of Witwatersrand, Johannesburg.

Netterberg, F. 1978. *Calcrete wearing courses for unpaved roads.* The Civil Engineer in South Africa, Vol 20, No 6, pp 129-138.

Newill, D. 1961. *An investigation of the linear shrinkage test applied to tropical soils, and its relation to the plasticity index.* Crowthorne: Road Research Laboratory, Research Note RN/4106/DN.

Paige-Green, P. 1989. *The influence of geotechnical properties on the performance of gravel wearing course materials.* PhD thesis, University of Pretoria, Pretoria.

Paige-Green, P. 1998. *Material selection and quality assurance for labour-based unsealed road projects.* Pretoria: Division of Roads and Transport Technology, CSIR, Contract Report CR-97/047.

Sampson, L.R. 1985. *A reproducibility study of Atterberg limits determined by the Casagrande methods and by an extended British cone penetrometer method.* Pretoria: National Institute for Transport and Road Research, Technical Report RP/5.

Sampson, L.R., Ventura, D.F.C. & Kalombo, D.K.

1992. *The linear shrinkage test: justification for its reintroduction as a standard South African test method.* Pretoria: South African Roads Board. Research Report 91/189.

Senadheera, S.P., Jayawickrama, P.W & Ashek Rama, A.S.M. 1996. Use of hydrated flyash as a flexible base material. In: *Issues in Geotechnical Engineering*, Washington: National Academy of Sciences (TRB 1546), pp 53-61.

Semmelink, C. 1991. *The effect of material properties on the compactability of some untreated roadbuilding materials.* Ph D thesis, University of Pretoria, Pretoria.

Southern African Transport Coordinating Committee (SATCC). 1997. *SATCC Standard Specifications for Roads and Bridge Works.* Maputo: SATCC, (Division of Roads and Transport Technology, CSIR, Draft Contract Report CR-97/103)

Sridharan, A & Prakash, K. 1998. *Mechanism controlling the shrinkage limit of soils.* Geotechnical Testing Journal, Vol 21, No 3, pp 240-250.

Texas Department of Transport (Texas DoT). 1953. *Standard specifications for construction of Highways, Streets and Bridges.* Austin: Texas DoT.

Texas Highway Department. 1953. *Soil testing procedures.* Austin, Texas:Texas Highway Department.

United States Engineering Office (USEO).1942. *Soils Manual.* Galveston, Texas: US Engineering Office,.

Van Rooyen, M. 1967. *Intercorrelation of engineering properties of soil.* Proc 4th Reg Conf Africa on Soil Mechanics and Foundation Engineering, Cape Town, pp 203-208..

Verhaege, B.M.J.A, Theyse, H & Vos, R.M. 1997. Emulsion-treated bases: a South African perspective. *Proc 10th AAA International Flexible Pavements Conference,* Perth, Australia.

Wall, GD. 1959. Observations of the use of lineal shrinkage from the liquid limit as an aid to control in the field. *Proc 2nd Regional Conference for Africa on Soil Mechanics and Foundation Engineering,* Lourenco Marques, pp 175-180.

Weston, DJ. 1978. *A comparison of the Casagrande cup and the BS cone methods for determining the liquid limit of soils and discussion of some of the uses of Atterberg limits.* Pretoria: National Institute of Transport and Road Research, Report RS/5/78, CSIR.

Geotechnics for Developing Africa, Wardle, Blight & Fourie (eds) © 1999 Balkema, Rotterdam, ISBN 90 5809 082 5

Use of the Osterberg Cell for pile load evaluation

A. L. Parrock – *ARQ Specialist Engineers, Pretoria, South Africa*

H. de Wet – *Portnet, Johannesburg, South Africa*

W. F. Hartley – *Ground Engineering, Johannesburg, South Africa*

J. A. Hayes – *Loadtest, Gainesville, Fla., USA*

ABSTRACT: The piling contract for the Richards Bay Dry Bulk Jetty was awarded to the LTA-Interbeton-DIC Joint Venture (LID-JV) and provided for the installation of 1 200 and 1 800mm diameter vertical piles. As part of the verification process for the load carrying capacity, two working piles of the jetty were tested to in excess of 1.5 times the specified working load.

The piles, permanently cased reinforced concrete, were installed from floating barges at sea level, through approximately 20m of water, between 17 and 43m of alternating bands of sand, silt and clay, and socketed into a low strength cretaceous layer.

In their tender submission, the LID-JV proposed the use of Osterberg Cells to determine the pile load/deflection characteristics.

This paper describes the characteristics of cretaceous mudstone at Richards Bay, design predictions, the installation/load testing of piles using the Osterberg Cell and back analysis of the results obtained.

1 RICHARDS BAY DRY BULK JETTY

In order to increase the export capacity of the Richards Bay Harbour, a Dry Bulk Jetty was constructed within the harbour confines. The postulated design for the jetty founding called for permanently cased concrete-filled piles to a depth of 20m below sea-bed level, continuing to the required founding level as nominally reinforced uncased piles, constructed under a drilling mud. The piles were designed to transmit the full loads to the rock socket, with most of the resistance attributed to side friction in the socket. (Portnet 1997)

2 PILE SELECTION, ANALYSIS, DESIGN AND SPECIFICATION

The overall jetty dimensions are 300m in length, sufficient for the berthing of 65 000 tonne dead weight bulk vessels either side, and 65m in breadth for the accommodation of a conveyor gallery with bulk ship loaders on each side. The structure was therefore required to resist large loads applied both vertically and horizontally. The use of a gravity structure was excluded by the poor founding conditions, whilst the use of raked piles was deemed unsuitable in view of the 20m depth of water and the great depth to bedrock. The structural system comprises 4 rows of piles in the transverse direction at 21m centres, the inner rows having 1 800mm diameter piles, with 1200mm diameter piles in the outer rows. The piles are spaced at 10m in the longitudinal direction.

To resist mooring and berthing forces by piles acting in flexure without mobilising excessive sway, it was necessary to design the deck as a single unit without expansion joints. In order to accommodate the resulting temperature movements at the extremities of the structure, the piles needed to have pinned connections to the deck to yield adequate flexibility in the longitudinal direction. In the transverse direction, however, maximum flexural rigidity was required in order to limit sway movement under berthing loads. In this direction the piles therefore have moment fixity at the deck connection.

In the light of the relatively high bending

moments resulting from the structural constraints, and in consideration of the installation conditions, the use of structural steel casings was an obvious choice. Wall thicknesses of 22mm and 16mm were specified for the 1 800mm and 1 200mm diameter piles respectively. These thicknesses were dictated by driving requirements rather than structural capacity. These casings were driven generally to a depth of 40m below Chart Datum (CD), the mean sea water level. The bending moments below this level were not significant, requiring only nominal reinforcing. The 20m embedment was also adequate to ensure pile stabilities in their unsupported state before construction of the deck. Installation below this level was to be by excavation under drilling fluid.

Pile socket design was carried out in accordance with the method of Williams, Johnston and Donald (1980). Utilising a rock modulus of 450MPa and a rock strength of 2.6MPa, a socket length of 6m was found to be satisfactory, provided a moderate tolerance could be accommodated on settlement. The question of differential settlement was already problematical in the light of rapidly varying bedrock depths in the range -25m to -68m CD. Whilst the 21m deck transverse spans were able to accommodate moderate settlement, the very stiff, short longitudinal spans could not. Any allowance for even the slightest settlement resulted in a rapid escalation of reinforcement requirements. The economic option was to provide flexibility in the short spans by way of providing rotational capacity. Consequently, it was not necessary to specify a tight settlement criterion on the piles, which would have required a deeper socket. These were the motivating factors for the postulated pile design.

3 POSTULATED PILE DESIGN

The tender document for the Bulk Jetty piles instructed that "the following key service loadings and deflection parameters shall be adopted."

Cased piles 1800mm diameter
i)	Axial Load (kN) - less self weight	14 050
ii)	Applied Bending Moment (kN-m)	5 100
iii)	Applied Lateral Deflection (mm)	40
iv)	Max Permissible Axial deflection (mm)	20

Cased piles 1200mm diameter
i)	Axial Load (kN) - less self weight	6 800
ii)	Applied Bending Moment (kN-m)	2 400
iii)	Applied Lateral Deflection (mm)	40
iv)	Max Permissible Axial deflection (mm)	20

4 CONTRACT

The contract for the works was awarded to the LTA-Interbeton-DIC Joint Venture (LID-JV). Interbeton based in Rijswijk, Netherlands is an international contractor with extensive experience in marine works and piling. LTA and its geotechnical engineering arm GEL are a South African civil engineering company while DIC is an emerging South African development company.

5 CONSTRUCTION

The contract valued at more than R50-million, involved the installation of 116 piles in four rows, extending 300m from the existing quay wall into the harbour. The central two rows of piles were 1.8m diameter while the outer two rows were 1.2m diameter. A typical construction sequence was:

Driving: a 42m length of 22mm wall thickness casing weighing some 20 tonnes was floated to the driving barge and positioned using a 225t crane. The casing was driven to a depth of approximately 40m below sea level with a Delmag D100 diesel hammer.

Drilling: this was carried out from a second floating platform using a Casagrande RM21 drill rig mounted on a 90t crane drilling inside the steel casing to a depth up to 65m below sea level. The uncased length of hole in the cretaceous mudstone was supported by a degradable environmentally friendly vinyl polymer slurry.

Steel and Concrete: A 45t crane placed reinforcing steel and shaft concrete from a third floating platform. High slump concrete was batched on site in a 30m³/hour plant, pumped to the edge of the quay wall and transported to the piles in 10m³ floating spinner drums and placed through a tremie extending to the bottom of the pile. Figure 1 details the general construction set up on the site.

6 OSTERBERG CELL TEST PILES

In their tender submission, the LID-JV proposed the use of Osterberg Cells to determine pile load/deflection characteristics.

The first Osterberg Load Cell (also called the O-Cell) was used for pile load tests as early as 1988. Since that date, some 300 tests have been conducted in 10 countries on drilled shafts, barrette piles, driven pipe piles and driven precast concrete piles up to depths of 90m, diameters up to 3m and loads

Figure 1: Positioning casings prior to driving (background) - drilling under vinyl polymer slurry (foreground)

up to 133MN. The O-Cell when placed on or near the bottom of a drilled shaft and pressurized internally, applies an equal upward and downward load, thus enabling side shear and end bearing to be determined separately. There is no need for dead load nor hold down mechanisms to react against, as in conventional tests.

The O-Cell essentially consists of a specially designed hydraulic jack-like device capable of exerting very large loads at high internal pressures. If the cell is inserted at the base of the pile, a small amount of concrete is placed on the bottom of a drilled shaft after which the O-Cell is lowered into the hole which is then filled with concrete. A pipe welded to the top of the centre of the cell extending to above the ground *or water* surface acts as a conduit for applying fluid pressure to the previously calibrated cell. Inside the pipe is a smaller pipe connected to the bottom with an open end. It extends to the surface and emerges from the large pipe through an O-ring. This pipe acts as a tell-tale to measure the downward movement of the bottom of the O-Cell as the load is applied. After the concrete has reached its desired strength, the cell is pressurized internally creating an upward force on

the concrete pile shaft and an equal but opposite force in end bearing. As the pressure increases, the telltale moves downwards as the load in end bearing increases and the shaft moves upwards as the side shear on the shaft is mobilised. Upward and downward movement of the various components can either be measured directly from the tell-tales or inferred from strain gauges attached to the pile steel reinforcement. Based on a description given by Osterberg (1998).

Two O-Cells were installed within working piles; one in a 1 200mm diameter pile and the other in an 1 800mm diameter pile. Figure 2 depicts one of the O-Cells prior to installation at the base of the pile, with Figure 3 illustrating some of the instrumentation attached to the test pile prior to installation. Clearly visible are the strain gauges and the tell-tale rods.

7 DESIGN CALCULATIONS

To enable the optimal position of the cells within the cretaceous rock socket to be determined, it was essential that predicted load deformation curves be generated for the performance of the pile under load tests. This was determined as follows.

8 DESIGN PARAMETERS

The data contained in Volume 3 of the Tender Document (Portnet 1997) was used in the prediction of representative design values.

8.1 Depth to cretaceous

The data contained in the profiles of the boreholes sunk on site, was evaluated to determine the depth below chart datum at which the cretaceous layer occurred. The SPT values recorded with depth were input to a spreadsheet. The deepest point at which the SPT registered a value in excess of 100 was adjudged to be the start of soft rock within the cretaceous layer. This data depicted in bar graph format in Figure 4 indicates that the +100 SPT material starts at CD depths between 37 and 63m.

8.2 Strength of cretaceous

Uniaxial compressive strength (UCS) tests on undisturbed samples retrieved during the

Figure 2: O-Cell prior to installation

Figure 3: Strain gauges and tell-tale rods attached to reinforcement

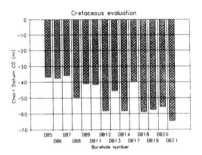

Figure 4: Chart Datum depth at which SPT > 100

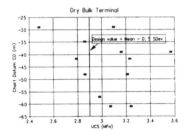

Figure 5: UCS vs CD depth

investigation, Portnet (1997), are depicted in graphical format in Figure 5.

As per the recommendations given by Simpson (1997), a representative value of UCS at 0.5 standard deviations below the mean was chosen as the design value.

8.3 Pile/cretaceous interface shear

8.3.1 A simple method of calculating the capacity of rock sockets for piles, involves estimating the undrained shear strength as 0.5 x UCS and using an α-factor (ratio of the shear strength of the insitu material to the adhesion to the side of the pile) of 0.6 to obtain the socket/rock interface shear. This translates into using a value equal to 30% of the UCS. (Knight 1998).

8.3.2 Figure 3.27 of Poulos and Davis (1980:41) indicates adhesion values which vary between 2 and 30% of the UCS for rocks of less than 10MPa strength.

8.3.3 Webb and Moore (1980), from tests conducted on sandstone with UCS' in the range 2.24 to 2.82MPa generated adhesion values which varied from 16.3 to 35.9% of the UCS.

Significantly, they demonstrated that for the *full* adhesion value to be generated, the ratio Length of socket : Diameter of socket, should exceed 3.

8.3.4 The recommendations contained in chapter 2.6.4d *Skin Friction and End Bearing of Grouted*

Piles in Rock of the publication by the American Petroleum Institute API RP2A (1987:51) state "The unit skin friction of grouted piles in jetted or drilled holes in rock should not exceed the triaxial shear strength of the rock or grout, but in general should be much less than this values based on the amount of reduced shear strength from installation."

8.3.5 Quoting from the results obtained from Webb and Allinson (1980), Brink (1985:45) states "In undrained triaxial tests average values, expressed in terms of total stress are c = 340kPa and ϕ = 45°".

8.3.6 Using the formulations contained in Hoek (1995) for the derivation of the Geological Strength Index (GSI), the Rock Mass Rating (RMR$_{76}$) as advocated by Bieniawski (1976), the attributes of the cretaceous layer as given in Portnet (1997) for boreholes 7,8,11 and 12, generated representative RMR$_{76}$-values at the (Mean-0.5*Standard Deviation) level (Simpson 1997) as depicted in Table 1.

For RMR$_{76}$-values > 18, the Geological Strength Index (GSI) = RMR$_{76}$. Using the formulations contained in Hoek (1995) and the spreadsheet HOEK95 encoded to calculate those formulations, the input parameters as contained in Table 2 were used to derive the Hoek Brown and Mohr Coulomb shear parameters of the cretaceous layer as detailed in Table 3.

At the envisaged depth of the O-Cells, the shear strength of the insitu material utilising the above formulations was calculated at 218kPa.

Based on the above formulations:

a) a lower bound adhesion value was postulated as: 15% of the lowest recorded UCS ie 15% of 2400kPa = 360kPa

b) a best estimate adhesion value was postulated as: 25% of the design UCS ie 25% of 2 900kPa = 725kPa

c) an upper bound adhesion value was postulated as: 35% of the design UCS ie 35% of 2 900kPa = 1 015kPa

8.4 *End bearing capacity*

As per the recommendations contained in table 2.6.4.1 Design Parameters for Cohesionless Siliceous Soils of API RP2A (1987:51), the material at a minimum CD of -43m (the bottom of a 6m socket) was conservatively assigned an N$_q$-value of 50.

Utilizing the buoyant density of 6.4kN/m^3 as per Table 3, the ultimate end bearing stress at -43m CD was calculated at 7 360kPa. This value is less than 9.58MPa, the recommended maximum for rock as specified in API RP2A (1987:51).

Table 1: Rock Mass Rating

Variable	Value
Strength - UCS = 2,9MPa	1
RQD = 42	8
Joint spacing = 0,8m	20
Joint condition - Openings = 5mm	6
Water	10
TOTAL	45

Table 2: Hoek's (1995) input parameters for derivation of shear strength

Variable	Value
Geological Strength Index - GSI	45
Mudstone - m$_i$	9
UCS	2,9

Table 3: Hoek Brown and Mohr Coulomb shear parameters

Variable	Value
m	1,262
s	0,002
a	0,5
c (kPa)	59
ϕ (°)	35,3
γ' (kN/m^3)	6,4

9 DESIGN

9.1 *Pile capacity in the transported materials above the Cretaceous*

Using the formulations as contained in API RP2A (1987) and the data from boreholes DB5, 6 and 7 of the geotechnical investigation (Portnet 1997), the ultimate capacity in shear of the material above the cretaceous, for the two pile sizes, was derived as per Table 4.

9.2 *Ultimate capacity of Cretaceous rock socket*

The ultimate capacity in shear and end bearing for the two pile sizes was derived for a 6m socket length as per Table 5.

The ultimate capacity in shear and end bearing

Table 4: Ultimate capacity (MN) in shear for material overlying Cretaceous

Pile diameter (mm)	Borehole Numbers		
	DB5	DB6	BD7
1 200	2,3	2,9	2,2
1 800	3,5	4,2	3,4

Table 5: Ultimate capacity (MN) in shear plus end bearing for Cretaceous rock socket of 6m length

Pile diameter (mm)	Condition		
	Lower bound	Best estimate	Upper bound
1 200	16,6	24,7	31,3
1 800	30,9	43,3	53,2

Table 6: Ultimate capacity (MN) in shear and end bearing for Cretaceous rock socket of 8,5m length

Pile diameter (mm)	Condition		
	Lower bound	Best estimate	Upper bound
1 200	19,9	31,6	40,8
1 800	36,0	53,6	67,5

Table 7: Factors of safety on predicted ultimate capacity for Cretaceous rock socket of 6m length

Pile diameter (mm)	Condition		
	Lower bound	Best estimate	Upper bound
1 200	2,4	3,5	4,5
1 800	2,2	3,1	3,8

Table 8: Factors of safety on predicted ultimate capacity for Cretaceous rock socket of 8,5m length

Pile diameter (mm)	Condition		
	Lower bound	Best estimate	Upper bound
1 200	2,8	4,5	5,8
1 800	2,6	3,8	4,8

for the two pile sizes was derived for an 8.5m socket length as per Table 6.

Many of the above capacities are in excess of the crushing strength of the concrete that was used in the manufacture of the pile and the loads in excess of this value are thus only theoretical.

For a working load of 7MN for the 1.2m pile and 14MN for the 1.8m pile, the factors of safety on ultimate load are detailed in Tables 7 and 8.

9.3 Position of O-Cell

The optimal position of the O-Cell will occur when the resistance of the material below the cell, at a given deflection limit, is equal to the resistance of the material above the cell.

The load deflection curve for the rock socket portion of the pile was assessed using the transfer functions as advanced by Everett (1991). Figures 6 and 7 detail typical predicted capacity in side friction and end bearing for 1.2 and 1.8m ϕ piles.

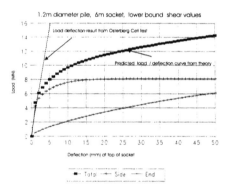

Figure 6:

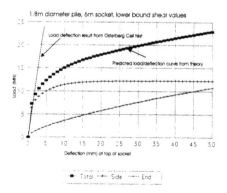

Figure 7:

Table 9: Predicted versus specified top of socket deflection under working load

Pile diameter (mm)	Deflections (mm)	
	Specified	Calculated
1 200	12	4
1 800	18	8

Table 10: Predicted versus specified top of socket deflection under 1.5 x working load

Pile diameter (mm)	Deflections (mm)	
	Specified	Calculated
1 200	20	13
1 800	30	31

The side friction is generated at a very low deflection levels while end bearing requires much larger movements of the base until the full load is developed.

From figures 6 and 7, the specified versus calculated deflections at the top of socket level, under working load, for the two pile types (1 200 and 1 800mm diameter piles) were derived as per Table 9.

Under 1.5 x the working load, the specified versus calculated deflections, at the top of socket level for the two pile types (1 200 and 1 800mm diameter piles) were derived as per Table 10.

9.4 Optimal position of O-Cell

Cognisance must be taken of the fact that end bearing and skin friction loads develop at differing deflection levels, in establishing where the O-Cell should be located within the rock socket. If the O-Cell is placed to close to the base, large deflections will be required until sufficient force is generated to shear the socket. If the O-Cell is placed to close to the top of the rock socket, there will be insufficient resistance above the cell to generate any load in the socket and base below.

Thus an iterative procedure was adopted whereby a location was chosen for the O-Cell and the resistance of the material above the cell was calculated as well as that below at a pre-determined deflection limit.

Utilising the above procedure and an arbitrarily chosen 20mm deflection limit, the optimal position of O-Cell for the 1.8m diameter pile and 8.5m long

socket was established as 4.7m from the top. However due to the fact that 3 pile diameters are required for the generation of full side shear, (See point 8.3.3), it was reasoned that this value should be increased to 5.4m (ie 1.8 x 3) leaving some 3m of socket below the cell. In this way it was known that the maximum shear would have been established in that portion of the socket above the cell. Even without the conclusions of the manipulations performed on the results of the strain gauges, this shear value would be the *maximum* value that could occur on the rock/socket interface below the O-Cell (as the L/d ratio for this portion is < 3) and correspondingly the *minimum* value of end bearing would be assessed.

This procedure would thus ensure that at the worst, conservative values would be derived for end bearing below the O-Cell and correct values for side shear above the cell. (As the L/d ratio for this portion of the socket = 3)

If the same position is adopted for the O-Cell in the 1 200mm ϕ tests, the condition of at least 3 diameters above the O-Cell would be satisfied (in fact 4.5) while below a value of L/d = 2.5 would be obtained. This latter value is close enough to the required limit of 3 not to warrant any further discussion.

10 INSTALLATION AND TESTING

Test piles D8 and B12 were installed as per the details given in numerical format in Table 11.

Clearing the base of the pile of loose material generated during the auger process was accomplished under slurry using a cleaning bucket.

Modulus values were back-calculated from the results of concrete cubes tests conducted at various ages.

The piles were tested via the O-Cell to 2.47 times the specified working load for pile D8 and 1.74 times for pile B12. Failure in neither side shear nor end bearing was induced in either of the test piles; test pile D8 deflecting a little over 6mm downwards at the base of the O-Cell and some 7.5mm upwards at the top of the O-Cell. The corresponding values for pile B12 were 6mm downwards and 4mm upwards.

Analysis of the movements recorded by the strain gauges and tell tales enabled the shear transfer in side friction and the end bearing stress at the tip of the piles to be assessed. The results of these calculations, which took the mass of the pile into account, are depicted in graphical format together with the relevant soil profiles in figures 8 and 9,

Table 11: Details of test piles installed

Pile Number	D8
Date cast	26 August 1998
Nominal shaft diameter (+0.55 to -38.03m CD)	1.22m
Nominal shaft diameter (-38.03 to -51.85m CD)	1.15m
Shaft modulus (+0.55 to -5.65m CD)	36.0GPa
Shaft modulus (-5.65 to -38.03m CD)	33.3GPa
Shaft modulus (-38.03 to -51.82m CD)	23.1GPa
Weight of shaft above O-Cell	918kN

Pile Number	B12
Date cast	6 September 1998
Nominal shaft diameter (+0.76 to -40.02m CD)	1.83m
Nominal shaft diameter (-40.02 to -65.35m CD)	1.79m
Shaft modulus (+0.76 to -5.06m CD)	25.4GPa
Shaft modulus (-5.06 to -40.02m CD)	24.0GPa
Shaft modulus (-40.02 to -65.35m CD)	14.5GPa
Weight of shaft above O-Cell	2 623kN

Table 12: Shaft friction values
deduced from O-Cell test

Chart Datum Depth Range (m)		Shaft friction derived from strain gauges (kPa)	
Test pile D8	Test pile B12	Test pile D8	Test pile B12
40.30-42.80	46.20-57.70	152	157
42.80-45.30	57.70-60.20	241	262
45.30-47.80	60.20-62.70	241	262
47.80-51.85	62.70-65.35	241	602

while Table 12 details the values obtained in numerical form.

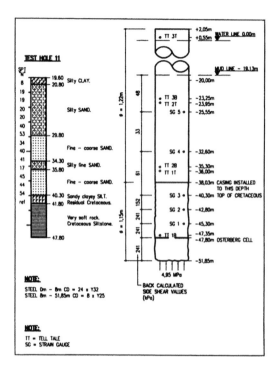

Figure 8: Pertinent data for test pile D8

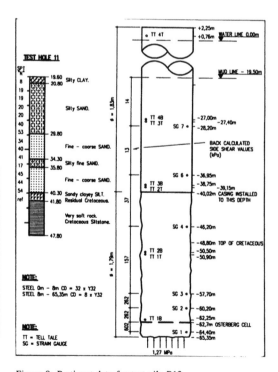

Figure 9: Pertinent data for test pile B12

11 CONCLUSIONS

Superimposed on the calculated values in Figures 6 and 7 are the reconstructed load-deflection curves at the top of socket level derived from the two tests conducted. These two tests indicated that:

- The Osterberg Cell provided a very elegant method of establishing the load capacity of large diameter offshore piles but very conservative load/deflection estimates are generated by the design process adopted.

- This verification process may well have been more useful in assigning revised working load capacities, if the load testing programme had been conducted prior to the installation of the working piles. This however was not possible due to the construction programme followed.

- More useful information would probably have been obtained if the socket lengths of the test piles were minimised and the O-Cell test conducted to limit modes of both side friction and end bearing. This however was rejected in the load test planning due to the tests being conducted on working piles.

12. REFERENCES

American Petroleum Institute 1987. Recommended Practice for Planning, Designing and Constructing Fixed Offshore Platforms. *API Recommended Practice 2A* (RP 2A. Seventeenth edition, April 1. 133pp.

ARQ 1997. *The use of "Inverted Buckets" as a foundation system for offshore structures.* ARQ document reference 674/2307 dated 22 September 1997.

Barton NR, Lien R and Lunde J 1974. Engineering classification of rock masses for the design of tunnel support. *Rock Mechanics* Volume 6 Number 4 pp 189-239.

Bieniawski, ZT 1973. Engineering classification of jointed rock masses. *The Civil Engineer in South Africa*, December, Vol 15 No 12. pp 335-343. "Errata" published in Vol 16 No 3 (March 1974) pp 119. "Discussion" in Vol 16 No 7 (July 1974) pp 239-254.

Bieniawski ZT 1976. Rock mass classification in rock engineering. Exploration for rock engineering, *Proceedings* of the Symposium, (editor ZT Bieniawski). Cape Town: Balkema, Vol 1 pp97-106.

Brink ABA 1985. *Engineering Geology of Southern Africa*, Volume 4 Post-Gondwana Deposits. Building Publications 332pp.

Everett JP 1991. Load transfer functions and pile performance modelling. *Proceedings* of the Tenth Regional Conference for Africa on Soil Mechanics and Foundation and Foundation Engineering and the Third International Conference on Tropical and Residual Soils. Maseru 23-27 September. pp 229-234.

Hoek E, Wood D and Shah S 1992. A modified Hoek-Brown criterion for jointed rock masses. Proceedings of the Rock Characterization Symposium, International Society for Rock Mechanics, Eurock'92 (editor JA Hudson). London: British Geological Society. pp 209-214.

Hoek E 1995. Strength of rock and rock masses. Extract from a book entitled *"Support of Underground Excavations in Hard Rock"* by E Hoek, PK Kaiser and WF Bawden. pp 4-16.

Knight K 1998. *Personal correspondence* with Alan Parrock.

Loadtest 1998a. *Preliminary data report on drilled shaft load testing* (Osterberg method). Test pile D-8, Dry Bulk Jetty, Richards Bay, South Africa.

Loadtest 1998b. *Preliminary data report on drilled shaft load testing* (Osterberg method). Test pile B-12, Dry Bulk Jetty, Richards Bay, South Africa.

Osterberg JO 1998. *The Osterberg load test method for drilled shafts and driven piles.* The first ten years. Dated 03/15/98. 17pp.

Portnet 1997. Port of Richards Bay. *Proposed Dry Bulk Jetty Geotechnical Report.* Protekon Design Offices.

Poulos HG and Davis EH 1980. *Pile Foundation Analysis and Design.* John Wiley and Sons. 397pp.

Simpson B 1995. Limit State Design in Geotechnical Engineering. *Proceedings* of the Conference by the Geotechnical Division of the South African Institution of Civil Engineers. October. Pretoria.

Webb DL and Moore LW 1980. The behaviour of piles. *Piling Panorama Symposium* presented by the Concrete Society of South Africa.

Webb DL and Allinson AJ 1980. Mechanical properties of the Cretaceous sediments of the Natal coast. *Proceedings* of the Seventh Regional Conference for Africa on Soil Mechanics and Foundation Engineering. Accra, AA Balkema, pp 57-66.

Williams AF, Johnston IW and Donald IB 1980. Structural Foundations on Rock, *Proceedings* of the International Conference. Sydney : Balkema pp 327-347.

Geotechnics for Developing Africa, Wardle, Blight & Fourie (eds) © 1999 Balkema, Rotterdam, ISBN 90 5809 082 5

The influence of pore water salt concentration on the measurement of soil suction using the filter paper method

A. M. Ridley
Imperial College of Science, Technology and Medicine, London, UK

E. O. Adenmosun
Posford Duvivier, Consulting Civil Engineers, Peterborough, UK

ABSTRACT: This paper presents a study of the influence of pore water composition on the suction measured using filter papers. Comparisons are made between the soil water retention characteristic of a silty soil with pure pore water and the characteristics of the same soil reconstituted with various salt solutions. The curves are represented in terms of total suction and matrix suction and are used to estimate the osmotic suction component over a wide range of moisture contents. A comparison is made between the osmotic suction derived from the measurement of total and matrix suction and the osmotic suction calculated from the concentration of salt in the soil.

1. INTRODUCTION

Total suction is a measure of the energy required to remove water from the pore water meniscus through evaporation into the vapour phase. It is related to the vapour pressure adjacent to the air-water interface through the relative humidity. Matrix suction is a measure of the free energy of the pore water and is a function of the capillary forces and surface absorption forces that act between the soil water and the soil particles. In a salt free soil the total suction and the matrix suction will be equal. However when the pore water contains dissolved salts the energy required to cause evaporation will be larger than the free energy of the pore water. Hence the vapour pressure and the relative humidity adjacent to the air-water interface are reduced and the total suction is increased. Under such circumstances the difference between the total suction and the matrix suction has frequently been termed the osmotic suction because of its dependence on the presence of dissolved salts.

The ability of a soil to retain and/or absorb water is by definition a measure of the soil suction. The relationship between the water content (gravimetric or volumetric) of a soil and the soil suction is known as the soil water characteristic curve. Theoretical aspects of the soil water characteristic curve have been well covered, but there are not many publications that consider the practical solution of simultaneously and accurately measuring the water content and soil suction. Moreover, for soils containing dissolved ions the soil water characteristic curve will be dependent on which component of suction is adopted and a method which can simultaneously measure the total suction, the matrix suction and the water content is desirable.

This paper will present measurements of total suction, matrix suction and gravimetric water content on samples of a clayey silt that was reconstituted using solutions of Sodium Chloride designed to give each sample a different initial osmotic suction.

2. SOIL TYPE

The soil used in this study was an artificial mix of two clays and a silt. The final composition consisted of

Table 1 Soil properties.

Sodium Chloride Concentration (M)	Liquid Limit (%)	Plastic Limit (%)
0	26.2	14.58
0.1	26.05	14.85
0.25	26.4	14.22
0.5	26.1	15.21

13% greater than 0.06mm particle diameter and 25% less than 0.002mm particle diameter. A predominantly silty material was chosen to reduce the likelihood of significant clay-salt-water interactions.

The soil was thoroughly mixed with pure water and three solutions of sodium chloride with concentrations of 0.1M, 0.25M and 0.5M to form four separate samples. The salt solutions have osmotic pressures of 459kPa, 1148kPa and 2298kPa respectively. Each sample was left to homogenise for 24 hours and used to determine Atterberg limits. At all stages of the test, solutions with identical salt concentrations were used to add moisture to each sample. The Atterberg Limit test results are shown in table 1. The variability is considered to be negligible. The specific gravity of the soil was 2.64.

3. SAMPLE PREPARATION

The soil was reconstituted with each solution of sodium chloride to a moisture content of 1.25 x the liquid limit, giving four new samples. After 24 hours each sample was poured into a separate consolidation mould. The mould was placed in a bath containing a sodium chloride solution with the same concentration as that used to prepare the particular sample. Drainage was maintained from both ends of the sample during consolidation. During the consolidation of the sample evaporation of the water in the bath and that on the top surface of the sample took place and it was necessary to replenish the lost moisture. To do this the volume of solution in the bath was returned to its original value using pure water. In this way the concentration of the salt solution remained unchanged during consolidation.

Each sample was consolidated in stages to a maximum vertical stress of 200 kPa. The loading sequence was 7.5 kPa, 25 kPa, 50 kPa, 100 kPa and 200 kPa. Each load was applied for a period of 24 hours prior to the addition of a subsequent load. Each sample was then unloaded in a similar sequence until the vertical stress was zero, when the sample was allowed to stand for 24 hours prior to extruding it from the mould. After consolidation each sample was circular in cross-section and approximately 25mm thick and 100mm diameter.

4. FILTER PAPER MEASUREMENTS

The moisture content of an absorbent material (such as laboratory filter paper) is related to suction in a similar manner to the moisture characteristic curve of a soil. Therefore, if a filter paper is allowed to absorb moisture from a soil specimen and the amount of moisture taken from the soil is small enough to have negligible effect on the soil suction then, when equilibrium is reached, the suction in the filter paper will be equal to the suction in the soil. A filter paper placed in intimate contact with the soil sample absorbs water through capillary flow and should equilibrate to the matrix suction of the soil. However a filter paper placed inside a sealed container without making contact with the soil absorbs moisture vapour and should equilibrate to a moisture content that is related to the vapour pressure in the container and hence the total suction in the soil. In this study Whatman No. 42 filter paper was used.

Duran (1986) presented a test arrangement that has been used to simultaneously measure total and matrix suction using calibrated filter papers. Each soil sample was sandwiched between two air dry filter papers and two perspex discs (of a slightly larger diameter than the soil sample) as shown in figure 1. One filter paper was in intimate contact with the soil sample and the other is separated from the soil sample by a wire gauze. The arrangement was wrapped in a protective seal (several layers of food wrap proved adequate for this), placed inside several plastic bags and stored in a temperature controlled environment for 7 days. It was then unwrapped and the masses of the soil and the filter papers were individually measured. The former was used to back-calculate the water content of the soil sample when the dry weight of the soil had been measured and the latter was used to establish the soil suction (total and matrix) from a relationship between the filter paper water content and soil suction. At this stage the volume of the soil sample could also be estimated using vernier callipers to measure the external dimensions of the sample in a number of places. The soil sample was then wrapped with two new filter papers and the procedure was repeated.

Using this procedure the quantity of water removed from the soil sample during each measurement is very small and the total time required to establish the

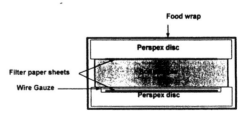

Figure 1 Apparatus for measuring soil suction.

400

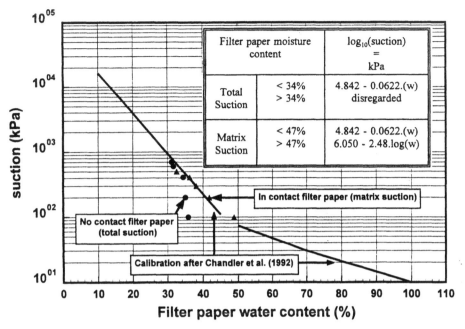

Figure 2 Filter paper calibrations for apparatus used in this study.

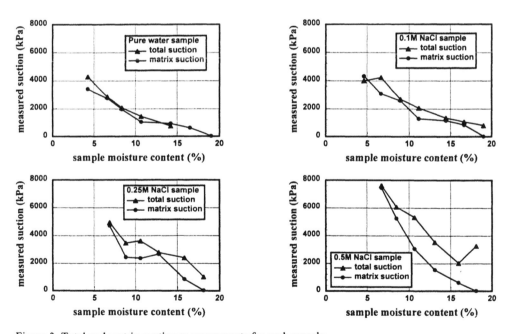

Figure 3 Total and matrix suction measurements for each sample.

complete soil water characteristic curve could be several months. However, the experiment can be accelerated by allowing the sample to dry slightly between each stage.

Ridley (1995) demonstrated the importance of calibrating the filter papers using an apparatus that is identical to the one that will be used during the

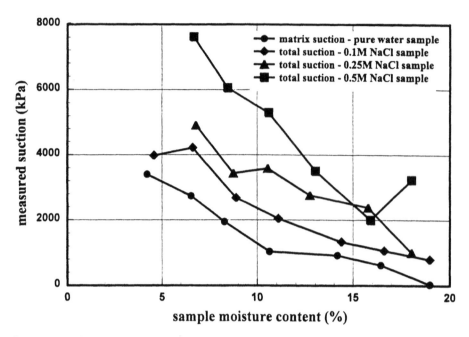

Figure 4 Total and matrix suctions

subsequent measurements. This was done for the apparatus that is used here and the results are presented in figure 2. Ridley pointed out that for filter paper moisture contents in excess of 34% the calibration for total suction measurements is relatively insensitive and produces poor accuracy. Therefore, in this study, only filter paper moisture contents less than 34% have been used in the measurement of total suction.

The calibration for measuring matrix suction measurements using this apparatus compares well with that due to Chandler et al. (1992) over a wide range of filter paper moisture contents. The calibrations that were adopted are presented in Figure 2.

5. RESULTS

The total and matrix suctions were measured on each sample at progressively lower moisture contents between 19% and 3%. The results are shown for each sample in figure 3.

It is notable that the matrix suction curves for each sample are different. If the dissolved salt in the soil was free to migrate into the filter paper, all four samples would have the same matrix suction curve. In addition, for the pure water sample and the 0.1M NaCl sample the difference between the measured total suction and the measured matrix suction is small at most of the sample moisture contents. At the higher

salt concentrations this difference is more significant.

Assuming therefore that the correct matrix suction - water content relationship for all four samples is that pertaining to the pure water sample, the graph can be replotted (figure 4) and show a more recognisable trend. Now the calculated osmotic suction (derived from the difference between measured total suction and the measured/inferred matrix suction) is seen to increase with both the increasing initial salt concentration and decreasing sample water content as each sample dries (figure 5).

Also shown on figure 5 are the theoretical values of osmotic suction calculated by assuming that the quantity of sodium chloride in each sample is unchanged from the initial value in the sample. At each stage of drying a revised salt concentration was determined from the adjusted moisture content of the sample and used to calculate the osmotic suction at each stage. Good agreement is obtained for the range of moisture contents investigated here.

6. CONCLUSIONS

Stage drying tests on samples of a clayey silt that was reconstituted with firstly pure water and then increasing concentrations of sodium chloride solution have shown that:

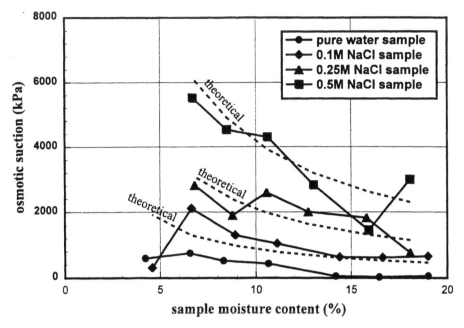

Figure 5 Derived and theoretical osmotic suctions

(a)   The filter paper method was not able to accurately measure matrix suction in the samples containing dissolved salts.

and

(b)   Osmotic suctions derived from the difference between the measured total and matrix suctions, compared favourably with those calculated from a theoretical consideration of the actual salt concentration in a sample at each stage. An implicit assumption in this comparison was that the matrix suction vs moisture content relationship for all of the samples was the same as that for the sample prepared with pure water.

REFERENCES

Chandler R.J., Crilly M.S. and Montgomery-Smith G. 1992. A low cost method of assessing clay desiccation for low-rise buildings. *Proc. Institution of Civil Engineers* **92**, No.2, 82-89.

Duran Gamarra A.J. 1986. Study of the effect of contact of the filter paper technique in the measurement of soil suction. *MSc thesis, University of London.*

Ridley A.M. 1995. Discussion on "Laboratory filter paper suction measurements" by Houston et al. *Geotechnical Testing Journal,* **18**, No. 3, 391-396.

Geotechnics for Developing Africa, Wardle, Blight & Fourie (eds) © 1999 Balkema, Rotterdam, ISBN 90 5809 082 5

A new approach to the determination of the expansiveness of soils

H. D. Schreiner
School of Civil Engineering, Surveying and Construction, University of Natal, Durban, South Africa

ABSTRACT: Previous correlations between soil index data and expansiveness have failed to apply basic principles of soil mechanics. This study applies to unsaturated soils relevant principles of soil modelling that have been in use in saturated soil mechanics for many years. The experimental method and the analysis that followed have produced an improved correlation between index test data and the measured expansion of the soils tested. This correlation is for use in assessing the expansiveness of soils and not for estimating the likely swell under field conditions.

1. INTRODUCTION

Accurate identification of expansive soils at the site investigation stage is essential for successful engineering design. Failure to do so could lead to failures and costly remedial works.

Attempts in the past to find a relationship between expansiveness and classification test data have proved unsuccessful, (Oloo et al 1987, Schreiner 1987). A problem common to most of these attempts has been the use of either compacted or undisturbed samples of natural soils, both of which introduce unknown stress histories and unknown microfabric. The microfabric has been shown to have major effects in testing for swell in the laboratory and can cause collapse to occur during swell testing (Schreiner and Burland 1991).
A clear distinction must be made between the intrinsic expansiveness of a soil, the potential swell on wetting from the *in situ* moisture content and the potential heave of the ground surface or of a road foundation. A pure montmorillonite clay, for example, will have a very high intrinsic expansiveness, but will have a zero potential swell if it is saturated and submerged i.e. if the suction is zero. The potential heave will be related to the potential swell, but account must be taken of overburden and foundation stresses, and initial and final suctions or moisture contents.

This paper reports a procedure developed from routine techniques in saturated soil testing whereby samples are prepared which are subjected to identical stress histories during sample preparation. The test sample is prepared from a slurry and which leaves a stable microfabric which is not subject to collapse on wetting. A small group of widely different soils has been examined using this procedure to allow study of their intrinsic expansiveness in terms of their index test data.

2. DEFINITIONS AND TERMINOLOGY

The definitions and terminology used in this paper follow the accepted usage in the literature. There are however, some cases where there is confusion in the literature or where concepts have not yet been adequately defined. The most significant of these the expansive property of a soil. There is still a significant lack of understanding of the relationship between swell and the fundamental properties of the soil. The most important definitions are those related to expansiveness and swell.

2.1 *Intrinsic expansiveness*

The ability of a clay mineral to adsorb and absorb water is an intrinsic property of the clay. It is not altered by the moisture content or suction which exists at a particular time. A mechanical analogy for the intrinsic expansiveness of a soil is the stiffness or load-displacement characteristic of a spring. Regardless of the load or displacement (suction or moisture content) applied, the stiffness (intrinsic expansiveness) will remain unchanged, i.e. the displacement (moisture content) is related to the applied load (suction) by the stiffness (intrinsic expansiveness).

Although the spring concept described is too simple to represent the full interaction between the intrinsic expansiveness, swell, suction and heave, it shows how the various components of the system interact. An attempt to model the full interaction would need to address as well as the applied stresses, previous stresses and suction, hysteresis and microfabric of the soil.

The intrinsic expansiveness of a soil is defined here as "a property of that soil resulting from its mineral composition and grading and its interaction with water. The intrinsic expansiveness determines the relationship between the change in volume (and corresponding change in water content), of a reconstituted sample, caused by a decrease in suction." The relationship between these parameters is not linear and, for natural soils, is dependent on stress history, fabric etc.

2.2 Swell

Swell of an intrinsically expansive soil may be described as the volume change that results from a change in moisture content or suction. For saturated soils only, the volume change of the soil will be equal to the volumetric moisture content change. In expansive soil engineering the case of complete saturation is rarely of importance in practical problems. In the partially saturated state the relationship between moisture change and soil volume change will be affected by factors such as the fabric and structure of the soil and hysteresis in the moisture content-suction relationship. In engineering usage it is more relevant to relate swell to the change in suction rather than to the change in moisture content.

Swell due to a unit change in suction and swell due to a unit change in applied stress may not be of equal magnitude, due to the non-equivalence of pore water pressure and applied stress in the unsaturated state. The relationship between the magnitudes of swell due to these two stress changes has not yet been determined, but is known to be non-linear and to vary with soil type.

2.3 Heave

Heave is the vertical displacement of a point in a soil mass or of a foundation that results from swelling of an intrinsically expansive soil. It is the parameter usually measured in the field and it is related to swell measured in the laboratory but the relationship

will be affected by the ability of the laboratory test to simulate the conditions and changes that prevail in the field.

A realistic simulation on an undisturbed sample would have to consider correctly at least:

a) the vertical stress due to overburden and imposed loads,
b) the lateral confinement,
c) representative initial and final suctions, and
d) previous maximum stress and suction.

2.4 Saturation

The term saturation is frequently misused with regard to expansive soils. A state of saturation describes a soil which contains solid particles and water but no air. A partially saturated soil contains solid particles and water and air. Neither term describes the state of stress or pore water pressure in the soil.

It has been common to describe the wetting stage of swell tests as a saturation stage. It is quite possible for saturated clay soils to have large negative pore pressures and it is also possible for initially partially saturated samples to remain partially saturated at the end of a laboratory swell test.

A correct description for this stage of the test is "soaking to zero suction or zero pore water pressure".

3. SOILS USED

Eight soil samples from sites in Sudan and Kenya were tested. The soils came from a wide range of climatic regions. Some of the soils are transported, some residual and some have been weathered *in situ* but may not be truly residual soils. An important improvement over soil samples used in previous correlations is the inclusion of red soils, yellow soils and soils which plot to the right of the A-line on the Casagrande Plasticity Chart. These soils would generally be considered to be non-expansive and Authors of previous correlations have limited their sample to those which are known, or expected, to be expansive.

All gravel sized material (i.e. larger than 2mm) was removed and discarded from the soils before index or volume change testing.

Plastic limit and cone liquid limit tests reported in Table 1 were carried out in accordance with

Table 1 Sources, classification test data and visual descriptions of soils used in the study.

Source	W_L	W_p	%- 425μ	Visual Description
1. Nairobi-Mombasa Rd. km 68 from Nairobi, Athi Plains, Kenya.	125	51	94	Very dark grey to black- slightly silty clay with occasional fine to coarse gravel. organic content high.
3. South Nyanza District near Aock-Muga School, Kenya	70	27	92	Grey to black clay with occasional sand and gravel. Organic content high.
5. Pile Test Site at Gezira University, Wad Medani, Sudan.	64	29	93	Dark brownish grey gravelly silty clay with calcareous granules. Organic content moderate.
6. Pile Test Site at El Fau Irrigation Scheme, Sudan.	71	37	98	Grey brown gravelly silty clay. Organic content moderate.
9. Cutting on Road C63 near Nairobi, Kenya.	68	37	100	Yellow to red, mottled yellow and white, stained dark red on joints, silty clay (Parent rock possibly volcanic ash). Organic content low.
10. Adjacent to Road C68 near Nairobi, Kenya.	84	42	100	Dark red clay. Organic content low.
12. Road A104 near Gilgil, Kenya.	48	28	83	Yellow brown clayey silt. Organic content moderate.
13. Road A104 16km from Nakuru toward Nairobi, Kenya.	81	48	83	Yellow to brown, stained black on joints, silty clay. Organic content low.

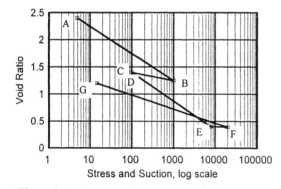

Figure 1 Diagrammatic representation of test procedure

B.S.1377, part 2, (BSI 1990). The classification data and brief descriptions of the eight soils used are given in Table 1.

The values quoted in Table 2 have been adjusted to represent the whole soil by:

W_L (whole soil) = W_L x (% -425μm)/100
W_p (whole soil) = W_p x (% -425μm)/100

The shrinkage limit, W_S, was determined by air drying a sample and measuring the mass and volume at intervals. This differs from the BS.1377, part 2 method (BSI 1990) in that the volume measurements of the circular disc shaped samples were made using vernier calipers and not with a mercury bath.

Specific gravities for the soils were determined from the measured volume, measured water content and mass of dry soil of the oedometer test samples and include any organic material present in the natural soil.

4. PRINCIPLES OF SOIL MODELLING

Engineers' abilities to understand saturated soil behaviour and predict deformations and stresses have been vastly improved since the use of reconstituted samples was introduced into research on the fundamental aspects of soil mechanics. A reconstituted sample is one which has been prepared by thorough mixing as a slurry (typically at a water content of 1.2 x W_L) and then consolidated to the required initial test conditions. Swelling back (or unloading) is used where over-consolidated samples

Table 2   Index test and swelling data for the soils tested

SOIL SOURCE	1	3	5	6	9	10	12	13
W_L(whole soil)	118	64	59	69	68	84	39	67
W_P(whole soil)	48	25	27	36	37	42	22	39
W_{SA}(whole soil)	13	11	13	10	23	25	23	20
W_{SB}(whole soil)	14	14	13	15	30	25	23	25
Plasticity Index	70	39	32	33	31	42	17	28
Shrinkage Index, USA	104	50	46	54	38	59	16	42
Shrinkage Index, UK	105	53	46	59	45	59	16	47
Sp gr, G_s, g/ml	2.60	2.61	2.78	2.79	2.68	2.77	2.44	2.66
$W_P - W_{SA}$	35	14	14	26	14	17	1	19
$W_P - W_{SB}$	34	11	14	21	7	17	1	14
Void ratio at W_L, e_L	3.08	1.67	1.64	1.93	1.82	2.33	0.95	1.78
Void ratio at W_P, e_P	1.25	0.65	0.75	1.00	0.99	1.16	0.54	1.04
Void ratio at W_S, e_{SL}	0.39	0.38	0.37	0.33	0.82	0.71	0.54	0.65
e^*_{100}	2.25	1.22	1.15	1.47	1.39	1.59	0.71	1.49
$e^*_{100} - e_{SL}$	1.86	0.84	0.78	1.14	1.39	1.59	0.71	1.49
Δe	1.32	0.45	0.60	0.79	0.14	0.35	0.015	0.53
$\Delta H/H_0$, %	85	33	41	55	9	19	1.5	32
Void ratio after swelling, e_{15}	1.75	0.84	0.98	1.22	0.97	1.14	0.57	1.22
Expansive strain, $(e_{15} - e_{SL})$x100/(1 + e_{SL})	98	33	45	67	8	25	2	35
C_S	0.170	0.074	.0100	0.145	0.054	0.064	0.025	0.082

List of Symbols

e_L Void ratio at the liquid limit
e_p Void ratio at the plastic limit
e_{SL} Void ratio at the shrinkage limit
e_{15} Void ratio after swelling under 15kPa vertical stress
Δe Change in void ratio during swelling under 15kPa vertical stress
G_s Specific gravity of all non-volatile components of soil
H   Sample thickness
Ho   Sample thickness before swelling
$\Delta H/H_0$ Vertical strain during swelling under 15kPa vertical stress
W_L Liquid limit
W_p Plastic limit
W_s shrinkage limit
PI   Plasticity index = $W_L - W_p$
SI   Shrinkage index = $W_L - W_s$

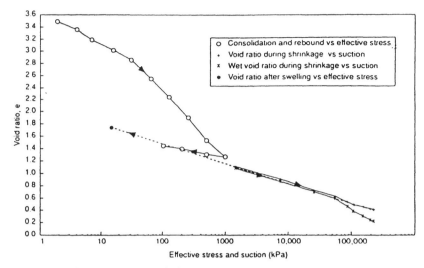

Figure 2 Volume change of soil from source 1.

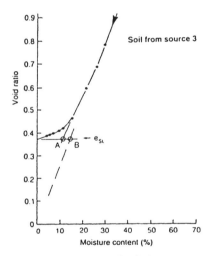

Figure 3 Shrinkage of soil from source 3.

are required. This technique permits the preparation of identical samples of a single soil type which can be subjected to various test procedures to examine the stress-strain behaviour of that soil. This avoids the inclusion in the soil model of the possibly unknown effects of fabric and stress history which may be present in undisturbed samples of a natural soil. It is for this reason that the correlation presented below must only be used for assessing the expansiveness of a soil and not for estimating the swell or heave, both of which are dependent on the above factors as well as the suction change and in-situ stresses.

In this study reconstituted samples of different soils have been subjected to the same test procedure. This permits comparisons to be made between the soil type, in terms of index test data, and the observed deformations.

5. DETAILS OF THE PROCEDURE

The procedure comprises three consecutive stages depicted schematically in Figure 1 and described in detail in Gourley & Schreiner (1993). These are:

(a) Compression, in an oedometer, under increasing effective stress, AB, and saturated swelling under decreasing effective stress, BC, to prepare a regular shaped soil sample which is strong enough to handle.

(b) Unconfined shrinkage after removal of the sample from the oedometer, DEF, with controlled moisture removal to a state drier at F than the shrinkage limit, at E.

(c) laterally confined swelling in an oedometer, FG, to zero suction under 15kPa vertical total stress.

During the unconfined shrinkage stage the suction was measured using filter paper. This technique has been widely used and the calibration curve favoured at the time that this research was done was that of Chandler and Gutierrez (1986). Subsequent work has shown that this calibration may not be appropriate for measurements in the high suction range (Schreiner 1988) and that the values may be overestimated due to a lack of contact between the water in the paper and that in the soil and consequently a new calibration is required (Gourley & Schreiner 1995). However, since the actual value of the suction has not been used in the derivation of the correlation given in this paper, it is considered acceptable to use the measured value to plot the data for diagrammatic purposes such as in Figure 2.

6. RESULTS

Table 2 lists the data related to the swelling after shrinkage together with various index values. Typical results are shown in Figure 2 for soil from source number 1. There are two lines on the figure that represent the shrinkage stage shown as DEF in Figure 1. The void ratio for the upper curve was determined from the overall sample volume and mass of dry soil. It thus represents the conventional void ratio. The void ratio for the lower curve, on the other hand, was determined from the moisture content and G_s using $e = w.G_s$. This void ratio therefore represents the voids filled with water. The difference between the two lines represents the air voids in the sample.

Figure 3 shows the results for the unconfined shrinkage of the soil from source 3 in terms of moisture content and void ratio. Points A and B represent the Shrinkage Limits obtained using the British (BSI 1990) and American (ASTM D427-83, 1992) standard methods. The former measures the void ratio directly during the test and used the actual shrinkage curve while the latter assumes full saturation until the shrinkage limit is reached. The two shrinkage limits are represented by W_{SB} and W_{SA} respectively.

Three measures of expansion or swelling have been chosen for comparison with the index data. These are the:

(a) change in void ratio, Δe, on soaking
(b) vertical strain, $\Delta H/H_o$, on soaking, in percent
(c) expansive strain on soaking, expressed as:

$$\varepsilon_{ex} = \{(e_{15} - e_{SL}) / (1 + e_{SL})\} \times 100, \%$$

The index data used for comparison with these measures of expansion are:

(a) Liquid limit, W_L
(b) Plastic limit, W_P
(c) Shrinkage limit, W_{SA} and W_{SB}
(d) Shrinkage index, $SI = W_L - W_S$ for both American and British cases
(e) Plastic limit - shrinkage limit for both cases.

(f) Void ratio at the liquid limit, e_L
(g) Void ratio at the plastic limit, e_P
(h) Plasticity index
(i) Compression Index, C_c
(j) Swell Index, C_s.

Initial attempts to find out whether or not there were correlations between any of these parameters and the measures of expansiveness described above were done graphically. This is a tedious process and was superseded by a statistical approach. This was done in two stages, first finding the most significant parameters and following this up with a regression analysis. The first part was done using the Spearman rank correlation test. This test does not in itself provide the correlation, merely indicates which of the parameters investigated is most significant for determining the correlation. Of the parameters investigated it was found that W_P- W_{SA} provided the highest level of significance, with W_P - W_{SB} being only slightly less significant. The highest level of significance in each case was found for the expansive strain, ε_{ex}. This measure of expansiveness was thus used in the remainder of the analyses.

The second stage comprised a linear regression analysis of each variable against the expansive strain, ε_{ex}. This produces a measure of the percentage

Table 3 Values of R^2 and t for the expansive strain, ε_{15}, and index test data

Parameter	R^2	t
W_L	58	2.87
W_P	32	1.66
W_{SA}	23	-1.32
W_{SB}	48	-2.39
SI, USA	77	4.44
SI, UK	76	4.42
W_P - W_{SA}	84	5.51
W_P - W_{SB}	90	7.32
e_L	58	2.88
e_P	30	1.62
PI	63	3.17
C_C	89	6.94
C_S	97	13.5

Table 4. Application of Equation 2 to soils 3, 6 and 9

Soil	W_L	W_P	PI	W_{SB}	ε_{ex}calc %	ε_{ex}meas %
3	64	25	39	11	38	33
6	69	36	33	10	60	67
9	68	37	31	23	4	8

of variation as R^2 and a level of significance as t. The results are presented in Table 3.

Here again W_P - W_{SA} is shown to be the best with W_P - W_{SB} only slightly less good. It is interesting to note that the PI shows a relatively poor correlation. Liquid Limit, Plastic Limit and Shrinkage Limit individually are even worse. The best results are found for C_C and C_S which are determined from oedometer tests and are thus, unfortunately, expensive to obtain.

The relationship between W_{SA} and ε_{ex} that was obtained from the regression is:

$$\varepsilon_{ex} = 3.0(W_P - W_{SA}) - 5.8, \% \qquad (1)$$

Further analysis, using the Plastic and Shrinkage Limits as separate variables in a single regression produced an even better correlation with $R^2 = 94$. The relationship from this analysis, where W_{SB} was found to give better results, is:

$$\varepsilon_{ex} = 32.5 + 2.4W_P - 3.9\,W_{SB}, \% \qquad (2)$$

This equation, eqn (2), should not be used for the prediction of actual swell or for the prediction of heave because these are dependent on many site or test specific factors such as vertical stress, change in suction, stress history etc. The correct application of this relationship is only in estimating the intrinsic expansiveness of the soil. Consider for example, soils 3, 6 and 9 in Table 2. Their index data is summarised in Table 4. Considering the Plasticity Indices, as is commonly done when assessing expansiveness, these soils appear to be similar. Liquid and Plastic limits are also of a similar magnitude. Superficially, therefore, an engineer would be justified in assuming that these soils would behave in a similar manner. However, the expansive strain, determined from the actual test data, shown in the last column of Table 4 shows that these are very different soils. The range of measured values of ε_{ex} in this study is about 0 to 100%, corresponding to "non-expansive" and "very expansive" soils, although these may not be the limiting values. The second last column shows the values of ε_{ex} that have been calculated using Equation 2. While not being identical the calculated values do rank the soils in the correct order, as defined by ε_{ex}meas, and indicate that soil 6 is significantly more expansive than soil 3 and very much more expansive than soil 9. The Plasticity Index does not do this and indicates that soil 3 should be the most expansive.

7. SWELL INDEX

The swell index for a soil is determined from the unloading stage of an oedometer test and is expressed as the change in void ratio per ten fold

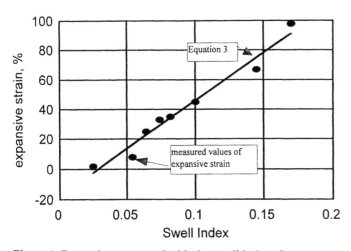

Figure 4 Expansion compared with the swell index, C_s.

reduction in vertical effective stress. It represents the slope of the unloading curve in a conventional semi-logarithmic plot of void ratio against vertical effective stress.

In the experimental procedure described in this paper, this parameter can be determined from stage BC shown in Figure 1. The swell indices for these soils are listed in Table 2.

These values can also be compared with the chosen measures of volume change. Of these parameters the best comparison is with ε_{ex}. The regression equation obtained between C_S and ε_{ex}, with R^2 of 97, is:

$$\varepsilon_{ex} = 644\ C_S - 18.4 \qquad (3)$$

The good comparison, shown in Figure 4, between the swell index determined from the saturated soil behaviour under effective stress and the swell on soaking indicates that both saturated soil swelling and un-saturated soil swelling are influenced by the same property of the soil.

8. RECOMMENDATIONS

It is important that geotechnical and materials engineers should have a reliable tool available for identifying expansive soils. Many attempts have been made in the past to provide such a tool by comparing swell data with one or more commonly determined soil index values. Each may have some validity for the particular set of soils and conditions from which it was derived but none has proved universally reliable. Reasons for the failures of past attempts have been discussed by Oloo et al (1987), Schreiner (1987) and Schreiner (1988) and centre around the soil microfabric and stress history as well as the absence of the application of soil mechanics principles.

The procedure followed in this study has avoided the problems inherent in the use of compacted or natural undisturbed samples. It is thus possible to compare soils purely on the basis of expansiveness without interference from microfabric and stress history. The results have confirmed several findings in Oloo et al (1987) even though that study was based on compacted samples. It has been found that expansiveness is not related to any single index test parameter.

A strong relationship was found between expansiveness and W_p and W_{SB} when used together. It must be cautioned that the data base for this study is extremely small, though it does include a wider

range of soil types than is usually the case in previous correlations between expansiveness and index test data. However, of importance in this respect is the inclusion of soils which plot to the right of the A-line on the Casagrande Chart and of red and yellow soils. These show that expansiveness is not restricted to soils which plot to the left of the A line nor to dark coloured soils.

On the basis of the results of this study the following recommendations can be made:-

(a) Liquid, plastic **and shrinkage** limits should be routinely determined for all samples at the site investigation stage. Liquid and plastic limits alone are inadequate for the determination of expansiveness.

(b) Expansiveness should be estimated from Equation 2, with the British shrinkage limit preferred..

(c) A strong correlation has been found between expansiveness and the swell index, C_S, derived from oedometer tests on reconstituted saturated samples. This shows that unsaturated expansive soils are not differnt from 'normal' saturated soils. Expansiveness can therefore be estimated from conventional oedometer tests.

9. REFERENCES

ASTM D4943-89. (1992). Standard test method for shrinkage factors of soils by the wax method. Annual Book of ASTM Standards, Vol. 04.08.

BSI. (1990). Methods of test for soils for civil engineering purposes; BS1377, part 2, classification tests. British Standards Institution, London.

Chandler, R.J. and Gutierrez, C.I. (1986). The filter-paper method of suction measurement. Geotechnique, Vol. 36, p265-68.
Gourley, C.S, and Schreiner, H.D. (1993). A new approach to the determination of the expansiveness of soils. Transport Research Laboratory Project Report PR/OSC/011/93.

Gourley, C.S, and Schreiner, H.D. (1995). Field measurement of soil suction. Proc. 1st Int Conf Unsat Soils, Paris, p601-608

Oloo, S.Y., Schreiner, H.D. & Burland, J.B. (1987). Identification and Classification of Expansive Soils. Proc. 6 Int. Conf. Expansive Soils. New Delhi, p 23-9.

Schreiner, H.D. (1987). The use of predictive

methods in expansive soil engineering. Proc. 9 African Reg. CSMFE, Lagos, p. 135-42.

Schreiner, H.D. (1988). Volume change of compacted highly plastic African clays. PhD. Thesis, University of London.

Schreiner, H.D. and Burland, J.B. (1991). A comparison of the three swell test procedures. Proc 10 African Regional Conference on Soil Mechanics and Foundation Engineering, Maseru, Lesotho.

Geotechnics for Developing Africa, Wardle, Blight & Fourie (eds) © 1999 Balkema, Rotterdam, ISBN 90 5809 082 5

A pilot investigation into the clay-sized fractions of some typical natural soils

D. J. Stephens
Civil Engineering Department, University of Natal, Durban, South Africa

J. N. Dunlevey
Geology Department, University of Durban-Westville, South Africa

ABSTRACT: Roads in under-developed regions require adequate in-situ material for their construction, but the normal test methods to indicate their suitability can be expensive and difficult to carry out. Attempts to reduce testing costs by using mathematical or graphical prediction models have been shown to be of limited reliability. Previous research on synthetic soils indicated that certain parameters could potentially be predicted using a modified Atterberg liquid limit value and the quantity of the 'clay' fraction present. However, these models could not be applied to natural soils, apparently due to the presence of non-clay ('inert') particles in the 'clay' fractions. A preliminary study was therefore undertaken to investigate the clay-sized fractions of some natural soils using X-ray diffraction analysis. The study indicated that a significant amount of non-clay mineral particles were present in many of the clay-sized fractions of the soils.

1 INTRODUCTION

The provision of roads in the under-developed regions of the world is vital to improving the economic potential and quality of life of disadvantaged communities. In designing and constructing these roads, it is necessary to quantify the capability of the in-situ material to carry the intended traffic volume, as well as identify suitable local sources of pavement construction materials. The normal method used to evaluate the strength of these materials is the California bearing ratio (CBR) test (eg NITRR 1986) , but this can prove problematic in under-developed regions as the test not only requires a large, representative, sample of the soil, but also has to be carried out with relatively specialised laboratory equipment.

Even in developed countries, attempts have been made over the years to reduce the cost of CBR testing by using mathematical or graphical CBR prediction models (Stephens 1990). Such techniques, utilising simple classification parameters, would obviously also be extremely useful in under-developed areas. However, the existing prediction methods have been shown to be unreliable (Stephens 1990). Subsequent research on compacted synthetic cohesive soil samples indicated that their CBR strength and swelling potential could be predicted by an equation using a modified Atterberg liquid limit value and the quantity of the 'clay' fraction present (Stephens 1992a; 1992b). The clay fraction was defined (according to current practice) as those soil particles with effective diameters smaller than 2 microns. The ensuing prediction models could,

however, not be applied to natural soils (Purnell and Day 1993; Wright et al 1993), the inconsistency being attributed to the presence of non-clay ('inert') particles in the 'clay' fractions, which had obscured the true amount of clay present (Stephens 1993).

In order to confirm this hypothesis, a study was undertaken to ascertain the mineralogical composition of the clay-sized fractions of some natural soils.

LABORATORY INVESTIGATION

2.1 *Soil types selected*

Soil samples for which the California Bearing ratio had recently been measured at a local soils laboratory were selected for investigation. Samples were selected from the available group to reflect soils typical of the KwaZulu-Natal region and thus soils from doleritic, granitic, lateritic, shale and sandstone origins predominated.

2.2 *Extraction of test samples*

The ASTM hydrometer soil test (ASTM 1972) was used to extract the clay-sized fraction from each sample, the material particles having effective diameter of less than 2 microns remaining in suspension in the hydrometer cylinder being decanted.

2.3 Preparation of test samples

Stephens and Dunlevey (1998) reported that the first trial suspensions extracted from the hydrometer test in this way produced a putty-like substance after the usual overnight drying. When subjected to X-ray diffraction these samples produced a pattern dominated by the broad peaks of poorly crystalline material thereby preventing a satisfactory mineralogical analysis.

The cause of this problem was shown to be the sodium hexametaphosphate $[Na_2O(PO_3)_{12-13}]$ dispersant used in the standard hydrometer test to separate the soil particles and prevent flocculation and settling of the suspension. The dispersant had been absorbed onto the surface of the individual particles to such an extent that on drying, it formed a coating that almost completely shielded the soil particles from the X-ray beam. Thus the combined effects of dilution and shielding removed virtually all indication of the mineral composition of the clay-sized material.

As the whole of the clay-sized fraction was required for investigation, the use of the dispersant was essential to prevent flocculation and settling. A procedure was therefore devised to remove the dispersant and recover the clay-sized fraction in a state suitable for the X-ray diffraction mineralogical analysis. Attempts to achieve this using high-speed centrifuging were unsuccessful and eventually a flocculant-based approach was adopted (Stephens and Dunlevey 1998). 100 ml of a one-molar Aluminium Chloride ($AlCl_3$) solution was added to the sample and the volume made up to 2 litres with distilled water. After standing for 48 hours the clear supernatant liquid was decanted to remove both sodium hexametaphosphate and aluminium chloride. Repeating this flocculation cycle 7 times produced a pure product containing less than 0,01% of dispersant.

2.4 X-ray diffraction of samples

The material collected from the above flocculation procedure was dried at 50°C to remove excess moisture while not causing dehydration of the phyllosilicate minerals. The aggregates of fine-grained materials formed by drying were crushed in an agate mortar and pestle to give a powder composed of composite grains less than 125 microns. Duplicate analysis were undertaken on each sample using surface and side-packed preparations in standard aluminium holders (Hutchison 1974; Klugg & Alexander 1974). The X-ray diffraction analysis of these preparations was undertaken using a Philips PW 1830 generator equipped with cobalt LFF X-ray tube (operated at 50kV and 30mA) and graphite monochromator, controlled by a Philips PW 3710 unit. Scans were undertaken from 3.00 to 60.00° 2 theta (34.2 to 1.8Å) with a step size of 0.01° 2 theta, and a counting time of 10 seconds at each step site.

Peak positions were derived from the scan raw data using the Philips APD and X'Pert (Vers. 1.1) software and mineralogical identification derived by comparison with data in the ICDD (1993) database. Although full pattern comparison was used in all cases, the presence of a peak at approximately 7° 2 theta (14Å) signifies the presence of smectite clay, and a peak at 14° 2 theta (7Å) indicates kaolinite. Sharp peaks at 24.2° 2 theta (4.2Å) and 31.1° 2 theta (3.3Å) are characteristic of quartz and minor peaks at 10.1° 2 theta (10Å) due to the presence of muscovite and sericite. The peaks produced by smectite and kaolinite clay minerals are significantly broader and less intense than those of quartz, due to the much less rigid structure of the phyllosilicate crystal structure and its greater ability of the lattice to accommodate ionic substitution.

3 RESULTS

The diffraction patterns obtained for each soil sample were compared with standard patterns for typical minerals (ICDD 1998). A good typical example of the diffraction patterns for the natural soils is the shale-derived soil shown in Figure 1. By comparing this diffraction pattern with the reference patterns in Figures 2 and 3 it can be seen that the soil contains two clay components, namely kaolin and montmorillonite. However, there is also a third set of peaks in the diffraction pattern for the natural soil, with the most notable occurring at approximately 30 degrees. This set of peaks is that of the mineral quartz which can be seen by comparison with the diffraction pattern for a commercial quartz powder in Figure 4. This powder had been processed undispersed through the hydrometer test ensuring a maximum particle size of 2 microns for direct comparison purposes.

Inspection of the diffraction patterns for the other four natural soils (figs 4 to 8) shows that quartz was present in all of the soil samples.

4 DISCUSSION

The diffraction patterns obtained for the five soil types all indicate the presence of a significant amount of quartz. Thus the initial hypothesis that the clay-sized fractions of a soil smaller than 2 microns, can, in certain circumstances, contain a significant quantity of non-clay minerals, has been shown to be valid.

The next phase of the research will involve extending these encouraging results to a larger sample of soil types, as well as attempting to quantify the amounts of quartz actually present so that the true clay fractions can be determined and hence the proposed prediction equations used with more certainty.

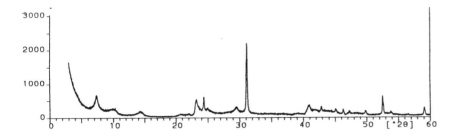

Fig. 1 X-ray diffraction pattern of clay-size fraction extracted from a typical shale soil.

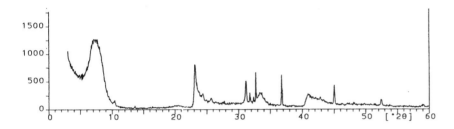

Fig. 2 X-ray diffraction pattern of a typical smectite (bentonite) clay.

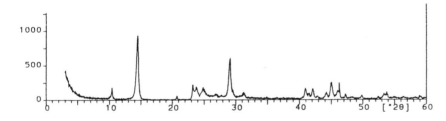

Fig. 3 X-ray diffraction pattern of a typical kaolinite (kaolin) clay.

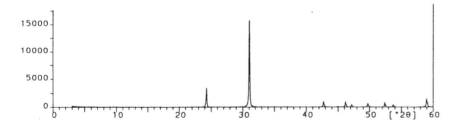

Fig. 4 X-ray diffraction pattern of clay-size fraction extracted from a commercial quartz.

417

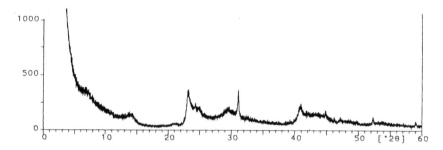

Fig. 5   X-ray diffraction pattern of clay-size fraction extracted from a typical dolerite soil.

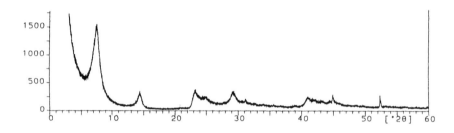

Fig. 6   X-ray diffraction pattern of clay-size fraction extracted from a typical granite soil.

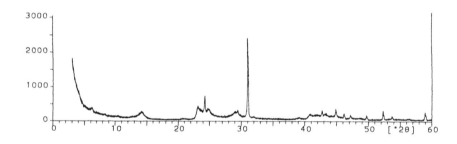

Fig. 7   X-ray diffraction pattern of clay-size fraction extracted from a typical laterite soil.

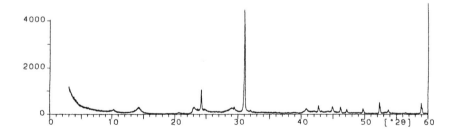

Fig. 8   X-ray diffraction pattern of clay-size fraction extracted from a typical sandstone soil.

5 CONCLUSION

The clay-sized fractions of some typical soil samples from KwaZulu-Natal in South Africa were examined using X-ray diffraction analysis and shown to contain non-clay particles (quartz), some in significant amounts.

6 ACKNOWLEDGEMENTS

The authors would like to thank Soils Design Laboratories (Natal) Pty Ltd for supplying the soil samples used in this investigation. The X-ray diffraction facilities of the University of Durban-Westville were used to analyse the soil samples.

7 REFERENCES

American Society for Testing and Materials. (1972). *Standard method for particle-size analysis of soils. Test designation ASTM 422-63 (reapproved 1972)*, Philadelphia: ASTM.

Hutchison, C.S. (1974). *Laboratory handbook of petrographic techniques*. New York: John Wiley & Sons.

International Centre for Diffraction Data (1993). *Mineral Powder Diffraction File Databook (Sets 1 to 42)*. Swarthmore, Pennsylvania: ICDD.

Klugg, H.P. & Alexander, L.E. (1974). *X-ray Diffraction Procedures for Polycrystalline and Amorphous Materials*. New York: John Wiley and Sons.

National Institute for Transport and Road Research (1986). *TMH1, Standard methods of testing road construction materials, 2nd edition*. Pretoria: CSIR.

Purnell R. & Day P.W (1993). Discussion on 'Variation of the California bearing ratio in some synthetic soils' and 'The use of a modified liquid limit for swell prediction'. *J. SAICE*, 35(2), Second Quarter, 1993:21-22.

Stephens D.J. (1990). The prediction of the California bearing ratio. *Civ. Engr. S. Afr.*, 32, 12, Dec. 1990: 523-527.

Stephens D.J. (1992a). Variation of the California bearing ratio in some synthetic soils. *Civ. Engr. S. Afr.*, 34, 11, Nov. 1992:379-380.

Stephens D.J. (1992b). The use of a modified liquid limit for swell prediction. *Civ. Engr. S. Afr.*, 34, 12, Dec. 1992:405-406.

Stephens D.J. (1993). Discussion on 'Variation of the California bearing ratio in some synthetic soils' and 'The use of a modified liquid limit for swell prediction'. *J. SAICE*, 35, 2, Second Quarter, 1993:21-22.

Stephens D.J. & Dunlevey J.N. (1999). The recovery of dispersed clay-sized suspensions for X-ray diffraction mineral analysis. *S. Afr. J.Sci.*, 94, (in press).

Wright, D.F., Clayton, R.A. & Smith, D.V. (1993). Discussion on 'The use of a modified liquid limit for swell prediction'. *J. SAICE*, (35), 2:21-22.

Geotechnics for Developing Africa, Wardle, Blight & Fourie (eds) © 1999 Balkema, Rotterdam, ISBN 90 5809 082 5

The use of embedded pressure cells to monitor geotechnical structures

J. P. van der Berg
Jones and Wagener, Rivonia, South Africa

G. Heymann
University of Pretoria, South Africa

C. R. I. Clayton
University of Surrey, Guildford, UK

ABSTRACT: The paper discusses the use of embedment pressure cells to monitor the stresses in a concrete/shotcrete structure. The performance of the cells was firstly evaluated under controlled conditions in the laboratory and it is shown that under these ideal conditions the cells performed satisfactory. Secondly the cells' performance was evaluated during an experiment simulating working conditions. A number of embedment cells were installed as part of the monitoring of a shotcrete tunnel lining. The results of the experiments, together with the results obtained from the cells installed as part of the monitoring of NATM tunnels are discussed in terms of the importance of installation details, factors affecting the measurements and the interpretation of the results. The investigation into the use of embedment cells shows that under the appropriate conditions, the monitoring of stresses can add valuable information to monitor the overall performance of a geotechnical structure.

1 INTRODUCTION

Geotechnical structures are monitored to evaluate their performance during construction, usually in order to confirm design assumptions. The monitoring may comprise either measuring the deformation of the structure and surrounding soils (i.e. strains and displacements) and/or the stresses within and on the structure. Methods and instrumentation for monitoring displacements and strains are well established and widely used. The performance of a structure is, however, evaluated typically in terms of stresses which makes interpretation of the deformation monitoring results difficult. It is therefore sometimes desirable to monitor stresses directly.

Of the parameters that may be monitored, total stress is possibly the most difficult to measure accurately. Total stress measurements can be divided into two categories. The first category is the measurement of normal total stresses in a medium such as a soil mass or a concrete mass. Embedment pressure cells are used for this purpose. The second category comprises the measurement of normal total stresses imposed by a soil mass on a structure such as a retaining wall, culvert, tunnel lining or shallow footing. Such stresses are measured using boundary pressure cells.

Two types of pressure cells are commonly used in practice. The first is a diaphragm type where the stress applied to the cell induces bending of a thin metal diaphragm. The magnitude of the bending is measured by electronic transducers such as bonded strain gauges, vibrating wire strain gauges (e.g. Thomas and Ward (1969)) or Hall effect gauges (Clayton and Bica (1993)). The second type of total stress cell is a hydraulic cell where a thin cavity between two circular or rectangular plates is filled with oil or mercury. Stresses applied on the cell are balanced by the pressure of the liquid inside the cell which is monitored by a pressure transducer. For a comprehensive discussion of design and installation of pressure cells see Dunnicliff (1988). The pressure cells discussed in this paper are of the hydraulic type. A typical hydraulic type pressure cell is shown in Figure 1.

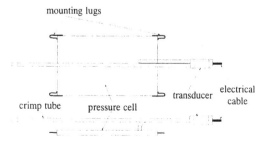

Figure 1 Embedment pressure cell.

During the construction of the NATM tunnels for an underground station at Heathrow Airport, the use and performance of embedment and boundary pressure cells were investigated. The paper describes two experiments conducted to investigate the performance of the embedment pressure cells under ideal and working conditions. The use of embedment pressure cells for the monitoring of the performance of geotechnical structures is then discussed with reference to the results of the experiments and the measurements obtained during the monitoring of the tunnel construction.

2 PRESSURE CELLS AS A MEASURING DEVICE

A major problem with pressure cells is that the presence of the instrument influences the distribution of the stress, which is the parameter being measured. When using pressure cells to measure total stress, the designer therefore has to consider the interaction of the cell and the surrounding medium under working conditions, as well as using appropriate calibration techniques.

In a medium which exhibits significant shear stiffness such as soil or concrete, the inclusion of the cell and the subsequent deformation of the cell disrupts the stress field in the vicinity of the cell. The stress measured by the cell may then be different from the stress in the absence of the cell. The *Cell Action Factor* (CAF) may be used to evaluate the performance of a total stress cell (Taylor (1947)). The CAF is defined as the ratio of the stress measured by the cell to the stress that would have prevailed in the absence of the cell. The nearer the CAF to unity, the more accurate the measurement of normal stress.

For embedment cells, the stress measured with the cell can be either higher of lower than the actual stress in the medium, depending on the relative stiffness of the cell and the medium. If the stiffness of the cell is higher than the stiffness of the surrounding material, the cell will over register (CAF > 1). On the other hand, if the cell stiffness is lower than the stiffness of the surrounding medium, the cell will under register (CAF < 1). In theory, to achieve a CAF of unity, the stiffness of the cell should be equal to the stiffness of the medium. In practice the stiffness of the medium depends on various factors, is time dependent and therefore varies significantly. To design a cell with the same stiffness as the medium is virtually impossible. However, Hvorslev (1976) showed that the sensitivity of the CAF to disparities between the stiffness of the cell and the medium diminishes if the aspect ratio (thickness/diameter) is reduced. In fact as the aspect ratio approaches zero, the CAF approaches unity. For this reason pressure cells are often manufactured as thin as practically possible.

The calibration of pressure cells is an aspect which needs careful consideration. Pressure cells are calibrated by the manufacturer usually in a pressure vessel by applying a fluid pressure to the cell. This fluid calibration is really only a calibration of the pressure transducer. To calibrate the pressure cells for working conditions, two options are available. Ideally, the cells should be calibrated under conditions similar to the operational conditions. If this option is not available, and rarely it is, the manufacturer's air or liquid pressure calibration and the CAF estimated from empirical or analytical information can be used to evaluate the cells' performance. Such information has been published by various authors (e. g. Coutinho (1953), Askegaard (1961), O'Rourke (1978), Selig (1980), Weiler and Kulhawy (1982), Clayton and Bica (1993)). It is however not possible to take the installation's sensitivity, for example, to temperature into account using the second approach.

Another important aspect of embedment pressure cells measurements is the practice of "crimping" the cells. It is common practice to re-pressurise the embedment cells, using the crimp tube (see Figure 1), at some stage after installation. Commonly referred to as "crimping", the purpose is to close any voids formed around the cell during the hydration of the concrete, to ensure adequate load transfer. During crimping, hydraulic fluid is forced into the cell by systematically flattening the crimp tube. If any voids exist, the pressure rise accompanying each pinch will be small. When the void is filled, the pressure rise with each pinch will increase dramatically. Crimping is continued until this difference in response is observed.

3 EXPERIMENTAL WORK

Embedment pressure cells were extensively used as part of monitoring the performance of the tunnel linings for an underground station at Heathrow Airport. Embedment cells were installed in the lining to measure the stresses in the shotcrete. As part of the monitoring programme a series of experiments was conducted to evaluate the performance and usefulness of the pressure cells for monitoring the performance of the tunnel lining. The experiments are described below.

3.1 *Performance under "ideal conditions"*

A series of tests on embedment pressure cells was conducted at the University of Surrey to evaluate the performance of the cells under "ideal conditions" (Clayton et al. (1998)). "Ideal conditions" are considered to be conditions similar to the field conditions, but without defects which may be introduced as a result of construction.

As part of the test series, the cells were first calibrated under conditions equivalent to a fluid calibration. This calibration was conducted in a pressure chamber with the face of the cells flush with a pressurized rubber membrane (Clayton et al. (1998)). The calibration technique was similar to that used by Selig (1980). Under these conditions the CAF for all cells were found to be close to 1. This is of course to be expected, since the calibration is similar to that used by the manufacturer.

The behaviour of the embedment pressure cells was evaluated under ideal conditions by casting two rectangular pressure cells (100mm x 200mm) with a full scale range of 20 MPa, into a concrete slab of dimensions 1m x 1m x 0.3m. Before casting, the pressure cells were positioned at one-third and two-thirds of the specimen height, staggered horizontally. The two cells were tied to a cage constructed from reinforcing meshes, which were identical to those used in the tunnel linings. A 25 MPa ready-mix concrete was poured into the formwork and vibrated to remove air voids. After 2 days the formwork was removed and the slab covered with moist rags during curing.

Zero readings of pressure and temperature were taken before casting, after casting and during the curing process. The temperature increased to about 32 °C and then decreased slowly over several days. After temperature equilibrium was reached, the slab was placed vertically in a 1000 kN loading frame. Hydraulic jacks were placed on bearing plates on the top of the slab. The slab was then subjected to a loading sequence (see Clayton et al. (1998) for details). It comprised monitoring the response of the cells as the vertical load on the slab was varied during a series of loading and unloading steps. The result obtained from a typical loading-unloading sequence to 1000 kN is shown in Figure 2.

From the figure it is evident that the cells performed exceptionally well. The CAF's for the two cells varied between 0.93 and 0.99.

3.2 *Performance under "working conditions"*

To investigate the performance of the embedded pressure cells under "working conditions", an experiment was conducted where four pressure cells were installed in two shotcrete slabs. The shotcrete slabs were constructed in the tunnel and closely simulated lining construction. Because the results of the experiment on the cells installed in the concrete slab in the laboratory showed that under ideal conditions the cells seemed to perform satisfactorily, it was hypothesised that a possible reason for not performing satisfactory might be installation effects.

A major difficulty with the installation of embedment pressure cells is ensuring that no voids are formed in "shadow zones" around the cell during shotcreting. A pressure cell situated along the shotcreting path could behave as a deflector and consequently a void may be formed in the resulting shadow zone. To investigate this hypothesis it was decided to pre-cast one of the cells in each of the two slabs in a tapered concrete block. Two cells were therefore embedded inside a tapered 30 MPa pre-cast concrete block, about 0.35m long, 0.25m wide and 0.17m deep, approximately a week prior to constructing the shotcrete slabs. A pre-cast pressure cell used in the experiment is shown in Figure 3.

The shotcrete slabs were constructed in the following way. Two plywood formworks, 1.0m high by 1.0m wide and 0.3m deep, were manufactured and

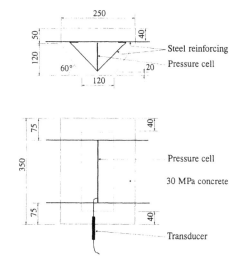

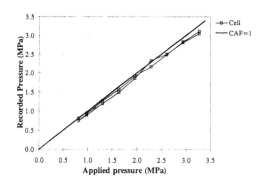

Figure 2   Result obtained from the ready-mix concrete slab.

Figure 3   Pre-cast pressure cell used for the experiment.

taken to the tunnel face. An initial layer of approximately 70mm of shotcrete was applied in the bottom of the formwork. This simulated the application of a sealing layer during tunnel lining construction. The mesh and pressure cells, one normal and one pre-cast cell, were fixed in each formwork, using the same technique used during installation of the cells in the lining. Instrument cables were tied to reinforcing bars and taken through holes drilled at the formwork sides. Crimping tubes were similarly tied to reinforcing bars and left protruding from the open 1.0m x 1.0m formwork surface. The cells were slightly staggered across the axis of the formwork to minimise the interference of the pre-cast cell on the other cell. The first layer of shotcrete was then applied. The second layer of mesh was fixed and the second layer of shotcrete was applied. After spraying each of the layers of shotcrete the loose material ("rebound") was removed in order to ensure that the "joints" between shotcrete applications did not form a weakness in the slab. Similar to the previous ready-mix concrete experiment, zero readings of pressure cells and thermistors were taken before shotcreting, after shotcreting, and during shotcrete curing.

The slabs were allowed to cure in the tunnel for about two days and were then transported back to the University of Surrey's laboratory for testing. The same testing set-up and essentially the same loading sequence used during the previous calibration test performed on the ready-mix concrete slab, were repeated for the shotcrete slabs.

The results obtained from both the slabs were remarkably consistent. Figure 4 shows the result obtained for a typical loading-unloading sequence to 1000 kN for a pre-cast and normal pressure cell in one of the shotcrete slabs. Both the pre-cast pressure cells and the "normal" cells recorded higher stresses than those applied. The cells pre-cast in concrete proved to be significantly stiffer than the surrounding shotcrete,

but being embedded in concrete did not have a desirable aspect ratio. As a result the cell action factor at the maximum applied pressure was 1.29 and 1.33 for the two pre-cast cells respectively. The normal cells performed remarkably well. The CAF at a maximum applied pressure of approximately 3.6 MPa was 1.08 for both cells. Repeating of the loading-unloading cycles showed that the results were highly repeatable.

4 FACTORS AFFECTING PRESSURE CELL PERFORMANCE

The experiments showed that the design of the pressure cells used in the experiments produced a CAF close to unity both under ideal conditions and working conditions. The experiments and the results obtained from the cells installed as part of the monitoring, however also indicated that to obtain an indication of the absolute stress in the structure being monitored, various other factors need to be considered.

Attention to the detail during the installation of the pressure cells was found to be very important. The most significant factor when installing the embedment cells was to ensure that the cells were securely fixed in position and could not vibrate during shotcreting. The nozzleman then had a key role to play to ensure that no voids were formed in shadow zones around the cell. Furthermore, in order to interpret the results properly, it is important to determine accurately the orientation and position of the cell within the lining, since the radial stresses will typically be only about 10% of the tangential stresses, and there are likely to be stress variations across the section as a result of bending moments. Due to the temperature sensitivity of the cells a base reading (including a temperature reading) must be taken once the cell is in position and prior to applying the shotcrete. It is desirable also to take a set of readings immediately after shotcreting.

It was found that the combination of the cells and the lining was temperature sensitive. Although a temperature correction is provided by the manufacturer as part of the calibration data, it corrects only for the transducer's temperature sensitivity and does not take into account pressure variations of the fluid in the cell due to temperature change. If a pressure cell in free air is subjected to temperature change, it will register a pressure change although no change in pressure is applied. This is due to the difference in coefficient of thermal expansion of the fluid in the cell and the steel plate capsule. When the cell is installed in shotcrete, with yet another coefficient of thermal expansion, the system's temperature sensitivity is further complicated. In order to be able to correct for temperature changes, it is therefore necessary to calibrate a cell installed in shotcrete and to subject the entire system to

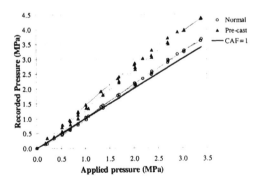

Figure 4 Loading-unloading sequence for one of the normal cells in the shotcrete slabs.

temperature changes. In practice this is not done and the assumption is made that temperature change during monitoring is small. Even this assumption may be incorrect, since seasonal temperature changes can be large in well-ventilated tunnels.

The initial response from a mercury filled embedment pressure cell installed in the tunnel lining, and the temperature measured with the pressure cell's thermistor, are shown in Figure 5. The pressure shown in the figure is after the manufacturer's temperature correction has been applied, and illustrates the temperature sensitivity of the cell. As the temperature increases due to hydration, the pressure increases correspondingly, and the trend is repeated as the temperature drops again. At least some of the pressure increase measured with the cell is therefore due to temperature change. If this effect is not calibrated for, it is not possible to monitor the stress change in the lining during hydration. The response during hydration can however be used to evaluate the soundness of the installation. Where the installation was poor, later confirmed by crimping the cell, the pressure increase during hydration was small. This indicates that the cell was able to expand more freely in the shotcrete, which could only happen if there were voids around the cell. Analysis of the pressure cell response during hydration can therefore be used to judge the soundness of the installation.

Furthermore, strains in the shotcrete which are not due to stress changes, for example shrinkage and creep, have an influence on the pressure cell readings. These strains cause changes in the fluid pressure in the cells which do not reflect stress changes in the lining. For example, using the base readings taken prior to constructing the shotcrete slabs, a stress of 800 kPa was measured in the unloaded slab in the laboratory. The actual stress in the slab was only the weight of the shotcrete, which was about 10 kPa at mid section. This high stress can therefore only be attributed to shrinkage of the slab.

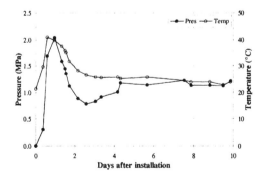

Figure 5 The initial response and temperature measured with a mercury filled pressure cell installed in a shotcrete lining.

The pressure cells installed during both experiments, and in the tunnel lining, were crimped in accordance with the manufacturer's recommendation. When crimping the cells, one of two responses was observed:

• A few cells installed in the tunnel linings showed a small pressure increase with each pinch, which continued until the entire crimp tube was flattened. This indicated the presence of voids around the cell, i.e. a poor installation.

• The remainder of the cells in the tunnel linings and the those installed during the experiments showed an immediate significant pressure increase with each pinch, indicating the absence of voids around the cell and therefore the necessity to crimp.

The crimping results therefore indicated that where the cell installation was sound, crimping was not necessary. Where crimping indicated a poor cell installation, the crimping did not succeed in changing the performance of the cell. Although crimping was apparently not necessary during this work, it was an effective way of judging the soundness of the installation.

5 CONCLUSIONS

The embedment pressure cells evaluated during this work performed well, both under ideal conditions and under working conditions. The pressure cell design is adequate to produce a CAF close to unity when subjected to stress changes. It may therefore be thought that provided cells are installed properly and calibrated under conditions similar to those encountered during monitoring, embedment pressure cells can give a good indication of the stresses in the structure being monitored.

In practice various factors influence the stress measured with the cell. This makes embedment pressure cell results difficult to interpret. In general this is due to the problem of compatibility of the cell and the medium. Factors such as temperature effects and strains in the medium due, for example, to shrinkage, register as pressure changes in the cell which do not reflect stress changes in the lining. To account for these effects is very difficult.

Installation is also important. Proper installation and the recording of the orientation and exact position of these cells, as well as the thickness of the lining, is difficult, but has a significant influence on the interpretation of the results. Careful records of crimping must be made.

In summary, the adequate CAF of the cells means that these cells can be used to obtain an indication of the stress variations in the lining, but factors resulting from the compatibility problem of the cell and the medium make it very difficult to measure the absolute stress in the structure.

ACKNOWLEDGMENTS

The authors would like to thank BAA plc, specifically the contribution made by the HEX Construction Director, Mr Chris Rust D'Eye. The funding of HEX and EPSRC is gratefully acknowledged. Colleagues from Mott MacDonald (particularly Dr David Powell), University of Surrey, HEX and Balfour Beatty have been instrumental in the success of the research and their contributions are appreciated.

REFERENCES

Askegaard, V. (1961). *Measurement of pressure between a rigid wall and a compressible medium by means of a pressure cell.* Acta Polytechnica Scandinavica, Civil Engineering and Building Construction Series, No. 11. Copenhagen: Technical University of Denmark.

Clayton, C. R. I and Bica, A. V. D. (1993). *The design of diaphragm-type boundary total stress cells.* Géotechnique, Vol. 43, No. 4, pp. 523-535.

Clayton, C.R.I., Bica, A.V.D., Hope, V.S. and Heymann, G. (1998). *The use of mercury-filled pressure cells for measuring stresses on and in shotcrete linings.* World Tunnel Congress, São Paulo, Brazil.

Coutinho, A. (1953). *Discussion on "Development of a device for the direct measurement of compressive stresses".* Journal of the American Concrete Institute, Vol. 25, pp.216/3-216/6.

Dunnicliff, J. (1988). *Geotechnical instrumentation for monitoring field performance.* Wiley and Sons, New York. ISBN0-471-09614-8, 577pp.

Hvorslev, M, J. (1976). *The changeable interaction between soils and pressure cells; tests and reviews at the Waterways Experiment Station.* Technical Report S-76-7. Vicksburg: US Army Engineer Waterways Experiment Station.

O'Rourke, J. E. (1978). *Soil stress measurement experiences.* Journal of the Geotechnical Engineering Division ASCE, Vol. 104, No. GT12. pp.1501-1514.

Selig, E. T. (1980). *Soil stress gage calibration.* Geotechnical Testing Journal ASTM, Vol. 3, No. 4, pp.153-158.

Taylor, D. W. (1947). *Pressure distribution theories, earth pressure cell investigations and pressure distribution data.* Vicksburg: US Army Engineer Waterways Experiment Station.

Thomas, H. S. H. and Ward, W. H. (1969). *The design, construction and performance of a vibration wire earth pressure cell.* Géotechnique, Vol. 19, No. 1, pp. 39-51.

Weiler, W. A. and Kulhawy, F. H. (1982). *Factors affecting stress cell measurements in soil.* Journal of the Geotechnical Engineering Division ASCE, Vol. 108, No. GT12. pp.1529-1548.

5 Methods of design and analysis
Les méthodes d'étude et de dimensionnement

Geotechnics for Developing Africa, Wardle, Blight & Fourie (eds) © 1999 Balkema, Rotterdam, ISBN 90 5809 082 5

Evaluation of risk factor for slope stability analysis

Atilla M. Ansal
Civil Engineering Faculty, Istanbul Technical University, Maslak, Istanbul, Turkey

Sonja Zlatović
Faculty of Civil Engineering, University of Zagreb, Croatia

ABSTRACT: Safety factors in geotechnical engineering, as in other fields of engineering, are utilised to assure sufficient safety against the uncertainty in the material properties and analysis methods. The values chosen for the safety factors depend on the available data and the engineering experience and may vary from one case to the other. In addition, safety factors do not reflect any information concerning the statistical distribution of the soil parameters and corresponding exceedance probabilities, thus yield no information about the risk of failure. The purpose of this study is to take into account the variability in the soil parameters to develop a relationship between safety factor and risk factor for the case of slope stability analysis. The risk factor is defined as the probability of slope failure. The proposed approach is demonstrated based on a slope stability analysis for a given embankment. The significance of the variability of soil parameters used in slope stability analysis was evaluated. A risk factor-safety factor relationship is developed assuming that the governing soil parameters are independent random variables. A multi-variant probability analysis is performed to obtain an optimum solution with respect to failure risk and additional cost of construction.

1. INTRODUCTION

An evaluation of slope stability in man made cuts and fills or in natural slopes requires a detailed geotechnical investigation composed of laboratory and insitu field tests with the main purpose of determining a feasible solution with respect to cost and safety. One possibility is to establish a relationship between additional cost and reduction in risk expressed in terms of risk factor.

Since there are various simplifying assumptions in all slope stability analysis and since there are uncertainties in the soil properties encountered in the soil profile, there is always a risk of instability or failure in all cases (Casagrande, 1969). The important issue is to know the level of risk under these conditions. Therefore, it is essential to perform a probabilistic analysis based on the statistical distribution of controlling parameters.

In this study slope stability analyses were conducted based on simplified Bishop method to determine the minimum safety factor, utilising commercially available chart programs such as Excel or Lotus, and Geoslope. These calculations were repeated for different shear strength parameters and for different slope configurations to evaluate the variation of the safety factor. Assuming the parameters used in the analyses are independent

random variables, a relationship was developed between safety factor and risk factor.

In evaluating the safety for any engineering application, both technical and social factors need to be considered. Since it is not possible to eliminate risk or since it becomes unfeasible to decrease the risk level after certain limits, it is required to define an acceptable risk level (Fell, 1994). The definition of an acceptable risk is a difficult issue and may change depending on economical, social and political conditions that prevail in the region. One possible approach is to compare different risk levels that people encounter during their life span and select a risk level in accordance with the observed probabilities of the events usually accepted as inevitable. The other alternative is to consider the probability of failure and additional cost of construction to increase the safety to find an optimum solution with respect to cost and safety.

2. SLOPE STABILITY ANALYSIS

Apart from the natural variability in soil properties, the uncertainties in slope stability analysis may arise due to the limitations of the site investigation and due to simplifying assumptions in the analyses. On the other hand, the values of the safety factors used depend on the experience and type of engineering structures and do not account for the variability in the related factors (Meyerhof, 1970; Cavounidis, 1987). Thus, the safety factor is not a measure of the risk level.

In a slope stability analysis, the safety factor would depend on (1) sliding or overturning forces (surcharge forces, earthquake forces, etc); (2) the properties of the soil layers located in the slope; and (3) the method of analysis adapted to determine the safety factor. The uncertainties related to these three factors lead to uncertainty in the calculated safety factor. One alternative is to determine the variability of the important parameters and evaluate risk factor using probabilistic procedures.

The simplified Bishop stability analysis (Graham, 1984) was adopted to evaluate slope stability based on a parametric study of an embankment. The embankment, with the geometry shown in Figure 1 was considered to be composed of a single homogeneous soil layer with unit weight, $\gamma = 20$ kN/m^3; cohesion, c = 20 kPa; and shear strength angle, $\phi = 30°$.

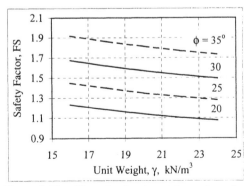

Figure 1. Selected slope profile with five different slope angles

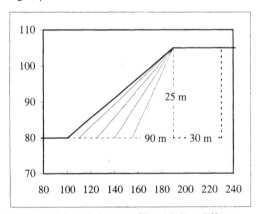

Figure 2. The effect of soil unit weight on the safety factor

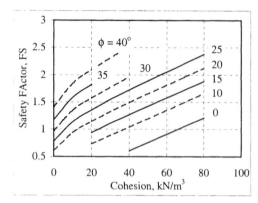

Figure 3. The effect of cohesion on the safety factor

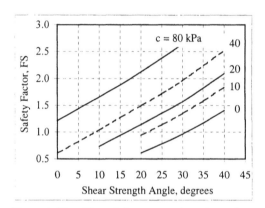

Figure 4. The effect of shear strength angle on the safety factor

The safety factor for the flattest slope was determined with the pore pressure coefficient $r_u = 0.5$. At this first stage a parametric study was performed by assuming each of the four parameters (γ_n, c, ϕ, r_u) as independent random variables to evaluate the effect of their variability on the safety factor. The results obtained are given in Figs. 2, 3, 4 and 5.

Figure 2 shows how an increase in the soil density leads to a decrease in the safety factor. The rate of decrease is not very significant and it is slightly nonlinear. The effect of a change in soil cohesion, as shown in Figure 3, and shear strength angle, as shown in Figure 4, are more significant.

The effect of the pore pressure coefficient, as shown in Figure 5, increases with an increase in the shear strength angle. Thus it may be justified to assume that the primary factors controlling the slope stability are cohesion, shear strength and the pore pressure coefficient. However, in performing effective stress analysis, cohesion term is normally neglected. Therefore, it would be more realistic to consider the variability of the shear strength angle

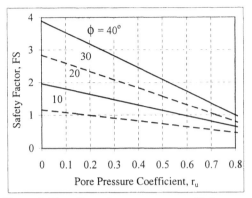

Figure 5. The effect of pore pressure coefficient on the safety factor

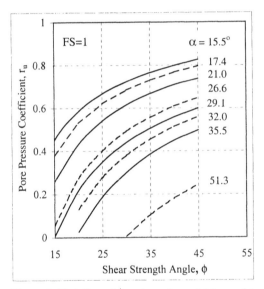

Figure 6. Variation of ϕ and r_u values yielding safety factor FS=1 for different slope angles

and the pore pressure coefficient as independent random variables.

Since the aim of this study is to determine the relationship between the construction cost and failure risk, the slope angle for the selected embankment was taken as a variable. The variation of the safety factor was determined with respect to the shear strength angle and pore pressure coefficient while the soil density and the cohesion were assumed constant (Ansal et al., 1991). The combination of values of these two variables that are yielding safety factor FS=1 for the eight slope inclinations are shown in Figure 6. It can be observed from Figure 6 that as the slope flattens, the values of pore pressure coefficient (r_u) yielding safety factor FS=1 increases.

3. FAILURE RISK IN SLOPES

In evaluating slope stability it would be essential to determine the failure risk based on statistical and probabilistic methods in addition to the conventional safety factor analysis (Vanmarcke, 1977; D'Anrea & Sangrey, 1982; Wu & Kraft, 1970). In the geotechnical engineering, there are uncertainties in the analysis performed to evaluate safety factor due to uncertainties in soil parameters used and due to simplifying assumptions in the analysis. If the statistical distributions of the related parameters are determined, it becomes possible to calculate the probability of being less or more and thus to obtain the relationship between the safety factor and failure risk or risk factor. In addition, such a relationship could also be used to evaluate the change in overall safety in case of deviations from initial assumptions and soil characteristics used in the analysis (Chowdhury & Xu, 1993; Chowdhury, et al., 1993, Christian, et al.,1994).

In determining failure risk levels for a slope, the statistical distribution of soil parameters and external forces should be known. Assuming that the variability of external forces is rather limited, excluding earthquake generated forces, the basic parameters that control the stability are the shear strength angle and pore pressure coefficient. The shear strength angle and pore pressure coefficient was assumed as the independent random variables that would be used to determine the risk level. The statistical distributions of these parameters were modelled by normal distribution function (Benjamin & Cornell, 1970, Lacasse & Nadim, 1996). The variability of these two parameters was evaluated and their effects were taken into account in determining the variation of risk factor with respect to the safety factor.

Shear strength angle and pore pressure coefficient was considered as independent random variables of the multi-variable normal distribution. Considering the characteristics of these two variables, the standard deviation of both parameters were assumed as 50% of the mean values used in the stability calculations. The overall failure risk is determined as the joint probability of being less for the shear strength angle and of being more for the pore pressure coefficient for their combinations yielding safety factor FS=1 for five different slope angles as shown in Figure 7. The overall failure risks for different slope angles are shown in Figure 7 with respect to shear strength angle. As can be observed, the set of parameters with safety factor of FS=1 does not yield equal failure risks. Therefore, in determining the safety factor - failure risk relationship, the maximum failure risks thus calculated were used to be on the safe side.

The safety factors obtained for different slope angles and corresponding maximum failure risks

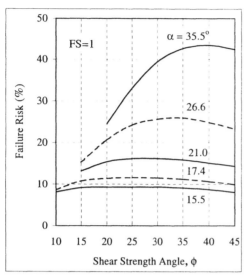

Figure 7. Variation of overall failure risk for safety factor FS=1 for different slopes

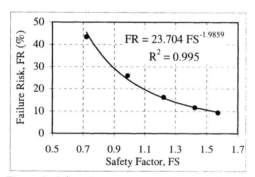

Figure 8. Safety factor versus combined failure risk for different slopes

were determined as shown in Figure 8. It was also possible to establish a correlation between safety factor and failure risk as shown in Figure 8 for the case of different slope angles with soil density 20 kN/m^3, cohesion 20 kPa, shear strength angle 30°, and pore pressure coefficient 0.5.

In this simple example, the failure risk corresponding to the safety factor of FS=1.5 is approximately 10 %. This risk value is dependent on the standard deviations selected for the shear strength angle and for the pore pressure coefficient. If these parameters can be determined more accurately, the failure risk may be smaller

One alternative to improve the stability is to flatten the selected slope. As shown in Figure 9, with the decrease in the slope angle, the failure risk decreases and the safety factor increases. In this situation the financial cost of the additional fill need

to be considered. In performing a cost analysis, half of the cross section area of the embankment for the steepest slope angle was used as the reference area with unit thickness.

The cost increase due to flattening of the slope was determined as the ratio of the additional fill area to this reference cross section area. Failure risks and safety factors corresponding to the cost increase ratios (CIR) were calculated. Since failure risk was determined by assuming shear strength angle and pore pressure coefficient as independent variables the cost increase ratio was calculated for these cases. Since the flattening of the slope requires additional fill costs, it is necessary to find an optimum point satisfying the safety of the slope for a minimum cost increase (Chang, 1989).

The change in the failure risk (FR) and cost increase ratio (CIR) with respect to increase in the safety factor due to the flattening of the slope is shown in Figure 10. The cost increase here is expressed as the ratio of the additional fill to the original fill.

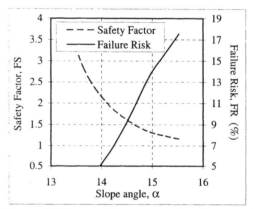

Figure 9. Slope angle versus safety factor and combined failure risk

Figure 10. Variation of failure risk and cost increase ratio with respect to safety factor

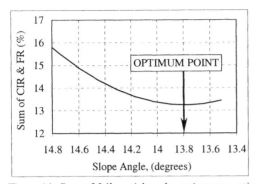

Figure 11. Sum of failure risk and cost increase ratio with respect to slope angle

An optimum point can be determined based on the variation of the sum of the failure risk and cost increase ratio since both were expressed in percentage. The sum of failure risk and cost increase has a parabolic variation with respect to slope angle (Figure 11) and also with respect to safety factor. Therefore the optimum point can be considered as the minimum point in these relationships.

In the simple slope example used in this study, the most efficient solution with respect to the cost of additional fill was obtained for the safety factor FS=2.5 corresponding to approximately 5% failure risk for the slope angle of 13.8°. This result may be different for different slope geometry, and different soil properties, as well as for different statistical distribution properties. However, it demonstrates that by using failure risk approach it would be possible to find a feasible solution. It also gives the engineer an idea about the failure risks that may be important depending on the importance of slope analysed. As in this example, 4 % failure risk may be considered very conservative and depending on the importance of the analysed slope, higher risk can be acceptable. However, one may not be aware of this risk if the analysis was conducted only based on the safety factor.

4. CONCLUSIONS

In evaluating the stability of slopes, it is essential to perform a parametric study concerning the variability of the factors entering the stability calculations. As shown in this study, such an approach would lead to more reliable evaluation concerning the overall stability of the investigated slope. In addition, by considering the statistical variability of the important parameters and by determining the failure risk based on probabilistic procedures, it would lead to better evaluation of the stability of the slope with respect to construction cost.

REFERENCES

Ansal, A.M., Uludağ, E. & Siyahi, B.G. 1991. Safety Factor, Failure Risk and Cost Relationship in Slope Stability Analysis. *Proc of the First Turkish National Landslide Symposium.* Black Sea Technical University, Trabzon, 129-138 (in Turkish)

Benjamin, J.R. & Cornell, C.A. 1970. *Probability, Statistics and Decision for Civil Engineers,* McGraw-Hill Book Com., NY.

Casagrande,A. 1969. Role of the "Calculated Risk" in Earthwork and Foundation Engineering. *ASCE Journal of Soil Mechanics and Foundation Eng.* 91(4):1-40

Cavounidis,S. 1987. On the Ratio of Factors of Safety in Slope Stability Analyses. *Geotechnique.* 37(2):207-210

Chang,H. 1989. Optimisation of Performance Reliability. *Computational Mechanics of Probabilistic and Reliability Analysis.* Ed. W.K.Liu, T. Belytschko, Elmepress International

Chowdhury,R.N. & Xu,D.W. 1993. Rational Polynomial Technique in Slope-Reliability Analysis. *ASCE Journal of Geotechnical Eng.* 119(12):1910-1928

Chowdhury,R.N., Tang,W.H. & Sidi,I. 1987. Reliability Model of Progressive Slope Failure. *Geotechnique.* 37(4):467-481

Christian, J.T., Ladd, C.C. & Baecher, G.B. 1994. Reliability Applied to Slope Stability Analysis. *ASCE, Journal of Geotechnical Eng.* 120(12):2180-2207

D'Anrea, R.A. & Sangrey, D. 1982. Safety Factors for Probabilistic Slope Design. *ASCE, Journal of Geotechnical Engineering.* 108(2):1101-1118.

Fell, R. 1994. Landslide Risk Assessment and Acceptable Risk. *Canadian Geotechnical Journal.* 31(2):261-272

Graham,J. 1984. Methods of Stability Analysis. *Slope Instability.* Ed.D.Brunsden & D.B.Prior. 6:171-215, John Wiley and Sons Ltd., London.

Lacasse, S. & Nadim, F. 1996. Uncertainties in Characterising Soil Properties. *NGI Report 514166-12.* Norway

Meyerhof,G.G. 1970. Safety Factors in Soil Mechanics. *Canadian Geotechnical Journal,* 7:348-355

Vanmarcke,E.H. 1977. Reliability of Earth Slopes. *ASCE, Jour.of Geotechnical Eng.* 103(11):1247-1265.

Wu,T.H. & Kraft,L.M. 1970. Safety Analysis of Slopes. *ASCE, Journal of Soil Mechanics and Foundations.* 96(2):609-630.

Geotechnics for Developing Africa, Wardle, Blight & Fourie (eds) © 1999 Balkema, Rotterdam, ISBN 90 5809 082 5

Résistance dynamique des pieux soumis à des actions horizontales

H. Ejjaaouani, V. Shakhirev & H. El Gamali
Laboratoire Public d'Essais et d'Etudes (LPEE), Casablanca, Maroc

J. P. Magnan
Laboratoire Central des Ponts et Chaussées (LCPC), Paris, France

RÉSUMÉ : Des essais de chargement horizontal par impact ont été réalisés sur trois pieux battus dans des sols argileux. Six autres pieux ont été testés sous des charges statiques dans des conditions géométriques identiques. Les essais dynamiques ont permis de définir le mécanisme d'interaction du pieu et du sol : les oscillations du pieu sous l'impact sont pratiquement élastiques, tandis que le sol garde des déformations résiduelles importantes dans sa partie supérieure. De ce fait, le pieu se comporte comme une poutre en console encastrée dans le sol à partir d'une certaine profondeur. Ce modèle de calcul a permis de déterminer les fréquences propres des oscillations du pieu et la valeur de la force statique équivalente à l'impact, qui sont en accord satisfaisant avec les résultats des essais.

ABSTRACT: Horizontal impact loading tests were performed on three piles driven in clayey soils. Six other piles were submitted to static loading in similar geometric conditions. The dynamic tests were used to define the conditions of interaction of the pile and the soil : the pile oscillations produced by the impact are essentially elastic, whereas the soil undergoes significant residual deformations in its upper part. Therefore the pile behaves as a cantilever beam embedded in the ground at a certain depth. This calculation model enabled the authors to determine the natural frequencies of the oscillations of the piles and the static load equivalent to the impact. The calculations were in good agreement with the experimental data.

1 INTRODUCTION

Pour la construction parasismique, on donne souvent la préférence aux fondations sur pieux, qui transmettent les charges dues à l'ouvrage et aux sollicitations sismiques à des horizons rocheux ou à des couches de sol plus compactes et plus résistantes, situées à une certaine profondeur sous la surface du terrain naturel. Dans ce cas, la technique de calcul des fondations est plus complexe car il faut vérifier la stabilité de la fondation sous des combinaisons de charges statiques et dynamiques verticales, horizontales et de moments (Hakimi et al., 1993a, 1993b, 1993c, 1993d ; Ejjaaouani et Chakhirev, 1995).

Dans tous les cas, on tente de ramener le calcul dynamique des structures à un calcul statique sous l'effet de charges statiques équivalentes aux actions dynamiques. La détermination des charges statiques équivalentes aux charges dynamiques est donc une question importante pour les projeteurs. Nous présentons ici une étude expérimentale consacrée à l'équivalence des actions statiques et dynamiques dans le cas des pieux soumis à des charges horizontales.

Nous n'avons pas connaissance de méthodes d'essais de pieux en place qui reproduisent de façon suffisamment proche les mouvements d'un pieu sous une charge horizontale sismique. Nous décrirons pour cette raison la procédure d'essai que nous avons adoptée avant de présenter et analyser les résultats des essais en vraie grandeur que nous avons réalisés sur des pieux battus dans le sol.

2 PRINCIPE DE L'ESSAI

L'exécution d'essais de chargement sismique en place sur des pieux présente des difficultés importantes au plan technique comme au plan économique. Au plan technique, deux problèmes se posent principalement :
- il faut trouver un mode de chargement qui imite le mieux possible la sollicitation sismique ;
- il faut pouvoir déterminer de façon fiable la valeur des impulsions transmises à la structure testée et leurs conséquences en termes de contraintes et de déplacements.

Parmi les techniques qui ont été proposées pour tester les pieux sous charge horizontale sismique, on peut citer :
- l'exécution d'essais de chargement à paliers de courte durée (moins de 5 minutes), qui reste assez éloignée des sollicitations sismiques, dont la durée se compte en secondes, voire en fractions de secondes ;
- le chargement des pieux par des explosions souterraines de puissance adaptée, à une distance du pieu qui permette de reproduire les paramètres de la sollicitation sismique désirée. Par exemple, les accélérations de la surface du sol près du pieu doivent au moins être égales à 2 m/s^2 pour une sismicité de 7 sur l'échelle de Richter, de 4 m/s^2 pour un niveau 8 et de 7 m/s^2 pour un niveau 9.

Lors des essais de chargement par explosion, on enregistre les composantes horizontales des vibrations à la surface du sol à proximité du pieu dans deux directions perpendiculaires afin de caractériser les ondes longitudinales (sous l'effet desquelles le sol subit des élongations et des compressions) et les ondes transversales (qui provoquent des distorsions et des rotations du sol). Mais ces essais ont pour but essentiel de déterminer l'influence de la sollicitation sismique sur la résistance du pieu aux charges verticales et le tassement supplémentaire qu'elle provoque et ne donnent qu'une idée approchée de la sollicitation horizontale imposée au pieu.

Nous avons donc recherché un type d'essai qui permette à la fois de déterminer précisément la quantité d'énergie ou la quantité de mouvement transmise au pieu par une sollicitation horizontale dynamique connue et de déterminer la réponse mécanique du pieu à ce chargement.

La solution retenue consiste à appliquer au pieu un impact horizontal connu, au moyen d'un marteau pendulaire, et d'enregistrer les vibrations qu'il provoque dans le pieu. La figure 1 montre le schéma de principe de l'essai.

3 MATERIEL D'ESSAI

L'équipement utilisé pour l'essai a été conçu pour remplir les conditions suivantes :
- appliquer une force d'impact horizontale pouvant atteindre 100 kN ;
- couvrir un domaine de fréquence de vibrations du pieu de 0 à 250 Hz ;
- atteindre un déplacement horizontal du pieu au niveau de la surface du sol correspondant au déplacement limite statiquement admissible, de l'ordre de u_o = 10 à 15 mm.

La pièce centrale du dispositif est un marteau pendulaire de masse 100 kg, suspendu par une barre tubulaire à un tripode, lui-même constitué d'éléments tubulaires. Le rayon de la trajectoire du marteau est de 1,5 m.

Pour mesurer la force d'impact réelle sur le pieu, on utilise un dynamomètre de type extensométrique, préalablement étalonné en laboratoire.

Les déplacements horizontaux du pieu testé sont mesurés par un capteur de déplacements de type inductif à circuit magnétique, possédant des paramètres magnétiques constants et fixé sur un repère fixe.

Le marteau est équipé d'un appareillage de type potentiométrique permettant de déterminer son inclinaison par rapport à la verticale. L'inclinaison maximale du marteau est de β = 70 degrés.

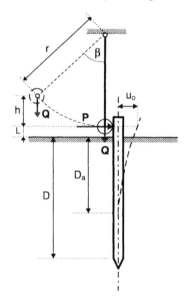

Figure 1   Schéma de principe de l'essai de chargement dynamique de pieu par impact horizontal

4 PROGRAMME EXPERIMENTAL

Les études décrites ici ont été réalisées sur un site expérimental où l'on trouve des sols argileux et argilo-limoneux de densité voisine de 2, avec un indice des vides de 0,78, une teneur en eau moyenne de 24%, un indice de plasticité un peu supérieur à 7 et un indice de liquidité de 0,3.

Les essais ont été réalisés sur des pieux battus de section carrée (30cm x 30cm), encastrés de D = 2m, D = 4m et D = 6m. La charge horizontale était appliquée à une hauteur de L = 0,25 m au-dessus de la surface du sol. La disposition des pieux sur le site expérimental est montrée sur la figure 2.

Trois des pieux (un de chaque longueur) ont été soumis à des charges d'impact au moyen du marteau pendulaire décrit ci-dessus, tandis que les six autres pieux étaient soumis par paires à des charges

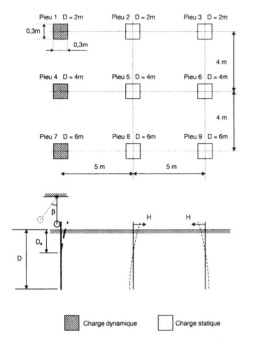

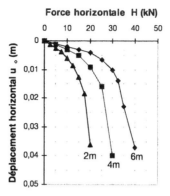

Figure 3 Déplacement horizontal des pieux sous charge statique

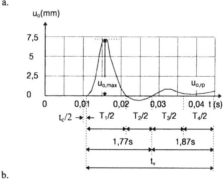

Figure 2 Disposition des pieux sur le site expérimental et principe des essais de pieux soumis à l'action de charges horizontales dynamiques et statiques

Figure 4 Oscillogramme de la force d'impact (a) et de l'amplitude des vibrations (b) pour le pieu de longueur 6 m et un lâcher du marteau d'un angle $\beta = 70$ degrés

statiques appliquées au moyen d'un vérin hydraulique à 0,25 m au-dessus de la surface du sol.

Pour les essais dynamiques, la force d'impact a été imposée en modifiant l'inclinaison initiale de la barre portant le marteau par rapport à la verticale, l'angle b prenant successivement les sept valeurs de 10 degrés, puis 20, 30, 40, 50, 60 et 70 degrés. La force d'impact et le déplacement horizontal du pieu ont été enregistrés au moyen d'un oscilloscope multicanaux.

Pendant chaque essai, on a déterminé les principales caractéristiques dynamiques de l'impact aux différentes étapes du chargement du pieu (amplitude et forme de l'impulsion du choc, durée de son action, forme des oscillations du pieu) et leurs interactions. Les longueurs des pieux ont été choisies pour couvrir les trois schémas de déformation classiques des pieux sous charge horizontale : pieu absolument rigide, pieu de rigidité finie et pieu souple.

Pour les essais statiques, réalisés deux semaines après le battage des pieux, on a attendu pour chaque palier de chargement la quasi-stabilisation des déplacements horizontaux du pieu, définie pour les sols argileux comme une vitesse inférieure à 0,1 mm/2 heures.

5 RESULTATS DES ESSAIS

Les essais de chargement statique horizontal ont permis d'établir la relation entre la charge horizontale appliquée H et le déplacement horizontal du pieu u_o au niveau d'application de la charge, soit à 0,25 m au-dessus de la surface du sol (Figure 3). On voit sur cette figure que les relations $H = f(u_o)$ sont non linéaires et que la résistance opposée par le pieu à l'action des charges statiques horizontales augmente sensiblement avec la profondeur d'encastrement du pieu.

Lors des essais dynamiques, on a enregistré la force d'impact et les oscillations du pieu au cours du temps, pour différentes valeurs de l'angle b de lancement du marteau. La figure 4 présente l'oscillogramme de la force d'impact et des déplacements d'un pieu, qui sont représentatifs des mesures effectuées dans tous les essais.

L'oscillogramme de la force d'impact (Figure 4.a) montre que la durée du choc est égale à $t_c = 0,0028$ s, pour un pieu de longueur 6 m et pour un angle de lancement du marteau de $\beta = 70$ degrés. L'analyse des autres oscillogrammes montre que les variations de t_c sont faibles et que presque toutes les valeurs sont comprises entre 0,0024 s et 0,0032 s. Mais il ne semble pas exister de relations entre t_c et les valeurs de l'angle β ou de la profondeur d'encastrement des pieux.

Le paramètre le plus important pour caractériser le comportement d'un pieu soumis à une charge d'impact est la valeur maximale de la force d'impact P_{max} et son intégrale au cours du temps, qui est l'impulsion S. La valeur de P_{max} dépend dans une certaine mesure de l'angle β (énergie de choc) et de la réaction du pieu et du sol, y compris les forces d'inertie).

L'examen de l'oscillogramme des déplacements du pieu (Figure 4.b) sous l'effet du choc décrit par l'oscillogramme de la force d'impact, montre que, par effet d'inertie, le début du déplacement du pieu ne coïncide pas avec le début du choc, mais est décalé d'un temps t à peu près égal à $t_c/2$, et qu'il correspond au pic de la force d'impact P_{max}.

La valeur maximale de l'amplitude de l'oscillation $u_{o,max}$ est atteinte après la fin du choc et les déplacements du pieu continuent sous l'effet des forces d'inertie. C'est pourquoi, lors des déplacements oscillatoires du pieu, la réaction du sol est mobilisée soit sur la face avant du pieu soit sur sa face arrière et la période T des oscillations n'est pas constante, mais varie quelque peu sous l'influence de la résistance à la flexion du pieu, de la réaction du sol et des forces d'inertie, qui changent de signe (Figure 4.b). Comme la résistance du pieu et du sol est assez élevée, le processus oscillatoire s'amortit très vite et la durée t_v des oscillations est comprise entre 0,03 et 0,04 s pour tous les pieux testés.

La période d'oscillation T et son inverse ω, fréquence des vibrations du pieu, dépendent de la longueur D du pieu et de la force d'impact déterminée par l'angle β. On voit sur la figure 5 que l'augmentation de l'angle β (c'est à dire de la force) et de la profondeur d'encastrement D du pieu fait diminuer sensiblement la fréquence ω des vibrations.

Les oscillations du pieu que provoque le choc sont pour l'essentiel élastiques. Les déformations résiduelles des pieux $u_{o,rp}$ (Figure 4.b) sont faibles et dépendent de la longueur du pieu et de la force

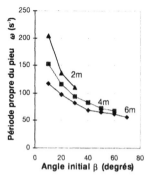

Figure 5 Variations de la fréquence des vibrations du pieu en fonction de l'angle de chute du marteau

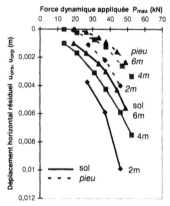

Figure 6 Évolution des déplacements horizontaux résiduels du pieu $u_{o,rp}$ et du sol $u_{o,rs}$ en fonction de la charge appliquée et de la longueur du pieu

d'impact (Figure 6). Dans l'expérimentation rapportée ici, elles varient entre 2 et 4 mm.

Les déformations résiduelles du sol entourant le pieu $u_{o,rs}$ sont plus importantes (Figure 6). Des vides se forment de part et d'autre du pieu et la valeur de la déformation résiduelle dans le sens de la force d'impact $u_{o,rs}^+$ est supérieure à celle qui apparaît dans l'autre sens $u_{o,rs}^-$ (Figure 7). Sous l'action d'ondes sismiques longitudinales et transversales, les contacts du pieu et du sol peuvent être totalement supprimés dans toutes les directions sur la longueur active D_a du pieu.

La formation d'un vide permanent entre le pieu et le sol a permis de déterminer expérimentalement la profondeur D_a de la zone active, dans laquelle le pieu subit des déplacements horizontaux importants. Elle vaut 1,3 m pour le pieu de 2 m, 2 m pour le pieu de 4 m et 2,5 m pour le pieu de 6 m.

Les valeurs mesurées de D_a correspondent à la zone des déformations résiduelles. Au-dessous, les oscillations élastiques du pieu sont égales à celles du sol et l'impact ne produit pas d'effets dangereux. Dans la partie inférieure du pieu, les particules du sol subissent des oscillations de faible amplitude et ne se déplacent pas dans le sens de la propagation des ondes, mais subissent des oscillations forcées autour de leur position d'équilibre, en modifiant l'énergie des ondes. Des conclusions analogues avaient été tirées par Stavnitser et Schekhter (1971) lors de leur étude expérimentale des oscillations horizontales d'un pieu soumis à des ondes sismiques.

On peut donc parler d'une certaine analogie du comportement des pieux soumis à des sollicitations sismiques et des pieux soumis à un impact, d'autant que la durée de ces sollicitations est du même ordre de grandeur que celle des actions sismiques.

C'est pourquoi, compte tenu de la forme des oscillations du pieu, on peut représenter son comportement sous un chargement par choc comme celui d'une console encastrée dans le sol à partir de la profondeur D_a sous la surface. Cette hypothèse est confortée par le fait que l'impact, qui dure une fraction de seconde, exerce sur le pieu une action locale et provoque un déplacement localisé du pieu, qui ne concerne pas toute sa longueur.

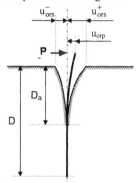

Figure 7   Représentation schématique des déformations résiduelles du pieu et du sol

6 ETUDE THEORIQUE ET INTERPRETATION DES RESULTATS

Considérons un pieu représenté par une console encastrée rigidement dans le sol à la profondeur D_a par rapport à la surface et étudions la forme des vibrations et la fréquence de ses vibrations propres. Pour déterminer la fréquence des vibrations propres de la console, on applique la méthode de Rayleigh. Cette méthode consiste à se donner la forme de la ligne élastique du pieu pour son premier mode de vibration, puis à déterminer les énergies potentielle et cinétique du système « sol-pieu » et enfin, sur la base de l'équation générale de la dynamique, à écrire l'égalité de ces deux énergies, d'où l'on déduit la valeur de la fréquence ω cherchée.

L'équation de la ligne élastique de la console correspond à un quart de sinusoïde :

$$u(z) = u_o\left[1 - sin\frac{\pi z}{2D_a}\right]$$

où u_o est le déplacement horizontal de la console au point d'application de l'impact. Cette équation vérifie les conditions aux limites du problème.

L'énergie cinétique est calculée au moyen de l'expression

$$E_c = \frac{m[v(z)]^2}{2}$$

où m est la masse du pieu par unité de longueur, $v(z)$ est la vitesse de déplacement du pieu à la profondeur z, égale à

$$v(z) = \omega\,u(z)$$

L'énergie cinétique totale $E_{c,tot}$ est obtenue par intégration sur la longueur libre D_a de la console, soit :

$$E_{c,tot} = \frac{m\omega^2}{2} \int_0^{D_a} [u(z)]^2\,dz$$

d'où

$$E_{c,tot} = \frac{m\omega^2 u_o^2}{2} \int_0^{D_a} \left(1 - sin\frac{\pi z}{2D_a}\right)^2 dz$$

Pour sa part, l'énergie potentielle totale des forces de retour élastique du pieu fléchi est donnée par l'expression

$$E_{p,tot} = \frac{1}{2EI} \int_0^{D_a} [M(z)]^2\,dz$$

où M est le moment fléchissant dans le pieu, égal à

$$M(z) = EI\frac{d^2u(z)}{dz^2}$$

EI est la rigidité de la section transversale du pieu.

On obtient finalement

$$E_{p,tot} = \frac{EI\,u_o^2\pi^4}{32\,D_a^4} \int_0^{D_a} sin^2\left(\frac{\pi z}{2D_a}\right)dz$$

En égalant les expressions de $E_{c,tot}$ et $E_{p,tot}$, on obtient finalement la formule de la fréquence propre des vibrations du pieu :

$$\omega = \frac{3,6654}{D_a^2}\sqrt{\frac{EI}{m}}$$

Cette expression est applicable quand il n'y a pas de sol dans la zone de flexion du pieu. Dans la réalité,

439

le sol qui entoure le pieu influence sensiblement le processus de vibration du pieu, qu'il ralentit en provoquant un amortissement très rapide des mouvements de vibration de l'ensemble « sol-pieu ».

La résistance du sol au processus d'oscillation du pieu est prise en compte en introduisant dans l'expression précédente une masse linéaire m_s complémentaire caractérisant la réaction q_s du sol et ses propriétés inertielles, répartie sur la longueur active du pieu. On a alors :

$$\omega = \frac{3,6654}{D_a^2}\sqrt{\frac{EI}{m+m_s}}$$

La masse linéaire m_s est liée à la résistance latérale q_s opposée par le sol au déplacement du pieu par la relation

$$m_s = \frac{q_s}{g}$$

avec

$$q_s = \frac{bku_o}{D_a}\int_0^{D_a}\left[1-sin\frac{\pi z}{D_a}\right]dz$$

  b - dimension transversale du pieu (m),
  k - coefficient de réaction du sol (kN/m^3),
  g - accélération de la pesanteur.
On obtient finalement pour ω l'expression :

$$\omega = \frac{3,6654}{D_a^2}\sqrt{\frac{EI\,g}{m+0,363bku_o}}$$

Considérons maintenant l'oscillation du pieu sous l'action d'une impulsion instantanée. L'impulsion correspond à l'impact du marteau, conformément au schéma de la figure 1. Sa valeur S est donnée par la formule

$$S = P\cdot t = m(u_1 - u_2)$$

avec
  P - force d'impact,
  t - durée de l'impact,
  m - masse du marteau (m=Q/g),
  u_1 et u_2 - vitesses du marteau avant et après le choc.

Si l'on admet que la vitesse du marteau après le choc est nulle (u_2=0), on obtient

$$S = mu_1 = \frac{Q}{g}u_1$$

La théorie du calcul dynamique des structures indique que l'impact S est lié à la charge statique équivalente P_e par la relation

$$P_e = \omega\,S = \omega\frac{Q}{g}u_1$$

Cette équation contient deux inconnues, P_e et u_1. On obtient leurs expressions en égalant l'énergie potentielle du choc dans le cas d'un marteau lâché de l'angle β et l'énergie cinétique au moment du choc :

$$E_p = mgh = mgr(1-cos\beta)\ ;$$

$$E_c = \frac{mu_1^2}{2}\ ;$$

d'où

$$u_1 = \sqrt{2gr(1-cos\beta)}$$

et

$$P_e = \omega Q\sqrt{\frac{2r(1-cos\beta)}{g}}$$

Finalement, l'expression de la force statique équivalente P_e est donc

$$P_e = \frac{3,6654}{D_a^2}Q\sqrt{\frac{2EIr(1-cos\beta)}{m+0,363\,bku_o}}\ .$$

En utilisant cette expression et les résultats des essais dynamiques, on peut déterminer la valeur de la force équivalente P_e pour différentes longueurs de pieux et différentes valeurs des actions extérieures dynamiques.

  Pour les calculs, les valeurs suivantes ont été attribuées aux paramètres :
  E = 30 GPa,
  Q = 1 kN,
  r = 1,5 m
  k = 22 MN/m^3,
  b = 0,3 m,
  m = 2,25 kN/m.
Les résultats des calculs et des mesures sont donnés dans le tableau 1.

L'examen de ce tableau montre que les efforts statiques (H, P_e) et dynamiques (P_{max}) sont liés par une relation que l'on peut écrire sous la forme approchée suivante :

$$P_e = H = 0,5P_{max}$$

La relation entre la charge statique équivalente P_e et la force dynamique maximale P_{max} est la plus stable et les écarts ne dépassent jamais 10% dans le cas considéré.

  La relation entre P_e et la charge statique H est moins stable pour les pieux de longueurs 4 m et 6 m et pour les faibles valeurs des déplacements horizontaux. Ainsi, pour des valeurs du déplacement horizontal u_o du pieu comprises entre 2 et 4 mm, l'écart entre H et P_e est de 20 à 25%. Dans tous les autres cas, la relation entre ces paramètres est satisfaisante.

Tableau 1 Comparaison des calculs et des mesures sur les pieux soumis à des charges statiques et dynamiques

a. Pieu de longueur D = 2 m

b degrés	u_o cm	P_e kN	P_{max} kN	P_e/P_{max}	H kN	H/P_e
10	1	13,9	27,5	1,98	13,4	0,96
20	1,8	18,5	56,1	1,95	17	0,92
30	2,5	22,4	45,2	2,02	18,5	0,82

b. Pieu de longueur D = 4 m

b degrés	u_o cm	P_e kN	P_{max} kN	P_e/P_{max}	H kN	H/P_e .
10	0,4	10	21	2,1	12	1,2
20	0,6	15,7	31,2	1,99	16	1,02
30	0,8	18,8	37,2	1,98	18,5	0,98
40	0,99	22,1	43,8	1,98	20,5	0,93
50	1,1	24	47,6	1,98	21,5	0,89
60	1,2	26,4	52,3	1,98	22	0,83

c. Pieu de longueur D = 6 m

b degrés	u_o cm	P_e kN	P_{max} kN	P_e/P_{max}	H kN	H/P_e
10	0,2	7,96	16,3	2,05	10	1,25
20	0,3	12,9	26	2,02	15	1,16
30	0,4	16,4	33,1	2,02	18,5	1,13
40	0,52	18,3	37,2	2,03	21,2	1,16
50	0,56	21,7	43,9	2,02	22,5	1,04
60	0,62	24,2	49,3	2,04	24,3	1,004
70	0,7	25,4	51	2,01	26	1,02

Il résulte de ce qui précède que, toutes les autres conditions étant égales, un pieu soumis à une charge horizontale dynamique sous forme d'un choc instantané peut supporter une charge deux fois plus grande que dans le cas d'une charge statique.

7 CONCLUSION

La modélisation des processus d'interaction dynamique entre pieux et sols soumis à des charges sismiques est généralement simplifiée sous forme d'un calcul statique de l'effet de charges statiques équivalentes. Les essais en vraie grandeur présentés ici ont permis de définir le mécanisme d'interaction du pieu et du sol et de mettre en évidence les principales lois de comportement du pieu sous l'effet de l'impact.

Les essais ont montré que les oscillations du pieu provoquées par un impact sont essentiellement élastiques. Les déformations résiduelles du pieu sont faibles et sont de l'ordre de quelques millimètres. Par contre, les déformations résiduelles du sol ont été importantes et étaient plusieurs fois supérieures aux déformations résiduelles du matériau du pieu. Les oscillations de la partie inférieure du pieu diffèrent très peu des oscillations du sol en l'absence de pieu. On peut ainsi représenter le comportement d'un pieu soumis à un impact comme celui d'une poutre en console encastrée dans le sol à une certaine profondeur à partir de la surface. Cette hypothèse est confirmée aussi par le fait que le choc,

qui dure quelques fractions de seconde, exerce sur le pieu une action locale qui provoque un déplacement local du pieu, qui ne concerne pas toute sa longueur.

Ce modèle de calcul du pieu a permis d'obtenir des formules de calcul des fréquences propres des oscillations du pieu et de la valeur de la force statique équivalente, qui sont en accord satisfaisant avec les résultats des essais.

REFERENCES BIBLIOGRAPHIQUES

Ejjaaouani H., Chakhirev V. (1995) Les problèmes de construction dans les régions sismiques. La Recherche Appliquée au LPEE, Bilan 1995.

Hakimi A., Acharhabi A., Cherrabi A. (1993a) Activités scientifiques et techniques marocaines pour la maîtrise des constructions vis-à-vis des séismes (1ère partie). La Revue Marocaine du Génie Civil, n° 44.

Hakimi A., Acharhabi A., Cherrabi A. (1993b) Activités scientifiques et techniques marocaines pour la maîtrise des constructions vis-à-vis des séismes (2ème partie). La Revue Marocaine du Génie Civil, n° 46.

Hakimi A., Acharhabi A., Cherrabi A. (1993c) Activités scientifiques et techniques marocaines pour la maîtrise des constructions vis-à-vis des séismes (3ème partie). La Revue Marocaine du Génie Civil, n° 47.

Hakimi A., Acharhabi A., Cherrabi A. (1993) Activités scientifiques et techniques marocaines pour la maîtrise des constructions vis-à-vis des séismes (4ème partie). La Revue Marocaine du Génie Civil, n° 48.

Stavnitser L.R., Schekhter O.Y. (1971) Les vibrations forcées sous l'action des ondes sismiques (en russe). Osnovaniya, fundamenty i mekhanika gruntov, Moscou, n°5/1971.

Geotechnics for Developing Africa, Wardle, Blight & Fourie (eds) © 1999 Balkema, Rotterdam, ISBN 90 5809 082 5

Analysis of progressive failure of earth slopes by finite elements

J.B.Lechman
Geomechanics Research Center, Colorado School of Mines, Fort Collins, Golden, Colo., USA

D.V.Griffiths
Division of Engineering, Colorado School of Mines, Fort Collins, Golden, Colo., USA

ABSTRACT: Slope stability analysis is traditionally performed using limit equilibrium methods that have remained essentially unchanged for decades. While giving generally conservative estimates of safety factors, the traditional methods give no indication of progressive failure or how the yield spreads. This paper will describe some finite element analyses of slope stability using a relatively simple elasto-plastic, Mohr-Coulomb soil model. Within the finite element model however, gravity can be applied in different ways. For example, a "gravity turn on" procedure can be used where an initially weightless slope is instantaneoulsy subjected to self-weight loads. Alternatively, the slope can be "built-up" using an embanking procedure that creates the mesh one lift at a time, or similarly, by an excavation procedure. The different loading strategies and their influence on yeilding of the slope is highlighted in the paper through the use of contour plots of the failure criterion, which indicate the spread of yield within the slope and hence the location and shape of the potential failure surface. The influence of the loading strategies and dilation on the ultimate slope factor of safety is also examined.

1. INTRODUCTION

Prior to the advent of modern computing techniques, methods for determining the stability of slopes were, by necessity, a matter of making various assumptions to allow for the solving of equations of static equilibrium. The results of these limited, simplified analyses are presented in the form of charts of stability numbers (e.g. Taylor 1937, Bishop and Morgenstern 1960, Spencer 1967, Janbu 1967, Cousins 1978) from which the factor of safety can be determined based on the soil's strength properties and slope geometry. Common to all methods from which the charts are derived is the assumption that the slope can be divided into slices. After this initial step, further assumptions are made in order to solve the problem of static indeterminacy created by the side forces acting on the failing mass slice. While the variety of assumptions in dealing with these side forces are in generally good agreement with regards to overall safety factor, they may produce large discrepencies in the distribution of stresses throughout the failure mass (Whitman and Bailey 1967, Wright, et al. 1973). Cousins (1978) also lists several restictions to the use of these charts which deal with matters ranging

from pore pressure, to limited slope angle range, to limited slope geometry, to limited or the lack of information about the failure surfaces. These charts are expedient and relatively easy to use, giving conservative values for the safety factor; however the aforementioned problems with limit equilibrium methods coupled with advances in computational ability beg the question of the possibility of more robust methods for analyzing slope stability.

One such method is the finite element method. The advantages of the method over limit equilibrium methods are stated by Griffiths (1996) as:

1. No assumption needs to be made in advance about the shape or location of the failure surface. Failure occurs "naturally" through the zones within the soil mass in which the soil shear strength is unable to sustain the applied shear stresses.

2. Since there is no concept of "slices" in the finite element approach there is no need for assumptions about slice side forces.

3. If realistic soil compressibility data is available, the finite element solutions will give

information about deformations at working stress levels.

4. The finite element method is able to monitor progressive failure up to and including overall shear failure.

This method has shown itself to be in good agreement with the various charts and proves to be very robust in that complications arising from the geometry of the slope and material property variations can be easily managed by it (Zienkiewicz et al. 1975, Griffiths 1980, 1989, Matsui and San 1992, Griffiths and Lane 1997). Advances made in refining the method and its applications may well prove to be a defining step in the maturation of soil mechanics. This paper examines the fourth item listed above via contour plots of the failure criterion as it is affected by gravity loading and dilation angle.

2. METHOD OF ANALYSIS

2.1 Finite element method used

The program is based closely on Program 6.1 in the text by Smith and Griffiths (1988)–the main difference being the ability to model more realistic geometries and better graphical output facilities. The programs are for 2-d (plane strain) analysis of elastic-perfectly plastic soils with a Mohr-Coulomb failure criterion. The programs use 8-node quadrilateral elements with reduced integration (4 Gauss-points per element) in both the stiffness and stress redistribution phases of the algorithm. A gravity 'turn-on' procedure generates nodal forces which act in the vertical direction at all nodes. These loads are applied in a single increment, or in percentage increments which generate normal and shear stresses at all the Gauss-points within the mesh. These stresses are then compared with the Mohr-Coulomb failure criterion. If the stresses at a particular Gauss-point lie within the Mohr-Coulomb failure envelope then that location is assumed to remain elastic. If the stresses lie on or outside the failure envelope, then that location is assumed to be yielding. Global shear failure occurs when a sufficient number of Gauss-points have yielded to allow a mechanism to develop.

The analysis is based on an iterative Modified Newton-Raphson method called the Viscoplastic algorithm (Zienkiewicz et al 1975). The algorithm forms the global stiffness matrix once only with all nonlinearity being transferred to the right hand side. If a particular zone within the soil mass is yielding, the algorithm attempts to redistribute those excess stresses by sharing them with neighboring regions that still have reserves of strength. The redistribution process is achieved by the algorithm generating self-equilibrating nodal forces which act on each element that contains stresses that are violating the failure criterion. These forces, being self-equilibrating, do not alter the overall gravity loading on the finite element mesh, but do influence the stresses in the regions where they are applied. In reducing excess stresses in one part of the mesh however, other parts of the mesh that were initially 'safe' may now start to violate the failure criterion themselves necessitating another iteration of the redistribution process. The algorithm will continue to iterate until both equilibrium and the failure criterion at all points within the soil mass are satisfied within quite strict tolerances. Convergence is achieved in a global sense by observing the change in nodal displacements from one iteration to the next. Convergence is satisfied when this change is less than 0.1%.

If the algorithm is unable to satisfy these criteria at all yielding points within the soil mass, 'failure' is said to have occurred. Failure of the slope and numerical non-convergence occur together, and are usually accompanied by a dramatic increase in the nodal displacements. Within the data, the user is asked to provide an iteration ceiling beyond which the algorithm will stop trying to redistribute the stresses. Failure to converge implies that a mechanism has developed and the algorithm is unable to simultaneously satisfy both the failure criterion (Mohr-Coulomb) and global equilibrium.

2.2 Soil model

An elastic-perfectly plastic (Mohr-Coulomb) model has been used in this work consisting of a linear (elastic) section followed by a horizontal (plastic) failure section.

The soil model used in this study consists of six parameters as shown in Table 1.

The dilation angle ψ affects the volume change of the soil during yielding. In this simple model ψ is assumed to be constant which is unrealistic, but will serve to show the affect of dilation on the final failure region.

The parameters c' and ϕ' refer to the cohesion and friction angle of the soil. Although a number of failure criteria have been suggested for use in representing the strength of soil as an engineering material, the one most widely used in geotechnical practice is due to Mohr-Coulomb. In terms of principal stresses and assuming a compression-negative sign convention, the criterion can be written as follows:

$$F = \frac{\sigma_1' + \sigma_3'}{2} \sin \phi' - \frac{\sigma_1' - \sigma_3'}{2} - c' \cos \phi' \quad (1)$$

where c' and ϕ' represent the shear strength parameters of the soil and σ_1' and σ_3' the major and minor principal effective stresses at the point under consideration. The failure function F can be interpreted as follows:

$F < 0$ inside M-C envelope (elastic)
$F = 0$ on M-C envelope (yielding)
$F > 0$ outside M-C envelope (yielding) and stresses must be redistributed

The unit weight γ assigned to the soil is important because it is proportional to the nodal loads generated by the gravity turn-on procedure.

In summary, the most important parameters in a finite element slope stability analysis are the unit weight γ which is directly related to the nodal forces trying to cause failure of the slope, and the shear strength parameters c' and ϕ' which measure the ability of the soil to resist failure.

2.3 Determination of the factor of safety

The Factor of Safety (FOS) of a soil slope is defined here as the factor by which the original shear strength parameters must be reduced to bring the slope to the point of failure. The factored shear strength parameters c'_f and ϕ'_f, are therefore given by:

$$c'_f = c'/FOS \tag{2}$$

$$\phi'_f = \arctan(\frac{\tan \phi'}{FOS}) \tag{3}$$

This method has been referred to as the 'shear strength reduction technique' (e.g. Matsui and San 1992) and allows for the interesting option of applying different factors of safety to the c' and $\tan \phi'$ terms. In this paper however, the same factor is always applied to both terms. To find the 'true' factor of safety, it is necessary to initiate a systematic search for the value of FOS that will just cause the slope to fail. This is achieved by the program solving the problem repeatedly using a sequence of user-specified FOS values.

However, when gravity is applied incrementally, the strength parameters are held constant and trail safety factor are held constant as loading is increased (compare Figures 1 and 4 or 2 and 3).

2.4 Visualization technique

For each successive trial safety factor or gravity load increment, the program described above gives a value of the failure criterion at each Gauss-point at convergence or failure (F). The coordinates of each Gauss-point and the corresponding value of F are then written to an output file which can be used in a commercial software package to produce contour plots which show the yielding regions of the slope. Of interest in the plots is the affects of an associated flow rule ($\phi = \psi$) on the spread

of yield through the slope. It is also interesting to compare the results with traditional methods, namely those reported by Cousins (1978) which were arrived at by a modified Taylor method. The charts produced by Cousins are of particular interest because they not only give the stability factor but also the depth that the failure surface extends into the foundation of the slope. The finite element results can be compared to these results on the basis of overall factor of safety, or, when the failure criterion is contoured, the two methods can be compared on the basis of the location of the failure surface.

3. RESULTS

For the purposes of this paper, a simple, dry slope composed of a soil exhibiting friction and cohesion was analyzed. In general, this simplification is not necessary in the finite element method. However, when comparing the results to traditional methods, simplification is necessary on the grounds that limit equilibrium methods typically require this in order that the problems may be solved with ease. The choice of geometries corresponds with specific points on Cousins' charts at which the depth factor (D, the amount by which the height of the slope should be multiplied by to give the maximum height of the failure surface) can be determined explicitly. It should be noted that the definition of D in the context of Cousins' paper differs from other definitions of D. Cousins' paper uses D as a multiplier of the height of the slope (H) to give the maximum depth through which the failure surface passes and not as a multiplier of the height to give the depth to a solid base. Defined this way, a depth factor equal to 1 (D=1) implies a toe failure, where D=1.5 implies a failure surface that extends into the foundational material a distance equal to half the height and does not necessarily pass through the toe of the slope.

Furthermore, gravity loading methods are examined as they relate to the spread of yield and the shape and location of the yielding zone at failure.

Also, an examination of the influence of dilation on the spread of yield and the final state of the yielding zone at failure is contained here. The analyses in this paper have, by virtue of the variability of the amount of actual volume change, been limited to two simple cases. The cases considered are where $\psi = 0$ and $\psi = \phi'$. When the angle of dilation equals the internal angle of friction ($\psi = \phi'$) the flow rule is "associated"; thus direct comparisons to classical plasticity can be made. Some results of the analysis are as follows.

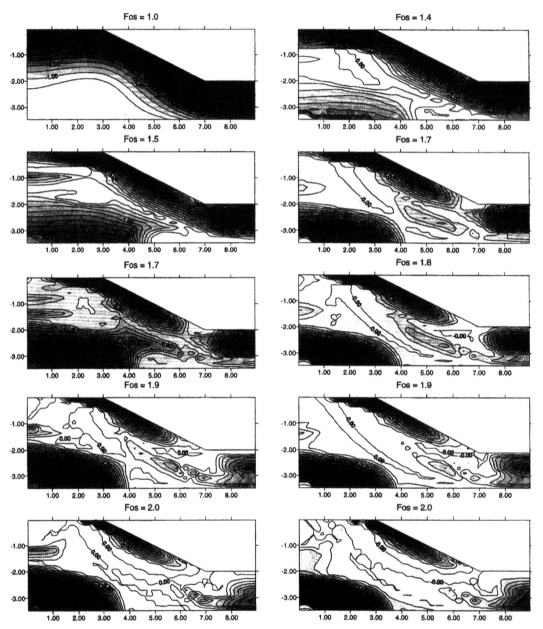

Figure 1. Embankment with a 2:1 slope, $c'/(\gamma H) = 0.15$, $\phi' = \psi = 16.7°$, and gravity applied instantaneously. Plots show contours of F at various FOS values. White regions are yielding $(F = 0)$.

Figure 2. Embankment with a 2:1 slope, $c'/(\gamma H) = 0.15$, $\phi' = 16.7°$, $\psi = 0°$, gravity applied instantaneously. Plots show contours of F at various FOS values. White regions are yielding $(F = 0)$.

Figures 3 and 4 show zones of yielding for an embankment in which the soil properties remain constant and gravity is appled gradually.

4. CONCLUSIONS

Conclusions drawn from the investigation can be summarized as follows:

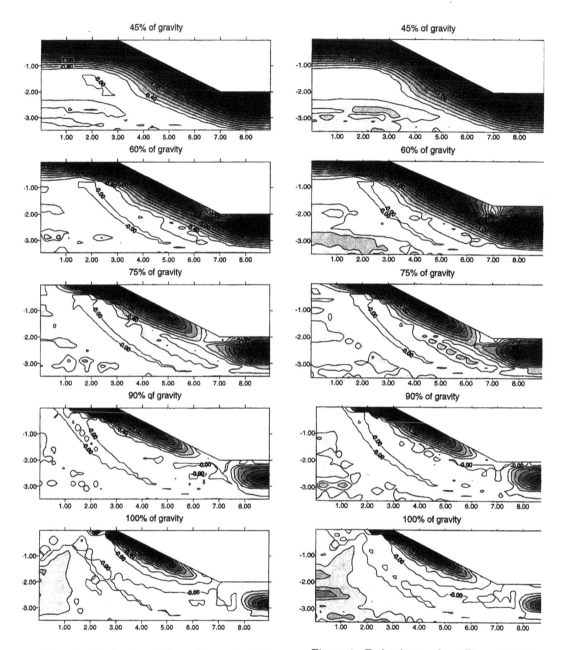

Figure 3. Embankment from Figure 2. Plots show contours of F at $FOS = 2$ with gravity applied incrementally. White regions are yielding ($F = 0$).

Figure 4. Embankment from Figure 1. Plots show contours of F at $FOS = 2$ with gravity applied incrementally. White regions are yielding ($F = 0$).

4.1 *Comparisons to Cousin's results*

1. Agreement was good for overall factor of safety. For the slope in the previous figures,

Cousins' charts give a factor of safety of 2.03 compared to a factor of safety of 2.0 for the finite element analysis.

2. Cousins' charts also give a depth factor (D) of 1.25 which is approximately equal to that indicated by the finite element analysis. In general the finite element analysis showed more "transitional" regions between the explicit depth factors of $D = 1.0$, $D = 1.25$, $D = 1.5$ considered by Cousins. That is to say the bottom of the failure surface was not necessarily at a depth exactly equal to that implied by Cousins' depth factors. This is to be expected. As stated earlier, a priori knowledge of the failure surface is not required in finite element analysis as it is in limit equilibrium analysis; therefore the failure surface is allowed to develop "naturally." Although the two failure surfaces are not arrived at by the same means, the results indicate that Cousins' charts agree well with finite element results.

4.2 Effects of Dilation

1. Overall factor of safety did not vary significantly (less than 5%), for the same slope geometry and soil strength parameters, between the two cases.

2. At failure, the yielding zones of the two cases appear to be equivalent in size and shape (See Figures 1 and 2, $FOS = 2.0$).

3. Although the failure zones were similar at failure, the spread of yield could vary significantly between the two cases. It may be argued that the associated flow rule gives a more "satisfying" failure in that the spread of yield seems to occur in a more intuitive sense.

4. Under normal gravity conditions for a given slope geometry and soil strength parameters (represented by a factor of safety equal to one in this analysis), regions within the slope may already be experiencing stress conditions that promote yielding of the slope material. These regions were exceptionally large in slopes with relatively small safety factors (less than 1.75). Usually, these regions of yield appeared in the $\psi = 0$ case, and corresponded with a deep failure surface. For the associated flow rule case, this tendency was not as prominent indicating that the stress redistribution algorithm was more "effective" here.

4.3 Effects of applying gravity incrementally

Reults of this analysis are shown in figures 3 and 4. Here the trial safety factor was set to the value that caused failure in the analysis where gravity was applied all at once. The shear strength parameters were factored accordingly, and held constant. Gravity was then applied incrementally untill failure occurred at 100%.

1. Overall factor of safety was virtuall unchanged. For both dilation cases.

2. The general shape and location of the yielding zones appear to be equivalent at failure while some differences can be seen at other trial factor of safety values.

3. For both cases considered, a "deep" failure mechanism developed first, but apparently was resisted by frictional forces along the rigid foundation below. In general, when cohesion "dominates" the soils ability to resist shearing, a deep failure is expected. This is particularly evident in Taylor's chart for undrained clays ($\phi' = 0$), but is also appearent in Cousins' charts for frictional/cohesive materials.

REFERENCES

[1] A.W. Bishop and N.R. Morgenstern. Stability coefficients for earth slopes. *Géotechnique*, 10:129–150, 1960.

[2] B.F. Cousins. Stability charts for simple earth slopes. *J Geotech Eng, ASCE*, 104(2):267–279, 1978.

[3] D.V. Griffiths. *Finite element analyses of walls, footings and slopes*. PhD thesis, Department of Engineering, University of Manchester, 1980.

[4] D.V. Griffiths. Advantages of consistent over lumped methods for analysis of beams on elastic foundations. *Comm Appl Numer Methods*, 5(1):53–60, 1989.

[5] D.V. Grifiths. Fe-emb2 and fe-emb1: Slope stability software by finite elements. Technical Report GRC-96-37, Colorado School of Mines, 1996. Geomechanics Research Center.

[6] D.V. Grifiths and P.A Lane. Slope stability analysis by finite elements. Technical Report GRC-97-47, Colorado School of Mines, 1997. Geomechanics Research Center.

[7] N. Janbu. Slope stability computations. In *Soil Mech. and Found. Engrg. Rep.* Technical University of Norway, Trondheim, Norway, 1968.

[8] T. Matsui and K-C. San. Finite element slope stability analysis by shear strength reduction technique. *Soils Found*, 32(1):59–70, 1992.

[9] I.M. Smith and D.V. Griffiths. *Programming the Finite Element Method*. John Wiley and Sons, Chichester, New York, 2nd edition, 1988.

[10] E. Spencer. A method of analysis of the stability of embankments assuming parallel interslice forces. *Géotechnique*, 17(1):11–26, 1967.

[11] D.W. Taylor. Stability of earth slopes. *J. Boston Soc. Civ. Eng.*, 24:197–246, 1937.

[12] R.V. Whitman and W.A. Bailey. Use of computers for slope stability analysis. *J Soil Mech Found Div, ASCE*, 93(SM4):475–498, 1967.

[13] S.G. Wright, F.H. Kulhawy, and J.M. Duncan. Accoracy of equilibrium slope stability analysis. *J Soil Mech Found Div, ASCE*, 99(SM10):783–791, 1973.

[14] O.C. Zienkiewicz, C. Humpheson, and R.W. Lewis. Associated and non-associated viscoplasticity and plasticity in soil mechanics. *Géotechnique*, 25:671–689, 1975.

Geotechnics for Developing Africa, Wardle, Blight & Fourie (eds) © 1999 Balkema, Rotterdam, ISBN 90 5809 082 5

Design and construction of the asphaltic concrete core at Ceres Dam

G. A. Jones & A. C. White
SRK Consulting, Johannesburg, South Africa

W. J. Schreuder
VKE Engineers, Cape Town, South Africa

ABSTRACT : The new Ceres Dam is unique in Southern Africa in having an asphaltic concrete core. This is particularly appropriate for the site since the core provides the required flexibility to sustain Maximum Credible Earthquakes of magnitude 7.0 and high differential settlements due to variable founding conditions. In addition the asphalt core is cost-effective because of local unavailability of clay materials and because it allowed all-weather construction.

The asphaltic concrete design is similar to that for road pavements but the emphasis is on minimum permeability - achieved by dense mixes (<2% air voids), and maximum flexibility-achieved by using bitumen contents of about 7%.

Rigorous monitoring of the computer controlled batch plant ensured that the mix produced as correct. Hence the core laying proceeded according to specification and within a tight schedule.

1 INTRODUCTION

The new Ceres Dam was constructed in 1997/98 for the Ceres Municipality and the Koekedouw Irrigation Board and is intended for domestic and irrigation water supply for these authorities. It replaces an old concrete gravity arch dam at the same site which no longer had sufficient capacity and had also suffered earthquake loading during the 1969/70 period when severe damage to houses had occurred at the nearby town of Tulbagh.

The earthquake potential and variable founding conditions, together with the tight curvature of the dam wall in a narrow valley necessitated a flexible core, hence the choice of asphaltic concrete.

The new dam was designed by Steffen Robertson and Kirsten in conjunction with Steinhobel Keller and Chantler. Specialist advice was provided by Professor Höeg from the Norwegian Geotechnical Institute (NGI) on both the seismic and the asphalt core design and construction. VKE Engineering assisted with the asphalt concrete design and construction supervision.

Grinaker Construction were awarded the tender for the construction of the dam and ancillary works. A joint venture of Colas and KOLO from Norway were the nominated specialist sub-contractors for the asphalt core, the latter having considerable international experience in such construction.

The detailed site selection was constrained by the presence of the existing dam which already utilised the optimum site and had to be maintained in operation. The final solution entailed an upstream storage coffer dam and the old dam wall, after partial demolishing, being contained within the downstream wall of the new rockfill dam.

The dam characteristics are given in Table 1.

Table 1. Dam characteristics

Characteristic	Value
Crest Level	646 m
Full supply Level	642 m
Maximum Height	60 m
Crest Length	275 m
Crest Width	7 m
Upstream Slope	1V : 1.6H
Downstream Slope	1V : 1.4 H
Capacity	17.2 mill m^3
Safety Evaluation Discharge	710 m^3/s

2 SITE DESCRIPTION, GEOLOGY AND SEISMOLOGY

The dam is located about 4 km west of the town of Ceres in the valley of the Koekedouw River which lies between the Witsenberg and Skurweberg mountain ranges. Figure 1.

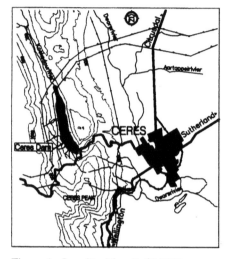

Figure 1 : Locality Plan (1:50 000)

Both the Witsenberg and Skurweberg ranges have been subjected to major folding. The resulting geology is complex and the rocks of the Nardouw Sandstone Formation of the Table Mountain Group comprise thickly bedded and weathered felspathic sandstone and shales, quartzitic sandstones and massively bedded quartzites. The rocks are generally highly jointed with regular bedding plane joints and closely spaced cross joints with numerous fault zones.

The river course approximately follows the trough of a syncline at the dam site but the presence of major fold hinge zones significantly complicates the geology. In the river course itself, under the centre of the dam wall, the relatively thin recent river sediments are underlain by highly weathered extremely friable soft sandstone such that for much of this zone careful triple tube drilling resulted in the recovery of no more than loose sand. The implication of this is that whereas the steep flanks of the valley comprise sound rock, albeit extensively jointed and steeply dipping, the centre section consists of very soft rocks to depths of up to 35 metres, hence the potential for significant differential settlements under the wall has to be recognised.

Recent seismic events in the area (1969/70) indicated that a specific investigation of the local seismicity was necessary. Professor Knight of Knight, Hall, Hendry carried out this investigation. This resulted in the dam site being assigned a Hazard Class IV (extreme) rating and the associated Risk Class IV rating, using ICOLD Bulletin 72/1989. The dam was therefore designed to withstand the Operating Basis Earthquake (OBE) with defined acceptable factors of safety. (Meintjes and Jones, 1999). The dam must also withstand the Maximum Design Earthquake (MDE) with acceptable damage ie. damage to the structure can be tolerated provided no catastrophic flooding occurs. For a high hazard and risk dam the MDE is defined as equal to the Maximum Credible Earthquake (MCE).

The seismic loading parameters for these design earthquake criteria are given in Table 2.

Table 2. Seismic loading parameters

Category/ Magnitude	Horizontal			Vertical		
	A	V cm/s	D cm	A	V cm/s	D cm
OBE 6.5	0.24g	27	14	0.13g	16	8
MDE 7.0 (=MCE)	0.35g	42	21	0.21g	25	12

Reservoir induced seismicity (RIS) was also considered, but since this was estimated as not exceeding magnitude 6.5 (ie. equivalent to the OBE) no further measures were required to handle any RIS.

The above loading parameters are those which are used for pseudo static design; they are severe criteria. It was therefore necessary to carry out full dynamic analyses from which estimates were made of the deformations of the dam wall and hence of the core under the most severe but realistic earthquake loads.

These estimates became a vital component in the design of the core, which in addition had to accommodate the differential settlement induced strains.

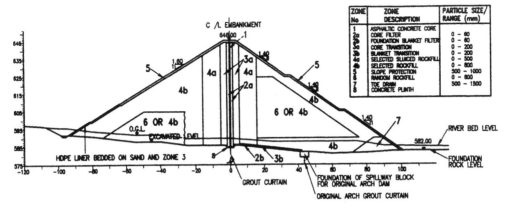

ZONE No	ZONE DESCRIPTION	PARTICLE SIZE/ RANGE (mm)
1	ASPHALTIC CONCRETE CORE	
2a	CORE FILTER	0 - 80
2b	FOUNDATION BLANKET FILTER	0 - 80
3a	CORE TRANSITION	0 - 200
3b	BLANKET TRANSITION	0 - 200
4a	SELECTED SLUICED ROCKFILL	0 - 800
4b	SELECTED ROCKFILL	0 - 800
5	SLOPE PROTECTION	500 - 1000
6	RANDOM ROCKFILL	0 - 800
7	TOE DRAIN	500 - 1500
8	CONCRETE PLINTH	

Figure 2 : Cross section through dam wall

3 ASPHALT CORES

Asphalt cores used in modern dams are generally thin vertical membranes positioned at the dam wall centreline. See Figure 2. The asphalt mixes are similar to continuously graded road mixes. The emphasis, however, is changed from mixes which have high strengths and resilient moduli to those having minimum permeability and, particularly where flexibility is required, as at Ceres, low moduli. The required characteristics are achieved by using higher bitumen contents and also higher filler contents than would be used for road mixes. In other respects typical dam core asphalts are very similar to road asphalts both in manufacturing and laying techniques. Unlike road asphalts, however, where occasional sub-standard batches could be tolerated or readily replaced or repaired, the asphalt and its laying in dam cores must be controlled with a very high level of confidence.

The experience of many asphalt core dams in Norway and elsewhere has led to the overriding philosophy that once the mix has been designed for the particular application the emphasis should be on simple but tight control of the manufacturing and laying. These can only be achieved by using good quality equipment and experienced personnel.

Although at this site flexibility of the core is a critical criterion, of equal importance is the permeability. This can be directly measured in the laboratory using adaptations of conventional soil testing equipment, but the difficulties in such testing are considerable, largely because of the very low range of permeabilities being measured. An indirect measure is preferred which in this case is the air voids content. Many tests have been conducted of the relationship between air voids and permeability and these are summarised in Figure 3 (Höeg, 1993)

It is generally understood that the design permeability of the asphalt should not be greater than about 10^{-9} cm/sec hence from Figure 3 the air voids should not exceed about 3%. The asphalt mix design to achieve this relatively low air voids (compared with road pavement mixes) entails using aggregate gradings complying with a Fuller's curve gradation – Figure 4 - and high bitumen contents. In practice a range of mix designs and bitumen contents are tested in the laboratory and conventional Marshall tests are conducted in which a target maximum air voids of 2% is set to allow for differences between laboratory and field conditions.

A range of mixes could be designed to achieve the air voids and hence permeability criteria having a range of bitumen contents and hence stiffnesses. For any particular dam the required stiffness will determine the bitumen content. Hence at Ceres Dam, because of earthquake and differential settlement considerations, the design bitumen content was raised to 7% and a 150/200 penetration grade bitumen selected. A high bitumen content necessitates the use of a filler which may be cement, lime or crushed aggregate the choice of which depends on the workability and the costs.

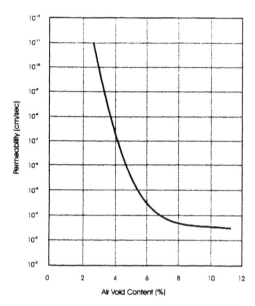

Figure 3: Air voids v permeability

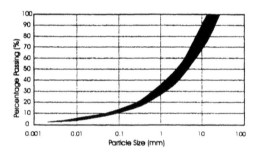

Figure 4: Aggregate grading envelope

The design of the mix for Ceres Dam are discussed in the following section.

4   ASPHALTIC CONCRETE DESIGN

Because of the local geology, seismology and topography the choice of rockfill for the dam wall as opposed to earthfill or concrete became clear. The combination of rockfill and asphalt core gave the most earthquake tolerant structure as well as providing other significant advantages. These are that at this site no suitable clay core material existed within economical haul distances and also that construction of an asphalt core can continue in all but the wettest weather. Since the construction programme was constrained to fit into well defined seasons, significant weather delays could have delayed completion by one year which would have very severely affected water supply to Ceres for irrigation.

The need for about 640 000m^3 of rockfill for the dam wall and 5000 m^3 of high quality aggregate for the concrete spillway, intake tower and other works and a further 5 000 m^3 for the asphalt core led to a search for aggregate sources within the dam basin. Fortunately an adequate source was located close to the dam wall although selection to separate the higher quality material was necessary.

This consisted of sandstones and quartzitic sandstones of the Nardouw Sandstone Formation of the Table Mountain Group. The drilling and laboratory testing showed that by appropriate selection a sufficient quantity of satisfactory aggregate for the asphalt core was available. In practice, however, because of programming constraints the site aggregate, which was located at lower elevations in the quarry, was not available for the early asphalt core laying and a commercial source was used.

No specifications have been set for asphalt core aggregates other than those used for pavements.   Unquestionably these are adequate but they may well be unnecessarily high since the asphaltic concrete in dam cores is certainly not subject to the same stresses and environmental influences as in road pavements.   Nevertheless it is considered prudent to use road specifications for dam cores and allow some latitude if the former specifications are not met.   In general the latitude would be with regard to the aggregate strength characteristics rather than shape or absorption since the latter could adversely influence the workability and stiffness which are more important considerations than strength.

Specifications for the asphaltic concrete aggregate and the site test results are shown in Table 3 and in Figure 4.   The aggregate grading should follow the Fuller curve within fairly close tolerances.

For asphalt cores the emphasis of bitumen specification has been pragmatic and standard road pavement bitumens have been used in

Table 3. Asphaltic concrete aggregate specification and test results

Test	Specification	Test Results
ACV	Max 29%	27%
10% FACT		
Wet	82.5 kN	102 kN
Dry	110 kN	121 kN
Wet/Dry	Min 75%	84%
L.A. Abrasion		37%

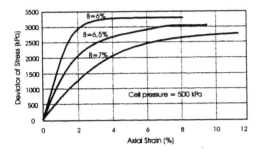

Figure 5: Stress-strain relationship at different binder contents

Norwegian practice, generally of penetration B60 or 65 compacted at a temperature of 160° – 180°C. Grade B180 has also been used at 140° – 155°C. For Ceres it was considered that 150/200 was more suitable because it would provide greater flexibility of the core.

Triaxial tests were performed on various mixes at the Institute for Transport Technology, Stellenbosch. Typical results are summarised in Tables 4 and 5. The former shows in particular how the stiffness

Table 4. Triaxial test results

Binder %	Sample No.	σ_3	Vertical Strain at Failure ϵ_V max	Secant Modulus at 1 % Axial Strain E_{sec} MPa
6.0	61	500	4.2	201
	62	750	5.8	194
	63	1000	3.9	223
6.5	657	500	5.9	115
	652	750	7.0	130
	653	1000	6.2	148
7.0	77	500	6.9	97
	79	750	14.6	80
	71	1000	11.8	65

represented by the secant modulus is very dependent on the bitumen content and this is also illustrated in Figure 5.

Table 5 shows that the Ceres mixes are similar to those used for Norwegian asphalt cores but less stiff which was a required attribute for inter alia the earthquake loading.

Table 5. Comparison Ceres (C) and Norwegian (N) tri-axial test results

Origin	Bitumen	%	Axial Stress at Failure kPa	Secant Modulus at 1 % Strain MPa
N	B180	5.9	4300	140
N	B60	8.0	4400	110
C	150/200	6.0	4739	223
C	150/200	6.5	4645	169
C	150/200	7.0	4861	78

5 CORE CONSTRUCTION

The asphaltic concrete core begins at the base of the dam wall on a reinforced concrete plinth or foundation. At Ceres the plinth is a standard width of 4 m under both the centre section and on the flanks. The plinth was also used as a platform for the foundation grouting.

The plinth, which had an average thickness of 0.75 m, varied from 0.4m up to a maximum of 1.5 m to accommodate changes of ground level. Due to the steepness of the flanks, in places the plinth was at a slope of up to 45 ° which was considered to be the maximum allowable; steeper natural sections were trimmed to this slope.

The plinth is a transition between the grout curtain and the asphalt core and provides a convenient way to start core construction: the plinth should not be viewed as a structural element and must be designed to accommodate differential settlements. At Ceres, because of the poor foundation material in the centre river bed section, this provision was essential and resulted in close spacing of the joints in the plinth and in particular emphasis on the adequacy of the water stops at the joints.

The interface between the core and the plinth is achieved by hand spreading a layer of mastic approximately 1.5 m wide and about 10 mm thick after the concrete has been thoroughly cleaned and acid washed to ensure a good bond. The mastic can be applied some time prior to the core placing because the latter will reheat the mastic to form a bond between the two.

The asphaltic concrete core at Ceres has a constant width of 0.5 m. Higher dams have wider cores, up to 1.0 m, and may be tapered to economise on core quantities.

The asphalt is laid with a conventional road paving machine adapted for the purpose. The paver simultaneously lays the core together with 1.5 m wide filter/transition layers on each side of the core in 0.2 m thick compacted lifts. The filter layers act as shuttering for the asphalt core. They comprise a graded rock which complies with a grading specification providing a filter for the aggregate in the asphaltic concrete and the rockfill shell surrounding the filter/transition zones. (ICOLD, 1992).

$$d_{100\,core} \geq d_{10\,trans} \text{ and } d_{100\,trans} \geq \frac{1}{4}d_{10\,shell}$$

The transition material, of max size 60 mm, should comprise hard crushed rock complying with the following requirements:

$$d_{50} > 10 \text{ mm and } d_{15} < 10 \text{ mm.}$$

Since it is not practicable to operate the paving machine for runs of less than about 30 m length which must be nearly horizontal, the initial few layers at the base of the dam were hand placed to a width of one metre using temporary shutters.

The main shell of the dam must be raised in conjunction with the progress of the core and rules have been developed to control the differences in level between the core and transition zones and the shell.

Compaction of the core is achieved by three self propelled smooth vibratory rollers travelling behind the paver as soon as the asphalt has cooled sufficiently to support the rollers.

For most valley dams the required rate of core production, which has to maintain the rate of rise of the rockfill, is low at the beginning of the project but becomes high towards the end when the crest length is at a maximum, as is the rate of rise. The capacity of both the asphaltic concrete production and laying must

therefore be sufficient to give the required end rate.

Placing at Ceres averaged two layers per day, each 0,2 m thick, with a maximum of three layers per day. Placement of more than three layers per day was not acceptable since the compaction of a fourth layer overlying three still hot underlying layers could not be achieved. The core, 61 m high, was placed in eight months.

In practice the core placing is carried out in cycles of nine to fourteen days of laying followed by five days for testing. At Ceres fourteen cycles were carried out.

6 QUALITY CONTROL

The quality assurance approach more closely follows a method specification than an end result specification in that the emphasis is on the manufacturing and laying process. Essentially if the asphaltic concrete mix is correct - and this is checked both by laboratory testing and trial sections, a similar approach to that adopted for concrete – then provided the paving equipment is adequate and the specified compaction process followed, the end result will be satisfactory. Frequent checks are made of the core by drilling 100 mm diameter cores and testing these for air voids, density and bitumen content. The emphasis of these tests is on the air voids since these give a measure of permeability. (Figure 3). A difficulty with coring to obtain samples is that it cannot be done when the mix is hot since the samples distort and the test results become inaccurate. Time must be allowed for the asphalt to cool.

A result of this is that the test cores are only taken at the top of each cycle at which stage the core and shell may be 5 m higher than the previously tested layer. If a defective layer should be identified, major disruptions to the construction programme and hence high costs would be incurred to correct the defect. This again emphasises the need for a method based specification for the paving and detailed checking of the mix production.

No defective layers were identified during the whole of the asphalt core production. This is ascribed both to the high level of quality control throughout the process and to the experienced personnel and also to the incorporation of trial sections into the contract. This was important because it was necessary initially to use an aggregate from a commercial source, Brewelskloof, before the site aggregate became available. The change to the latter required a higher bitumen content to achieve the required low air voids. At the change of aggregates some higher air voids (>2%) were recorded for a short period before the appropriate adjustment to the bitumen content was made.

It was also noted that for the site aggregate in particular that the percentage bitumen added into the mix consistently exceeded the measured bitumen content extracted from the cores by about 0.2%. This was due to absorption by the aggregate and it was decided to enrich the mix to account for this apparent loss to ensure that the long term flexibility was not impaired.

Table 6 indicates the different characteristics of the Brewelskloof and Site mixes.

Table 6. Comparison of bitumen content, densities and air voids for Brewelskloof and Site Aggregate mixes

	Bitumen %	Density	Air Voids %	
			Plant	Core
Cycle 1 – 5 (Brewel)	6.58	2.466	0.33	1.20
Cycle 6 – 14 (Site)	6.98	2.348	0.44	1.61

It would be ideal if the core could be tested immediately or soon after laying, as with roadworks and considerable experimentation has been undertaken worldwide to achieve this. At Ceres Troxler density tests were undertaken to provide comparative data with the measured core densities.

The results are summarised in Table 7.

Table 7. Comparison of Troxler and core densities

	Troxler	Core
Maximum	2.518	2.498
Minimum	2.208	2.320
Average	2.360	2.404
Std Dev	0.083	0.063

It can be seen that the Troxler and core densities are very similar and that the former on average under-measures the density by about 0.04. Whilst this may appear to be a small amount it is significant in terms of air voids. At the time of construction there was insufficient data to place confidence in the Troxler results but certainly sufficient for them to give a preliminary indication that a layer was very probably satisfactory.

Further investigation and analyses of these results is being undertaken and it may well be that in future reliance can be placed primarily on Troxler type testing backed with less frequent coring.

7 CONCLUSIONS

The new Ceres Dam is believed to be the first in the southern hemisphere to utilise an asphaltic concrete core. It follows what has recently become conventional Norwegian asphalt core practice both in design and construction.

The site at Ceres is particularly suitable for an asphalt core since it has a high seismic risk classification and in addition has a potential differential settlement problem due to relatively soft foundation material in the central section with steeply sloping rock flanks. The asphalt core provides the required flexibility to accommodate these strains. Clay material for a conventional core was not readily available and even if it had been the construction programme which necessitated working through rainy periods would have been severely jeopardised.

The asphalt core proved to be cost effective with the overall tendered price being similar to those for conventional clay cores.

The dam started storing water in July 1998 and deformation and seepage measurements show both to be lower than was conservatively estimated for design purposes.

It is concluded that dam designs currently being considered should consider the asphaltic concrete core option taking particular note of the advantages of simplicity and all-weather construction. Where seismic loading is high or large differential settlements are possible, the use of a flexible asphalt core should probably be amongst the first options considered.

8 ACKNOWLEDGEMENTS

The design and construction team acknowledge with gratitude the inputs particularly from those who are not members of the staff of SRK or SKC viz Professor Knight and Professor van Schalkwyk, The Institute for Transport Technology, Stellenbosch, Department of Water Affairs and Forestry and many others.

Grinaker Construction are to be congratulated on their enterprise in undertaking the construction with their sub-contractor Colas and in dealing professionally with the inevitable snags of the novel construction.

Extensive use was made of Norwegian expertise through the appointment of Professor Höeg of the Norwegian Geotechnical Institute as specialist advisor and KOLO of Oslo as both nominated sub-contractors in joint venture with Colas South Africa and as consultants on asphaltic concrete core construction.

9 REFERENCES

Höeg, K., 1993, *Asphaltic Concrete Cores for Embankment Dams,* Norwegian Geotechnical Institute, Oslo

ICOLD, 1992, *Bituminous Cores for Fill Dams,* ICOLD, Bulletin 84, Paris

ICOLD, 1983, *Seismicity and Dam Design* ICOLD Bulletin 45, Paris

ICOLD, 1989, *Selecting Seismic Parameters for Large Dams* ICOLD Bulletin 72

Meintjes, H.A.C & G.A. Jones, 1999, *Dynamic Analysis of Ceres Dam.* Proc 12th Reg. Conf. Africa Soil Mech. Fdn. Engng, Durban

Geotechnics for Developing Africa, Wardle, Blight & Fourie (eds) © 1999 Balkema, Rotterdam, ISBN 90 5809 082 5

Influence de la rotation des axes principaux des contraintes et/ou des déformations sur la rhéologie des sols et des matériaux routiers

N. Laradi, S. Haddadi & F. Kaoua
Institut de Génie Civil, Université des Sciences et de la Technologie Houari Boumediène, Alger, Algérie

RESUME : Dans cette étude nous avons cherché à comprendre et maîtriser l'orientation des sollicitations sur un échantillon de sol et d'un enrobé bitumineux d'analyser des aspects du comportement des matériaux granulaires encore mal connus. Car les essais dits « en axe «, c'est à dire qui ne peuvent faire évoluer les directions principales que par sauts de 90 degrés, ne sont pas suffisants pour étudier les effets de la rotation des axes principaux ou encore cerner complètement ceux de l'anisotropie.

Notre approche expérimentale a deux objectifs : dans une première étape, mieux comprendre des phénomènes très présents in-situ tels que la rotation des axes principaux des. Dans une deuxième étape, elle doit permettre d'améliorer les modèles rhéologiques existants. L'expérience intervient alors à deux niveaux : En amont pour formuler ou rectifier les hypothèses de base et en aval pour valider les prévisions théoriques le long de chemin de sollicitations les plus variés. Cette approche contribuera certainement à promouvoir la géotechnique au service du développement de l'Afrique.

1 INTRODUCTION

Les routes, les barrages, les ponts, les tunnels, les plates-formes de forage en mer... tant d'édifices, autant d'interactions des structures avec le sol. En effet, l'ensemble de ces ouvrages fait intervenir les matériaux terrestres. L'homme a toujours, avec plus ou moins de bonheur, cherché à domestiquer cet environnement de manière à construire le mieux possible, le plus sûr et le moins cher. Or une conception et réalisation correcte des ouvrages en génie civil passe d'abord par une bonne connaissance du comportement mécanique et rhéologique des matériaux mis en jeu, et donc des sols en particulier.

C'est dans ce contexte que se situe notre travail, car si beaucoup a été fait pour comprendre le comportement des sols, des zones d'ombre subsistent. Nous citerons, entre autres, le rôle de la rotation des axes principaux des contraintes et/ou des déformations ou encore les effets de l'anisotropie des enrobés bitumineux utilisés dans la géotechnique routière, qui ne vérifient jamais les hypothèses prises en compte dans le calcul et le dimensionnement à savoir : milieu homogène, continu et isotrope. En effet les matériaux sont souvent caractérisés par une anisotropie de leur propriétés physiques et mécaniques, dues à une orientation privilégiée de leur structure, de leur fissuration. Avec le progrès de l'informatique et des méthodes numériques, notre capacité actuelle à calculer a dépassé nos possibilités d'analyser, de mesurer, de comprendre et de caractériser les propriétés constitutives de mécanique des sols.

En plus des études sur la rotation des axes principaux de contraintes et/ou de déformations qui sont présentés dans les premières parties, nous présentons dans cette étude, la caractérisation de l'anisotropie d'un enrobé bitumineux routier.

2 EXEMPLES DE SITES SOUMIS A DES SOLLICITATONS ROTATIONNELLES

Il nous semble important d'illustrer par quelques exemples l'importance et l'omniprésence in situ du phénomène qui a motivé ce travail : la rotation des axes principaux de sollicitations.

La réorientation des directions principales des contraintes et/ou des déformations est un aspect commun à tous les sols affectés par des travaux d'ingénierie ou des phénomènes naturels comme les séismes. Nous avons regroupé dans la figure 1 des exemples extraits de la littérature.

Le premier (fig.1 A) représente un élément de sol sous-marin qui peut être le support de constructions off shore comme les plates-formes de forage. Sous l'effet de la houle, un point quelconque de ce massif voit les directions principales de contraintes changer constamment de direction.

Le second exemple (fig.1 B) illustre la réorientation des directions principales ou cours d'un creusement de tunnel.

Sous une fondation (fig.1 C) les directions principales de contraintes changent suivant la position du point considéré par rapport à la surface d'appui. Cette variation dans l'espace peut être aussi temporelle s'il s'agit d'un chargement cyclique type trafic routier ou ferroviaire par exemple.

Nous pouvons citer aussi l'exemple des noyaux de barrage (fig.1 D). Selon la charge hydraulique appliquée, les contraintes principales changent d'ampleur et de direction.

Enfin, la figure 1 E représente un cas particulier : celui des édifices tournants tels que les grands radars. Là aussi, le sol est soumis à des sollicitations de type rotationnel.

Ces quelques exemples pris parmi tant d'autres montrent l'importance de la rotation des axes principaux in situ. Or les modèles rhéologiques existants, notamment les modèles élasto - plastiques n'en tiennent pas compte en tant que tel : La charge n'étant exprimée qu'en fonction des valeurs principales des contraintes, une modification de leur orientation ne va engendrer aucune déformation plastique.

3 EFFETS DE LA ROTATION DES AXES PRINCIPAUX

Il est très difficile de déterminer les effets de la rotation des axes principaux de contraintes, car aucun type d'essai existant ne permet réellement de dissocier cette rotation des autres paramètres. En effet même si des essais comme le cylindre creux avec pressions intérieures différentes permettent de faire tourner les axes principaux, ils n'arrivent pas à maintenir, au même temps, les valeurs des contraintes principales constantes. Au mieux, ils peuvent garder la pression moyenne constante, (Kharchafi, 1988).

Néanmoins, en s'appuyant sur les essais à pression moyenne constante, (Kharchafi, 1988) et en confrontant, par la suite, ces résultats avec ceux obtenus par d'autres auteurs, nous essayons de cerner le comportement des sols pulvérulents quand ils sont soumis à une rotation continue des axes principaux de contrainte, même si celle-ci s'ajoute à une variation d'autres paramètres comme le coefficient b (b=$(\sigma_2 - \sigma_3)/(\sigma_1 - \sigma_3)$).

4 ETUDE DE LA ROTATION DES AXES PRINCIPAUX EN DEFORMATION PLANES

Des travaux, visant à cerner les effets de la rotation des axes principaux de contraintes et utilisant des appareils de cisaillement ont été publiés par ailleurs, (Ishihara et Towhata, 1985), (Arthur, 1979). Au cours de ces essais, l'état de déformations est plan et donc l'état de contraintes mal défini.

Nous citons, dans ce cadre, l'étude menée par Arthur (1979) à l'aide de l'appareil de cisaillement directionnel (figure 2). En contrôlant les trois contraintes σ_x, σ_y et τ_{xy} exercées sur les faces verticales de l'éprouvette cubique de sable (fig.2 A), Arthur pouvait imposer des rotations cycliques des axes principaux de contraintes, tout en maintenant le rapport R = σ_1/σ_3 donc l'angle de frottement ϕ, presque constant (fig.2 B). La figure 2 C résume les résultats obtenus pour deux séries d'essais avec deux amplitudes d'angle de rotation θ différents : 40° et 70°, réalisés à une fréquence de 7,5 cycles/minute. Pour chaque amplitude θ, plusieurs valeurs de R(ou ϕ) ont été adoptées (fig.2 C). On voit d'après ces courbes, représentant les déformations principales majeures ε_1 cumulés en fonction du nombre de cycles que :

Pour une même amplitude des cycles θ, les déformations majeures cumulées augmentent avec l'angle de frottement ϕ mobilisé.

Hormis l'essai à ϕ=33°, toutes les déformations engendrées par des cycles d'amplitude θ =70° sont supérieures à celles provoquées par des cycles à θ = 40°.

La déformation se stabilise au bout d'environ 25 cycles en ce qui concerne les essais effectués avec une amplitude θ = 40°. Alors que pour les cycles à θ=70°, les déformations du sol continuent d'augmenter même après 75 cycles excepté le cas où l'angle de frottement mobilisé n excède pas 33 degrés.

Par ailleurs Roscoe (1967) avait essayé, grâce à sa boite de cisaillement d'étudier l'évolution des directions principales de contraintes et de déformations au cours d'un essai de cisaillement simple. Les principaux résultats de ces essais, réalisés respectivement sur deux échantillons de sable de Leighton Buzzard lâche (e_0=0,68) et dense (e_0=0,53), sont résumés sur la figure 3 A et B.

D'après ces courbes qui sont à interpréter avec précaution à cause de tous les problèmes d'uniformité des contraintes relatifs à ce type d'essai, Roscoe (1967) souligne deux points importants: D'abord la concordance presque parfaite des directions principales de contraintes et des incréments de déformation (α_σ =αd_ε), spécialement après le point de contractance maximale (e_0). Ensuite, le fait que les directions des incréments de contraintes et des déformations ne coïncident

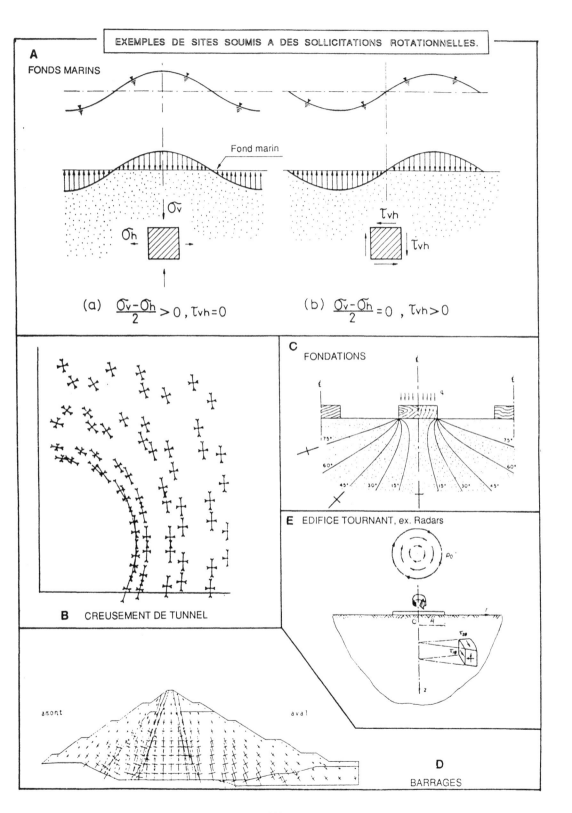

EXEMPLES DE SITES SOUMIS A DES SOLLICITATIONS ROTATIONNELLES.

A
FONDS MARINS

Fond marin

$\widetilde{\sigma}_v$

σ_h

τ_{vh}

τ_{vh}

(a) $\dfrac{\widetilde{\sigma}_v - \sigma_h}{2} > 0 \, , \, \tau_{vh} = 0$

(b) $\dfrac{\widetilde{\sigma}_v - \sigma_h}{2} = 0 \, , \, \tau_{vh} > 0$

B CREUSEMENT DE TUNNEL

C FONDATIONS

E EDIFICE TOURNANT, ex. Radars

D BARRAGES

amont

aval

461

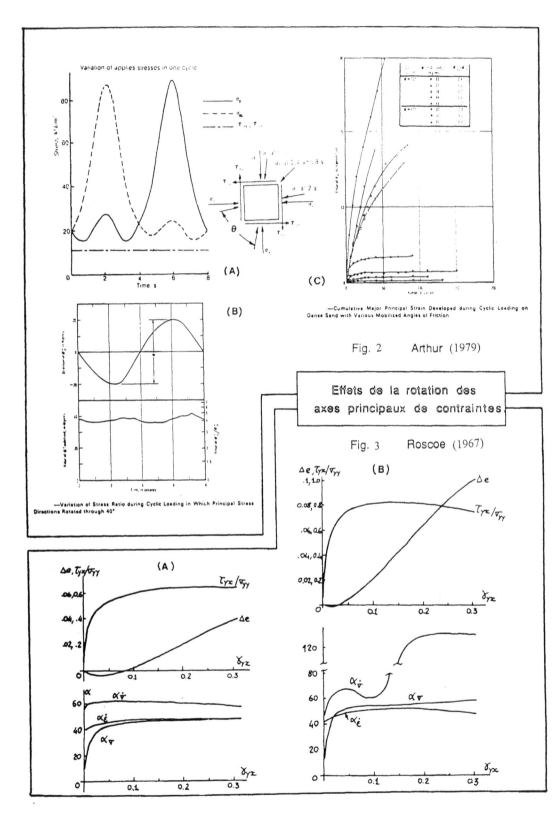

Variation of applied stresses in one cycle

(A)

(B)

(C)

—Cumulative Major Principal Strain Developed during Cyclic Loading on Dense Sand with Various Mobilized Angles of Friction

Fig. 2   Arthur (1979)

—Variation of Stress Ratio during Cyclic Loading in Which Principal Stress Directions Rotated through 40°

Effets de la rotation des axes principaux de contraintes

Fig. 3   Roscoe (1967)

462

presque jamais ($\alpha d_\sigma = \alpha d_\varepsilon$) et ne sont relativement proches que pour les faibles valeurs des déformations.

5 ETUDE DE L'ANISOTROPIE DES ENROBES BITUMINEUX :

Cette étude (Laradi et Haddadi, 1999) a pour but de caractériser l'anisotropie d'un enrobé bitumineux de type routier, qui est liée aux techniques de mise en place de ces matériaux, de voir son influence sur les paramètres intrinsèques et son comportement réel sur une route.

Deux campagnes d'essais (compression et traction simple en grande déformation (quelque 10^{-2}) ainsi que des essais triaxiaux de révolution en petite déformation (quelque 10^{-5})ont été réalisé sur l'anisotropie des enrobés bitumineux ; Nous avons réalisés nos essais à différentes vitesses de déformation axiale constantes (0,05 %/mn, 1 %/mn, 4 %/mn et 16 %/mn suivant trois directions (I, II et III).

L'examen des résultats expérimentaux met en évidence d'une part l'existence d'une forte anisotropie et d'un autre coté l'influence des paramètre intrinsèques sur le comportement rhéologique des enrobés bitumineux (Laradi et Haddadi, 1999 b), à savoir : la teneur en bitume, la densité du béton bitumineux, la pression de confinement, le temps et la température (Laradi et Dermèche, 1997).

L'ensemble des résultats ont mis en évidence l'existence d'un domaine de déformations réversibles du matériau (Laradi, 1990), où les valeurs des modules de rigidité correspondant au sens I sont supérieurs à ceux correspondant au sens II et III, ce qui met en évidence une influence de l'anisotropie. Et les mêmes observations sont valables pour les coefficients de Poisson .

Pour les essais en petites déformations (Laradi 1995), une première analyse de l'ensemble des résultats montre que :

- Il existe une symétrie de révolution autour de l'axe I, il résulte donc qu'une confection des éprouvettes par compactage oedométrique peut donner les mêmes résultats qu'une confection à l'aide d'un compacteur à plaque,

- Il existe une anisotropie entre la direction de compactage I et les directions perpendiculaires, cette anisotropie est probablement une anisotropie due à la forme des grains et à l'orientation des plans de contact entre particules.

Nous avons également retenu une quantité importante d'informations (Laradi et Haddadi, 1999 a):

La similitude de comportement avec l'ensemble des matériaux possédant un squelette granulaire :
Le comportement fortement visqueux et irréversible du matériau.
L'aspect non linéaire du comportement du matériau, ce qui est contradictoire avec les hypothèses classiques concernant le comportement des enrobés bitumineux.
La symétrie de révolution existante autour de l'axe I entre les directions longitudinales et transversales, perpendiculaires à la direction de compactage.
Existence d'une anisotropie entre la direction de compactage et les autres directions.
La confirmation que le critère de rupture est anisotrope.
La contractante initiale puis la dilatante lors des sollicitations en compression simple.
L'évolution des modules d'Young initiaux et contraintes de rupture en fonction de la vitesse de sollicitation.
La contrainte de relaxation est indépendante de la vitesse de sollicitation.
Le matériau est plus résistant dans la direction 1 de compactage pour la compression, et il est plus résistant dans les autres directions dans le cas de la traction.
Il semble que plus la vitesse de sollicitation augmente, plus le matériau est dilatant en compression et plus contractant en traction. Le temps court ne nous a pas permis de traiter à fond, toutes les informations sur la quantité importante des essais que nous avons fait, il convient donc de traiter ces informations qui sont facilement accessibles par un traitement informatique.

Pour ce qui est des essais, nous avons réalisé des essais « dans les axes » ($\theta = 0°$ et $\theta = 90°$) .Nous pensons que les seules résultats obtenus à l'aide de ces essais « dans les axes ($\theta = 0°$ et $\theta = 90°$) sont insuffisants pour la prévision des variations des caractéristiques mécaniques, il faut donc penser à faire des essais « hors axes » et réfléchir à leurs interprétations et analyses et à leurs réalisations qui sont très difficiles pour ne pas dire « impossibles ».

6 CRITERE DE CONTRAINTES MAXIMALES

6.1 *Résultats expérimentaux*

Les résultats montrent que les points représentant les contraintes maximales , dans les axes contrainte axiale - contrainte radiale, sont alignées si l'on considère les essais à une vitesse de déformation axiale (ε_1) et une température (θ) constantes. Pour cela, les essais ont été réalisés à une température de 23°C et une vitesse de déformation axiale de 1%/mn et ont été spécialement analysés pour étudier l'évolution de la contrainte maximale des enrobés.

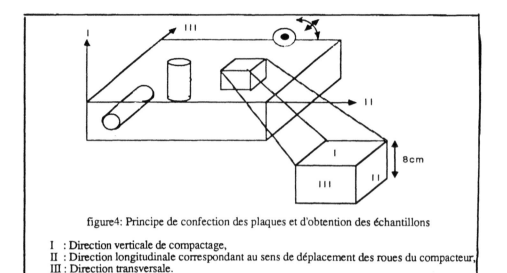

figure4: Principe de confection des plaques et d'obtention des échantillons

I : Direction verticale de compactage,
II : Direction longitudinale correspondant au sens de déplacement des roues du compacteur,
III : Direction transversale.

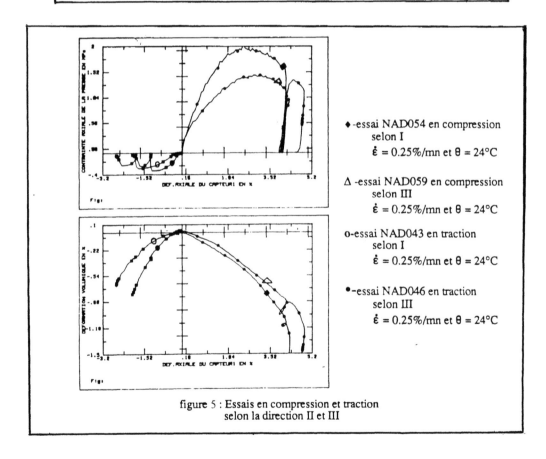

♦ -essai NAD054 en compression
selon I
$\dot{\varepsilon} = 0.25\%/mn$ et $\theta = 24°C$

Δ -essai NAD059 en compression
selon III
$\dot{\varepsilon} = 0.25\%/mn$ et $\theta = 24°C$

o-essai NAD043 en traction
selon I
$\dot{\varepsilon} = 0.25\%/mn$ et $\theta = 24°C$

•-essai NAD046 en traction
selon III
$\dot{\varepsilon} = 0.25\%/mn$ et $\theta = 24°C$

figure 5 : Essais en compression et traction
selon la direction II et III

Afin de présenter une formulation pour un critère de contrainte maximale, nous supposons les deux hypothèses suivantes :

H1: Le critère a une forme conique, et il est divisé en deux parties : une, lorsque les contraintes principales sont positives, l'autre quand au moins

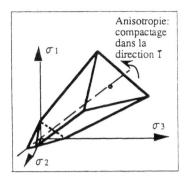

Fig. 6 Représentation dans l'espace des contraintes principales
(σ_1 - σ_2 - σ_3).

une des contraintes principales est négative (soit quand certaines facettes sont soumises à une tension).

H2: Le critère est isotrope. En effet lors de leur fabrication, les enrobés bitumineux sont fortement compactés selon une direction verticale. Afin d'évaluer si ce compactage induit une anisotropie une série d'essais a été réalisés sur des échantillons cubiques, dans différentes directions matérielles Laradi, 1995, Laradi et Haddadi, 1999).

6.2 *Essais de caractérisation de l'anisotropie*

Lors de la mise en place des enrobés, trois direction peuvent être distinguées : la direction de compactage, la direction de roulement du compacteur et la direction perpendiculaire aux deux précédentes (fig.5). Des essais de compression et d'extension simples à vitesse de déformation constante, ont été réalisés sur des éprouvettes cubiques découpées dans des plaques d'enrobés obtenues à l'aide du compacteur de plaques type LPC. Les éprouvettes cubiques, de compacité C_p=93,6%, ont été sollicitées selon les trois directions I, II et III, correspondant respectivement au sens de compactage, de roulement des roues du compacteur, et perpendiculairement aux deux précédents.

Une analyse plus détaillée des résultats est présentée dans par ailleurs (Laradi, 1990 et Laradi, 1995 et Laradi et Haddadi 1999). Nous ne présentons dans cette étude, que les résultats relatifs aux contraintes maximales.

Nous remarquons que la direction I semble correspondre à un axe de symétrie de révolution pour le matériau. Les valeurs des contraintes de ruptures obtenues en traction (signe opposé) et en compression pour les différentes vitesses de déformation sont reportées sur la figure 7.

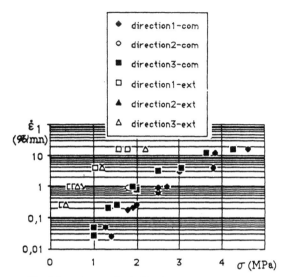

Fig. 7 Contrainte maximale (Mpa) en compression et en traction simple

Nous constatons sur cette figure que l'écart relatif entre les valeurs absolues des contraintes maximales lors des sollicitations dans la direction I et la direction II (ou III) est compris entre 10 et 25%. Le signe de l'écart est inversé entre la compression et la traction : plus grande résistance dans le sens I en compression et moindre résistance dans ce même sens en traction.

Les écarts observés indiquent une anisotropie du milieu; néanmoins, compte tenu de leurs valeurs relativement faibles, en première approximation, que le matériau possède un critère de contraintes maximales isotrope. Ce qui confirme l'hypothèse formulée, pour améliorer la modélisation.

7 CONCLUSIONS

Comme nous l'annoncions en introduction de ce chapitre, la difficulté à comprendre le rôle des divers paramètres (rotation des axes principaux, anisotropie…) dans le comportement des matériaux ou des sols réside dans l'impossibilité d'isoler complètement leurs influences respectives au cours d'un essai. Preuve en est la diversité des conclusions auxquelles aboutissent les auteurs répertoriés, et qui sont souvent contradictoires.

Néanmoins, d'après nos résultats expérimentaux, nous avons pu mettre en évidence quelques aspects importants concernant le comportement avant la rupture des matériaux granulaires. Nous avons ainsi mis en évidence les effets de l'anisotropie induite par un premier chargement ; le comportement des matériaux grenus au cours des chemins de sollicitation à rotation continue des axes principaux.

Le comportement des matériaux granulaires a beaucoup été étudié à travers le monde (exception faite pour les enrobés bitumineux). Néanmoins si des aspects de comportement sont à peu près bien cerné, beaucoup reste à faire pour maîtriser complètement le rôle de l'anisotropie et la rotation des axes bitumineux. Surtout que ces derniers présentent un grand intérêt pour l'Algérie qui investit beaucoup d'argent pour la construction en travaux publics (Routes, Barrages, etc...)

8 REFERENCES

Arthur J.R.F., Chua K.S, Dunstan T. (1979): " Dense *Sand Weakned by continuous principal stress direction rotation* ». Geotechnique Vol. 29 N° 1, pp. 91-96.

Ishihara K. , Towhata I. (1985): « *Effects of rotation of principal stress directions on cyclic reponse of sand* ». Soils and fondation, vol. 25, N° 3, pp. 73-84

Kharchafi M (1988): « *Contribution à l'étude du comportement des matériaux granulaires sous sollicitations rotationnelles* » Doctorat, Ecole Centrale Paris, Ecole Nationale des Travaux Publics de l'Etat –Lyon, p. 314.

Laradi N. (1990): « *Modélisation du comportement des matériaux traités aux liants hydrocarbonés* » Thèse de Doctorat, Ecole Nationale des Travaux Publics de l'Etat – Institut Nationale des Sciences Appliquées – Lyon, p. 253.

Laradi N. (1995): « *Anisotropie des Enrobés Bitumineux* ». 6ème Congrès Arabe de la Mécanique des Structures. Damas, Syrie, pp. 253-258.

Laradi N. & Dermeche Y. (1997): « *Influence de la température sur le module complexe des mélanges bitumineux algériens* ». A.A. BALKEMA Eds. Roterdam, 5th International Rilem Symposium on Mechanical Tests for Bituminous Materials, Lyon, France .

Laradi N. & Haddadi S. (1999): « *Etude de* l'aniso-*tropie des Enrobés Bitumineux* ». Eurobitume Workshop, Luxembourg 03-06 june. Paper N°001, 7p.

Laradi N. & Haddadi S. (1999 a): « *Carctérisation rhéologique des bitumes purs* ». Eurobitume Workshop, Luxembourg 03-06 june. Paper N°002, 8p.

Laradi N. & Haddadi S. (1999 b): « *Influence de la dureté du bitume, de la fréquence et de la température sur les mélanges bitumineux* ». Eurobitume Workshop, Luxembourg 03-06 june. Paper N° 004, 6p.

Roscoe K.H. (1967) : « *Principal axes observed during simple shear of sand* ». Procedings of the Geotechnical conference Vol. 1. pp.321-237.

Geotechnics for Developing Africa, Wardle, Blight & Fourie (eds) © 1999 Balkema, Rotterdam, ISBN 90 5809 082 5

Essais de cisaillement d'un grès sur un appareillage de petite et de grande dimension

Mamba Mpele
Ecole Polytechnique de Yaoundé, Cameroun

J.J. Fry
Electricité de France

RESUME : L'enrochement de grès qui fut utilisé pour la construction du barrage de Vieux - Pré en France; a été soumis à des tests de cisaillement à la petite et à la grande boîte de Casagrande. Les résultats obtenus montrent que: les valeurs d'angle de frottement à la rupture données par les deux appareillages sont différentes. Toutefois lorsqu'on introduit le facteur d'échelle Ψ défini par le rapport entre la dimension caractéristique L_0 de l'échantillon et le diamètre maximum d_{max} des grains constitutifs du matériau testé: nous constatons que les différences observées entre les angles de frottement données par la grande et la petite boîte de cisaillement pour les valeurs de Ψ inférieures à 20 disparaissent dès que ce paramètre devient supérieur ou égal à 20.

ABSTRACT : The sandstone rocfills used during the construction of Vieux-Pré's dam in France is submitted to a small and large-scale shear box test in oder to evaluate their strength. The result we obtained show that the internal friction's angle of tested sandstones given by two apparatus are differents. If we introduce the scale factor parameter Ψ, define as the ratio the characteristic size of tested specimen L_0 with the maximum size d_{max} of sandstone particles which are the constituent element of sample; we remark that the difference of the angle given by both shear box machines observed when Ψ is inferior to 20 does no longer exist when its superior or egal to 20.

1 INTRODUCTION

L'angle de frottement interne et le module de cisaillement des matériaux grossiers sont parmi les paramètres fondamentaux qui intéressent le géotechnicien. La détermination de ces paramètres exige la construction des appareillages de grande dimension qui nécessitent la mobilisation des moyens financiers lourds qui ne sont pas à la portée des tous les laboratoires de mécanique des sols. Dans cet article nous nous intéressons à l'angle de frottement à la rupture d'un enrochement de grès obtenu avec la petite et la grande boîte de Casagrande.

2 APPAREILLAGE

La petite boîte de Casagrande que nous avons utilisée pour tester les échantillons de 100mm de coté sur 50mm d'épaisseur a une forme carrée. Cet appareillage classique de cisaillement direct à vitesse contrôlée se rencontre dans tous les laboratoires de mécanique des sols et l'on ne peut par conséquent insister sur sa description.

2.1 *Appareillage de grande dimension*

L'appareillage de grande dimension utilisé pour faire des essais de cisaillement sur des matériaux grossiers et qui est représenté par la figure1, a un encombrement total de 2m85 x 3m32 x 2m42. Son poids total à vide est de 30KN et le volume utile occupé par l'échantillon à tester est de 1m20 x 1m20 x 1m00. Ce dispositif d'essai de cisaillement est constitué de deux demi - boîtes: La demi - boîte inférieure (Bi) logée dans une fouille rectangulaire en béton de 0m90 de profondeur et de section égale à 3m40 x 3m40. Elle est immobile et repose sur un socle en béton par l'intermédiaire des IPN 360. Le fond de cette demi - boîte est tapissé par des madriers jointifs de 15cm d'épaisseur et qui jouent le rôle des pierres poreuses. La demi - boîte supérieure (Bs) est mobile et glisse sur la demi - boîte inférieure grâce à des guides munis des rouleaux de 2cm de diamètre. Dans sa partie supérieure se trouve un plateau amovible (P) sur lequel sont fixées les extrémités inférieures des 6 vérins hydrauliques pouvant développer chacun un effort de 250KN. Les extrémités supérieures de ces vérins sont fixées à un plateau en acier qui se trouve

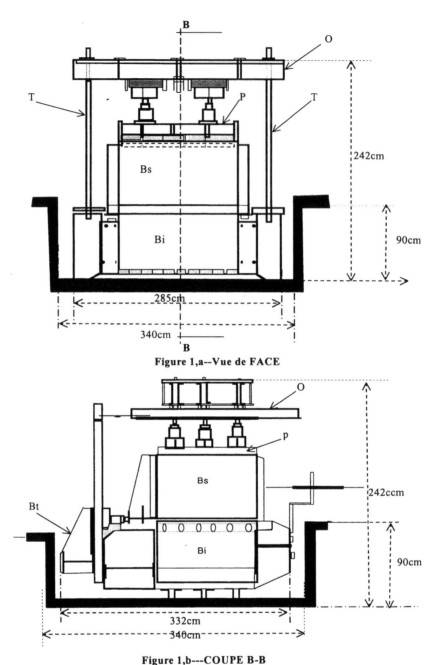

B

O

T

P

T

242cm

Bs

Bi

90cm

285cm

340cm

B

Figure 1,a--Vue de FACE

O

p

242ccm

Bt

Bs

90cm

Bi

332cm

340cm

Figure 1,b---COUPE B-B

Figure 1, Shèmas de la grande boîte de Casagrande

en contact avec des poutres de l'ossature supérieure (O) par l'intermédiaire des roulements logés à l'intérieur des guides qui n'autorisent que des mouvements de translations horizontaux du système demi - boîte supérieure et plateau amovible.

L'ossature supérieure est solidaire à la demi - boîte inférieure immobile grâce à un système de 6 tirants (T) de 40mm de diamètre. C'est ce système de tirant qui permet d'appliquer la contrainte verticale de consolidation désirée sur l'échantillon testé lorsqu'on on déclenche leur dispositif de mise en marche.

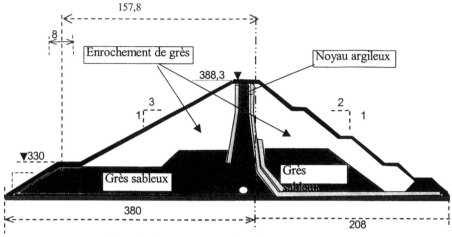

Figure 2 : Coupe transversale du barrage de Vieux-Pré

Le dispositif de cisaillement est constitué par 2 vérins hydrauliques pouvant développer chacun un effort de 500KN. Ceux-ci sont fixés au bâti inférieur grâce à une butée (Bt) et à la demi - boîte supérieure grâce à des appuis rigides. Ces appuis transmettent au niveau du plan de cisaillement des efforts tangentiels développés par les vérins et qui provoquent le déplacement de la boîte supérieure.

Les efforts de cisaillement et de consolidation sont mesurés par des capteurs de force ; les déplacements verticaux et horizontaux par des capteurs inductifs. L'ensemble de ces capteurs sont reliés à une table traçante et à une centrale de mesure pour traitement. C'est avec ces deux appareillages de cisaillement que l'enrochement de grès a été testé.

3   LE MATERIAU: LE GRES DE VIEUX PRE

Le grès de Vieux-Pré est un matériau friable à ciment siliceux qui fut utilisé pour la construction du barrage de Vieux-Pré en France. Il s'agit d'un barrage en enrochement à noyaux central argileux ayant une hauteur de plus de 50m et dont les recharges amont et aval sont constituées d'enrochement de grès. (voir figure 2). La qualité du matériau est variable suivant son altération et sa fracturation importante. Une altération plus forte pouvant se traduire par un pourcentage de la fraction sableuse plus élevée atteignant parfois 54% de matériau de dimensions inférieures à 2mm. La courbe enveloppe de ce grès est donnée par la figure 3. Les essais d' identification effectués ont confirmé le caractère tendre de cette roche dont les Los Angeles obtenus étaient de l'ordre de 98! C'est avec cet enrochement que nous avons fabriqué des échantillons qui ont été soumis à des essais de

cisaillement simple à la grande et à la petite boîte de Casagrande.

4   CONFECTION DES ECHANTILLONS

Les échantillons testés ont été obtenus par la méthode d'écrêtage directe de la courbe granulométrique naturelle du grès de Vieux-Pré. Les coupures 0/12.5mm, 0/5mm, 0/2.5mm et 0/0.63mm ont été testées avec la petite boîte de Casagrande de 100mm et à des contraintes de consolidations de 0.2MPa et 0.5MPa. Par contre les fractions 0/150mm, 0/63mm, 0/20mm et 0/5mm ont été testées avec la grande boîte de cisaillement de 1200mm et pour des valeurs de contraintes de consolidation identiques aux précédentes. La figure 4 nous montre les courbes granulométriques de ces échantillons. La valeur moyenne de la densité sèche et de la teneur en eau des échantillons testés étant constantes et égales à 19.30KN/m3 et 5%.

5   MISE EN PLACE DES ECHANTILLONS

La mise en place des échantillons à l'intérieur des boîtes de cisaillement s'est faite en trois couches (la deuxième couche étant cisaillée à mi-hauteur) et selon le principe suivant: connaissant le volume final et la densité apparente finale de l'échantillon, on calcule la masse totale Mt du sol à prélever. Le tiers de cette masse qui constitue la masse de chaque couche est mise à l'intérieur de la boîte puis compacté jusqu'à ce que sa hauteur finale soit égale au tiers de la hauteur totale de l'échantillon. L'on procède ainsi à la mise en place des autres couches. Le compactage de chaque couche se faisant à l'aide

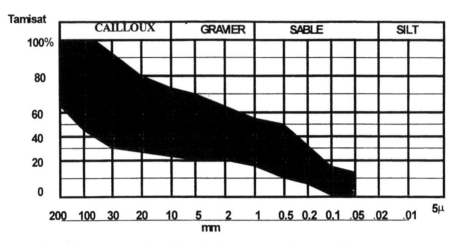

Figure 3: Fuseau granulométrique de l'enrochement du grès de Vieux-Pré.

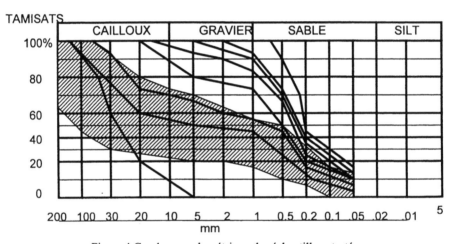

Figure 4:Courbe granulométrique des échantillons testés

d' une petite dame manuelle pour les essais de cisaillement à la petite boîte ou avec une pilonneuse pour les tests à la grande boîte.

Ces échantillons ont été cisaillés à une vitesse constante égale à 2mm/mn. Pendant le cisaillement on a mesuré la force de cisaillement F_T les déplacements horizontaux ΔL et la force de consolidation verticale F_v. La contrainte verticale σv et tangentielle t ainsi que l'angle de frottement à la rupture φ_0 ont été obtenus à partir des relations (1), (2) et (3).

$$\sigma_v = \frac{F_v}{S_0} \tag{1}$$

$$\tau = \frac{F_T}{S_0(1 - \frac{\Delta L}{L_0})} \tag{2}$$

$$Tan(\phi_0) = \frac{\tau_{rupt}}{\sigma_v} \tag{3}$$

Où $S_0 = L_0 x L_0$ est la surface initiale de cisaillement de l'échantillon (L_0=100mm pour la petite et 1200mm pour la grande boîte). ΔL désigne le déplacement de l'échantillon correspondant et τ_{rupt} sa contrainte tangentielle à la rupture.

PRESENTATION DES RESULTATS

6.1 Essais à la petite boîte

Les courbes contraintes tangentielles en fonction des déformations $\Delta L/L_0$ concernant les fractions 0/12.5mm, 0/5mm, 0/2.5mm et 0/0.63mm qui sont

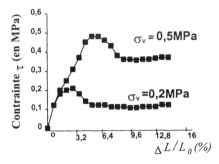

Figure 5-a: Coupure: 0/5mm

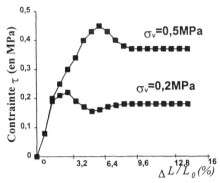

Figure 5-c: Coupure 0/12,5mm

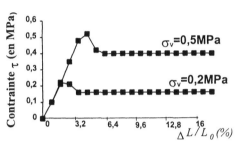

Figure 5-b: Coupure 0/2,5mm

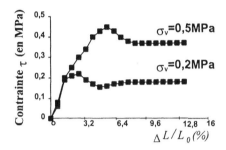

Figure 5-d: Coupure 0/0,63mm

Figures 5: Essais de cisaillement à la petite boîte de Casagrande ou boîte de 100 : courbes contraintes tangentielles t en fonction de $\Delta L/L_0$

représentées par les figures 5 montrent dans tous les cas, l'existence des pics. En outre on remarque que les contraintes tangentielles à la rupture augmentent lorsque que la contrainte verticale croît. Par contre les résultats de la figure 7 montrent que : l'angle de frottement à la rupture diminue lorsque la contrainte de consolidation augmente.

6.2 Essais à la grande boîte de 1200mm

Les courbes contraintes tangentielles en fonction des déformations $\Delta L/L_0$ des fractions 0/5mm, 0/20mm, 0/63mm et 0/150mm (testées avec la grande boîte) ; montrent que si les pics se sont adoucis pour les fractions 0/20mm, 0/63mm et 0/150mm ; ils ont par contre disparu sur la fraction 0/5mm. De plus dans la figure7 on note que l'angle de frottement φ_0 est une fonction de la dimension maximale d_{max} des grains constitutifs de l'enrochement et de la contrainte de consolidation σ_v.

Cet angle en effet diminue lorsque la contrainte verticale augmente et ce quel que soit la valeur de d_{max}. Par contre il augmente quand d_{max} croît. De plus quel que soit les valeurs de d_{max} et σ_v les valeurs d'angle de frottement données par le petit appareillage sont supérieures à celles données par le grand appareillage.

De plus si l'on examine les courbes de la figure 8 qui donnent les angles de frottement obtenus avec les deux appareillages en fonction du facteur d'échelle Ψ défini par la relation (4).

$$\Psi = \frac{L_0}{d_{max}} \qquad (4)$$

Où L_0 désigne la dimension caractéristique des échantillons qui est égale à 100mm ou à 1200mm lorsqu'ils sont testés dans la petite ou la grande boîte de cisaillement de Casagrande et d_{max} la dimension maximale des grains constitutifs des enrochements. On constate également ici que l'angle de frottement diminue quand la contrainte augmente et lorsque le facteur d'échelle croît. De

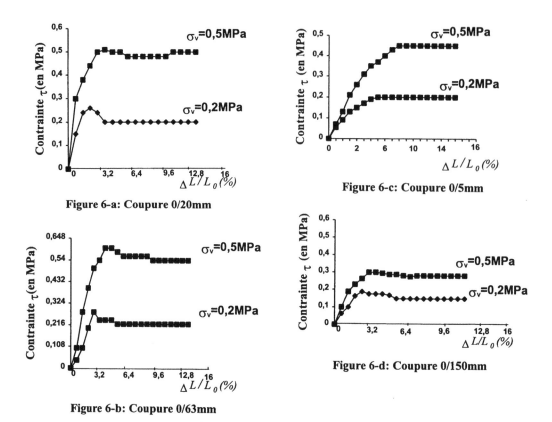

Figure 6-a: Coupure 0/20mm

Figure 6-c: Coupure 0/5mm

Figure 6-b: Coupure 0/63mm

Figure 6-d: Coupure 0/150mm

Figure 6: Essais de cisaillement à la grande boîte de Casagrande : courbes contraintes tangentielles t en fonction de $\Delta L/L_0$.

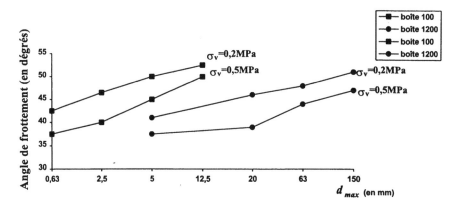

Figure 7 : Angle de frottement en fonction de la dimension maximale des grains d_{max}

plus pour les valeurs de Ψ supérieures à $\Psi_{min} = 20$, on observe que les angles de frottement donnés par les deux appareillages sont du même ordre de grandeur. Ce résultats confirment les travaux de R.K. RATHEE (1981) qui réalisa des essais similaires sur un matériau sableux. En dessous de cette valeur Ψ_{min} les différences s'accentuent. Ces très fortes variations des angles de frottement à la rupture observées quand Ψ est inférieur à 20 peuvent s'expliquer par le fait qu'étant dans le domaine où les dimensions des particules ne sont plus négligeable par rapport aux

472

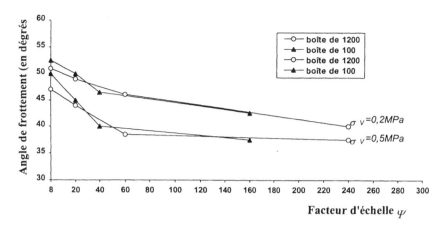

Figure 8 angle de frottement en fonction du facteur d'échelle $\Psi = L_0 / d_{max}$

dimensions de l'appareillage, il apparaît des phénomènes de comportement locaux au niveau du plan de cisaillement qui se manifestent par la rupture des granulats se trouvant de part et d'autre de la surface de rupture et qui ; perturbent très fortement le comportement global des échantillons. Les valeurs d'angles obtenues dans ces conditions sont loin de représenter le comportement global des matériaux

7 CONCLUSIONS

A la lumière de ces résultats expérimentaux, on peut dire que l'angle de frottement à la rupture de l'enrochement du grès de Vieux-Pré augmente lorsque d_{max} croît. Les angles obtenus avec le grand appareillage semblent plus faibles que ceux donnés par le petit appareillage. Mais lorsqu'on introduit le facteur d'échelle Ψ les angles de frottement obtenus avec les deux appareillages sont du même ordre de grandeur lorsque les valeurs de ce paramètre deviennent supérieures à 20. L'application concrète de cette conclusion est de donner la possibilité de déterminer en laboratoire les angles de frottement des matériaux grossiers à l'aide du petit appareillage lorsque le paramètre Ψ est convenablement choisi.

BIBLIOGRAPHIE

FRY J.J, FLAVIGNY E. MAMBA MPELE, 1989. Classification et propriétés des enrochements: cas d'un grès. Proceedings of twelfth international Conference on soil Mechanics and Foundation engineering Rio de Janeiro Août 1989.
MAMBA MPELE 1989. Résistance au cisaillement des enrochements et matériaux grossiers: applications aux calculs des barrages. Thèse présentée à l'Université de Lille I, soutenue le 1er Déc. 1989 pour l'obtention du Grade de Docteur.
MILTON E. Harr, 1981. Mécanique des milieux formés des particules. Presses Polytechniques Romandes.
RATHEE RK 1981. Shear strength of granular soil sand its prediction by modelling technique. Journal C.I Vol. 62. September 1981

Geotechnics for Developing Africa, Wardle, Blight & Fourie (eds) © 1999 Balkema, Rotterdam, ISBN 90 5809 082 5

Dynamic analyses of the new Ceres Dam

H.A.C. Meintjes & G.A. Jones
SRK Consulting, Johannesburg, South Africa

ABSTRACT: A 60m high rockfill dam with an asphalt core has been constructed to replace an existing concrete dam near Ceres in an area with some earthquake activity. The asphalt core – the first of its kind in southern Africa – has been designed to provide the flexibility required to accommodate not only differential settlements due to variable founding conditions but also to take account of the pseudo-static and dynamic behaviour of the rockfill under dynamic loading conditions. The STABL program was used for the pseudo-static analysis, and the FLUSH program for dynamic analyses. A Newmark procedure was used to calculate the displacements through the asphalt core under dynamic loading conditions, using the program Slip developed by NGI.

1. INTRODUCTION

The new Ceres dam is in an area of high seismicity and as a result has been designed using both pseudo-static and dynamic analyses according to ICOLD guidelines. The dam wall is unique in South Africa in that it has an asphaltic concrete core in a zoned rockfill shell.

The maximum height is 60m and the crest length 275m and width 7m. The upstream and downstream slopes are 1:1.6 and 1:1.5 respectively, the latter including two 3m wide berms.

The dam replaces a 20m high concrete gravity dam on the same site which, as well as having insufficient capacity, had also deteriorated from a combination of alkali-aggregate reaction and earthquake loading.

The new dam was constructed in 1996/98 for the Ceres Municipality and Koekedouw Irrigation Board and designed by Steffen, Robertson and Kirsten. Specialist input on the local seismology was provided by Prof Knight of Knight Hall Hendry and on seismic design by Prof Höeg of the Norwegian Geotechnical Institute.

The main contractor was Grinaker Construction and the asphaltic concrete core construction was undertaken by a joint venture between Colas (South Africa) and KOLO (Norway). The latter also provided specialist advice on the design and construction of asphalt cores.

2. MATERIAL PROPERTIES

The rockfill was quarried from a site immediately upstream of the dam wall and comprised sandstones and quartzitic sandstones of the Nardouw Formation of the Table Mountain Group. The rockfill properties were determined by appropriate laboratory testing and are represented by a cohesionless material with a normal stress dependent friction angle varying from 40° to 50°. This design line is shown in Figure 1, which also shows Marsal's and Leps' data.

For analytical purposes, the dam wall was divided into zones each having a friction angle dependent on the normal effective stress level within that zone – typical values are shown in Table 1.

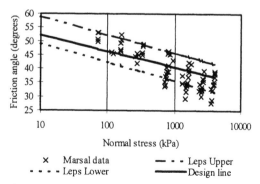

Figure 1. Rockfill strength model

Table 1. Material properties

Normal Effective Stress kPa	Bulk Density kg/m³	Saturated Density kg/m³	Friction Angle °	Material
	2620	2650	40	Bedrock
	2350	2450	40	Asphalt
<18	2025	2245	52	Rockfill
18 – 75	2025	2245	48	Rockfill
75 – 156	2025	2245	46	Rockfill
156 - 343	2025	2245	44	Rockfill
343 - 750	2025	2245	42	Rockfill
750 - 1200	2025	2245	41	Rockfill

Note: The cohesion intercept was taken as zero for the embankment and as 100kPa for bedrock.

After the completion of the analyses, the dam was divided into different material zones. Dividing the wall into zones is an approximation for a continuous depth dependent profile. The sensitivity of stability analyses to this approximation was checked by carrying out detailed analyses using multiple zones. These showed the influence is negligible. Similarly strength models having both constant cohesion and depth dependent cohesion, in conjunction with appropriately reduced angles of friction, were tested. These too showed no significant differences in stability analyses from the zero cohesion, depth dependent friction model. The approach followed, complied with the method used by Leps for the derivation of the strength parameters of Figure 1 (1970).

3. SEISMOLOGY

The dam is located in a seismically active area as evidenced by the Tulbagh and Ceres earthquakes of 1969/70, of magnitude 6.3 which caused considerable damage to buildings. The area is in the Cape Fold Belt, which comprises parallel mountain chains with extensive faulting.

The structural axes trend north-northwest between Clanwilliam and Ceres and generally east-west from Ceres eastwards. A zone of syntaxis is formed where these meet west of Ceres in the area where the dam is located.

Intraplate movements continue to cause earthquakes in this region. Two significant faults occur nearby: these are the Worcester and Groenhof faults.

The Worcester fault is a normal fault with a dip of 70-75° to the south with a throw of about 3km and the Groenhof fault, much closer to the dam is generally less well defined.

It is presumed that any earthquake activity would be caused by post-rift adjustment and would therefore tend to occur along existing faults and be most severe in areas of greatest stress such as the zone of syntaxis close to Ceres.

The seismology report recommended that the design parameters given in Table 2 should be used for seismic loading for the Operating Base, Maximum Design and Maximum Credible Earthquakes – OBE, MDE and MCE (Knight, 1995).

The majority of the stability analyses were carried out using only horizontal accelerations. The vertical accelerations could however have had an influence on the factors of safety since the recommended ratio of the vertical to horizontal accelerations was 0.6.

For example, therefore, the vector sum of the horizontal and vertical accelerations coefficients where the former is 0.12 (MCE) is 0.135. Stability analyses were therefore carried out with coefficients of up to 0.135 to check the sensitivity of the factors of safety to this increase. Typical results are given in Table 3.

From the detailed results it was concluded that it is sufficient to consider only the earthquake coefficient of 0.112 for horizontal acceleration since the addition of the vertical accelerations has no significant influence on the stability analyses results.

4. EARTHQUAKE LOADING

4.1 Pseudo-static loads

In pseudo-static analyses earthquake loads are represented by additional loads obtained by increasing static loads using earthquake coefficients which are associated with specific factor of safety criteria.

The earthquake coefficients are:
No earthquake. EQ Coefficient   = 0.0
Maximum Design. EQ Coefficient   = 0.1
Maximum Credible. EQ Coefficient = 0.12

Figure 2 shows a relationship between earthquake magnitudes and coefficients which can be used to estimate the earthquake coefficient for any magnitude earthquake (Seed, 1979). Hence for Ceres, where the MDE is 6.5 the coefficient is 0.1

Table 2. Seismic loading parameters

Category	Magnitude	Horizontal Values			Vertical Values		
		A_n (g)	V_h	d_H (cm)	A_v (g)	V_v (cm/s)	d_V (cm/s)
OBE	5,3	0.188	14.5	7.2	0.071	8.7	4.3
MDE	6.5	0.244	27.3	13.6	0.134	16.4	8.1
MCE	7.0	0.353	47.2	20.6	0.212	25.3	12.4

Table 3. Influence of vertical acceleration coefficients on stability

Case No.	Earthquake coefficient EQ_{hor}	Earthquake coefficient EQ_{ver}	Upstream analysis (FOS)	Downstream analysis (FOS)
1	0.116	0.070	1.289	1.403
2	0.116	-0.070	1.408	1.453

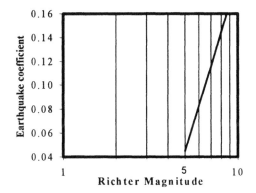

Figure 2. Selection of seismic coefficient

and for the MCE (7.0) is 0.116. The associated minimum factor of safety for the MCE is 1.15.

4.2 Dynamic loads

The relatively high seismic profile at Ceres necessitated dynamic analyses. Ideally, actual earthquake records at Ceres would have been used for this, but unfortunately such records which were available, did not provide adequate information for the purpose.

Two suitable records were obtained from the international database, namely the Nahanni and the Friuli earthquakes which contained significant energy in a wide range of frequencies since this was required to simulate the nature of the hypothetical Ceres earthquake.

The peak acceleration at Friuli was 0.345g whereas that at Nahanni was 0.163g: the latter was therefore scaled up to 0.35g to match the Ceres MCE estimated horizontal acceleration – Table 2.

The computer program FLUSH was used since it is internationally accepted for dynamic analyses for dams. At some stages, FLAC was also utilised and gave similar results.

5. DESIGN CRITERIA

5.1 Static and Pseudo-static Criteria

Design criteria for stability analyses for dams are universally expressed as factors of safety derived from limiting equilibrium analyses. Whilst this approach could be criticised in that the influence of the variability of the inputs is not explicitly addressed, it remains the norm at present. The sensitivity to variations in the input parameters is covered by simple checks carried out by re-running the analyses with a limited range of parameters, generally on the lower side of the selected normal or average design parameters, and assessing by judgement whether a lower bound factor of safety is acceptable.

Generally the method of analysis is not specified in conjunction with the design criteria. The analytical method used for Ceres dam was the conventional limiting equilibrium approach using STABL, in which a variety of failure surfaces can be tested, e.g. circular or non-circular and sliding blocks, and the positions of these can be random or specifically defined.

The internationally accepted static and pseudo-static factors of safety are given in Table 4.

5.2 Dynamic Criteria

The assignment of design criteria for dynamic analyses takes a different form from that for static and pseudo-static analyses. The latter are expressed as factors of safety against failure (Table 4), which have been established on the basis of long experience rather than theoretical considerations.

Dynamic analyses essentially result in estimated deformations of the dam wall. These are then compared with differential strains which are judged to be acceptable for the particular structure; for example at Ceres dam the narrow asphalt core may be assumed to be the critical feature and the acceptable strains within it under earthquake loading determine the limiting criteria.

Table 4. Design criteria factors of safety against sliding

Design Condition	Reservoir Level	Sliding in Slope	Minimum Factor of Safety
End of construction and first impoundment	Empty	Upstream	1.3
	Critical level (min. stability)	Downstream	1.5
Steady state seepage	High flood water level	Downstream	1.5
Rapid draw-down	Critical level (min. stability)	Upstream	1.3
Seismic loading	FSL +EQ	Up and down	1.15

Large vertical and horizontal displacements could rupture the 0.5m wide asphalt core if they result in large differential strains over short distances. On the other hand large horizontal displacements which are evenly distributed over a wide zone will be readily tolerated by the flexible asphalt core. In a rockfill dam with a flexible core, it is anticipated that the displacements will not be restricted to a narrow zone (the basic assumption in the limit state analysis), but rather be distributed over a wider zone.

The output from the displacement analysis is given as a displacement along an earthquake induced slip surface and it is this displacement which should be compared with the asphalt core width.

There are no recognised criteria to relate the width of the core to the allowable calculated earthquake induced slip displacement through the core. The risk of failure of the core through earthquake induced slip displacement is a function of the width of the core, the elevation of the slip surface passing through the core and the slip displacement. More slip displacement through the core could be tolerated closer to the full supply level of the reservoir than deeper down, where, if the core was sheared, erosion of the rockfill might take place, depending on the specific core conditions and the hydraulic gradients. Taking account of the above factors, it is considered that the displacement through the core should definitely not exceed 50% of the core thickness i.e. 250mm, but rather that the displacements for the design case should be limited to less than 100mm.

6. STATIC AND PSEUDO-STATIC ANALYSES

Static and pseudo-static analyses were carried out for different flood levels, coefficients of seismic acceleration and locations of failure surfaces. For the upstream slope (v:h) of 1:1.6, the down stream slope (v:h) varying between 1:1.4 and 1:1.5, and the crest width varying between 7m and 12m, the cases for analysis that were considered, are:
a) Flood levels: rapid draw-down, full supply level (FSL), regional design flood (RDF) and safety evaluation flood (SEF).
b) Horizontal seismic acceleration coefficients: 0.0, 0.1 and 0.12.

c) Failure surfaces:
i. Deep seated failure surfaces from crest to toe.
ii. Higher failure surfaces (from crest intersecting the core) which should provide critical conditions under earthquake loading.

Typical factors of safety of the most critical failure surface obtained for the upstream slope are listed in Table 5 (upstream and downstream slopes of 1:1.6 and 1:1.5 respectively). The start and end positions for the generation of failure surfaces were selected so that failure through the dam foundation was allowed. However, as the dam foundation is considered to have some cohesion (more than 50kPa), the generated failure surfaces did not intersect the dam foundations, but in all cases extended above the foundations and passed through the rockfill.

A series of static and pseudo-static analyses was also carried out for different flood levels and coefficients of seismic acceleration, with the range within which the toe of the failure surface emerged varied in a controlled way in order to assess:
a) the stability of the embankment as the height of the dam increases during construction;
b) the stability of the upper part of the embankment for possible steepening of embankment slopes higher up the embankment;
c) the influence of the crest width.

The analyses show that at any height the factors of safety are similar to those set out in Table 4. Steepening the slopes closer to the crest was therefore not recommended.

A further series of analyses was carried out in which the stability of the upstream slope was considered to assess the effects of rapid draw-down on the dam wall. Two different crest widths were considered, with no seismic acceleration. The most critical factors of safety which were obtained are listed in Table 5: these comply with the design criteria given in Table 4.

Table 5. Upstream rapid draw-down analyses

Crest Width (m)	Factor of Safety	Failure Surface Type
7	1.822	Random
7	1.685	Circle
9	1.825	Random
9	1.683	Circle

7. DYNAMIC ANALYSES

7.1 Procedure

Dynamic analyses were undertaken using the results of the static and the pseudo-static analyses. The FLUSH software was used for the analyses. The following procedure was used:

a) For a set of rockfill material properties, a FLUSH analysis and a pseudo-static stability calculation were undertaken. The output of the FLUSH analysis is a dynamic record of accelerations. The output of the stability calculation is a yield surface and various acceleration coefficients for different analysis cases.

b) A spectral analysis programme was used to calculate the pseudo-spectral velocity, and the spectral acceleration, from the FLUSH output an acceleration-time record.

c) Using the failure surface geometry together with selected nodal dynamic histories, displacement slip analyses were done using the Newmark method (1965), to assess the relative displacement at the failure surface.

The Newmark method comprises a calculational procedure based on the dynamic displacement of a block on a rough surface. When the dynamic acceleration of the block exceeds the yield acceleration (the friction of the block on the rough surface), the block will displace for the time period when the dynamic excitation is larger than the inertial resistance. Similarly, when the dynamic excitation is in the opposite direction, then displacement can occur in the opposite direction.

The Slip program developed by NGI, calculates the algebraic sum of the slip displacements, taking account of their relative directions. Instead of using a block on a rough surface, the Slip program is based on an assessment of the dynamic inertia of a circular failure surface of the rockfill embankment (see Table 6, and Figure 5).

A sophistication incorporated was to accurately calculate the dimensions of the failure surfaces used for the slip analysis, to distinguish between the different crest widths of 7m or 9m (Table 6).

7.2 Discussion of Results

Typical results of the frequency analyses are presented in Figures 3 and 4. These results confirm the previous findings where it was observed that the earthquake frequency content is modified within the dam. As part of the summary of dynamic analyses the amplification ratio has been determined as the ratio of the largest crest acceleration to the dam base excitation. For the design case the amplification ratios are 1.42 and 0.73 for the Friuli and Nahanni earthquakes respectively. In many of the analyses undertaken the response spectra for the dam crest show a shift in frequency content of the excitation motion to typically between 1.2 Hz and 2.0 Hz.

The high frequency content of the Nahanni excitation is then damped in the embankment (Figure 4), whilst the lower frequency (in the range of 2hz) is amplified within the embankment. Similar principles apply to Figure 3.

Many analyses were undertaken to assess, for the embankment geometry and stiffness, the relative shift in frequency at the dam crest for the two selected earthquake excitations. Also an assessment was made of the sensitivity of the slip displacements to the rockfill stiffness and density. It was concluded that the rockfill density variation between 2000 and 2200 kg/m³, has little effect on the displacements calculated.

For the two cases of embankment crest widths, 7m and 9m, the respective upstream and downstream cut-off (yield) seismic acceleration are summarised in Table 7. The cut-off acceleration is defined as the threshold acceleration above which permanent displacements start to take place (i.e. the earthquake acceleration corresponding to a factor of safety of one).

Table 6. Properties of slip failure surfaces

	Crest 7m Upstream	Crest 7m Downstream	Crest 9m Upstream	Crest 9m Downstream
Circle centre (m)	X=100.1 Y=201.3	X=112.3 Y=170.0	X=100.0 Y=201.4	X=112.3 Y=170.0
Circle radius (m)	80.1	50.2	80.1	50.3
Circle angle (°)	39.2	65.2	40.4	66.8
Moment of inertia of surface (m⁴)	55254.1	76887.8	60199.0	81825.0
Centroid of the surface (m)	X=134.9 Y=134.5	X=134.4 Y=132.3	X=134.4 Y=134.3	X=134.0 Y=132.2
Area of the surface (m²)	339.8	454.6	352.9	468.5

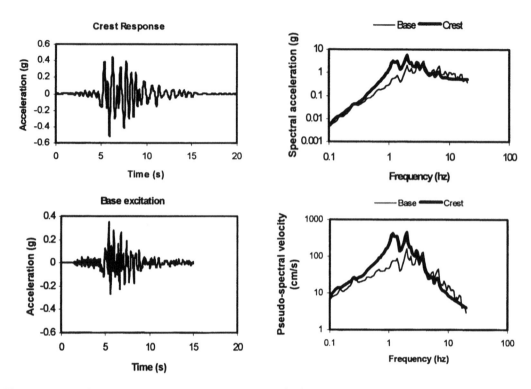

Figure 3. Dynamic analysis using the Friuli earthquake excitation

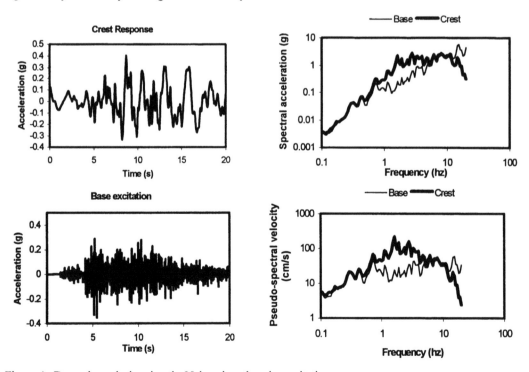

Figure 4. Dynamic analysis using the Nahanni earthquake excitation

Table 7. Summary of cut-off (yield) seismic accelerations (k)

Crest Width (m)	Downstream Slope (g)	(FoS)	Upstream Slope (g)	(FoS)
	0.31	1.00	0.218	1.00
7	0.23	1.15	0.17	1.15
	0.16	1.32	0.155	1.20
	0.32	1.00	0.222	1.00
9	0.24	1.15	0.175	1.15
	0.16	1.34	0.155	1.22

Table 8. Summary of nonlinear dynamic analyses

Rockfill Properties (kg/m³)	Earthquake	Crest Width (m)	Slip (mm) for FOS=1
Non-linear G, γ_{up}= 2404' γ_{dn}= 2195	Friuli	9	10
		7	12
	Nahanni	9	0
		7	0
Non-linear G, γ_{up}= 2245' γ_{dn}= 2404	Friuli	9	13
	Nahanni	7	14
G= 145MPa, γ_{up}= 2404' γ_{dn}= 2245	Friuli	9	0
	Nahanni	7	0
G= 145MPa, γ_{up}= 2245, γ_{dn}=2025	Friuli	9	0
	Nahanni	7	0
G= 150MPa, γ_{up}=2245, γ_{dn}= 2025	Friuli	9	0
	Nahanni	7	0

The values of seismic acceleration for both upstream and downstream were determined for a factor of safety of 1.15, to assess the sensitivity of the factor of safety to seismic acceleration (Table 6). In this calculation, the same failure surface was used for the factors of safety of 1.0 and 1.15; an example is shown in Figure 5.

A summary of the properties of the different failure surfaces is shown in Table 6.

For the various appropriate failure surfaces the dynamic slip displacements through the core of the dam are summarised in Table 8. The calculated slip displacement is very sensitive to the selection of the dynamic stiffness of the rockfill, together with the selected non-linear behaviour thereof. Some of the analyses showed slip displacements up to 300mm, if a linear dynamic stiffness strain level relationship is used.

For the earthquake histories used in the analyses, the calculated slip displacements through the asphalt core are typically less than 20mm. This is considered acceptable as the core can tolerate at least five times more slip displacement without failure occurring.

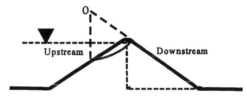

Figure 5. Upstream critical slip circle for crest width of 7m

Table 9. Summary of stability analyses

Condition	Reservoir	Slope Considered	Required FoS	FoS Obtained Crest width = 7m D/s slope = 1.5
Static and Pseudo-static Analyses				
E.C	Empty	Upstream	1.3	Not critical
	Critical level	Downstream	1.5	Not critical
Steady State	RDF	Downstream	1.5	1.51
Rapid D/down	Critical	Upstream	1.3	1.16
Seismic	FSL	Upstream	1.15	1.16
(EQ coeff 0.12)	FSL	Downstream	1.15	1.25
Dynamic Analyses			Factor of safety and cut-off (yield accelerations)	Displacement (mm)
Permanent displacement through the core on the critical failure surface for design earthquake conditions	FSL	Upstream	FoS=1.0 (k=0.22g)	12
		Downstream	FoS=1.0 (k=0.31g)	0

Larger slip displacements can be calculated when the time period -of when actual slip acceleration exceeds the yield acceleration - increases. This can happen under two conditions: the frequency content of excitation decreases, or if the earthquake amplitude increases.

8. CONCLUSIONS

The results of both the static and pseudo-static analyses and the dynamic analyses show that the dam designed with an upstream slope of 1:1.6 and a downstream slope of 1:1.5, with a crest width of 7m, satisfies all the relevant internationally accepted criteria for dam design. The results of the analyses are summarised in Table 9. These meet the design requirements.

9. ACKNOWLEDGEMENTS

The background and technical assistance provided by F Nadim and the guidance by K Höeg of NGI is much appreciated. The permission to write this paper, granted by Koekedouw Irrigation Board and the Ceres Municipality, is kindly noted.

REFERENCES

Barton, N. & B. Kjærnsli 1981. Shear strength of rockfill. *ASCE, Proc. of ASCE, Vol. 107, GT7, pp. 873 - 892, July.*

FLAC 1995. Fast Lagrangian Analysis of Continua. Version 3.3 Itasca Consulting Group, Inc Minneapolis, Minnesota, USA.

Høeg, K. 1993. *Asphaltic Concrete Cores for Embankment Dams.* ISBN82-546-0163-1, Oslo.

Høeg, K., T. Valstad, B. Kjærnsli & P.M. Johansen 1993. Norwegian experience with embankment dams. *Proc. Maintenance of older dams - accidents, repairs and safety assessment.* Chambery, France, September.

International Commission On Large Dams 1989. Selecting Seismic Parameters for Large Dams-Recommendations. *ICOLD Bulletin No. 72.*

Jones, G.A., A.C. White & W. Schreuder 1999. Design and construction of asphaltic concrete core at Ceres Dam. *12th Reg. Conf. Africa Soil Mech. Fdn. Engng.,* Durban.

Kjærnsli, B., T. Valstad & K. Høeg 1992. *Rockfill Dams - Design and Construction.* ISBN82-75598-012-11, Trondheim.

Knight Hall Hendry & Associates 1995. *Geologic and seismologic investigation of the Ceres Dam site.* Second report.

Leps, T.M. 1970. Review of shearing strength of rockfill. *Journal of SM and Foundations Division, ASCE, Vol. 96. No. SM4.* Paper 7394. July, pp.1159-1170.

Lysmer, J., T. Udaka, C.F. Tsai & H.B. Seed 1975. *FLUSH - A computer program for approximate 3-D analysis of soil-structure interaction problems.* Earthquake Engineering Research Centre, Report EERC 75-30, University of California, Berkeley.

Makdisi, F.I. & H.B. Seed 1978. Simplified procedures for estimating dam and embankment earthquake-induced deformations. *Journal of Geotechnical Engineering, ASCE, Vol. 104, No. GT7.*

Newmark, N.M. 1965. Effects of earthquakes on dams and embankments. *Geotechnique, Vol. 15, No. 2.*

Norwegian Seismic Array 1993. Troll-Mongstad Pipeline - *Earthquake criteria update.* Final report - Contract No. T.183539 to Statoil, 4 November.

Seed, H.B. 1979. Considerations in the earthquake-resistant design of earth and rockfill dams. *Geotechnique, Vol. 29, No. 3,* pp. 215-263, Sept.

STABL5 1998. Joint Highway Research Project Informational Report JHRP-88/19 *User Guide for PC.* Purdue University, West Lafayette, USA.

US Committee on Safety Criteria for Dams 1985. *Safety of Dams: Flood and Earthquake Criteria.* National Research Council, National Academy Press, Washington DC, USA

Valstad, T., P.B. Selnes, F. Nadim & B. Aspen 1991. Seismic response of a rockfill dam with an asphaltic concrete core. *Water Power and Dam Construction, April.*

Geotechnics for Developing Africa, Wardle, Blight & Fourie (eds) © 1999 Balkema, Rotterdam, ISBN 90 5809 082 5

Utilisation des paramètres de forage pour le calcul de la capacité portante d'un pieu foré

B. Melbouci
Institut de Génie Civil, Université Mouloud Mammeri, Tizi-Ouzou, Algérie

J.C. Roth
Labo LPMM, Université de Metz, France

RESUME: Nous avons montré dans cet article que l'outillage Starsol-Enbesol fournit des diagraphies instantanées, dont l'utilisation est prometteuse pour la connaissance des caractéristiques des sols au cours de forage. En effet, le rôle de la tarière peut être interprété comme celui d'un véritable pieu. Il est intéressant d'essayer de déduire des paramétres de forage (couple et vitesse d'avancement) la capacité portante d'un pieu réel.

La réalisation d'un nombre conséquent d'essais de forage avec mesure de ces deux paramètres, a fourni les données expérimentales permettant de proposer une méthode de calcul de portance des pieux forés.

ABSTRACT : We have shown in this paper that the new Starsol- Enbesol equipment provides an instantaneous log which proved very promising in determining characteristics of soil during drilling. In fact, the continuous flight auger could be compared to a real pile. From the drilling parameters: torque and advancment rate, consequently is possible to compute the bearing capacity of a real pile.

Testing a sufficient number of bored piles after measuring the two parameters during drilling has provided the subsequent data. This work proposes a method to compute a bearing capacity of piles bored by continuous flight auger.

1 INTRODUCTION

L'appareil faisant l'objet de notre travail se compose de deux parties:

-la Starsol : outillage de forage à tarière continue (Fig. 1)

-l'Enbesol : centrale d'acquisition de paramètres.

Cet outillage a été développé par SOLETANCHE essentiellement pour améliorer la mise en œuvre des pieux forés du point de vue qualité, fiabilité et performances. La réalisation de ces pieux nécessite deux phases importantes: la phase de forage et la phase de bétonnage.

Dans la phase de forage, seul le couple et la vitesse d'avancement sont enregistrés. Dans la phase de bétonnage, c'est la pression de bétonnage et le profil de bétonnage qui sont enregistrés.

Ce procédé de forage à la tarière évite le remaniement du sol, l'effondrement des parois de forage et assure un bon contact en pointe et le long du fût (béton-sol). Dans ce document, on exploitera uniquement la phase de forage qui nous permettra de valider les paramétres de forage liés à la réponse du terrain (vitesse d'avancement et couple de rotation) (Fig. 2).

- Vitesse d'avancement : la variation de cette vitesse traduit la résistance du sol avec l'outil de forage. Elle est inversement proportionnelle à la compacité ou la consistance des sols.

- Couple de rotation : il peut renseigner sur l'adhérence du terrain à l'outil. Sa variation donne une idée de la compacité du matériau traversé.

2 PRINCIPE DE REALISATION DES PIEUX

Ce procédé, dérivé des techniques de reconnaissance, consiste à utiliser en guise de tubage, l'outil lui même qui, en l'occurrence, est une tarière continue dont l'âme creuse est munie d'un tube plongeur intérieur coaxial pouvant coulisser sur 1.2m et lui confère l'une de ses qualités majeures, qui est la possibilité de conduire le bétonnage en maintenant la partie basse du tube plongeur noyée dans le béton, d'où élimination du risque de désamorçage.

Le forage est réalisé en vissant dans le sol la tarière (Fig. 1) dont la longueur est au moins égale à celle du pieu à exécuter. En fin de vissage, le béton est injecté par l'axe creux de la tarière qui alors progressivement arrachée du sol sans aucune rotation en assurant un bon contact entre le pied de la tarière et

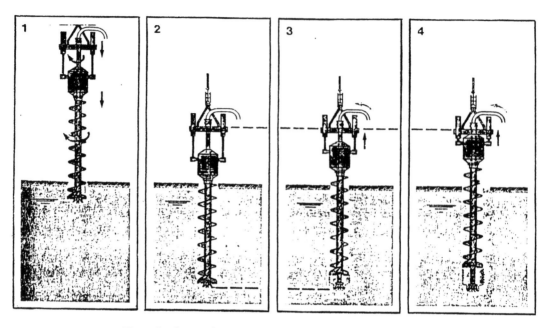

Figure 1 : phasage des opérations de mise en œuvre d'un pieu Starsol

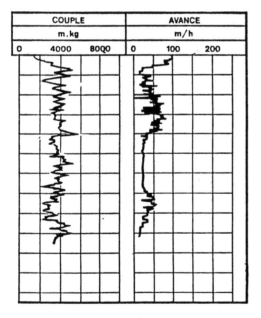

Figure 2 : exemple d'enregistrement des deux paramètres (vitesse d'avancement et couple de rotation).

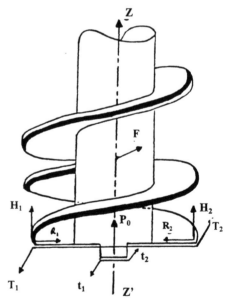

Figure 3 : efforts appliqués sur la tarière.

le béton mis en place, de manière à éviter de découvrir les parois du forage entre tarière et béton.

Le tube plongeur et la possibilité d'injecter sous pression, assurent une qualité de contact béton-terrain supérieure à celle des pieux forés tradition-nels. Cette disposition assure aussi, une excellente qualité de pointe du pieu en supprimant le risque de décompression du sol en forage avant bétonnage.

Après la phase de bétonnage et après nettoyage des déblais, une armature est mise en place par battage.

3 SOLS TRAVERSES

La réalisation des pieux est envisageable dans tout sol de module pressiométrique inférieur ou égal à 100MPa.

4 METHODE DE CALCUL DE PORTANCE

Cette méthode repose sur la résolution d'un système de trois équations, deux d'entre elles caractérisent les efforts qui se développent dans la tarière, la troisième concerne le bilan énergétique global.

La mise en équation d'un tel système nécessite la détermination d'un certain nombre d'hypothèses. En effet l'analyse des efforts dépendra essentiellement de l'influence de trois facteurs intervenant sur le comportement de la tarière en pénétration:
- de la nature du sol (couches rencontrées, répartition des déformations pendant le forage et le mode de travail de l'outil).
- de la géométrie de la tarière (obliquité de l'hélice et le diamètre de la tarière).
- du frottement tarière-sol.

La modélisation du comportement de la tarière et la mise en équation des efforts nécessitent le choix d'un certain nombre de paramètres et d'hypothèses définies comme suit :
- la tarière et le sol sont en contact continu,
- la tarière est animée d'un mouvement de rotation de vitesse angulaire « w » et d'un mouvement de translation , de vitesse d'avancement « V »,
- la pénétration de la tarière dans le sol est constituée sous le seul aspect du cisaillement et sans déplacement du sol : (cisaillement en pointe et frottement de contact tarière-sol),
- l'effet du sol est caractérisé par des contraintes tangentielles à l'hélice et à l'âme de la tarière pour le frottement latéral et par des réactions verticales pour le terme de pointe (Fig. 3),
- les frottements latéraux au niveau de l'hélice et de l'âme sont de même type. Les forces de frottement sont inclinées suivant l'angle d'inclinaison de l'hélice (ψ),
- la résultante de toutes ces forces de frottements est la somme des frottements sur le tube et de part et d'autre de la tôle hélicoïdale. Elle est positionnée par rapport au centre de gravité:

$$F = \Sigma f \qquad (1)$$

- la résistance de pointe globalise tous les effets de cisaillement à ce niveau : celle ci constitue la première équation du système,
- la deuxième équation est obtenue en considérant que la somme algébrique des couples est nulle,
- La troisième équation traduit le bilan énergétique global.

Le système d'équation ainsi obtenu , s'écrit :

$$P - R_p - F \sin \Psi = 0 \qquad (2)$$
$$C_m - C_f - C_T(1 + d/di) = 0 \qquad (3)$$
$$E_m - F.L - C_T.2\pi (1 + d/d_i) - R_p.a = 0 \qquad (4)$$

Avec :
Ψ : angle d'obliquité de l'hélice
C_m et C_f : sont respectivement le couple moteur et le couple de frottement de l'hélice,
C_T : couple résistant au niveau de la pointe,
d et d_i : sont respectivement les diamètres de la tarière et du tube,
E_m : énergie mécanique,
P : poids de la tarière,
a : enfoncement de la tarière pour un tour.

Les solutions de ce système donnent le frottement latéral « F » et la résistance de pointe R_p.

$$F = \frac{2\pi \, C_m \, - \, (E_m - P.a)}{\dfrac{2\pi \, df}{\cos\Psi} - L + a \sin \Psi} \qquad (5)$$

$$R_p = P - F \sin \Psi \qquad (6)$$

où
df et L sont respectivement le bras de levier de F et la longueur par tour de l'hélice au point d'application de F.

5 EVALUATION DE LA PORTANCE DU PIEU FORE:

Ces formules sont basées sur la séparation de la charge portante en résistance de pointe et en frottement latéral.

$$QL = Q_l{}^p + Q_l{}^f \qquad (7)$$

5.1 Terme de pointe

La résistance de pointe est définie par la formule suivante:

$$Q_l{}^p = R_p / \sin\Psi \qquad (8)$$

Dans le cas d'un sol homogène, la résistance de pointe est obtenue au premier tour par la formule (8) ci dessus. Dans le cas d'un multicouche, la résistance de pointe est donnée par l'équation (8) de la dernière couche au dernier tour.

5.2 Terme de frottement latéral

Dans le cas d'un sol homogène, les calculs sont effectués pour un tour de la tarière, le frottement latéral est déterminé par la formule (7). Pour déterminer le frottement latéral le long de la couche, il faut multiplier « F » par le nombre « n » de tours que la tarière a effectué pour traverser la couche considérée.

$$Q^f_L = F * n \qquad (9)$$

Dans le cas d'un multicouche, le frottement latéral est déterminé au niveau de chacune d'elle et le frottement latéral total est la somme des frottements mobilisés dans les différentes couches.

5.3 Remarques sur le domaine d'utilisation de la résistance de pointe et du frottement latéral

En pointe de la tarière, on estime que la présence de la contre hélice sur un demi-tour facilite la pénétration de la tarière, on considère donc que le domaine de la résistance de pointe se limite à ce dernier tour de l'hélice. Quant au frottement latéral, il est supposé régner sur toute la longueur de la tarière à l'exception de l'extrémité inférieure sur une hauteur de un pas, c'est à dire, sans la présence de la contre hélice.

6   DETERMINATION DU COEFFICIENT DE SECURITE:

Pour affiner notre méthode et étendre son application, il est indispensable de comparer et vérifier nos résultats par rapport aux données fournies par des essais de chargement. Car , il est vrai que seule la multiplication des essais de chargement en vraie grandeur, réalisés sur fondations dûment instrumentées selon un mode opératoire normalisé offrira la possibilité de mieux cerner la réalité. Malgré le faible nombre d'essais dont nous disposons, il nous a paru possible de nous faire une idée globale de la validité de la méthode proposée en comparant pour l'ensemble des pieux étudiés la charge limite calculée et la charge limite réelle.

Pour les capacités portantes totales, le rapport entre valeur expérimentale et valeur de calcul est égal au coefficient de sécurité F_1.

Tableau 1: comparaison des forces portantes limites (en KN)

Sites	mesures expérimentales QL	méthode starsol QLc	F1=QL/QLc
Toulouse	4301	3578	1.2
Strasbourg	2480	1964	1.26

Où   QL : est la charge limite mesurée aux essais de chargement

QLc : est la charge limite calculée par la méthode Starsol.

Lorsqu'on applique les coefficients de sécurité habituels (3 pour la pointe et 2 pour le frottement latéral) pour obtenir les charges nominales et que l'on calcule les coefficients de sécurité réels Fr , on obtient pour les différents pieux :

Tableau 2 : comparaison des forces portantes aux états limites de service en (KN)

Sites	mesures expériment QF	méthode Starsol QN	Fr=QF/QN
Toulouse	2750	1712	1.61
Strasbourg	1250	950	1.31

Où   QF et QN sont respectivement la charge de fluage et la charge nominale.

Si l'on était descendu jusqu'à 2 pour le coefficient de sécurité de la résistance de pointe, on aurait trouvé comme coefficient de sécurité Fr :

Tableau 3 : forces portantes Starsol proposées en (KN)

Sites	mesures expériment QF	méthode Starsol QN	Fr=QF/QN
Toulouse	2750	1789	1.54
Strasbourg	1250	982	1.27

On estime donc que le coefficient de sécurité réel est compris entre 1.27 et 1.61. Pour plus de sécurité nous adopterons un coefficient de sécurité uniforme égal à 2 pour le terme de pointe et le frottement latéral.

La réduction du coefficient de sécurité de 3 à 2 du terme de pointe peut s'expliquer par le bon contact pointe sol et au contrôle continu et systématique réalisé pendant le forage.

7 CONCLUSION

L'étude de synthèse sur la portance effectuée à partir des résultats obtenus après réalisation d'un certain nombre d'essais de chargement, a permis de vérifier que la méthode Starsol conduit à des portances bien calées sur la réalité.

Présentant en outre , l'avantage de reposer sur un essai, réalisable pratiquement dans tous les types de sol, avec une exécution particulièrement soignée

(curage parfait du fond de forage et un contact pointe sol de bonne qualité), la méthode s'avère fiable et suffisamment représentative.

Par contre , nous sommes tout à fait conscients que le nombre de pieux instrumentés analysés (essais de chargement) est assez faible et que les conclusions proposées sont susceptibles d'être modifiées dans l'avenir en fonction des expérimentations futures. De ce point de vue, la multiplication des essais de chargement de pieux instrumentés est infiniment souhaitable.

BIBLIOGRAPHIE:

[1] Bustamante M.
   -Compte rendu d'essai pieu « STARSOL » Colombes - F.A.E.R 1 -05-02-5 Mai 1985
   -Compte rendu d'essais de chargement cité judiciaire de clermont ferrand -sept 1985-Note technique.
   -Compte rendu d'essais de chargement de pieux « Starsol » sur le Zac cevennes-F. A. E .R
   1-05-02-6 Mars 1986.
[2] Bustamante M.1987. Etude expérimentale du procédé de fondations profondes STARSOL. Performances et évaluation de la capacité portante. Bul Liaison Labo. P. Et Ch., N°149, Mai-Juin1987,pp29-36.
[3] Bustamante M., Gouvenot D.1981. Incidence des conditions d'exécution et du délai de repos sur le comportement et la portance des fondations forées-Revue Française de Géotechnique ,N°16, Août 1981 , pp17-33
[4] Melbouci B.1988. Détermination de la capacité portante des pieux réalisés avec le dispositif STARSOL-ENBESOL-Thèse de Docteur Ingénieur- INPL Nancy (France) Fev 1988
[5] Schlosser F. , Blondeau F. , guilloux A.1985 Pieu foré injecté au tube plongeur. Analyse des performances du procédé. Terrasol Bureau d'ingénieur Conseils en géotechnique.

Geotechnics for Developing Africa, Wardle, Blight & Fourie (eds) © 1999 Balkema, Rotterdam, ISBN 90 5809 082 5

Nonlinear finite element analysis of load-deformation of soil-culvert systems

Yahia E-A. Mohamedzein
Building and Road Research Institute, University of Khartoum, Sudan

ABSTRACT: A finite element based methodology has been developed for the analysis of soil-culvert interaction. An elastic plastic work hardening cap soil model and a large deformation formulation are included in the computer program. The possibility of yielding of the culvert is accounted for by using a Von Mises yield criterion. The compaction and construction processes are also simulated. Predictions with this program agreed with the results of field, laboratory, and full scale tests. Prime locations for yielding of the culvert are at the spring line and shoulders; shear failure develops in the soil in the neighborhood of the spring line and above the upper half of the culvert. Yielding of the culvert, soil shear failure, and large buckling deformations(individually or combined) were identified as the causes of failure of soil-culvert systems. Minimum soil cover of 1 to 2m above the crown is required to protect flexible buried culverts from failure under the influence of heavy live loads.

1 INTRODUCTION

Modes of failure that can lead to the complete collapse of soil-culvert systems consist of large buckling deformations, soil shear failure and yielding of the culvert wall. Thus, to accurately determine the performance limit of culverts, it is required to trace the load-displacement of the system up to failure. A method to achieve this must use an appropriate soil model, and model correctly the nonlinear behavior of culvert such as yielding and large strains. A finite element computer program was developed in this study to simulate the mechanisms of the most relevant modes of failure.

2 THE COMPUTER PROGRAM

The program is a plane strain finite element program for analysis of soil-structure interaction problems with special emphasis on soil-culvert systems (Mohamedzein 1989). A typical mesh used in the program is shown in Figure 1. The program has the following features and capabilities:(1) A soil library which includes the elastic plastic cap model with Drucker-Prager failure surface (Chen & Baladi 1985), as well as Duncan-Chang, modified Duncan (Duncan et al. 1978), variable modulus, and linear elastic soil models, (2) Large deformation formulation to account for the effects of large strains associated with conditions leading to failure of soil-culvert systems, (3) Simulation of both construction sequences and compaction loads, (4) Yielding of the culvert using a Von Mises yield criterion, (5) Interface elements to simulate the interaction between the culvert and the soil, (6) Trench and embankment installations simulated; mesh is automatically generated based upon the dimensions of the culvert , and (7) The modified Newton with Aitken acceleration scheme or BFGS iteration methods to solve for equilibrium equations (Bathe 1982). The predictions of the program were compared with the results of field, laboratory and full scale tests and were found to agree well with the experimental results (Mohamedzein 1989, Mohamedzein 1992, Mohamedzein & Chameau 1996).

3 PERFORMANCE LIMITS

Performance limit (failure) is identified whenever one or more of the following conditions are satisfied during the analysis:

1. The displacements increase rapidly even for small load increments.

2. The stiffness matrix becomes non positive definite.

3. Convergence problems arise even for small load increments.

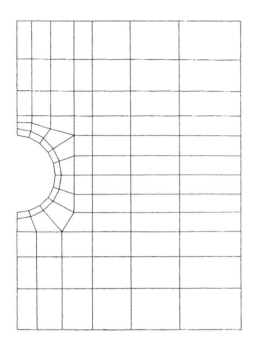

Figure 1. Typical finite element mesh (After Katona et al 1976).

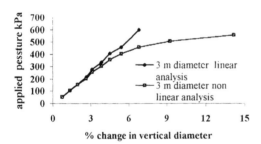

Figure 2. Typical load-deformation curve.

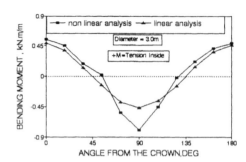

Figure 3. Distribution of the bending moment.

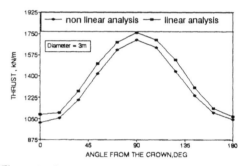

Figure 4. Distribution of the thrust.

A typical load deformation curve for culverts is shown in Figure 2 for a corrugated steel culvert of 3m diameter. Similar curves were obtained for culverts with diameters of 1.5m and 6m. The figure clearly indicates the effect of nonlinearity on culvert performance. For nonlinear analysis the culvert failed when the percentage change in vertical diameter (i.e. crown deflection)was 21.7, 20.8 and 13.3% for the 1.5m, 3m and 6m diameter culverts, respectively. In all cases analyzed here, failure occurred as a result of soil failure, culvert yielding and/or large deformations.

Typical graphs for the distribution of the bending moment and thrust are shown in Figures 3 and 4, respectively. The figures again show the different culvert response when nonlinear analysis is performed.

Figure 3 shows that the maximum negative bending moment occurs at the springline (90° from the crown), while the maximum positive bending moment occurs at the crown and the invert (180° from the crown). It should be emphasized that the shape of the bending moment diagram around the culvert is not unique since it depends on soil type and compaction and culvert type and size among other factors .The distribution shown in Figure 3 is typical for a loose soil and a large diameter culvert. Figure 4 shows that the location for maximum thrust is the springline. Unlike the bending moment ,the shape of thrust diagram is the same for all cases and is not sensitive to soil type, compaction, culvert type and size.

4 FACTORS AFFECTING FAILURE LOAD

There are a number of factors which affect the load-carrying capacity of soil-culvert systems, e.g. culvert size, culvert shape, degree of soil compaction, culvert yielding, height of soil cover, live loads, bedding, construction details, and installation procedures. In this section the effects of the following factors are considered: culvert yielding, degree of soil compaction, height of soil cover, and live loads.

4.1 Yielding of the Culvert Wall

Large deformation analysis of three corrugated steel culverts with diameters of 1.5,3.0 and 6.0m were performed with and without accounting for culvert yielding. The results of the analyses are given in Table 1. In the yielding analysis, once all the culvert sections have yielded, the load-carrying capacity of the whole system is impaired. For the non-yielding analysis, the system fails as a result of large deformations. Yielding of the culvert decreases the load-carrying capacity of the soil-culvert systems by 10 to 30% and the effect is more significant for large size culverts. In all cases yielding occurred first at the spring line and the shoulders. Thus special considerations should be given to increase the resistance to yielding at these locations. Special features such as rib stiffeners can be used.

4.2 Degree of Soil Compaction

Design codes (e.g. AASHTO Design Code 1983) usually specify well selected backfill to be placed around the culvert, however, the extent of the selected backfill is not always known, and the compaction requirements are not always satisfied. Two steel culverts with diameters of 3 and 6m were analyzed twice, first using a loose sand and then a dense sand as backfill materials. Each problem was analyzed using the cap model and large deformation formulation, and yielding of the culvert was also accounted for. For the 3m diameter culvert in a dense sand, failure occurred as a result of partial yielding of the culvert wall (about 75% of the culvert sections) and large deformations, however, for the other cases soil failure, culvert yielding, and large deformations all contributed to the failure of the soil-culvert system. The soil elements that reached a failure condition were located near the spring line and above the crown and shoulders. The results of these analyses are summarized in Figure 5 and Table 2.

Table 2 shows that a dense sand increases the load-carrying capacity of the soil-culvert systems by 40 to 60%. Also, the table indicates that the pressure carried by the 6m diameter culvert is less than that carried by the 3m diameter culvert. The pressure deflection curves for the two culverts (Figure 5) are similar in trend to the test results reported by Watkins and Moser (1969). The deflection curve of a culvert buried in a dense soil is steeper than that for the same culvert buried in a loose soil. The computed limiting crown deflection (Table 2) is about 6% for the dense sand , while it ranges from 11 to 14% for the loose sand. This indicates that the 5% limiting deflection recommended by the current design methods is a reasonable limit for dense backfill, however ; it is smaller than the values computed herein for loose backfill. With less than 50% of yielded sections of the culvert, the computed crown deflections were 8 to 13% for loose sand and the applied loads are in the range 30 to 50% of the failure loads. Thus, by limiting deflection to about 10%, less than 50% of the culvert sections would yield and the culvert will retain its load carrying capabilities.

4.3 Soil cover and live loads

A large number of buried culverts have failed under the influence of live loads due to inadequate soil cover (Selig et al 1977). To reduce the effect of live loads, design codes specify a minimum soil cover. For example, the AASHTO Design Code (1983) specifies minimum soil covers of 0.61 to 1.22m (2 to 4ft) for culverts with diameters greater than 4.5m (15ft). These specifications are applicable to only a few cases of buried culverts, and thus analytical methods are required to compute the allowable live loads for a given soil cover and culvert diameter. Two corrugated steel culverts with diameters of 1.5m and 6m, respectively, were analyzed. An elastic-plastic cap soil model was used to represent a dense sand backfill, and yielding of the culvert and large deformations were included. Live loads were applied over soil covers of 0.3 to 3m above the crown . These live loads are assumed to be carried by two wheels on one axle to simulate the action of an AASHTO HS-20 truck weighing 32000lb (142.4 kN). Because the live loads are concentrated loads, they are represented by equivalent line loads using the Boussinesq elastic theory.

The failure load is given as a function of soil cover above the crown in Figure 6. For soil covers less than 1m, the two culverts failed at small live loads; then the failure loads increase rapidly with the soil cover specially for the 1.5m diameter culvert. For a soil cover of 1.5m, this 1.5m culvert can sustain live loads up to 56000lb (249.2 kN). The 1.5m diameter culvert can withstand the HS-20 truck for

Table 1. Effect of yielding of the culvert on failure pressure.

Dia. (m)	Yielding Analysis Pressure (kN/m²)	Nonyielding Analysis Pressure (kN/m²)
1.5	695	771
3	448	648
6	275	379

Table 2. Effect of soil compaction.

Dia. (m)	Soil Type	Failure pressure (kN/m²)	%change in vertical diameter
3	dense	634	6.2
6	dense	475	6.4
3	loose	448	13.7
6	loose	303	11.4

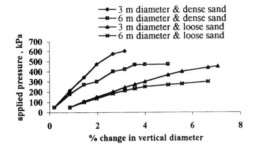

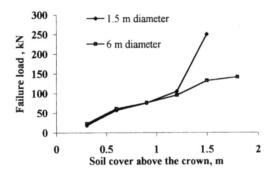

Figure 5. Effect of soil compaction.

Figure 6. Effect of soil cover on live load.

increased soil cover under traffic loads, and decreased culvert size. Although these results were obviously expected they have been clearly quantified with the present analyses.

Minimum soil covers of 1.22 to 1.83m are required to guard against failure of steel culverts with diameters of 1.5 to 6m when AASHTO HS-20 type trucks are considered in the design.

Crown deflections as large as 20% were computed in this study. The 5% limiting crown deflection recommended by design codes can be unjustified restriction specially where loose backfill is used. With less than 50% of yielded sections of the culvert, crown deflections in the range 8 to 13% were computed for loose backfill, and the applied loads are only 30 to 50% of the failure loads. Thus, the limiting crown deflections for loose backfill could be specified as 10%. For dense backfill, the limiting crown deflection of 5% is reasonable.

soil covers greater than 1.22m, while the minimum soil cover is 1.83 m for the 6m diameter culvert. These results are similar to those obtained from experiments (e.g. Bacher & Kirkland 1986, Temporal & Johnson 1988).

Although the analyses performed here are for a dense sand backfill, the computed minimum soil covers are higher than those recommended by AASHTO Design Code (1983). This suggests that the AASHTO specifications may not be sufficient for cases where buried culverts are exposed to significant live loads.

5 CONCLUSIONS

A finite element based model was developed for prediction of load-deformation characteristics of soil-culvert systems. The program models the non-linear behavior of soil culvert-systems (both material and geometric nonlinearity) and accounts for installation procedures.

A parametric study was performed to investigate the effects of degree of soil compaction, culvert yielding, height of soil cover, and live loads on the load-carrying capacity of soil culvert systems. With regard to these factors, the failure of buried culverts can be minimized with increased yield stress of the culvert material, increased degree of soil compaction,

REFERENCES

AASHTO 1983. Standard specifications for highway bridges. 13th Edition.

Bacher, A. E. & D. E. Kirkland 1986. Corrugated steel plate structures with continuous longitudinal stiffeners: live load research and recommended design features for short-span bridges. Transportation Research Record 1087, TRB, Washington, D.C., USA, 25-31.

Bathe, K. J. 1982. Finite element procedures in engineering analysis. Prentice Hall Inc.

Chen, W. F. & G. Y. Baladi 1985. Soil plasticity: theory and implementation. Elsevier Science Publishers.

Crabb, G. I. & D. R. Carder 1985. Loading tests on buried flexible pipes to validate a new design method. TRRL Research Report No. 28, Crowthorne, Berkshire, UK.

Duncan, J. M., Byrne, P., Wong, K. S., & P. Mabry 1978. Strength, stress-strain and bulk modulus parameters for finite element analyses of stresses and movements in soil masses. Report No. UCB/GT/78-02, University of California, Berkeley, Calif., USA.

Katona, M. G., J. M. Smith, R. S. Odello, & J. R. Allgood 1976. CANDE: A modern approach for structural design and analysis of buried culverts. Engrg. Report, User Manual, System Manual, Reports FHWA-RD-77-5, 77-6, 77-7, U. S. Naval Civil Engineering Laboratory, port Hueneme, Cal., USA.

Mohamedzein, Y. E-A. 1989. Nonlinear finite element analysis of soil culvert interaction. Thesis presented to Purdue Univ., West Lafayette, Ind., USA, in partial fulfillment of the requirements for the degree of Docator of Philosophy.

Mohamedzein, Y. E-A. 1992. Predicting the

performance limit of soil-culvert systems.
Proceedings of the 9th ASCE Engineering
Mechanics Conference, College   Station,
Texas, USA, pp. 908-911..

Mohamedzein, Y. E-A. & J-L. Chameau 1996.
Elastic plastic finite element analysis of soil
culvert interaction. Sudanese Engrg. Journal,
Vol. 44, No. 34, pp. 16-29.

Selig, E. T., Abel, J. F., Kulhawy, F. H. & W. E.
Falby 1977. Review of the design and
construction methods of long-span, corrugated-
metal, buried culverts. Report FHWA-RD-77-
131, Federal Highway Administration, Washington D.C., USA.

Temporal, J. & P. E. Johnson 1988. Loading tests on
two long span corrugated steel culverts. Research
Report No. 141, TRRL, Crowthorne, Berkshire
UK.

Watkins, R. K. and A. P. Moser 1969. The structural
performance of buried corrugated steel pipes.
Report, Engineering Experimentation Station,
Utah State University, Logan, Utah. USA.

Geotechnics for Developing Africa, Wardle, Blight & Fourie (eds) © 1999 Balkema, Rotterdam, ISBN 90 5809 082 5

Osmotic pressures and effective stresses

A. D. W. Sparks
Department of Civil Engineering, University of Cape Town, South Africa

ABSTRACT : It was shown by the author in 1961 that stresses in clays with osmotic pressures can be described by an effective stress equation for partly saturated sands. This paper amplifies the connection between osmotic pressures and this equation. A method is also provided for predicting the soil-suction curve from a grain size analysis of the soil.

1. INTRODUCTION

The fundamental stress equation (Sparks 1961) was of the following form :-

$$\sigma_X = \sigma'_X + \alpha . u_A + \beta . u_W - \gamma . T \dots (1)$$

where

- σ_X = the total stress in the x direction (with respect to atmospheric pressure).
- σ'_X = the normal effective stress in the x direction.
- α = the proportion of the cross-sectional area occupied by air.
- β = the proportion of the cross-sectional area occupied by water.
- T = the Surface tension force (approx 73×10^{-6} kN / m).
- γ = the length of the cut air-water boundary in the unit projected area.
- u_A = the air pressure (with respect to atmospheric pressure; usually zero).
- u_W = the freewater pore water pressure (this is usually negative in the equation 1).

The effective stress σ'_X contains the nett effect of all repulsive and attractive forces between the solid soil grains, and the effect of the osmotic pressure , and the transmitted portion of applied and self-weight stresses. This is shown in equation (10) of this paper.
When using equation (1), the effective stress is found as the unknown quantity. The pore water pressure u_W corresponds to the pore pressure which is found by connecting an engineering device such as a manometer tube or pressure gauge to the (slightly salty) free pore water.

2. THE BISHOP CHI FACTOR

The relationship to the Bishop Chi factor χ can be illustrated by the following equations.
Factors λ and η were defined (equations 4.15 to 4.23 in Sparks 1961) as follows :-

$$\sigma_X = \sigma'_X + \lambda . u_A + \eta . u_W \dots (2)$$
$$\sigma_X = \sigma'_X - \eta . (u_A - u_W) + u_A (1 - A_{xr}) \dots (3)$$
$$\text{and } \eta = \beta + \gamma . T /(u_A - u_W) \dots (4)$$

where A_{xr} = total of the projected contact areas between solid particles on to the x-y plane as a fraction of the total x-y section area. (The value of A_{xr} is usually zero for a moist clay).

It is therefore seen from equations (2) and (3) that η was defined in a similar manner to the value of χ as used by Bishop and Blight. However the value of η is obtained from the values of β and γ as described by equation (4).
A problem arises when using the χ factor for a non-wetting soil such as Talc (wetting angle greater than zero) because the value of $(u_A - u_W)$ approaches zero while the degree of saturation is still less than unity, and γ is positive. In other words χ or η tends to plus or minus infinity as the degree of saturation approaches the condition when u_A equals u_W. The author prefers to use equation (1) rather than equation (2).

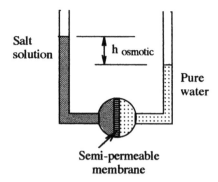

Fig. 1 The usual osmotic experiment

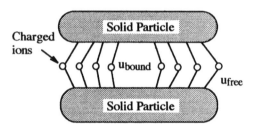

Fig. 2 - Simplified model - Ions attached to rubber-like bands. Salt content is higher at centre than in the free water.

3. EFFECTIVE SOIL PARTICLES

The concept of the effective particle was used in the derivation of stress equations for saturated or partly saturated soils (Sparks 1961, 1963). The effective soil particle can be defined as the solid particle plus the surrounding micelle water layer (i.e double layer). An effective particle can also include several clay particles which are close together and which have overlapping double layers. (e.g. the core of the effective particle can consist of parallel packing , or a domain type structure).

Lambe (1965, p 55) states that the thickness of the double layer is the distance from the solid particle required to neutralize the nett charge on the particle, i.e. it is the distance over which there is an electrical potential. Sridharan and others (1982) showed that particles in parallel packing move closer together when the salt concentration is increased in the pore water. It seems that the larger concentration of available cations permits the neutralization over a shorter distance of the nett charge on the solid particle. i.e. a greater salt concentration will cause thinner

micelle layers of adsorbed cations.

We can conclude that an effective clay particle, which might include several stacked solid clay particles, will become smaller in total size if extra salt is added to the water surrounding the clay particle.

4. OSMOTIC PRESSURE

A student is introduced to the concept of osmotic pressure via a diagram such as Figure 1 which shows that at equilibrium the water level in the standpipe containing a salt solution, is higher than the water level in the tube containing pure water. The stand-pipe pressure measurement system shown in Figure 1 can be termed an "engineering measurement system".

It was suggested that the average water pressure u_{bound} across the area of contact between the effective particles is greater than the value of the pressure u_{free} in the free capillary water (Sparks 1961). The author believes that the osmotic pressure in saline water becomes evident when the salt ions are unable to move towards the unbound water because these salt ions are restrained either by a semi-permeable membrane or because the ions are held in the adsorbed water double layer by electrical charges on the solid particles. Figure 2 illustrates this principle.

In other words when these water pressures are both measured by engineering pressure gauges or by levels in piezometer tubes, the water pressure in the held saline water u_{bound} is greater than the water pressure u_{free} in the adjacent unbound water.

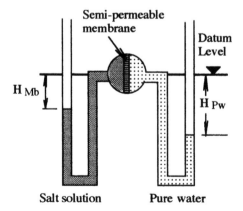

Fig. 3 Greater suction tension in the case of the pure water.

5. MEASURING PRESSURES

5.1 Engineering or Matric water pressures

The values ubound and ufree are the true engineering water pressures (corresponding to matric water pressures) which should be used in pressure or force equations. These values could in theory be measured by engineering pressure gauges or by levels in piezometer tubes.

5.2 Vapour pressures and Total water pressures

An "artificial" water pressure is estimated when the water pressure is calculated from vapour pressure measurements. This "artificial" water pressure is called the "total" water pressure. The author refers to this "total" water pressure as an "artificial" water pressure because it cannot be solely measured in a salt solution by using an "engineering pressure measurement system.".
This "total" water tension is important because it is useful in the understanding of the wilting tension adjacent to plant roots in saline soils. i.e. this "total" water tension which the pore fluid within the plant must overcome is equal to the negative matric tension u in the unbound soil water plus an osmotic tension due to the salt concentration in the soil water adjacent to the roots.

5.3 Water heads and water tensions.

All water pressure heads are measured upwards from datum, but water tensions are measured downwards from the level of the soil sample.
Within the micelle layer, the osmotic pressure probably increases closer to the solid particles (i.e. the water level in a standpipe from this held water will be higher as one moves the sensor tip of the standpipe closer to the solid particles).

5.4 Five effects due to saline solutions.

a.) The equilibrium water vapour pressure over a saline solution will be lower than that over pure water - if all other factors are the same. In fact condensation readily occurs on a concave meniscus on a saline solution because of the curvature of the concave surface and because of the salinity of the pore water (see Figure 4).

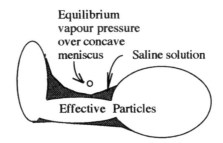

Equilibrium vapour pressure over concave meniscus — Saline solution — Effective Particles

Fig. 4 - Equilibrium vapour pressure is lowered because of saline solution and because of concave meniscus (associated with -ve u).

b.) Because of the bound salts and soil particles which act as an semi-permeable membrane, the water will rise to a higher level in an engineering standpipe if the lower end of the stand-pipe is in the more saline water (when compared with the rise from the less saline water in the soil).

c.) In the example in section (b) above, the water suction associated with the level of the saline water in a stand-pipe is lower than that associated with the level of the water for the stand-pipe from the less saline water within the same soil.

d.) The suction head measured downwards from the sample to the level of the water level in the standpipe from the pure water is known as the "total suction" and is greater than the "matric suction of the saline water" which is measured from the sample down to the water level in the standpipe which emanates from the saline water. It is seen from section 5.4(a) above, that a system which measures the vapour pressure which exists over a saline solution will provide the "total suction" and not the "matric suction of the saline water"; the latter being the pressure or suction to be used in engineering equations of static equilibrium (as in Section 6 of this paper). A saturated clay sample located in the air space above a super-saturated salt solution in a closed container, will reach equilibrium corresponding to a "total suction" determined by the vapour pressure of the super-saturated salt solution.

e.) The addition of salt to a clay soil decreases the void ratio, the thickness of the adsorbed layers and the diameters of the effective soil particles. The Liquid Limit will also decrease. Salt addition will change the "spring constant" between particles and hence the matric suction.

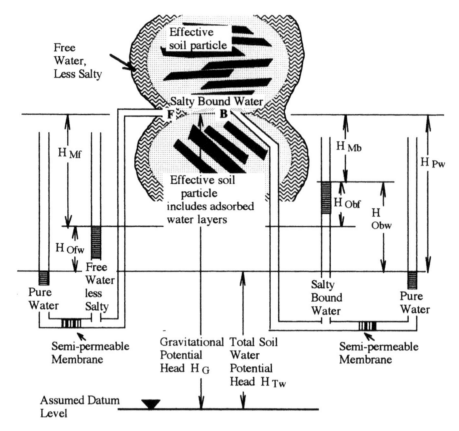

Note that if the Datum Level is moved up to the level of the soil sample, then the Total Soil Water Potential Head H_{Tw} will be numerically equal to the Head H_{Pw}.

Fig. 5 - Equilibrium levels in standpipes connected to different water zones.

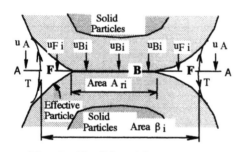

Fig. 6 - If solid particles are not in contact

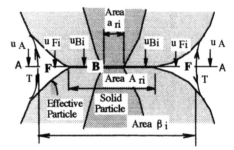

Fig. 7 - Solid particles in contact (Sparks 1961, p 4-50)

6. EQUATIONS RELATING TO FIG. 5

For the free-water in Figure 5, it follows that the following equation can be written :-

Total Soil Water Potential Head
$$H_{Tw} = H_G + (-H_{Mf}) + (-H_{Ofw}) \quad ......(5a)$$

i.e. Total Head = Gravitational Head + Matric Tension + Osmotic Tension + + ... (5b)

498

(the ++ indicates that other terms might be added in future , e.g. Temperature effects. Similar terms will also be added to the other side of the equation).

The Total Head and the Gravitational Head are measured with respect to the Datum Level.

In equation 5(b) , the Matric Tension and the Osmotic Tension are actually negative quantities. But they are merely indicated by positive distances in Figure 5.

The Total Head is the height from the datum level to the level of the pure water in the standpipe, while the matric head plus the osmotic head in equation (5) refer to the slightly salty free water and are negative when measured downward from the level of point F.

The point F in the free water is at the same horizontal level as the point B in the bound water. Hence it follows that :-

$$(H_{Mf} + H_{Ofw}) \; \text{FreeWater}$$
$$= (H_{Mb} + H_{Obw}) \; \text{Bound Water}$$
$$= H_{G} - (H_{Tw}) \; \text{Pure Water}$$
$$= H_{Pw} \quad \text{Pure Water} \qquad ....(6)$$

At all positions which are at the same level, the numerical sum of the Matric Tension Head and the Osmotic Head is constant, even though the Osmotic Head will vary from point to point, due to variations in the salt content.

The pressure heads in Figure 5 agree with the opinion expressed by the author (Sparks 1961, p 4-50) that the average water pressure across the area of contact between the effective particles is greater (i.e. less negative) than the value of the pressure in the free capillary water. This is due to the higher salt concentration at the area of contact between the effective particles. Observe the levels in the two salt water tubes (Figure 5).

Figure 6 illustrates the ith point of contact between the effective particles. The average pore water pressure in the bound water is equal to u_{Bi} and the pore water pressure in the free pore water is equal to u_{Fi}. The matric pressure of the free water is equal to ($u_{A} - u_{Fi}$).

7. STRESS EQUATIONS INCLUDING OSMOTIC PRESSURES

With reference to Figures 6 and 7 , the average water pressure at the Micelle contact zone B between the two effective particles is u_{Bi} , where

$$u_{Bi} = u_{Fi} + \Delta u_{osmotic} \qquad .... (7)$$

and $\Delta u_{osmotic}$ equals a difference in osmotic pressure due to the difference in ion concentration between the region F in the free water (slightly salty) and the region B in the zone of contact (very salty) between the effective particles. In fact $\Delta u_{osmotic}$ corresponds to the head difference H_{Obf} in Figure 5.

The pressure u_F in the free capillary water is the normal engineering water pressure (matric water pressure = $u_F - u_A$) measured by direct engineering devices such as a pressure gauge or the water level in a manometer tube. When this water pressure is multiplied by the area on which it acts, then one obtains a real engineering force which can be used in stability equations.

Indirect methods using vapour pressure measurements measure the vapour pressure measurements over the pure water surfaces in the two outer tubes of Figure 5, and hence they provide the Total Soil Water Potential Head H_{Tw} above a datum level or the Total suction head H_{Pw} (as in Figure 5). Hence the heads calculated from vapour pressure readings provide the levels of the water in the outer two tubes in Figure 5 , and these must be corrected by an osmotic head H_{Ofw} to provide the Matric Pressure Suction Head H_{Mf} (Figure 5) which equals the engineering pressure in the free capillary water.

The salt ions in the free and bound water can cause at least four effects , as follows :-

a.) They can cause an osmotic pressure difference $\Delta u_{osmotic}$ defined by equation (7),

b.) They can cause a change $\Delta F_{Ri \, osmotic}$ in the Repulsive Forces between effective particles. (e.g. due to electric charge effects)

c.) They can cause a change $\Delta F_{Ai \, osmotic}$ in the Attractive Forces between effective particles. (e.g. due to electric charge effects). This can be combined with the forces in (b).

d.) They can cause a change in the overall radius of the effective particles. (i.e. an increase in salt concentration decreases the thickness of the adsorbed water layers).

Consider a surface A-A which is approximately horizontal (in the x and y directions). The surface

Total Stress $\sigma_z = \sigma'_z + \alpha.u_A + \beta.u_{Fw} - \gamma.T$ ......(9) (1957, 1963)

$\sigma_z = (\sigma'_{Pz} + \sigma'_R + \sigma'_{Os} - \sigma'_A) + \alpha.u_A + \beta.u_{Fw} - \gamma.T$ ....(10) (1961)

| Effective Stress due to solid contacts, if any | Effective Stress due to repulsive forces acting at a distance | Effective Stress due to osmotic pressure difference | Effective Stress due to attractive forces acting at a distance | Pore air pressure in the continuous phase | Pore water pressure in the free pore water even if it is salty. (usually -ve for partly sat. soils) | Surface Tension Force in the air water interface. |

Fig. 8 - Components of the stress equation for Clays.
Note that equation (10) transforms into equation (9).

A-A is larger than that shown in Figure 6. It can be considered to consist of steps with horizontal portions which pass through between the effective particles.

Consider a portion of this surface A-A so that the plan area of this portion (on the x-y plane) is equal to unit area (one square metre). Let there be n points of contact within this portion of the surface A-A. For simplicity we will consider the air phase to be continuous and its pressure is u_A. (e.g. with respect to vacuo pressure).

Within this square metre (of surface A-A) let the total area occupied by water (free and bound) be equal to β. And the total area occupied by the air phase within this square metre is equal to α.

Within each area β_i there is an internal area $A_{r\,i}$ shown in Figure 6 and Figure 7 in which one can assume that there is a larger pore pressure u_{Bi}. Let the total area on which the larger water pressure acts be A_r when considering this square metre of the area A-A. The total area within this square metre which is occupied by solid to solid point contacts is a_r and it will be small, but it will increase as the soil dries out (Figure 7).

The surface tension force T (approx 73 x 10 $^{-6}$ kN / m) acts across the surface A-A at every air-water meniscus. The total length of these air-water menesci which are cut by this portion of the surface A-A is equal to γ metres.
The total compressive force across this portion of the surface A-A is equal to the total pressure σ_z

It follows that, the total pressure σ_z is equal to the sum of all the compressive forces across this one square metre portion of the surface A-A. i.e.

σ_z = (Sum of the normal components of the solid-to-solid contact forces if they exist)
+ (Sum of repulsive forces acting at a

distance between effective particles)
- (Sum of attractive forces acting at a distance between effective particles)
+ (Area A_r x Δu osmotic i.e. the difference $u_B - u_F$ only acts on A_r)
+ (Area β x u_F i.e. as though pressure u_F of Free Water acts across whole β)
+ (Area α x u_A)
- (γ x T) ....(8)

This is the same as equation (10) of this paper (Sparks 1961, or eqn. 4-97) which simplifies to equation (9).

The effective stress is the algebraic sum of compressive point loads and mutual repulsions between the grains or effective particles in this one square metre of the surface A-A. This effective stress controls the shear strength.

The first four terms of the right hand-side of the equation (5) can be combined to provide this effective stress σ'_z.

In other words one can consider the third term (osmotic pressure difference effect) to be part of the effective pressure.

Should it be necessary to estimate the value of the osmotic pressure in a salt solution, a method based on electrolyte conductivity was presented by the U.S. Dept of Agriculture (Fredlund & Rahardjo, 1993, Fig 4.85). The electrical conductivity of the pore water containing dissolved salts is measured. The following formula seems to fit the results :-

Π (kPa) $= 30 \times (\rho_{25})^{1.084}$ ....(11)

where Π = osmotic pressure (kPa)
ρ_{25} = electrical conductivity at 25° C of the pore fluid (millimohs / cm)

The equation (8) has been re-written as equation (10) in Figure 8. Equation (10) transforms into equation (9), because we can regard the effective stress due to osmotic pressure as contributing to the repulsive forces between the effective soil particles. It will also be noticed that in equation (9) one can use the pore water pressure u_{Fw} in the free water. If the total stress is expressed with respect to vacuo pressure then the values of u_A and u_{Fw} must also be expressed with respect to vacuo. However it is more usual to express the pressures on both sides of the equation with respect to atmospheric pressure, or with respect to u_A. The value ($u_{Fw} - u_A$) is often called the matric pressure.

8. METHOD OF USE OF EQUATION (9)

Equation (9) is also the same equation which was derived for sands in which there are no osmotic pressures (Sparks 1961).
When using the above equation (9) for sands or for clays , it is useful to attempt to estimate all the parameters and pressures, in such a manner that the effective stress σ'_z is the only unknown value.
Curves were derived for estimating the factors α, β, and γ at different degrees of saturation.
 These curves for sand were published (Sparks 1963, 1998). In the case of clays it is useful to use the concept of the Degree of Saturation of the Free Water, which is equal to the volume of the free water divided by the volume of the free pores. Curves can be entered with the value of the Degree of Saturation of the Free Water in order to obtain the factors α , β, and γ for clays.

9. ESTIMATING THE MATRIC PRESSURE

It is necessary to estimate the value of the matric water pressure ($u_{Fw} - u_A$) which is required for insertion into the stress equations (9) and (10) in Figure 8. The author has derived two methods for estimating this "soil-suction" curve.

9.1 The "Soil-Suction" Curve.

The "soil-suction" curve is a vital curve which for example provides the relationship between the Degree of Saturation and the Matric Suction, or between the Water Content and the Matric Suction e.g. for a sample which is drying out

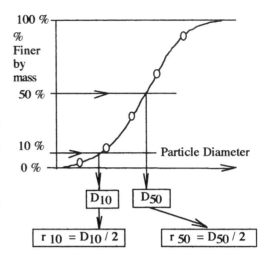

Fig.9 - Use of Grain Size Distribution Curve.

monotonically. (The matric suction is the water pressure which can be measured by engineering pressure guages or engineering stand-pipes). The Matric Suction is therefore a logical pressure to use in derivations such as described in equation (10) of this paper which are based on the equilibrium of forces across a section in the soil.

9.2 "Soil-Suction" curve based on Liquid Limits

The author has published a method for estimating the Matric Suction from the value of the present water content for a clay with a known Liquid Limit (i.e. for a clay which has been drying). The equations of these curves were presented (Sparks 1998) and also at this conference. Although these curves have been tested mainly in the region of saturated conditions (e.g. water content greater than 18 %) it seems that they might apply to lower degrees of saturation. However for partly saturated conditions the author recommends the method in the next section.

9.3 "Soil-Suction" curve from grain sizes

The following method (Sparks 1998, Vol 2) is based on using two grain sizes, namely D50 and D10 to estimate the "Soil-Suction" curve.
This method has provided excellent agreement with measured soil-suction curves for sands, silts and clays (Anderson & Wiklert 1972) .
In the case of clays it is useful to consider the D10 size to have a size limit which corresponds to the size of "an effective clay particle" which consists of several solid particles (e.g. its

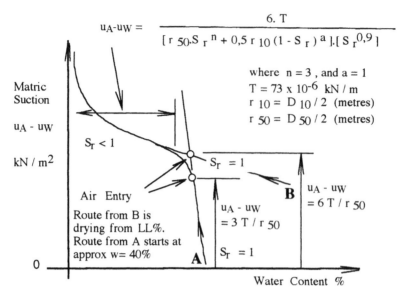

$$u_A - u_W = \frac{6 . T}{[\, r_{50}.S_r{}^n + 0,5\, r_{10}\, (1 - S_r)^a \,].[\, S_r{}^{0,9} \,]}$$

where $n = 3$, and $a = 1$
$T = 73 \times 10^{-6}$ kN / m
$r_{10} = D_{10} / 2$ (metres)
$r_{50} = D_{50} / 2$ (metres)

Matric
Suction

$u_A - u_W$

kN / m^2

$S_r < 1$

$S_r = 1$

Air Entry
Route from B is
drying from LL%.
Route from A starts at
approx w= 40%

A

$S_r = 1$

$u_A - u_W$
$= 3\, T / r_{50}$

B $\begin{array}{l} u_A - u_W \\ = 6\, T / r_{50} \end{array}$

0

Water Content %

Fig. 10 - Soil Suction Curve derived from D_{10} and D_{50} grain sizes.

diameter might be about 0,6 µm for most clays, and this is discussed in section 10).

In this method the grain size Diameters D10 and D50 are used to find the corresponding grain size radii r10 and r50 and these radii are inserted into the equation shown on Figure 10 to provide the soil-suction curve for the partly saturated region where the Sr value is Sr = 2.7 w / 0.5 (approx) and w is the water content (as a ratio)

10. THE EFFECTIVE SIZE OF FINE CLAYS

It is important to note that many solid particles combine in the form of layered "books" to form an effective clay particle. The values of α , β, and γ which should be used in equation (9) at a particular degree of saturation are determined by the size of the pores between the effective clay particles.

Natural clay particles congregate to form effective soil particles which are larger than those found by hydrometer grading of deflocculated samples. It must also be remembered that the size of the effective soil particle diminishes as the clay dries out.

In natural clays, coagulation of particles occurs during deposition, and subsequent leaching of weathered chemicals takes place and increases the free pore diameters, and hence the air entry can take place at lower suctions than in remoulded laboratory samples of the same

clay. In many clays large vertical effective pressures can cause horizontal layering of the plate-like clay particles, as in shales.

10.1 *Mercury Porosimetry to measure clay pores*

The hydrometer grain sizing method can measure dispersed clay particles down to a size of approximately 1 µm. But even at this size different percentages of sizes can be obtained for different deflocculating agents.

Mercury porosimetry can be used to measure the sizes of the fine pores in clays down to a pore size of about 0.01 µm which would correspond to an effective grain diameter of about 0.06 µm.

Sridharan (1971) found that in hydrometer tests Kaolin had more than 50 per cent of the grain sizes finer than 1 µm, but most of the pore diameters were 0.1 µm or larger. In Illite the pores were of the same magnititude but were larger than those in Kaolin. Perhaps the value of D_{10} in Figure 9 can be found in hydrometer tests conducted without the use of a deflocculant.

11 CONCLUSIONS

This paper derives the equation for effective stresses in partly-saturated clays with osmotic pressures and shows that the equation is similar to the equation for partly-saturated sands. The

Find water content or normal S_r

↓

From Grading Curve estimate the Matric Suction ($u_A - u_W$). Hence we know u_W.

↓

Estimate the degree of Saturation of the Free Pore Water = S_{rf} = (Vol free water) / (Volume of free pores)

↓

Enter curves or formulae with the value of S_{rf} in order to estimate α, β and γ for entry into formula (9).

↓

If the external total stress σ_z is known (e.g. $\sigma_z = 0$ if atmospheric pressure) then the effective stress σ'_z is the only unknown and can be found from equation (9)

Fig. 11 - Steps in the calculation of the effective stress of a partly saturated clay.

same methodology applies, but certain curves and formula must be modified.

It was pointed out (Sparks 1961) that the effective stress for shear strength calculation will not be identical to that used for volume changes. This paper does not ignore the Chi-factor method used by Blight and others, but it can provide an insight to these other methods. In particular this paper sheds light on the role which grain-sizes (Figure 10) can play in modifying the curve for Chi-factor versus degree of saturation. The latter curve is known to vary for different soils.

Future research in this topic will involve the following equation

$$\chi = \eta = \beta + \gamma . \, T \, /(\, u_A - u_W \,) \, \dots\dots\dots\dots(\, 12 \,)$$

Use of this equation will permit the use of measured values of χ in order to derive better functions for β and γ in terms of the degree of saturation. It has been shown in Figure 10 that the values of ($u_A - u_W$) depend on the grain size distribution and the degree of saturation. Hence equation (12) will permit predictions of χ for soils with different grain size distributions.

12. ACKNOWLEDGEMENT

The author wishes to acknowledge the encouragement which he has received from Prof. Kenneth Torrance from Carlton University, Ottawa. However, any errors in this paper are the responsibility of the author.

13 REFERENCES

1. Anderson S., Wicklert P. Water-holding capacity of Swedish Soils (in Swedish), *Grundforbattring*, Vol 25, 1972, No 2-3, pp53 -143, Inst. of Soil Science, Dept of Agr, Hydrotechnics, Uppsala, Sweden.
2. Bishop A.W.,Blight G.E.,Some Aspects of Effective Stresses in Saturated and Partly Saturated Soils, *Geotechnique*, Sept. 1963, pp177-197
3. Fredlund. D. and Rahardjo, *Soil Mechanics for Unsaturated Soils*, Wiley, 1993.
4. Lambe T.W., Whitman R.V., *Soil Mechanics*, Wiley, 1969
5. Sparks A.D.W. *Stresses in Partially Saturated Soils*, M.Sc. , Wits, 1961
6. Sparks A.D.W. Theoretical considerations of stress equations for partly saturated soils. *3rd Regional Conf for Africa, ISSMFE,* Salisbury Rhodesia,1963,Balkema
7. Sparks A.D.W. "Stresses in Unsaturated Soils", *Proc. of the Second International Conference on Unsaturated Soils,* Vol 1 & Vol 2, Beijing, Aug 1998.
8. Sridharan A., Altschaeffl A., Diamond S. Pore Size Distribution Studies, *Proc. of the Geotechnical Div.,SM5, ASCE,* May 1971.
9. Sridharan A. and Jayadeva A., Double Layer theory and compressibility of clays, *Geotechnique,* Vol 32, No 2, 1982, p 133-144

Geotechnics for Developing Africa, Wardle, Blight & Fourie (eds) © 1999 Balkema, Rotterdam, ISBN 90 5809 082 5

A unified theory for clays, for students and practice

A. D. W. Sparks
Department of Civil Engineering, University of Cape Town, South Africa

ABSTRACT: A mathematical model is presented which simulates the volume changes in clays due to pressure changes, or due to wetting of the clay. The model incorporates the initial water content and the Liquid Limit. It also permits the prediction of undrained or drained strength.

1. INTRODUCTION

1.1 Expectations of Geotechnical Engineers

Geotechnical engineers require a Model for Heaving clay which shows the following :-

a.) When wetted, clays with high Liquid Limits swell more than clays with low Liquid Limits.
b.) The lower the initial water content prior to wetting, the greater the percentage swell.
c.) Swelling is mainly controlled by a change in the pore water suction values. The percentage swell should increase if the final pore water suction (after wetting) approaches zero.
d.) Swelling is reduced if the soil swells against a high overburden pressure, or against a high foundation pressure.

In addition, Geotechnical Engineers would like to know the following values :-

e.) The swelling pressure if the clay is wetted while volume change is prevented.
f.) The final water content after wetting.
g.) The approximate shear strength of the clay at different water contents.
h.) The approximate drained value of the Young's Modulus at different water contents.
i.) The Compression Index C_c at different water contents.

This model provides all of the above.

2. ISOTROPIC CONSOLIDATION

2.1 Consolidation by Desiccation

A remoulded sample of saturated clay can be located above a supersaturated salt solution within a desiccator. After a few weeks the water content of the moist clay will reach a constant value which is consistent with the pore water suction caused by the low equilibrium vapour pressure of the particular salt solution. (Fig 1. shows two drying curves.)

Similar curves in Figure 2 are also for isotropic consolidation but they were derived from void ratio versus pressure curves for two-dimensional consolidation (see section 2,7 in this paper for this alternative method).

The simplified suction curves in Figure 3 are less accurate than the suction curves in Figure 2, but these simplified curves were used to derive the formulae listed in Table 1 (Sparks 1998).

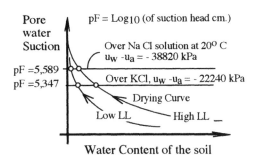

Fig. 1 - Clays drying over salt solutions

2.2 A simple concept

The value of the negative pore pressure (-u) in a soil sample can be estimated with difficulty by measuring the water content of a filter paper (Whatman's No 42) which is in contact with the soil. The concept introduced in this present model, is that one can directly use the water content of the particular clay soil to estimate (-u).

2.3 Consolidation due to Average Stress

It has been assumed in this model that one can enter the compression curve by using the average effective stress (p'_{av}) instead of an isotropic stress (p') , in order to find the corresponding water content (w%). Or one can enter with w% to find p'_{av} or to estimate the value of -u .

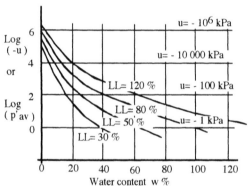

Fig. 2 Pore pressure (-u) or average effective stress (p'_{av}) versus water content w%

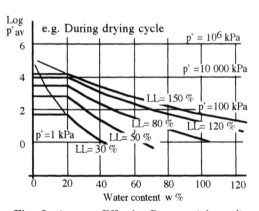

Fig. 3 Average Effective Pressure ($p'_{av}= p'$) due to (-u). For curves use formula (1).

2.4 Definitions and units used in Formulae.

p'_0 = Initial average effective pressure (kN/m^2) at w_0 % .

w_0 % = Initial percentage water content (e.g. prior to wetting).

LL % = Liquid Limit of Clay (expressed as a percentage).

p'_v = Overburden pressure at end of wetting process (kN/m^2).

p'_f = Final value of equivalent average effective pressure after wetting (kN/m^2).

= (2.3 . p'_v - u_f) for isotropic soil.

= (0.4 . p'_v - u_f) for weathered shale which expands normal to the layers.

(Note that the value of u_f is negative, and therefore increases p'_f.)

p'_f = (p'_v) if there is no heave or settlement on wetting (and u_f = 0).

u_f = Pore pressure after wetting (Usually zero, or depends on water table).

D_0 = The original overall thickness of the soil layer (metres).

2.5 Original formulae based on Figure 3.

These formulae (Sparks 1998) are shown in Table 1. These long formulae are better shown in a wide table.

2.6 To find swelling pressures

a.) To estimate Swelling Pressure (volume change prevented) :-

One can assume that that p'_v = p'_0 if volume change is prevented (a wider range can exist).

Enter equation (1) with the original water content w_0 % and with the known Liquid Limit LL% to find the swell pressure p'_0 which equals p'_v .

b.) To estimate Water Content at which there is no volume change (for a certain loading) :-

Enter equation (1) with p'_0 = p'_v and find the water content w_0 %.

TABLE 1. Formulae based on Figure 3.   (see also Sparks 1998).

To describe the curved portion of Log p'o versus wo%   (i.e. curves in Figure 3) :-

$$\log_{10} p'_o = \left[\frac{16 \cdot LL \%}{(1.5 \; LL\% + w_o \%)} - 5.6 \right] \quad \dots\dots\dots\dots\dots\dots\dots (1)$$

To estimate the Compression Index C c :-

$$C_c = \frac{(15 \; LL\% + w_o \%)^2}{600 \cdot LL \%} \quad \dots\dots\dots\dots\dots\dots \dots\dots\dots(2)$$

To estimate the Swelling Index C_s :-

use $C_s = C_c / n$ (e.g. n = 4) $\dots\dots\dots\dots\dots\dots\dots\dots(3)$

To estimate the Heave (e.g. metres) :-

$$\frac{\text{Heave}}{\text{(metres)}} = \frac{(1.5 \; LL\% + w_o \%)^2}{(n \cdot 600) \cdot LL \%} \cdot \frac{(\text{Log}_{10} p'_o - \text{Log}_{10} p'_f)}{1 + 0.027 \; w_o \%} \cdot D \dots\dots (4)$$

(where n has the same value as in equation 3 .)

Water content at which there is no volume change (if average confining stress = p'o) :-

$$w_o \% = \left[\frac{16 \cdot LL \%}{(\text{Log}_{10} p'_o + 5.6)} - 1.5 \; LL\% \right] \quad \dots\dots\dots\dots\dots\dots(5)$$

This will also provide the water content at the level of the Water Table.

To estimate the undrained shear strength of the clay (especially if $w_o \% > 20 \%$) :-

$$\Phi' = 35^o - 17^o \cdot \log_{10} \left[\frac{LL\%}{20} \right] \quad \dots\dots\dots\dots\dots\dots\dots\dots (6)$$

$$s_u = c_u = m \cdot p'_o \cdot \text{Sin} (\Phi') \quad \dots\dots(\text{ e.g. } m = 0.8)\dots\dots\dots\dots (7)$$

where the value of p'o is obtained from equation (1)

To estimate Young's Modulus E' for settlement (i.e. for slow drained loading) :-
(for monotonic compressive settlement, especially if $w_o \% > 20 \%$)

$$E'_{compr} = 117.5 \cdot p'_o \cdot \left[\frac{(1 + 0.027 \; LL\%) \cdot 16 \; LL\%}{(1.5 \; LL\% + w_o \%)^2} \right] \quad \dots\dots\dots\dots (8)$$

Or one can use :- $E'_{compr} = 0.74 / m_v = \dfrac{0.74 \times 2.3 \; (1 + 0.027 \; w\%) \cdot p'_{vert}}{C_c} \dots\dots(9)$

where $p'_{vert} = 2 \cdot p'_o$ and p'o is found from equation (1)

Approx rule:- $E'_{compr} = 35 \cdot s_u$ (if $w\% > 40 \%$); $E'_{compr} = 40 \cdot s_u$ when $w\% = 20\%$ ..(10)

$E'_{\text{Slow cyclic drained}} = E'_{\text{Slow unloading}} = n \cdot E'_{compr}$ ...(n as in eqn 3.) $\dots\dots\dots$ (11)

Young's Modulus (undrained) = $E_u = (250 \text{ to } 500) \cdot s_u$ $\dots\dots$(ex Bjerrum)$\dots\dots$(12)

2.7 Alternative method to find curve DJ (Fig 5.)

Curve ABC in Figure 4 is the usual one-dimensional consolidation curve as found in an oedometer cell. The curve DEF in Figure 4 is for isotropic consolidation in which p'_0 is the average effective pressure.

Measurements by the author suggest that the compression curve AB can be moved by about 0,3 of a log load cycle in order to estimate the position of DE (the formula is shown in Fig.4.). The curve axes are then rotated so that curve DE in Figure 4 yields the curve DEJ in Figure 5.

In Figure 5, the swelling takes place along the unloading curve from E to F due to an unloading of the p'_0 value (which includes -u which is changed by wetting). At F the clay approaches a degree of saturation of 95% to 100%. The value of the "heave" yields the new void ratio at F, and the new water content can be found from $w=e/G$ where w is a ratio. As the clay again dries out in the field, its state moves up from point F (Fig.5) to rejoin the curve EJ.

The curve DEJ is vital to the clay behaviour.

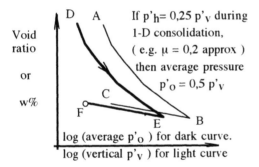

If p'_h= 0,25 p'_v during 1-D consolidation,

(e.g. μ = 0,2 approx)

then average pressure

$p'_0 = 0,5\ p'_v$

log (average p'_0) for dark curve.

log (vertical p'_v) for light curve

Fig. 4 Compression Curves

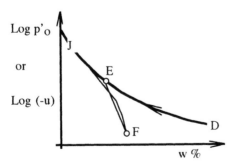

Fig. 5 Soil-Suction Drying Curve DJ

2.8 Discussion of the formulae in Chart 1

All the formulae except (6), (7) and (11) are derived from formula (1) which is based on Figure 3. It is tempting to derive "improved" formulae based on Figure 2. This can be done by multiplying the right-hand side of equation (1) by a multiplier M of the form M=(0.9+0.003 LL%).

This leads to higher values of p'_0 which can lead to unrealistic high shear strengths in equation (7). Equation (6) could also be changed to give lower values of Φ' at high Liquid Limits as found by Simons(1963), but one would also still need to use a lower factor other than 0,8 in equation (7). However the values for Φ' from equation (6) are very close to the values in Terzaghi, Peck and Mesri,(Fig 19.7, 1996) and equation (6) appears to be realistic.

The formulae given in Chart 1 will be upgraded if necessary, but they seem to be sufficient.

2.9 The Swelling Index C_S

The value of the "Heave" is greatly affected by the value used for C_S. In a laboratory oedometer sample the "heave" on wetting can be underestimated due to the downward frictional shear stresses between the expanding soil and the inner surface of the oedometer ring. These shear stresses do not occur on site where general heave takes place over a large plan area. Henkel (1959) used triaxial tests to find compression and swelling due to all-round pressures for London Clay and Weald Clay.

Henkel's curves show that the value of "n" in formula (3) in the present paper can be 2,0 (for Weald Clay) or 2,6 (for London Clay). It seems that future research should be directed at the variability of the value of "n" in equation (3). Another factor is that wetting tests in the field do not yield the full "heave" within one year, and several years may be necessary to obtain the full heave due to wetting which changes (- u_f).

Comparisons between the percentage expansions in the field and published theories can be misleading if the field values have not yet reached final equilibrium values. Shales with horizontal bedding will expand more in a vertical direction than in a horizontal direction.

For normal use, the present author suggests the use of a value of 4 for "n" in formula (3), but when estimating an upper limit for the "heave" of a clay, a value of "n" equal to 3 can be used.

2.10 *The Swelling Pressure for confined samples*

Consider a saturated clay sample in which the average effective pressure p'_o will be equal in value to the negative suction pressure (-u).

If this sample can be trimmed to fit exactly into a confined container, there will be no pressure between the inner surface of the container and the soil sample while the negative suction (-u) exists in the sample.

If free water (at pressure u = 0 = atmospheric pressure) is made available to the sample, the sample would like to swell because its effective stress will have been reduced to zero. The sample can only be kept at its original volume by the imposition of an external stress from the container onto the soil surface, and this new confining stress must still be equal to p'_o. This confining stress is known as the swell pressure and it will be equal to (-u). Hence the curves in Figure 2 or Figure 3 provide the swell pressure after wetting of the confined sample from the original pre-wetting water content w%.

Equation (1) in Table 1 provides the formula for these curves, and hence the value of p'_o in equation (1) is the value of the swell pressure when the confined sample is wetted from a pre-wetting original water content of w%.

2.11 *Warning.: Clays mixed with other soils.*

If the soil contains clay and less active components, the standard tests require the Liquid Limit Test to be performed only on the fine portion of the soil sample. Preliminary calculations of "heave" in equation (4) will be based on this value of the Liquid Limit.

It is important to estimate the proportion of the material which is likely to be active, because the preliminary calculated heave from equation (4) will be too great, and it should be reduced by multiplication by the proportion of the active ingredient. The values of C_c (equation 2), and the water content from equation (5) must also be reduced. The value of E in equations (8), (9) and (10) will be increased. The swell pressure given by equation (1) will probably remain unaltered as the proportion of the active minerals is altered, but it will fall rapidly for very low clay contents. Great care must be taken when looking at published results. A research worker might have tried to correlate Liquid Limits with "heave" without realizing that the Liquid Limits were measured on only a fine portion of the soil.

3 COMPARISONS WITH OTHER RESEARCH

3.1 *The Compression Index C_c*

According to Terzaghi and Peck (p66 , ref 2) the value of the Compression Index C_c can be estimated by the following formula :-

$$C_c = 0,009 (LL\% - 10) \quad \dots\dots\dots\dots\dots(14)$$

Obviously this value was intended to be used over the normal working range for vertical effective pressures. We might assume that the average effective pressure of p'_o might be about 100 kPa.

The formulae given in this paper are based on LL% values obtained by falling cone, whereas equation (14) was derived for LL% values from the Casagrande Cup and groove method.
The author has found that LL% values from the cone are approximately 8% to 10% higher than values from the Casagrande Cup experiment.
Table 2 shows a comparison between values found from equation (2) and equation (14).
For the same clays, the values of C_c from the Skempton / Terzaghi equation (2) are extremely close to the values of C_c from equation (14).

3.2 *Measured Expansions at Vereeniging*

Collins and Jennings provided measured field values of swell, and values measured from double-oedometer expansion tests (Collins,1958).

The formula (4) in Table 1 of this paper, was applied at each level in the soil profile to yield the swell for the soil layer at each level. This calculation table was published (Sparks 98).

The method using formula (4) produced a total heave of 8,3 cm, whereas the measured heave for this site at Leeuhof in Vereeniging was 7,9 cm.

TABLE 2 . Comparison of C_c Values
at p'_o = 100 KN / m^2

LL% (Cone)	LL% (Cup)	C_c from (2) (Using Cone)	C_c from (14) (Using Cup)
50 %	45%	0.37	0.32
75 %	67.5	0.55	0.52
100 %	90	0.74	0.72
125 %	112.5	0.92	0.92
150 %	135	1.11	1.12

3.3 Research by Burland.

Instead of using the water content w% as a major variable, Burland (ref 5.) uses the void index I_V where

$$I_V = (e - e^*_{100}) / (e^*_{100} - e^*_{1000}) \quad............(15)$$

and

e^*_{100} = void ratio at p'_V = 100 kPa

e^*_{1000} = void ratio at p'_V = 1000 kPa

Burland showed that a single unique compression curve is obtained for different soils, when I_V is plotted against log (p'_V). If the void ratio e_L at the liquid limit is known, then

$$e^*_{100} = 0.109 + 0.679 e_L - 0.089 e_L^2 + 0.016 e_L^3$$

and $C_c = (0.256 e_L - 0.04) = (e^*_{100} - e^*_{1000})$

The Burland method requires one to estimate the void ratio for example at the liquid limit or at p'_V = 100 kPa in order to predict the void ratios at other values of p'_V.

One can use the approximation log(p'_O) = log($0.5p'_V$) together with equation (1) in Table 1 (Sparks, Beijing, 1998) to estimate the water content and hence the void ratio corresponding to different values of p'_V. These are presented in Table 3.

Figure 6 shows the dimensionless curves for soils during normally consolidated compression, together with the curve CD derived from Formula (1) in Table 1 (values in Table 3)

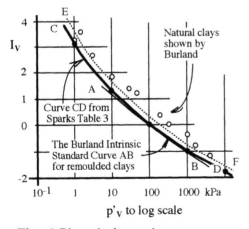

Fig. 6 Dimensionless e-p' curves.

TABLE 3 : Void Ratio and Pressure values derived from equation (1) for plotting in Fig. 6.

p'_V	log (p'_O)	e for LL=80	I_V	e for LL=150	I_V
1 kPa	-0.31	3.293	3.14	6.17	3.14
10	0.69	2.254	1.32	4.22	1.31
100	1.69	1.500	0	2.81	0
1000	2.69	0.929	-1	1.741	-1
10000	3.69	0.480	-1.79	0.900	-1.79

The thick curve CD based on the values from Table 3 lies on the Burland curve AB (Fig. 6).

Using the Burland formula (15) together with the e values from Table 3, we obtain two superimposed dotted curves in the position EF (Fig. 6) i.e. one curve for each Liquid Limit value. It is obvious that curve CD from the formula (1) in Table 1 agrees well with the Burland curve for remoulded clays, but the curve CD from formula (1) also lies midway between those for remoulded clays and natural clays (if the latter are not cemented or highly sensitive).

The writer's curves are based on the Liquid Limit cone apparatus, but if the Burland curves are based on Liquid Limits from the percussion cup, then the Liquid Limits used by the present writer will be slightly higher than Burland's values (e.g. 8%). The values of e_L, C_c, and e^*_{100} must be altered before making comparisons.

3.4 Remoulded Soils versus Natural Deposits

Burland concluded that extra bonding between the grains in natural deposits, causes the "heave" of such natural deposits to be less than half the heave of remoulded soils. However the bonding increases the strength of natural deposits when compared to remoulded samples.

In this model in equations (3) and (7) the present author uses the following :-

a.) For natural soils, use n=4, and m = 0.8

b.) For remoulded soils use n=3, and m= 0.6 ; but to predict swell in oedometer tests use n = 4.5 to account for side friction within the oedometer.

3.5 The shear strength formula

The clay is precompressed at a pressure p'_c. The stress path for a quick undrained triaxial test will follow the curve AB in the M.I.T. stress plot shown in Figure 7. The precompression pressure

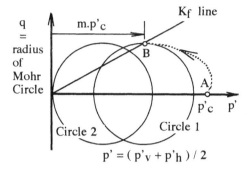

Fig. 7 Undrained Effective Stress Path AB

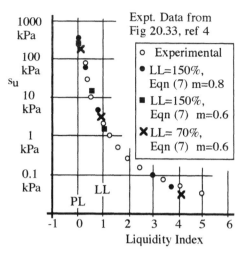

Fig.8 s_u at High Water Contents

p'_c is equal to $(-u)$ plus the smaller average effective stress $(p'_v + p'_h + p'_h) / 3$ which is caused by the overburden effective stress p'_v.

The value of $(-u)$ is equal to the value of p'_o from equation (1) in Table (1). Hence one can estimate the value of p'_c. For practical purposes one can often assume $p'_c = p'_o$ because the initial overburden effects are small compared to the value of $(-u)$.

The undrained shear strength s_u is given by :-

$$s_u = m. \, p'_c . \, Sin \, (\, \Phi' \,) \quad ....(16)$$

The writer prefers to use m= 0.6 for remoulded clays and m= 0.8 for uncemented natural clays. Cemented clays will have higher shear strengths. It seems that m= 0.8 can be used to find a lower shear strength limit for natural clays which are not cemented, nor sensitive (Norwegian clays). The writer has used unconfined compression tests (Circle 2) in order to check the formula (16) for air-dry remoulded clays.

3.6 *Shear strength at high water contents*

Figure 8 shows remoulded undrained shear strengths of clays at water contents greater than their Plastic Limits.

An interesting fact from formula (1) in this paper is that at the Liquid Limit, the value of p'_o will be the same for all clays and it is independent of the value of the Liquid Limit. Formula (1), at the Liquid Limit reduces to :-

$$\log (p'_o \text{ kPa}) = 0.8 \quad \text{i.e.} \quad p'_o = 6.3 \text{ kPa} \quad ....(16)$$

The following formulae are also used :-

LL% = 10 + 1.4 PI% = 3.5 PL% - 25 ....(17)
PL% = 7 + 0.286 LL% = 10 + 0.4 PI% ....(18)
PI% = LL% - PL%
Liquidity Index = (w% - PL%) / (LL% - PL%)

Values of the undrained strength s_u for two different clays have been calculated by using equation (7) in Table 1, and these have been plotted in Figure 8. These show excellent agreement with measured experimental values.

3.7 *Shear strength at lower water contents*

a.) A slurry of London Clay was consolidated under an isotropic pressure of 1000 kPa to a water content of 29 % and an undrained triaxial test provided a shear strength of 220 kPa. (Terzaghi, Peck, Mesri,1996, p144). (LL% = 67.5% in cup test, or LL%=73 for cone test).
Equation (5) in Table 1 shows that for p'_o= 1000 kPa, the value of the water content is 26.3 %. Using m=0.6 for remoulded clays, equation (7) yields s_u = 0.6 p'_o Sin (25.44^o) = 250 kPa. These values are acceptable.
b.) Henkel found a Liquid Limit of 43% for Weald Clay. This will probably be equal to a Liquid Limit of 48% when using the cone method. He found that an isotropic pressure of p'_o = 120 lb / in^2 = 827.4 kPa caused the clay slurry to consolidate to a water content of 18.4 % and the shear strength was 35 lb/in^2 = 241 kPa.

Equation (5) predicts that for a Liquid Limit of 48% and an isotropic pressure of 827.4 kPa the water content after consolidation will be 18.2 %. Using m=0.6 for a remoulded clay, equation (7) provides :-

$s_u = 0.6 \times 827.4 \times \mathrm{Sin}\,(28.5^o) = 236$ kPa

In this case both the predicted water content and the predicted shear strength are identical to the measured values.

The equations in Table 1 were derived before the above measured values were discovered in the literature. It is therefore even more gratifying to find such agreements between measured and predicted values.

However various factors such as fissures, cemented bonds or extra-sensitive structures will cause deviations between the predicted and measured soil strengths.

4. ADVANTAGE OF WATER CONTENTS

4.1 *Water Contents are easy to measure*

Water contents are easier to measure than other parameters such as void ratios and suction pressures. A mathematical model based on water contents is more useful than more complex models.

4.2 *Drained and Undrained Shear Strengths*

The shear strength of a normally consolidated clay is determined by its water content. Geotechnical engineers are aware of the curve which relates water content to undrained shear strength (Fig 9). However, it is not well known that the points for shear in a drained test will also fall on this curve providing the water content at failure is used as the horizontal abscissa. Henkel

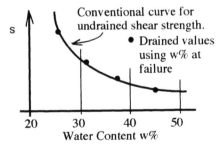

Fig. 9 Drained or Undrained strengths for clay fall on the same curve.

showed that for normally consolidated clays, the strengths for undrained and drained shear tests will fall on the same curve as in Figure 9.

The water content at failure can be used in equation (1) to find p'_o which can be substituted into equation (7) to provide the drained strength.

5. A PHILOSOPHICAL MATTER

Consider for example a saturated clay layer located above the water table. It is possible for this clay layer to dry out to a water content which is drier than that which would be caused solely by the overburden pressure (state A, Fig 10(a)). This clay would be defined by most engineers as "precompressed by desiccation".

However if one takes into account the negative pore pressure , the clay may actually be normally consolidated, i.e. at condition X in Fig. 10(b).

Another insight relates to the swell due to wetting. Engineers who use the double oedometer

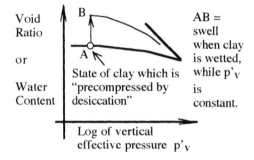

Fig. 10 (a)   The historical outlook

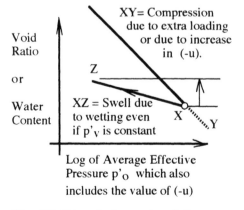

Fig. 10 (b)   A clearer description.

test to estimate heave, are accustomed to working with the diagram in Figure 10(a). The vertical intercept AB is used to calculate the heave. This vertical intercept is identical to the vertical component due to the change XZ in Figure 10(b). It is important to be able to estimate the final average effective stress p'_f at Z after wetting or unloading.

6. ESTIMATING THE VALUE OF p'_f

The value of p'_0 in the heave equation (4) is usually very large, and it dominates over the final value p'_v which is part of p'_f. The pore pressure u_f is also included in p'_f. A minor change in the formula (25) for the calculation of p'_v has little effect on the calculated percentage heave in equation 4.

a.) Stresses during vertical compression:-

If an elastic medium is compressed vertically while horizontal strain is prevented, then the change $\Delta\sigma'_h$ in the horizontal stress is related to the change $\Delta\sigma'_v$ in the vertical stress as follows:-

$$\Delta\sigma'_h = \Delta\sigma'_v \cdot \mu /(1 - \mu) = K_0 \cdot \Delta\sigma'_v \quad ......(19)$$

The value of K_0 is assumed to be $(1-\sin\Phi)$ or one often assumes that $\mu = 0.27$ for drained soils. However based on compression measurements, the author uses the following simplified expression during vertical compression :-

$$\Delta\sigma'_h = 0.25 \cdot \Delta\sigma'_v \quad ....(20)$$

which corresponds to $\mu = 0.2$

Equation (20) leads to :-
$$\Delta p'_0 = (1+0.25+0.25)\Delta\sigma'_v / 3 = 0.5 \cdot \Delta\sigma'_v \quad ...(21)$$
The final value of $p'_f = (-u_f) + 0.5 \cdot p'_v \quad ....(22)$
where p'_v is the effective pressure due to the effective weight of the overburden.

Equation (22) will apply on the virgin-consolidation curve XY (Fig 10b).

b.) Stresses during Swelling due to wetting :-

(i) Non-shaly soils (weathered igneous rocks):

If a soil swells vertically due to wetting, it is convenient to assume that the increase in the horizontal effective stress $\Delta\sigma'_h$ is much larger than the increase $\Delta\sigma'_v$ in the vertical effective stress. Some might wish to assume passive conditions in a swelling soil which is constrained horizontally. (In this respect the equations in Table 1 include estimates of c, Φ' which can be used). However the author has preferred to assume that during swell due to wetting the following apply to non-shaly layers in weathered igneous rocks or lacustrine deposits :-

$$\Delta\sigma'_h = 3 \cdot \Delta\sigma'_v \quad ....(23)$$

Equation (22) leads to
$$\Delta p'_0 = (1+3+3)\Delta\sigma'_v / 3 = 2.3 \cdot \Delta\sigma'_v \quad ....(24)$$
The final value of $p'_f = (-u_f) + 2.3 \cdot p'_v \quad ....(25)$
where p'_v is the effective pressure due to the effective weight of the overburden

(ii) Stresses in weathered shales during swelling

In the case of weathered shales with horizontal bedding, it seems that most of the swelling takes place normal to the bedding planes.
   The horizontal stress $\Delta p'_h$ might be between $0.1\Delta p'_v$ and $\Delta p'_v$ during swelling due to wetting. For weathered shales ,the writer uses the lower value , hence :-

$$p'_f = (-u_f) + 0.4 \cdot p'_v \quad ....(26)$$

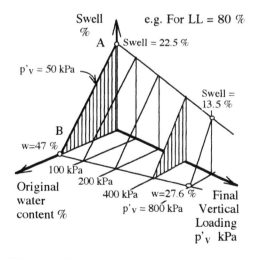

Fig. 11 Using equation (4).

513

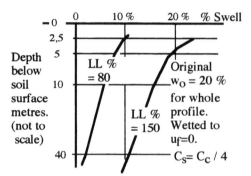

Fig. 12 Results from equation (4).

7. PARTLY SATURATED CLAYS

The original paper (Sparks 98) discussed the fact that the void ratio tends to be constant as the clay dries beyond the shrinkage limit (approx w = 16 %). However it was still assumed that if the soil is very dry (e.g. w% = 2%) it will swell more than if it has a higher water content (e.g. w% = 14%). No special corrections are made in the simplified version in this paper for heave from the dry partly saturated condition (low values of w_O). When measuring shear strengths, the end bearing surfaces of dry triaxial specimens must be planar as prominences cause brittle failures and lower the measured shear strengths.

8. THE FORM OF THE HEAVE EQUATION

Results from the heave equation (4) can be plotted in the forms shown in Figures 11 and 12. Figure 11 shows the percentage heave which results from wetting a clay (to u_f=0). Figure 12 shows the percentage heave at each level in a clay which is originally at a water content of 20 percent and is soaked (to u_f= 0). In practice the values of w_O %, LL %, and u_f vary at each level in a soil. The method copes with these variations.

9. A SIMPLIFIED EQUATION

Curve AB (Fig.11) is slightly concave.
The following empirical equation approximates equation (4) for most clays, if $C_s = C_c / 4$.

% Swell
$$= 0.02(LL\%^{1.36}) - 6.65 \, Log_{10}((p_v - 0.43u_f)/100)$$
$$- 0,37 \, (w_O\% - 20) \, ...(27)$$

The final pressures p_v and u_f are in kPa units.

10. CONCLUSIONS:-

The Liquid Limit values can be regarded as the factors which distinguish between different types of clays. The model provides estimates of the volume changes arising from wetting of the clay from any water content. The user can specify the final distribution of the water suction pressures, and for each distribution the user can obtain an estimate of the heave of the soil surface.

The model provides reasonable values for the void ratio which corresponds to a certain pressure system. The model also provides a method for estimating settlements of saturated clays in which pore water suctions exist.

Estimates can be made of the minimum shear strengths which can be estimated in undrained or drained shear tests for remoulded clays and for uncemented natural clays.

The cohesion of a clay is determined mainly by the negative pore suction pressure (Section 3.7). In other words the negative water suction pressure seems to be more important than electrical charges between the colloidal particles.

REFERENCES :-

1. Sparks A.D.W. Stresses in Unsaturated Soils, *Proc. of the Second International Conference on Unsaturated Soils* , Beijing, Aug 1998.
2. Terzaghi K., Peck R.B. *Soil Mechanics in Engineering Practice* , Wiley, 1948.
3. Henkel D.J., The relationships between the strength, pore-water pressure, and volume-change characteristics of saturated clays. *Geotechnique* , Sept. 1959, p119.
4. Terzaghi K.,Peck R.B.,Mesri G., *Soil Mechanics in Engineering Practice* ,Wiley,96.
5. Burland J.B., On the compressibility and shear strength of natural clays.,*Geotech* ,1990
6. Collins L.E.., Some observations on the movement of Buildings on Expansive Soils.. *Proc.S.A.I.C.E.*, Sept 1957 & June 1958.
7. Simons N.E.,Fundamental shear strength parameters of two South African soils, *3rd Reg.Conf.S.M.& F.E.*,Salisbury, 1963, p 207

6 Soil improvement
Amélioration des sols

Geotechnics for Developing Africa, Wardle, Blight & Fourie (eds) © 1999 Balkema, Rotterdam, ISBN 90 5809 082 5

The design and construction of a dynamic soil replacement foundation in Mauritius

G. P. Byrne, E. A. Jullienne & E. A. Friedlaender
Frankipile, Bramley, South Africa

ABSTRACT: A unique soil improvement foundation solution was successfully used to support heavily loaded on-grade unreinforced concrete surface beds for a large warehouse complex in the Mer Rouge area of Port Louis Harbour, Mauritius. The choice of material parameters, the design, and the construction of the on-grade solutions as well as the settlement performance of a full scale loaded test slab are described. The predicted versus measured slab settlement performance is compared and the key parameters and construction techniques are discussed. These include soil parameters, load distribution, rock column spacing and soil raft geometry.

1 INTRODUCTION

A tender for the construction of a 19,000sq.m Dry Store, 7,000sq.m Cold Store, 4,000sq.m Transformation Building and an Administration Building together with roads and hardstand areas was awarded on an alternative soil improvement foundation solution in October 1996. The alternative soil improvement solution was accepted by the Client on the Dry Store and Transformation Building. The specified suspended reinforced concrete ground floor supported on piles was maintained on the Cold Store and Administration Building. An alternative of driven cast-in-situ piles with an enlarged base was accepted as a more economical alternative to the specified driven precast concrete piles for these two structures. An important issue in the design and construction of the heavily loaded surface beds was the stringent total and differential settlements specified to ensure the satisfactory operation of the high bay racking systems. To verify the calculated settlement performance of the "untried" soil improvement system, the project engineers required a fully loaded, full-scale test slab to be constructed and monitored to satisfy the Client that the substantial cost savings offered by the alternative would not expose them to large maintenance and operating costs. The entirely satisfactory performance of this test ensured the successful on-schedule completion of the project in February 1998.

2 GEOTECHNICAL INVESTIGATION

From available information it was expected that the site would be underlain by dredged coral hydraulic fill, lagoonal deposits and weathered / decomposed Basalt bedrock at relatively shallow depths.

The field investigation comprised the drilling and sampling of rotary cored boreholes and Dynamic Probe Superheavy Penetration Tests (DPSH). Standard Penetration Tests (SPT) were done at 1.5m intervals in the upper horizons and double tube 55mm diameter rotary cored samples were recovered in the hard horizons during the rotary core drilling. The DPSH tests were carried out by continuously driving a 50mm diameter solid cone using an SPT hammer and recording the number of blows required to achieve 300mm of penetration. Shallow test pits were excavated to expose and sample the dredged fill horizon.

3 SITE GEOLOGY

3.1 *General Geological History*

The island of Mauritius is part of the Mascarine group of islands which were formed by Volcanic activity 8 Million years ago. Three distinct phases of Volcanic activity have been identified by Simpson (1951) and McDougall (1988). The geology of Mauritius indicating areas underlain by the three volcanic phases is shown in Figure 1. The site, to the north of Port Louis, shown in Figure 1 is underlain at relatively shallow depth by highly weathered late lavas of the Younger Basaltic Series. These lavas are characterised by 0,2 Million year old Doleritic Flow Basalts extruded from vents to form the plains topography of the island. Recent changes in sea level and a phase of erosion has resulted in the deposition of a thin horizon of soft silty deposits in the Mer Rouge lagoonal environment. The reclamation of the Mer Rouge was facilitated by the dredging of a shipping and berthing channel and the coral sand hydraulic fill presently occupies the surficial horizon.

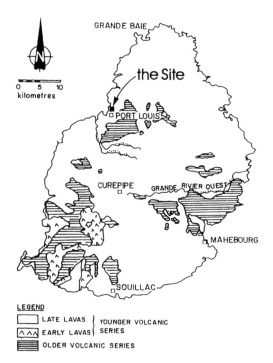

Figure 1.

3.2 Detailed Ground Conditions and Material Properties

Figure 2 shows the geology of the site as interpreted from the borehole profiles. It is of note that large variations to Basalt bedrock occur on the site and that the geological section shown in Figure 2 represents relatively uniform and shallow bedrock depths under the first phase of the site development.

The upper fill horizon comprises loose, medium and coarse coral sands with SPT's in the range 5 to 10. This horizon is free draining and is easily compacted.

The underlying lagoonal deposits are predominantly silty with varying quantities of clays, silts and fine sand included in the soil skeleton. These recent deposits are normally consolidated with average SPT 'N' values of 5 in the range 2 to 8.

The lagoonal deposits have been deposited directly onto Basalt bedrock generally but are underlain by residual Basalt under localised areas of the site. These residual soils comprise clayey silts and are stiff in consistency. SPT 'N' values of between 20 and 40 were recorded in this horizon.

Figure 2 shows the SPT 'N' profile recorded on Phase 1 of the project. Basalt bedrock, considered a rigid boundary, was recorded at elevations of between 4.0m and 9.0m below M.S.L. Existing ground level on the site before the construction phase was measured at + 2.2m M.S.L. on average and the phreatic surface was recorded at an elevation of + 0.75m M.S.L.

4 FOUNDATIONS OPTIONS AND CONSTRUCTION

Three foundation solutions and combinations thereof were investigated at tender stage and comparison of cost and performance made for each solution. The three solutions considered were (A) Piled Foundations (B) Dynamic Replacement combined with Dynamic Compaction of the granular upper layers (C) Dynamic Compaction of the granular upper horizon. From the preliminary analysis it was clear that Options A and B would meet the stringent differential settlement criteria but that there was a risk of the specified total settlements being exceeded if large areas of heavy loading were applied to Option B. Option C presented severe risk of exceeding both the specified total and differential settlements but provided the most economical solution. Option B offered substantial savings on the piled ground floor structure specified in the tender documents and calculations showed that satisfactory performance could be anticipated for the Dynamic Replacement solution. Details of this unique solution are given below in Figure 3.

4.1 Construction details and methodology

The soil improvement solution was carried out in two phases and comprised an initial phase of Dynamic Replacement rock columns driven through the soft lagoonal deposits and founded on stiff residual Basalt or Basalt bedrock. Rock columns 1.2m nominal diameter supporting the surface beds were installed on a 3.75m x 5.0m grid to suit the main structural grid. An initial pre-excavated crater was formed through the existing coral sand and rock fill placed in the crater was driven through the lagoonal deposits up to 7m thick by heavy tamping using high penetration 12 and 14 tonne pounders. Drop heights of between 15m and 20m were used. Rock columns

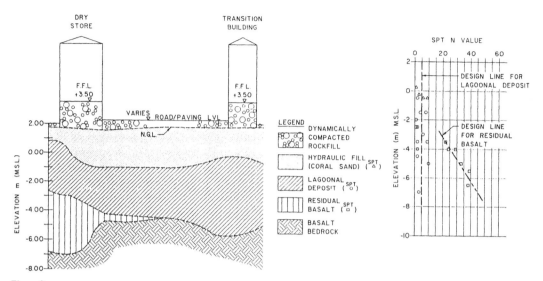

Figure 2.

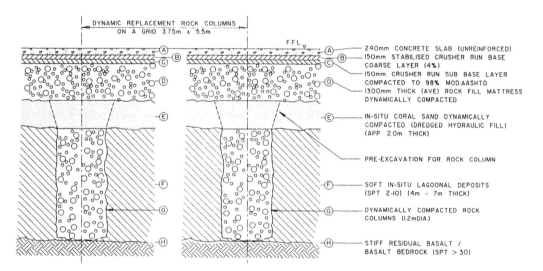

Figure 3.

were also placed beneath the main structural columns located on a 25m x 10m grid and provided support for the shallow spread foundation solution used to support the main structural frame.

On completion of the Dynamic Replacement phase, a rock fill "blanket" comprising basalt cobbles and boulders 0.3m to 1.0m in diameter was placed beneath the building footprint. An average rock fill thickness of 1.3m was required to achieve a finished floor level of + 3.5m M.S.L. The rock fill and the underlying approximately 2m thick hydraulic coral sand fill were Dynamically Compacted using lower energy "ironing" pounders.

The earthworks terrace was completed by placing a 150mm thick crushed stone sub-base layer followed by a 4% cement stabilised crushed stone base coarse layer. A 240mm thick unreinforced concrete powerfloated surface bed was placed in 22.5m x 22.0m panels with toggle joints on far sides and sawn joints at 3.50m centres in one direction and 4.4m centres in the other direction.

Figure 3 shows the typical geometry of the "as built" slab on-grade solution. The typical rock column plan geometry is shown in Figure 4.

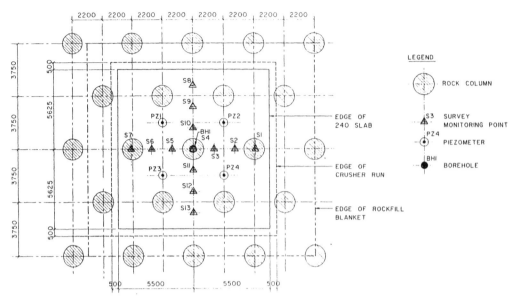

Figure 4.

5 DESIGN AND SETTLEMENT ANALYSIS

The Warehouse floors were required to accommodate a uniformly distributed load (U.D.L.) of 50kPa and rack loads of 60kN per leg which could be increased to 90kN per leg in the longer term. Flexibility was required for the Warehousing operations and these concentrated loads had to be accommodated anywhere on the slab. Maximum total vertical floor movements of 10mm were specified and a differential settlement of 5mm between any two points 3m apart were required.

The floor slab was designed to meet the required criteria using three methods of analysis. Lambe (1973) in the 13th Rankine Lecture stated that there are many techniques for making predictions of stress and deformations but the application of these techniques has limitations. These include the difficulty of determining accurately the field situation, the difficulty of accurately predicting appropriate pre and post compaction soil parameters and the modelling of the rock column / soil raft system. This could have been done with a three dimensional finite element analysis incorporating variable soil parameters and time dependant effects but a simplified analysis was chosen using relationships for rock column diameter, column spacing and soil raft thickness given by Varaksin (1981). The settlement of the foundation system and the design of the concrete surface bed was carried out using the following three simplified methods of analysis :

1 The maximum total vertical settlement of the slab was estimated on the most onerous load case (50kPa U.D.L.) using a method of analysis for piled embankments proposed by Hewlett & Randolph (1996).

2 The maximum differential settlement of the slab was estimated on the most onerous racking leg load configuration using a layered elastic settlement method of analysis.

3 The unreinforced jointed concrete surface bed was designed to carry the rack leg loads and heavy duty forklift truck wheel loads using procedures outlined by the British Concrete Society (1994) and recommendations of Marais and Perrie (1993) for the design and construction of concrete surface beds.

5.1 Design Parameters

The design soil stiffness parameters used for the settlement analysis are given in Table 1 below. The drained stiffness values for the unimproved in-situ horizons was based on the penetration test results (S.P.T.) using empirical correlations derived for similar soils by Webb (1974).

The estimated stiffness of the compacted in-situ coral sand, rock fill and crushed stone base coarse horizons was derived from plate tests carried out using similar methods of compaction and material types together with correlations developed by Stroud (1971) for overconsolidated granular soils. The stiffness and capacity of the compacted rock columns was also based on previous results of large diameter plate tests. An equivalent modulus value was assumed for the rock column / lagoonal deposit horizon in the layered elastic analysis.

520

Table 1.

Horizon	Average SPT 'N'	Drained Modulus Ev (MPa)
Concrete Surface Bed	-	15 000
Stabilised Base	-	150
Sub Base	-	75
Rockfill	-	50
Coral Sand (Compacted)	25	50
Lagoonal Deposits	5	5
Replaced Horizon	-	11.5
Rock Column	-	75

5.2 Uniformly Distributed Load

The maximum total vertical settlement of a large uniformly loaded area was estimated assuming the Dynamically Compacted in-situ coral sands and rock fill forming a soil arch between the rock columns. The load transfer to the supporting rock columns was based on the methodology described by Hewlett & Randolph (1988). It was assumed the rock columns would transfer this applied load to the rigid boundary in end bearing only on Basalt bedrock / very stiff residual Basalt.

The total settlement of the surface was estimated by summating the elastic compression of the upper compacted horizons within the soil arch and the elastic shortening of the rock columns. The settlements induced by the rock fill placed on the coral sands was ignored in the calculation since it was assumed to have taken place before the placing of the concrete surface bed. The calculation model used in the analysis is shown in Figure 5. The predicted total settlement using this method of analysis was 15.5mm.

5.3 Rack Leg Loads

The maximum differential settlements induced on the concrete surface bed by combinations of varying leg loads was estimated using the computer programme VDISP. The programme calculates the surface settlement of a selected area and load pattern using a Boussinesque (1886) stress distribution and moduli outlined in Table 1. A typical settlement contour for a uniformly loaded rack bay is given in Figure 6. Maximum differential settlements of 6mm were calculated for the most adverse load combinations.

5.4 Surface Bed Design

The concrete surface bed was designed in accordance with the recommendations and methodologies given above. For ease of construction and the achievement of high levels of serviceability, an unreinforced jointed 240mm thick concrete pavement was chosen. Subgrade moduli for the base and sub-base horizons used in the analysis were derived from plate load tests.

6 FULL SCALE TRIAL

6.1 Test Slab Construction, Loading and Monitoring

The construction of a 20m x 20m test area identical to the solution accepted for the Dry Store and Transformation Building surface beds, commenced in November 1996.

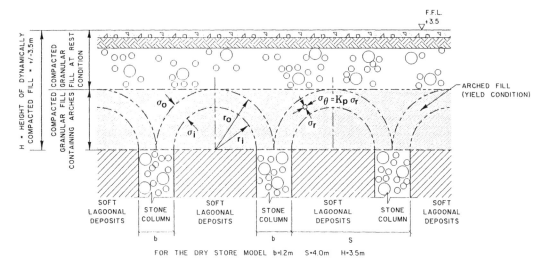

Figure 5.

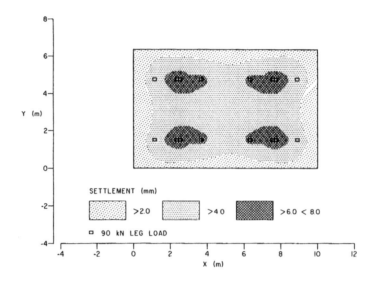

Figure 6.

Prior to the installation of the rock columns a borehole was installed centrally in the test area and four number piezometers were installed in the lagoonal deposit at a depth of 4.5m below finished slab level. Details of the test area are shown on Figure 4. The borehole was wash bored and S.P.T. tests carried out at 1.0m intervals. The depth to Basalt bedrock was 6.5m below existing ground level and 1.5m of rock fill was placed over an area approximately 25m x 25m. The crusher run base coarse layers and 240mm thick concrete surface bed were constructed in accordance with the details shown in Figure 3. Dynamic Probe Superheavy Tests were carried out before and after Dynamic Replacement. Three 1.2m diameter plate load tests were carried out on top of the Dynamically Compacted rock fill blanket as well as the top of the Dynamic Replacement stone columns. These results are given in Figure 9.

The 240mm thick test slab was loaded in two stages. The first stage of loading comprised the uniform placement of 410 tonnes of steel ballast giving a uniform surcharge pressure of 33kPa. This load was maintained for a period of 11 weeks.

The final design surcharge load of 50kPa was achieved by placing a further 230 tonnes of ballast and this load was maintained for a period of 4 weeks.

Precise levelling points were installed on the slab as shown in Figure 4. These points were levelled at 2 hourly intervals throughout the duration of the test.

6.2 Test Results

The average measured drained modulus values of the compacted rock fill and replacement rock col-

Table 2.

Elevation (m) M.S.L	SPT 'N' before Compaction	SPT 'N' after Compaction
+ 1.0	6	-
+ 0.5	4	42, 39, 26
- 0.5	-	15, 15, 17
- 1.0	5	-
- 1.5	-	21, 15, 17
- 2.0	1	-
- 2.5	-	22, 16, 17
- 3.0	3	-
- 3.5	-	8, 6, 6
- 4.0	23	-
- 4.5	-	25, 29, 41
- 4.9	Ref.	-

umns was 65MPa and 85MPa respectively. The DPSH tests carried out showed a marked densification of the coral sand fill horizon while a slight increase in penetration resistance was noted in the lagoonal deposit approximately 1 week after compaction. An increase in shear strength and stiffness was indicated by this result but no corresponding increase in pore pressure during and after compaction was noted in the piezometer readings. The pre and post compaction penetration test results are tabulated in Table 2 below. Total settlements recorded on the slab were an average of 5mm in the range 2mm to 9mm. The test slab settlement profile as well as the time versus settlement response are given in Figures 7 and 8 respectively.

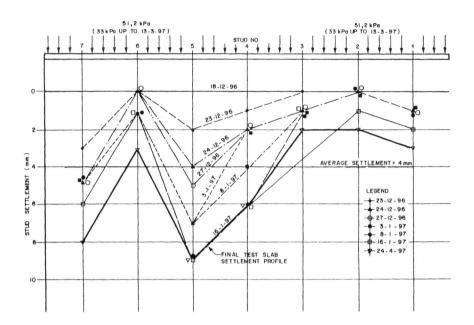

Figure 7.

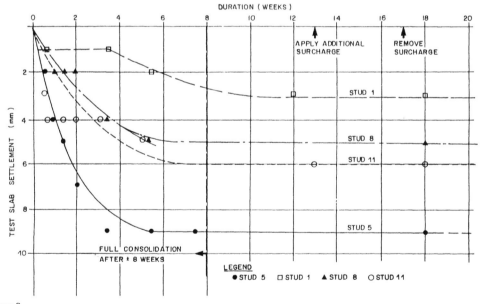

Figure 8.

7 CONSTRUCTION, MONITORING AND CONTROLS

The adequate settlement performance of the foundation system was dependant on the thickness and densification of the fill horizons as well as the load capacity and settlement performance of the Dynamic Replacement rock columns. These important aspects of the operation were carefully monitored as follows.

7.1 Fill Horisons

To ensure an overall soil raft thickness of 3.0m was achieved beneath the building footprints a 25m x

523

25m grid of DPSH tests as well as shallow test pits was carried out prior to commencement of soil improvement. In this way both the thickness and precompaction density of the in-situ coral sand horizon was verifed and the elevation of the surface of the lagoonal deposit recorded over the area of compaction. The post compaction stiffness of the rock fill "blanket" and coral sand was monitored by large diameter plate tests carried out on a 50m x 50m grid at the surface of the fill horizon as well as a depth of 1.5m. A levelling grid before and after the "ironing" phase of compaction provided a detailed record of settlements induced by the compaction process.

An average settlement of 250mm in the range 100mm to 400mm was recorded on the approximately 3m thick compacted horizon. The average measured drained modulus for the fill horizon was 35Mpa in the range 20 – 80MPa.

7.2 Rock Columns

The applied energy together with the volume of rock used at each compaction point was carefully recorded. The penetration depth of the rock columns was verified by rotary borehole drilling through installed rock columns in representative areas of the site. The load capacity and stiffness of the as-built rock columns was measured on approximately 5% of the compaction points using large diameter (1000mm) plate load tests loaded to a maximum stress of 500kPa.

7.3 Full Scale Load Test

A full scale load test was carried out in a completed area of the Transformation Building underlain by a thick localised horizon of lagoonal deposit (approximately 9.0m). After completion and curing of the relevant 11.5m x 11.5m section of floor slab, the area was loaded uniformly with 650 tonnes of steel billets and the movements of the surface bed monitored in an identical method to that outlined for the full scale trial noted in 6 above. The trial settlements recorded were less than 6mm and the differential settlement between any monitoring point 3.0m apart was less than 3mm.

8 DISCUSSION

The entirely satisfactory load capacity and settlement performance of the soil improvement solution adopted for the project vindicated the decision to opt for this "untried" solution in order that the substantial cost savings effected with the Dynamic Replacement system would ensure the economic viability of the project. The success of the project once again highlighted that the combination of the use of readily available and economical construction materials in combination with innovative construction techniques providing the end user with "value engineering".

8.1 Soil Parameters

The choice of appropriate soil parameters is arguably the most difficult part of any normal foundation design procedure. The predicted and measured settlement performance clearly indicated the necessity of measuring soil stiffness in-situ using appropriate test methods. The choice of large in-situ plate load tests to both predict and monitor the stiffness of heterogeneous horizons which include particles up to 1.0m in diameter was both practically feasible and representative of the horizons tested.

8.2 Load Distribution

The choice of calculation model to enable the prediction of load and stress distribution within the foundation structure together with appropriately chosen soil parameters is fundamental to the accurate assessment of foundation performance. The simplified analysis methods chosen provided a safe but reasonable prediction of settlement behaviour. The difficulties of isolating post compaction soil stiffness effects from load distribution effects makes the accurate prediction of the field situation unattainable. The increased stiffness of the fine grained lagoonal deposits as shown by the post-compaction S.P.T.'s is likely to be the single largest factor in contributing to the difference between the predicted and measured settlement behaviour. The increase in stiffness of the lagoonal deposit would result in a reduction of load transmitted to the rock columns with a consequent reduction of settlement within the column. It is of interest to note that settlements estimated by the consultants at tender stage using vibro replacement analysis methods predicted total settlement in excess of 40mm under the 50kPa U.D.L. loading.

8.3 Rock Column Spacing and Soil Raft Geometry

The spacing of the soil replacement compaction points together with the thickness of the soil / rock raft are key variables in providing a cost effective solution. Procedures and the design of soil replacement solutions are given by Varaksin (1981) and can be used as an initial guide for estimating purposes. Testing of the actual field situation is essential in defining the optimal geometry and applied energy for the soils encountered. The thickness of the soil / rock raft will also influence the spacing of the supporting replacement columns.

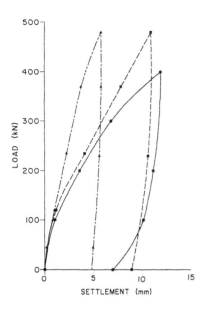

———•——— 1200mm DIA. PLATE ON ROCK COLUMN.
TEST SLAB AREA

——·—▲—·—— 1000mm DIA. PLATE ON ROCK COLUMN D16.
DRY STORE

——·—■—·—— 1000mm DIA. PLATE ON
ROCKCOLUMN B5. TRANSFORMATION BUILDING

Figure 9.

9 CONCLUSIONS

The relative accuracy of the predicted settlements using simple theoretical models to analyse a complex physical soil replacement problem once again emphasises the importance of choosing soil parameters that are not only accurate but representative of the field situation. The satisfactory performance of the foundation also emphasises the importance of using established and proven principles of design and construction in developing new concepts or techniques for a particular application where stringent performance is required.

REFERENCES

British Concrete Society Technical Report N⁰ 34 (1993). *Concrete Industrial Floors on the Ground.* Cement and Concrete Institute.

Hewlett, W.J. Randolph, M.A. (1988). *Analysis of Piled Embankments.* Ground Engineering Vol. 21, N° 3

Lambe, T.W. (June 1973). *Predictions in Soil Engineering.* Geotechnique Vol. 23 N° 2.

Marais, L.R. Perrie, B.D. (1993). *Concrete Industrial Floors on the Ground.* Cement and Concrete Institute R.S.A

McDougall, I. Chamalaun, E.H. (1969). *Isotopic Dating and Geomagnetic Studies - Volcanic Rocks from Mauritius, Indian Ocean.* Geological Society of America Vol. 80 N° 8

Simpson, E.S.W. (1951). *The Geology and Mineral Resources of Mauritius.* Dissertation for a Master of Science Degree, University of Cape Town

Stroud, M.A. (1974). *The Standard Penetration Test – its applications and interpretations.* Proceeding I.C.E. Conference on Penetration Testing. U.K. Birmingham.

Varaksin, S. (1981). *Recent Developments in Soil Improvement Techniques and their Practical Applications.* Soils N° 38/39

Webb, D.L (1974). *Penetration Testing in South Africa.* State of the Art Report. European Symposium on Penetration Testing 1. Stockholm.

Geotechnics for Developing Africa, Wardle, Blight & Fourie (eds) © 1999 Balkema, Rotterdam, ISBN 90 5809 082 5

Liquéfaction et consolidation des sols sableux sous l'action des charges dynamiques

M.S. Diané
Centre de Recherche Scientifique Conakry-Rogbanè, Guinée

S. Diané
Direction Nationale du Génie Rural, Conakry, Guinée

RESUME : Le présent article traite de l'efficacité du compactage des sols peu cohérents saturés sous l'action des charges dynamiques. L'expérimentation a porté sur l'étude des particularités du processus de la destruction de la structure, de la liquéfaction et de la consolidation des sols sableux sous l'action de l'explosion et du choc. La mise en œuvre de techniques expérimentales conséquentes et l'étude des phénomènes se produisant dans lesdits sols ont permis:
- d'élucider la nature du processus de la consolidation des sols peu cohérents saturés sous l'action des charges dynamiques,
- d'établir des relations empiriques entre les déformations volumiques et le degré de compacité desdits sols;
- d'élaborer une méthode de prévision du degré de compactage des sols;
- d'élaborer une méthode de détermination du coefficient de perméabilité desdits sols au cours de leur passage à l'état de liquéfaction complète sous l'action des chocs.

ABSTRACT : This paper deals with the technical feasability of compacting cohesionless and saturated soils by dynamic loading. The experimentation has focussed on the investigation of the particularities of the destruction process of the structure, the liquefaction and the densification of the sandy soils by explosion and shocks. From the result:
- the nature of the consolidation process of cohesionless and saturated soils by dynamic loading is clarified;
- empirical relationships between deformations and the degree of densification of the aformentioned soils are established;
- a method of estimation of the compaction degree of soils is elaborated;
- a method of determination of the soils coefficient permeability by their transfer in full liquefaction state is elaborated.

1. INTRODUCTION

Les dernières décennies du siècle sont caractérisées par une large utilisation des sols peu cohérents, en particulier, des sables fins et limons, pour l'acquisition de nouveaux territoires par remblayage en vue du développement des villes et de la construction des entreprises industrielles. Pour ce faire on a le plus souvent recours aux zones submersibles des cours d'eau et peu profondes des océans, aux terrains marécageux, etc. Dans plusieurs cas le remblayage des sols s'effectue sous l'eau, ce qui conduit à la formation d'une structure très meuble et peu stable des sols peu cohérents saturés. Pour assurer la stabilité de leur structure, éliminer la possibilité de leur passage à l'état de liquéfaction et diminuer les tassements susceptibles de se produire sous les ouvrages qu'ils supportent, il s'avère nécessaire de les consolider ,en particulier, par voie de compactage. Pour de vastes étendues de terrains cons-

titués de sols meubles de remblais, l'une des méthodes de compactage les plus efficaces est l'utilisation de l'explosion de charges d'explosifs relativement petites. C'est pourquoi l'étude du processus du compactage des sols peu cohérents, meubles et saturés sous l'effet de chocs ou d'explosions avec l'élaboration des méthodes d'évaluation de la densité de leur gisement et de prévision du degré de leur compactage sous l'action de charges dynamiques est un sujet d'actualité.

De nombreux auteurs à travers le monde (Forin 1952 Ivanov 1965, 1969, Cid 1966, 1977, Murayama 1958, Prugh 1963) se sont intéressés au problème de la consolidation des sols sans cohésion sous l'action des charges dynamiques.

L'une des principales particularités du développement de la méthode de consolidation des sols par explosion sur le plan international (Damitio 1970-72, Dembiski 1980, Donchev 1980, Ivanov 1965, Kisielova & al, 1980, Kisielova, Osiecimski, 1980, Menard,

TABLEAU I Composition granulométrique du sable de l'île Vassilievski (Saint Petersbourg).

Masse volumique ρ g/cm³		Densité ρ$_s$, g/cm³	Composition granulométrique(%) Diamètre des fractions (mm)									
max	min		5-2	2-1	1-0,5	0,5-0,25	0,25-0,1	0,1-005	0,05-0,01	0,01-0,005	0,005 0,002	< 0,002
1,95	1,49	2,67	0,9	5,3	3,8	34,4	39,0	11,3	3,4	0,3	1,0	0,6

1974, Menard & Broise, 1975, Simon, 1980) est l'élargissement considérable de son domaine d'utilisation et en particulier le large diapason des types de sols saturés: des enrochements aux sables, limons et autres résidus industriels, loess et même aux argiles. C'est pourquoi les auteurs ont souvent adopté le terme de "sols peu cohérents" dans le travail, généralisant ainsi les différents types de sols cités plus haut. Ceci a paru possible en raison du fait que, comme l'ont montré les résultats de nos recherches, la nature du phénomène de liquéfaction dans ces sols sous l'action des charges dynamiques (explosion, choc) est pratiquement identique. Par ailleurs, il convient de remarquer que toutes les études effectuées dans le présent travail ont été réalisées avec des sols à structure remaniée ou des sols de remblai, c'est-à-dire qu'il n'a pas été tenu compte de la stabilité de structure et des liaisons de cimentation caractéristiques des sols naturels. C'est pourquoi dans cette étude on limite le domaine d'application des recommandations pratiques aux seuls sols des remblais et ballasts.

2. PROGRAMME EXPERIMENTAL

Le but principal de l'expérimentation est l'étude des particularités du processus de la destruction de la structure, de la liquéfaction et de la consolidation des sols peu cohérents saturés sous l'action de l'explosion et du choc. Elle a porté sur du sable fin des nouveaux quartiers de Saint Petersbourg (île Vassilievski).

La composition granulométrique dudit sable est donnée dans le tableau I

2.1.Expérimentation sous l'action des chocs au laboratoire :

L'expérimentation sous l'action des chocs a été effectuée sur une table vibrante sur laquelle est installé un récipient métallique de hauteur 14,6 cm et de diamètre intérieur 15,5 cm (Fig. 1). Le choc est provoqué sur le côté de la table par une masse de type pendulaire. Ses oscillations horizontales sont assurés par des dispositifs spéciaux. Celles-ci sont enregistrées sur un enregistreur NZZ-1M par l'intermédiaire d'un capteur magnétique à induction de type 217. Le principe de travail de celui-ci consiste en la transformation du déplacement mécanique de l'extrémité du transformateur en tension analogique proportionnelle à ce dernier. Un exemple d'oscillogramme de la table est donné sur la Figure 2. On a fait varier graduellement l'angle d'incli-

naison du pendule par rapport à la verticale de 5, 10, 15, 20, 25 et 30°. La valeur des déplacements maxima de la plate-forme de la table a varié de 0,3 à 3,4 mm. La durée du premier déplacement était de 0,02 - 0,025 secondes. On a mesuré la pression interstitielle sur la hauteur et au voisinage du fond du récipient à l'aide d'un capteur à faible débit. Pour cela on a installé des piézomètres ayant à leur extrémité des filtres pour empêcher le passage des particules du sol. Chaque choc doit être effectué seulement après la stabilisation de la consolidation due au choc précédent, ce qui se reconnaît facilement par la baisse totale de la pression interstitielle jusqu'à sa valeur initiale.

L'intensité du choc a été évaluée par la valeur de l'angle d'inclinaison du pendule par rapport à la verticale et la valeur conventionnelle de l'accélération des oscillations lors du premier déplacement :

$$\eta = A\omega^2$$

où A- amplitude du déplacement de la table; au moment du choc, sa valeur était proportionnelle à celle de l'angle d'inclinaison du pendule (Fig. 3); η- fréquence des oscillations de la table pour la première période d'oscillation.

Le sable ,à l'état sec, a été versé lentement et par couches successives dans le récipient de façon à obtenir la structure la plus meuble possible. Lors du chargement du récipient métallique une attention particulière avait été accordée à l'uniformité de la compacité initiale. Dans tous les essais on a procédé à la détermination du tassement de la surface du sol sous l'action des charges. La détermination de la variation de la pression interstitielle provoquée par le choc a été effectuée à l'aide de capteurs à membrane. Cette pression a été mesurée en trois points disposés selon la hauteur du récipient à partir du voisinage du fond et sous l'action de chocs successifs d'égale intensité

2.2. Expérimentation sous l'action de l'explosion au laboratoire :

L'expérimentation sous l'action de l'explosion a été effectuée dans un bac métallique (Fig. 4) de hauteur 118 cm et de diamètre 151 cm. Les actions dynamiques ont été provoquées à l'aide de l'explosion d'une substance explosive de 1,5 g placée à la profondeur de 30 - 40 cm permettant la réalisation d'une explosion camouflée. L'essentiel de la méthode fut l'utilisation de manomètres à faible inertie pour les mesures de la pression interstitielle et une méthode électrométrique pour la mesure de la porosité des sables. Le sable a été

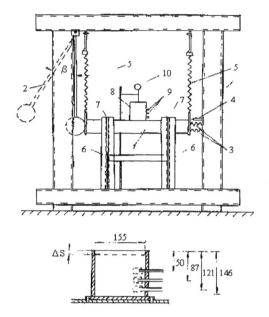

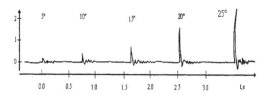

Fig. 2: Oscillogrammes de déplacements horizontaux de la table sous l'action de chocs d'intensité diverse.

Fig. 1- table à chocs : 1 plate-forme 2 pendule; 3 amortisseur: 4 capteur: 5 ressorts: 6 supports de la sable; 7 poteaux; 8 récipient; 9 piézomètres; 10 indicateur de mesure du tassement du sol.

versé dans le bac de la même manière que dans le récipient de l'expérimentation sous l'action des chocs. Dans le but de comparer les essais avec différentes compacités initiales du sol on a déterminé la valeur du tassement de la surface, c'est-à-dire le rapport $\Delta\delta_n/\Delta\delta_{n+1}$ (voir le tableau II).

Au bout de 20 à 25 secondes après l'explosion, des jets d'eau surgissent à la surface du sol . La partie essentielle du tassement se produit pendant les 5-7 premières minutes et au bout de 30 minutes environ le tassement est pratiquement terminé. Il s'est amorti progressivement avec l'augmentation du nombre

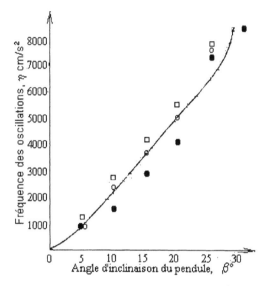

Fig.3: variation de la fréquence des oscillations de la table en fonction de Δ.

d'explosions (Fig. 5) pour atteindre sa valeur optimale après les 3 à 4 premières explosions (tableau II).

TABLEAU II Valeurs du tassement du sable de l'île Vassilievski après l'explosion.

N° de explosions	Tassement moyen				$\frac{\Delta\delta_n}{\Delta\delta_{n+1}}$	ρ_s g/cm^3	Degré de densité relative I_D	Variation du degré de densité relative ΔI_D
	Absolu		Relatif					
	S, mm	δ, %	$\Delta\delta$,mm	$\Delta\delta$, %				
0	-	-	-	-	-	1,559	0,187	-
1	22,90	2,22	22,90	2,22	-	1,595	0,279	0,092
2	35,61	3,47	12,71	1,25	1,78	1,614	0,326	0,047
3	46,15	4,52	10,54	1,05	1,19	1,632	0,369	0,043
4	55,62	5,47	9,47	0,95	1,10	1,648	0,407	0,038
5	62,42	6,16	6,80	0,69	1,37	1,659	0,433	0,026
6	69,12	6,84	6,70	0,68	1,01	1,677	0,456	0,023
7	73,91	7,33	4,79	0,49	1,39	1,679	0,477	0,021
8	77,05	7,65	3,14	0,32	1,53	1,684	0,489	0,012
9	79,80	7,95	2,75	0,30	1,07	1,689	0,499	0,010
10	82,75	8,23	2,95	0,28	1,07	1,694	0,511	0,012

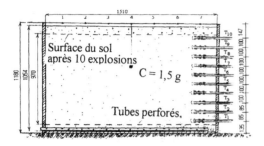

Fig. 4: Bac pour l'essai des sables à explosion: T_1 à T_{10} -points de mesure de la pression interstitielle.

2.3 Expérimentation in situ :

Pour une comparaison des résultats ainsi obtenus avec ceux des recherches in situ, des essais ont été effectués sur le terrain par Kroutov (1978). Le sol de remblai était caractérisé par une teneur élevée en particules limoneuses (jusqu'à 98%). Le niveau des eaux souterraines était situé à la profondeur de 0,1-0,2 m. Les couches supérieures étaient pratiquement saturées d'eau en raison des remontées capillaires. Les études ont été menées dans une zone de grande profondeur remblayée hydrauliquement sous l'eau. Pour ce faire, des explosions profondes de type camouflé ont été

provoquées avec des charges de masse 0,5-5,0 kg placées dans le sol à la profondeur de 2-5 m. Ainsi la couche de sol à étudier atteignait 7-8 m et englobait toute la zone du remblai hydraulique qui atteignait 4-7 m de rayon. Pour la mise en place de la charge explosive on a creusé des forages avec des carottiers de diamètre 108 mm qui ont été retirés avant le chargement et le remplissage du trou par du sol. A titre d'exemple on donne ci-dessous la description de l'explosion de la charge de 4,8 kg, effectuée sur une zone où l'épaisseur du sable atteignait 4,8 m. Au moment de l'explosion il est apparu une petite éruption à la tête du forage.

On n'a pas observé d'élévation de la surface du sol. 2 à 3 minutes après l'explosion commença l'expulsion de l'eau interstitielle sous l'action de la surpression. La zone d'expulsion de l'eau atteignit un rayon de 20 - 25 m. A certains endroits, à la distance de 3-5 m de la tête du forage, on a observé la formation d'un "gonflement" par suite de la surpression de l'eau et la création de chemins de filtration concentrés. Ensuite après 5-10 minutes commença l'augmentation croissante des griffons et

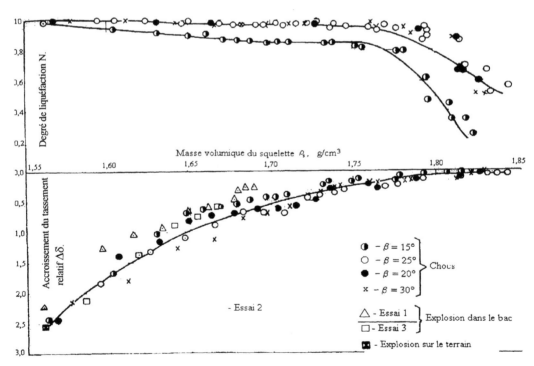

Fig. 5: variation du degré de liquéfaction N et de l'accroissement du tassement relatif $\Delta\delta$ en fonction de la masse volumique du squelette du sol.

TABLEAU III Résultats des explosions camouflées des nouveaux quartiers de Saint Petersbourg.

Explosions	Nombre de charges	Profondeur de compactage m	Tassement moyen du sol 1 jour après l'explosion		Tassement moyen du sol 4 jours après l'explosion		Tassement moyen du sol 24 jours après l'explosion	
			Absolu cm	Relatif %	Absolu cm	Relatif %	Absolu cm	Relatif %
1ère série	20	7	13	1,8	18	2,6	-	-
2ème série	12		8	1,1	11	1,6	-	-
3ème série	16		6	0,8	-	-	14	2,0
Total			27	3,8			40	5,7

TABLEAU IV Effet de l'explosion camouflée sur les sols remblayés des nouveaux quartiers de Saint Petersbourg.

N° explosion	Epaisseur de la couche de sol, m	Masse De l'explosif kg	Profondeur de la charge, m	Rayon d'action de la charge, m	Tassement moyen, cm	Profondeur de consolidation calculée, m	Tassement relatif de la couche, %	Rayon maximum m
1	4,8	5,0	4,8	5	9,8	7,2	1,4	10
2	4,2	5,0	4,5	5	11	6,8	1,6	12
3	3,3	4,0	4,2	5	10	6,3	1,6	13
4	3,4	3,0	3,8	4	9,8	5,7	1,7	10
5	4,3	1,5	3,0	3	9,8	4,5	2,2	10
6	3,2	1,5	3,0	3	6,7	4,5	1,5	7
7	3,8	0,5	2,0	2	8,2	3,0	2,7	9
8	3,4	0,5	2,0	2	4,8	3,0	1,6	7

flèches d'entraînement du sable. On a observé une sortie intense de l'eau pendant 1-1,5 heure. Par suite de l'explosion il est apparu une zone de tassement du sol atteignant 11-12 m de rayon. Le tassement moyen dans un rayon de 5 m était de 9,8 cm et le tassement relatif, en considérant que la profondeur de destruction possible de la structure est d'environ 7 m, atteignit 1,4%.

Au même endroit il a été procédé à une seconde explosion de charge de poids 5 kg à la profondeur de 4,8 m. On a observé le même phénomène que pour la première explosion. Au cours de la deuxième explosion tous les geysers n'ont pas commencé à travailler et 50% environ des griffons n'ont pas réapparu. Le rayon de répartition des geysers a diminué de 10 - 6 m. Le tassement moyen du sol après deux explosions a atteint 16 cm et la valeur relative du tassement 2,2%. Il convient de remarquer que la partie essentielle du tassement (80%) s'est produite au bout de 24 heures après l'explosion. Les paramètres des explosions profondes sont données dans les tableaux III et IV.

3. ETUDE ANALYTIQUE

3.1. Destruction de la structure et consolidation des sols peu cohérents sous l'action des chocs :

Comme on le sait, la possibilité de la destruction de la structure du sol non cohérent est déterminée par l'intensité de la charge dynamique, la densité relative et l'état de contrainte initial du sol. L'influence de ces facteurs se reflète largement sur les résultats des essais

de consolidation par chocs. Ainsi il est établi que la consolidation ou l'ameublissement du sol non cohérent lors de la destruction de sa structure est déterminé par la valeur du déplacement relatif des grains. Indépendamment de la compacité initiale du sol, on peut considérer que pour de faibles déplacements réciproques la structure du sol se raffermit, et pour de grands déplacements elle s'ameublit. Ceci revient à dire qu'il existe une valeur critique des déplacements réciproques des particules qui provoque l'ameublissement du sol.

On considère que cette valeur dépend de la dimension et de la forme des particules du sol ,ainsi que de leur compacité. Plus les particules sont fines et le sol compact, plus la valeur du déplacement relatif critique est petite. Pour une compacité maximale les déplacements critiques doivent tendre vers zéro (Ivanov ,1968).

Les essais ont permis d'établir la relation entre le tassement relatif δ ou son accroissement $\Delta\delta$, la densité ρ_s ou le degré de densité relative I_D, le degré de liquéfaction N, le nombre de chocs n et l'intensité de l'action des chocs. On examine ci-dessous les principales lois observées.

La particularité caractéristique de tous les essais est pratiquement l'élévation spontanée de la pression de l'eau interstitielle au moment du choc jusqu'à sa valeur maximale, ce qui témoigne du passage simultané de toute la masse du sol dans l'état de liquéfaction. Par suite du traitement des oscillogrammes obtenus au niveau des trois points considérés le long de la couche de sable (Fig. 6), on a construit l'épure de répartition des pressions maximales dans

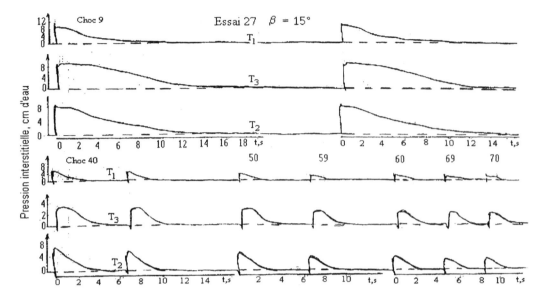

Fig.6: oscillogramme de la pression interstitielle dans le sable sous l'action d'une serie de chocs.

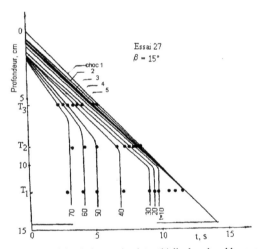

Fig. 7 répartition de la pression interstitielle dans le sable sous l'action d'une série de chocs.

l'eau interstitielle après chaque choc (Fig.7). Le caractère triangulaire de l'épure des pressions correspond à la valeur de la pression interstitielle maximale, c'est-à-dire correspondant au poids de la couche de sol en suspension, ce qui témoigne du passage du sol dans l'état de liquéfaction totale sur toute l'épaisseur de la couche. Celui-ci est suivi du dépôt de ses particules, c'est sa consolidation. Les courbes d'enregistrement de la variation de cette pression indiquent la présence, à chaque instant, d'une frontière entre la partie consolidée du sable et la partie se trouvant encore dans l'état de liquéfaction complète.

Par ailleurs, dans tous les essais il a été observé le passage du sol dans l'état de liquéfaction complète sous l'action du 1er au 20-30ème choc (Fig. 8). Cela a été confirmé par la valeur du degré de liquéfaction qui est égale à l'unité, ainsi que par le graphique de la Figure 6. Par ailleurs les valeurs de l'accroissement des tassements relatifs obtenues lors de l'action des chocs s'inscrivent sur la courbe Degré de liquéfaction–Accroissement du tassement relatif (Fig. 9).

L'analyse des résultats des données expérimentales relatives à l'action des chocs a montré que pour chaque degré de densité relative du sol, à l'exception de la valeur maximale ($I_D= 1$), correspond une intensité critique qui provoque la destruction de la structure et la consolidation du sol. Les déformations volumiques augmentent avec l'augmentation de l'intensité du choc, mais jusqu'à une valeur limite au-delà de laquelle une augmentation ultérieure de celle-ci a peu d'effet sur le degré de consolidation. La déformation limite dépend essentiellement de la valeur de la densité relative initiale et diminue avec la diminution de celle-ci. Ainsi, sous l'action de chocs isolés chaque degré de densité relative initiale a ses déformations volumiques maximales.

La relation $I_D = f(\Delta I_D)$ (Fig. 10) peut bien être exprimée par l'équation:

$$I_D = \frac{a}{a + bI_D} \tag{1}$$

où a, b et c - Coefficients empiriques dépendant du type du sol: pour le sable considéré on a:
a = 0,10, b = 1,33 et c = 0,1 ;

532

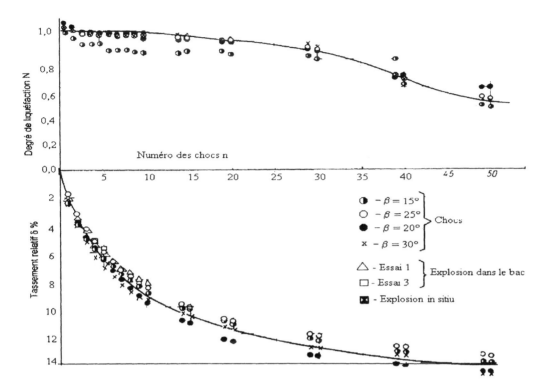

Fig. 8: variation du degré de liquefaction N et du tassement relatif □ du sable sous l'action d'une serie de chocs.

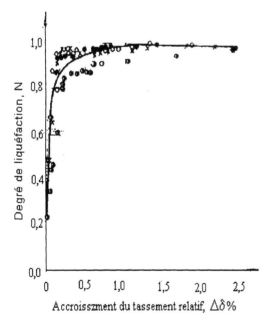

Fig. 9: variation de l'accroissement du tassement relatif en fonction du degré de liquéfaction du sable C.

sachant que:

$$I_D = \frac{e_{max} - e}{e_{max} - e_{min}} \; ; \Delta I_D = \frac{\Delta e}{e_{max} - e_{min}} \text{ et } \Delta\delta = \frac{\Delta e}{1 + e}$$

la relation entre la déformation limite $\Delta\delta_{lim}$ et la variation limite du degré de compacité ΔI_{dlim} peut être présentée sous la forme

$$\Delta\delta_{lim} = \frac{e_{max} - e_{min}}{1 + e} \Delta I_{D\,lim} \qquad (2)$$

où e_{max}, e_{min} et e sont les indices des vides du sol dans les états de densité maximale et minimale ainsi que dans l'état considéré respectivement et la relation $\Delta\delta = f(\rho_s)$ (Fig. 5) peut s'écrire:

$$\Delta\delta = \frac{\rho_{s\,max} - \rho_s}{b' \rho_s - a'} \qquad (3)$$

où ρ_{smax} et ρ_s - densités du sol sec dans les états compact et considéré respectivement et a', b' et c' – coefficients; pour le sable de l'étude on a: a' = 0,50 g/cm^3, b'= 0,10 et c '= 4,20.

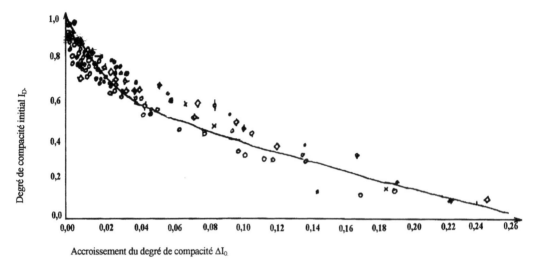

Fig. 10 variation limite de l'accroissement du degré de compacité ΔI_0 en fonction du degré de compacité initial I_0.

3.2. Processus de la consolidation des sols peu cohérents saturés sous l'action de l'explosion :

La particularité des effets de l'explosion ,comme celle des effets du choc, est son instantanéité, une grande amplitude des oscillations de l'onde de choc au moment de l'application de la charge et des amplitudes relativement faibles pour les oscillations suivantes. Le caractère des ondes provoquées par l'explosion est identique à celui des ondes du choc. Les ondes de l'explosion se transmettent aux particules du sol se trouvant dans l'eau interstitielle et entraînent la destruction de la structure du sol sableux.

Les résultats des mesures de la pression dans l'eau interstitielle ont montré que le sol meuble est passé dans l'état de liquéfaction complète sur toute l'épaisseur de la couche. Ceci est confirmé par le caractère de l'épure des pressions et l'équivalence des valeurs de la pleine pression interstitielle et du poids mort de la couche du sol en suspension. Après le passage du sol dans l'état de liquéfaction complète, suit le dépôt de ses particules. L'épure de répartition de la pression dans l'eau interstitielle pour différents intervalles de temps après l'application de la charge d'explosion est identique à celle obtenue dans le cas des chocs .

Les essais in situ ont montré que les sables limoneux et les limons saturés sont susceptibles de passer en état de liquéfaction complète sous l'action des ondes d'explosion. Le caractère du phénomène de liquéfaction et le dépôt ultérieur des particules dans les essais in situ et au laboratoire avec l'explosion est le même que dans les essais avec l'impact dans les sols considérés. La différence réside essentiellement dans le temps de l'accomplissement du processus de liquéfaction et les conditions de reflux de l'eau interstitielle. Il est essentiel de remarquer que les va-

leurs des tassements relatifs obtenues sous l'action de l'explosion (in situ et au laboratoire) et des chocs sont identiques.

4. PREVISION DE LA DENSITE RELATIVE DES SOLS PEU COHERENTS SATURES

Les études de laboratoire sur le compactage des sols sous l'action du choc et de l'explosion, ainsi que les données des essais d'explosion in situ permettent de proposer une méthode de prévision du degré de compactage desdits sols. Ladite méthode est basée sur la représentation physique du processus de la consolidation des sols peu cohérents saturés par suite de l'action de chocs isolés décrits plus haut (§ 2 et 3). Elle se base sur le fait que le degré de consolidation du sol peu cohérent saturé après la destruction complète de sa structure et sa liquéfaction dépend uniquement de sa densité relative initiale et non de l'intensité du choc. Pour faire un pronostic sur le degré de compactage du sol, on procède de la manière suivante:
- pour chaque choc on détermine le degré de liquéfaction par la relation: $N = p/p_{max}$, où
- p est la pression interstitielle maximale mesurée;
- p_{max} est la pression maximale possible lors de la liquéfaction complète du sol.
- on détermine le tassement relatif de la couche de sol par la relation: $\delta_i = S_i/h_{i-1}$ où :
S_i est le tassement de la couche de sol dû au ième choc;
h est l'épaisseur de la couche de sol.
- ayant la valeur du tassement et la masse volumique initiale du squelette, on détermine la masse volumique du squelette du sol après le ième choc par la relation:

$$\rho_{s,i} = \frac{\rho_{s,i-1}}{1-\delta_i} = \frac{\rho_{s,i-1}h_{i-1}}{h_{i-1}-S_i} \qquad (4)$$

- Dès lors on peut construire les graphiques N= f(ρ_s) et $\Delta\delta$ = f(ρ_s) (Fig.5). Par ces graphiques il est possi-ble de prédire le degré de compactage du sol après une série d'explosions ou évaluer celui-ci dans les conditions naturelles sur la base des résultats de sondages par explosion sur le terrain. Pour cela l'intensité du choc doit être au moins égale à la valeur limite entraînant la liquéfaction complète de toute la couche de sol. La condition sine qua non de l'utilisation de la méthode proposée est l'obtention de la liquéfaction complète de la couche de sol, c'est-à-dire N = 0,9-1,0. Ainsi, on peut résoudre beaucoup de problèmes pratiques à l'aide des travaux de laboratoires réalisés dans le cadre de l'étude, en particulier les suivants:

4.1. Prévision du degré de densité relative des sols peu cohérents :

Par l'intermédiaire du graphique du type de la Figure 5 pour une densité donnée ρ_s,0, on détermine le tassement relatif après la première explosion (ou premier choc) δ_s,1 et, par la formule, (4) la masse volumique du squelette correspondante ρ_s,1; ensuite par la valeur de celle-ci et le même graphique on détermine le tassement relatif δ_s,2 après la deuxième série d'explosions, on détermine ρ_s,2 et ainsi de suite pour toutes les séries d'explosions nécessaires. Ayant la valeur projetée de la densité du squelette du sol $\rho_{s,k}$, de la condition $\rho_{s,n} \geq \rho_{s,k}$, on détermine le nombre nécessaire de séries d'explosions n. Par la valeur du tassement relatif enregistrée au cours de l'essai de compactage sous l'action du choc on détermine le tassement du sol après chaque série d'explosions par la relation: $S_i = \sigma_i h_i$.

Le nombre nécessaire de séries d'explosions pour la valeur donnée du tassement S_k se détermine à partir de la condition:
$$(\delta_1 + \delta_2 + ... + \delta_n) h_{comp} = S_k \qquad (5)$$

4.2. Détermination de la masse volumique du squelette des sols peu cohérents :

La méthode proposée ci-dessus peut être utilisée pour la résolution du problème inverse du précédent. A cet effet, on procède de la manière suivante: connaissant, par les résultats d'essais d'explosion réalisés in situ, le tassement du sol S, on détermine le tassement relatif moyen de la couche de sol par la relation: δ = S_{comp}/h_{comp}.

Ensuite on procède aux essais de compactage ,par chocs, d'échantillons du sol et par le graphique obtenu (Fig. 5) et la valeur δ on détermine la masse volumique du squelette du sol dans les conditions naturelles. Par ailleurs, ayant déterminé pour l'échantillon de sol les valeurs limites de la masse volumique du squelette dans les états compact et meuble on obtient facilement le degré de compacité I_D. Ainsi, sur la base des essais du sol par explosions in situ et chocs au laboratoire, on peut approximativement déterminer la compacité du sol peu cohérent saturé, ce qui constitue l'un des problèmes les plus complexes des levés géologiques. Cette méthodologie avait été utilisée sur le territoire remblayé des nouveaux quartiers de la ville de Saint Petersbourg (voir tableau IV) et dans le bassin de déchets industriels de Madinéoulé.

Pour déterminer la densité, on a construit le graphique du sondage par explosion de toute la couche de sol (Fig. 11a) en fonction du tassement relatif. Ensuite, sur la base de la méthodologie proposée et utilisant la relation Tassement relatif-Densité du sol (Fig. 5) ,on a construit le même graphique (fig. 11b) d'après les valeurs obtenues du poids volumique du squelette, ce qui permet de distinguer nettement les couches les plus denses du sol.

Une comparaison de la valeur de la densité des couches supérieures du sol donnée par le graphique de la Figure 11 (1,58-1,59) et celle obtenue par la méthode électrométrique (ρ_s=1,52-1,57 g/cm^3) montre que celles-ci sont assez voisines, ce qui confirme la validité de la méthodologie proposée.

Pour l'évaluation de la stabilité dynamique de la structure et la possibilité du passage dans l'état de liquéfaction des sols du remblai hydraulique du bassin de déchets industriels de Madinéoulé , il a été effectué en 1979 un sondage par explosion. A cet effet on avait utilisé des charges explosives profondes camouflées de masse 1-5 kg disposées à la profondeur de 2-4,5 m (tableau V). L'étendue des zones de consolidation, évaluée selon les données de mesure du tassement des repères , a atteint 6-11m.

A l'aide des graphiques du sondage par explosion et de l'essai de compactage par choc (Fig.5), on a construit celui de la variation du poids volumique du squelette en fonction de la profondeur (Fig. 11). Les résultats du sondage par explosion témoignent de

TABLEAU V Paramètres des explosions camouflées profondes du territoire remblayé des nouveaux quartiers de Saint Petersbourg.

N° essais	Poids de la charge, kg	Profondeur de la charge m	Tassement moyen cm	Profondeur de compactage m	Tassement relatif, %	Masse volumique, g/cm^3
1	3	3,2	40(42)	8	5,55	1,38
2	1	2,5	30	6,2	4,84	1,41
3	5	4,5	58	11,4	5,38	1,39

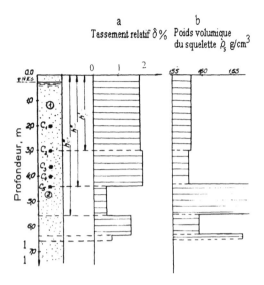

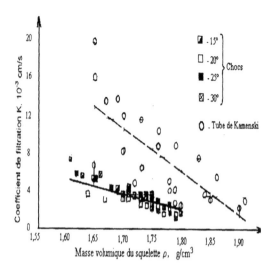

Fig. 11: graphique de sondage par explosions d'une couche de sable: a- variation du tassement relatif en fonction de la profondeur; b- variation de la masse volumique du squelette en fonction de la profondeur ; h_{cl} – profondeur de compactage par la charge c_l.

l'homogénéité des sols de remblai vis-à-vis de leur compacité. Par prélèvement d'échantillons à structure intacte, on a obtenu la masse volumique du squelette du sol de 1,42-1,67 g/cm3, ce qui est très proche des données du sondage par explosion et des essais de compactage par choc.

4.3. Détermination du coefficient de perméabilité du sol :

Les recherches effectuées ont montré que la nature de la destruction de la structure, du passage dans l'état de liquéfaction et de la consolidation ultérieure des différents types de sols peu cohérents saturés sous l'action des chocs est identique. La différence essentielle est le délai pendant lequel le sol demeure dans l'état de liquéfaction (ou le temps de consolidation).

Dans les conditions d'un problème plan (cas du sol) la vitesse de déplacement de la frontière entre les couches liquéfiée et consolidée du sable est déterminée par la vitesse de l'expulsion de l'eau interstitielle. La consolidation du sol s'effectue sous l'influence de son poids propre. C'est pourquoi la vitesse de consolidation est déterminée essentiellement par la perméabilité du sol. A partir des oscillogrammes de la Figure 7, on peut déterminer le temps pendant lequel une couche de sol demeurera dans l'état de liquéfaction. Le temps nécessaire pour atteindre la frontière de liquéfaction à la surface du sol correspond au temps plein pendant lequel le sol demeure dans l'état de liquéfaction (t_l). Ce sont les couches supérieures du

Fig. 12: comparaison des valeurs du coefficient de filtration du sable obtenues d'après les données des essais de chocs à l'aide du tube de Kamenski.

sol qui demeurent le plus longtemps dans cet état . Ce temps diminue avec l'augmentation de la densité du sable, aussi bien en raison de la diminution de la zone de destruction que par suite de la faible consolidation du sol et, par conséquent de la diminution de la quantité d'eau expulsée.

En conclusion, on peut affirmer que le coefficient de perméabilité du sol peut être déterminé, à partir des essais de consolidation par chocs, par la formule ci-dessous :

$$r_1 = K_1 \frac{\gamma_s}{\gamma} t \frac{1 - n_1}{n_1 - n_2} \qquad (6)$$

où n_1 et n_2 sont les porosités du sol dans l'état de liquéfaction et après compactage respectivement à condition de s'assurer du passage du sol dans l'état de liquéfaction complète.

Ainsi on a: $K = \dfrac{n_1 - n_2}{1 - n_1} \dfrac{h}{t_l} \dfrac{\gamma}{\gamma_s}$ $\qquad (7)$

considérant que: $n_1 = 1 - \dfrac{\rho_{s,1}}{\rho_s}$, $n_2 = 1 - \dfrac{\rho_{s,2}}{\rho_s}$ $et \dfrac{\gamma_s}{\gamma} \approx 1$

la relation (7) prend la forme:

$$K_i = \frac{\rho_{s,i} - \rho_{s,i-1}}{\rho_{s,i-1}} \frac{h_{i-1}}{t_{li}} \qquad (8)$$

où ρ_s - poids volumique du squelette du sol;

   h - épaisseur de la couche du sol;

   t_l - temps mis par la couche de sol dans l'état de liquéfaction;

   i - numéro de la charge dynamique (choc).

Ainsi, on a établi la possibilité de la détermination du coefficient de perméabilité des sols sableux par de mesure de leurs déformations et des pressions correspondantes dans l'eau interstitielle. Sur la Figure12, on compare les valeurs du coefficient de perméabilité des sols étudiés obtenues par la méthode proposée et celle du tube de Kamenski.

Il convient de faire remarquer que dans plusieurs cas, par exemple, dans les sables très compacts la méthode proposée peut s'avérer être pratiquement la seule possible et la plus juste. Par ailleurs elle est très simple et ne nécessite aucun appareillage spécial et peut être utilisée aussi bien au laboratoire que dans les conditions de chantier.

5. CONCLUSION

L'étude du phénomène de la liquéfaction et de la consolidation des sols non cohérents saturés sous l'action du choc et de l'explosion a montré que sa nature est identique dans les deux cas de sollicitation.

Par ailleurs il ne dépend que de la composition granulométrique desdits sols laquelle détermine la durée de leur maintien dans l'état de liquéfaction. L'étude a permis d'obtenir les résultats suivants:

Il a été établi qu'en cas de passage du sol dans l'état de liquéfaction complète, l'augmentation de l'intensité du choc n'a pas d'effet sur la variation ultérieure de sa porosité, mais peut accroître sa zone de destruction de la structure et sa compacité .

On a établi des relations empiriques entre la déformation volumique et le degré de compacité initial du sol.

On a élaboré une méthode de détermination du tassement relatif des sols en fonction de la variation de leur densité sous l'effet de chocs successifs.

Il a été élaboré une méthode de prévision du compactage des sols peu cohérents sous l'action de l'explosion d'après les données de leur essai sous l'action des chocs, En conséquence, on peut approximativement déterminer le nombre de séries d'explosions nécessaires à leur consolidation effective.

Il a été proposé une méthode d'évaluation de la compacité des sols peu cohérents saturés d'après les résultats de leur sondage par explosions et essai par chocs.

Il a été élaboré une méthode de détermination du coefficient de perméabilité des sols peu cohérents saturés par voie de leur passage à l'état de liquéfaction complète sous l'action de chocs.

POSTFACE:

Cette étude est le fruit d'un séjour de deux ans à la chaire de Mécanique des Sols et Fondation de l'Institut Polytechnique de Saint Petersbourg.

Feu Petr Léontiévitch IVANOV, alors Chef de la dite chaire dirigeait, depuis 1951, les recherches en laboratoire et sur chantier sur la consolidation des sols sans cohésion par explosion. Dans cette optique, il a largement contribué à la diffusion de cette technique, dans le monde par l'entremise, entre autres, de la cinquième conférence internationale de mécanique des sols et des travaux de fondation (Paris 1961) et du Colloque International sur le Compactage (Paris1980).

Grâce à la grande compréhension du Pr. P.L. IVANOV, le premier auteur eut l'occasion de réaliser cette étude autant en laboratoire que sur la base des rapports de recherches effectuées, ces dernières années par la chaire de mécanique des sols de l'Institut Polytechnique de Saint Petersbourg.

REFERENCES BIBLIOGRAPHIQUES:

Cid H.B. § Docker J.Q. (1977), Stabilisation of potentially liquefiable sand deposits using gravel drains, *J. Geotech. Engng. Div.*, ASCE, (103), GT7, p. 757- 768.

Cid H.B., Lac K.L. (1966), Liquefaction of saturated sands during cyclic loading, *J. Soil Mech. and Found. Div.*, ASCE, vol. 92, p. 543-552.

Damitio C. (1970-72), La consolidation des sols sans cohésion par explosion, Paris, *Construction*, (25), 3, p. 100-108, 9, p. 292-302, (2), 7/8, p. 264-271, (27), 3, p. 90-97.

Dembiski N. § al (1980), Compactage des fonds marins sableux à l'explosif, Col. Int. sur le compactage, vol. 1, Paris, Avril, p. 301-305.

Donchev P.(1980) , Compaction of loess by saturation and explosion, *Proc. Int. Conf. on compaction*, Paris, Avril, p. 313-317.

Florin V.A. (1952), Le phénomène de la liquéfaction et les procédés de consolidation des sols sableux, meubles et saturés. (en russe), Annale de l'Accadémie des Sciences de l'URSS, N° 6, pp. 824-833.

Ivanov P.L.. (1965), Compaction of cohesionless soils by explosives, - *Proc. 6th ICSMFE*, Montréal, vol. 3, p. 325 - 329.

Ivanov P.L. (1969) Liquefaction et consolidation des sols sans cohésion des ouvrages hydrotechniques sous l'action des charges dynamiques (en russe), thèse de doctorat, Léningead, 1969, 376 p.

Kisielova N., Osiscimski R. (1980), ,A method of underwater compaction of cohesionless soils, Coll. Int. sur le compactage, Paris, Avril, vol. 1, p. 455-460.

Kroutov A.P. (1978), Foundations in tamped pits with enlarged bottom, *Osnov. Fund. Mech. Grunt.*, (20), 3, p. 306.

Menard L. ,(1974), La consolidation dynamique des sols de fondations, *Revue des Sols etFondations*, N° 320, p. 79-82.

Menard L. § Broise Y. (1975), Theoretical and practical aspects of dynamic consolidation, *Geotechnique*, (15), 1 , p. 3-18.

Prugh B.I. (1963) ,Densification of soils by explosive vibrations, *Proc. of the ASCE*, March, vol. 89, N° 001, p. 79-100.

Simon A. (1980) ,Comparaison de l'efficacité de trois procédés de compactage d'une couche de sol de grande épaisseur sur remblai hydraulique, Coll. Int. sur le compactage, Paris, Avril, vol 1, 363-368.

Geotechnics for Developing Africa, Wardle, Blight & Fourie (eds) © 1999 Balkema, Rotterdam, ISBN 90 5809 082 5

Grouting of dispersive dam foundations

W. F. Heinz & P. I. Segatto
Rodio South Africa (Pty) Limited, Johannesburg, South Africa

ABSTRACT: Dispersive soils, their occurrence, identification and their judicious utilization for embankment dams and other applications have become part of the geotechnical "toolbox" of contractors and consulting engineers alike.

Southern Africa has its fair share of dispersive soils and much research has been done during recent decades to improve the identification process and to develop adequate preventive and remedial techniques. Early identification and the introduction of appropriate elements at design stage have gone a long way to solve or prevent potential problems with dispersive soils. South African geotechnical engineers and contractors are sensitized to potential problems associated with these soils.

Most research has focused on potential embankment problems resulting from the use of dispersive materials. Practically no publications are available in the field of grouting of foundations, which are dispersive.

The authors endeavour to present the most important aspects of dispersivity as they relate to foundation engineering with special reference to dam foundations. Little or no experience is available where dams have been founded on dispersive soils; the precautions and controls with respect to dispersive foundations during construction and during operation is the subject of this paper.

An important and interesting case study is used to illustrate the typical problems with dispersive residual granites used as foundation for a dam.

1 INTRODUCTION

Dispersivity or dispersive soils first appeared as possible ground engineering hazard in mid-1960. Several dam failures, notably in Australia, led to a more intensive investigation into the reasons for these failures; the conventional soil mechanics tests such as Atterberg limits, etc. did not provide any answers. However, it was found that clays rich in sodium cations were often structurally unstable, easily dispersed and, therefore, highly erodible. Furthermore, it was found that these dispersive clays eroded rapidly in moving water; even in stagnant water these soils would slake rapidly. In the agricultural field, dispersive soils had been recognized for many decades already as dispersivity has important repercussions for agriculture.

Today it is generally accepted that earth embankment dams constructed using dispersive soils often for economic reasons, will operate satisfactorily and safely for many years if the design is correct and the appropriate precautions are taken during construction and operation.

Practically all publications and experience accumulated during previous decades relate to dam embankments: failures, identification, remedial construction methods, precautions, etc. No information exists in the literature on similar aspects relating to dispersible soils in dam foundations.

In this publication the author discusses the effect of dispersive foundation soils on the typical drilling and grouting activities required for dam construction and recommends certain new techniques under these conditions.

2 DISPERSIVITY, DISPERSIBILITY, DISPERSIVE SOILS

2.1 Definition

Dispersive Soil – structurally unstable soil that readily deflocculates or disperses in water into its constituent particles such as sand, silt or clay; normally highly erodible.

Dispersiblity of a soil as it affects the engineering characteristics of construction materials or foundations is a complicated interaction between water AND soils with specific material properties such as clay content, clay type, type and concentration of certain ions and the concentration of free ions in the soil. In addition the type and concentration of certain ions in the water affect the dispersibility of the soil.

Although dispersivity and its effects as well as controls are better understood today, the complicated interaction between potentially dispersive soils and water seems to indicate the use of several different identification tests for positive and reliable identification.

2.2 Identification

Several tests have been developed and improved during recent years. Today field and laboratory tests can identify reliably the dispersivity of a sample. It is important to note that the conventional soil mechanics laboratory tests such as Atterberg limits, grain size analyses and others are not able to determine the dispersivity of a soil; simply because dispersivity is a physical manifestation of chemical and mineral – dependent physico – chemical properties of soils and the soils interaction with water.

Identification will commence with the first site visit of a potential construction site. Dispersive soils exhibit typical erosion patterns such as jagged formations, deep erosion gullies (dongas in South Africa) and tunnels and piping. Washed-out fans of light colour and turbidity of stored water are indications of the presence of dispersive soils.

The high salinity of the soil often results in a stunted growth of vegetation or the appearance of certain species, which are less affected by the high salinity such as certain acacia species as well as the mopane trees (Colophospermum mopane).

Although these surface "signals" may indicate the presence of dispersive soils, the absence of such indicators is no guarantee that the foundation soil will not be dispersive simply because the upper layers may not be dispersive at all, while lower horizons may be highly dispersive.

Originally, it was believed that dispersive soils occur mainly as alluvial clays, as slope wash, lake bed deposits, loess deposits and flood plain deposits (Sherard et al, 1972). However, particularly in Southern Africa, dispersive soils have been associated with residual granites, granodiorites, mudstones and sandstones as well as alluvial deposits. Dispersivity is found in clayey, silty and sandy soils.

Earlier studies had indicated that dispersive soils were mainly found in arid and semi-arid regions, however, formations in more humid climates with well-established vegetation can also be dispersive.

Hence, identification based on climatic conditions, geographic location and geological origin is not reliable.

In addition the high variability of the degree of dispersivity in any dispersive foundation soil makes an assessment of the seriousness of the problem difficult. Also where free salts are found in the pore water in equilibrium with the in situ soil, dispersivity may be minimal. If this equilibrium is disturbed the potentially dispersive soil may in fact disperse rapidly. Therefore, although identification tests are well developed and will detect dispersibility reliably, the overall geotechnical engineering is an important aspect of the design and construction of dams. Where dispersive material is used for the construction of an embankment dam, the appropriate precautions required are by now well known. In cases where these precautions have been taken, satisfactory results have been achieved. For the purpose of this publication dispersive soils also refer to highly weathered rocks.

2.3 Field and Laboratory Tests

It is beyond the scope of this paper to describe in detail the various tests that have been developed for the identification of dispersivity, nevertheless a short description of the more important tests will be given to highlight some useful characteristics of these tests:

The Crumb Test
The crumb test is a simple field test; it gives a good indication of the potential erodibility of clay soils. A soil specimen of about 3.5cm³ is placed in about 250ml distilled water. The degree of turbidity (colloidal cloud) in the water is an indication of the degree of dispersivity of the specimen.

The Double Hydrometer Test
The grain size distribution is determined using the standard hydrometer test where the sample is dispersed in distilled water by a chemical dispersant and by mechanical agitation. A second test is done without chemical dispersion or mechanical agitation. The difference between the two curves is a measure of the potential dispersivity of the sample.

To quantify the result the percentage finer than 5 microns without chemical dispersion and mechanical agitation is divided by the same percentage with dispersion and mechanical agitation. The resultant percentage is an indication of the potential dispersivity of the samples.

The percentage dispersion does not correlate to the amount of sodium present in the soil, but the test is reasonably reliable in indicating problem soils.

A dispersion percentage below 15% is an indication of non-dispersive soil, while over 30% is considered moderate; severe problems have been experienced where the percentage exceeds 50.

Pinhole test
The pinhole test simulates directly the dispersibility of clay soils as water flows through a small pinhole in the soil specimen. The dispersion of clay colloids is observed by cloudiness in the discharge effluent. For dispersive soils the measured flow rate increases as the pinhole diameter enlarges. The pinhole test as modified by the USBR is regarded as the most reliable test in the USA (Sherard et al, 1976).

Chemical tests
Chemical tests refer to the determination of cation concentration in the soil and pore water. As sodium seems to be the dominant cation related to dispersibility of soils the ESP test is the most important test to determine potential dispersibility.

ESP = exchangeable sodium percentage

$$= \frac{\text{exchangeable Na} \times 100}{\text{CEC}}$$

CEC = cation exchange capacity

$$\text{SAR} = \text{sodium absorption ratio} = \frac{\text{Na}}{(0.5\,(\text{Ca+Mg}))^{\frac{1}{2}}}$$

Soils with ESP larger than 15% are regarded as dispersive (Harmse, 1990), however, soils with ESP as low as 5% have been found to be dispersive (Van der Merwe et al, 1985). Harmse, (1980, 1988) has proposed a procedure for identification of dispersive soils by chemical testing as follows:

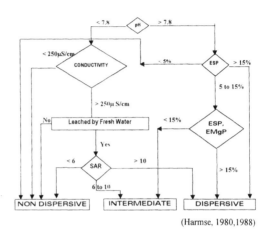

(Harmse, 1980,1988)

2.4 Occurrence

As mentioned earlier dispersive soils occur not only in arid and semi-arid regions but also in more humid climates. In Southern Africa these soils seem to be more predominant in areas with rainfall less than 850mm per year. (Bell and Maud, 1994) Rainfall in these areas occurs as flash floods with considerable erosion.

Dispersive soils have been derived from the sedimentary rocks of the Molteno Formation, the Beaufort Group, the Ecca Group and the Dwyka Group, which are all part of the Karoo sequence; the Witteberg Group, the Bokkeveld Group and the Table Mountain Group which belong to the Cape Supergroup; and the Malmesbury Group; the Nama Group; and the Kirkwood formation and Sunday River Formation which are assigned to the Uitenhage Group of the Cretaceous system. Dispersive clays also occur in all granites and in soils found over the granodiorites in the Swaziland Basement Complex. In the Cape Province soils are largely derived from granites or Malmesbury mudrocks and can possess dispersive characteristics even where the ESP values are less than 5. (Elges, 1985; Bell and Maud, 1994)

Dispersive clays can develop under the following circumstances in Southern Africa: (Elges, 1985)

- Low-lying areas where the rainfall is such that seepage water has high SAR (Sodium absorption ratio) values; especially in regions where the N-values are higher than two. Soils developed on granites are especially prone to the development of high ESP values in low-lying areas.
- In areas where the original sediments contain large quantities of illite and other 2:1 clays (montmorillonite, vermiculite) with high ESP values, dispersive soils will generally occur. This in particular is the case with the mudstones and siltstones of the Beaufort Group and the Molteno Formation in regions where the N-values are higher than two. Soils in the low-lying areas of the above formations are, virtually without exception, dispersive.
- In the more arid parts of the country where the N-values exceed 10 the development of dispersive soils is generally inhibited by the presence of free salts despite high SAR values. Highly dispersive soils can develop should the free salts with high SAR values be leached out.

The N-value was introduced and defined by Weinert (1980) and is equal to 12Ej/Pa where Ej is the evaporation during the hottest month and Pa is the annual precipitation.

3 EFFECT ON EMBANKMENT

3.1 Experience

Experience from dam embankment failures as a result of dispersivity, subsequent investigations and corrective actions or precautions may hold some lessons for foundation treatment.
Most dam failures have occurred in small earth dams, which typically are not as well engineered and designed as large important dams (Donaldson, 1975). Some dams may have failed due to dispersibility in the past but these failures were not recognized as such, as dispersive soils had not been recognized as a potential hazard in the dam construction field at the time. (Wagener et al, 1981)

Causes of dam failure and corrective action and/or relevant precautions are listed below and may have some relevance for dispersive foundation soils.

1. Piping – Most failures of dam embankments seemed to have been initiated by piping although no failure as a result of piping in a foundation has been reported.
2. Dispersive erosion in embankments commences typically in zones of higher permeability. These zones often occur:
   a. around rigid (concrete or rock) structures in the embankment where embankment material was not compacted sufficiently.
   b. where differential settlement has opened fissures.
   c. where compaction is inadequate
   d. where desiccation has caused cracks in the embankment.
   e. by hydrofracturing.

However, where designs have been done carefully e.g. adequate and effective filters have been installed and all normal precautions have been taken during construction such as proper compaction (2% wet of optimum of say 98% standard proctor) no failures of any significance have been reported.

4 EFFECT ON FOUNDATION

4.1 Experience

Typically the occurrence of piping through deep foundations is very infrequent; almost all reported piping problems in dispersive soils were embankment related.

Furthermore, the weight of the embankment may assist in closing fissures. Also after impounding when seepage is accelerated, the initial water moving through the foundation will probably not cause dispersibility, as it will be in equilibrium with the in situ formation.

Experience has shown that dispersive soils have many construction related problems: severe caving during drilling, rapidly progressing deterioration of the subsurface formation as drilling progresses, erratic and unreliable water test and grouting results, low permeability but anomalous high takes of grout, problems with the installation of casing etc. Some of these problems may cause construction delays and possibly quality related problems.(See Annexure: Case Bokaa Dam).
The main objective related to foundation improvements is seepage control and strength rehabilitation:

The main activities are drilling, water pressure testing and grouting.

4.2   Drilling

Drilling can be rotary such as diamond core drilling or rotary percussion, rarely rotary tricone is used, except for the installation of casing only in highly weathered dispersive formations. (air core drilling is only an option for very soft formations such as loose alluvia).

Present State-of-the-Art as specified most frequently by consulting engineers requires the following types of drilling:
1. Diamond core drilling with water flushing
2. Percussion (destructive) drilling with water flushing
3. Percussion (destructive) drilling with air flushing
   a. with top hammer
   b. with down-the-hole-hammer

Diamond drilling is specified to obtain cores for reliable strata identification and to obtain cores for various laboratory tests (frequently N-size is specified).

Water flushing is required as this is the best technique for drilling and water as drilling fluid has many functions in the drilling process (Heinz, 1994).

Percussion with water flushing is specified to avoid blocking fissures, which should be left open for subsequent grouting.

In general, percussion is specified as it is faster and more economical especially for blanket grouting but also for curtain grouting.

As contact with water causes dispersion the introduction of water into dispersive foundations is problematic. The dispersion of soils by water is a very rapid process as is shown by the fact that most embankments that failed, failed on first wetting or filling. Hence, it must be assumed that any type of drilling with water in dispersive foundations must be detrimental to the foundation. It also makes a reliable analysis of subsurface conditions difficult as an intrusive process such as drilling changes the results, which are being investigated.   Particularly where drilling is accompanied by water loss, dispersion may take place without being detected.

Drilling with water in dispersive soils will result in caving as the surrounding foundation will react with water and disperse.   Therefore, the typical specification which consulting engineers use i.e. that drilling will only be allowed with clean, potable water must be revised wherever dispersive soils may be encountered.

An enormous wealth of knowledge on drilling in difficult formations has been accumulated in the oil industry. Practically all drilling in the oil fields is done in sedimentary rocks.   Where difficult swelling, slaking or dispersive formations are expected, the following tests are made:
1. Clay mineral analysis, cation exchange capacity and exchange cations.
2. Balancing salinity – a test to determine the salinity required to balance the in situ activity of sub-surface formations e.g. shales.
3. Swelling measurement.
4. Dispersion tests.

The drilling muds are then selected and designed on the basis of these test results.

Hole stability is achieved by:
a) "Protecting" the formation against water: several polymer based drilling fluids can achieve this, hydrophilic shales are controlled in this manner.
b) Using a saline drilling fluid which is in balance with the in situ cation concentration (or slightly higher). The in situ ground water cation concentration will be in equilibrium with the surrounding formation and hence should not cause dispersion.

Therefore, drilling should be air percussion where possible. Where diamond core drilling is required, the most economical solution is probably drilling with a well designed saline solution.

4.3   Water Pressure Testing

Water pressure tests are an integral part of normal dam grouting techniques. Typically the Lugeon test is used which requires the isolation of a certain stage of a borehole by packers. Water is then pumped into this section at increasing and decreasing pressures. The result is expressed in Lugeon values (=litres per metre per minute). Where the maximum pressure of 10 bars cannot be reached (e.g. inadequate volume of water, hydrofracturing) the maximum test pressure is reduced; the result is then normalized to 10 bars.

Pumping water into dispersive foundation soils can be detrimental, especially at higher pressures and high velocities.

Also it is not logical to test at ten bars when the height of the dam is only 30m. The simple normalization calculation to 10 bar is misleading if the maximum test pressure that can be achieved is only say 0.5 bar.

The simplest precaution in dispersive foundation soils would be to use balanced saline water for the test. Either water from a borehole on the site is used or the in situ salinity is determined and salts including small concentrations of $CaSO_4$ are added to the test water to obtain the required salinity. The latter method seems more feasible as dispersive formations are often rather impermeable. The addition of polymers of very low viscosity can also be considered particularly as the Lugeon test is really a before and after test.

The other most important aspect of the water test is the pressure to be used.

The water test pressure should be in some way related to the water pressure at operation of the dam. The water test pressure should also be related to the grouting pressure used for the foundation, despite the fact that for various reasons Lugeon values and grout takes normally do not correlate.

Between the two extreme philosophies of grouting exemplified by the European: "rather fracture to achieve some penetration" and the Australian (Houlsby, 1990): "don't hurt the patient", the Engineer has to find a proper way on site to solve the problem within the context of the specific foundation condition. This is possibly more important for dispersive foundation soils than for any other type of dam foundation.

Therefore, for drilling and water pressure testing we would recommend the following procedure where dispersive foundation soils have been encountered:

1. Determine dispersibility of the foundation material including the degree and extent of the dispersivity.
2. Determine the in situ salinity of the ground water. The ground water should be stagnant or slow moving.
3. Use water with the in situ balancing salinity and some $CaSO_4$ to do water tests as well as for drilling.

4. Determine hydrofracturing pressure.
5. Operate the water tests at approximately 20 – 30% below the determined hydrofracturing pressure.

4.4 Grouting

In essence the task of the grouting engineer and/or contractor is threefold:
1. To determine the pressure at which the grouting material is pressed into the foundation.
2. To design a grouting material that will match the subsurface requirements i.e. to reduce the permeability to the specified level, economically.
3. To design and implement, on the basis of the specific site conditions, an efficient, effective and economical grouting procedure.

On grouting pressure, the world is still divided, in simple terms, into the "high pressure grouters", mainly European and South African practice and the "low pressure grouters" mainly American and Australian practice. For example European engineers often specify grouting pressures at $1kg/cm^2$ per m depth (Heinz, 1987) the Australian specification (Houlsby, 1990) usually states 1lb/sq.inch per ft depth; the difference is a factor of 4. More recent grouting techniques have been developed by Lombardi. Lombardi proposed the GIN principle (Grouting Intensity Number). The most important principles of this method are: (Lombardi and Deere, 1993)
1 A single "stable" grout mix for the entire grouting process (water:cement by weight of 0.67 to 0.8) with superplasticizer to increase penetrability.
2 A steady low to medium rate of grout pumping gradually increasing pressure as the grout penetrates further into the rock fractures.
3 The monitoring of pressure, flow rate, volume injected, and the penetrability in real-time by PC graphics; and
4 The termination of grouting when the grouting path on the displayed pressure versus total volume (per metre of grouted interval) diagram intersects one of the curves of limiting volume, limiting pressure, or limiting grouting intensity as given by the selected GIN hyperbolic curve (a curve of constant pressure times volume, $p \cdot V$, a measure of energy expended.)

In general dispersive soils are quite impermeable with predominantly fine fissures. Nevertheless in some foundations isolated zones with large grout takes were found. (Bokaa Dam, Vaalkop Dam). In the case of the Bokaa Dam one control borehole absorbed 3435kg after primary and secondary grout curtain holes had been drilled and grouted. At the Vaalkop Dam 3 boreholes accepted 62000kg. The major part of the Bokaa Dam foundation absorbed 4 to 6 kg/m. (See Annexure)

It is possible that these isolated takes may be due to pseudokarstic phenomenon resulting from suffosion which is a process of undermining of transported or residual soils by mechanical and chemical action of underground water. The phenomenon is closely associated with residual granites. Suffosion can lead to piping and is the main process responsible for the development of collapsible grain structures in residual granite soils (Brink,1979). Collapsible and dispersive soils are both associated with residual granite in the Southern African region. While the former is found predominantly in areas of annual water surplus, the latter is more readily found in areas of less rainfall. However, overlaps of these areas exist. Therefore, dispersive and collapsible soils may be found on one site.

On the basis of experience to date on dispersive materials and the present State-of-the-Art of grouting dam foundations, the following recommendations are presented:

4.4.1 Grouting Pressure

Grouting pressures should be as high as possible to achieve reasonable penetration in as short a time as possible; the grouting pressure should be redefined after appropriate tests on site have been made.
For example shallower horizons may require very low pressures whereas fissured granites below the weathered granites may require proportionately higher pressures than the shallower horizons i.e. a simple formula for the pressure linearly related to depth may not be adequate.
Hydrofracturing should be avoided, as it is difficult to grout all fissures that have been opened by fracturing. In dispersive soils this is of particular importance.

4.4.2 Grouting materials

As a result of the high sensitivity towards water, grouting materials should have a low water content. First choices would be normal OPC

(Blaine 3500cm²/g), rapid hardening cement (RHC) (Blaine 4500cm²/g) and microfine cement (Blaine 12000 to 15000cm²/g). Typically OPC is specified under normal conditions but as RHC is reasonably inexpensive in South Africa, RHC blended with pozzolanic products (South African coal mines produce a PFA of excellent quality) is often used and is an excellent grouting material. However, PFA binds the excess calcium hydroxide which is set free by progressive hydration. In dispersive soils this free calcium hydroxide may reduce the dispersibility potential, hence it may be better not to use PFA under these circumstances.

With "thicker" grouts, superplasticizers should be used as is recommended by Lombardi (Lombardi et al, 1993). Gypsum is often added to cement to achieve a) flash setting to combat lost circulation b) gelling or thixotropic properties and c) expansion properties in the set cement (very low 0.3%). About 4 to 10% of gypsum may be added to reduce the effect of dispersibility of the subsoil. Cement grouts containing salts help to protect shale reactions from sloughing and heaving during cementing and hence may reduce dispersibility of foundation soils. Because of the Na-sensitivity in dispersive soils it is probably better to use KCl for salt cements (Smith, 1987). Long term stability and durability of the cement grout should also be considered.

The above mentioned procedures are proven techniques from oil field cementing technology, nevertheless, each case must be considered on its own merits, and based on the specific subsurface conditions.
Where cement products may be too coarse to achieve any notable penetration in fine fissures sodium silicate is often the most economical alternative. However, there is still some doubt as to the permanency of sodium silicate particularly under conditions of frequent wetting and drying. Shrinkage can result in an increased residual permeability. Sodium silicate has many uses including as deflocculants, detergent bleach etc. In addition sodium salts may be exuded from silicate gels. Also the quality control of the sodium silicate process is difficult especially at the interface between grout material and soil. (Karol, 1990). No experience is available of sodium silicate grouting of dispersive soils. With the above characteristics sodium silicates do not seem to have the right properties for grouting dispersive foundation soils.

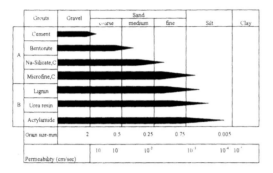

Grouts	Gravel	Sand			Silt	Clay
		coarse	medium	fine		
A	Cement					
	Bentonite					
	Na-Silicate,C					
	Microfine,C					
B	Lignin					
	Urea resin					
	Acrylamide					
Grain size-mm	2	0.5	0.25	0.75	0.005	
Permeability (cm/sec)		10 10	10²	10¹	10⁴ 10⁷	

GROUT PENETRABILITY   A: Suspensions B: Solutions

If the use of cement products, OPC to microfine, has been exhausted chemical grouting using acrylamides should be applied. However, costs are usually high. A combination of cement products and chemical grouts, except sodium silicate, may provide the most economical solution. See also the attached drawing for an indication of grout penetrability.

4.4.3 Grouting Technique

The grouting technique refers to the "how" to grout and place the grouting material where it is needed. In general this refers to the split spacing method, tube-à-manchette (TAM) (Heinz, 1983), Multiple Packer Sleeved Pipe system (MPSP) (Bruce, 1991), the rate of thickening if required and the variation of the W:C ratio as required etc.

The grouting technique is not only important from the perspective of the quality of the end product but it may have a significant impact on the construction programme. While grouting activities may only be 2 to 5% of the entire cost of a dam project, delays of main contractor activities due to grouting may be very costly. (See Annexure: Case Bokaa Dam). As dispersivity adds an additional "hazardous" dimension to the grouting process and hence to the entire project it is imperative that identification of these potential problems is made early on in the construction process so that the construction team including its experienced geotechnical subcontractor can address these potential problems speedily in order to avoid expensive delays and still achieve an acceptable end product.

For a given pressure the maximum distance of penetration is determined by the yield strength (as defined by the Bingham model) while the viscosity governs the rate of flow i.e. the time to get there. (Lombardi and Deere, 1993)

For a given material and type of fissures high pressures will increase penetration distance, increase the hole spacing which is desirable in dispersive foundations but may cause hydrofracturing. Low pressures will decrease the penetration distance, reduce the hole spacing which is not desirable in dispersive foundations and will probably not hydrofracture which is desirable in dispersive foundation soils. Therefore, in general, the Australian method of grouting with very low pressures typically leads to very close spacing of grout holes. Hence in dispersive foundation soils the judicious choice of grouting pressures for various horizons is extremely important.

The best 'all round' method in difficult formations where severe dispersion and/or caving occurs is the tube-à-manchette (TAM) method. A further development of this system is the multiple packer sleeved pipe system (MPSP). These techniques are described in detail elsewhere (Heinz, 1983; Bruce, 1991). In short, the holes to be grouted are drilled to their final depth in one operation, a sleeved pipe (sleeves = manchette = non-return valves) is installed and the formation is grouted through these pipes. The system has many advantages, however, it is rather slow hence generally more costly than the standard grouting techniques. The pipe and sleeves are permanently installed hence contribute to the costs; the MPSP technique is similar to the TAM method.

The most significant advantage of these methods is that drilling is minimized in each hole, which is an important factor in dispersive subsoils and, of course, collapsing or caving soils can be treated effectively. Without the effective treatment of caving soils, drilling and redrilling may significantly impact on the entire construction programme. While these techniques may be more costly, invariably expensive delays are avoided.

The decision to grout upstage or downstage should be made in favour of the method that guarantees least contact with drilling water, assuming that no satisfactory drilling fluid can be found. A comparison of the two methods is given below (Bruce, 1991). It seems that upstage grouting with packers at the top may be the most effective method in dispersive soils.

4.5 Special Techniques

Several techniques which have been used to prevent seepage in dam foundations also may have some

	Downstage	Upstage
A D V A N T A G E S	1. Ground is consolidated from top down, aiding hole stability and packer seating and allowing successfully higher pressures to be used with depth without fear of surface leakage.	1. Drilling in one pass
		2. Grouting in one repetitive operation without significant delays
		3. Less wasteful of materials
	2. Depth of the hole need not be predetermined: grout take analyses may dictate changes from foreseen, and shortening or lengthening of the hole can be easily accommodated.	4. Permits materials to be varied readily
		5. Easier to control and programme.
		6. Stage length can be varied to treat 'special' zones
	3. Stage length can be adapted to conditions as encountered to allow 'special' treatment.	7. Often cheaper since net drilling output rate is higher.
D I S A D V A N T A G E S	1. Requires repeated moving of drilling rig and redrilling of set grout: therefore, process is discontinuous and may be more time-consuming	1. Grouted depth predetermined
		2. Hole may collapse before packer introduced or after grouting starts, leading to stuck packers and incomplete treatment
	2. Relatively wasteful of materials and so generally restricted to cement based grout	
	3. May lead to significant hole deviation.	3. Grout may escape upwards into (non-grouted) upper layers or the overlying dam, either by hydrofracture or bypassing packer. Smaller fissures may not then be treated effectively at depth
	4. Collapsing strata will prevent effective grouting of entire stage, unless circuit grouting method can be deployed	
	5. Weathered and/or highly variable strata problematic	4. Artesian conditions may pose problems.
		5. Weathered and/or highly variable strata problematic
	6. Packer may be difficult to seat in such conditions	

distinct advantages when used in dispersive soils. Some of the more recent proven techniques are discussed.

4.5.1 Diaphragm or slurry trench wall

Very effective though rather costly especially where only certain lower horizons require impermeabilization. Possible problems: The effect of a Na-montmorillorite slurry on highly dispersive soils has not been researched. Probably a high bentonite concentration would be required for the slurry which could then give problems when the slurry is displaced by the concrete i.e. inclusions of soil or slurry in the concrete.

4.5.2 Jet Grouting

Extremely effective and cost effective especially where only certain deeper horizons require impermeabilization. Dispersivity is irrelevant with this technique.

4.5.3 Crystallization Process

As this process is less known in dam foundation engineering, a more detailed description of the process will be presented here. Further detailed information can be found in Ziegenbalg and Crosby, 1997 and Ziegenbalg and Holldorf, 1998.

Many naturally occurring processes are known which lead to closure of seepage paths. These processes are the result of recrystallization processes of coupled dissolution and precipitation processes. Normally nature requires many years to produce satisfactory sealing. However, if the crystallization process can be controlled and accelerated a very exciting grouting technique is created.

For example gypsum precipitation occurs immediately when a solution containing Na_2SO_4 is mixed with a solution containing $CaCl_2$. However, the addition of an inhibitor allows a mixing process without spontaneous crystallization.

Ziegenbalg and Crosby (1997) have shown that it is possible to prepare large volumes of $CaSO_4$ oversaturated grout solution under field conditions. Depending on the degree of supersaturation and the relative inhibitor content, the solutions lead to the artificial precipitation of gypsum within two to eight hours. In this way micro fissures can be sealed.

The method is particularly interesting for the sealing of dispersive foundation soils because:
1. The grouting material is a solution capable of penetrating fine fissures which typically occur in

dispersive soils at least in the Southern African Region and
2. The process precipitates gypsum, which is the favoured and proven material to inhibit dispersibility of soils.

5   CONCLUSION

Dispersivity is a complex interaction between relatively pure water and soils with specific chemical characteristics. Dispersivity in earth embankments can be identified and controlled by including certain design features, during construction and operation.

Practically no information or data exist with respect to the control of dispersivity in foundation soils particularly for dam foundations where grouting is required.

Based on the available knowledge and experience worldwide as well as experience from the grouting of several dams on dispersive soils in the South African Region, the authors come to the following conclusions and recommendations

1. On dispersive soils, which can be identified, all activities that introduce water into the foundation such as drilling and grouting must be limited to a minimum.
2. The most important activity to utilize water is drilling – rotary and percussion. To minimize the effect on potentially dispersive soils, drilling water designed with a balancing salinity and the inclusion of some $CaSO_4$ should be used for all activities such as drilling and water testing.
3. Similarly, grouting procedures should be designed on the basis of the effect of the grouting material on dispersivity. Grouting materials with relatively low water cement ratios, "salt cements" or chemical grouts should be used. Sodium silicate should only be used after careful evaluation.
4. Techniques such as the TAM or MPSP have distinct advantages when grouting dispersive foundation soils; especially in the face of collapsing or severely caving conditions.
5. Other alternative techniques to the standard drilling and grouting such as slurry trenches or jet grouting have been applied with great success. However, whereas slurry trenches are costly, jet grouting is an excellent and economical technique, especially where only selected horizons require impermeabilization.
6. A very exciting and promising new grouting technique to impermeabilize dispersive soils is the controlled crystallization of oversaturated grout solutions, especially $CaSO_4$ in the foundation. As a solution, the technique holds great promise for sealing micro fissures and to reduce dispersivity by gypsum precipitation.
7. No simple test can be relied on to identify dispersive foundation soils. The ESP test is still the most reliable test to identify dispersive soils in the Southern African region. By early identification of dispersive soils the required precautions and controls can be introduced timeously. In this way, safe and economic structures can be built even if foundation soils should be found to be dispersive.

Finally it seems abundantly clear that intricate or complex foundation conditions such as the drilling and grouting of dispersive foundation soils requires the nomination of specialized and experienced geotechnical and grouting contractors if unnecessary delays are to be avoided and an economical solution of high quality is to be achieved.

6   REFERENCES

Bell, F.G. and Maud, R. R (1994) - *Dispersive Soils: a review from a South African perspective,* Quarterly Journal of Engineering Geology, Vol 27.

Brink, A.B.A (1979-1985) - *Engineering Geology of Southern Africa,* Pub. by Building Publications, 4 Vol

Bruce, D.A. (1991) - *Grouting with the MPSP Method at Kidd Creek Mine,* Ontario, Ground Engineering.

Bruce, D.A. and Shirland, J.N (1985) - *Grouting of completely weathered granite with special reference to the construction of the Hong Kong Main Transit Railway,* Proc. 4th Int. Sym., IMM, Brighton.

Clarke, M.R.E (1987) - *Mechanics, Identification, Testing and Use of Dispersive soils in Zimbabwe.*

Darley, H.C.H. and Gray, G.R. (1988) - *Composition and Properties of Drilling and Completion Fluids,* 5th Ed, Gulf Publishing, Houston.

Donaldson, G.W (1975) - *The occurrence of dispersive soil piping in central South Africa.* Proceedings of the Sixth Regional Conference for Africa on Soil Mechanics and Foundation Engineering, Durban.

Elges, H.F.W.K (1985) - *Dispersive Soils in South Africa.* State-of-the-Art, The Civ. Eng. in S. A.

Gerber, F.A & von Maltitz Harmse, H.J (1987) - *Proposed procedure for identification of dispersive soils by chemical testing,* Die Siv. Ing. in Suid Afrika.

Heinz, W.F. (1995) - *Grouting the Future,* Proc. Int. Drill '95. Annual Conf. ADIA, Australia

Heinz, W.F. (1994) - *Diamond Drilling Handbook,* 3rd Ed, Published by SADA.

Heinz, W.F. (1993) - *Extrem tiefe Injektionsschürzen,.* Proc. Int. Conf. On Grouting in Rock & Concrete. Balkema.

Heinz, W.F. (1987) - *The Art of Grouting in Tunnelling*, SANCOT seminar, Nov.

Heinz, W.F. (1983) – *Tube-à-Manchette: Description and Applications*. Proc. Grouting Symp, Jhb.

Houlsby, A.C. (1990) - *Construction and Design of Cement Grouting*, John Wiley & Sons Inc., NY.

Karol, R.H. (1990) - *Chemical Grouting*, Marcel Dekker, New York, 2nd Ed.

Lombardi, G. and Deere, D. (1993) - *Grouting Design and Control using the GIN Principle*, Water Power and Dam Construction.

Sherard, J.L., Decker, R.S., Ryker, N.L. (1972) - *Hydraulic Fracturing in Low Dams of Dispersive Clay*, Proc. Spec. Conf. On Performance of Earth and Earth-Supported Structures, ASCE.

Sherard, J.L, Dunnigan, L.P., Decker, R.S. (1976) - *Identification and Nature of Dispersive Soils*, Journal Geotechnical Division, ASCE.

Smith, D.K. (1987) - *Cementing*, Monograph, SPE.

Von M. Harmse, H.J., Gerber, I.A. (1988) - *A Proposed Procedure for the Identification of Dispersive Soils*, Proc. 2nd Int. Conf. On Case Histories in Geotechnical Engineering, St. Louis.

Von M. Harmse, H.J. (1990) - *Report on Dispersive Soils from Bokaa Dam Botswana*, Eurosond report by Prof. Von M Harmse of the Univ. of Potch.

Von M. Harmse, H.J. (1980) - *Dispersiewe Grond, Identifikasie en Stabilisasie*, Ground Profile, no.22.

Wagener, F., Von M. Harmse, H.J., Stone, P., Ellis, W. (1981) - *Chemical Treatment of a Dispersive Clay Reservoir*, 10th International Conference on Soil Mechanics and Foundation Engineering, Stockholm.

Weinert, H. H. (1980) – *The National Road Construction Materials of South Africa*, Academica, Cape Town.

Ziegenbalg, G, Holldorf, H. (1998) - *The directed and controlled crystallization of naturally occurring minerals – a new process to seal porous rock formations and to immobilize contaminants*, IAH Syposium, Quebec.

Ziegenbalg, G, Crosby, (1997) - *An overview of a pilot test to reduce brine inflows with controlled crystallization of gypsum at the IMC Kalium K2 brine inflow*, Mineral Resources Eng., Vol. 6, No. 4.

ANNEXURE - Case Bokaa Dam

The Bokaa Dam is situated approximately 35 km NNE of Gaborone, Botswana. The dam is part of the Metse motshlaba Scheme. Dam construction commenced on January 1989 and was completed during 1993; the dam is situated on the Metsemotshlaba River.

The dam is an earth embankment dam, it is 1700m long, its maximum height is 30m. The spillway is on the left embankment and is approx. 500m long. The pumphouse is centrally situated close to the old riverbed.

Dispersivity in the foundation soil was not investigated at design stage, although during construction the dispersive foundation soils caused significant problems and subsequent delays

According to a report by Prof. Von Maltitz Harmse (Harmse, 1990) some of the fissures filled with clay could be classified as highly dispersive. The soil samples investigated could be prone to piping. ESP values as high as 60.6% where encountered on the site. For example the average ESP for seven samples between 1m and 7m depth on Pattern E was 30%.

The dam foundation is a typical residual granite foundation. Residual granite foundations as they occur in the Southern African region, are possibly one of the most difficult formations for dam foundations.

Typically the formation has the following characteristics: (see drawing below) (Brink, 1979)
1. Troughs of decomposition adjacent to tors of unweathered rock.
2. The presence of small to large core-stones within the residual granite soil (see drawing).
3. Abrupt changes from highly weathered to unweathered rock.
4. Moderate to slightly weathered rock – weathering is mainly along joints; joints may be void, partially or completely clay filled.
5. Irregularly dispersive in degree and extent.

For grouting engineering purposes, this soil profile translated into the following characteristics:
1. Low permeability in the weathered to highly weathered horizons.
2. Anomalous high grout takes in some stages.
3. Low grout absorption over the entire grout curtain.
4. Rapid deterioration of soils and hence caving in boreholes, flushing out of weathered soils.
5. Erratic water pressure test and grouting results. After grouting of primary and secondary and sometimes tertiary boreholes with decreasing absorption, sudden relatively high grout takes in the final test and control boreholes.
6. Difficulties in installing casing in the rapidly caving (dispersing) formation.
7. Difficulty in defining base rock due to the presence of core-stones; makes it difficult to define depth to which casing must be installed.
8. Due to variability of degree of dispersibility it was difficult to assess the extent and seriousness of dispersivity.

Due to the above-mentioned problems with the foundation, long delays were caused as the Engineer blamed the Geotechnical contractor for incorrect grouting procedures. The Engineer did not realize the extent and seriousness of the dispersivity in the foundation soil. The complete drilling and grouting programme of 22,000m percussion and 3000m diamond drilling was completed within 6-7 months. Approximately a similar amount of work was requested

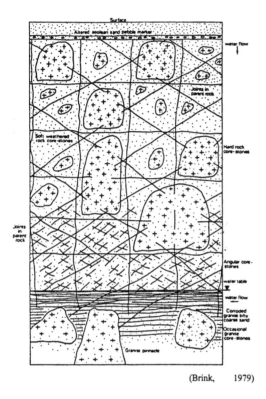

Surface
Altered aeolian sand pebble marker
water flow
Joints in parent rock
Soft weathered rock core-stones
Hard rock core-stones
Joints in parent rock
Angular core-stones
water table
water flow
Corroded granite silty coarse sand
Occasional granite core-stones
Granite pinnacle

(Brink,   1979)

by the Engineer and completed in an additional 18 months with little or no improvement in the total grout absorption. A claim of the main contractor against the client resulted. The client recovered a large amount of this claim in a lengthy arbitration case against the Engineer.

Geotechnics for Developing Africa, Wardle, Blight & Fourie (eds)© 1999 Balkema, Rotterdam, ISBN 90 5809 082 5

Geosynthetics – An alternative method of soil improvement

W. P. Hornsey & G. M. James
Kaytech, Durban, South Africa

ABSTRACT: Internationally, geosynthetics have become established in geotechnical engineering practice as a useful means to stabilize soils over poor ground and in walls, steep slopes and embankments. This paper attempts to highlight the positive contribution geosynthetics can make to the design of these structures. Although geosynthetics perform various functions in a wide range of applications, this paper focuses on the reinforcing and separation functions. In the 20 years since geotextiles were first designed into near vertical retaining walls, widespread utilization of this method has been realized, however conservatism in design prevails.

1 INTRODUCTION

Throughout the world, and in southern Africa in particular, the demand for improved services and structures is constantly increasing, while funds available for their construction and maintenance is decreasing at an alarming rate. In recent years informal, urban settlements have been expanding into areas where development has been avoided due, in part, to the underlying soil conditions as well as the scarcity and expense of suitable natural materials commonly used in conventional construction. The modern engineer is therefore faced with the challenge to design and construct more economical structures with limited resources. Experience has shown that geotextiles can make a significant contribution towards this end.

The use of modern geosynthetics for soil improvement in South Africa dates back to 1970 when a continuous filament non-woven needle punched polyester geotextile (Bidim) was first imported from France. The only product available at that time was a $340g/m^2$ fabric which was used as a "one size solves all problems" product and the pioneering design methods used a great deal of engineering judgement. Since that time the range of geosynthetic products has grown considerably and so too has knowledge and understanding of their strengths and weaknesses.

The field of geosynthetics has grown to such an extent that the 6th International Geosynthetics Conference held in Atlanta, GA, USA attracted over 1800 delegates representing 61 countries to access the 160 technical papers accepted for publication. The papers covered a wide range of topics, such as drainage, reinforcing, waste management and separation to name a few. These conferences show not only the depth of interest in the subject but they also encourage scientific research into the design and function of geosynthetics.

This paper concentrates on Reinforcement and Separation, which are areas not usually (until recently) associated with geosynthetics, but they do have a marked influence on geotechnical engineering designs.

Combining good geotechnical engineering practice with the latest, engineered geosynthetic products and advanced technical knowledge can produce cost effective, safe and durable structures.

2 REINFORCING

2.1 *History*

Geosynthetic Reinforced Soil Structures (GRSS) have been in existence for centuries. The earliest remaining examples of soil reinforcement are the ziggurat of Agar-Quf in Iraq and the Great Wall of China. The Agar-Quf ziggurat, located 5 km north of Baghdad, was constructed of clay bricks varying in thickness between 130 and 400mm, reinforced with woven mats of reed laid horizontally on a layer of sand and gravel at vertical spacing varying

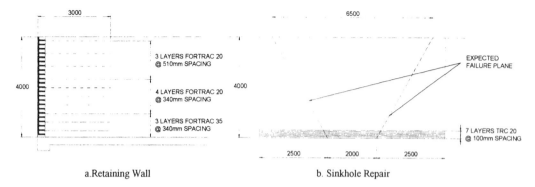

a.Retaining Wall   b. Sinkhole Repair

Figure 1: Reinforcing Applications

between 0.5 and 2.0m. Today the Agar-Quf structure is 45m tall compared to its original height believed to have been over 80m. It is estimated to be over 3000 years old.

Possibly the best modern example of the durability of GRSS was highlighted during the recent Kobe earthquake in Japan where, in comparison with reinforced concrete retaining walls, geogrid reinforced soil walls performed very well. In particular, the geogrid reinforced soil walls that were constructed in 1992 at Tanata did not collapse despite the fact that the site was located in one of the most severely shaken and seriously damaged areas. Based on these experiences, many damaged embankment slopes and conventional retaining walls were replaced with geogrid reinforced soil retaining walls [10]. This can be attributed to the inherent flexibility of GRSS, which are not adversely affected by movement e.g. settlement, vibrations etc.

2.2 Applications

The term reinforcing, in this section, coveres the strengthening of walls and embankments and the somewhat unusual application of sinkhole stabilisation and rehabilitation.

Walls can have front faces, which vary from 70° to 90° with a number of facings such as concrete panels, blocks, gunite etc, Figure 1a. This diversity makes this system suitable for a wide range of engineering and landscaping applications.

Embankment reinforcing allows the construction of slopes at steeper angles than the angle of repose of the fill material. This is of particular importance in areas where property boundaries pose a restriction on the height of an embankment.

Sinkhole stabilisation is of importance when constructing roads and structures in areas that are prone to subsidence, e.g. over old mine workings, Figure 1b.

All the mentioned applications are ideally suited to labour based construction methods because of the ease of handling of the geotextile and the simple method of placement. No heavy or sophisticated equipment is required. In these times of high unemployment and government support of black empowerment it would appear that geosynthetic reinforcement can be used as an incentive to proceed with the construction of a scheme before other schemes, which do not have a high level of employment opportunities, are initiated.

2.3 Design

At present there is only one Code of Practice available in South Africa for the design of reinforced soil structures and that is the British Code of Practice BS8006 (1995). The purpose of geosynthetic reinforcing is to convert a mass of loose soil into a homogeneous monolith and hence the structure can be designed as a mass gravity structure. A number of factors over and above the usual geotechnical considerations must be considered when designing a GRSS these include:

- Geosynthetic polymer e.g. polyester or polypropylene
- Geosynthetic type e.g. geotextile or geogrid
- Geosynthetic structure e.g. woven multi-filament, PVC coated grids etc.
- Geosynthetic strength i.e. stress/strain tensile behaviour
- Soil friction angle vs. geosynthetic to soil friction behaviour

This information is then used to design the GRSS.

A correct design should cover at least the following potential areas of failure (Fig. 2):

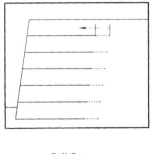 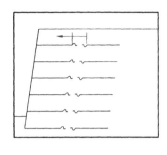 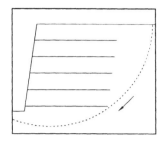

Pull Out   Tensile Overstress   Slope Stability

Figure 2: Design Considerations

- Pull out resistance – where the embedment length is insufficient, tensile forces overcome the soil/geosynthetic friction and the geosynthetic is pulled out of the soil.

- Tensile overstress – there are insufficient layers of geosynthetic reinforcing and the geosynthetic ruptures.

- Overall slope stability – the geosynthetic reinforcing does not intersect the critical failure plane and global failure occurs.

2.4 *Construction*

Construction of a GRSS is relatively simple, with no special training required for the labour installing the geosynthetic, but it is imperative that the geosynthetic is installed at the correct spacing, length, direction and that compaction of the backfill is adequate. Site supervision is no less critical in the construction of earth reinforced structures than reinforced concrete structures where a resident engineer would not allow a concrete pour to go ahead without first checking the reinforcing. It is therefore surprising to find that when back analysis of failed GRSS are carried out the results show that the geosynthetic reinforcing was either the incorrect grade, in the wrong position or was left out completely. Gassner and James (1998) cover this topic extensively. They show that both consultants and contractors do not regard GRSS in the same light as a conventional reinforced concrete structure. This is wrong because the economic ramifications of failure are equal for both systems. This attitude has greatly harmed the reputation of GRSS in the past. Both consultants and contractors must understand that savings are incurred through construction methods and the materials used, not by lowering standards!

2.5 *Conclusion*

Experience has shown that geosynthetic reinforced soil structures are better or at least equal to conventional reinforced concrete structures at 1/3 to 1/2 of the costs. Why then persist with these uneconomical forms of construction?

3 SEPARATION

3.1 *History*

Separation in road construction can be traced back to the Roman road builders who used animal skins laid out over swampy areas to prevent contamination of fill material. In the South African context one of the first uses of geosynthetic separators was during the construction of the N2 South freeway along the southern coast of Kwa-Zulu Natal where large embankments were built over deep alluvial deposits (hippo mud) at the numerous river crossings. The inclusion of these separation layers greatly reduced the fill requirements and allowed the blanket drains below the embankments to operate efficiently.

3.2 *Function*

As shown in Figures 3 & 4 the function of a geotextile separator is to:

- maintain the integrity of the fill material
- allow the dissipation of excess pore water pressure in the existing material while limiting soil migration and
- allow the release of pore water pressure through and within the plane of the geotextile.

Research world wide on the subject of separation has resulted in the following conclusions:

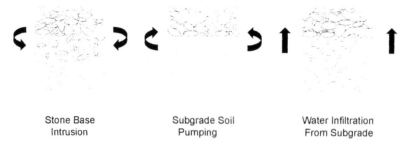

| Stone Base | Subgrade Soil | Water Infiltration |
| Intrusion | Pumping | From Subgrade |

Figure 3: Typical Reasons for Deterioration of Fill

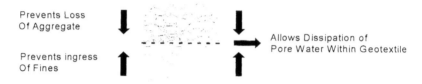

Prevents Loss
Of Aggregate

Prevents ingress
Of Fines

Allows Dissipation of
Pore Water Within Geotextile

Figure 4: Role of Geotextile Separator

- Needle-punched nonwovens have the best overall long-term drainage performance[6&11]
- Drainage of sub-grade improves with increased geotextile thickness[11]
- The use of a geotextile separator results in the reduction of porewater pressure, and hence increases the consolidation, in the sub-grade[7]
- The use of a geotextile eliminates base/sub-grade intermixing, if the geotextile survives the installation and placement operations[1,2,8&11]
- From a construction point of view, the use of a geotextile greatly enhances the initiation of construction of the embankment by providing a more stable working platform and allows the subsequent filling operation to be carried out more rapidly[8]
- Greater densities are achieved in the fill material when using a geotextile separation layer[12]
- Thick geotextiles are most effective in reducing fines pumping[4]

The U.S. Department of Transportation, Federal Highway Administration (FHWA) published Table 1 below, which shows the estimated aggregate thickness loss as a function of sub-grade strength.

3.3 Application

Geotextile separation layers are used predominantly in the construction of the following structures:

1. Embankments for roads including soil platforms for housing, commercial and industrial developments.
2. Road and runway layer works
3. Railways below ballast and in layer works

3.4 Design

With these varied applications so will requirements of the geotextile vary. The following factors should be considered when including a separation layer in a design:

3.4.1 Survivability

The performance of the geotextile separation layer is directly related to the amount of damage it incurs during construction. This is because the geotextile is subjected to higher stresses and loads during this stage than at any other time during the life of the structure. Therefore if the geotextile survives the installation/construction stage, it will have sufficient strength to perform adequately as a separator for the design life of the structure. Hence the first step in the design process is to ensure that the geotextile is not damaged to such an extent that it can no longer function adequately as a separator.

A geotextile with good conformability (high elongation) is less susceptible to puncture and damage. Therefore the full stress-strain curve (not only the elongation at failure) of the geotextile must be considered when choosing the generic makeup of the geotextile, as each method of manufacture will result in a different stress-strain curve. In some cases geotextiles with similar elongation at failure may have vastly different stress-strain relationships leading up to failure, which will influence their susceptibility to damage. Factors such as strength of sub-grade and size/shape of fill material, type of compaction equipment used and cover thickness should be considered when deciding on a suitable geotextile.

Various Road and Railway authorities around the world base the survivability requirements of the geotextile on Burst Strength, Grab Strength, Trapezoidal Tear Strength and Puncture Resistance.

3.4.2 Soil Retention

The soil retention requirements are similar to that of a subsoil drain:

- The larger pores in the geotextile must be smaller than the larger soil particles to prevent piping, and
- The smaller pores in the geotextile must be larger than the smaller soil particles so that clogging and blinding will not occur.

It should be noted that particle movement takes place under slightly different circumstances to that of a subsoil drain. In this instance the vibrations caused by vehicular traffic will cause the movement of soil particles, the presence of water in the soil will encourage this movement. Flow conditions may be either unidirectional (static) e.g. in embankment separation or multidirectional (dynamic) e.g. in rail track separation.

To determine the soil retention requirements,

for the mechanical filter stability of geotextiles, a full grading and hydrometer analysis as well as the plasticity index (PI) of the soil to be filtered are required. The retention requirements are expressed in terms of Apparent Opening Size (AOS) of the geotextile. The AOS or O_{95} of a geotextile indicates the approximate size of the largest particle, which will pass through the geotextile. Where flow conditions are expected to be unidirectional the geotextile AOS value will be determined by wet sieving (O_{95W}) while for multidirectional flow conditions the hydrodynamic sieving method (O_{95H}) is used.

3.4.3 Permeability

Geotextile separators should allow the drainage of excess porewater in two planes:

- Transverse where the water is able to flow horizontally so as to reduce the amount of ground water infiltration into the imported layer. This is known as transmissivity, and
- Normal, where the water is able to flow vertically through the geotextile into the free draining imported layer above. This is known as permitivity.

The reduction in porewater pressure in the sub-grade due to the drainage capacity of the geotextile increases the bearing capacity of the sub-grade and hence the life of the road. In separation the transverse permeability (transmissivity) of the geotextile is of the utmost importance in maintaining the strength of the imported layers. This characteristic allows the ground water to flow in the plane of the geotextile into the longitudinal subsoil drains, thereby reducing the amount of water that would normally percolate into the imported layers.

The principle of all permeability criteria is that as long as the permeability of the geotextile (k_g) is greater than the permeability of the soil (k_s) the flow of water will not be impeded at the soil / geotextile interface.

3.4.4 *Anti-Clogging*

In order to function adequately as a separator for the life of the road the geotextile must be highly porous so that the permeability of the geotextile is not significantly reduced should some of the pores become blocked by soil particles.

3.4.5 *Durability*

Ultraviolet degradation due to exposure to sunlight occurs in all polymers to some greater or lesser extent. Polyester and polyethylene are the more resistant polymers[9]. If the application requires the geotextile to be exposed to sunlight for an extended period of time allowances must be made for losses in strength.

3.5 *Conclusion*

Geosynthetic separators can result in considerable savings in materials, time and cost.

4 SUMMARY

The use of geosynthetics to improve soil structures is an established fact. Since 1979 six International Conferences have been held around the world on Geosynthetics. Design approaches and procedures have been published and are readily available through these proceedings. Education is of vital importance in making engineers, contractors and developers aware of the benefits of designing with geosynthetics. Most of the internationally established experts in the field of geosynthetics are academics running design schools within civil engineering departments of recognised tertiary education institutions. This paper serves to highlight only two of the many applications of geosynthetics, namely reinforcing and separation, in an effort to convince the civil engineering fraternity that geosynthetics are not a means to an end but an end in themselves.

Flexibility and ease of installation allows one to construct Geosynthetic Reinforced Soil Structures

(GRSS) in very difficult conditions like steep topography and seemingly inaccessible areas. Since 1986 South Africa has seen an increasing number of GRSS (some in excess of 10m height) being constructed. Examples of failed structures and reasons as to why they failed are mentioned or referred to in this paper. These failures can help engineers gain knowledge of the limitations that are placed on such structures by poor design, poor material selection, bad installation and insufficient education and training of contractors.

The most under-utilised application for geosynthetics in South Africa is in the area of sub-grade separation and basal reinforcement. Through this application previously inaccessible areas, i.e. informal settlements, may be improved for development. The challenge is made to engineers to be more innovative with their designs and break out of the conventional design mould and offer prospective clients cheaper and workable solutions.

Good performance after a successful 28-year presence in South Africa coupled with good performance begs the question: "How long before geosynthetics (geotextiles), although established practice, become standard?"

REFERENCES

Austin D.N, Coleman D.M, 1993, *A field evaluation of geosynthetic reinforced haul roads over soft foundation soils*. Proc. Geosynthetics '93 Conference, Vancouver, Canada, 65-80.

Brorsson I, Eriksson L, 1986, *Long term properties and their function as a separator in road construction*. Proc. 3rd Int. Conference on Geotextiles, Vienna, Austria, 93-98.

Gassner F.W, James G.M, 1998, *Failure of two fabric reinforced segmental block walls in South Africa*. Proc. 6th Int. Conf. on Geosynthetics, Atlanta, Georgia, USA, 565-572.

Glynn D.T, Cochrane S.R, 1997, *The behaviour of geotextiles on glacial till sub-grades*. Proc. Geosynthetics '87 Conference, New Orleans, USA, 27-36.

McGown A, Andrawes K.Z, Pradhan S, Khan A.J, 1998, *Limit state design of geosynthetic reinforced soil structures*. Proc. 6th Int. Conf. on Geosynthetics, Atlanta, Georgia, USA, 143-184.

Metcalf R.C, Holtz R.D, Allen T.M, 1995, *Field investigations to evaluate the long-term separation and drainage performance of geotextile separators*. Proc. Geosynthetics '95 Conference, Nashville, Tennessee, USA, 591-967.

Nishida K, Nishingata T, 1994, *The evaluation of separation functions for geotextiles*. Proc. 5th Int. Conf. on Geotextiles, Geomembranes and Related Products, Singapore, 139-142.

Olivera A, 1982, *Use of non-woven geotextiles to construct a deep highway embankment over swampy soil.* Proc. 2nd Int. Conference on Geotextiles, Las Vegas, USA, 625-630.

Raumann G, 1982, *Outdoor exposure tests of geotextiles.* . Proc. 2nd Int. Conference on Geotextiles, Las Vegas, USA, 541-546.

Tatsuoka F, Koseki J, Tateyama M, 1997, Earth Reinforcement, *Performance of reinforced soil structures during the 1995 Hyogo-ken Nanbu Earthquake.* Rotterdam: Balkema, 973-1007.

Tsai W.S, Savage B.M, Holtz R.D, Christopher B.R, Allen T.M, 1993, *Evaluation of geosynthetics in a full scale road test.* Proc. Geosynthetics '93 Conference, Vancouver, Canada, 35-48.

Werner G, Resl S, Montalvo R, 1989, *The influence of non-woven geotextiles on the capacity of the fill material.* Proc. Geosynthetics '89 Conference, San Diego, USA, 345-350.

Geotechnics for Developing Africa, Wardle, Blight & Fourie (eds) © 1999 Balkema, Rotterdam, ISBN 90 5809 082 5

Geosynthetic reinforcement of soft clay under highway embankments

I. M. Ibrahim & H. M. A. Salem
Faculty of Engineering, Zagazig University, Egypt

ABSTRACT: Soft clay as weak subgrade soil, under highway embankments should be treated to support pavement structure and traffic loads. Reinforcing such weak soil with geosynthetics is considered simple and economic but efficient treatment. This paper is devoted to present the effect of reinforcing soft clayey soils with two different geosynthetic types (geotextile and geogrid) on the soil characteristics represented by its shearing and bearing resistance. Triaxial and California Bearing Ratio are adapted in this research. It is found that geosynthetic reinforcement of soft clay highly improves its shearing and bearing resistance. It is also found that geogrid improves shearing resistance significantly, while its improvement to bearing resistance is moderate compared to the geotextile. Therefore, it is recommended to use geogrid reinforcement where higher shearing resistance is required such as prevention of slope stability failure. While, geotextile is recommended under road base to increase soil bearing capacity and reducing vertical settlement due to traffic loads.

1. INTRODUCTION

Treatment of soft clayey soils is essential when used as a subgrade under highway embankments. Methods of soil treatments are presented in detail by NCHRB [9], Sinacori et al [12], and Arman [4]. In brief, these methods are : (1) using barms or flatter slopes in order to increase embankment slope stability, (2) using light-weight fill material for road embankments, (3) supporting roadways and embankments by piles, (4) replacing weak soil by suitable fill, (5) Stabilizing the weak soils by chemicals, compaction, or preloading, (6) reinforcing weak soils by ground anchors, nailing, rootpiles and geosynthetics. Geosynthetic reinforcement of weak soil is considered the simplest and the most economical but efficient treatment method [1]. The effect of geosynthetic reinforcement on the strength of clayey subgrade soils was studied by Abd El- Hady [1] and Al- Taher and Salem [2] through conducting a set of CBR tests using different geosynthetic types. They found that CBR values are highly increased because of the geosynthetic reinforcement and it enables the soft soil to be safely compacted using lower compacting efforts. Broms [5] conducted a set of triaxial tests on dense sand to determine the effect of geotextile reinforcement when it placed at various locations from the triaxial specimen height. He found that most beneficial effects are obtained by placing the geotextile layer at mid or one third of the height.

Where the fabric interrupts potential shear zones and has influence of increasing the overall strength.

This study is devoted too evaluate the improvement in both shearing and bearing resistance of soft clayey subgrade soil as a result of reinforcing it with two different geosynthetic material (geotextile and geogrid). This is carried out through conducting an extensive experimental program using both triaxial and CBR tests.

2. INVESTIGATED MATERIALS

The investigated soil is soft clayey soil which is subgrade soil of Balotha- Port Foad road 240 km north east of Cairo, Egypt. A series of standard identification tests were conducted according to AASHTO [3], BS: 1377 [8], and Egyptian Code [6] on the investigated soil in order to determine its engineering characteristics. A summary of the tests and the soil characteristics are shown in Table (1).
Two different geosynthetic types are used in this investigation. The first is Typar 3857 geotextile (T1) and the second is Netlon CE 121 geogrid (N). The technical characteristics of these two types as provided by the manufactured companies are illustrated in Table (2).

Table (1): Different Engineering Characteristics of the Investigated Subgrade Soil.

Test	Specification	Determined Soil Character Value
1- Atterberg Limits:		
-Liquid Limit, %	AASHTO: T-89	43
-Plastic limit, %	AASHTO: T-90	24
-Plasticity index, %	AASHTO: T-91	19
2- Percentage passing No.200 sieve,%	AASHTO: T-27	98.5
3- AASHTO Soil classification	AASHTO: M-145	A-7-6
- Unified Soil classification		CL
4- Specific gravity	AASHTO:T-100	2.769
5-Standard Proctor compaction	BS:1377	
-Max. dry density, ton/m³		1.53
- Opt. moisture content, %		23.2
6- Organic matter content, %	Egyptian Code	11.12
7- Shear strength of natural soil, kPa		< 14

Table (2): Technical Data of the Different Investigated Geosynthetic Types (*After their Companies' Technical Reports*).

Property	Geosynthetic Name	Polyfelt TS 700 (P1)	Polyfelt TS 500 (P2)	Typar 3857 (T1)	Typar 3337 (T2)	Netlon CE121 (N)
1-	Weight, gm/m²	280	140	290	110	730
2-	Thickness (under pressure of 2 kN/m²), mm	2.6	1.4	0.78	0.45	3.3
3-	Strip tensile strength, kN/m	18.0	9.2	16.0	4.0	7.68
4-	Elongation at max. load, %	50-80	50-80	33.0	25.0	20.2
5-	Cost, L.E. /m²	6.60	3.20	10.00	3.75	4.00

3. EXPERIMENTAL PROGRAM

The experimental program consists of two sets of tests, triaxial and CBR sets. Each set includes evaluation tests and reinforced soil tests as follows:

3.1 Triaxial tests

Under three different confining pressure of 100 kPa, 200 kPa, 300 kPa, three groups of UU triaxial tests are conducted on compacted identical sample sizes and densities. First group of tests is conducted on non-reinforced soil samples have a nominal dimensions of 50.4 mm diameter and 105 mm height. This group is carried out in order to evaluate the shearing resistance, determine the strain – stress characteristic curve, and define the failure surfaces of non-reinforced soil samples. Second group is conducted on reinforced soil samples with Netlon CE 121 geogrid, while third group is conducted on reinforced soil samples with Typar 3857 geotextile. In second and third groups, samples are reinforced with circular geosynthetic layers in the mid height of the samples at or close to the failure surface of the non-reinforced samples.

3.2 California Bearing Ratio (CBR) tests

Two series of unsoaked and soaked CBR tests are conducted in this investigation, according to BS:

Group	Test No.	Nature of sample		Geosynthetic type		Surcharge load (kg)			
		Nonreinforced	Reinforced	T1	N	10	25	35	50
G1	1	*					*		
	2		*				*		
	3		*				*		
	4		*	*			*		
	5		*				*		
	6		*		*		*		
G2	7	*				*			
	8		*	*		*			
	9		*		*	*			
	10	*						*	
	11		*	*				*	
	12		*		*			*	
	13	*							*
	14		*	*					*
	15		*		*				*

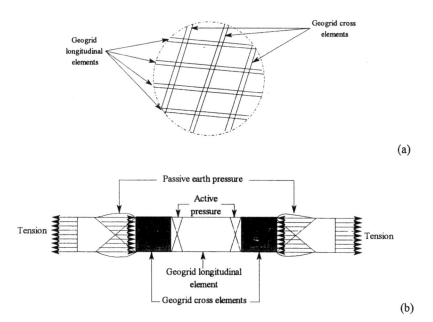

(a)

(b)

Figure (1): Sections in Triaxial Sepcimen at the Reinforcement Position Showing Passive Resistance to the Tensile Force Resulted from Shear, (a) Cross Section, (b) Longitudinal Section. [Not to Scale]

1377 [6] and Egyptian Code [8] on both non-reinforced and reinforced soil samples. Table (3) summaries the conditions of the CBR tests that have been achieved in this investigation, including the sample nature, type of geosynthetic and applied surcharge.

4. RESULTS AND DISCUSSION

The results of triaxial and CBR tests conducted in this investigation are presented and discussed in this section. The effect of geosynthetic reinforcement type under different confining pressure on the

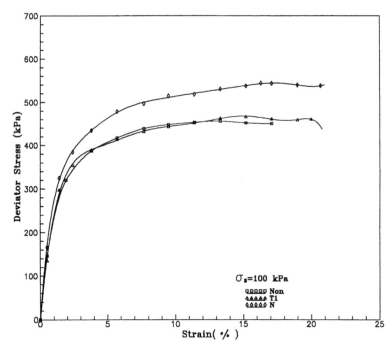

Figure(2): Stress–Strain Curves for UU Triaxial Test,
with & without (T1 & N) Reinforcement Types

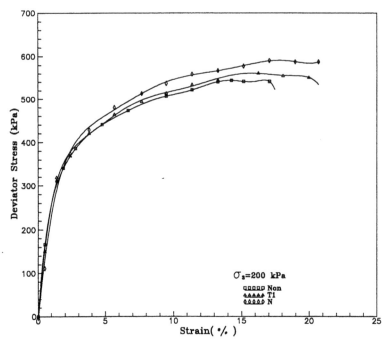

Figure(3): Stress–Strain Curves for UU Triaxial Test,
with & without (T1 & N) Reinforcement Types

shearing resistance is discussed. The effect of geosynthetic reinforcement type under different surcharge load on the bearing resistance of the investigated soil for both soaked and unsoaked conditions is also discussed.

4.1 Effect of geosynthetic reinforcement on triaxial test results

Figures (2) through (4) show the stress-strain characteristic curves obtained from UU triaxial tests conducted under confining pressure of 100 kPa, 200 kPa, 300 kPa for non-reinforced and reinforced soil samples. Reinforced soil samples are reinforced with two different geosynthetic types (T1 geotextile and N geogrid).

4.1.1 Effect of geosynthetic type on soil shear strength

Figures (2) and (3) illustrate the increase of the deviator stress due to geosynthetic reinforcement especially at high strains and with N geogrid type. The figures indicate that the initial modulus, the failure strain, and the maximum deviator stress are highly increased due to soil reinforcement with geosynthetics in general and with N geogrid type in particular. Therefore, N geogrid type is more effective in improving soil shear strength than T1 geotextile. This phenomenon can be explained as follows: The pore openings of N geogrid type are large enough to allow better penetration of soil grains into the pores. Which let the soil below and above the geogrid form continuous medium in discrete areas (geogrid pores). When the soil is sheared in the plane of the geogrid layer, a tension force is generated in the geogrid due to good interlock and friction between soil particles and geogrid. This tension force causes a slight movement, due to strain, in the geogrid. The movement represents a compression state in the soil which provides a passive earth pressure against the shearing direction due to the presence of geogrid cross members, as illustrated in Figure (1). This passive pressure in addition to friction and/or adhesion increase soil shearing resistance and raise the failure strain of the reinforced soil.

T1 geotextile has small pore openings into which soil particles may not penetrate, thus geotextile layer separates the triaxial soil specimen into two parts. Therefore, a negligible amount of interlocking developed. The interface shear strength is therefore, gained mainly from the friction and/or adhesion the soil and the geotextile, which is small compared with the case of geogrid. As the friction between the soil and the nonwoven geotextile is considered to be equal to the soil friction itself [10], no significant increase in the deviator stress is achieved, while the failure strain increases. Therefore, since geotextile

such as T1 have low interface shear strength, higher anchorage lengths are required, compared with geogrid when they are used in soil reinforcement. While, because of their small opening sizes, their separation and filtration functions are better than those for geogrid. On the other hand, geogrid such as N type have good anchorage because of the higher interlocking and friction interface forces while its separation and filtration functions are not fulfilled.

Therefore it recommended to use geogrid in soil reinforcement where high shearing resistance and good anchorage are required as in the case of prevention of slope stability failure. While it is recommended to use geotextiles in soil reinforcement where high bearing resistance and good filtration are required with the ability of increasing the anchorage length as in the case of road construction on soft soils.

4.1.2 Effect of Confining Pressure on Geosynthetic Reinforcement Efficiency

Figure (5) illustrates the relation between the confining pressure and the maximum deviator stress for three test cases. The figure indicates that the rate of increasing the maximum deviator stress due to N geogrid reinforcement is much higher with lower confining pressure. Such rate decreases gradually with increasing the confining pressure. While, the rate of increasing the maximum deviator stress due to T1 geotextile reinforcement is much relatively small, and it is nearly constant with all confining pressure values. Figure (4) indicates that at higher confining pressure (of 300 kPa) the differences in the maximum deviator stress in the three test cases are insignificant. Therefore, if a high confining pressure is to be applied, there is no much need for geosynthetic reinforcement.

As confining pressure simulates the surcharge load which the soil subjected to in the field, and it is direct proportional to road thickness, i.e. low confining pressure means low road thickness, the geosynthetic effect decreases with surcharge load increase.

4.2 Effect of Geosynthetic Reinforcement Type on CBR Values

Figure (6) and (7) summarize the load-penetration relationship of different reinforcement types for both soaked and unsoaked cases. The samples have the same conditions of group (G1, see Table 2). These Figures indicate that for any load value, reinforcing the investigated soil reduces its penetration value significantly. This means that the reinforced soil sample can withstand loads of higher values than those for non-reinforced soil samples. This true for both soaked and unsoaked conditions.

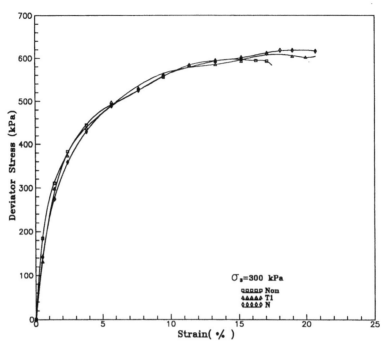

Figure(4): Stress—Strain Curves for UU Triaxial Test,
with & without (T1 & N) Reinforcement Types.

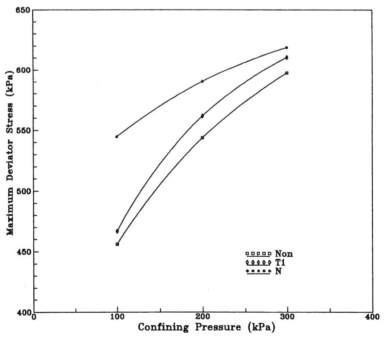

Figure(5): Effect of Confining Pressure on the Maximum
Deviator Stress.

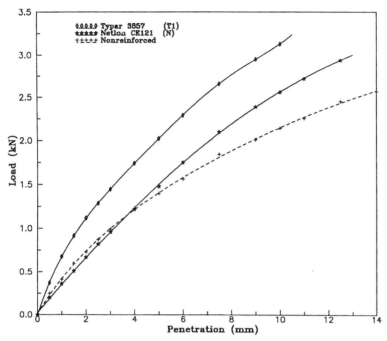

Figure(6): Load—Penetration Curves for The Unsoaked CBR
Test With Different Reinforcement Types.

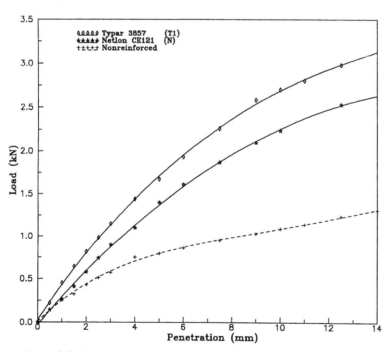

Figure(7):Load—Penetration Curves for The Soaked CBR
Test With Different Reinforcement Types.

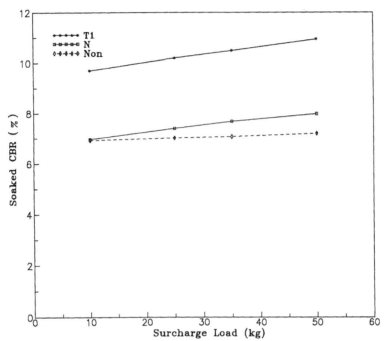

Figure(8):Relationship between Surcharge Load and Soaked
CBR Values for Nonreinforced,and Reinnforced
Soil Smples with T1 and N Geosynth.Types.

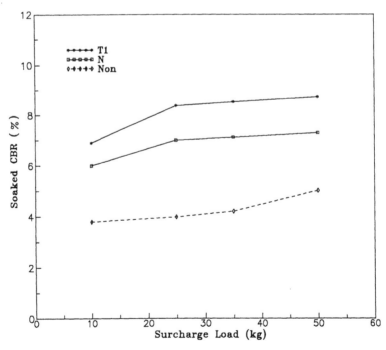

Figure(9):Relationship between Surcharge Load and Soaked
CBR Values for Nonreinforced,and Reinnforced
Soil Smples with T1 and N Geosynth.Types.

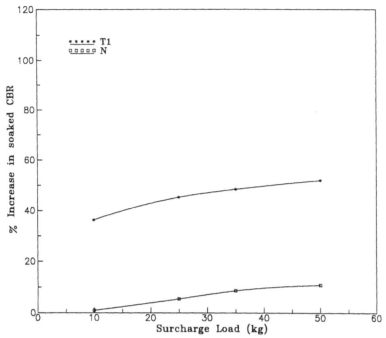

Figure(10):Relationship between Surcharge Load and
The Percentage Increase in Soaked CBR
Values,w.r.t Nonreinforced Case.

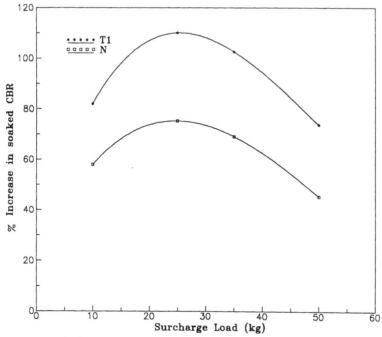

Figure(11):Relationship between Surcharge Load and
The Percentage Increase in Soaked CBR
Values,w.r.t Nonreinforced Case.

The figures also show that T1 geotextile type achieves the highest improvement in the soil bearing resistance (CBR value), while N geogrid type has little effect on increasing the CBR values for both soaked and unsoaked conditions. This may be due to the large size of its pore openings that cause the geogrid to sink through the soil when a vertical load is applied from the CBR plunger.

4.2.1 Effect of Varying Surcharge Load on CBR Values

Figures (8), and (9) show the relationship between surcharge load values and the CBR values for reinforced and non-reinforced soil sample under both soaked and unsoaked conditions. The figures illustrate that CBR values increase as surcharge load increases for reinforced samples significantly, while it moderate for non-reinforced samples, especially under the soaked conditions. Figures (10), and (11) show the percentage increase in the CBR values for reinforced soil sample with N geogrid and T1 geotextile relative to non-reinforced soils sample under both soaked and unsoaked conditions. The figures indicate that the percent increase of the CBR value is increasing as the surcharge load increases under the unsoaked conditions, while it approaches a maximum value at certain value of surcharge load and then decreases as surcharge load increases. They also show that the percent increase for soil reinforced with T1 geotextile type is higher than that reinforced with N geogrid type under both soaked and unsoaked conditions. For soaked conditions, the percent increase of CBR value of the soil reinforced with T1 geotextile type reaches as high as 110% CBR value of non-reinforced soil. While the maximum percent increase of the CBR value for the soil reinforced with N geogrid is about 70% CBR value of non-reinforced soil. For unsoaked conditions, the rate of the percent increase of CBR value is almost steady (0.50 % per unit increase in surcharge load) for the reinforced soil with both types of geosynthetic. Although the rate of percent increase is steady for both cases of reinforcement, the initial percent increase for T1 geotextile type is more significant than that for N geogrid type.

CONCLUSIONS

In order to evaluate the improvement in both shearing and bearing resistance of soft clay subgrade soils due to reinforcing it with two different geosynthetic types (geogrid and geotextile), an extensive experimental program using both triaxial and CBR tests was carried out. Analyzing and matching the test results, the following conclusions can be drown:

1. Reinforcing weak subgrade soils such as soft clay will any type of geosynthetic highly improves soil shearing and bearing resistance, especially at low confining pressure,

2. Maximum befits of geotextile reinforcement in bearing resistance is achieved by reducing road thickness above geotextile layer to its optimum value,

3. It is recommended to use geogrid in soil reinforcement where high shearing resistance and good anchorage are required as in the case of prevention of slope stability failure in road embankment,

4. It is recommended to use geotextiles in soil reinforcement where high bearing resistance and good filtration are required with the ability of increasing the anchorage length as in the case of road construction on soft soils,

5. It is recommended to use larger triaxial cell that suite larger specimen size in order to determine the effect of the sample size on the results obtained.

REFERENCES

[1] Abd EL-Hady, G. M. 1990. 'Use of Geotextile in Subgrade Reinforcement" Master Degree Thesis, El-Mansoura University,.

[2] AL-Taher, M. G. & Salem, H. M. A. 1995. " Improving The Characteristic of Weak Soils Using Geosynthetic Reinforcement" Proc. 2nd Int. Conf. on Engineering Research, Suez Canal University, Port Said, Egypt, 1995.

[3] American Association OF State Highway AND Transportation Officials (AASHTO) " AASHTO Standard and Specifications", 1974.

[4] Amran, A. 1978. "Current Practice in the Treatment of Soft Foundations" Soil Improvement report by Committee on Placement and Improvement of Soils, Geotechnical Eng. Division. ASCE, , pp. 30 – 51.

[5] Broms. B. B. 1977. "Triaxial Tests with Fabric-Reinforced Soils" C. R. Coll Ins. Soils Test, Paris, Vol. 3, 1977, pp. 129- 133.

[6] BS 1377, " Methods of Test for Soils for Civil Engineering Purposes" British Standards Institution, 1975.

[7] Colllios, Delmas, Goore & Giroud 1980. " The Use of Geotextiles for Soil Improvement" ASCE National Convention, Portland, Oregon, 1980, pp. 53- 73.

[8] Egyptian Code, " Egyptian Code for Soil Mechanics and Foundation Design" Part ii, 3rd. edition, Cairo, Egypt, 1993.

[9] NCHRP synthesis of highway practice 147, "Treatment of Problem Foundations For Highway Embankments"; Transportation Research Board, National Research Council, Washington D. C. 1989, pp. 12-36.

[10] Resl, S. & Werner, G. 1986 " The Influence of

Nonwoven Needlepunched Geotextiles on the Ultimate Bearing Capacity of the Subgrade" Proc. 3rd Int. Conf. on Geotextile, Vienna, Austria, 1986, pp. 1009- 1012.

[11] Salem, H. M. A. 1995. "The Use of Geotextile in Highway Construction over Weak Soils" Master Degree Thesis, Zagazig University.

[12] Sinacori, Hofmann, & Emery 1952 " Treatment of Soft Foundations For Highway Embankments" HRB Proc. 31st Annual Meeting, Highway Research Board, National Research Council, Washington D. C. 1952, pp 601 - 621.

Geotechnics for Developing Africa, Wardle, Blight & Fourie (eds) © 1999 Balkema, Rotterdam, ISBN 90 5809 082 5

Compaction grouting-application as a method of ground improvement for construction on dolomite

S. Makura & A. Schulze-Hulbe
Africon Consulting Engineers, Pretoria, South Africa

L. Maree
Department of Civil Engineering, University of Pretoria, South Africa

ABSTRACT: Compaction grouting has been successfully applied to improve subsurface conditions for construction of a building on ground characterised by sub-surface karst features common in dolomitic areas. This paper discusses the method of compaction grouting, the design procedures used and their implementation on site as well as the verification of the procedures with respect to ground improvement. It highlights the combined use of rational and empirical design methods for the purposes of drawing up a grouting programme and estimating of quantities.

1 INTRODUCTION

Following subsidence of an existing structure, which was founded on a well compacted soil mattress within an area classified to have a moderate to high risk of dolomite related instability, alternative methods of soil improvement were investigated for the new adjacent chipping plant building. The proposed structure is known to exert considerable dynamic as well as static loads on the soil profile.

The geological profile at the site comprises an upper layer of sandy clayey chert rubble fill, underlain by stiff residual shale and sandstone of Karoo origin. This sequence is encountered at varying depths between 4 m and 12 m. Underneath this formation is the residual dolomite, which consists of weathered chert and manganiferous clay (wad). It extends to the highly irregular dolomite bedrock at depths of between 35 and 50 m. The water level is found at about 14 m depth from the surface. The residual dolomite layer is prone to subsurface erosion by water percolating through the soil profile from the surface and removing the often loosely compacted highly erodible wad matrix which occurs between the gravel to boulder sized chert fragments and transporting it into cavernous areas occurring lower down in the residual dolomite profile. It is also prone to collapse type settlement that can be triggered by dynamic as well as static loads and this can result in the formation of sinkholes and dolines. Potential trigger mechanisms that could lead to instability on the site are the ingress of water into the soil profile, a fall in the ground water table and vibrations caused by heavy machinery to be installed in the chipping plant.

Several founding solutions were considered and these included piling and the use of a soil raft/mattress. The piling option using percussion drilled pre-cast driven piles extending into bedrock (auger drilling refuses at shallow depths on chert boulders) has been considered. It is not considered a practical solution, as the average depth to bedrock dolomite is about 35 m below the surface. This option is likely to be very expensive and would still not eliminate risk due to subsidence related movement within the dolomitic residuum. This is because the long slender piles in dolomite residuum become susceptible to buckling when the shaft support is partially removed as a result of erosion of the residuum. The soil mattress option was also considered. For a mattress to comply to the anticipated requirements it would need to be of very high quality and have considerable thickness and adequate stiffness. It would need to adequately dissipate the heavy loads of the proposed structure and bridge subsidence movements of up to 300 mm occurring over areas extending between 8 m and 10 m in diameter as have occurred at the adjacent existing building. To accommodate such subsidences and allow deflections of an order of magnitude small enough to allow unhindered operations in the proposed building was not considered to be easily obtainable in an unreinforced soil mattress.

Following extensive evaluation of the options discussed above it was decided to employ compaction grouting as a preventative measure.

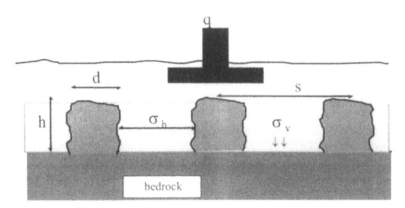

Figure 1: Section across treated area

2 COMPACTION GROUTING

Compaction grouting is achieved by the injection of non-plastic thick cement grout introduced under pressure in a borehole at the desired location to form a bulb of material, which displaces and compresses the residual dolomite. This procedure has been likened to blowing up balloons in the soil with hydraulic pressure. Compaction grouting does not depend on grout entering the void space in the soil, but rather on the displacement and compaction of soils resulting from intrusion of a mass of thick grout. In principle compaction grouting compacts and stabilises inherently unstable soft and porous areas within the residual dolomitic soil profile, thereby restricting the nature and magnitude of the ground movements, reducing and perhaps eliminating the risk of damage to structures. This is attained through,

• significant reduction in the soil permeability which in turn reduces the risk of sub-surface erosion due to the ingress of water.
• the filling of any cavities which may occur as well as an increase in the stiffness of the soft or loose material occurring in the residual soil profile.
• the increase in the lateral pressure caused by the injection of the grout results in an increase in the density, shear strength and bearing capacity of the residual dolomite.

The design philosophy and rationale of this solution is described below. A detailed analysis of the uptake and distribution of the grout in the residual dolomite formation is also discussed. Verification of the effects and improvements due to compaction grouting have been undertaken.

3 DESIGN

Due to the relatively recent development of the technique, and the general nature of grouting, definite design methods and procedures for estimating costs have until recently not been well developed. The success of past projects has often depended on trial and error procedures, and the amount of experience of the engineer and the contractor. Therefore the design approaches used in compaction grouting are largely based on experience. Common practice has shown that this type of design is characterized by borehole spacing and stage length, both of which are dependent on the anticipated radius of compaction at each injection point.

According to Graf (1992) the radius of compaction is thought to be a function of the following. (a) The restraining pressure of the soil, which is directly proportional to weight of soil above the bulb. (b) The total radial force and hence the surface area of the grout bulb and (c) the grout pressure at the bulb, which for practical purposes is the grout pressure at the collar. The effective radius of compaction of each grout hole will vary according to soil type, its moisture and its relative density. It is also influenced by the grout injection rate and the resulting pressure. Based on experience it has been established that in most cases significant densification of soil will occur within a radius of 0.3m to 1.8m from the interface of the soil and the grout (Warner, 1992). It is however recognized that some effect will be experienced at greater distances from the point of grout injection. Borehole spacing is therefore dependent on the soil type and soil conditions, the depth at which the grouting is carried out, the improvement requirements and experienced judgement of the engineer.

The basic guidelines used for the empirical design of the ground improvement at the chipping plant were based on previous experience of the engineers and the contractor gained from grouting in similar materials in the West Rand. The primary boreholes were placed at 6 m centres, with provisions to incorporate secondary and tertiary boreholes when and if required. With regards to grout injection points down the hole, a standard 1 m stage length was adopted. Due to the limitations of cost the up-stage grouting method was chosen as it provides a more cost effective alternative with adequate control for application. Conditions for the end point refusal were predefined with consideration for the anticipated goal of the grouting. The refusal criterion, which was also based on experience, included attaining a consistent pumping pressure in excess of 3 MPa, or pumping a volume of 2 m^3 of grout at each point of injection.

This design was fine-tuned using a new theory and rational design for compaction grouting after Schmertmann and Henry (1992). The theory uses shear enhancement along grout columns, due to lateral stress and density increase, to support the soil above and between the grout columns. Laboratory tests have been conducted to help support the design theory. This design approach is considered a breakthrough and a major improvement to the experience-based empirical design discussed above. The combined use of the two approaches provides a more rational design process that can also be used for the purposes of cost estimation and verification of completed projects. The rational design can be explained with the aid of the cross section in Figure 1.

Consider that high pressure injections of low slump grout form a geometric pattern of columns each of a height h. Each column helps to displace and hence compact the soil in area s^2 between adjacent columns and increases the lateral pressure σ_h, and soil friction τ at the column/soil boundary. The resultant density and stress increases result in a potential column / soil shear force T and a top bearing pressure R. By varying suitable combinations of column spacing s, height h and average diameter d until the overburden soils can be entirely supported by T and R, the bearing capacity of the soil can be varied. Any surcharge q generates an additional ΔT and ΔR in the soil.

The design analysis aims at achieving a certain factor of safety F, using only effective forces and stresses. Depending on the critical conditions, total forces and stresses may also be used (Atkinson and Bransby, 1978). The following equations were used to define the various forces and stresses considered in evaluating the factor of safety.

The weight requiring support is expressed according to equation (1).

$$F_v\!\downarrow = (\gamma_z + q)\, s^2 \dots\dots\dots\dots\dots\dots\dots\dots\dots(1)$$

(Where $\gamma =$ unit weight of the soil)

The forces available for support $F_v\!\uparrow$, are vertical upward forces, which consist of T and R plus ΔT and ΔR from any surcharge and are described by the following equations: (2) to (7).

$$T = \tau \pi dh \dots\dots\dots\dots\dots\dots\dots\dots\dots\dots\dots\dots(2)$$

$$\tau = \sigma_h \tan \phi_g \dots\dots\dots\dots\dots\dots\dots\dots\dots\dots(3)$$
(Where ϕ_g = soil/grout column friction angle)

$$\sigma_h = [K_o + 0{,}5\,(K_g - K_o)]\,(\sigma_v) \dots\dots\dots\dots\dots(4)$$
(Where K_o and K_g denote the earth pressure coefficient before and after grouting respectively.)

$$\Delta T = K_o\, q \tan \phi_g\, \pi dh \dots\dots\dots\dots\dots\dots\dots(5)$$

$$R = \sigma_{vc}\, \pi d^2 /4 \dots\dots\dots\dots\dots\dots\dots\dots\dots(6)$$
(Where σ_{vc} = grouting pressure at top of columns)

$$\Delta R = \theta q\, \pi d^2 /4 \dots\dots\dots\dots\dots\dots\dots\dots\dots(7)$$
(Where θ = stress concentration factor on top of columns (after Boussinesq theory)

The support for the imposed weight is given by

$$F_v\!\uparrow = T + \Delta T + R + \Delta R \dots\dots\dots\dots\dots(8)$$

The factor of safety is therefore given by:
$$F = \frac{F_v\!\uparrow}{F_v\!\downarrow} = \frac{T + \Delta T + R + \Delta R}{(\gamma_z + q)s^2} \dots\dots\dots\dots(9)$$

This design concept provides a rational basis for the design of grouted soils that support the overlying materials between and above the soil columns. The method is best suited for uniform materials such as wad where near cylindrical shaped grout bulbs will be formed on grouting.

Whilst this approach provided a very important design method it is rarely feasible to devise a programme in advance that turns out exactly as predicted. Particularly when considering the nonuniform nature of the material presented by alternating chert and wad layers. Maximum flexibility was therefore maintained by integrating the empirical and rational methods to allow for the refining of the design, as drilling data became available for every hole completed. This exercise was done to produce a thorough programme so that the cost of the works was less likely to be substantially increased.

Table 1: Trial and error results

d	3	3	3	1.5	1.5	1.5	1	1	1
h	25	15	5	17	15	10	15	20	25
FOS (s=1.5 m)	99	63	35	29	26	17	16	21	28
FOS (s=4 m)	13	8	4.9	4.4	3.6	2.5	2.2	3	3.9
FOS (s=6 m)	6.1	3.9	2.1	1.8	1.3	1.1	1	1.3	1.7
FOS (s=10 m)	2.2	1.4	0.8	0.6	0.6	0.4	0.4	0.5	0.6

FOS= factor of safety

Using the method described above and applying the parameters indicated in Table 1, several factors of safety were calculated. From this variation of parameters it was clear that the dimensions and specifications suggested from the empirical approach could attain a factor of safety of 1,8 which is acceptable. Due to the limitation of the non-uniform nature of dolomite residuum application of the factors of safety calculated here were applied as a rough guide only aided largely by the empirical approach.

3 IMPLEMENTATION

As mentioned above the upstage grouting technique was employed for the grouting process which entailed the following steps:-

- drilling of the borehole to a predetermined depth of 25 m, which was later revised to a depth of 26 m to cater for the spoil accumulating at the bottom of the borehole.
- installation of a tremie pipe to the bottom of the borehole;
- lifting the tremie pipe to the required height and;
- pumping grout until refusal pressures were attained or a maximum of 2 m^3 of grout was pumped.
- repeating the lifting of the tremie pipe in one meter stages and pumping until the base of the covering layer (about 8 m depth) was reached.

Proportions of the grout mix were determined in accordance with ASTM C938 and then modified to attain the required slump. The original grout mix comprised 40 kg of water (7.8%),: 25 kg of cement (5.1%): 420 kg of sand (86.6%) to attain a target 28 day strength of 2 MPa. Slump tests were periodically undertaken and randomly at the request of the engineer to ensure that the grout complied with the specification of between 25 and 50 mm slump. The importance of the slump is to minimise grout leakage and segregation as well as to avoid hydraulic fracturing of the soil often associated with thin grouts. Variations of the amount of water used in the mix were decided upon depended on the moisture content of the sand to maintain the water cement ratio. Cube tests, which were conducted, confirmed that the required grout strength was attained.

The pumping pressure of the grout was closely monitored during the operation using a pressure gauge installed at the collar of the borehole being grouted. The maximum pressures, which were used to determine refusal, take into consideration the pressure losses as a result of friction in the pipelines and the overburden pressure. Pressure guidelines in the range between 1,5 MPa and 3 MPa were used when pumping the grout, taking into consideration the two aspects mentioned above.

As a guide the maximum volume injected per stage in each borehole was initially set at 2 m^3 as calculated from the configuration decided upon from Figure 1. This volume was later revised to 2,7 m^3 following results of the test grouting which revealed that after pumping 2 m^3 of grout the pumping pressures were significantly less than the anticipated refusal pressure. However when the volume was increased to 2,7 m^3 pumping pressure in the order of 80% to 90% of the recommended refusal pressures were attained indicating significant improvement. This additional grout is thought to have found its way into voids, which needed to be filled up prior to the process of densification.

The grouting was undertaken in rows of primary boreholes at six meter centres, ensuring that a distance of at least 20 m was maintained from the drilling rig to minimise the adverse effect of the air pressures used for drilling. Constant monitoring of the drilling, the geology and analyses of the grouting results was maintained to decide on the requirement and location of secondary boreholes. Criteria for deciding on the location of secondary boreholes were largely dependent on the local geology and the grout takes recorded in adjacent boreholes. From the results observed in the test grouting conducted in the initial rows, it was decided that if the adjacent primary boreholes each recorded grout takes in excess of $15m^3$ without attaining refusal pressures below the 20 m depth, a secondary borehole would be located equidistant from the primary boreholes. This was done to ensure that no zones of soft residual dolomite remained untreated. The secondary boreholes also served a further function of assessing the effectiveness of the compaction grouting carried out in the primary boreholes. Effectiveness using secondary boreholes was measured according to the decrease in grout takes relative to the primary boreholes.

4 ANALYSIS OF GROUT UPTAKE AND DISTRIBUTION

Of the 4 771 m drilled during this programme 528 m were drilled in 23 secondary boreholes, whilst the rest were in the 179 primary boreholes. The engineer on site decided upon the positions of secondary boreholes after analysis of the grout takes recorded in the adjacent boreholes using the criteria described above.

On average four boreholes were drilled per day and three were grouted per day. Chip samples could only be collected for 60% of drilled meterage on average. There was a widespread loss of sample below the water table and in areas where air loss occurred in the cavities associated with the leached dolomite residuum

Based on previous experience of compaction grouting contracts it was assumed that unless a cavity or a wad filled cavity was intersected during drilling then grout takes would be relatively low or in the region of the measurements predicted at the tendering stage.

Results of the actual grout takes were plotted on the borehole layout plan shown in Figure 2 and zoned according to the total grout takes recorded for each borehole. It is interesting to note that the zoning pattern shown on Figure 2 implies a configuration of typical subsurface dolomite morphology, viz. wad filled cavities and irregular dolomite rockhead as is normally found in the West Rand (Brink 1981).

The distribution of boreholes considered to have 'low grout takes' is coincident with areas where Karoo conglomerate rock extends to depths of between 16 m and 20 m. This formation does not contain voids and the residual soils are relatively less compressible when compared to dolomite residuum. Similar grout take patterns were also recorded in areas where the dolomite residuum largely comprises

abundant chert rubble with minor or no wad. This material is distributed across the site, but is more prevalent at the western and southern extremes of the treated area.

Areas in which boreholes recorded total grout takes between 12 m^3 and 20m^3 have been classified in the zone of 'moderate grout takes'. The geological profiles of the boreholes drilled in this zone indicate that the Karoo rocks are generally not as thick as in the zone of low grout takes. In addition the underlying dolomite residuum comprised alternating subhorizons of abundant chert with minor wad and subhorizons of abundant wad with minor chert. The latter sub horizons were responsible for up to 50% of the grout takes in this zone and are usually located below 20 m from the ground surface.

A third zone defined by areas where boreholes recorded grout takes in excess of 20 m^3 has been classified as a zone of 'high grout takes'. Grout takes of up to 33 m^3 were recorded in this zone. The typical geological profile in this zone comprised a surfcial cover of fill, transported material and Karoo conglomerate up to 10 m thick underlain by a wad dominated dolomite residuum. Drilling in this zone was characterised by extensive sample and air loss, which are indicative of the presence of cavernous ground. Over 75% of the grout takes recorded in this zone are confined below 20 m depth measured from the ground surface.

Twenty three secondary boreholes were drilled and grouted. The function of the secondary boreholes was twofold. Firstly to ensure that adequate compaction and filling of cavities and wad filled cavities was attained in areas where the specified grouting pressures were not reached in primary boreholes and secondly, as a measure to evaluate the effectiveness of the grouting completed in the adjacent primary boreholes.

5 VERIFICATION AND CONCLUSION

A variety of methods are available for measuring the performance of grouting either directly or indirectly. These methods are used to determine a change in some property of the subsurface after grouting or to detect the presence of grout. Such methods are generally indirect and require interpretation of the desired information from some other measured property. The intrusive methods can either collect samples for viewing or other analysis, or measure some in-situ properties. Some methods can be used while grouting is underway and others are sensitive to disturbances caused by ongoing construction and can only be used when all operations have stopped.

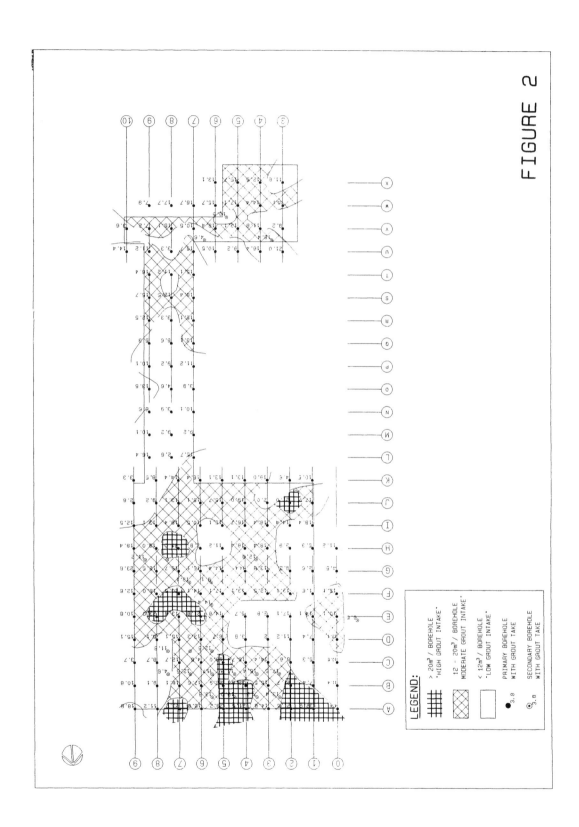

FIGURE 2

Due to the budget constraints and the lack of experience in conducting and interpreting results obtained from techniques such as seismic and other geophysical remote sensing in compaction grouting, information obtained from drilling and grouting of secondary boreholes was used to verify the effectiveness of compaction grouting.

Analyses of the secondary boreholes show that the grout takes below 20 m are in the order of 80% to 90% of those recorded in the adjacent primary boreholes. This is particularly true in the zone of 'high grout takes' and indicates that slightly less than the desired compaction was attained below 20 m in this zone. However significant decreases of up to 95% in the intake of grout were recorded between depths of 12 m and 20 m in the same boreholes, which indicates that considerable compaction has been attained at these depths.

Compaction grouting has become an acceptable method of ground improvement over the past decades. It is still a relatively more expensive option for most conventional ground densification treatments. However in instances such as that discussed in this paper it provides a viable alternative to the more commonly used methods. The combined use of rational and empirical approaches in design has contributed significantly towards acceptance of this method as a ground improvement alternative in development of sites underlain by dolomite.

6 REFERENCES

ASTM C938 1991, *Standard practice for proportioning grout mixtures for preplaced-aggreagate concrete* Philadelphia: American Society for Testing and Materials.

Bowen R 1981, *Grouting in Engineering Practice"* 2nd Edition; Applied Science Publishers Ltd, London.

Brink A.BA 1981. Engineering Geology of Southern Africa, Silverton, Pretoria.

Brown D.R. and Warner J. 1973. *Compaction Grouting.* Journal of the Soil Mechanics and Foundations Division, ASCE, Vol. 99, No. SM8, Proc. Paper 9908, August 1973.

Byle M.J. and Borden R.H. 1995. *Verification of Geotechnical Grouting.* Geotechnical Special Publication No 57, ASCE, October 23-27, 1995.
Nonveiller E (1989) *"Grouting Theory and Practice"*, Developments in Geotechnical Engineering, 57 Elsevier Science Publishers.

Graf ED (1992), *Compaction Grout* 1992, Grouting, Soil Improvement and Geotsynthetics, RH Borden ed., ASCE Geotechnical Special Publication No 30, pp 275-287.

Schmertmann J.H. and Henry J.F. 1992. *A Design Theory for Compaction Grouting.* Grouting, Soil Improvement and Geosynthetics, R.H. Borden ed., ASCE Geotechnical Special Publication No. 30 pp 275 to 287.

Stilley AN 1982, *Compaction Grouting for Foundation Stabilisation* Proceedings of the Conference on Grouting in Geotechnical Engineering, WH Baker ed., ASCE pp 923-937.

Vesic, AS, 1972 *Expansion of Cavities in Infinite Soil Mass*, Journal of the Soil Mechanics and Foundations Division, ASCE, Vol. 98 No. SM3, Proc. Paper 8790 pp 265-290.

Warner J 1982, *Compaction grouting – the first thirty years,* Proceedings of the Conference on Grouting in Geotechnical Engineering, WAH Baker ed., ASCE pp 694-707.

Warner J. 1992. *Compaction Grout; Rheology vs Effectiveness.* Grouting Soil Improvement and Geotsynthetics, RH Borden ed., ASCE Geotechnical Special Publication No 30 pp 229-239.

Warner J. and Brown D.R, 1974 *Planning and Performing Compaction Grouting,* Journal of the Geotechnical Engineering Division, ASCE Vol. 100, No. GT6, Proc. Paper 10606, June 1974, pp 653-666.

Geotechnics for Developing Africa, Wardle, Blight & Fourie (eds) © 1999 Balkema, Rotterdam, ISBN 90 5809 082 5

Soft soil treatment using stone columns

Y. E-A. Mohamedzein, A. M. El Sharief, M. A. Osman & H. A. E. Ismail
Building and Road Research Institute, University of Khartoum, Sudan

ABSTRACT: The Construction of a high approach embankment (6.5 to 10.25m high) on soft soil was facilitated by the installation of stone columns. The soft soil consisted of highly plastic clay and silt with an undrained shear strength as low as 4 kPa. The thickness of the soft soil ranged from 2.5 to 8.5m. The stone columns were installed through the soft soil by vibro replacement techniques. The stone columns were anchored about 0.5m into the underlying dense to very dense sand, sandstone or mudstone. The quality of the constructed stone columns was evaluated by insitu dynamic penetration tests. The overall performance of the treated soil under the embankment load was evaluated using insitu instrumentation such as settlement plates.

1 INTRODUCTION

Stone columns are used to increase the load carrying capacity of soft clays. The stone columns carry a significant portion of the applied load especially during the initial stages of construction. The stone columns also act as drains and increase the rate of consolidation of clays and hence increase the shear strength of clays. Stone columns have been used to support different structures on soft clays, e.g. highway embankments, buildings and storage tanks (Poorooshasb & Meyerhof 1997, Holtz, 1989, Som, 1995).

Analytical methods have been developed to analyze the problem of stone column-soil interaction (e.g. Poorooshasb & Meyerhof 1997, Poulos & Davis 1980, Balaam & Booker, 1981). The validation of analytical methods requires model testing and monitoring of structures as built. This paper summarizes a case history of a high bridge approach embankment constructed on soft silty clay alluvial deposits in Omdurman, Sudan. To the authors best knowledge this is the first site in Sudan where stone columns technology has been used.

2 PROJECT DESCRIPTION

The New White Nile Bridge is intended to link the two cities of Khartoum and Omdurman. The bridge is about 700m long and crosses the main permanent stream course of the White Nile River. An additional approach embankment of about 500m long is proposed to cross the seasonal flood plains on the Omdurman side. The embankment is 6.5 to 10.25m high and about 30m wide at the crest. The side slopes range from 1:1.5 to 1:2.

3 SOIL PROFILE

The site investigation consisted of the drilling of 9 boreholes and 28 cone penetration tests (CPT). The soil consists of alternating layers of highly plastic silt and medium to highly plastic clay. These formations extend from the ground surface to depths ranging from 2.5 to 8.5m. The formations are soft with undrained shear strength as low as 4 kPa and natural moisture content well above the plastic limit and approaches the liquid limit in few cases. The measured properties of the soft soils are shown in Table 1. Below the soft deposits dense or very stiff sandstone, mudstone or sand were identified.

4 DESCRIPTION OF STONE COLUMNS

Stability and settlement analyses have shown that the soft deposits cannot support the proposed embankment. The factor of safety against slope failure is less

Table 1. Properties of Soft Soil

Property	Range	Average	Standard Deviation	Coefficient of variation (%)	No. of observations
Thickness (m)	2.5 to 8.5	6.0	2.2	36.7	9
Natural moisture content(%)	20 to 64	47	10.8	23.0	44
Liquid Limit(%)	42 to 65	54	6.4	11.8	39
Plasticity Index(%)	15 to 40	26	5.0	19.4	39
%passing sieve#200	80.2 to 99.7	94.2	6.57	6.9	16
Undrained shear strength (KN/m2)	4.2 to 65.1	17.5	17.25	98.6	13
Long term shear strengthϕ(deg)	27 to 32	29.25	2.2	7.5	4
OCR(%)	1 to 2.22	1.42	0.54	37.9	7
Cc	0.37 to 0.927	0.541	0.189	35.0	7
eo	1.177 to 2.325	1.674	0.343	20.5	7
tip resistance* (KN/m2)	200 to 1607	784	377.8	48.2	20
Skin friction*(KN/m2)	45 to 160	91.3	33.2	36.4	20

*From CPT soundings

Table 2. Physical properties of stone columns and embankment

Region	Thickness of soft soils (m)	Length of stone columns (m)	Spacing (m)	Average Proposed Embankment height (m)
Ia	7.5 to 8.5	8.5	2.25	10.278
Ib	7.5 to 8.5	8.5	2.8	9.07
II	5.5 to 7.0	7 to 8.5	3.25	7.5
III	4 to 5.5	4.5 to 6	3.5	7
IV	2.5 to 4	3 to 4	3.8	6.5 to 6.7

than 1 (about 0.79). The consolidation settlement for a 6-m high embankment is greater than 60 cm. This settlement was estimated using one dimensional consolidation theory.

Among all options for soil treatment and modification methods, the owner elected to use stone columns to modify the soft clay. Table 2 lists a summary of the physical dimensions of the stone columns. A total of 2376 stone columns were constructed. All stone columns were 91 cm in diameter. The length of an individual stone column ranged from 3 to 8.5m. The spacing between stone columns varied from 2.25 to 3.8m. The total length of all stone columns installed was 14365m. The majority of stone columns were constructed using natural gravel. In some stone columns a mixture of natural gravel and crushed stone was used. Construction of stone columns took about 4 months (from Jan. 26, 1998 to May 30, 1998).

The method used for construction of stone columns followed standard practice (Holtz 1989). The borehole was advanced to the appropriate depth using displacement techniques. Water pressure of 400 to 600 kPa was used to displace the soil. The hole was cleaned twice before gravel was poured in layers

(about 50 cm in thickness) into the hole. Each gravel layer was well compacted. The quality of compaction was checked by measuring the electric current transmitted to the vibroflotation machine. Further confirmation of the quality of stone columns was made by conducting sounding (SPT) tests at selected stone columns.

After installation of all stone columns, the ground was leveled to the design grade elevation. A sand cushion was then spread on the leveled ground. The sand cushion was about 50 cm in thickness and was placed in layers and compacted. The sand cushion provides an upper drainage layer to drain water away from clay.

After the placement of the sand cushion construction of embankment was started. By the end of July 1998 about 4m of embankment material was placed all over the site. Construction of embankment was stopped to allow for construction of abutment of the bridge. Embankment construction will resume in 1999.

5 QUALITY ASSURANCE, INSTRUMENTATION AND RESULTS

Six settlement plates were installed about 0.5m below the design subgrade level and before the construction of the sand cushion. Steel pipes connected to the settlement plates were extended through the sand cushion and embankment to allow for regular reading of the movement of the settlement plates. Typical time history of settlement as recorded by the settlement plates is shown in Figure 1. The figure indicates for a 4m high embankment the primary consolidation took about 150 days and the amount of settlement was 20 cm. Using one-dimensional consolidation theory the estimated settlement for a 4m

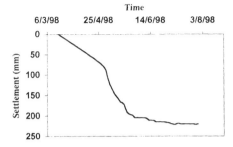

Figure 1. Settlement curve.

high embankment without stone columns is about 40 to 50 cm. Thus the performance ratio is about 0.4 to 0.5 (the performance ratio is the ratio of the settlement of treated soil to that of the untreated soil under identical surcharge (Poorooshasb & Meyerhof 1997)).

It is impressive to compare the measured results with the prediction of the analytical procedure developed by Poorooshasb & Meyerhof (1997). In that study a chart showing the performance ratio versus spacing parameter is used to estimate the performance ratio from the spacing parameter and the properties of stone column material (e.g. gravel). The spacing parameter is defined as the distance between adjacent rows divided by the diameter of the stone column. In this manner the spacing parameter defines the influence of the stone column. In this project the spacing parameter is about 2. The properties of the granular material (gravel) can be reasonably estimated as : $\phi = 41°$, strain at failure, $\varepsilon = 5\%$. Thus using the chart in the above study, the performance ratio (PR) is about 0.4.

The amount of settlement can be computed from the equation:

$$PR = \delta\, E\, /\, (H*UDL) \qquad (1)$$

where δ = settlement

E = Young's Modules of soil (recommended value for soft soils by Poorooshasb and Meyerhof (1997) is E = 1000 kPa)

H = Thickness of Soft Soil

UDL= Uniformly distributed applied load.

For the case considered here H= 8.5m, E=1000 kPa and UDL = 80 kPa. Based on these values δ =27.2 cm. This value is in reasonable agreement with the measured value. Better estimate can be obtained if the value of the Young's Modulus of the soft soil and the properties of the gravel (e.g. ϕ, ε) are properly evaluated.

Cone Penetration Tests (CPT) were performed before and after the installation of the stone columns. The latter CPTs were performed about 15 to 30 days after construction of the stone columns. Comparison between CPT results before and after installation of stone columns is shown in Figure 2. The figure indicates no improvement in soft soil shear strength at most depths. It should be noted, however, that significant improvement in shear strength is not expected at this stage of construction because the loading process by radial pressure from stone column is only significant on the immediate vicinity of the stone column. Also the time allowed for consolidation is too short.

Undisturbed soil samples were taken from boreholes before and after installation of stone columns and tested in the laboratory. The results of consolidation tests are shown in Figure 3. The figure indicates that the compressibility (Cc/1+eo) of the soil decreases from 0117 to 0.091. This apparent improvement is attributed to the relatively long waiting period (about 40 days) between installation of stone columns and drilling of the second borehole.

6 CONCLUSIONS

The paper presents a case of an approach embankment constructed on soft soil deposits reinforced by stone columns. Based on the one dimensional consolidation theory the estimated consolidation settlement for a 4m embankment, without stone columns, is about 40 to 50 cm. The measured settlement of the 4m embankment after the installation of the stone columns is 20 cm. The stone columns reduced the

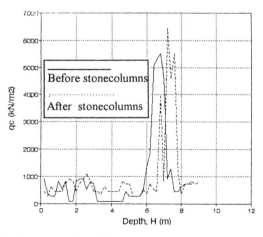

Figure 2. Results of CPT.

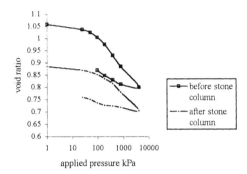

Figure 3. Consolidation curves.

consolidation settlement of the soft soil by about 40 to 50%. The amount of the shear strength gain in the soft soil depends on time allowed for consolidation after the construction of the stone columns. Decrease in the compressibility of soft soil was noted after a relatively long period of time following the stone column installation.

ACKNOWLEDGMENT

The owner of this project is the ministry of Engineering Affairs, Khartoum State, Sudan. The stone column contractor was Jilin International Company (JIC) of China. Dar Consultants of Khartoum and Building and Road Research Institute (BRRI) of University of Khartoum supervised the work. The help of all these parties is gratefully appreciated.

REFERENCES

Balaam, N.P. & J. R. Booker 1981. Analysis of rigid raft supported by granular piles. International Journal of Numerical and Analytical Methods in Geomechanics, Vol. 5, No. 4, pp. 397-403.

Holtz, R. D. 1989. Treatment of problem foundations for highway embankments. NCHRP, Synthesis of Highway Practice No. 147, TRB, Washington, D. C., USA.

Poorooshasb, H. B. & G. G. Meyerhof 1997 Analysis of behavior of stone columns and lime columns. Journal of Computers and Geotechnics, Vol. 20, No.1, pp. 47-70.

Poulos, H. G. & E. H. Davis 1980. Pile Foundations: analysis and design. John Wiley and Sons.

Som, N. 1995. Consolidation settlement of large diameter storage tank foundations on stone columns in soft clay. Proceedings of the International Symposium on Compression and Consolidation of Clayey Soils, Hiroshima, Japan, pp.654-658.

Geotechnics for Developing Africa, Wardle, Blight & Fourie (eds) © 1999 Balkema, Rotterdam, ISBN 90 5809 082 5

Behaviour of Con-Aid treated fine grained Kenyan soils

J. N. Mukabi
Construction Project Consultants Incorporated, Nairobi, Kenya

P. A. Murunga
Construction Project Consultants Incorporates, Malindi, Kenya

J. H. Wambura & J. N. Maina
Materials Testing and Research Department, Ministry of Public Works and Housing, Kenya

ABSTRACT: The engineering characteristics of soils such as black cotton in Kenya are known to be complex and problematic in the sense that their behaviour cannot be satisfactorily interpreted on the basis of conventional concepts that characterise soils in temperate regions. In most cases therefore, they have been disregarded and discarded as useless construction materials. In this study, the effects of a liquid chemical stabilizer, Con-Aid, on some of the problematic microstructural related properties and engineering behaviour of black cotton soil are examined and an attempt made to analyse the concomitant characteristics.

1 INTRODUCTION

Conventional road construction methods are usually expensive particularly for developing countries. As a consequence, infrastructure development is grossly curtailed both in terms of the quality of work, and length of road construction and related maintenance. On the other hand , a vast portion of inhabited areas in Kenya for example, are dominated largely by tropical soils of residual expansive clay characteristics. These soils are known to be highly susceptible to pore grain moisture~suction variations characterised mainly by swell due to wetting and shrinkage due to drying. This precipitates large reduction or gain in strength respectively. A group of such soils have also been known to show changes in plasticity characteristics upon drying; attributed mainly to the alteration of the clay minerals on dehydration and aggregation of fine particles to form larger particles due to the presence of cementing agents and/or pore fluid~grain particle interaction. It is considered that due to the drying, large suction pressures are developed, and hence small aggregates are drawn together to form large aggregates (Pandian et. al. 1993). Newill (1961) and Frost (1967) identified gibbsite and allophane found in such soils that develop over volcanic materials as the secondary minerals that are most susceptible to drying. This contributes to the moisture~suction variations in such soils. Prevalence of such changes in the post construction stage can be extremely detrimental to the stability and bearing capacity of the structure, mainly related to deformation and cracking. Due to such character-

istics, expansive soils are usually rejected as construction materials despite the fact that they occur widely in arid regions. It is usually desired therefore, that such in-situ material be economically stabilized to contain the moisture-suction condition and the related variables, in order to effectively adopt them for construction.

Consequently, this study aims at exploring one of the possibilities of so doing by examining the effect of Con-Aid on the basic engineering behaviour of two types of black cotton soils sampled from two different borrow pits along the Tana Basin Road Development Project along the Malindi~Garsen (B8) road, currently under construction in Kenya.

2 MATERIALS TESTED AND TESTING PROGRAMME

2.1 Materials tested

The black cotton soils adopted in this study were dark grey and plain grey in colour. They were sampled from borrow pits located at km91+900 to 92+200 (sample A) and km217+800 RHS (sample B) respectively. This area is predominantly of very flat terrain with elevations of between 30~70m above mean sea level. Table 1 is a summary of the basic properties of the two soils.

The stabilizing agents used were hydrated lime with pozzolanic reaction capability and Con-Aid CBR plus, a surface active agent which supposedly takes advantage of the ion exchange capacity of fine grained soil minerals to change a material from the hydrophilic (water absorption) nature to a hydrophobic (total water repellent) character thereby achieving higher strength and stabilization as well as aiding in compaction.

Table 1	Average values	
Soil parameter	Km91+900 ~ 92+200 (A)	Km217+ 800 RHS (B)
Liquid limit, LL (%)	84	65
Plasticity index, PI (%)	37	34
Natural Moisture Content, ω_c %	20.7	18
Linear shrinkage, LS (%)	18.5	17
Percent passing 0.475mm	84	75
Specific gravity, G_s	2.456	2.51
Compaction dry density γ_d (kg/m³)	1495	1577
Compaction moisture content OMC (%)	23.6	18.0
Maximum dry density M.DD (kg/m³) (AASHTO T99)	1378	1665
Optimum moisture content O.M.C (%) (AASHTO T99)	24.5	19.7

2.2 Testing program

Upon sampling and determining basic in-situ parameters, both soils were thoroughly air dried and crushed into small fine particles prior to storage under virtually constant room temperature conditions. For soil A, one batch of untreated and three batches of Con-Aid, Con-Aid +2% lime and Con-Aid +5% lime treated materials were prepared.

Soil B was prepared in five batches of untreated, Con-Aid, 3%, 4% and 5% lime treated. The lime treated materials were initially thoroughly mixed in a dry state before addition of water, whereas predetermined quantities of Con-Aid were diluted with clean water (pH<8) in optimum ratios and thoroughly stirred by using a mixer prior to adding the solution to the soils and subjecting the paste to further mixing. Subsequent to recording the texture and physical properties of the batched materials, moisture contents were monitored as the bulk materials were cured with water for at least 24 hours, re-mixed for homogeneity whilst moisture content variations and Atterberg Limits were measured for close monitoring purposes. Based on these results, either the curing process would be continued or compaction carried out. Optimum states to achieve MDD~OMC relations that would ensure fairly similar moulding properties were then determined. The necessary parameters were varied as required and standard compaction, CBR and unconfined compression tests performed. The detailed testing method is described in the Con-Aid R&D Proposal by Mukabi (1998).

3 DISCUSSION OF TEST RESULTS

3.1 Plasticity characteristics

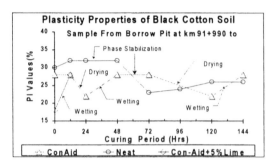

Fig. 1 Effect of Curing period on drying and wetting cycles in relation to PI characteristics.

The test results depicting the combined effects of curing period and drying/wetting cycles on the plasticity characteristics of untreated and treated samples are shown in Fig. 1. It can be seen from this figure that there was an immediate drop in the plasticity index upon treatment of the material from 30 to 28 for Con-Aid and 30 to 18 for Con-Aid + 5% lime. The drastic drop in PI of the Con-Aid + 5% lime treated material may be as a result of a relatively high development of capillary stresses. This can be partially confirmed from the fact that upon wetting after the first 12 hours of curing, the PI increased from 18 to 28. This increase may be reciprocal to the high suction stresses created by capillarity. This shows that the particle aggregation may not have acquired a stable state due to the short curing period. It can also be observed that the plasticity characteristics of the untreated material are erratic whereas those of the Con-Aid treated material show steady recovery to the initial plasticity states in both upper and lower bounds through wetting and drying and are little influenced by the curing period for up to 144 hrs.

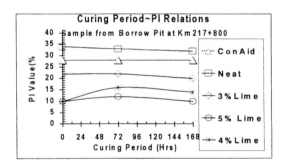

Fig. 2 Effect of curing period on the PI characteristics.

Fig. 2 shows the effect of curing period on the PI characteristics of sample B material that was stored under quasi-constant room temperature. The materials treated with 4% and 5% lime show the most significant drop in PI, from 34 to 10 prior to curing, followed by 3% lime, Con-Aid treated and with the untreated material exhibiting the highest PI even after 168 hrs of curing under air-dried room conditions. Again, it can be noted that after the initial drop in PI, the plasticity characteristics of the Con-Aid treated material become insensitive to the period of curing.

To simulate the effects of rapid drying on the PI characteristics of Con-Aid treated material, a comparison of air and oven (110^0C) dried untreated and treated material was carried out. Fig. 3 is a plot of these results. Whereas the untreated material is characterised by a significant reduction in PI with curing period upon rapid drying, the plasticity characteristics of the Con-Aid treated material are neither influenced by rapid drying effects nor curing period and maintain the same magnitude of PI (28) as in previous cases.

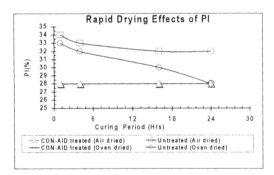

Fig. 3 Rapid drying effect vs. curing period PI characteristics.

Since Plasticity Index (PI) is a measure of the plasticity of a soil in terms of the amount of water which must be added to change a soil from its plastic limit to its liquid limit and the liquid limit, which was observed to be practically constant for Con-Aid treated material, represents the water content at which a soil has such a small shear strength of definite value, it would imply that for Con-Aid treated material, the suction stress that would induce the moisture variation to reduce the material strength to its minimal may be constant under otherwise similar conditions. The linear shrinkage was also observed to be constant and insensitive to curing period after an initial drop from LS = 16% to LS = 14%. This may mean that the water content that is just sufficient to fill the pores when the soil is at it's minimum volume upon

drying, would be otherwise constant for Con-Aid treated material.

From these results it may also be deduced that Con-Aid did not evaporate upon relatively high temperature (110°C) heat effects. This is probably due to the fact that it had penetrated into the clay mineral almost immediately in this case. This may also indicate that Con-Aid's action on the absorbed water and clay particle dispersion had already occurred.

It may seem therefore that Con-Aid has the effect of initially reducing the plasticity index of a material and subsequently stabilising it such that it is only influenced by gradual drying upon phase stabilisation of the microstructure and pore geometry and that this occurs without exceeding the magnitude of its initial PI notwithstanding the hysterisis of wetting and drying. This may imply a stable Con-Aid vs. water interaction and that the interpretation, relation and characterisation of variables related to plasticity against those of moisture-suction changes may therefore be achieved much more easily for Con-Aid treated material as opposed to the untreated one.

3.2 Compaction characteristics

Figs. 4a~4e show the time effects on the compaction characteristics for untreated, Con-Aid, 3%, 4% and 5% lime treated materials respectively. In this case, treatment was carried out and materials cured in bulk prior to being subjected to single compaction. The respective unsoaked CBRs are also indicated. The following observations can be made from these results: -

a) The compaction characteristics of Con-Aid treated material are steady and consistent with curing period in comparison to those of the other materials tested. The time dependent density-moisture content relation may therefore be empirically characterised by the following formulae proposed in this study.

$$\beta_{ct} = \Delta(\gamma_d)_t / \Delta(\omega_c)_t \bullet e^{(\delta\gamma - \delta\omega)t} \qquad (3.1)$$

where β_{ct} is the proposed compaction time factor and $\delta_\gamma, \delta_\omega$ are constants determined from density~time and moisture content~time curves respectively.

$$\phi_{ct} = \Delta(\gamma_d)_t / \Delta(\omega_c)_t = \{(\gamma_d)_{t2} - (\gamma_d)_{t1}\} / \{(\omega_c)_{t1}$$
$$- (\omega_c)_{t2}\} \qquad (3.2)$$

$$\delta_\gamma = \Delta\gamma_d / \Delta_t = \{(\gamma_d)_t - (\gamma_d)_i\} / (t_t - t_i) \qquad (3.3)$$

$$\delta_\omega = \Delta\omega_c / \Delta_t = \{(\omega_c)_i - (\omega_c)_t\} / (t_t - t_i) \qquad (3.4)$$

From these parameters, time t_c (hrs) required to achieve a maximum dry density to suit the design standards may be calculated based on the results of

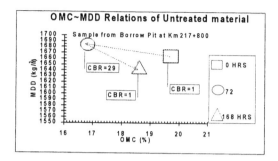

Fig. 4a Time effects on compaction characteristics of untreated material

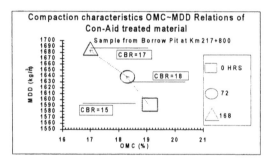

Fig. 4b Time effects on compaction characteristics of Con-Aid treated material.

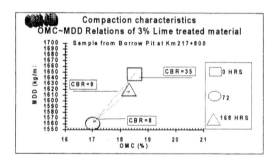

Fig. 4c Time effects on compaction characteristics of 3% Lime treated material.

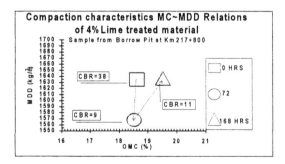

Fig. 4d Time effects on compaction characteristics of 4% Lime treated material.

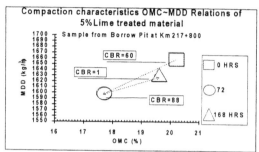

Fig. 4e Time effects on compaction characteristics of 5% Lime treated material.

compaction tests performed on 0-day curing period in order to save testing and consequently design as well as construction time.

An example of this calculation is as follows:-

$$(\gamma_d)_{req} - (\gamma_d)_i = \delta_\gamma t / \beta_{ct} \quad (3.5)$$

where $(\gamma_d)_{req}$ is the required design maximum dry density to be achieved during compaction and $(\gamma_d)_i$ is the initial MDD determined on the day of moulding the specimens in the laboratory. Given that $\delta_\gamma = 0.565$ (determined in this study), $\beta_{ct} = 1.265$ and the required and initial maximum dry densities are $(\gamma\delta)_i = 1800 kg/m^3$ and $1590 kg/m^3$ respectively, then,

$$t_c = \Delta\gamma_d / \delta_\gamma \beta_{ct} = 210/0.565 \times 1.265 \quad (3.6)$$
$$= 470 \text{ hrs (plotted in Fig. 16)}$$

This means that if a bulk layer of similar material treated with Con-Aid is cured continuously for about 20 days and subsequently compacted, a maximum dry density of $1800 kg/m^3$ may be achieved.

To calculate the corresponding optimum moisture content, the following method may be applied:
From equation 3.2

$$\phi_{ct} = \Delta(\gamma_d)_t / \Delta(\omega_c)_t$$

$$\therefore \Delta(\omega_c)_t = \Delta(\gamma_d)_t / \phi_{ct} \quad (3.7)$$

Incorporating the time element functions

$$\Delta(\omega_c)_t = \{(\omega_c)_i - (\omega_c)_f\} \delta_\omega / \beta_{ct} \quad (3.8)$$

Substituting $\Delta(\omega_c)_t$ from equation (3.7)

$$\Delta(\gamma_d)_t / \phi_{ct} = \{(\omega_c)_i - (\omega_c)_f\} \delta_\omega / \beta_{ct} \quad (3.9)$$

Rearranging equations (3.6) and (3.7),

$$(\omega_c)_t = (\omega_c)_i - (t_c / \phi_{ct}) \delta_\omega / \beta_{ct} \quad (3.10)$$

$t_c = 470$ hrs, $(\omega_c)_i = 19.2\%$, $\delta_\omega = 0.655$, $\phi_{ct} = 44$ (determined from this study).

Substituting these values,

$(\omega_c)_f = 19.2 - (470/44) \times 0.655 / 1.265 = 13.5\%$

Hence the corresponding optimum water content would be 13.5% (plotted in Fig. 17). $\omega_c \sim \gamma_d$ values calculated on the basis of the proposed formulae are plotted in Fig. 18. A good agreement between the calculated values and those determined experimentally can be seen.

b) Although the rate of strength development is minimal for the Con-Aid treated materials, it can be significantly noted that a steady reduction in the OMC and gain in MDD is apparent. On the contrary, the other materials seem to exhibit erratic characteristics. In other words, the compaction characteristic curve of the Con-Aid treated material seems to shift steadily with time from the wet zone to the dry zone of high density and low moisture content.

The effect of curing period on moisture content movement for the five different materials is shown in Fig. 5. Con-Aid treated material shows a virtually linear and consistent reduction in the optimum moisture content with time. The other materials show a tendency of a decrease in the OMC up to a curing period of 72 hrs, followed by an increase.

On the other hand, a plot of the curing period vs. maximum dry density depicted in Fig. 6 shows a steady and virtually linear increase in density with time for the Con-Aid treated material and a slight increase and subsequent reduction for the untreated

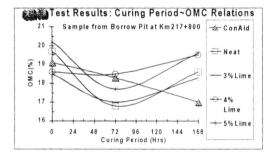

Fig. 5   Curing period effect on moisture content.

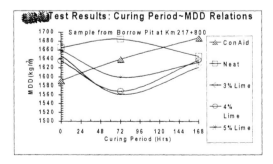

Fig. 6   Curing period effect on maximum density movement.

material. The characteristics observed in both Figs. 5 and 6 may be attributed to the arbitrary effect of drying and lime treatment on the pore geometry, particle orientation, and inter-particle contact associated with meniscus effect due to one or a combination of the following reasons: -

(i) For the first 72 hrs curing period, lime may have undergone a chemical reaction with pore water in the soil causing volumetric increase through swelling which initially translated into the reduction in density $(\gamma_d)_{max}$. If this reaction is inversely proportional to the lime content (since part of the higher lime content would contribute to particle aggregation or hydration), then the characteristic exhibited in Fig. 6 would justify this theory. As can be seen, the largest reduction in $(\gamma_d)_{max}$ was by the 3% lime stabilized material $\{(\gamma_d)_f - (\gamma_d)_i\} = 86 kg/m^3$ while the lowest was by the 5% lime treated material whose $\Delta\gamma_d = 60 kg/m^3$. The subsequent increase in $(\gamma_d)_{max}$ can be attributed mainly to the particle aggregation, hydration and to some extent carbonation.

(ii) The autogeneous interaction of hydration and drying shrinkage.

The compaction characteristics may therefore lead to the following observations: -

(a) The inconsistency between strength development and densification may be due to the different curing effects on the bulk materials and specimens; since compaction tests were performed on bulk materials and strength tests on specimens that had been initially compacted and cured in moulds for the same period as that of the bulk.

(b) The erratic behaviour of the other materials may also be due to inhomogeneity of the specimens prompted by capillarity. This may have caused water to travel more rapidly in small pores where the pull is stronger than in the large pores. From the observations made of the steady and consistent behaviour of Con-Aid treated material, this would essentially mean that Con-Aid may have, through the hydrophobic and hydrophillic action, concentrated around the clay mineral/water layer interfaces and stabilised the pore geometry of the material or its capillary pore system and that this action is highly time dependent.

(c) The fact that the initial maximum density of Con-Aid treated material is achieved at a higher OMC than for the other materials treated with lime means that Con-Aid's effect on the capillary pore system may be weaker than that of the lime. However, the magnitude of its effect develops with curing time and compaction effort (compaction effort of AASHTO T180 on Con-Aid treated material showed that it gained in maximum dry density (MDD) from 1448 to 1640 kg/m³ and reduced in optimum moisture content (OMC) from 25.3% to 18.6%) almost immediately.

3.3 *Strength characteristics*

The effect of curing period on the unsoaked CBR strength of Con-Aid treated, lime treated and untreated material for samples A and B is shown in Figs. 7 and 8 respectively, while the normalised (CBR/γ_d) characteristics of the latter are depicted in Fig. 9.

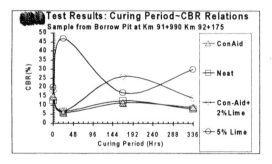

Fig. 7   Effect of curing period on unsoaked CBR

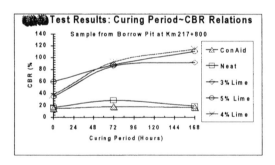

Fig. 8   Curing period Vs. CBR relations.

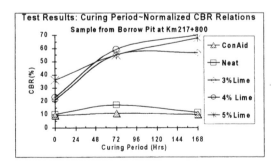

Fig.9 Curing period effect normalized CBR(CBR/γ_d)

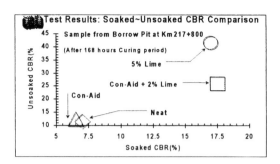

Fig. 10   Soaked vs. unsoaked CBR.

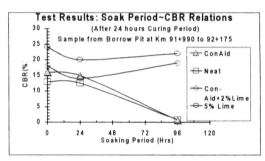

Fig. 11   Effect of soaking period on soaked CBR.

It can be seen from Fig. 7 that whereas the 5% lime treated material exhibits significant gain in CBR strength within the first 24 hrs, the Con-Aid treated, untreated and material treated with a mixture of Con-Aid + 2% lime all show a drop in CBR strength to virtually the same value. On the contrary, the 5% lime treated material shows a drastic drop in CBR strength from 47 to 17% between a curing period of 24 and 168 hours. On the other hand, the Con-Aid + 2% lime treated material gains from 6.5 to 26% within the same period whereas the other two show very mild gains in CBR strengths during this period.

Fig. 8 clearly shows that the lime treated materials have much larger strengths than the Con-Aid treated material. It can also be observed that Figs. 8 and 9 exhibit similar characteristics. This implies that the effects of density variations were minimal and did not have direct bearing on the anomaly in CBR strengths. In both cases, it can be seen that the Con-Aid treated material exhibits virtually no gain in strength over the curing period. This could possibly be attributed to its weak effect on capillarity and persistent interaction with water in the absence of continuous compaction.

A comparison of soaked and unsoaked CBR and the effect of soaking period on soaked CBR in relation to the behaviour of untreated, Con-Aid, Con-

Aid + 2% and 5% lime treated materials are depicted in Figs. 10 and 11 respectively. The 5% lime treated material shows the highest ratio (2.44) of the unsoaked/soaked CBR, followed by the Con-Aid treated (1.92), while the Con-Aid + 2% treated material has the least unsoaked/soaked CBR strength ratio of 1.49. This shows that the effect of soaking is more apparent for lime and Con-Aid treated materials. For lime treated material this may be attributed to the effect of the magnitude of the capillary pore system while for the Con-Aid treated material it is considered to be possible higher particle aggregation due to drying. This particle aggregation may lead to the creation of larger voids ratios which create pores for more water accommodation thereby causing volumetric increase.

For the untreated material, the difference in unsoaked and soaked CBR is considered to be purely due to the weak capillary stresses created by gradual drying and possibly due to slight inter-particle attraction resulting from the same.

The material treated with a mixture of Con-Aid plus 2% lime shows the best-soaked and unsoaked strength characteristics. This may imply that the interaction of the two stabilisers can complement each other in the strength development (ref. to Fig. 11) related to the components of capillary pore system, inter-particle attraction, pore geometry and moisture-suction matrix stabilisation. This may advantageously be applied in curtailing significant differences in soaked and unsoaked CBR strengths. This compound effect of Con-Aid + Lime may be explained as follows: -

(a) The initial unsoaked CBR strengths of Con-Aid (CBR = 15%), Con-Aid + 2% (CBR = 18%) compare well with that of 5% lime (CBR = 20%) treated material implying that all stabilizers may have displaced some absorbed water molecules through the ionic exchange. Their reaction with absorbed water within the capillaries is assumed to take place faster since it is not physically bound to the clay minerals.

(b) The next phase of strength reduction of Con-Aid +2% lime treated material within the following 24 hrs curing period is due to the fact that the initial process could have induced autogeneous relative humidity gradients within the clay particle structure so that the transfer of water molecules from the large surface area of calcium silicate hydrated into empty capillaries, whilst the Con-Aid's continued reaction with the absorbed water molecules was taking place simultaneously.

(c) The subsequent phase of strength development between 24 and 168 hrs curing period may have been prompted by the continuous process of

adsorbed water molecule dispersion by Con-Aid. This water could have aided in hydration and the depositing of calcium carbonate ($CaCO_3$) in the voids within the clay minerals resulting into the increase in strength.

(d) Other reactions that may be cited as complementary are :-

- reaction with colloidal silica and colloidal alumina minerals
- Carbonic acid interaction
- Hydration/dehydration cycles etc.

It can also be noted that the soaked CBR for Con-Aid + 2% and 5% lime treated materials are virtually the same while those of pure Con-Aid treated and untreated materials are quite similar in both cases.

The similarity of the soaked CBRs of Con-Aid + 2% and 5% lime treated materials may be due to the fact that the capillary stresses developed by the drying effect of lime were still concentrated within the small pores creating high suction stresses which caused the rapid travel of water to those pores as is evident in the drastic drops in soaked CBR in the first 24 hrs from Fig.11. Since the rates in drop of strength are virtually the same for these two materials, it may be deduced that about 2% of the lime may have carried the small pore component of capillarity. Similarities between soaked and unsoaked CBR strengths of the Con-Aid and untreated materials indicates that Con-Aid reaction in displacing water molecules and creating strength development components is practically non-existent within the span of this curing period or that the pores created by the dispulsion of water molecules by Con-Aid are larger contributing to the vitiation of strength as may be evident from Fig. 11.

Fig. 12 compares the unconfined compression strength of five materials over a curing period of 336 hrs. It can be seen that the magnitude of compressive strength is highest for the 5% lime treated material and that it increases with increasing lime content. The untreated material exhibits the lowest strength

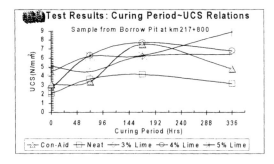

Fig. 12 Effect of curing period on unconfined compression strength.

overall. It can be noted however, that although the UCS of the untreated material (UCS = 5.1 N/mm^2) is higher than that of Con-Aid treated (UCS = 3.11 N/mm^2) material initially, this trend inter-changes with Con-Aid treated material exhibiting almost the highest compressive strength (UCS=7.5N/mm^2) at the 168 hrs curing period but later dropping to UCS= 4.8N/mm^2 at the 336hrs curing period. Despite the fact that it is different in magnitude, this trend of the Con-Aid treated material is analogous to that of the effect of curing period on the CBR strength development depicted in Fig.7. This supports the concept that was earlier mentioned concerning components related to particle movement and aggregation, dispersion, capillarity pore system, etc.

3.4 Swell characteristics

The effect of curing period and soaking period on the swell characteristics of Con-Aid treated, lime treated and untreated material are depicted in Figs. 13~15. Fig. 13 shows that the highest swell differentials were exhibited during the 0-hrs curing periods for all materials and that after 168 hrs of curing, all materials show a tendency of reduction in their swell behaviour. A comparison of the untreated and Con-Aid treated material shows an initial similar value of 6% swell. However, with time, a difference of 2.5% ensues whereby the untreated increases and treated one reduces after the materials had been cured for 72 hrs. After an initial increase in swell of up to 5%, the material treated with a mixture of Con-Aid + 2% lime shows a continuously steady decrease to an approximate 0% value after a curing period of 336 hrs. This may mean that interaction of Con-Aid and lime highly stabilises the moisture-suction characteristics with increasing curing period based on the physical and chemical reactions earlier explained and that for such treated material this variable can be contained after about 400hrs of curing. It can also be noted from the initial increase in

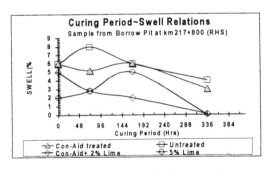

Fig. 13 Effect of curing period on SWELL characteristics.

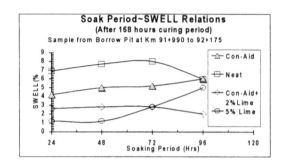

Fig. 14 Soaking period effect on swell characteristics after 168 hours curing period.

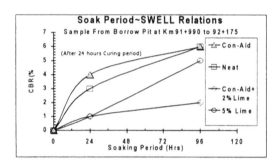

Fig. 15 Soaking period effects on swell characteristics after 24 hours curing.

swell that overall the swell characteristics of the Con-Aid treated material do not vary drastically.

From the soak period-swell relations shown in Fig. 14, it can be seen that:-

(a) Swell characteristics of Con-Aid + 2% lime treated material exhibit the best swell behaviour especially after the initial swell induced in the first 24 hrs with + 0.2% at 72 hrs and a reduction (- 0.8%) after 96 hrs of soaking period. This reduction may be associated with non-uniform water migration or meniscus effect upon reaching a certain limiting value of saturation.

(b) Except for the 5% lime treated material, all the others show drastic swelling characteristics within the first 24 hrs of curing predominantly attributable to their interaction with the capillary pore system. The drastic increase in swell after the first 48 hrs of soaking the 5% lime treated material can be associated with the fact that the capillary pore system reached a limiting state causing the water to flow into the larger pores, the consequence of which was swelling. This may also be associated with its hydration reaction with the pore water. This implies that even after 168 hrs of curing, it is only the Con-

Aid +2% lime treated material that achieves a steady moisture-suction state. The swell characteristics of the same materials cured after only 24 hrs depicted in Fig. 15, further confirm the foregoing discussion.

(c) After a certain period of curing, the swell/shrinkage characteristics with time of these materials may be characterised by a linear relationship (ref. Fig. 13).

3.5 Overall Comparison of results

Although Con-Aid treated black cotton soil shows fairly good plasticity and compaction characteristics, the CBR strength development characteristics, especially under soaked conditions, are discouraging. On the other hand, the unconfined compression strength development as well as the swell characteristics with time for the Con-Aid treated black cotton soil are fairly appreciable when compared with those of the untreated black cotton soil.

A comparison of the unsoaked CBR and unconfined compression strength development and OMC~MDD compaction characteristics with time for the Con-Aid treated black cotton soil shows that, whereas the compressive strength development with time compares consistently well with that of maximum density with time (refer to Figs. 4b and 12), that of the unsoaked CBR strength does not. This may be attributed to the scale and inhomogeneity factor in relation to loading method, i.e. compression of mass and localised penetration. Another factor that would be contributing to these anomalies is likely to be the curing effect related to the quantitative bulk factor, specimen size and compaction effects.

4 CONCLUSIONS

Based on the results obtained from this study, the following conclusions may be made: -

(a) Con-Aid showed a good effect on the plasticity characteristics of the black cotton soils tested

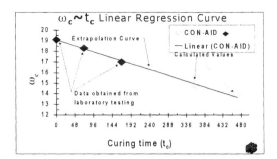

Fig. 17 ω_c~t_c Linear Regression curve.

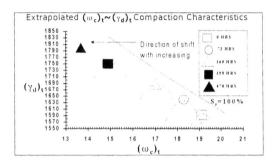

Fig. 18 Extrapolated $(\omega_c)_t$~$(\gamma_d)_t$ Compaction characteristics.

related to reduction and stabilisation despite variations in temperature and wetting/drying cycles.

(b) The density-moisture content characteristics of Con-Aid treated black cotton soil were the most consistent compared to those of lime treated and untreated. Con-Aid may therefore have the effect of improving workability in terms of compaction and densification.

(c) The CBR strength development characteristics of Con-Aid treated black cotton soils did not show a tendency to increase within the curing period tested in this study.

(d) Con-Aid may have an effect of containing swelling after ample curing.

ACKNOWLEDGEMENTS

The authors wish to acknowledge the contribution of Mr. Toda, Mr. Kimura, CPC staff members, Prof. Tatsuoka, Dr. Njoroge, Mitsui & Co., CON-AID PTY Ltd., Prof. Savage, Eng. Leou, as well as Ruth Mukabi and Silvester Mukabi of Revtech Ltd.

REFERENCES

CPC Consultants. Tana Basin Road Development Project, phase II. Materials Report Vol. 3

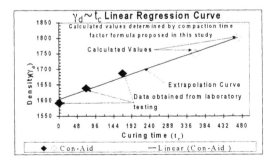

Fig. 16 γ_d~t_c Linear regression curve.

Frost, R.J. (1967). The importance of some tropical soils. Proc. South East Asian Conf. On Soil Engrg., 1. 43-53

Mukabi, J.N. & Tatsuoka, F. (1995). Effects of Swelling and Saturation of Unsaturated Clay in Laboratory Testing. Symposium on Saturated Soil Behaviour and Applications.

Mukabi, J.N. (1995). Deformation Characteristics at Small Strains of Clays in Triaxial Tests. PhD. Thesis. University of Tokyo

Mukabi, J.N. (1998). Con-Aid Research and Development Proposal.

Newill, R.J. (1961). A Laboratory Investigation of Two Red Clays from Kenya. Goetechnique. 11(4) 302~318.

Pandian, N.S., Nagaraj, T.S. & Sivakumar Babu. G.L. (1993). Tropical Clays, Part II. Engineering Behaviour. J. Geotech. Engrg., ASCE.

Savage, P.F. & Leou, J. (1998). Guidelines on Use of CON-AID Liquid Chemical Stabilizer.

Savage, P.F. (1998). Some Experiences on the Use of Con-Aid; A Water-Soluble Ionic Additive. University of Pretoria.

Geotechnics for Developing Africa, Wardle, Blight & Fourie (eds) © 1999 Balkema, Rotterdam, ISBN 90 5809 082 5

Properties and performance of lime kiln dust for stabilisation of soft clay

H. R. Nikraz
School of Civil Engineering, Curtin University of Technology, Perth, W.A., Australia

ABSTRACT: This paper is concerned with the influence of additives such as lime byproduct - lime kiln dust and lime on the stabilisation of fine-grained soil commonly found in the Narrogin area of Western Australia. The effect of lime kiln dust or lime on the engineering properties of the fine-grained soil, with the clay fraction of 41 percent of the weight of the soil are discussed. It was found that adding lime or lime kiln dust greatly improved the engineering properties of the soil. Lime kiln dust exhibited the same characteristics as the lime, however, it did not have the same degree of effectiveness.

1 INTRODUCTION

Problems associated with unstable subgrade soils such as expansive clays and soft clayey silts are reported in many parts of the world. In Western Australia the Westrail Narrogin division has had a long history of problems attributed to the existence of an extensive, weak clay subgrade. The use of XU wagons on the track, each exerting a 19 tonne axle load along with the poor clay formation caused failure of the subgrade. This meant that several days each week had to be allocated to maintenance alone.

The failure of the subgrade to support the applied loads led to extensive damage to the track and its components. Along the section of track, wooden sleepers had subsided deep into the ballast. In addition, severe splitting was common in both the longitudinal and traverse directions. Sleeper damage in turn caused distortion to the rails. Drifting of carriages was commonly experienced with even the cow-catcher at times making contact with the rails. As a result, train speeds were limited to 20 km/h to avoid derailment.

A number of possibilities were considered to rectify the problem. These included: (a) undertrack sledding and metal ballasting utilising geotextile at the ballast/formation interface; (b) replacement of subgrade; and (c) stabilisation with lime. Previous studies in Narrogin clay have shown that pure lime stabilisation is effective but expensive (Jujnovich, 1993 and Moore, 1992). With this in mind, an attempt has been made to examine the feasibility of using a lime by-product, lime kiln dust (LKD), as an alternative admixture for pure lime. The proper use of LKD can reduce the relative high cost of pure lime stabilisation in the order of 85 percent.

Lime is the most extensively used stabilizing agent for fine grained soils in different parts of the world. The role of physico-chemical reactions between lime and clay minerals present in the soils, which are responsible for stabilizing the soils are well documented (Diamond and Kinter, 1965; Davidson et al, 1965; Stocker, 1972 and Ingles, 1972). Improvements in the engineering characteristics of lime-soil mixtures are attributed to three basic reactions: (a) cation exchange; (b) flocculation and (c) aggregation or pozzolanic reactions. The addition of lime to a soil supplies an excess of divalent calcium cations and results in the replacement of monovalent cations by calcium cations. The addition of lime to the soil increases the electrolyte content of pure water and as a result of cation exchange, flocculation and agglomeration of the clay particles can occur.

This paper describes an investigation into the effect of the addition of LKD to a Narrogin clay on the engineering properties of the clay.

2 PROPERTIES OF MATERIAL TESTED

The clay studied in this investigation was from areas along the railway line in the Narrogin district, Western Australia. The consistency limits and mechanical analysis are given in Table 1. The clay is defined as sandy clay according to unified classification systems.

Both the lime and LKD were supplied by Cockburn Cement Limited, Western Australia. Properties of lime and LKD are given in Table 2 and 3, respectively. By nature, LKD is a variable material, but Table 3 states the normal ranges of various compounds. LKD is produced during the quicklime production when the calcium carbonate is calcined in rotary at 1100°C. The lime dust is collected via electrostatic precipitation and transported within the process to be pelletised for agricultural use or stored in a soil ready for dispatch as a very cost effective neutraliser.

3 TEST PROGRAMME AND TECHNIQUES

A series of laboratory tests were carried out on selected formation soils recovered along the rail to examine the feasibility of using LKD as an alternative admixture for lime. Furthermore, parallel testings with lime were also conducted to gain data on the relative effect of the two additive materials and a comparative evaluation was made mainly on the basis of two criteria, namely: (1) reduction in plasticity index; and (2) strength gain.

The soil samples were prepared in accordance with AS 1289 A2, 'Preparation of disturbed soil samples for testing'. After sifting a soil sample through a 4.75 mm sieve, it was mixed with the required quantities of LKD and lime. Water was then added to bring the mixture to the required moisture content. Mixing was continued until, by visual inspection, a uniform mixture was achieved. The additives content was defined by the ratio of the weight of the additive to the dry weight of the natural clay, expressed as a percentage.

To ascertain the engineering properties of the clay, the following laboratory testing was carried out:

1. The consistency limits of the clay treated with between 0 to 15% lime or LKD were tested after 48 hours curing time.

2. Compaction tests were performed on the clay samples treated with between 0 to 15% lime or LKD. All specimens were compacted according to the stand proctor method, and the effect of lime or LKD content on dry density was evaluated.

3. California Bearing Ratio (CBR) tests were conducted according to the procedure in AS1289 F1.1, 'Determination of the California Bearing Ratio of a Soil'. The clay samples were treated with 5% lime or LKD. The CBR tests were then conducted on both soaked and unsoaked specimens at 28 days. Soaking proceeded for four days prior to the test. For each natural and treated clay, four separate CBR tests were conducted and the mean values were recorded.

4. Unconfined compressive strength tests (UCS) were performed to observe the effect of the additives (0 to 15% lime or LKD) and curing time on the strength development of the treated clay. The initial curing time used was 48 hours before moulding and between 1 to 28 days after moulding into specimens of 50 mm in diameter and 100mm in height. The mixed material was placed in the mould in four layers. The amount of soil placed in a layer was determined from the moisture-density relation for the particular mixture. Compaction to standard proctor density was achieved by compressing the required mass into the given volume one hour at a time.

5. Consolidated undrained triaxial compression (CU) tests were performed on specimens of natural and treated clay with 6% lime or LKD, compacted to their optimum moisture contents and maximum dry densities. The CU tests were constructed mainly to observe the effect of the additive (6% lime or LKD) on the shear strength parameters of the treated clay. Tests were conducted according to the procedure in Bishop and Henkel (1962).

Table 1. Properties of the Clay.

Property	Value
Liquid limit (%)	40.00
Plastic limit (%)	16.00
Plastic Index (%)	24.00
Shrinkage limit (%)	13.00
Percent gravel size	1.00
Percent sand size	38.00
Percent silt size	20.00
Percent clay size	41.00
Specific gravity	2.67

Table 2. Hydrated Lime Composition.

Composition	Percentage by weight
Calcium hydroxide	80-90
Magnesium hydroxide	5.5 - 6.5
Silicon Dioxide	3 - 7
Aluminium oxide	0.3 - 0.5
Iron oxide	0.1 - 0.3

Table 3. Properties of Lime Kiln Dust.

Property	Lime Kiln Dust
Percent passing on sieve size (µm):	
45	100
16	80
10	40
Chemical Analysis (%):	
SiO_2	0.1 - 2.0
Al_2O_3	0.1 - 0.3
Fe_2O_3	0.1 - 0.6
C_aO	20.0 - 40.0
MgO	0.8 - 4.0
Na_2O	1.0 - 2.0
K_2O	0.1 - 0.5
SO_3	1.0 - 4.0
$CaCO_3$	40.0 - 71.0
$MgCO_3$	5.0 - 11.0
Cl	0.3 - 0.6

6. Swelling tests were performed in the oedometer on specimens of natural and treated clay with between 0 to 15% lime or LKD. For the purpose of investigation, swelling potential is defined as the percentage. Swell of the laterally confined sample, 102mm in diameter and 51mm in height at 96 hours soaking under 6.9 kPa surcharge after being compacted to the maximum density at optimum moisture content of treated or untreated sample, as the case may be.

4 TEST RESULTS AND DISCUSSION

Consistency limits tests indicated that increasing lime or LKD content reduced the plasticity index of the clay (see Figure 1). This may assist comminution for effective mixing and compaction by improving workability of the clay. Figure 1 also shows that

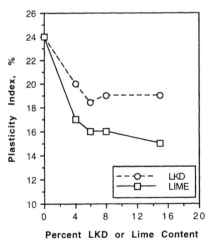

Figure 1. Effect of LKD and Lime on Plasticity Index

within the range of percentage of LKD used for tests, 6 percent LKD seems to be the optimum percentage.

Compaction tests indicated that increasing lime content reduced the maximum dry density but increased the optimum moisture content of the mixture (see Figure 2). The same trend existed for the application of the LKD. While optimum moisture contents vary greatly, the LKD mixtures exhibited lower densities than the lime mixtures for the same added amount. Compaction curves for both lime or LKD treated clay were generally flatter. This makes moisture content less critical and reduces the variability of the density produced.

The CBR tests conducted on the clay with 5% lime produced very high CBR values. CBR values of 135% and 76% were recorded for unsoaked and soaked samples respectively. This was a significant improvement from the initial recorded CBR values of 4% for both soaked and unsoaked of the natural clay. The clay treated with 5% LKD also returned high CBR values of 77% unsoaked and 47% soaked for the clay. The CBR tests indicated that LKD was approximately 60% as effective when compared to lime.

The UCS tests indicated that both additives significantly improved the strength of the clay after 48 hours curing (see Figure 3). The graph plots of additive content versus strength indicated a sharp increase in strength for additive contents up to approximately 6.5% and 8% for lime and LKD respectively. Progressing from these points, the rate of strength increase diminished with increasing additive content and the graphs tended to flatten. The UCS after 1, 2, 7 and 28 days curing time are presented as a function of LKD contents in Figure 4. Generally, the UCS increased with the curing period. For example, after 28 days curing, an 8% LKD treated clay attained a UCS 2.4 times greater than that of the untreated clay.

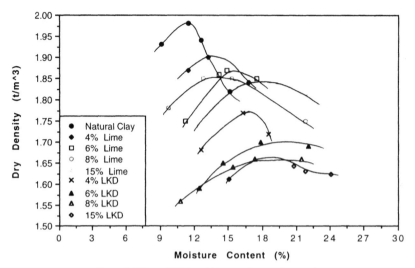

Figure 2. Effect of LKD and Lime on Proctor Compaction.

The influence of the additives on the effective strength parameters determined from CU tests. The strength parameters changed from $\phi' = 19°$, $C' = 185$ kPa for the untreated clay to $\phi' = 26$, $C' = 168$ kPa, and $\phi' = 24$, $C' = 175$ kPa for treated clay with lime and LKD respectively. The treated clay increased the frictional property and marginally decreased the cohesion property of the clay.

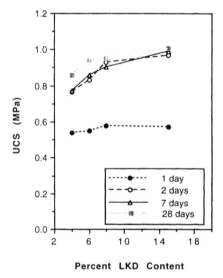

Figure 4. Influence of LKD Content on the Unconfined Compressive Strength of the Clay

The results of the swelling tests conducted on untreated clay as well as clay treated with lime and LKD are shown in Figure 5. For example, it can be seen from Figure 5 that the addition of 15% LKD decreased the swelling potential of the natural clay by a factor of about 7. Comparable results were obtained when the clay samples were treated with lime (see Figure 5).

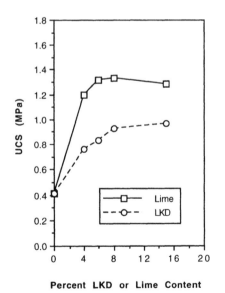

Figure 3. Effect of LKD and Lime. The Unconfined Compressive Strength of the Clay.

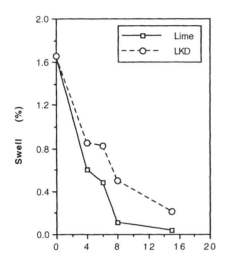

Figure 5. Effect of LKD and Lime on Swelling Potential

5 CONCLUSIONS

The most common method of chemical stabilisation employed for stabilising clay soils is lime stabilisation. The high strengths obtained from lime stabilisation may not always be required, however, and there is justification for seeking a cheaper additive which may be used to satisfactorily alter the clay soil properties in such cases.

From the results of a series of laboratory tests carried out on subgrade clay soils recovered from areas along rail formation in Narrogin, Western Australia, it was found that the addition of LKD to the tested clay soils generally results in decreased plasticity index, maximum dry density and swelling potential, while it increased soil strength as indicated by UCS, CU and CBR tests. Although it exhibited the same characteristics as the lime did, it did not have the same degree of effectiveness. It was, however, preferred for use over the lime for the reason that it would still significantly increase the strength of the soil, but would cost far less in its application than if lime was used.

REFERENCES

Bishop, A.W. & D.O. Henkel 1962. *The Measurement of Soil Properties in the Triaxial Test.* 2nd ed., London: Edward Arnold, p. 227.

Davidson, L.K., T. Demirel. & R.L. Handy 1965, Soil pulverisation and lime migration in soil-lime stabilisation, *Highway Research Record* No. 92.

Diamond, S. & B.E. Kinter 1965. Mechanisms of soil-lime stabilisation: An interpretative review. *Highway Research Record* No. 92, p. 83.

Ingles, O.G. & J.B. Metcalf 1972. *Soil Stabilisation.* Sydney: Butterworth.

Jujnovich, A. 1993. *Lime Stabilisation of Soft Clay.* B.E. final year project, Curtin University of Technology, p. 102.

Moore, T. 1994. Private communication.

Standard Association of Australia. *Methods of Testing Soils for Engineering Purposes.* AS1289

Stocker, P.T. 1972. *Diffusion and Diffusion Cementation in Lime and Cement Stabilised Clay Soils,* Special Report No. 8, Australian Road Research Board.

Geotechnics for Developing Africa, Wardle, Blight & Fourie (eds) © 1999 Balkema, Rotterdam, ISBN 90 5809 082 5

Analysis of the stability of reinforced hydraulic engineering

K. Sh. Shadunts
Kuban State Agrarian University, Krasnodar, Russia

Mado-Abary Haruna
Republic of Niger

ABSTRACT: The construction of earth dams in order to create the ponds which are necessary for irrigation is included into the program of the development of economy of Niger. In the first place it is supposed to build Kandagy dam. Modern earth constructions become more effective when they are reinforced by geotextiles. We have worked out the methods of analysis of stability of the hydraulic engineering embankments reinforced with the help of a set of flexible and rigid elements on the grounds of computer modelling.

1 INTRODUCTION

The design of the construction of Kandagy dam on the river Niger in Republic of Niger envisages an improvement of water supply of the population, generation of electrical energy, river discharge regulation, utilisation of irrigation in farming agriculture. It is possible to increase economical efficiency of the hydraulic engineering construction at the expense of erection of a dam made of reinforced soil. Such technology allows to reduce significantly a volume of earthworks at the expense of an increase of steepness of the slopes. The area occupied by the construction is reduced.

Flexible elements are usually used as reinforcement in the practice of construction of reinforced soil dams (Ter-Minassian 1976). They resist stretching stresses very well. Slope failure character analysis has shown that reinforcement accepts such stresses if it is arranged in the upper part and the lower one of a dam. In the middle part where shearing efforts take place by slope displacement, flexible reinforcement arranged in the horizontal position is ineffective.

It is possible to arrange geotextiles along the inclined areas oriented orthogonally relative sliding surface (Shadunts & Eshchenko 1990), but it makes soil compacting technology more complicated.

Construction components equistrength principle has been assumed as a basis of reinforced embankment construction optimisation. Effective arrangement of geotextile in the soil layers being consolidated has been searched in the process of the investigations in order to determine height of the embankment and optimum length of reinforcement.

Construction of an earth structure (application for invention in Russia № 97119444 "Embankment" priority dated 26.11.1997) reinforced with the help of flexible and rigid elements has been worked out with the arrangement of geotextiles in the upper part and the lower one of the embankment and rigid reinforced concrete beams in the middle part (Figure 1).

2 PHYSICAL AND MATHEMATICAL MODELLING OF REINFORCED EMBANKMENT

An influence of reinforcement on stability and slope failure mechanism has been investigated on the grounds of physical and computer modelling, because analytical calculations of the structures made of composite material are rather complicated.

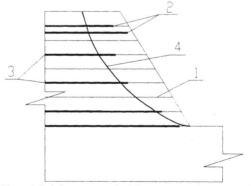

Figure 1. Reinforcement methods beam suggested: 1 - layer of soil, 2 - geotextiles, 3 - reinforced concrete beam, 4 - slip failure surface.

The slope models in a tray with a transparent wall were prepared using equivalent materials in accordance with the methods worked out at the institute of mechanics of Moscow State University. Soil equivalent material was chosen in such a way that its characteristics fulfil the conditions of similarity on conversion to nature and correspond to the international standard for reinforced soil structures BS 8006. Sand mixture with 3% of cup grease according to weight met these requirements; this mixture had the following characteristics: cohesion was 0.001 MPa, angle of internal friction was 31°, specific gravity was 1.86 g/cm³, heterogeneity factor of granulometric composition was 6.7, weighed average diameter of particles was 0.31 mm.

Here are the additional terms of modelling which have allowed to exclude ambiguity of test results (Jewell & Milligan 1985; Schlosser & Elias 1978):

$$I_d^H = I_d^M \qquad (1)$$

$$\frac{D_{cp}^H}{L^H} = \frac{d_{cp}^M}{l^M} \qquad (2)$$

where I_d^H and I_d^M are the dam material compactness grades of natural object and model; D_{cp}^H and d_{cp}^M is a size of a mesh of natural and model reinforcing geotextiles; L^H and l^M are the linear sizes.

Strength of a grid for the models was chosen in such a way which helped to eliminate a rupture under influencing loads, and mesh length prevented a slip of soil along reinforcement. Kapron grid with meshes of 1 x 1 mm was used. Rigid elements were modelled with the help of the bars of 10 x 10 x 250 mm.

Stresses were created in the model due to continuous growth of load on the horizontal area. The deformations of the orthogonal grid of the lines arranged on the lateral surface were registered. Consequent series of tests of non-reinforced embankment, the one reinforced only with the help of flexible elements arranged on different levels, only with the help of rigid elements and a combination of both elements were carried out.

The experiments proved an advantage of the reinforcement method being suggested. The deformations of the slip proved to be less than in non-reinforced embankment (nearly two times as much) and than in an embankment reinforced only with flexible elements.

The failure process began with the formation of cracks of a fracture in the upper area of the embankment. Then recesses were formed near the lower edge due to soil loss. When the deformation zones reached critical dimensions, failure took place

along slightly curved arc with the convexities facing downwards. Reinforced slip failure took place within one stage. Displacement surface is arc-shaped and convex.

For numerical calculation of the earth construction reinforced with flexible and rigid elements the boundary element method was used which allowed to take into account geometrical and physical heterogeneity, non-linearity of properties of materials of reinforced embankments.

Strengthening influence of reinforcing elements can be evaluated due to a change of stress-strained state of the embankment under load as compared with non-reinforced one.

Non-linear deformation of strengthened massive can be considered as a result of mutual displacements of linearly deformed soil and non-linearly deformed reinforcement.

$$\begin{cases} \sigma_{ij}^{rp} = 2G^{rp}(\varepsilon_{ij}^{rp} + \dfrac{\mu}{1-2\mu}\delta_{ij}(\varepsilon_{ij}^{rp} + \varepsilon_{jj}^{rp})) \\ \sigma_i^{ap} = \sigma(\varepsilon_i^{ap}) \qquad i...j = X...Y \\ \sigma_{ij}^{ap} = \sigma_{\max}^{rp} ...\forall i \neq j \end{cases} \qquad (3)$$

where σ_{ij}^{rp} is stress in i soil layer; G^{rp} is soil humidity degree; ε_{ij}^{rp} is modulus of deformation of i soil layer; δ_{ij} is strain; σ_i^{ap} is stress in i layer of reinforcement; $\sigma_{\max}^{rp}$ is maximal stress in soil; μ is Poisson's ratio.

A solution of a set of equations (3) is carried out on the grounds of theory of plasticity assuming that displacements take place in each point of the surface of a contact of soil and reinforcement until tangential stresses reach the limit determined by soil strength criterion. When the limit is reached, tangential stresses are reduced up to permissible values

$$\tau = \tau_{np}; \qquad \forall \, \tau > \tau_{np} \qquad (4)$$

where τ is current value of maximal stresses, and τ_{np} is maximal permissible value of tangential stress in an element according to strength condition.

A series of calculations of embankment stability has been carried out when strength characteristics and deformation ones accepted according to the recommendations of standard BS 8006. For sand soils modulus of deformation is E = 60 MPa, Poisson's ratio $\mu = 0.30$; cohesion C = 18 kPa, angle of internal friction $\varphi = 25°$; specific gravity of soil $\gamma = 19$ kN/m³, contour interval of slopes 1:1. Height of the embankment was assumes equal to 15 m.

When numerical calculation of stress - strained state of the embankment under an increasing load acting on the main area was carried out with the help of boundary element method, a magnitude corresponding to a transition of load - settlement dependence graph from a linear part to non-linear one was assumed as critical stress value.

A packet of programs of COSMOS/M complex worked out by Structural Research and Analysis Corporation, Santa Monica, California, on the grounds of elastic plastic model of soil deformation suggested by Droocker and Prager allows to determine critical load for each variant of reinforcement of a computer model of the embankment. The results being obtained have shown that a construction reinforced with the help of flexible reinforcement in the upper part and the lower one and with the help of rigid reinforcement in the middle part resists maximal load.

Imitation calculations allow to optimise a position and the dimensions of reinforcing elements. A search of the rational arrangement of rigid elements has been carried out under the condition of soil "non-forcing through" between them taking bridging into account. The best strengthening was reached when a force corresponding to the beginning of forcing through was equal to shearing strength in vertical planes passing through the axes of holding elements.

3 CONCLUSIONS

1. When the investigations on the models of the embankments reinforced with the help of flexible elements (geotextile) and rigid ones (reinforced concrete beams) are carried out, it is necessary to provide additional criteria (besides the main ones) which take into account discreteness of soil and the peculiarities of its contact interaction with reinforcement.

2. As a results of the experiments it has been proved that maximal efficiency of reinforcement is reached when flexible elements as well as rigid ones are introduced into soil.

3. As COSMOS/M program has been used, a computation with the help of boundary element method has allowed to solve a task concerning optimal distribution of geotextile according to the height of the embankment taking into consideration its shortage in Niger. Flexible reinforcement significantly increases stability if it is arranged in the upper layers and the lower ones of soil.

4. Efficiency of computation and design of the constructions made of composite materials on the grounds of computer modelling has been proved.

REFERENCES

Jewell, R.A. & G.W.E. Milligan 1985. Interaction between soil and geogrids. *Proc. conf. of polymer grid reinforcement:* 18-30. London.

Schlosser F. & V. Elias 1978. Friction in reinforced earth. *Proc. ASCE Symp. on Earth Reinforcements:* 735-763. Pittsburgh.

Shadunts, K.Sh. & O.Yu. Eshchenko 1990. Embankment construction method. Author's certificate of the USSR No. 1562407, cl. E02D 17/20. *Bulletin of invent.* 17: Russia.

Ter-Minassian, W. 1976. Reinforced earth as material for a new type of composite dams. *Problems of Geomechanics.* 7: 119-123.

Geotechnics for Developing Africa, Wardle, Blight & Fourie (eds) © 1999 Balkema, Rotterdam, ISBN 90 5809 082 5

Impact compaction trials: Assessment of depth and degree of improvement and methods of integrity testing

J. H. Strydom

Africon Engineering International (Pty) Limited, Pretoria, South Africa

ABSTRACT: Impact compaction trials were performed at Kriel on a residential site overlain by compressible windblown sands. The purpose of the trials was to assess the depth and degree of improvement attainable with various compactors of varying energy levels and included a vibratory compactor and four impact compactors. In addition, the trials had to indicate the most sensible ways of measuring the effectiveness of compaction.

Several in situ and laboratory tests were performed to characterise the in situ materials in terms of their salient engineering properties before and after compaction. The latter included sections that were subjected to 20, 40 and 60 passes of the compaction equipment.

The trials have indicated that on this site significant improvement of the soil was attained by the impact compactors to depths of 1,5 m as opposed to only about 0,3 m by the vibratory compactor. Tests which have proved worthwhile in characterising the virgin material and measuring the effectiveness of the compactive effort included level surveys; DCP tests; plate bearing tests, indicator tests, moisture content tests, oedometer tests and laboratory compaction tests. The specific value of some these tests are highlighted in terms of their ability to reflect different aspects of improvement during the compaction process.

1 INTRODUCTION

The technique of compacting soils and rockfill with a roller compactor that imparts a significant impact in the rolling process, has been around for several years. Mainly three, four and five-sided compactors have been used and several compaction trials or results of projects on which the equipment had been used, have been described in the literature by (Pinard et al (1988), Barret and Wrench (1984), Scheckel (1985), Pinard and Strydom (1998). Various aspects, like site charactersisation and the measurement of relevant engineering parameters before and after the compaction process, are addressed in these documents, but to date no specific procedures have been suggested that represent a formalised approach to impact compaction.

Landpac is a South African based contractor that specializes in the development of impact compaction equipment, as well as their application in the compaction process. As part of their ongoing technology development program they performed extensive trials which comprised the treatment of the in situ soil on a site near the town of Kriel in South Africa with various types of compaction equipment. During the compaction trials they varied the number of compaction passes within each section treated with specific equipment. The broad objectives were to demonstrate the effectiveness of impact compaction in absolute terms, but also relative to conventional vibro compaction. To demonstrate the effectiveness of impact compaction, the depth and degree of compaction had to be evaluated. This also implied establishing norms for charactersing a site, as well as the most appropriate parameters and type of integrity tests to measure the conditions of the soil before and after compaction.

A range of tests was consequently selected to satisfy the above requirements. This involved the following:

- Characterising the soils across the site via indicator and moisture content tests. Maximum dry density, optimum moisture content and CBR tests were conducted in the laboratory to assess compaction properties.
- Measurement of settlement in the compacted areas and to assess its significance as a measure of degree of compaction.
- Assessing the stiffness of the in situ material both in the virgin and compacted state. This was done via plate bearing tests to obtain static moduli and via falling weight deflectometer (FWD) tests to obtain dynamic moduli. A limited number of consolidation and in situ density tests were also done to assess the compressibility characteristics of the in situ materials.
- Measuring the consistency (strength) of the virgin and compacted soil via hand DCP equipment to limited depths and via a heavy mechanical DCP apparatus to larger depths.

2   SITE DESCRIPTION

The site is situated on the outskirts of the town of Kriel in the Thubelihle township in Mpumalanga. The topography comprises mild slopes and vegetation consists only of veld grass.

Geologically the site is underlain by sandstone of the Vryheid Formation of the Ecca Group of the Karoo Sequence sedimentary rocks. Bedrock depth varies from 2 m in places to more than 6 m, but generally speaking bedrock depth is well in excess of 3 m at the site.

The soil profile is fairly simple and uniform. It consists almost entirely of a redbrown clayey fine sandy silt aeolian deposit. On inspection the material is moist and distinctly pinholed and reworked, indicating that it is potentially, highly compressible. It is generally soft or soft to firm but stiff at surface due to desiccation.

3   COMPACTION TRIALS

The compaction trials involved treatment of 5 designated sections of the site, each with a specific compactor.

The compaction equipment used were the following:
- a 12-ton vibro compactor
- a 4-sided single drum compactor of which the compaction energy is ± 10 kJ
- two 5-sided dual drum Landpac Impact Compactors of respective energies 10 kJ and 15 kJ
- a 3-sided dual drum Landpac Impact Compactor of which the energy is rated at 25 kJ.

Each section comprised 4 lanes of which the degree of compaction was varied. The first lane was not compacted and represented virgin soil, while the other 3 lanes were subjected to 20, 40 and 60 passes respectively.

4   SCHEDULE OF TESTS

Test pits were dug at representative positions across the site to depths of about 3 m to allow inspection and recording of the soil profile, as well as sampling for laboratory tests.

A wide range of in situ and laboratory tests was performed, a summary of which appears in Table 1.

A detailed description of the tests and their results are covered in a report by Strydom and Hefer (1997).

5   MATERIALS CHARACTERISATION

5.1   *Indicator Test Results*

Samples were taken at various positions across the site and submitted for foundation indicator tests. Representative soil samples of each of the five test sections were taken and a total of 13 indicator tests were done to determine the type of soils present and the degree of uniformity of the soil properties across the site.

The results generally indicate that the hillwash material across the site comprises fine sandy clayey silt of low to moderate plasticity and without exception classifies as A-6(3) or A-6(5) material according to the Unified classification. The results confirm the high degree of uniformity of the soil indicator properties over the entire site, as all the parameters show limited variation.

Table 1: Test summary per compactor type at various levels of treatment (0, 20, 40 and 60 passes).

TEST	VIBRO				10 kJ				15 kJ				25 kJ				4-SIDED			
	0	20	40	60	0	20	40	60	0	20	40	60	0	20	40	60	0	20	40	60
Level surveys	20	20	20	20	20	20	20	20	20	20	20	20	20	20	20	20	20	20	20	20
Visual settlement																1				
Indicator	1	1		2		1		1		1		1	1			2			1	1
CBR				1						1										1
Moisture content		5	1	5		5	1	5		5	1	5	5	1	1	5		1	5	5
Hand DCP	3	3	3	3	3	3	3	3	3	3	3	3	3	3	3	3	6	6	6	6
Heavy DCP	3	3	3	3	3	3	3	3	3	3	3	3	3	3	3	3	3	3	3	3
Plate bearing	1			3		1	2		1		1	2	3	1	1	7	1			2
In situ density	3			3									4			11				
Consolidation													3			3				
FWD Deflection measurements	10	10	10	10		10	10	10		10	10	10	10	10	10	10		10	10	10

The results are summarised in Table 2.

Table 2: Indicator test results.

Statistical Parameter	Clay and silt (%)	Sand (%)	GM	LL (%)	PI (%)	LS (%)
Mean	47	53	0.62	23	11	5.4
Standard deviation	5.9	6	0.07	3	2.6	1.5

Legend   LL   =   Liquid limit
          LS   =   Linear shrinkage
          PI   =   Weighted plasticity index
          GM   =   Grading modulus

5.2   Moisture Content Results

The moisture content of the in situ soils was regarded very important since it generally has a marked influence on the stiffness and compactability of a material. This is especially the case with finer grained soils that are generally slightly or moderately plastic.

56 Soil moisture content samples were taken at various levels from all five test sections from each of the 20, 40 and 60-pass lanes, as well as the 0-pass lane in the 25 kJ test section. The moisture content envelope against depth appears in Figure 1.

It is evident from the above figure that the optimum compaction moisture content (OMC) of the in situ soils, which is between 10 and 11 % according to Section 5.4 below, is generally lower than the prevailing moisture content of the in situ soils, except over the upper ± 0,5 m of the soil profile. The significance of this phenomenon is discussed in Section 6.5 where it is shown that the degree of

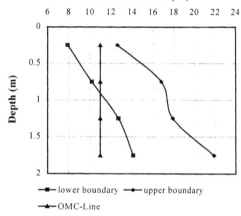

Figure 1: Moisture content envelope.

saturation varies between 38 and 58 %. These moisture contents are substantially higher than the 2,4 to 4 % reported in the trials by Pinard et al (1988) or the 6 to 9 % reported by Barret and Wrench (1984).

5.3 Compaction Test Results

Compaction tests were performed to establish the basic compaction parameters of the soil against which some of the soil properties could be gauged. In addition, the compaction tests were done to establish whether the compaction properties of the soils across the site were fairly similar.

Three samples were taken from different positions and the salient parameters were as follows:

• Optimum moisture content: 9.7 to 11.1%
• Maximum dry density: 1958 to 2024 kg/m^3
• CBR at 93% Mod AASHTO 15 to 25

The results indicate that the in situ material has a fairly high maximum dry density and a moderate optimum moisture content, while the CBR values at densities typically specified in the field are moderate.

6   RESULTS OF TESTS ON COMPACTED SECTIONS

6.1   Settlement Results

6.1.1   Level survey

From the levels taken during the compaction trials, significant conclusions can be drawn. These results are summarized in Figure 2 as the cumulative settlement versus number of passes.

When one considers that the settlement must in some way be related to compaction of a certain degree and to a certain depth, the following trends are evident from the above figure:

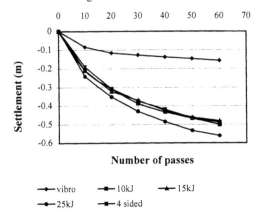

Figure 2: Settlement vs number of passes.

- The 12-ton vibro compactor only achieved a settlement of 150 mm as opposed to 470 to 500 mm of the other impact compactors, except the 560 mm that was achieved by the 25 kJ impact compactor. The 10 kJ, 15kJ and 4-sided compactor have yielded remarkably similar results.
- After 60 passes the settlement induced by the impact compactors does not seem to have leveled of entirely. While this may be said of the vibro compactor as well, the increase in settlement beyond 20 passes achieved by this compactor is very limited.
- Although the settlement induced by the 25 kJ compactor is only about 10 % more than that of

the other compactors, it represents significant additional induced stiffness to achieve the additional settlement. It can be likened to "strain hardening" that has taken place as soil density increased.

Settlements achieved in the trials performed by Pinard et al (1988) varied between 115 and 200 mm after 60 passes with a 25 kJ impact compactor. The variation is ascribed to material variation. The material is broadly similar to that at the Kriel site with very similar void ratios. However, the overall much lower magnitude of settlement is attributed to the low soil moisture content (2.5 to 6 %) which, in conjunction with the cementing effect of iron oxides and salts, gave rise to high soil suction. These effects cause significant interparticle bonding and consequently difficulty in breaking down the soil structure, as well as re-arranging the particles into a denser packing. Barret and Wrench (1984) report settlements of 340 mm after 60 passes with a 25 kJ compactor on aeolian sands near Middelburg. The moisture content of these soils varied between 6 and 9 % and the void ratios between 0.9 and 1.05. They mention significant additional settlement after further treatment some time later and ascribe the additional settlement to be most likely due to an increase in soil moisture after summer rains.

6.1.2   Visual settlement indicator

A visual indication of the depth of soil that was influenced by compaction was obtained in the 60-pass lane of the 25 kJ compactor. A narrow strip of soil (0.6 m) was excavated across the 60-pass lane to about 3 m depth before compaction was conducted. The soil was subsequently backfilled and nominally compacted to a density of about 85 to 90%, deemed to be the density of the in situ material. This was done in 300 mm layers, except the top layer which was 500 mm thick. Each layer was separated from the one above by a thin horizontal marker layer of lime. Lime was chosen simply because of its color contrast with the redbrown in situ soil.

After compaction was completed, the strip of soil was exhumed to reveal an elevational view of the marker layers. A view of the exhumed section appears in Figure 3.

Volume-wise the marker strip of soil was small in relation to the surrounding virgin soil and compacted to a density close to that of the surrounding soil. It would therefore be fair to assume that the behavior of the marker strip and the surrounding soil would be very similar.

Figure 3: A view of the visual settlement section indicating degree of settlement with depth

From the above figure it is evident that:

- Intense shearing of the soil has taken place in the transition zone between the virgin soil and the compacted soil. This is evident from the intense curvature of the marker lines in this area.
- Below 1.5 m depth the influence diminishes rapidly and beyond 2 m depth the marker lines are almost horizontal, indicating little if any compaction influence beyond this depth.
- The settlement of each marker layer can roughly be measured, and hence the cumulative settlement below that depth. This allows an estimate of the settlement within each marker layer.

The estimated settlement within each marker layer is summarized below.

0 – 500 mm:	75 mm
500 – 800 mm:	100 mm
800 – 1100 mm:	125 mm
1100 – 1400 mm:	100 mm
1400 – 1700 mm:	75 mm
1700 – 2000 mm:	50 mm
2000 – 2300 mm:	25 mm
Total	550 mm

The total settlement agrees well with the average 560 mm settlement obtained via the level survey.

It is also interesting to note that the maximum unit settlement is not at the surface. Even though the surface layer is 500 mm thick as opposed to the 300 mm of the lower marker layers, the settlement of the top layer is about 75 mm as opposed to 100, 125 and 100 mm of the next three layers. The implication is that the highest degree of compaction is between about 0.5 and 1.5 m below surface. The largest strain in the third marker layer is an incredible 42 % (125 mm settlement within a 300 mm layer).

6.2 Hand DCP Test Results

Hand DCP soundings were conducted at three different positions on each test section and in each lane after 0, 10, 20, 30, 40, 50 and 60 passes of the various compactors. These tests were valuable in qualitatively comparing the stiffness of the soil with depth in the various sections. Since the driven cone causes shear failure of the soil, the DCP blow count is not a direct measure of stiffness, but it stands to reason that the penetration rate should be in some way related to stiffness. One should however be mindful that a relationship between the blow count and stiffness is not unique for all soil types.

The results of DCP soundings are known to be sensitive to variations of moisture content of the soil, especially in fine-grained soils. Although the moisture content increases somewhat with depth (see Section 5.3), this trend is the same across the site. This implies that over the same depth range the soil moisture content does not vary by more than 4 to 5 % across the site. In addition, over the depth range (± 2 m) that the DCP tests were done, the moisture content increases on average by 7 %. Figure 4 reflects the DCP results for the 25 kJ and vibro compactors after 30 and 60 passes respectively.

It is interesting to note that there is still a slight difference in consistency between the virgin soil co-passes and the vibro treated soil at depths of 800 mm and more. However, there is no difference in consistency below 400 mm depth between a 30 and 60 pass treatment of the vibro compactor. The difference in consistency between the vibro compactor and the 25 kJ compactor is significant and at 2 m depth the latter compactor still reflects an improvement beyond a 30 pass treatment.

6.3 Plate Bearing Tests

Vertical plate bearing tests were conducted, since stiffness is the intrinsic and most important engineering parameter in terms of assessing the effectiveness of compaction.

607

Penetration Rate (mm/blow)

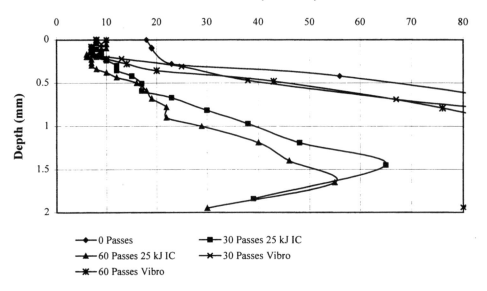

Figure 4: Variation in DCP blow count with depth: Vibro compactor versus 25 kJ compactor.

Tests conducted at surface did not allow accurate reflection of the stiffness of the materials at larger depth, due to a crust of desiccated material (± 300 mm thick) which shielded the underlying materials from the imposed plate stress.

Consequently, 8 tests were conducted at various depths below surface employing a 400 mm plate loaded to a stress of 780 kPa. Tests included an unloading-reloading loop at about halfway of the stress range.

Figure 5 shows a typical depth vs modulus plot for the results of the 25 kJ and vibro compactors where E is the secant modulus on virgin loading over the stress range 0 – 400 kPa.

The results indicate the significant difference in stiffness below 0.4 m depth to about 1.2 m between the impact compactor and the vibro compactor. Since the vibro compactor is only effective to depths less than 0.4 m, the results for this compactor below 0.4 m in Figure 5 essentially represents that of virgin soil.

6.4   Consolidation Tests

No collapse potential tests were conducted, since the collapse potential values would be low at the relatively high in situ moisture contents and

therefore not of much significance. A reasonable estimate could have been obtained had the samples been air dried to a moisture content of say about 6%. Six consolidation tests were done on undisturbed samples of material from the 25 kJ test section. The undisturbed samples were taken at depths of 0.75 m, 1.5 m and 2.25 m in each of the 0-pass and 60-pass lanes.

According to the moisture content results over the upper 1.5 m of the profile, the moisture content ranges typically between 12 and 18 %.

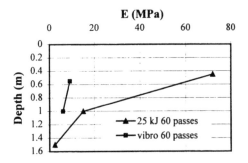

Figure 5: Typical modulus results from vertical plate bearing tests.

At these moisture contents and assuming an average void ratio of 0.84 for the virgin soil, the degree of saturation varies between 38 % and 58 %. This implies that although the moisture content is above OMC, it is still well below saturation (zero air voids) and hence would not affect the compaction process adversely as a result of excess pore pressure as the soil densifies. However, one would expect that as the soil reaches a high degree of densification over about the top 1 m of the profile, a slightly higher density and stiffness could be achievable, had the soil been at OMC.

Figure 6 and 7 reflect the above-mentioned trends.

It is worth noting that the consolidometer dry densities concur with the observations of the visual indicator experiment, ie that beyond about 2 m depth there is no significant influence of the 25 kJ compactor (Figure 6).
In Figure 7 the moduli are shown over the loading range 10-100 kPa and 100-200 kPa.

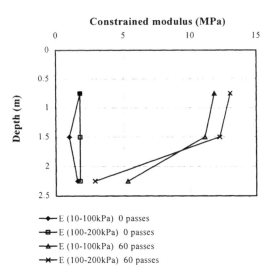

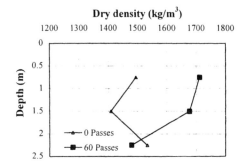

Figure 6: Consolidometer dry density results, 25 kJ compactor.

In addition, according to Figure 7 the constrained moduli do not differ significantly over the stress ranges indicated. The modulus of the virgin soil hardly changes over the entire depth range in question. Above 1.5 m the moduli seem fairly consistent, but reduce significantly below this depth.

6.5   In Situ Density Tests

In situ density tests via the sand replacement method were done at various depths in the 0 and 60-pass lanes of each of the vibro and 25 kJ compactor sections.

Figure 7: Consolidometer constrained moduli 25 kJ compactor.

According to Figure 8, which represents a typical comparison of in situ density vs depth for the 25 kJ and vibro compactor, the 25 kJ compactor has a much greater effect on the density and depth of compaction than the vibro compactor.

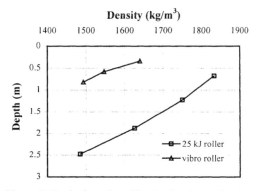

Figure 8: Typical results of in situ density vs depth.

The densities achieved with the 25 kJ compactor vary almost linearly from about 75 % relative density at 2.5 m depth (where virtually no effect was achieved) to about 93 % at 0.5 m depth. The relative density at 1 m depth was 90 %.

Pinard et al (1984) achieved similar results after 60 passes of the 25 kJ roller, albeit somewhat lower. In two trial areas the densities at 0.5 and 1 m depths were respectively in the ranges 90 to 92.5 % and 85 to 86 %. Barret and Wrench (1984)

609

achieved densities of only 73 and 77 % after 60 passes at respective depths of 0.5 and 1 m, which appear to be anomalous considering surface settlements of up to 340 mm.

6.6 *Falling Weight Deflectometer (FWD) Measurements*

Due to constraints of length, these results will not be discussed in detail here, save to state that:

- The trials proved that these tests can sensibly be done on a subgrade material to determine its dynamic modulus. In other words its application is not limited to roads with high stiffness surfacing and upper layers.
- The back calculation of dynamic moduli with depth using currently available software requires the identification of "layers" in the subgrade of similar consistency or stiffness. This can be done via DCP tests but is obviously less well defined as opposed to a layered pavement system.

7 COMPARISON OF TESTS AND TRENDS ESTABLISHED

From the description of the various tests it is evident that certain comments can be made on the value of the tests performed and certain trends that were established. Recommendations can also be made on how to employ tests optimally. This can be summarized as follows:

- Characterization of the site, especially in terms of material type and moisture content is important since both can have a profound affect on the compaction results, if conditions vary significantly across the site. In this case conditions have been highly uniform in terms of material type, basic engineering properties, compaction characteristics and moisture content.
- The level survey (settlement) results indicate a gradual increase versus number of passes and indicate a significant reduction of effect beyond 20 passes for the vibro compactor and a lesser reduction beyond 40 passes for the other compactors. Level surveys remain a good qualitative and valuable measure of the compaction effect achieved. It is an important and simple indicator of the optimum number of passes for a specific type of compaction equipment on a certain type of material. It can be fruitfully used in conjunction with other tests to tailor the compaction process.

- DCP soundings have also proved themselves a valuable means to obtain qualitative and comparative results of consistency. They have the advantage of providing a continuous profile of consistency cheaply and quickly as opposed to in situ densities and plate bearing tests. They have indicated that after 60 passes the improvement achieved by the vibro compactor is limited to the upper 0.5 m of the profile, while the effect of the 25 kJ compactor extends to 2 m depth. Unfortunately DCP tests do not provide a direct measure of soil stiffness and in addition, the results are quite sensitive to changes in soil moisture. The 2 m hand DCP, in conjunction with the mechanical DCP, yields satisfactory results.
- In situ density tests have limited value, despite being in good agreement with the results of specifically the DCP and consolidation tests. In addition, they have to be conducted as sand replacement tests if conducted below ground level, since nuclear density tests may not be reliable in excavations due to backscatter. Sand replacement tests are time consuming and cumbersome to do in test holes, cause large disturbance of the compacted area and do not provide a direct measure of stiffness, nor do they provide continuous results with depth, unless conducted at close depth intervals (which is costly and impractical)
- Plate bearing tests are simple but fairly time consuming to conduct and provide a static stiffness which is of high significance from a compressibility/settlement point of view in the case of building structures. When conducting plate bearing tests, account should be taken of crusting effects during testing at the surface. The plate bearing test reflects the average stiffness of a volume of soil to a depth of about 1.5 times the plate diameter. Hence to obtain good definition of stiffness with depth, a fair number of tests have to be conducted and the test depths judiciously chosen. It requires no laboratory work and test results can be processed very quickly using available spreadsheets.
- Constrained moduli obtained from the consolidation tests are in fair agreement with the plate bearing moduli but somewhat lower. Calculations taking into account the void ratio of the virgin soil and the prevailing soil moisture content indicate a degree of saturation of less than 60 %. Although the in situ soil moisture is above compaction optimum, it is well below full saturation and did therefore not affect the compaction process adversely.

Consolidation tests also have significant value since they provide both a measure of static stiffness and at the same time involve measurement of the in situ density. In other words, if consolidation samples can be extracted in a simple and effective manner, it supersedes the need to do conventional in situ density tests.

8 CONCLUSIONS

The value of appropriate tests like indicator, moisture content and compaction tests are highlighted to characterise a site, over and above proper soil profiling.
The effectiveness and depth of impact compaction can be captured by a variety of tests which focus on different aspects. A combination of these like level surveys, DCP, plate bearing and consolidation tests yield complimentary parameters which together provide a sound overall picture of the effect achieved. Test procedures can be adapted to represent a system that not only advances site characterisation but also integrity testing.

All the tests have indicated that at moisture contents close to or somewhat above optimum, compaction of predominantly fine sandy and silty soils is achieved far more effectively and to much greater depths in the case of impact compactors as in the case of vibro compaction.

REFERENCES

Barret, A. J. & Wrench, B. P., (1984). Impact rolling trial on 'collapsing ' aeolian sand, *Proc. of the 8th African Regional Conference on Soil Mechanics and Foundation Engineering, Harare, 1984.*

Pinard, M. I., Ookeditse, S. & Fraser C., (1988). Evaluation of impact roller compaction trials on potentially collapsing sands in Botswana, *Proc Annual Transportation Convention, Pretoria: Vol 3B, Paper 7.*

Pinard, M. I. & Strydom, J. H., (1998), Innovative Compactive Techniques for roads and airfields using high energy Impact Compactors, *Proc. Second International Conference on Unsaturated Soils, Beijing China, 27 – 30 August 1998.*

Scheckel, D. M., (1985). Impact rolling of collapsing sand subgrades for low maintenance roads, *Proc Annual Transportation Convention, Pretoria: Vol 5B, Paper 3.*

Strydom, J. H. & Hefer, A., (1998). Report on the trials at Kriel to assess the effectiveness of impact compaction and to establish appropriate methods of integrity testing. *Africon report 50444AE/G1/98 to Landpac, January 1998.*

7 Case studies

Etudes de cas

Geotechnics for Developing Africa, Wardle, Blight & Fourie (eds) © 1999 Balkema, Rotterdam, ISBN 90 5809 082 5

Lavumisa Dam, Swaziland – Interpretation and analysis of foundation seepage and grouting

G. N. Davis
Council for Geoscience, Pretoria, South Africa

H. F. W. K. Elges
Department of Water Affairs and Forestry, Pretoria, South Africa

ABSTRACT: Steadily increasing foundation seepage recorded upon filling of the recently completed Lavumisa Balancing Dam was cause for some concern. Attempts to reduce seepage by means of a partial clay blanket did not have the desired effect. Rather than grouting of the entire centre-line, a contact zone between the founding lava and intrusive dolerite was identified as a possible seepage path and grouted. In spite of extending the grout curtain, continued monitoring of the seepage revealed little overall reduction in seepage flow rates. These foundation grouting results are discussed in the light of this apparent failure in achieving a sealing of the foundation. The predominantly fine fractures are evidently not groutable. In such cases a clay blanket is a better option, provided specific design criteria are adhered to.

1 INTRODUCTION

The Lavumisa Balancing Dam is located just within the southern border of the Kingdom of Swaziland, approximately 4 km north-west of the Lavumisa-Golela border post.

The dam was constructed in 1993 - 1994 to compensate for inundation of part of Swaziland which results from filling of the Pongolapoort Dam in South Africa. Water from the Pongolapoort Dam is pumped to the Lavumisa Dam via two pumping stations and a 8,25 km rising main.

The dam comprises a zoned earth embankment with a 300 m long side-channel spillway chute on the right flank (Figure 1). Other statistics are presented in Table 1. A typical embankment section is shown in Figure 2.

The area of interest is underlain by basalt of the Lebombo Group which was subsequently intruded by numerous N-S striking dolerite dykes. The basalt lavas dip towards the east at low to moderate angles

Table 1. Lavumisa Balancing Dam statistics

Crest length	578 m
Maximum wall height	12,3 m
NOC level	RL 220,3 m
Full Supply Level	RL 218,2 m
Total capacity at FSL	373 330 m³

(approximately 20°), while the younger dolerite dykes are roughly perpendicular to this and dip very steeply to the west. A number of faults are recognized (Morgan 1951, Rossouw, 1951).

2 INVESTIGATION SYNOPSIS

Three dam sites were initially considered; the favoured site being located in a small valley roughly at the position of an old farm dam. All three sites were investigated.

At the selected site eighteen auger holes were drilled at regular intervals in the proposed basin. Results of laboratory testing are summarized in Table 2. Sampling revealed higher clay contents in low-lying areas, presumably from deposition within the old farm dam, yet the mean clay content was only 14,6 % (after discarding the highest and lowest values). In addition, seven holes were drilled for purposes of in situ permeability testing (DWA 1993). Results varied between 10^{-3} - 10^{-4} cm/s whereas the alternative sites yielded permeabilities of approximately 10^{-2} cm/s.

Additional investigations conducted at this favoured site included the drilling of six shallow rotary diamond boreholes as well as the excavation of several test pits along the proposed spillway (Davis 1993). The boreholes along the proposed centre-line produced extremely poor core recoveries, suggesting small core-stones within a matrix of highly to completely weathered material which was largely washed away during the drilling process.

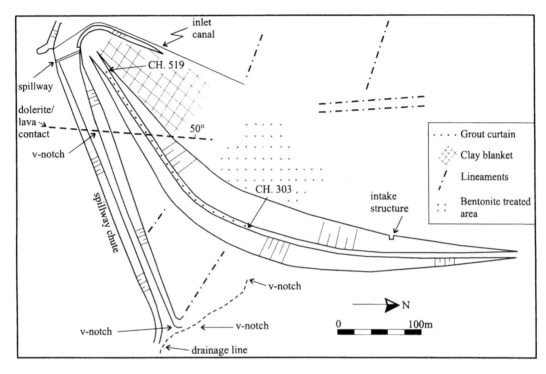

Figure 1. Structure layout

Water pressure (lugeon) tests were not possible in the circumstances.

Construction material investigations considered sources of impervious core material as well as general fill, filter and concrete materials (DWA 1993). Investigation of potential sources of rip-rap material included the drilling of additional boreholes with laboratory testing focussing on the question of durability (Davis 1993).

3 DEVELOPMENT OF SEEPAGE PROBLEM

Only minor seepage (less than 1 l/s) was initially recorded within the spillway excavation, but at the time reservoir levels were low. Subsequent monitoring revealed a worrying trend; with the reservoir maintained at higher levels (RL 214 - 215), a steady increase in seepage was observed from less than 1 l/s to almost 5 l/s, measured at the end of the spillway chute. Of prime concern were possible long term effects of this seepage on the founding materials; while the loss of this pumped and therefore expensive water also had certain financial implications.

At roughly one third distance along the spillway seepage was seen to flow from the chute floor. Other seepage exit points were recognized at intervals along the chute. Importantly, observed seepage flows comprised clear water. The lower reaches of the chute were obscured by dense reed growth thus preventing closer observation.

A limited clay blanket located immediately upstream of the embankment had no noticeable effect in reducing seepage losses, neither did bentonite treatment of selected areas within the basin (Figure 1). Earlier exploratory boreholes were also relocated and backfilled. Further options considered at that stage included bentonite treatment of the entire basin, and foundation grouting.

Table 2. Summarized laboratory test results

	Min.	Max.	Mean
Clay %	3,9	81,1	18,4
LL %	12,3	92,8	54,2
PL %	20,3	39,2	30,0
PI	13,4	57,2	27,8
LS %	6,3	20,0	11,8
Permeability cm/s	$1,3 \times 10^{-8}$	$2,1 \times 10^{-7}$	$4,6 \times 10^{-8}$

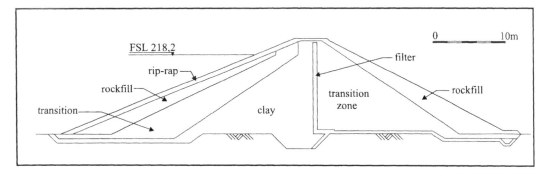

Figure 2. Embankment section

4 INVESTIGATION OF SEEPAGE

As a first approach, a series of percussion boreholes were drilled along the embankment toe to allow observation of piezometric levels. Initial observations revealed a marked depression in an area roughly aligned to the seepage exit point in the spillway chute. This suggested the presence of a more-pervious zone within the foundations. Subsequent monitoring recorded a 5 - 9 m rise in water levels together with the disappearance of this depression, even though reservoir levels were essentially the same. It was assumed that increased seepage had in effect 'flooded' this more-pervious zone.

4.1 Chute geology

The excavated spillway chute provides the only opportunity for detailed post-construction observation of the geological conditions.

Several dolerite dykes are seen to intersect the lava. These intrusive bodies are typically narrow (2 - 16 m wide). Although mostly N-S striking, with westerly dips between 50° - 60°, some variation is evident with dykes also striking in NE or NW directions while dipping at angles between 45° - 70°.

Jointing within the dolerite is well-developed and very closely spaced. The most prominent joint set possesses the same attitude as the intrusive body, with joints dipping steeply (50° - 60°) in a westerly direction. These joints are continuous across the width of the spillway chute, possibly extending into the dam basin. The joints are smooth and planar, and surfaces as a rule are stained with thin clayey infill (thickness less than 1 mm). Two sets of sub-vertical joints are also recognized, striking in westerly and north-easterly directions, respectively.

While the dolerite intrusions are moderately weathered, the basalt lava is more variably weathered with alternating zones of highly weathered friable rock and medium hard rock. In contrast to the dolerites, the lavas are more massive

in appearance with variable jointing. Joints are however typically stained and smooth.

The prominent seepage point coincided with a contact between the basalt lava and a major intrusive dolerite dyke dipping steeply (50° - 60°) in a westerly direction (Figure 1). Closer observation revealed the contact to comprise narrow, alternating horizons of basalt and dolerite.

Rather than grouting the entire centre-line, this dolerite-lava contact was targeted. The orientation of this recognized seepage path, and therefore the point of intersection beneath the dam foundation, was confirmed by excavation of a trench within the dam basin.

5 FOUNDATION GROUTING

Initially, primary grout holes at 6 m centres were drilled between CH. 387 - 453, and grouted, followed by secondary and tertiary holes where necessary.

Holes were grouted downstage. Casings were installed within the clay core with a packer being installed at the base of this casing. The shallow depths and the associated risk of damaging the clay core necessitated the use of low grout pressures. Near-surface grouting was conducted at 100 kPa, increasing to 200 or 300 kPa, or even 400 kPa for deeper stages. Generally a 2:1 water cement mix (by mass) was used, occasionally starting with a 3:1 mix, thickening to 1:1 where excessive grout takes were recorded. Ordinary Portland cement was used, together with pulverized fly ash (pfa). A high speed / high shear mixer was used, thus ensuring a high quality grout. Although joint measurements in the spillway chute and exploratory trench indicated steeply-dipping and sub-vertical joints were dominant, grout holes were drilled vertically both to avoid unnecessary disturbance of the clay core as well as for ease of operation. All grout stages were water pressure tested prior to grouting.

617

With no apparent reduction in seepage flows this curtain grouting was extended as follows;

- Within the area of the contact zone (CH. 438 - 453) the grout curtain was deepened to a maximum depth of 45 m beneath the core trench.
- The curtain was extended in the direction of the right flank (to CH. 519). Only primary holes were drilled and grouted here, to a depth of 15 m into the foundation.
- The curtain was also extended towards the central portion of the embankment. Primary holes only were drilled between CH. 303 - 387, extending to depth 20 m below the core trench.

5.1 Results

Water pressure testing of the grout stages showed an even spread of water acceptances from very low values (< 2 lugeons) to very high values between 30 - 50 UL. Only 5 % of the stages recorded losses greater than 50 UL. In contrast; the majority of the grout takes (> 60 %) may be classified as very low, after Deere 1976, with grout takes of less than 12,5 kg/m. The number of stages recording grout takes greater than this show a marked reduction. Only three stages yielded high grout takes (> 200 kg/m).

No real pattern can be discerned for the respective lugeon values and grout takes. High WPT values were recorded at apparent random intervals along the grout curtain. Between CH. 387 - 453 the stages yielding high WPT values were not restricted to primary or even secondary holes. Tertiary holes commonly recorded high values; suggesting limited grout penetration from the flanking holes. Almost all the WPT values greater than 50 UL are broadly associated with the dipping contact zone originally targeted for grouting. Very low WPT values were however also recorded within the contact zone, indicating the potential seepage path is not uniformly pervious. A further zone reflecting a concentration of high WPT values is located between CH. 387 - 411. Only primary holes were drilled in areas between CH 303 - 387 and CH. 459 - 519 and, considering the random distribution of significant WPT values, more pervious zones are not readily identified in these areas.

Because so few grout stages recorded high takes, the apparently random distribution is not unexpected. No grout takes greater than 100 kg/m are associated with the relatively pervious contact zone. A few stages recording takes greater than 100 kg/m are however associated with the zone between CH. 387 - 411 which also yielded the relatively high proportion of high WPT values. As with the case of the WPT values, the high grout takes were not restricted to primary holes but were rather predominantly associated with the secondary holes.

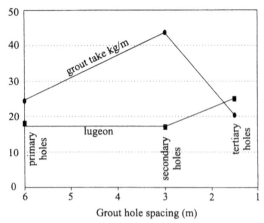

Figure 3. Summarized water and grout takes (upper stages to depth 15 - 20 m)

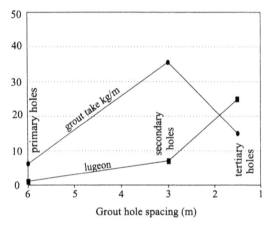

Figure 4. Summarized water and grout takes (lower stages 15+ m)

Because all stages were water pressure tested prior to grouting it is possible to briefly consider the relationship between lugeon value and grout take. For a number of reasons, correlation between lugeon value and grout take is generally weak, as noted previously by other authors (Houlsby 1990, Ewert 1992). A few stages yielded the combination of high WPT values and large grout takes indicative of open joints, while many stages yielded low WPT values and small grout takes. A number of stages recorded low WPT values but significant grout takes, suggesting the opening of fractures due to the grouting pressures. Most commonly, high WPT values were associated with small grout takes indicating very fine fissures which are not groutable (Ewert 1992).

Only between CH. 387 - 453 were primary holes followed by intermediate secondary and tertiary

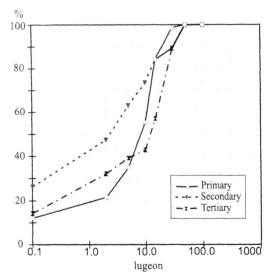

Figure 5. Frequency distribution of water acceptances

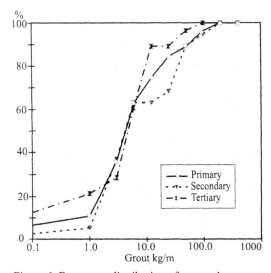

Figure 6. Frequency distribution of grout takes

values may be severely distorted by the occasional unusually high lugeon values.

With the apparent failure of the grouting to significantly reduce the seepage flows, further illustrated by relative frequency distribution plots after Ewert, 1998 (Figures 5 & 6), the decision was taken to suspend grouting.

6 ONGOING SEEPAGE MONITORING

Two additional v-notches were installed, located immediately below the prominent seepage exit point in the spillway, and downstream of the toe drain outlet, respectively. Until that time, it had been assumed that most seepage measured at the end of the spillway chute originated from this upper dolerite - lava contact.

However, at relatively low reservoir levels (RL 214) the upper seepage point associated with the dolerite - lava contact contributes only 20 - 30 % of the spillway seepage, which in turn comprises 82 % of total seepage. At higher reservoir levels (say RL 217) this seepage point contributes roughly 60 % of the measured spillway seepage which still represents approximately 80 % of total seepage.

Other observations from the ongoing seepage monitoring (Figure 7) include the following;

• With gradual emptying of the reservoir due to this seepage and evaporation, a similar gradual decrease in seepage flow rates was observed.
• Upon filling of the reservoir by pumping an immediate response is noted in the form of increased seepage flows. Similarly, an immediate decrease in seepage is recorded following releases from the reservoir.
• In terms of contributions to total seepage, most measured seepage is intercepted by the spillway chute. Seepage collected by the toe drain is significantly less.
• Toe drain seepage is noticeably less-responsive to fluctuations in reservoir level. This is particularly evident with gradual lowering of reservoir level where the toe drain seepage remains more-or-less constant.

7 DISCUSSION

The continued seepage flows monitored after completion of foundation grouting would appear to point to the failure of the grouting programme. This is further reflected in the plot of mean lugeon values (Figures 3 & 4) which show steadily increasing WPT values from primary through secondary to tertiary holes, indicating that closure was not achieved. Total post-grouting seepage flows, measured in excess of 6 l/s, might be considered to

holes. Only in this area can the effects of closure therefore be observed. Considering the upper and lower stages of the respective grout holes separately (Figures 3 & 4), it is evident that the respective water acceptances show a steady increase from primary through secondary to tertiary holes. The corresponding grout takes increase from primary to secondary hole spacing but decrease for the tertiary holes, suggesting wider openings may have been grouted but not the finer fissures. Although it may be assumed that closure has not been achieved, the limited number of data sets implies these mean

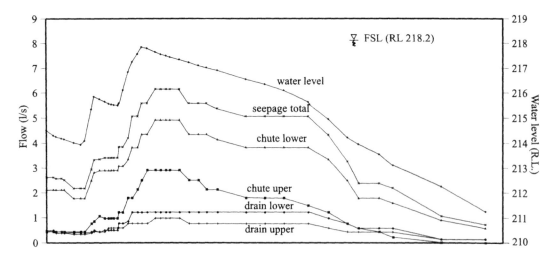

Figure 7. Post-grouting seepage monitoring (March 1998 - February 1999)

further underline the apparent failure of the grouting in sealing the foundation.

For meaningful comparison however, seepage flow measurements from the same position and for similar reservoir levels should be considered. Pre-grouting seepage flows were only recorded at the end of the spillway chute, where losses approaching 5 l/s were recorded for reservoir levels between RL 214 - 215. Post-grouting seepage flows between 2 - 3 l/s were measured at the end of the spillway chute, at similar reservoir levels. It would therefore appear that foundation grouting reduced these seepage losses by approximately half.

It should however be borne in mind that combined spillway and toe drain post-grouting seepage losses in excess of 6 l/s are recorded for a reservoir level approaching the full supply level (RL 218,2).

Large scale aerial photographs (1:3000) provide a possible explanation for this seepage. Two lineaments representing potential seepage paths are recognized. The more-prominent E-W striking lineament follows the natural valley, intersecting the centre-line and coinciding with the seepage area within the spillway chute. The N-S striking lineament roughly corresponds to the main dolerite-lava contact. The two lineaments intersect within the dam basin.

Although there is a clay lining within the basin this blanket only covers portion of the right flank between the embankment and the inlet canal (Figure 1), extending a maximum of 80 m from the embankment heel. The bentonite-treated areas, notably within the central part of the basin, are similarly limited in extent. It is unlikely that a blanket of such limited extent would effectively cut

off the seepage - unless the actual point of ingress was itself targeted. No such area was identified within the basin, although the soil cover in places is almost non-existent.

An issue to consider here is whether the Lavumisa Dam foundations were in fact groutable. According to Ewert (1992), WPT values greater than 20 UL indicate groutable conditions. As some 35 % of grout stages recorded WPT values greater than this, it might be expected that a similar proportion of these grout stages would have recorded significant grout takes. However only 16 % of stages recorded moderate to high takes of 50 kg/m and above.

The joint separation, or width of fracture opening, is of critical importance when considering the groutability of the rock mass, particularly if seen in the context of grout particle size.

The dolerite within the spillway chute, as well as within the trench excavated in the basin, was observed to be very closely jointed. Jointing of the basalt lava was more variable, although still closely spaced in places. As mentioned, the joints are generally tight but typically stained, with a thin clay infill (width up to 1 mm) observed within the trench. Conditions at depths greater than exposed within the 6 m deep chute are not known with certainty, but at worst may be expected to be similar. Greater joint separation is unlikely.

Rock mass permeability, fracture spacing and separation are all closely related. For a given permeability of say 20 UL, a joint separation of 0,15 mm is calculated for a spacing of one joint / metre; reducing to a separation of 0,1 mm for four joints / metre (Fell et. al. 1992, after Louis 1969). Considering therefore the closely to very closely jointed nature of the Lavumisa founding rock mass, mean fracture separations much less than 0,1 mm must be assumed.

There appear to be conflicting views on the groutable widths of fractures. Houlsby (1990) considers a limit of 0,5 mm for normal grouting, or 0,4 mm for high quality grouting. Fell et. al. (1992) consider the limiting fracture width to be three times the maximum particle size of the grout in suspension. For conventional, rather than specialized micro-fine cement this would imply minimum groutable fracture openings between 0,15 - 0,18 mm. With the exception of occasional wider openings, it would appear that the fine fractures at Lavumisa are therefore not groutable. This is borne out by the high percentage of grout stages which recorded high WPT values but only small grout takes.

The observed pattern of apparently random WPT values and grout takes, particularly for the tertiary holes, may imply that grout penetration from the respective primary and secondary holes was less than 1,5 m; hence no reduction in WPT values for these tertiary holes was recorded. Conversely, the marked increase in tertiary WPT values, compared to the secondaries, but with reduced grout takes, might suggest the development of fine cracks during grouting of adjacent primaries and secondaries; cracks too fine for grout penetration.

The somewhat unfavourable grout hole orientation with respect to the predominant sub-vertical or steeply-dipping joint sets has been mentioned. As a result, intersection of all possible seepage paths cannot be assumed, particularly in areas where grout holes were not drilled closer than a 6 m primary hole spacing.

8 CONCLUSIONS

Viewed solely in terms of the desired grouting criteria, namely the sealing of foundation seepage, grouting at Lavumisa cannot be considered successful.

Considering comparable seepage flows at similar reservoir levels would however appear to suggest seepage reduction by approximately half - at least in certain areas. Significant seepage nevertheless still occurs, with losses in excess of 6 l/s at high reservoir levels.

Indications are that the generally fine fractures within the foundation were not groutable. In cases like this, installation of a clay blanket might be considered the better option. A discontinuous clay blanket will however have little or no effect in preventing seepage. The partial blanket placed at the Lavumisa Balancing Dam was too patchy and, considering the variable thickness of 250 - 300 mm, too thin. It is essential that such a clay blanket should have a minimum thickness of 1 m, placed in 3 - 4 layers to ensure good compaction, and extend into the basin for a distance of at least 12 - 20 times the maximum head of water. Also important is competent site supervision in order to ensure adherence to design specifications.

Further options for reducing the seepage which might be considered at this stage include bentonite treatment of the entire basin or perhaps chemical grouting of the fine cracks, but these are invariably very expensive. Pumping of the seepage back into the reservoir might also be considered. Currently, the favoured approach is to continue monitoring of seepage flows, as well as reservoir levels. Evidence of internal erosion or steadily increasing seepage flows would be of particular significance. Perhaps, with time, seepage will be reduced to acceptably low levels by natural sedimentation within the reservoir.

ACKNOWLEDGEMENTS

The Director General of the Department of Water Affairs and Forestry and the Director of the Council for Geoscience are thanked for permission to publish.

REFERENCES

Davis, G.N. 1993. Lavumisa Balancing Dam: Swaziland. First Engineering Geological Design Report. Unpublished *Council for Geoscience Report* 1993 - 0116.

Davis, G.N. 1997. Lavumisa Balancing Dam: Swaziland. First Engineering Geological Maintenance Report. Unpublished *Council for Geoscience Report* 1997 - 0294.

Deere, D.U. 1976. Dams on Rock Foundations - Some Design Questions *in* Rock Engineering for Foundations and Slopes. *Conference II.* Boulder, Colorado.

DWA. 1993. Lavumisa GWS: Materials Investigation for Balancing Dam. (in afrikaans). *Department of Water Affairs Report* no W450/04/MJ01.

Ewert, F-K. 1992. The Individual Groutability of Rocks. *Water Power and Dam Construction.* Jan 1992.

Ewert, F-K. 1998. Permeability, Groutability and Grouting of Rocks Related to Dam Sites. Part 4. Groutability and Grouting of Rocks. *Dam Engineering.* Vol VIII. Issue 4.

Fell, R., MacGregor, P. & Stapledon, D. 1992. *Geotechnical Engineering of Embankment Dams.* Balkema.

Houlsby, A.C. 1990. *Construction and Design of Cement Grouting: A Guide to Grouting in Rock Foundations.* John Wiley & Sons, Inc.

Louis, C. 1969. A Study of Groundwater Flow in Jointed Rock and its Influence on the Stability of Rock Masses. *Imperial College Rock Mechanics Report No. 10.*

Morgan, R.P.E. 1951. The Geology of the Gollel Area. Unpublished *Geol. Surv. S. Afr. Report* 1951 - 0059.

Rossouw, P.J. 1951. The Geology in the Area of Pongola and in North-east Zululand. (in afrikaans). Unpublished *Geol. Surv. S. Afr. Report* 1951 - 0097.

Geotechnics for Developing Africa, Wardle, Blight & Fourie (eds) © 1999 Balkema, Rotterdam, ISBN 90 5809 082 5

The methodology used to prove the clay reserves at Maguga Dam, Swaziland

K. Le Roux
Maguga Dam Joint Venture, Swaziland & SRK Consulting, Johannesburg, South Africa

H. A. C. Meintjes
SRK Consulting, Johannesburg, South Africa

ABSTRACT: At Maguga Dam borrow areas for clay core material were proved by the implementation of a simple methodology that identified the potential borrow areas, delineated the potential clay core deposits to determine the available volume and thirdly, implemented a detailed laboratory testing programme. The testing programme determined the engineering behaviour of the material and consequently the suitability for use as clay core material. This paper describes how this methodology was successfully implemented at Maguga Dam.

1. INTRODUCTION

Maguga dam is currently under construction near Pigg's Peak in Swaziland and will ultimately supply water for agriculture to the downstream Swaziland and South African users. A small hydroelectric power station is planned to supplement the Swaziland national grid. Maguga Dam will be 115m high and retain a reservoir of 332 000 000 m³.

The development of Maguga Dam has been considered since the early 1980's and a total of four geotechnical investigations have been undertaken since 1985. The most recent and final investigation conducted in 1996/1997, formed part of the detailed design phase of the dam project.

Several dam design options were considered for detailed design, these being: an asphalt core rockfill dam, a concrete face rockfill dam and a clay core rockfill dam. All three of these options were included in the tender documentation and the winning tender was for a clay core rockfill dam.

One of the objectives of the detailed design geotechnical investigation was to obtain an estimate of the quantity of available core material in close proximity to the dam site. Ideally, the core material must be situated within the basin below full supply level to minimise the impact of the construction of the dam on the environment.

2. SITE GEOLOGY

Maguga Dam is situated in the north west of Swaziland, approximately 15km South of Pigg's Peak (see Figure 1). This area is underlain by a granite batholith known in Swaziland as the Lochiel Hood Granite. The batholith intruded to form a sub-horizontal hood over the Ngwane gneiss and Swaziland Supergroup about 3000 to 3200 million years ago. In the vicinity of the dam the Ngwane gneiss, the oldest rocks in Swaziland at approximately 3500 million years, occur as minor inliers or xenoliths.

Several younger granitic plutons are exposed in Swaziland, namely the Mswati Granite and the Nkalangeni Granite Pluton aged at about 2600 million years ago. In addition to these granitic intrusions there are several mafic intrusions that occur throughout the area. The dykes vary in orientation from northwest to North-north west.

These mafic intrusions vary in age and have, for the purposes of this paper, been divided into two categories

a. karoo age mafic intrusions (about 1800 million years old)

b. pre-karoo mafic intrusion which are generally metamorphosed to greenschist facies. A pre-karoo gabbro dyke 100-150m wide occurs on the left bank of the Komati river at the dam site and a pre-karoo dolerite dyke forms the natural gorge along which the Mangunga stream flows on the right bank of the river. A Karoo age dolerite sill forms the hilltop to the north west of the left flank and 7km north of the dam site. The colluvial and residual material derived from these mafic intrusions provide suitable core material for the clay core rockfill dam.

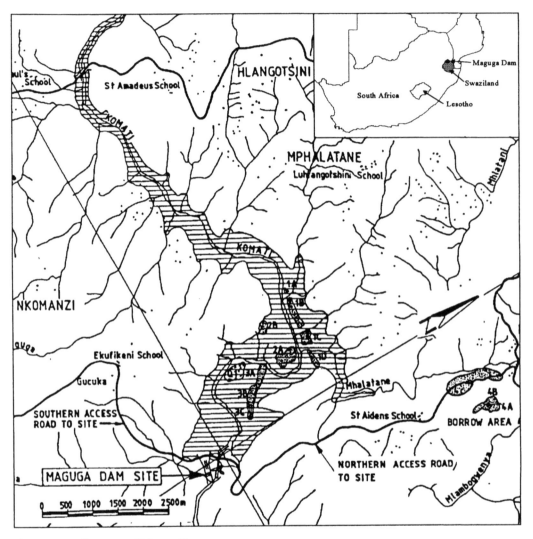

Figure 1: Locality maps of Maguga Dam

3. INVESTIGATION METHODOLOGY

The methodology that was implemented in this project to prove the feasibility of a clay core dam is not new. It involves a sequence of logical and practical steps based on sound geological and geotechnical practice. This methodology is flexible and can be adapted to any similar geotechnical project.

There are 3 major objectives to a proving a clay core borrow area. All three of these objectives have significant technical and economic implications in the feasibility of the project. These essential objectives are:

- sufficient volume
- suitable quality
- close proximity

In order to achieve these 3 objectives a three-phase investigation programme was implemented. The three phases can be described as follows:

a) Identification of the potential borrow areas through the study of available literature, aerial photograph interpretation and the confirmation of the research through surface mapping.

b) Delineation of the borrow areas in three dimensions from surface and subsurface investigation. The calculation of the available volume from the delineated geometry of the clay deposit.

c) Proving, through an extensive sampling and laboratory testing programme, the suitability of the material in terms of it's engineering properties.

The single most critical aspect of this methodology is the existence of an information feed-back loop that enables an evaluation of the project as data and information become available. The assessment of the data is critical to on-site decisions regarding further investigation requirements.

The methodology implemented at Maguga Dam is illustrated in Figure 2 below.

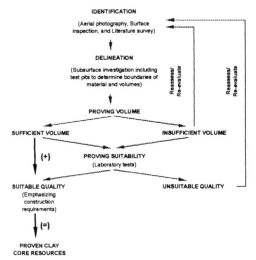

Figure 2: The methodology implemented at Maguga Dam.

The figure illustrates that the evaluation of the data is an ongoing process that fine-tunes the investigation as more knowledge of the clay deposit is gained. The evaluation of the data may lead to the need to identify further potential borrow areas, or a need to redefine the boundary of the borrow area thereby changing the volume.

The Maguga Dam project is a good example of how this methodology can be successfully implemented. All of the above phases will be discussed in more detail with reference to the Maguga project below.

4. MAGUGA DAM CASE STUDY

The objectives of the geotechnical investigation for borrow material at Maguga Dam are summarised below:

• To prove an available volume of at least 1.5 times the required 1 000 000 m³ of core material.

• The material must have enough plasticity to ensure easy working of the material and prevent cracking of the core due to differential settlement. In addition the material, when compacted to the construction dry density, must be of low permeability (1×10^{-10} m.s^{-1}) and have sufficient shear strength. This will ensure the long-term stability of the dam.

• Ideally the borrow area must be situated close to the dam site and, preferably, be within the dam basin.

4.1 Identification of Borrow Areas

The identification of potential borrow areas at Maguga Dam was achieved by following three steps:

• Obtaining information from previous geotechnical work in the area.

• Aerial photograph interpretation.

• Surface inspection of possible borrow areas.

Since 1985 three geotechnical investigations have been conducted at this site prior to the final detailed design investigation conducted by Maguga Dam Joint Venture (MDJV, 1997).

Only the first investigation (M.J. Mountain, 1986) considered the clay core rockfill dam option feasible. The second and third investigations (Sir Alexander Gibb and Partners, 1990 and George Orr and Associates, 1996) concentrated on the feasibility of concrete dams. Insightful information was gained with regards to the location and engineering properties of potential clay core borrow areas.

The MJ Mountain investigation (1986) identified the gabbro dyke on the left bank of the Komati River and a dolerite sill about 7km north east of the site as potential borrow areas.

An extensive aerial photographic analysis of the area using large scale (1:500) confirmed the occurrence and locality of the gabbro dyke and dolerite sill identified in the MJ Mountain (1986) investigation. In addition a large dolerite sill north west of the dam site was identified. Colluvial material located down slope, and derived from, this dolerite sill was considered a potential source of clay material.

Surface inspections of the identified borrow areas confirmed that the soil did exhibit different characteristics to the residual granite exposed over most of the site. The test results conducted on the borrow material later proved that it was non-dispersive.

Generally, the soil in the borrow areas are redder than elsewhere on the site. The borrow material has a higher fines content and an increased plasticity in comparison to the surrounding residual granite. The borrow areas were selected from non-dispersive clay sources.

The aerial photographic analysis identified a few disruptions in the continuity of the gabbro dyke. These disruptions were caused by alluvium deposited in drainage gulleys formed on small faults across the dyke. This indicated that the MJ Mountain (1986) investigation over-estimated the available volume of core material in this borrow area.

In total 4 borrow areas were investigated in detail in the MDJV (1997) detailed design investigation. The borrow areas are:

• Borrow Area 1 - colluvial clay on the left bank of the Komati River about 3km upstream from the dam position.
• Borrow Area 2 - colluvial clay and reworked residual gabbro on the right bank of the Komati River about 3km upstream from the dam position.
• Borrow Area 3 - colluvial and residual sandy clay derived from the underlying gabbro dyke. The gabbro dyke is exposed on the left bank of the Komati River at the dam site and extends to the northwest for 1.5km.
• Borrow Area 4 - colluvial and residual clayey silt derived from the underlying dolerite sill. The Karoo age dolerite sill is exposed about 7km east of the dam position.

Once the approximate positions of the borrow areas were fixed on site, the detailed subsurface investigation to determine the volume of the potential borrow areas could proceed.

4.2 Delineation of the Borrow Areas

The most practical and reliable way to define the geometry of a clay deposit is to excavate test holes that provide "windows" of information on which the volume of suitable material can be assessed. Representative samples of different material occurring in the borrow area are taken for detailed laboratory testing.

Generally the more test pits excavated, the more information regarding the variability of the clay deposit is known. Consequently, a more accurate assessment of the total volume and variation in engineering behaviour can be made. A total of 144 test pits were excavated in 4 borrow areas in the detailed design (fourth) stage of investigation at Maguga Dam (MDJV, 1997).

Positioning of test pits to prove a borrow area is controlled by two major objectives:
• The accurate determination of the boundaries of the clay reserve in three dimensions.
• An assessment of the variability of the engineering behaviour of the proposed core material.

The total available volume of suitable core material is calculated from the three-dimensional geometry and variability within the borrow area.

Borrow Area 3, the gabbro dyke, was initially identified as the primary source of clay by the first phase geological investigation (MJ Mountain, 1986). Detailed investigation of this borrow area during the detailed design phase (MDJV, 1997) proved a total of only 530 000 m^3 of clay was available. This is not sufficient volume for the construction of the dam, as the volume required for the dam core is approximately 1 000 000 m^3.

To supplement the material available from Borrow Area 3, Borrow Areas 1, 2 and 4 were investigated. The results of the investigation are summarised below.

4.2.1 Borrow Area 1

A total of 37 test pits were excavated to an average depth of 5m to determine the geometry and engineering behaviour of the clay deposit.

Borrow Area 1 comprises colluvial material derived from a dolerite sill situated upslope of the borrow area. The colluvial clay overlies granite that forms the bottom boundary of the borrow area. Alluvial sand deposits associated with drainage channels that traverse the borrow areas divide it into sub-areas. The field description of the suitable core material is: slightly moist red brown, stiff to very stiff sandy clay, colluvium with occasional (<5%) boulders (SAIEG, 1976).

The sub-areas were found to vary in terms of engineering behaviour, and thickness of suitable core material. Table 1 summarises the variation between the sub-areas. The occurrence of boulders and gravel within the suitable core material reduces the volume of available material. The variability of the location and percentage of bouldery material impedes the efficient exploitation of the clay deposit. However the boulder layer generally occurs at the base of the exploitable horizon, and careful excavation would ensure efficient exploitation of the clay deposit.

4.2.2 Borrow Area 2

A total of 34 test pits were excavated to determine the geometry and engineering behaviour

Table 1: Variations in plasticity and percentage of boulders between the sub-areas of Borrow Area 1.

Sub-area	Layer Thickness (m)	Plasticity	Boulders (%)
1a	2.9	-	>5
1b	1.8	13	5
1c	2.7	23	10
1d	2.0	19	5-10

of the clay deposits. The test pits were concentrated near the estimated boundaries and consequently only 11 of these encountered suitable core material.

Borrow area 2 is divided into 2 distinct sub-areas (2a and 2b) on account of their different origins. Sub-area 2a comprises colluvium derived from mafic intrusions, and metamorphosed mafic rocks that have intruded into the granite.

The field description of the suitable core material is: moist red brown, stiff to very stiff sandy clay, colluvium with occasional (<5%) boulders (SAIEG, 1976).

Sub-area 2b comprises colluvial and reworked residual gabbro material derived from the underlying gabbro dyke. This gabbro dyke is the northern extension of the gabbro dyke that forms Borrow Area 3. The field description of the suitable soil is: moist reddish brown stiff sandy clay, colluvium and residual soil derived from the underlying gabbro (SAIEG, 1976).

4.2.3 Borrow Area 3

The colluvial and residual material derived from the underlying gabbro dyke is considered the primary source of core material. A total of 45 test pits of which 23 encountered suitable core material were excavated. Several test pits were excavated close to the gabbro dyke boundary to determine the extent of the gabbro dyke and consequently the horizontal limits of the borrow area.

The borrow area is relatively homogenous and is described in terms of a generalised profile in Table 2 below.

Table 2: Generalised soil profile of Borrow Area 3. (SAIEG, 1976).

Range of Depth (m)	Description
0.0 – 1.7	Reddish brown CLAYEY SILT. Colluvium derived from completely weathered gabbro, grano-diorite and dolerite.
0.5 – 2.00	Reddish brown SANDY CLAY. Residual soil derived from completely weathered gabbro.
0.9 – 6.00	Green grey SANDY CLAY. Residual soil to completely weathered gabbro

The reddish brown residual and colluvial material was identified, in the field, as being suitable core material. The suitability of the completely weathered green grey gabbro was questioned in the field, on account of its low plasticity. Laboratory results confirmed that the completely weathered green grey gabbro could only be used in the core if it was mixed with the overlying red brown highly plastic soil.

Sub-division in this borrow area is based primarily of thickness of suitable material which ranges from 2.7m to 5.0m. The subdivision of Borrow Area 3 is shown in Figure 1.

4.2.4 Borrow Area 4

This borrow area comprises colluvial and residual red-brown clayey silt derived from the underlying Karoo age dolerite sill. Only sub-areas 4a, 4b and 4c (Figure 1) were investigated in detail. This is because the remainder of the dolerite sill has a very shallow weathering profile and could not be exploited as a borrow area.

A total of 28 test pits were excavated to determine the depth and variability of the suitable clay deposit. One of the primary sources of variability is the percentage of boulders occurring within the exploitable material. The percentage of boulders ranged from about 5% in sub-areas 4a and 4b to about 10% in sub-area 4b.

4.2.5 Estimated Quantity of Core Material

The total estimated volume of available material is summarised in Table 3 below.

Table 3: Summary of estimated available volume of clay core material at Maguga Dam.

Borrow Area	Colluvium Top layer	50% Colluvium 50% Res. Soil
Area No. 1:		
Subarea A	20 000	
Subarea B	200 000	
Subarea C	200 000	
Subarea D	50 000	
Sub-Total	470 000	
Area No. 2:		
Subarea A	240000	
Subarea B		100 000
Sub-Total	340 000	
Area No. 3:		
Subarea A		210 000
Subarea B		130 000
Subarea C		90 000
Sub-Total	430 000	
Area No. 4:		
Subarea A	235 000	
Subarea B	450 000	
Subarea C	560 000	
Sub-Total	1 245 000	
Total	2 485 000	

The total estimated volume of core material far exceeds the required volume of 1 000 000 m^3. The volume of material available in the basin (Borrow Areas 1, 2 and 3) is 1 240 000 m^3. This is sufficient to construct the clay core dam but does not fulfil the requirement of proving at least 1.5 times the required volume of material. The haulage distance and the added environmental considerations in developing a borrow area outside of the reservoir render Borrow Area 4 the least attractive of the borrow areas. Consequently it was recommended that the material within the reservoir be exploited first and the material from Borrow Area 4 be kept in reserve to be exploited if required.

4.3 Laboratory Testing

The objective of a laboratory testing programme is to confirm the field assessment of the proposed clay core material. The laboratory test results quantify the engineering properties and design parameters of the core material, needed to design the dam.

An extensive and comprehensive testing programme is an essential part of any geotechnical investigation of dam building materials. This is particularly pertinent when, as in the case of Maguga Dam, the subsurface investigation indicates variable material. The testing programme will indicate whether this variability will have a significant influence on the placement technique or indeed viability of the clay deposit.

One of the difficulties encountered in the testing programme at Maguga Dam was that the proposed borrow material is classified as tropical soils. The material was classified as exhibiting tropical soil behaviour on account of the mode of weathering. Most of the gradings displayed bi-modal behaviour. Tropical soils typically have higher strength, higher permeability and lower density than soils with similar gradings from temperate climates.

4.3.1 Tropical Soils

Once the proposed borrow areas were identified as consisting of tropical soils all the testing conducted on the soil had to conform to the requirements for testing tropical soils. Tropical soils do not behave as expected when using the standard laboratory testing methods.

The following considerations are essential when testing the engineering behaviour of the tropical soil (Uehara, 1982):

• The type of dispersing agent affects the gradings determined by hydrometer analysis.
• The typical USCS is not fully appropriate as the method of testing has a marked influence on the Atterberg limits. The cementing clusters are also not taken into account.
• Cementation of clay particles into clusters and aggregates are responsible for:

- High liquid limit but low plasticity index.
- High void ratios
- High shear strength
- Low compressibility
- Low Proctor dry density with high optimum moisture content.
- Sometimes higher permeability than expected from grain size distribution.

The density of tropical soils cannot be directly related to strength. Compaction test results can vary significantly (OMC ± 4%) using different sample drying methods - the maximum allowable temperature is 60° C. The tropical soils encountered in this project were wetter than optimum, a common characteristic of tropical soils. The drying of the soils at higher temperature can lead to the aggregation or cementation of particles forming clusters. Drying was therefore done with care and repetitive Procter compactions were done.

4.3.2 Testing Programme

Selected samples taken from the four borrow areas were tested to prove their suitability as core material and to determine design parameters. One of the main objectives of the testing programme was to identify areas containing high plasticity material needed for placement in small zones where maximum differential settlement is expected. This includes the areas with a significant change of gradient, the material placed at the base of the core trench and the zone between the clay core, and the concrete structure of the spillway.

The following indicator tests were conducted to describe the soil characteristics:
• Grain size analysis by wet sieving to determine the sand and gravel fraction,
• hydrometer test to determine the silt and clay fraction,
• Atterberg limits to determine the plasticity and linear shrinkage index,
• natural moisture content to enable a comparison with the optimum moisture content. This allows for the planning of the necessary moisture control of the construction material.
• Specific gravity to calculate the void ratio of the compacted material.
• Dispersivity testing using primarily the double hydrometer method and some pinhole tests.

The following tests were conducted on selected samples from each borrow area to determine the design strength parameters and the density moisture relationship. These results were used to prepare the technical specifications regarding the construction of the clay core.

• Standard Proctor tests to determine the density-moisture relationship and the maximum dry density of core material that can be achieved by compaction under optimum moisture conditions. Both MOD AASHTO and Procter tests were conducted on the samples.
• Consolidated drained (CD)-triaxial tests to analyse the stress strain behaviour of remoulded cohesive soil to obtain the deformation modulus. The shear strength for different load stages under compacted and drained conditions. This information is required for the stability of analysis of the dam during operation.
• Consolidated undrained (CU)-triaxial tests to analyse the pore water pressure development during loading and shearing under undrained conditions. This information is necessary for the stability analysis of the dam at the end of construction and after seismic loading.
• Permeability tests in the triaxial cell to confirm the required low permeability coefficient of the core material.

4.3.3 Laboratory Test Results

Assessments of the laboratory results highlighted the variability of the clay reserves. Not only did the engineering properties of the core material vary from borrow area to borrow area but variation was also observed within each borrow area, in both the vertical and horizontal directions.

The variability between the individual borrow areas was anticipated from the field observations as the parent material and mode of deposition is different in each of the borrow areas.

The laboratory test results, engineering properties and consequent exploitation concerns are discussed briefly for each borrow area below.

4.3.3.1 Borrow Area 1

Colluvial doleritic material deposited on the western flank of the Komati River comprises Borrow Area 1. The alluvium associated drainage channels that flow into the Komati River and the Komati river terrace are not suitable as core material because of the dispersive nature of the alluvial clay.

The borrow area can be divided into 4 sub-areas on account of location and variation in engineering properties. The relative positions of the sub-areas are indicated in Figure 1.

A summary of the parameters of Borrow Area 1 is presented in Table 4.

It was concluded, as was observed in the field, that sub-area 1a and 1b contain less plastic material, with a smaller fines content than sub-areas 1c and 1d. To achieve relatively homogenous material it will be necessary to mix the material in the vertical

profile and at the stockpiles with material from other sub-areas within Borrow Area 1. In addition, material from sub-area 1c can be placed in areas of the core where maximum differential settlement is expected due to the high plasticity index of 23.

Table 4: Parameters of Borrow Area 1.

Parameters	Design	Range
Fines content<0.075mm	66	26-92
Clay content (%)	33	9-61
Liquid limit	42	22-65
Plasticity index	21	10-36
Specific gravity (kN/m^2)	27	26.4-27.1
Proctor dry density (kN/m^2)	17.1	16.8-18
Optimum moisture content (%)	20	16-19.8
Natural moisture content (%)	21.7	20-22.9
Effective friction angle (°)	30	-
Effective cohesion (kPa)	10	-
Pore water A-value	0.75	0.63-0.87
Permeability (m/s)	5x10^{-9}	4x10^{-9} - 4x10^{-9}

Figure 3 presents a summary of the compaction results showing the maximum, minimum and average PI of the material. The optimum moisture content varies from 12% to 28% and the maximum dry density from 1450 to1920kg/m^3.

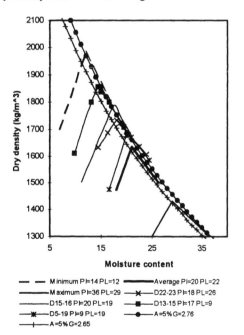

Figure 3: Summary of compaction results for Borrow Area 1

4.3.3.2 Borrow Area 2

Borrow Area 2 is situated on western bank of the Komati River and is divided into 2 sub-areas 2a and 2b. Sub-area 2a comprises colluvial material derived from predominantly mafic greenstone xenoliths and doleritic intrusions. Sub-area 2b comprises residual gabbro, the north-western extension of the gabbro dyke that forms Borrow Area 3.

Indicator tests clearly show the anticipated difference in the two sub-areas: Sub-area 2a has moderate plasticity with 52% of the particles <0.075mm and sub-area 2B has a high plasticity with 90% of the particles <0.075mm. The compaction properties of Borrow Area 2 are indicated in Figure 4.

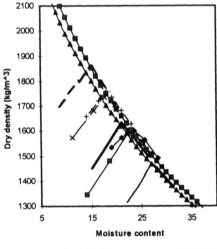

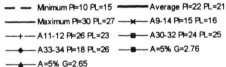

Figure 4. Summary of compaction results for Borrow Area 2.

The compaction properties vary significantly from an optimum moisture content of 27% and a maximum dry density of 1465 (kg/m³) to an optimum moisture content of 15% and a maximum dry density of 1860 (kg/m³).

A summary of the parameters of Borrow Area 2 is presented in Table 5.

It was concluded that sub-area 2a with moderate plasticity and favourable sand content was suitable for use in the construction of the clay core. The material is workable under construction conditions and a relatively high compacted density is achieved.

The clayey soil in sub-area 2b has a relatively high plasticity and does not allow compaction to a high degree. This material is best suited to zones of the core where maximum differential settlement is expected and therefore a highly plastic material is required.

Table 5: Design parameters for Borrow Area 2.

Parameters	Design	Range
Content of fines<0.075mm	58	24-90
Content of clay (%)	32	9-53
Liquid limit	42	25-57
Plasticity index	21	10-30
Specific gravity (kN/m²)	27	26.4-27.6
Proctor dry density (kN/m²)	16.3	14.5-18.6
Optimum moisture content (%)	21	16-26
Natural moisture content (%)	21.7	19-26
Effective friction angle (°)	30	-
Effective cohesion (kPa)	10	-
Pore water A-value	0.83	0.77-0.87
Permeability (m/s)	$7x10^{-11}$	$6x10^{-11} - 9x10^{-11}$

4.3.3.3 Borrow Area No. 3

Borrow Area 3 comprises completely weathered and residual gabbro overlain by colluvium derived from gabbro. An XRD analysis revealed that there is no significant difference in the mineralogical composition of the residual and colluvial materials in this borrow area.

The design parameters are relatively similar across the borrow area with the only significant variation occurring vertically. The residual material varies from highly to completely weathered gabbro. The colluvium and residual soil have a higher fines content and a correspondingly higher plasticity index than the highly to completely weathered gabbro.

Table 6 summarises the plasticity values, Table 7 summarises the design parameters of mixed material, and Figure 5 the compaction results.

The colluvial and residual soil can be used directly for the construction of the dam core. The completely weathered gabbro does not meet the requirements of an appropriate core material as it does not have enough plasticity. If the colluvial, residual and completely weathered gabbro is mixed that the material is suitable core material.

By applying an appropriate vertical excavation method it will be possible to mix the soil column to obtain a relatively homogenous soil. Ideally a 50:50 ratio of colluvium to weathered gabbro should be used.

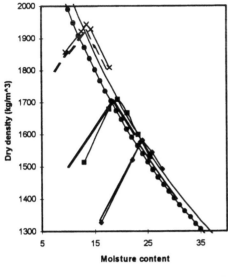

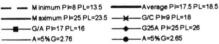

Figure 5. Summary of compaction results for Borrow Area 3

Table 6: Summary of plasticity values for Borrow Area 3.

Material	Parameter	Average	Range
Colluvium	LL	44	21 – 56
	PI	20	13 – 25
Completely	LL	29	26 – 31
weathered gabbro	PI	17	14 - 22
Mixing 50% coll.	LL	36.5	23.5 – 43.5
And 50% gabbro	PI	18.5	13.5 – 23.5

Table 7: Design parameters for Borrow Area 3.

Parameters	Average	Range
Fines content<0.075mm (%)	60	50-70
Clay content (%)	38	18-40
Liquid limit	36	23-44
Plasticity index	18.5	13.5-24
Specific gravity (kN/m^2)	26.8	26.4-27.1
Proctor dry density (kN/m^2)	18.5	17.0-19.5
Optimum moisture content (%)	15	13.5-19.5
Natural moisture content (%)	22.5	20.2-24.0
Effective friction angle (°)	33	31-38
Effective cohesion (kPa)	20	17-25
Pore water A-value	0.54	0.48-0.75
Permeability (m/s)	1×10^{-9}	1×10^{-9} - 1×10^{-9}

Additionally sub-area 3c should be mixed with sub-areas 3a and 3b, to ensure it is suitable as dam core material.

The natural moisture content of the mixed material will be close to optimum and must be taken into account when conditioning the material for construction.

4.3.3.4 Borrow Area 4

Borrow area four is comprised of residual dolerite and doleritic colluvium derived from a Karoo aged dolerite dyke.

XRD analyses showed that the mineral composition was significantly different from the material exposed in the other borrow areas. The predominant minerals are gibbsite and hematite that typically characterise laterized bauxite.

On account of the typical behaviour of tropical soils the determination of the design parameters was not easy. Every precaution was taken to ensure reliable results were achieved from the laboratory testing. The special consideration in testing tropical soils is summarised in section 4.3.1 above. A summary of the design parameters of Borrow Area 4 is presented in Table 8 below.

Table 8: Summary of the design parameters of Borrow Area 4.

Parameters	Design	Range
Fines content <0.075mm (%)	83	57 – 93
Clay content (%)	20	9 – 35
Liquid limit	47	41 – 61
Plasticity index	32	28 - 45
Specific gravity (kN/m^2)	30.7	29.7 – 33.0
Proctor dry density (kN/m^2)	15	13.2 – 17.3
Optimum moisture content (%)	29.5	22.5 – 39
Natural moisture content (%)	31.7	20.1 – 40.1
Effective friction angle (°)	35	
Effective cohesion (kPa)	10	
Pore water A-value		
Permeability (m/s)	1.2×10^{-9}	

It was concluded that the residual and colluvial material of Borrow Area No. 4 is suitable for the dam core. Care should be taken in the working of the material because of the high natural moisture content, low quartz content and the small percentage of coarse particles.

5. CONCLUSIONS

The geotechnical investigation proved a total of almost 2 500 000 m^3 of suitable core material in close proximity to the dam site. This is more than twice the required volume for the clay core rockfill dam. The success of this geotechnical investigation is significant because the dam is situated in a granite terrain that does not produce suitable dam core material.

The success of this investigation can be attributed to the implementation of the simple three phased methodology that can be summarised as:
- Identification
- Delineation
- Suitability

The use of the feed back loop system "fine tuned" the investigation so that any potential problem in the investigation could be corrected immediately and prevent any delays at a later stage.

This methodology has proved successful in the proving of material deposits but it is simple and flexible enough to be implemented in many other aspects of geotechnical investigation.

6. ACKNOWLEDGEMENTS

The authors would like to thank the Komati Basin Water Authority (KOBWA) and Maguga Dam Joint Venture (MDJV) for permission to publish this paper.

7. REFERENCES

Gibb, Sir A and Partners 1992. *Review and Feasibility Study for Komati River Basin Development in Swaziland. Phase II Geotechnical Investigation and Maguga Dam Site.* Factual Report March 1992, Interpretative Report May 1992.

George, Orr and Associates 1996. *Maguga Dam-Report on the Third Phase Geotechnical Investigations.*

Jennings, J.E., Brink A.B.A. and Williams, A.A.B. 1973. Revised Guide to Soil Profiling for Civil Engineering Purposes in Southern Africa. *Trans. S. Afri. Instu. Civil Engrs.* 15: 3-12.

Mountain M.J. and Associates 1986. Komati River Development Maguga Dam, *First Phase Feasibility Study-Engineering Geotechnical Investigations.*

Skempton, A.W., 1953. Soil Mechanics in Relation to Geology. Proc. Yorkshire Geol. Soc., 28: 33-62.

Uehara, G. 1982. Soils Science for the Tropics. Engineering and Construction in Tropical and Residential Soils. *Proc. ASCE Geot. Eng. Div. Spec. Conference.* 1982: 13-26.

Geotechnics for Developing Africa, Wardle, Blight & Fourie (eds)© 1999 Balkema, Rotterdam, ISBN 90 5809 082 5

The stratigraphy and engineering geological characteristics of collapsible residual soils on the Southern Mozambique coastal plain

C. L. McKnight
McKnight & Associates, Sandton, South Africa

ABSTRACT: From detailed geotechnical investigations for a smelter complex on the outskirts of Maputo in southern Mozambique, a composite stratigraphic column has been compiled for the upper Cretaceous and overlying Neogene marine sediments. It is proposed that the 6m thick diagnostic Pectenid-rich calc-arenite of Miocene age be termed the Matola Formation.

A Pliocene marine transgression reworked reddened aeolian dune sands with deposition offshore to form a tabular fine sandstone. Subsequent exposure and weathering during the Pleistocene has resulted in a 15-20m thick potentially collapsible residual soil profile. Collapse settlement potential of up to 13% occurs at saturation loads of 1000kPa. As a result, initial pile design was revised to allow for this geotechnical hazard. The critical degree of saturation for the red soil mantle is approximately 30%, which corresponds to a moisture content of 6%-6,5%.

1 INTRODUCTION

Detailed geotechnical site investigations for a large aluminium smelter complex on the outskirts of Maputo in southern Mozambique has identified a 15-20m thick mantle of red, iron oxide cemented, medium and fine sandy residual soils derived from extensive chemical weathering of Neogene marine sediments. Extensive laboratory testing of undisturbed samples was applied to quantify a severe collapse settlement potential within the upper 10-12m of the sequence. Design of preliminary pile founding solutions for the complex had to be re-evaluated to account for this potentially hazardous geotechnical problem. Collapse settlement potential of up to 13% at saturation loads of 1 000kPa was determined using the one-dimensional single odometer test of Jennings and Knight (1975) and verified by insitu vertical plate load testing.

In addition, the pile load test programme was modified to include determination of pile performance at working load when subjected to shaft and toe saturation. Byrne (in press) reviews load test results and pile deflection analysis.

Rotary core drilling at both the smelter site, the approach bridge over the Rio Matola and at the proposed harbour silos and jetty in the estuary, some 20km to the south, has enabled a concise local stratigraphic section to be compiled. Figure 1 shows the locality of the study area in southern Mozambique.

The diagnostic 6m thick *Pectenid*-rich calc-arenite and bioclastic limestone unit that underlies the residual soil mantle bears strong resemblance to and similar stratigraphic position as the Monzi and Sapolwana Formations of the Uloa Subgroup in northern KwaZulu-Natal. It is proposed that this biostratigraphic marker that may be a lateral equivilant of the Tembe/Santaca Formations (Dingle *et al*, 1983) be named the Matola Formation in the southern Mozambique area.

The objective of this paper is to present the local stratigraphy as defined by rotary core drilling results and to outline the engineering geological parameters of the thick red soil mantle. A genetic link between particle size distribution and critical degree of saturation has enabled the field of collapse potential to be defined in terms of insitu soil moisture content.

2 BACKGROUND

Relevant literature for the interpretation of stratigraphic and engineering geological data has been reviewed. Description of Cainozoic geology along the Southern Mozambique coastal plain is provided by Frankel (1972), Flores (1973), Förster (1975) and Dingle *et al* (1983). King (1972) studied the geomorphology of the coastal plain and linked the evolution of marine platforms to epeirogenic movements in the interior of southern Africa. Partridge

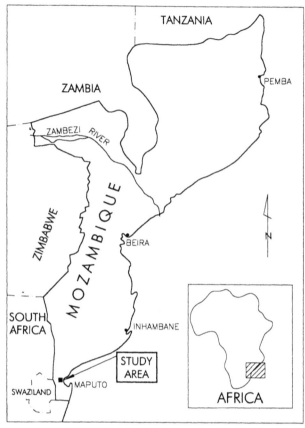

Figure 1. Locality of the Study Area

and Maud (1987) and Partridge (1998) also give detailed accounts of geomorphic evolution of southern Africa throughout the Cainozoic period.

Early geological work on 'Berea Red Sand' (Maud, 1968) exposed along the KwaZulu-Natal coastline was carried out by McCarthy (1967) and Maud (1968) and more recently Jermy and Mason (1983) described red soils encountered during construction of the Glenwood Tunnel in Durban. Cooper and McCarthy (1998) provided an updated description of the stratigraphy of the Uloa Subgroup that was published in SACS (1980).

Discussion on the engineering properties of 'Berea Red Sand' is provided by Brink (1984) and state of the art understanding of collapse potential test results by Schwartz (1985). Rogers *et al* (1994) has carried out a detailed bibliographic review of the mechanisms of collapse potential throughout the world.

3 GEOLOGY

A detailed composite section of the stratigraphy from the study area is shown on Figure 2. The sequence of coastal marine sediments has been subdivided into the following sections, from eldest to youngest:

3.1 *Mesozoic bedrock*

The bedrock in the southern Mozambique area consists of interlayered eastward dipping basalt and rhyolite of the late Mesozoic Lebombo Group, Karoo Supergroup. The resistant felsic volcanic strata form the topography 30km to the west of the study area. A detailed account of lithostratigraphy of the Lebombo Group is presented by Cleverly and Bristow (1983). The uppermost basalt flows of the Movene Formation outcrops approximately 15km to the west near Boane and typically weather to dark brown and black clayey vertisols.

3.2 Upper Cretaceous

In the southern Mozambique area, argillaceous deep water marine sediments of the Grudja Formation overlie the Karoo volcanics (Dingle et al, 1983). From approximately 12m below sea level at the smelter site, the upper 10m of a dark grey, massive fine sandy mudstone unit was intersected, with scattered external and internal bivalve casts. The upper 4-5m of the succession is moderately weathered to a yellow brown colour. This unit is unconformably overlain by 15m thick medium to fine-grained sandstone with a 300mm thick basal medium to small pebble, clast-supported polymictic conglomerate with numerous sub-rounded yellow brown siltstone clasts. Commuted shell fragments occur within intermittent coarse-grained, carbonate-cemented horizons.

Westward from the smelter site, outcrops of similar mudstone and siltstone occur along the Umbeluzi River floodplain and contain thin matrix supported, clayey, medium to small pebble conglomerate with abundant Baculites fragments (cephalopod). Cooper (pers comm) suggests that these lenses represent Senonian-aged debris flows. Further westward at Boane, feldspathic, medium- to coarse-grained sandstone exposed within a railway line cutting was described as continental sediment of Turonian age by Förster (1975). Off-white and pale red mottling of the bioturbated and cross-bedded sandstone suggests that these sediments may be southern equivalents of the Malonga Formation (formerly Malvernia Formation) as described by Botha and De Wit (1996). The sandstones conformably cap a sequence of completely weathered siltstone containing 4 distinct lenses of unfossiliferous matrix supported medium to small pebble conglomerate.

At the bridge site over the Rio Matola to the east of the smelter complex, a 4m thick zone of pale yellow grey laminated mudstone lenses with infilled mudcracks and sandy layers occur at the top of moderately weathered, medium to fine grained sandstone. Thin coarse-grained sandstone with scattered fine gravels often demarcates the base of individual units. A 6-10m thick siltstone marker horizon occurs at approximately 15m below sea level.

Drilling at the silo complex at the Matola Harbour revealed a similar package of yellow brown sandstone and siltstone extending to depths in excess of 60m below sea level. Below a depth of 50m, the sediments become grey in colour. Professor Siesser of Vanderbilt University, Tennessee, USA kindly carried out the identification of nannofossils from the 10-15m thick siltstone marker horizon. The population range confidently places this marine sedimentary succession as late Campanian-Maastrichtian in age (Upper Cretaceous).

Lateral and vertical transitions within the Upper Cretaceous show a marked facies change over 20km from deposition of deep water mudstone and siltstone with broad, sinuous off-shore fine sandy channels to shallow water coastal marine sands with encroaching sub-aerial fluviatile continental sedimentation. It can be concluded that a marine regression was occurring in the latter Cretaceous in southern Mozambique with the distinct possibility of arenaceous sedimentation across the K-T boundary. Figure 3 shows the local schematic facies relationships of the Upper Cretaceous at the smelter site.

The Grudja Formation of southern Mozambique can be correlated with the St Lucia Formation of northern KwaZulu-Natal that belongs at the top of the Hluhluwe Group of the Kazu Supergroup (Cooper and McCarthy, 1998).

3.3 Tertiary

3.3.1 Matola Formation

The predominantly fine grained arenaceous Upper Cretaceous marine succession is overlain by approximately 6m of shell-rich, coarse grained to gritty, well-cemented, hard rock bioclastic limestone and calcarenite. The basal clast-supported conglomerate is typically 200-300mm thick and contains predominantly rhyolite and dolerite clasts. Below the unconformity is 1-3m thick cross-bedded, relatively unfossiliferous medium grained sandstone that is usually much softer and friable. The top of the unit is marked by pale greenish yellow, glauconitic limestone up to 2m thick with off-white, hard, calcified nodules. At the base of the limestone is a thin grit horizon (10-15mm).

Diagnostic macrofossils identified in this zone are Pecten sp. and Amussium umfolozianum (King, 1953). Clearly, this stratigraphic unit bears strong correlation to the Monzi and Sapolwana Formations of the Uloa Subgroup (Cooper and McCarthy, 1998) and similar description as the Tembe/Santaca Formations (Dingle et al, 1983). It is proposed that this unconformity bound unit of Late Miocene age be called the Matola Formation. Sparry carbonate cement is common and gives the rockmass typical hard rock strength below the limestone marker horizon. Numerous small dissolution cavities are infilled with loose brown sand. The Matola Formation cuts across the older underlying Upper Cretaceous stratigraphy and outcrops along the lower slopes of the incised Rio Matola estuary near the bridge site. Stratigraphic equivalents have also previously been reported from Santaca, some 60km to the southeast (Soares and da Silva, 1970).

3.3.2 Ponta da Vermelha Formation

Overlying the Matola Formation is a thick package of predominantly medium to fine-grained decalcified and rubified sandstone that belongs to the Ponta da Vermelha Formation. The equivalent stra-

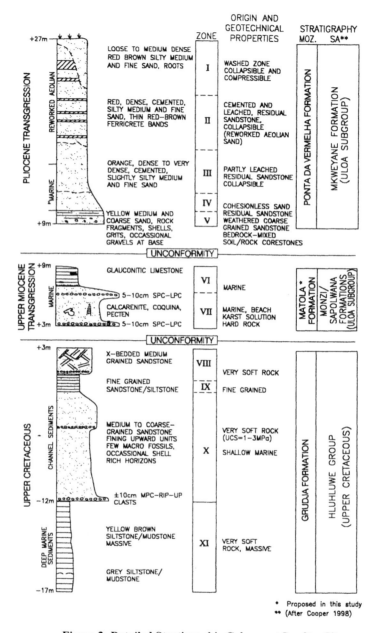

Figure 2: Detailed Stratigraphic Column at Smelter Site

tigraphic unit in the Uloa subgroup of northern KwaZulu-Natal is the Mkweyane Formation. Five distinct zones can be recognized and are described from bottom to top as follows:

Unit V: consists of a thin (1-2m) of buff coloured, coarse grained, shell-rich calcarenite with abundant casts of bivalves and gastropods and in particular the giant oyster Ostrea sp. Scattered grits and small pebbles occur on the erosional interface with the underlying Matola Formation.

Unit IV: consists of a completely weathered medium to coarse-grained sandstone that has deteriorated in places to a yellow-orange, dense, cohesionless sand (facies C of Jermy and Mason, 1983). In some areas, corestone remnants of hard rock calcarenite and sandstone form a mixed zone above Unit V.

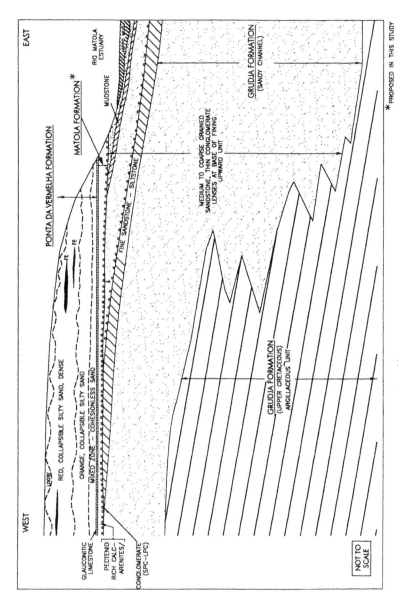

Figure 3: Schematic Cross-Section at Smelter Site

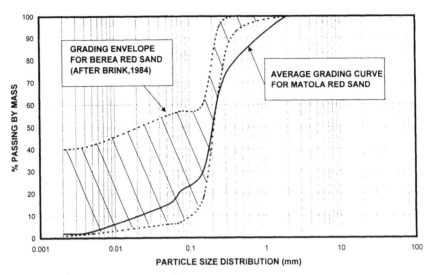

Figure 4. Average Grading Curve vs Berea Red Sand Envelope

Unit III: consists of pale reddish orange, dense, partly leached, iron-cemented, silty, medium and fine sand. Thin bands of hard, red brown ferricrete represent a falling water table during the late Quaternary period. Scattered, coarse sand particles and occasional pinhole voids occur. This unit is usually 10-15m below ground level and can be correlated with facies B of Jermy and Mason (1983).

Unit II: consists of a red brown to bright red (haematite rich), dense and cemented, silty, medium and fine sand that may contain broad zones of slightly more coarser grained material (remnant of former stratification). Hard ferricrete layers up to 300mm thick occur throughout. It is within this horizon that collapse settlement potential is the most severe. This zone can be correlated to facies A of Jermy and Mason (1983). Belderson (1961) and McCarthy (1967) have regarded these fine grained, ferruginous soils to be weathering products of aeolianite. This residual horizon is often referred to as "Berea Red Sand". Figure 4 shows an average grading curve for the Matola-Maputo sand in relation to the Berea Sand (Brink, 1984). Of note is the coarse fraction that occurs in the Maputo sand. Eriksson (1998) determined that the larger grains are quite angular and jagged in shape with variable sphericity and moderate to poor roundness. Recent research carried out at the Department of Geology at the University of the Witwatersrand has shown that the distinct population of coarse sand clasts consists of well rounded and poorly rounded k-feldspar and quartz. This 20% fraction may represent contamination of aeolian sand with immature marine sand during re-deposition offshore. It should be noted that the amount of feldspar does not change significantly with depth.

Unit I: consists of an irregular (up to 13m thick) zone of dark reddish brown, loose, compressible, slightly clayey, medium and fine sand that has originated from recent podsolic weathering of unit II (Maud, 1968). Note that there is no pebble marker at the base of this unit.

It is likely that the residual soils of the Ponta da Vermelha Formation in southern Mozambique represent shallow marine sediments that formed due to reworking of extensive aeolian sands during the major marine transgression of the early Pliocene (King, 1972). Cooper and McCarthy (1998) regard the Mkweyane Formation to be a blanket transgressive sequence at the base that oversteps the lower shell-rich Sapolwana and Monzi Formations. The upper fine-grained sand has generally been considered to be aeolian deposits of the regressive phase. However, the tabular nature of the deposit across the region in southern Mozambique, the gradational relationship with the underlying transgressive sedimentary strata and the coarse sand fraction suggest otherwise.

It is probable that in KwaZulu-Natal, the highlying old dune cordons were not reworked by the transgression and therefore preserve the macro-features of the aeolian depositional environment (linear ridge, large-scale planar cross-bedding and uniform grading).

3.4 *Quaternary*

The crests of broad interfluve areas around Maputo are gently undulating and punctuated by sinuous drainage courses with thin white alluvial sand deposits. Thin accumulations of white to pale brown, loose fine windblown sand has infilled hollows and formed indistinct ridges on the surface. Reworking of the older red soils by rain and wind has occurred throughout the Quaternary period. Insitu weathering without major lateral transportation has formed the compressible reddish brown soil described as Unit I (termed 'washed residual sandstone' during the geotechnical investigation). Major rivers in the region that are not filled with black esturinal clay (Hippomud) show a distinct north-south lineation and probably reflect the macro-tectonic framework of the southern Mozambique region.

Near the main tarred road along the smelter site access road, borrowpits have been worked for the 1-2m of perched alluvial gravels that overlies Cretaceous sandstone bedrock containing poorly preserved Brachiopod casts. Along the edge of the Umbeluzi floodplain to the east is 3-4m of layered alluvial gravel and sand deposits. At Boane, mottled Cretaceous sandstone is capped by ferruginised rudaceous deposits with individual rounded clasts up to 300mm in diameter. All of these raised alluvial deposits are terrace remnants of once high-energy rivers that eroded down in response to lowering of sea level during the Pleistocene glacial maxima.

Erosion of the red soil profile can be quite severe and in the Maputo area, stripping of Unit I has exposed a thick ferricrete horizon near the Cardoza Hotel (Ponta Vermelha) in the city center. Large borrowpits near the airport also indicate that the red residual soil horizon is continuous across the region, thickening eastwards to 30-40m and terminating at the fault scarp along the eastern coastline of the city.

4 WEATHERING

Evidence from scanning electron microscope research has shown that specks of primary titanium-enriched iron-oxide cement that infills minute indentations on quartz clasts appears to have been the original cement. This material appears to have been bevelled during a subsequent erosion event and suggests that the original dune deposits were 'reddened' prior to the Pliocene transgression. Hydrolysis of the heavy mineral suite occurred to form the initial cement.

Another unique feature of the red sand deposit is the large amount of secondary gel-like iron-oxide cement that always completely surrounds the sand-sized clasts. It is suggested that the source for this fine material may well be contributed to by volcanic dust derived from the volumous eruptions of basaltic lava and pyroclastic flows in the southern portion of the Mascarene ridge (Old Volcanic Stage of Mauritius and Reunion). According to McDougall and Chamalaun, (1969) these eruptions occurred approximately 6,8-7,8 Ma ago in the late Miocene (Messinian) and therefore prior to the main transgressive event. Fine ash could have been blown westward by anti-cyclonic weather patterns and accumulated along the eastern seaboard of southern Africa.

Since the regression of sea level during the glacial maxima of the Pleistocene, the poorly lithified and porous sandstone deposits of the Ponta da Vermelha Formation have been exposed. Subsequent weathering has produced a quartz-rich, red residual soil mantle that now blankets large areas around Maputo. Preservation of feldspar grains throughout the profile suggests that the intensity of weathering was low and only the iron-rich matrix was leached and oxidised.

The warmer Eemian period, some 135 000 years ago with a sea level stand approximately 5m above current sea level, may well have been responsible for the development of the Berea-type red residual soils along the eastern seaboard of southern Africa (Maud, 1998). This phenomenon has formed an iron-cemented regolith super-imposed a wide range of coastal sediments of Neogene age.

The removal (leaching) of soluble salts, iron hydroxide gels and colloidal clays by water percolation and trans-evaporation through the vadose zone is responsible for the formation of the metastable structure in the soil mass. Quartz grain contacts are coated in a layer of amorphous clay/iron-oxide. The precipitation of these coatings during desiccation has caused apparent overconsolidation of the soil. Prolonged desiccation has formed the bright red haematite content of the soil, thus slightly increasing the soil specific gravity and forming the hard, laterally discontinuous hardpan sheets. All original macro-structures such as cross-bedding, shells and bioturbation have been completely obliterated.

The hard ferricrete layers at variable depths within the Units II and III suggests that the water table was falling during the cooler and drier latter Holocene. Podsolic weathering has further increased the clay content within the upper 5-6m and formed smectite clays by addition of mobile cations to kaolinite. Variation in thickness of the softened soilmass was determined by standard penetration tests carried out on a grid of 25m. Figure 5 shows an isopach contour plot of N<30 blows per 300mm. Linear ridges of the older cemented soil are preserved as 'highs' within Unit I with deep zones probably indicating lines of preferential moisture migration.

At depth, circulation of mildly acidic groundwater at the bedrock/soil interface has been responsible for the development of fine karst dissolution features in the hard, carbonate-cemented bioclastic limestone and calcarenite of the Matola Formation.

5 ENGINEERING GEOLOGY

The red residual soils of the Ponta da Vermelha Formation represent reworked reddened aeolian sand

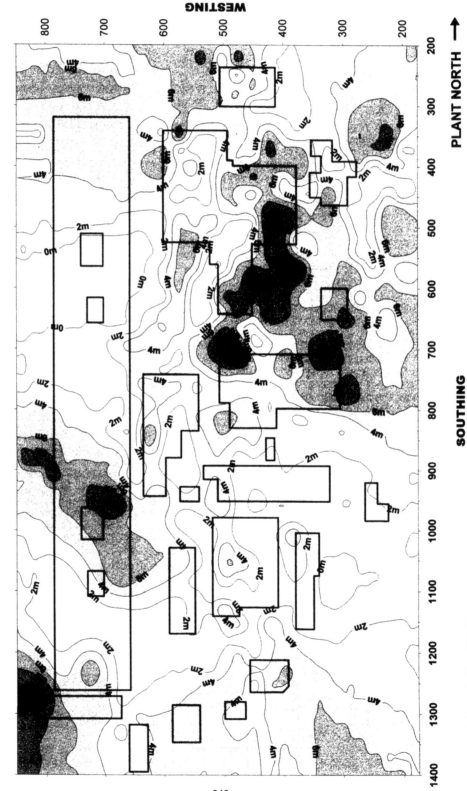

Figure 5: Isopach Plot of Unit 1

that was re-deposited on a wave cut platform during a major marine transgression during the Pliocene (King, 1972). During the Holocene, deep chemical weathering has produced a ferruginous-ferralitic soilmass that is metastable and iron-cemented.

Tactile profiling of large diameter trialholes and logging of boreholes immediately highlighted the possibility of a deep collapsible soil profile to 15-18m depth and so laboratory testing focused on the quantification and distribution of collapse settlement potential across the site.

The detailed geotechnical investigation for construction of the smelter complex included undisturbed soil sampling from large diameter trialholes, limited deep vertical plate load testing and collapse potential testing of test piles. The following section summarizes the principal engineering geological characteristics of the red soil mantle and shows the unique relationships that exist between collapse potential, particle size distribution and degree of saturation.

5.1 Particle Size/Atterberg limits

Statistical analysis of 79 particle size distribution/Atterberg limits tests are summarized in Table 1 below:

Table 1. PSD / Atterberg limits.

Parameter	$\bar{X}$	X_M	X_m	s	s/$\bar{X}$
L.L.	22	37	16	4.3	0.20
P.I.	1.0	5.1	0.3	0.6	0.6
Coarse Sand	14	18	11	1.92	0.14
Total Sand	81	89	49	4.6	0.06
Silt	15	48	9	4.6	0.31
Clay	2.7	5.8	0.4	1.6	0.59

L.L= Liquid Limit P.I.= Plasticity Index $\bar{X}$= Mean
X_M = Maximum X_m= Minimum s= Standard deviation
s/$\bar{X}$ = Coefficient of Variation

The variation of selected parameters is shown on Figure 6. The main points are:
- The total sand content remains fairly uniform with depth, decreasing slightly to 8m depth.
- Silt content increases with depth to 8-9m and then decreases.
- Clay content is quite scattered at shallow depth.

The red sands are slightly coarser grained than typical 'Berea Red Sand' with a lower clay content and a conspicuous 'coarse fraction' (see Figure 4).

Clay content over the upper 6-8m shows two populations, namely the 1-3% range corresponding to cemented material and the 5-6% range corresponding to the areas of washed residual sandstone where recent podsolic weathering and softening of the old profile has taken place.

Analysis of the particle size distribution data indicates that the 8m of the profile shows a gradual coarsening upward pattern. This may either reflect original stratification of the deposit or be due to the effect of eluviation during the formation of unit 1.

5.2 Physical parameters

A total of 230 laboratory tests carried out on undisturbed samples are summarized in Table 2 below:

Table 2. Physical parameters.

Parameter	$\bar{X}$	X_M	X_m	s	s/$\bar{X}$
Dry density (kg/m^3)	1696	2438	1429	107.4	0.06
Moisture content (%)	5.0	11.8	0.2	1.4	0.28
Void ratio	0.546	0.791	0.087	0.08	0.15
Specific gravity	2.61	2.66	2.56	0.02	0.01
Degree of saturation (%)	24	62	4	7.38	0.31
Porosity (n)	0.35	0.442	0.08	0.04	0.11
Air-voids (%)	26.7	33.3	6.4	0.04	0.00
Collapse potential (1000kPa)	5.3	13.2	0.0	3.41	0.64

Figure 7 shows the variation of dry density, moisture, void ratio and collapse potential with depth. It can be seen that soil density, and therefore void ratio remains constant with depth. However, there is quite a strong inverse correlation between magnitude of collapse and natural moisture content. Clearly the magnitude of collapse settlement decreases as moisture content increases.

5.3 Collapse potential

During the geotechnical investigation, collapse potential testing was carried out on undisturbed samples at soaking pressures of 200kPa, 500kPa and 1000kPa. A limited number of double odometer tests were carried out. It was noticed that to a depth of approximately 10m, the rate of collapse settlement was significantly higher than for the deeper samples. Figure 7 shows that the insitu moisture content is crucial in determining the amount of collapse settlement that will occur.

In order to determine the effect of moisture content variation, Figure 8 shows a plot of collapse potential against degree of saturation. The critical degree of saturation above which no collapse occurs is approximately 30% with serious collapse (>4%) at a maximum depth of approximately 12-13m.

Assuming that the average void ratio is 0.546 and the Specific Gravity is 2.6, the critical moisture content of the soilmass is 6.3%. The average insitu moisture content is 5%. Figure 9 shows the distribu-

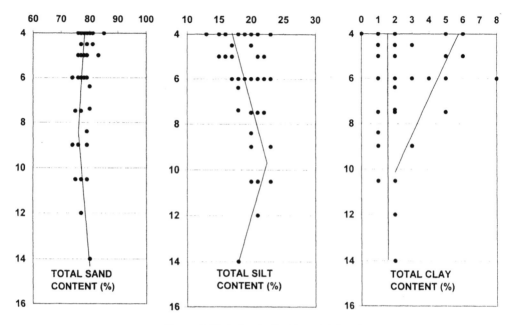

Figure 6: Variation in Grading with Depth

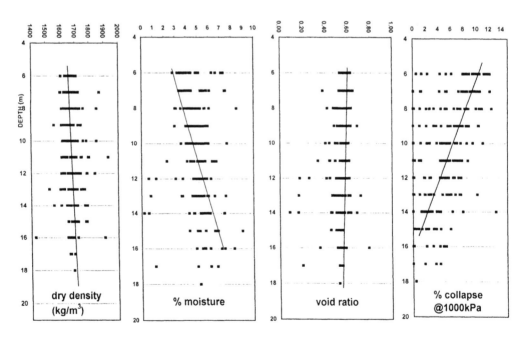

Figure 7: Variation in Physical Parameters with Depth

tion of insitu moisture content and that 83% of the samples tested had moisture contents below critical. The desiccated nature of the profile in a usually high rainfall area can only be attributed to the biomass and high trans-evaporation rate. During insitu profiling of large diameter trialholes, roots were noted

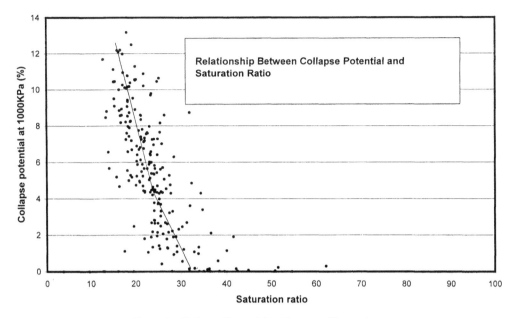

Figure 8. Collapse Potential vs Degree of Saturation

to depths in excess of 20m. Alteration of site boundary conditions by removal of the biomass during construction will certainly cause the ambient soil moisture content to rise and possibly the water table. Currently the water table occurs at a depth of approximately 3m above sea level with very little hydraulic gradient to the Rio Matola.

It is generally accepted that full saturation of the soil mass is not required to provide moisture conditions primed for collapse (Fredlund, 1979 and Mahmoud *et al*, 1995). Only sufficient water to soften the iron-clay cement buttresses and dissolve any soluble salt bridges is necessary for collapse to occur. Houston *et al* (1995) indicates that a saturation ratio of 50% is sufficient. In the Maputo area, degree of saturation ranges from 4-62%.

Figure 10 shows a graph of percentage soil passing the 0,075mm sieve plotted against the critical degree of saturation (taken from Schwartz, 1985). Of note is that above 75% passing the 0,075mm sieve, collapse settlement should not occur ie. clay-silt rich soils. However Schwartz (1985) warns that a high clay content may not necessarily mean that collapse will not occur. The data point provided by this study substantiates subdivision of the graph into the field of expected collapse settlement and the field of no collapse settlement at insitu moisture content. This plot can be used as a quick guide as to the likelihood of collapse settlement. The distance below the line of critical saturation can be roughly correlated to the potential magnitude of collapse.

Table 4 shows the broad range of collapse potential within Units II and III of the Ponta da Vermelha Formation:

Table 4. Distribution of collapse potential with depth

Zone	Depth range (m)	Collapse potential (%)
Unit II	0 - 12	4 - 13
Unit III	12 - 18	0 - 4

General trends across the site are that the magnitude of collapse settlement decreases with depth and that the zone of severe collapse occurs from below Unit I to an average depth of 8-10m. Although sampling of Unit I is problematical, both collapse and consolidation settlement can also occur in this layer.

Figure 11 shows a typical section at the aluminum smelter site with the distribution of collapse potential at soaking pressures of 1000kPa. The areas of 0-1% collapse are often well ferruginised and tending to hardpan ferricrete. The sandstone bedrock is usually very mixed and can contain appreciable amounts of very dense, cohesionless sand with slab-like hard rock corestones.

Results of pile tests on driven displacement piles with bulbous bases at approximately 6-7m depth and saturated under load showed rapid collapse settlements of up to 35mm. Detailed special trialpile test results from the smelter site are discussed by Byrne (in press).

Large diameter vertical plate load tests were also carried out depths of 8-10m with saturation at

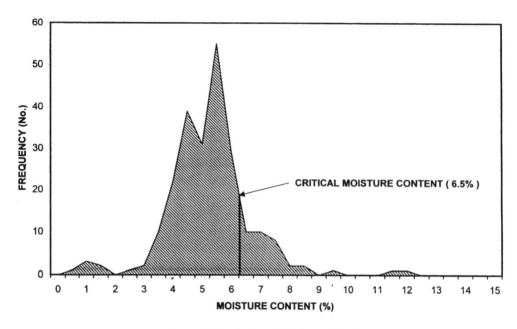

Figure 9: Distribution of Moisture Content

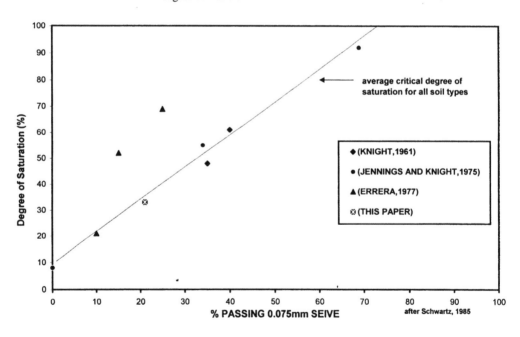

Figure 10: Critical Degree of Saturation vs Grading

500kPa. Collapse settlements of 30-35mm were measured in a single cycle test.

6 CONCLUSION

Detailed geotechnical investigations for a large smelter complex on the outskirts of Maputo in southern Mozambique has enabled the lithological

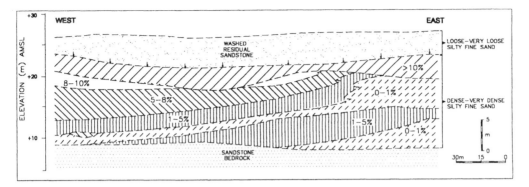

Figure 11. Typical Cross-Section

description of over 85m of composite stratigraphy.

In addition, extensive collapse potential testing on undisturbed samples with representative grading/Atterberg testing has allowed a unique opportunity to describe the engineering geological parameters that were critical for design of piled foundations.

Upper Cretaceous mudstones, siltstones and sandstones of the Campanian-Maastrichtian aged Grudja Formation are unconformably overlain by arenaceous and calcareous fossiliferous marine sediments of the Uloa Subgroup.

A diagnostic 6m thick sequence of Pecten-rich bioclastic limestone and calcarenite cuts across the underlying sandstone. This Miocene aged coastal marine deposit is the equivalent of the Monzi/Sapolwana Formations in northern KwaZulu-Natal. It is proposed that this unit be called the Matola Formation.

The upper Ponta da Vermelha Formation consists predominantly of red and orange ferruginous sandy residual soil (15-18m) that was originally reddened dune sand that were reworked during the Pliocene marine transgression. It may be possible that ancient soils were contaminated by dust derived from sub-aerial basaltic volcanic activity on the southern edge of the Mascarene ridge.

Chemical weathering and erosion of the exposed sediments during the Pleistocene has formed a highly collapsible residual soil horizon. Similarity exists to the Berea Red Sands in the Durban area (preserved above the 45m sea level stand).

The main engineering geological features in the stratigraphic profile at the smelter site are as follows:

- Compressibility of washed zone up to 13m thick
- Severe collapse settlement in red soils to 8m depth.
- Lower collapse settlement in orange soils to 15m depth.
- Cohesionless sand to 18m.
- Karst in hard Miocene calcarenite/conglomerate to 21m (Matola Formation)

- Very soft rock Cretaceous sandstone to 38m.
- Unweathered Cretaceous mudstone to 50m.

Tactile profiling of trialholes showed pinhole voids over the red, haematite stained, leached and metastable horizon. An average grading curve shows a distinct 20% coarse fraction consisting of angular clasts probably of marine origin. It is suggested that the Ponta da Vermelha Formation consists of reworked aeolian sand that was re-deposited by the early Pliocene transgression.

Dry density (1690kg/m^3) and void ratio (0.546) of the soilmass remains fairly uniform with depth. Plasticity Index is on average 1% and clay content varying from 6% in the recently weathered zone (Unit 1) to 2% in the older cemented soilmass.

The critical degree of saturation is 30% for 21% passing the 0,075mm sieve. This correlates to a critical moisture content of 6-6.5% above which collapse settlement will not usually occur. The profile at Matola is desiccated with 83% of the samples selected being dry of critical. The collapse potential mechanism is controlled primarily by moisture content and influenced by the grading of the material. At saturation pressures of 1000kPa, collapse settlement can be up to 13%. Severe collapse is usually confined to the upper 12m of the profile and therefore the revised piling solution was to take the pre-drilled driven displacement piles to safe depths of 15-18m below original ground level.

REFERENCES

Belderson, R. H. 1961. *The Size Distribution Characteristics of Recent Shallow Marine Sediments off Durban, South Africa.* M.Sc. thesis, (Unpubl.), Univ. Natal, Durban.

Botha, G.A. & De Wit, N.C.J. 1996. *Post Gondwanan continental sedimentation, Limpopo region, southeastern Africa.* Journal of Africa Earth Sciences Vol 23, No. 2: 163 - 187.

Brink, A.B.A. 1984. *Engineering Geology of Southern Africa.* Volume 4. Building Publications, Silverton: 332 pp.

Cooper, M.R. & McCarthy, M.J. 1998. *The Cainozoic palaeontology and stratigraphy of KwaZulu-Natal: Part 2. Stra-*

645

tigraphy of the Uloa Subgroup. Durban Museum Noviates. 23: 3 - 28.

Cleverly, R.W. & Bristow, J.W. 1983. *A Note on the Volcanic Stratigraphy and Intrusive Rocks of the Lebombo Monocline and Adjacent areas.* Trans. geol. Soc. S. Afr., 86, 55 - 61.

Dingle, R.V., Siesser, W.G. and Newton, A.R. 1983. *Mesozoic and Tertiary Geology of Southern Africa.* A.A. Balkema, Rotterdam.

Eriksson, P.G. 1998. *Report on Berea Red Sand samples from Maputo.* In: Report on the Results of the Detailed Geotechnical Investigation and Foundation Testing. Vol. 3, Frankipile - McKnight & Associates.

Flores, G. 1973. *The Cretaceous and Tertiary sedimentary basins of Mozambique and Zululand.* In: G. Blant (Editor), Sedimentary Basins of the African Coasts. 2nd part. Assoc. Afr. Geol. Surv., Paris. 81 - 111.

Förster, R. 1975. *Die geologische Entwicklung von Süd-Mozambique seit der Unterkreide und die Ammoniten-Fauna von Unterkreide und Cenoman.* Geologisches Jahrbuch. B12: 1 - 324.

Förster, R. 1975. *The geological history of the sedimentary basin of southern Mozambique, and some aspects of the origin of the Mozambique channel.* Palaeogeopgraphy, Palaeoclimatology, Palaeoecology. 17: 267 - 287.

Frankel, J.J. 1972. *Distribution of Tertiary sediments in Zululand and Southern Mozambique, Southeast Africa.* The - American Association of Petroleum Geologists. V 56, No. 12: 2415 - 2425.

Fredlund, D.G. 1979. *Appropriate Concepts and Technology for Unsaturated Soils.* Canadian Geotechnical Journal. No. 16: 121 - 139.

Houston, S.L., Mahmoud, H.H. & Houston, W.H. 1995. *Downhole Collapse Test System.* Journal of Geotechnical Engineering division, ASCE, Vol. 121. No. 4: 341 - 349.

Jennings, J.E. & Knight, K. 1975. *A guide to construction on or with materials exhibiting additional settlement due to collapse of grain structure.* Proceedings, 6th Regional Conference for Africa SM and FE Durban,.

Jermy, C.A & Mason, T.R. 1983. *A Sedimentary Model for the Berea Formation in the Glenwood Tunnel, Durban.* Trans-geol. Soc. S. Afr., 86, pp 117 - 125.

King, L. 1953. *A Miocene marine fauna from Zululand:* Trans. geol. Soc. S. Afr., 56: 59 - 91.

King, L. 1972. *The coastal plain of Southeast Africa: it's form, deposits and development.* Zeitschrift für Geomorphologie, N.F., 16 (3): 239 - 251.

Mahmoud, H.D., Houston, W.N. & Houston, S.L. 1995. *Apparatus and Procedure for an In Situ Collapse Test.* Geotechnical Testing Journal. Vol. 18. No. 4: 431 - 440.

Maud, R.R. 1968. *Quarternary geomorphology and soil formation in coastal Natal.* Zeitschrift für Geomorphologie. N.F., 16: 239 - 251.

Maud, R.R. 1998. *Report on the Geology and Geomorphology of the Mozal Site, Maputo, Mozambique.* In: Report on the Results of the Detailed Geotechnical Investigation and Foundation Testing, Vol. 3, Frankipile - McKnight & Associates.

McCarthy, M. J. 1967: *Stratigraphic and Sedimentological Evidence from the Durban Region of Major sea-level Movements since the Late Tertiary.* Trans. Geol. Soc. S. Afr. 48: 31 - 42.

McDougall, I. and Chamalaun, F.H. 1969. *Isotopic Dating and Geomagnetic Polarity Studies on Volcanic Rocks from Mauritius, Indian Ocean.* Geological Society of America Bulletin, v. 80, p.1419-1442.

Partridge, T.C. 1998. *Of diamonds, dinosaurs and diastrophism: 150 million years of landscape evolution in southern Africa.* S.Afr.J.Geol.,101(3) ,167-184.

Partridge, T.C. & Maud, R.R. 1987. *Geomorphic evolution of southern Africa since the Mesozoic.* S.Afr.J.Geol. 90 (2): 179 - 208.

Rogers, C.D.F., Dijkstra, T.A. & Smalley, I.J. 1994. *Hydro-consolidation and subsidence of loess: Studies from China, Russia, North America and Europe.* Engineering Geology. 37. 83 - 113.

Schwartz, K. 1985. *Collapsible soils.* Civil Engineer in South Africa. July 1985, 379 - 393.

Soares, A.F. & G.H. da Silva 1970. *Contribaicão para o Escado da Geologia do Maputó.* Rev. Ciênc. Geol., Univ. L. Marg. 3: 1 - 85.

South African Committee for Stratigraphy (SACS) 1980. *Stratigraphy of South Africa, Part 1* (Comp. L.E. Kent). Lithostratigraphy of the Republic of South Africa, South West Africa/Namibia, and the Republics of Bophuthatswana, Transkei and Venda, Handbk. Geol. Surv. S. Afr., 8, 690pp.

Geotechnics for Developing Africa, Wardle, Blight & Fourie (eds) © 1999 Balkema, Rotterdam, ISBN 90 5809 082 5

Case studies on the failure mechanism of embankment dams along the outlet conduit due to massive leakage

S.C.Ng'ambi
Department of Civil and Structural Engineering, Nottingham Trent University, UK

H.Shimizu, S.Nishimura & R.Nakano
Department of Land and Water Engineering, Gifu University, Japan

ABSTRACT: Many cases of embankment dam failure along the periphery of the conduit due to massive leakage have been reported in literature in past and recent times. The mechanism of failure is, however, not fully understood at the present time. Two case studies supplemented by FEM analysis are presented to elucidate the mechanism from the point of view of hydraulic fracturing and some measures to help cope with this problem are recommended for consideration in current dam design and construction.

1 INTRODUCTION

It is a common practice to install an outlet conduit underneath an embankment dam for water conveyance from upstream to down stream. For large dams, the conduit is designed as a tunnel, which is usually driven by conventional drilling and blasting methods and in the case of small low cost dams, the conduit is generally designed as a reinforced rigid culvert frame because tunneling is too costly. The embankment is often compacted against the culvert in a reliable fashion to ensure a good seal between the soil and the concrete structure.

Many past studies on buried conduits have confirmed that arching takes place around the culvert which increases the vertical earth pressures on the top of the culvert and reduces the horizontal earth pressures on the sides of the culvert. Because of the practitioners' interest in a more rational structural design of buried conduits, more attention has been paid to the vertical load imposed on the top of the conduit than on the sides.

In the case of culverts constructed underneath dams, however, the state of reduced horizontal stresses on the sides of the conduit could be critical when the stresses are lower than the reservoir water pressure, which by definition is a potential hazard of hydraulic fracturing. The current writers infer that hydraulic fracturing is the most probable cause of failure due to massive leakage along the out let conduits and this study programme was carried out to establish that inference. Two case studies supplemented by FEM analysis are presented to elucidate the mechanism of failure along the outlet conduit from the point of view of hydraulic

fracturing. Some measures to help cope with this problem are also recommended.

2 CASE STUDY SITE DESCRIPTION

2.1 *Gennaiyama dam site*

This is a nearly 13m high dam located in the south of Niigata Prefecture in Japan with a box shaped concrete outlet conduit installed underneath the dam. The embankment has two zones, an impervious Zone I and a pervious Zone II as shown in Figure 1.

Pressure meters were installed in the periphery of the culvert barrel as shown in Figure 2 to measure the earth pressures around the conduit at two locations, 19.5 m upstream in Zone I and the dam center in Zone II. Cross arms to measure the settlement were also installed at the dam center as shown in Figure 3.

2.2 *Embankment fill at Goi dam site*

This was a nearly 20m high embankment fill of material later on used for the impervious zone of Goi Dam located in Toyama Prefecture in Japan. A horseshoe shaped concrete conduit was constructed underneath the fill as a cover over a natural stream that flowed through the site. Similarly, pressure cells were installed on the periphery of the culvert as shown in Figure 4 to measure the earth pressure.

3 MATERIAL PROPERTIES

Zone I of Gennaiyama Dam is an impervious layer

of a mixture of river terrace deposit (d_{max}=150 mm) and pyroclastic loam (volcanic ash) by a volume ratio of 75% to 25% respectively and was compacted in 0.3m thick layers to unit weight γ_f=18 kN/m³. Zone II is a pervious or semi-pervious layer composed of sands, gravels, cobbles (d_{max}=300 mm) of river terrace deposit whose density after compaction was 21.5 kN/m³.

The embankment fill material at Goi dam site was a mixture of residual soil (M_0) originated from sandstone and mudstone, weathered mudstone (M_{1W}) and crushed fresh mudstone (M_{1F}) by a volume ratio of 1:1:1 respectively with an average wet unit weight (γ_s) of 16.7 kN/m³ after compaction. The physical properties of these materials are summarized in Table 1.

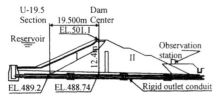

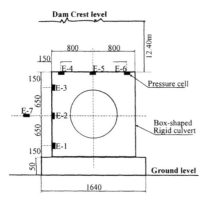

Figure 1. Cross section of Gennaiyama dam

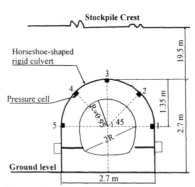

Figure 2. Cross section of the box shaped conduit

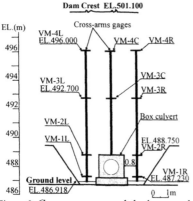

Figure 3. Cross-arms around the box conduit

Figure 4. Cross section of the horseshoe shaped conduit

4 MEASURED RESULTS AND OBSERVATIONS

4.1 *Earth Pressure*

The recorded earth pressures around the box conduit in Zone I of Gennaiyama dam are shown in Figure 5. The recordings were done from the start through to completion of the embankment build-up and the overburden pressure expressed as $\gamma_t \times$ h i.e. the wet unit weight (γ_t) of soil times the vertical height (h), is also shown.

The earth pressure recordings at the dam center in Zone II of Gennaiyama dam were discarded due to some erratic inconsistencies. This was deemed to be due to the large boulders and gravels (d_{max} = 30 cm) of Zone II with respect to the size (10 cm dia.) of the pressure cells.

The earth pressure recordings around the horse shaped conduit underneath the embankment fill at Goi dam site are shown in Figure 6.

4.2 *Settlement*

The embankment settlement above and adjacent to the box conduit recorded at the dam center of Gennaiyama dam is shown in Figure 7. The records show higher settlement at mid-height of the embankment (VM-3R = 52 mm), and smaller at both the top (VM-4R = 32 mm) and the bottom (VM-1R = 7 mm).

4.3 *Reliability of measured results*

It's apparent from the results presented in Figure 5 and Figure 6 that the earth pressure records show a consistent behavior i.e. following the increase in embankment height, which is an indicator that the earth pressure instrumentation at both sites functioned well and reliably. The settlement values given in Figure 7 are also typical of the settlement patterns normally obtained in any real embankment. When plotted, they give an oval shaped graph as shown subsequently.

Table 1. Physical properties of soils used for the embankments

Dam Site	Material	Group symbol	G_S	ρ_t (t/m³)	% of fines (−74μm)	Wn (%)	W_L (%)	W_P (%)	K (cm/s)
Gennaiyama dam	Pyroclastic loam	VH	2.66	1.4	90	110-130	86-152	61-73	$n\times10^{-7}$ ~10^{-8}
	River terrance deposit	GP-GC	2.80	2.0-2.1	3-8	12	47	35	$n\times10^{-2}$ ~10^{-4}
Goi dam	Residual Soil(M_0)	MH-SM	2.71	1.7-1.8	12	30-40	37-54	NP	$n\times10^{-5}$ ~10^{-6}
	Weathered mudstone (M_{1W})	-	2.72	1.9-2.1	50-70	21-29	60-80	35-40	-
	Fresh mudstone (M_{1F})	-	2.72	2.1-2.3	-	9-21	50	23	-

Note
G_S: Specific gravity of soil particles; W_n: Natural water content; W_L: Liquid limit; W_P: Plastic limit; K: Coefficient of permeability. For mudstone, W_n, W_L, and W_P were measured for powdered material and for river terrace deposit, were measured for material less than 420 μm whose Group symbol is ML.

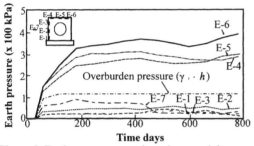

Figure 5. Earth pressure records on box conduit

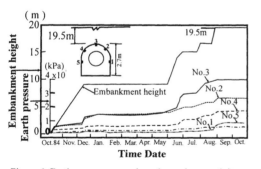

Figure 6. Earth pressure records on horseshoe conduit

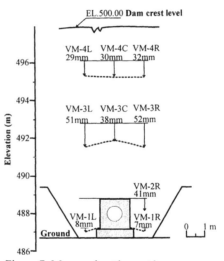

Figure 7. Measured settlement by cross-arms on box conduit

The ordinates of the earth pressure records measured by pressure cells E-1, E-2 and E-3 on the sides of the box shaped conduit (ref. Figure 5) are lower than the ordinates of the overburden weight $\gamma_t\cdot h$, whereas those for E-4, E-5 and E-6 on the top of the culvert are higher than those of $\gamma_t\cdot h$ (ref. Figure 5). A similar tendency is also observed on pressure records for the horseshoe shaped conduit (ref. Figure 6). Earth pressure recordings by pressure cells 1 and 5 on the sides of the conduit are lower than $\gamma_t\cdot h$ and the vertical earth pressure measured by pressure cell 3 on the top of the conduit is higher than $\gamma_t\cdot h$.

These results show that there is an increase in the nominal overburden pressure load on top of the conduit and a reduction in the horizontal earth pressures acting on the sides of the culvert, a phenomenon commonly known as arching or the transfer of stresses. This phenomenon is caused by the differential vertical displacement of the two soil columns (adjacent to the sides of the culvert and directly above the culvert) previously indicated by

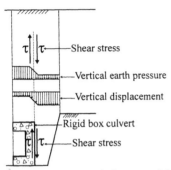

Figure 8. Conceptual diagram of the arching around a rigid box culvert underneath fill

Boundary conditions
BCDEFG: completely fixed (assumed rigid)
AB, HG: vertical movement only is allowed (in Y-direction)

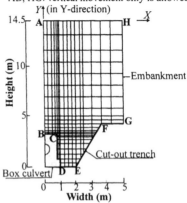

Figure 9. FEM mesh around box conduit

the cross arm recordings in Figure 7. The soil column on the sides of the conduit has more settlement than the adjacent soil column directly above the conduit (VM-3R = 52 mm and VM-3L = 51 mm > VM-3C = 38 mm respectively) creating a differential movement between the two columns. This movement results in the transfer of stresses from the column of soil on the sides of the culvert to the adjacent column of soil directly above the culvert crown increasing the vertical earth pressure load on top of the conduit and reducing the horizontal earth pressure on the sides of the conduit, as schematically described in Figure 8.

5 NUMERICAL ANALYSIS BY FEM

Not withstanding the reliable instrumentation data obtained at both sites, plane stress build-up analysis by FEM was performed.

5.1 Model Description

The structure is assumed to be symmetrical and thus only a half-section was modelled in the analyses and taken as a two-dimensional body of unit thickness under plane strain conditions in the upstream-down stream directions. The finite element configuration used in analyses at dam center of Gennaiyama dam is shown in Figure 9 which contains 291 elements and 331 nodal points. All elememts were four-node isoparametric and were hypothesized the linear shape function. The nodal points along the boundary of the cut-out trench and the periphery of the concrete culvert were completely fixed, whereas those on both vertical boundaries of the embankment were constrained to move only in the Y-direction.

5.2 Parameters and type of analysis

UU triaxial tests were carried out on remoulded soil samples of Zone I material of Gennaiyama dam and the embankment fill for Goi dam to obtain the E (elastic modulus) and ν (poisson ratio). The overburden load above the conduit crown was used to determine the appropriate σ_3 (minor principal stress). Secant modulus at E_{50} was taken from the stress-strain curve of the UU test, to give the appropriate value of E and the corresponding strain ε was in the order of 0.4%.

Considering that strain falls in the range less than ε_{50}, application of linear elastic finite element analysis using E_{50} was warranted. The parameters E and ν employed in the analyses for Zone I material of Gennaiyama dam and the embankment fill for Goi dam are given in Table 2. Build-up analyses (placement of successive layers) of linearly elastic material were performed with constant values of E and ν. Twenty-one layers of varying thickness (smallest at the bottom of the conduit and progressively increased towards the top of the embankment) were employed in the analyses.

6 RESULTS OF NUMERICAL ANALYSIS

The analyses were first performed for U-19.5 (Zone I) of Gennaiyama dam around the box shaped conduit and for the embankment fill of Goi dam around the horseshoe shaped conduit since the soil parameters (Table 2) for use in analysis were available from laboratory tests.

Table 2. Parameters employed in build-up linear analyses

Case	Elastic modulus E ($\times$ 100kPa)	Poison ratio ν	Unit weight γ_t (kN/m^3)
Zone I (Gennaiyama dam)	58	0.35	18
Embankment fill (Goi dam)	54	0.30	16.7

Stress ratios $\sigma_V / (\gamma_t \cdot h)$ which is the ratio of the vertical stress on top of the conduit to the overburden pressure and $\sigma_H / (\gamma_t \cdot h)$, the ratio of the horizontal stress on the sides of the conduit to the overburden pressure, were calculated. In order to verify the accuracy and reliability of the FEM analyses, the computed stress ratios by FEM analysis were compared with those calculated from the measured results. Both results (analysed and measured) agreed fairly well in both cases of the box shaped and the horseshoe shaped conduits as shown in Figure 10 and Figure 11 and tabulated in Table 3 and Table 4 respectively.

Table 3. Stress ratio values for Figure 10

Location	Stress Ratio	
	Measured	Analysed
E-1	0.35	0.4
E-2	0.5	0.55
E-3	0.4	0.45
E-4	1.5	1.3
E-5	1.4	1.2
E-6	1.6	1.3

Table 4. Stress ratio values for Figure 11

Location	Stress Ratio	
	Measured	Analysed
1	0.2	0.1
2	0.8	0.7
3	1.3	1.3
4	0.5	0.65
5	0.25	0.1

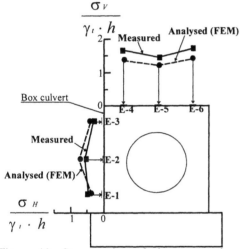

Figure 10. Stress ratios of the measured and analysed earth pressures on the box conduit at 19.5m upstream of Gennaiyama dam

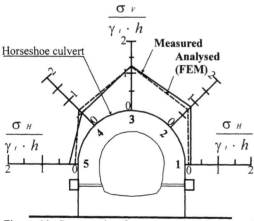

Figure 11. Stress ratios for measured and analysed earth pressures on horseshoes culvert underneath the embankment fill for Goi dam site

As mentioned earlier on, the measured earth pressure readings at the dam center (Zone II) of Gennaiyama dam were discarded because the large soil boulders of Zone II material caused the pressure cells to give erratic recordings. It was impossible too to carry out reliable laboratory tests on the same material (due to the large boulders) to obtain the soil parameters for performing the FEM analyses as was done for Zone I material at U-19.5 of Gennaiyama dam and for the embankment fill at Goi dam.

In order, therefore, to obtain the stress ratios around the conduit at the dam center, the settlement data given by cross arm recordings at the dam center was used to perform back analysis by FEM. A reference value of settlement measured by the cross arm was picked and employed in the back analyses. In this particular case, the VM-3 were chosen because the measured settlement values at these points gave a distinct differential settlement pattern (ref. Figure 7).

Iteration back analyses were carried out in reference to the point VM-3 by adjusting the E value employed in each back analysis until the value of settlement obtained by back analysis coincided with the measured value at the same point as shown in Figure 12. The values of settlement for other points VM-1, VM-2 and VM-4 obtained by back analysis were then compared with measured values at those points respectively and both the measured and analysed values agreed fairly well as shown in Figure 12. At that point, the stress ratios $\sigma_H / (\gamma_t \cdot h)$ and $\sigma_V / (\gamma_t \cdot h)$ around the box conduit at the dam center of Gennaiyama dam were finally obtained as shown in Figure 13 and tabulated in Table 6. The ratio $\sigma_H / (\gamma_w \cdot h)$ which is the horizontal earth pressure on the sides of the conduit to the reservoir water pressure is also shown in Figure 13.

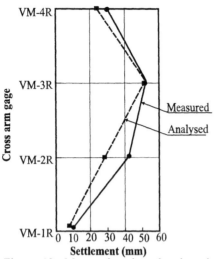

Figure 12. Measured and analysed settlement at Gennaiyama Dam Center

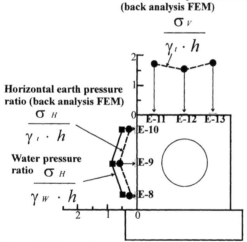

Figure 13. Reservoir water and earth pressure ratios obtained by back analysis at Gennaiyama Dam Center

Table 6. Tabulated values for figure 13

Location	Stress Ratio	
	$\sigma_H/(\gamma_t \cdot h)$	$\sigma_H/(\gamma_w \cdot h)$
E-8	0.25	0.45
E-9	0.5	0.8
E-10	0.3	0.5
	$\sigma_V/(\gamma_t \cdot h)$	
E-11	1.755	
E-12	1.5	
E-13	1.75	

The analyses showed that the vertical stress ratio $\sigma_V/(\gamma_t \cdot h)$ is higher than 1 in the range of 1.5 ~ 2.0 and the horizontal stress ratio $\sigma_H/(\gamma_t \cdot h)$ is less than 1 in the range of 0.1 ~ 0.6, which indicates, as stated previously, some severe arching (transfer of stresses) around the buried conduit. These figures of stress ratio values agree well with values recorded in other past studies on buried conduits, confirming further that the results of the current study are reliable e.g. Marston recorded $\sigma_V/(\gamma_t \cdot h)$ = 1.92 (Trollope et al. 1963) and the Japan Highway Corporation recorded $\sigma_V/(\gamma_t \cdot h)$ = 1.60 (Design Manual 1992).

The results shown previously also point out that the arching takes place on both the box shaped and horseshoe shaped conduits. Although the dome shape of the upper part of the horseshoe shaped conduit should supposedly reduce the arching on the sides of the conduit compared to the sharp-edged box shaped conduit, the effect is negligible, as the calculated stress ratios clearly show. The reason the horseshoe shape conduit has been adopted in several designs of buried conduits mainly for its structural merit on bending moments over other shapes.

The results also show that the effect of arching on the sides of the conduit is more critical at the bottom of the conduit with $\sigma_H/(\gamma_t \cdot h)$ < 0.4 which, again, is similar for both box or horseshoe shaped conduit.

Comparison of the horizontal stresses on the sides of the conduit with the water pressure taking the highest reservoir level showed that (ref. Figure 13) the ratio $\sigma_H/(\gamma_w \cdot h)$ is less than 1 in the range of 0.45 ~ 0.8 and also larger than the ratio $\sigma_H/(\gamma_t \cdot h)$. This condition is deemed highly critical and vulnerable to hydraulic fracturing as discussed subsequently.

7 MEASURES TO MITIGATE THE ARCHING EFFECT

Structural designers of buried conduits have recommended in past years to place compressible materials such as straw or some geo-textile material in the fill directly above the crown of a buried conduit to reduce the added vertical load on the top of conduit due to arching. Whilst this idea has been useful especially in highway embankments, it cannot be used for the embankment dam for the reason that such material would permit water to leak through them. When such materials rot, it may also jeopardize the structural reliability of the embankment.

Instead, an impervious but compressible material such as very soft clay would be plausible although it has never been used before. A numerical simulation by FEM was carried out with a layer of soft clay (thickness, t=1m) placed on top of the crown of the buried conduit. In each simulation, the E of soft clay was reduced i.e. making the layer softer with respect to the E of embankment material around the conduit-$E_{fill} : E_{soft\,clay}$.

The result showed that the adverse effect of arching on the sides of the conduit is considerably reduced by this method as shown in Figure 14. The horizontal stress ratio $\sigma_H / (\gamma_t \cdot h)$ approaches 1 (diminishing arching effect) as the simulated clay layer gets softer. The long-term consolidation process (including secondary consolidation) and the changing elastic modulus of soft clay with time may, however, require a very rigorous analysis to determine the most appropriate thickness of the soft clay layer to be placed above the conduit. The very soft clay (nearing slurry) may also give difficulties with workability.

σ_H : Horizontal stress on conduit side walls

$\gamma_t \cdot h$: Earth pressure due to overburden

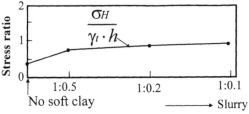

No soft clay ——————→ Slurry

Elastic Modulus Ratio ($E_{fill} : E_{soft\,clay}$)

Figure 14. Effect of placing a compressible layer of soft clay in the section above the culvert to reduce the arching in the fill adjacent to the conduit

8 DISCUSSION

The reduced horizontal earth pressures on the sides of the conduit, which are lower than the reservoir water pressure makes the embankment along the outlet conduit a potentially hazardous zone to hydraulic fracturing. It has been established through both experimental and theoretical approaches that hydraulic fracturing is induced in a soil mass when the fluid pressure exceeds the horizontal stress by a value equal to the tensile strength of the soil (Vaughan 1971; Bjerrum et al. 1972; Penman et al. 1981; Mori et al. 1987; Atkinson et al. 1994; Nishimura et al. 1998). Since typical values of tensile strength of soil are very small, it is considered that only the acting horizontal stress is available to resist hydraulic fracturing. It's now believed that hydraulic fracturing would occur in soil mass on any plane where the minor principal stress is lower than the fluid pressure. Current assessments of dam safety against the risk of hydraulic fracturing are thus geared to comparing the total stresses in the embankment with the reservoir water pressure.

Based on the results obtained in this study, the writers believe that the leakage along the outlet conduits is an outcome of hydraulic fracturing. Many reported cases of dam failure along the outlet conduit due to massive leakage are said to have occurred when the reservoir had reached full capacity, e.g. during heavy rains or floods. To mention just a few of the reported cases:

The Mississippi and Oklahoma failures; the failure in all the three Mississippi low dams for flood control occurred directly adjacent to conduits and seven out of 15 piping holes observed in the 11 failed Oklahoma low dams, were located at or near the plane of the conduit (Sherard 1972). All these failures occurred during heavy storms and the reservoirs are most likely to have been filled to full capacity.

In Yamaguchi prefecture in Japan, a total of 23 dams failed during 8 days of continuous heavy rainfall (max. intensity = 250 mm/day) of the June 1985 rainy season. 16 of the 23 failed due to piping and 11 of the 16 were cases of piping particularly along the outlet conduit (Yamada et al. 1986). The process of failure of one of the 11 dams called Hasuike Reservoir, was photographed by an observer and the pictures showed that the reservoir water level had nearly reached the crest when piping leakage that led to the failure of the dam occurred. The failure of the other dams was not photographed but judging from the same rainfall conditions, the other reservoirs too must have reached their highest water levels before failure. According to the results of this study, the conditions along the outlet conduit become more vulnerable to hydraulic fracturing when the reservoir is at or near full capacity as the water pressure at that stage highly exceeds the horizontal earth pressure on the sides of the conduit adjacent.

At this juncture, the writers wish to point out that the problem of reduced stresses adjacent to the outlet conduit, discussed so far, is not the initial and only cause of hydraulic fracturing along the conduit but rather a condition that makes the zone around the conduit more vulnerable. It cannot be conclusively proven at present that a dam with a smooth uniformly homogeneous face of the impervious zone, has any risk of hydraulic fracturing even when the water pressure becomes higher than the total stresses at any given plane. Past studies suggest that hydraulic fracturing requires the presence of a discontinuity, such as an existing crack, a borehole, or a zone of loose soil through which water pressure can penetrate and act by wedging action. No instances of hydraulic fracturing have been observed in the absence of such discontinuities (Jaworski et al. 1981). There is substantial evidence from past studies and filed investigations of embankment dams, that embankment cracking is inevitable even for dams designed and constructed under accepted standards during the following circumstances: an earthquake – even at low magnitude an earthquake would induce cracking for any type of cohesive soil; shrinkage due to over-drying as shown in Figure 15, especially in arid regions when the reservoir is empty or construction is suspended for long spells; and mechanical failure due to compressive tensile strain or severe arching and differential settlement as shown hypothetically in Figure 16 and Figure 17 respectively (Sherard 1973). These investigations have shown that cracks formed by an earthquake or over-drying can be as deep as five meters into the impervious zone of the embankment. Analytical studies have also shown that designs for a narrow core or a core trench with steep sides or an uneven foundation with a jagged rock or an embankment with a rigid structure buried underneath like the box shaped outlet conduit discussed previously, can cause arching and differential settlement in the embankment capable of inducing micro internal cracks. Although cracking due to differential settlement may be avoided by good design and construction techniques, Casagrande 1950 made a statement based on two dam failures due to leakage that clay, including that compacted at optimum moisture content, does not necessarily possess the ability to follow even small differential settlements without cracking. These evidences have recently led to the conclusion that hydraulic fracturing in fill dams is, in fact, initiated through existing cracks in the embankment. Even though cracks cannot be seen by engineers on some embankment dams which may seem to be free of cracking, it is plausible to conclude that cracks as narrow as hairline exist in any embankment. As water pressure can easily penetrate into very narrow or hairline cracks, it is considered that reservoir water enters pre-existing cracks on the reservoir side of the embankment and induces crack advance from the tip of the existing crack when the critical stress intensity at the crack tip is reached. The crack then propagates steadily and when it eventually accesses the low stress zone around the outlet conduit, a drastic failure occurs leading to massive leakage, as schematically shown in Figure 18. Suffice it to say, therefore that, in retrospect, the mechanism of failure due to leakage, along the outlet conduit or any other region of the embankment with conditions of very low reduced total stresses, involves at the initial stages the propagation of the existing cracks in the embankment. Using the fracture mechanics approach, the writers conducted a study to measure the fracture toughness of cohesive clay soil used for the impervious zone of a dam (Ngambi et al. 1998). Fracture toughness is regarded as a material property that indicates the material's resistance to fracturing i.e. propagation of an existing crack in the material. The study showed that the fracture

toughness of soil is very small in the range of 0.3 ~ 2.0 kPa·m$^{0.5}$ which is negligible to resist hydraulic fracturing of an existing crack hence supporting the inference that an existing crack is easily extended by water pressure. Starting with the premise therefore that cracks exist in any embankment dam and water can easily penetrate into cracks, it would be said that a dam with low stresses within the impervious zone especially where further stress reduction occurs by either arching or any form of transfer of stresses, has a potential hazard of hydraulic fracturing, and such is concluded to be the failure mechanism of dams along the outlet conduit due to massive leakage.

Figure 15. Embankment cracking due to drying

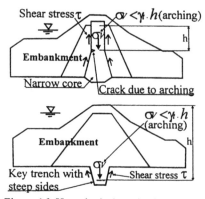

Figure 16. Hypothetical cracks due to arching

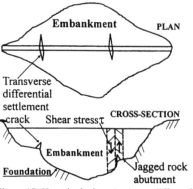

Figure 17. Hypothetical cracks due to differential settlement

654

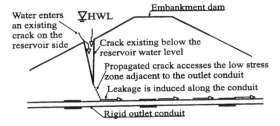

Figure 18. Schematic diagram of sequence of crack extension leading to failure along the outlet conduit

σ_N : Normal stress on culvert wall

$\gamma_t \cdot h$: Earth pressure; $\gamma_w \cdot h$: Water pressure

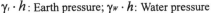
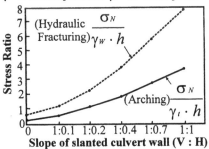

Figure 19. Effect of slanting the walls of the culvert to increase the normal stress acting on the side walls

9 SOME COUNTERMEASURES

In old time practice, seepage-cutoff-collars were designed on the sides of the buried conduits underneath a dam to elongate the seepage path and drop the seepage potential. With the development and use of modern compaction equipment such as heavy rubber-tyred rollers, current designs often exclude the seepage-cutoff-collars as they make impossible good compaction against the conduit walls.

9.1 *Slanting culvert side walls*

In order to compact the fill directly against the culvert walls in a reliable fashion, with such equipment, some past investigators have proposed slanting the conduit walls at a slope of 1H : 8V~ 10V (Sherard 1984). This technique not only improves the seal at the soil-conduit interface but also increases the normal principal stress on the slant wall surface which could be advantageous to reducing the risk of hydraulic fracturing. Therefore several simulation analyses by FEM were carried out for various slopes of slant walls to obtain the normal stress σ_N. The stress ratios $\sigma_N / (\gamma_t \cdot h)$ which is normal stress on the slanted wall to the overburden pressure showing the arching and $\sigma_N / (\gamma_w \cdot h)$, the normal stress on the slanted culvert wall to the water pressure showing the safety factor against hydraulic fracturing were computed whose graphs are shown in Figure 19. Both graphs rise with further increase in slant slope. It was significant to see that the critical slope where $\sigma_N / (\gamma_w \cdot h) = 1$ is nearly 1: 0.1 and beyond that slope gives a safety factor more than 1.2 against hydraulic fracturing.

9.2 *Filter*

As filter material is cohesionless and collapsible, it would work as a crack stopper of a propagating crack by hydraulic fracturing. To confirm it, a laboratory test was conducted whereby a hydraulic fracture crack was induced in a clayey soil specimen

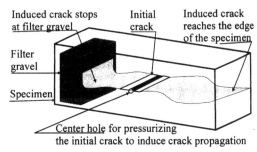

Figure 20. A hydraulic fracture test in a specimen with a filter zone

with a filter gravel zone at one end. The induced hydraulic fracture crack propagated in both directions from the initial slit but stopped and dispersed at the filter gravel whilst the other crack in the opposite direction without the filter gravel zone reached the edge of the specimen as schematically shown in Figure 20.

The filter zone cannot prevent the leakage i.e. the loss of water from the dam but will consequently eliminate the possible catastrophic damages to the dam, environment, human life etc. by safely dispersing the concentrated leakage. In case of the outlet conduit, a filter should be placed around the downstream portion of the conduit. The use of filter zones such as chimney drains, especially in homogenous dams of low to medium height is, therefore, a matter of necessity rather than preference as is the case under the current state-of-the-art of fill dams.

9.3 *Impervious blanket cover*

At another study site (Ngambi et al. 1998) where massive leakage of the dam foundation was believed to have occurred through desiccation cracks on the surface of the reservoir base, the reservoir was emptied and a blanket of an impervious layer of clay was placed over the reservoir surface. Figure 21 shows the leakage levels before and after the blanket was laid. The blanket layer sealed the existing cracks on the reservoir side of the dam from water penetration thus preventing crack propagation which would lead to leakage and failure.

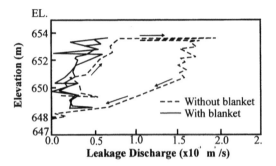

Figure 21. Leakage levels before and after placing the blanket cover of impervious clay

10 CONCLUSIONS

The main conclusions are summarized as follows:
(1) Arching takes place around the outlet conduit underneath an embankment and severely reduces the horizontal earth pressure on the side walls of the conduit in both box shaped or horseshoe shaped conduits. In the case of dams, this situation creates a potential failure hazard along the conduit with regard to hydraulic fracturing.
(2) When the reservoir level is at full capacity, the water pressure may exceed the horizontal earth pressure on the conduit sides, which is a basic condition for hydraulic fracturing to occur.
(3) Since most dam failure due to massive leakage along the conduit is reported to occur when the reservoir is at or close to full capacity, the failure is most likely caused by hydraulic fracturing.
(4) The arching and reduction of stresses around the conduit simply makes the conditions more favourable for hydraulic fracturing and aggravates the failure mechanism but is not necessarily the initiating mechanism too.
(5) Massive leakage that occurs along the outlet conduit or any other part of a dam is initially induced through existing cracks in the dam when water pressure enters such cracks and triggers crack propagation within the dam. When the propagating crack accesses a low stress zone such as around the conduit, hydraulic fracturing occurs drastically as a failure evidenced by massive leakage.
(6) Slanting the conduit walls at a slope of 1V: 0.1H provides a safety factor of at least 1.2 against

hydraulic fracturing along the conduit.
(7) The following recommendations are made to current dam designers and practitioners:
(i) every dam design should include downstream filters and in the case of conduits, a filter wrapped around the downstream portion of the outlet conduit is desirable.
(ii) covering the reservoir surface and part of the upstream slope face with a blanket layer of impervious soil is necessary when cracks are observed.
(iii) avoid designs that promote arching and differential settlement in the embankment.

REFERENCES

Brown, C. B. 1967. Forces on rigid culverts under high fills, Journal of structural division, Proc. *ASCE*, 93, No.ST5: 195-215

Casagrande, A. 1950. Notes on the Design of Earth dam. *Journal of Boston Society of Civil Engineers*, Vol. 37.

Jaworski, G. W., Duncan, J. M. and Seed, H. B. 1981. Laboratory study of hydraulic fracturing. *ASCE*, Vol. 107, GT6: 713-732.

Ngambi, S., Nakano, R., Shimizu, H. and Nishimura, S. 1996. Cause of leakage along the outlet conduit underneath a low fill dam with special reference to hydraulic fracturing, *JSIDRE*, No. 35: 35-46.

Ngambi, S., Shimizu, H., Nishimura, S. & Nakano, R. 1998. A fracture mechanics approach to the mechanism of hydraulic fracturing in fill dams, *JSIDRE*. No. 195: 47-58.

Ngambi, S., Shimizu, H., Nishimura, S. & Nakano, R. 1998. Laboratory testing for fracture toughness of clayey volcanic ash loam soil. *Proc. Int'l Symposium on Problematic Soils, IS-TOHOKU '98*: 141-144.

Ngambi, S. 1998. A study on the mechanism of leakage of fill dams- mainly from the view point of hydraulic fracturing. In partial fulfilment of the degree of Ph D, Gifu Univeristy, Japan.

Penman, A. D. M., Charles, J. A., Nash, J. K. T. L. and Humphreys, J. D. (1975). Performance of Culvertunder Winscar Dam, Geotechnique 25, No. 4: 713-730.

Sherard, J. L. 1973. Embankment Dam Cracking. Embankment Dam Engineering, *Casagrande Volume*: 271-353.

Sherard, J. L. 1986. Hydraulic fracturing in embankment dams, *ASCE*, Vol. 112, No. GT10: 905-927.

Trollope, D. H., Speedie, M. G., and Lee, K. I. 1963. Pressure measurements on Tullaroop Dam Culvert,Proc. *4th Australia and New Zealand conference on soil mechanics and foundation.*

Geotechnics for Developing Africa, Wardle, Blight & Fourie (eds) © 1999 Balkema, Rotterdam, ISBN 90 5809 082 5

Failure and reconstruction of a small earth dam in residual granites

R.J.Scheurenberg
Knight Piésold Consulting, Rivonia, South Africa

ABSTRACT: A 4m high earth embankment was constructed out of compacted residual granite soil to form a reservoir to contain spills of waste water at a sewage treatment works outside Johannesburg. No spillway was provided. The dam was overtopped and a 15m wide break formed. Repair of the dam incorporated a drop inlet spillway with an outlet pipe and a pressure relief drain pipe. The residual soil was again used for the embankment earthfill. Precautions were taken in the design of the various components of the works and in the specification of the construction procedures because of the high erodibility of the soil. Despite all this, on first filling of the reservoir a piping failure occurred through the embankment. Investigation of this second failure revealed a number of deficiencies in the detailing and in the execution which could have contributed to the failure.

1 THE SITUATION

The spills dam at the Driefontein Waste Water Treatment Works is provided to capture sewage flowing into the site, which cannot be accepted by the treatment plant for any reason, and hold it temporarily until it can be pumped back into the plant.

The original dam is approximately 140m long along the hillside by 30m wide across the hillside which falls to the west. The basin was formed by excavating into the east side of the hill and constructing a 4m high earthfill embankment above original ground level on the west side.

The underlying geological formation is granite, which is decomposed to a silty sand down to depths of several metres. The earthfill in the embankment is a residual granite obtained from excavations nearby. As the dam is less than 8m high and holds a relatively small volume of water, it is classified as of "low" safety risk.

At some time early in 1997 the dam overtopped and a breach about 4m wide at the bottom and 15m wide at the top formed in the west wall, approximately one third of the length from the south corner (Figure 1).

Knight Piésold was appointed by the Wastewater Subcluster of the Greater Johannesburg Metropolitan Council to design remedial measures for reconstructing the dam and ensuring its proper functioning in future, to prepare specifications, drawings and tender documents for the reconstruction and to su-pervise the contract for carrying out the necessary works.

The main contractor for various constructions at the treatment-works site appointed a sub-contractor to carry out the earthworks for the spills dam. Earthfilling of the breach was practically completed on 9 December 1997. The dam first held water on 12 December 1997. In the early hours of 3 January 1998 a tunnel was formed by erosion of the earthfill by water flowing through the repaired breach section (Figures 2 and 3).

2 DESIGN CONSIDERATIONS

The design had to take into account two main considerations:

2.1 *Hydraulic aspects*

A spillway would decant excess water from the pond at a level to prevent overtopping. A 600mm diameter pipe through the wall would convey the water from a 900mm diameter drop inlet spillway at the heel of the embankment inside the dam, upstream of the repaired breach.

A 160mm drainage relief pipe from under the concrete tank to be constructed at the north east corner of the spills dam basin would also have to exit through the repaired breach in the wall.

Figure 1. Downstream (west) face of dam after original failure – general view

Figure 2. Downstream face of dam after failure showing concrete pipes washed out piped breach and dropped at outlet slab

Figure 3. Downstream end of "pipe" tunnel

2.2 Geotechnical aspects

The residual granite soil was readily available to be used as fill for reconstructing the dam wall. Although this material had a suitable grading and plasticity for the purpose, it was known to be easily erodible and possibly potentially dispersive.

Even though dispersiveness tests had not been done at the design stage, it was realised that particular attention would need to be paid to the selection, placing and compaction of the earthfill to ensure its integrity, even if dispersive, and prevent the flow of water through it any any significant rate. Although the embankment would not need to be 100% watertight for reasons of retaining water, nevertheless seepage should be minimised and zones or layers of more pervious material should not be permitted.

The results of laboratory tests for potential dispersiveness carried out on a total of seven samples of the residual granite soil (after the failure occurred) are summarised in Table 1.

Table 1. Potential dispersiveness test results

Sample location	Crumb test	Double hydrometer	Pinhole	Chemical test, ESP
New fill from side of cut downstream of tunnel	1	3	1	1
Old fill from side of cut inside tunnel	1	3	1 or 2	1
New fill from upper layer in cut	1	2	1 or 2	1
New fill from upstream face of embankment	1	2	1	1
Fill from around relief drain pipe	1	3	1	1
Fill from layer above relief drain pipe	1	3	1	1
Old fill or *in situ* decomposed granite	3	3	1	1

Dispersiveness grade: 1 – Non-dispersive
2 – Intermediate
3 – Dispersive

Such discrepancies between different tests on the same materials are common. The pinhole test, which is regarded by some authorities as the best model of likely behaviour in the field (Reference USBR Test Designation 5410, has indicated that the material tested is non-dispersive to possibly slightly dispersive, while the US Soil Conservation Services double hydrometer test results (Reference USBR Test Designation 5405 suggest that the material is definitely dispersive. The chemical tests all plot in the completely non-dispersive region (low Exchangeable Sodium Percentage and low Cation Exchange Capacity) of the graph proposed by Gerber and Harmse.

Despite the inconclusive results obtained from the potential dispersiveness tests, design precautions were clearly called for because of the known easy erodibility of the residual granite soils.

3 SPECIFICATIONS

3.1 Drawings

Figures 4, 5 and 6 show details of the breach and construction requirements as issued to the Contractor.

3.2 Earthfill material

The impervious fill in the emergency spills dam wall and the bedding cradle material for the pipes was specified to be the same, sourced from selected borrow and meeting the following requirements:

Max. particle size	37,5mm
Finer than 0,425mm	50% max.
Finer than 0,075mm	20-40%
Finer than 0,002mm	5-20%
Plasticity index	6 to 18
Grading modulus	1 to 1,8

Such material is considered suitable for impervious earthfill in a small dam such as this, which is meant to hold water only occasionally.

3.3 Compaction control

The dam wall was specified to be backfilled and compacted in 150mm layers to at least 93% of modified AASHTO maximum density at OMC-0+4%. The same fill material was to be thoroughly rammed into the trench around each of the two pipes through the wall to achieve at least 93% mod. AASHTO maximum density, both in the bedding cradle of compacted impervious earthfill and the carefully compacted earthfill blanket up to 300mm above the pipe.

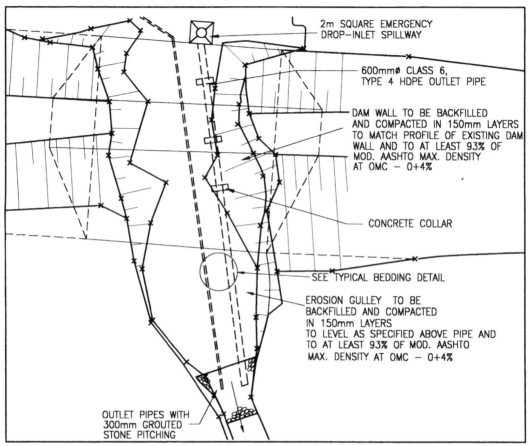

Figure 4. Plan of breach in dam wall

3.4 *Pipe materials*

For the 600mm diameter outlet pipe, HDPE Class 6 Type 4 was originally specified. When difficulty was experienced with timeous delivery to site of this pipe, it was agreed that 600mm diameter concrete spigot and socket pipe, Class 100D with a 30mm sacrificial layer, could be used instead.

For the 160mm diameter drainage relief pipe, flexible corrugated HDPE pipe was specified. The corrugations on the outside of the pipe would inter-lock with the surrounding soil to add resistance to any water flow that might tend to occur along the length of the pipe. This kind of pipe has been used successfully in roadside drainage and for un-derdrains in tailings dams.

3.5 *Pipelaying*

The working drawing showed each of the two pipes (160mm dia. and 600mm dia.) as being placed in its own separate pipe trench, of width 300mm outside

the pipe on both sides. Three concrete collars of 1500 x 1500 x 500m were shown around the 600mm diameter pipe and the construction method for this pipe was specified in detail.

1. Backfill to 500mm above pipe crown.
2. Excavate pipe trench, lay pipe and backfill.
3. Excavate collars and cast.
4. Complete backfill and reconstruct dam wall.

No collars were shown around the 160mm diameter pipe, but in all other respects the construction proce-dure was required to be the same as for the 600mm diameter pipe.

3.6 *Evaluation*

In principle, these specifications and details were considered appropriate for obtaining an homogene-ous impervious fill which would satisfactorily pre-vent undue seepage through any part of the dam wall. If properly carried out, these methods should

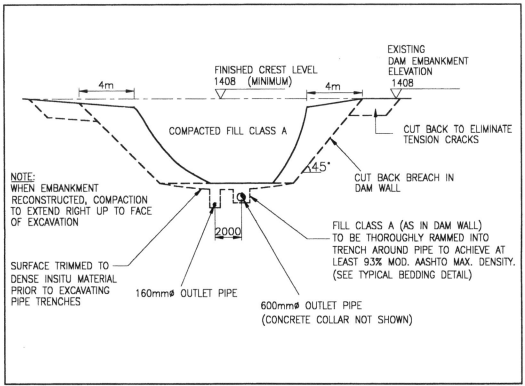

Figure 5   Section through breach in dam wall and temporary excavation
                    required for reconstruction

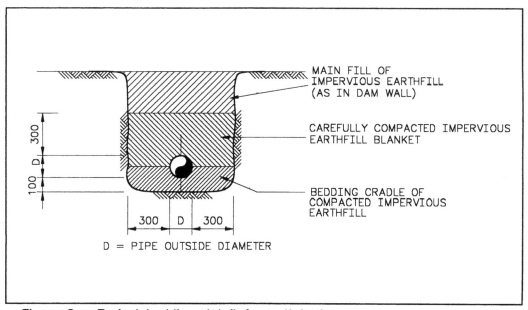

Figure 6   Typical bedding detail for outlet pipes

661

also have been adequate in the presence of potentially dispersive soils.

4 CONSTRUCTION DEFECTS

On this project an all-too-common problem was experienced in that both the Engineer's and the Contractor's site representatives were inexperienced and did not appreciate the importance of careful attention to detail and to the requirements of the specifications. As a result, various construction defects were afterwards noted:

4.1 *Earthworks quality control*

Insufficient tests were carried out to confirm that the earthfill material being placed complied with the grading and plasticity criteria, or that the specified compaction densities were achieved throughout. Careful checking of the few test results that were presented would have revealed a number of cases where the specifications were not met. Also of concern was the inability to identify from the test report sheets the positions where the tests were done.

After the failure the conclusion was reached that the compaction of the fill in the breach area and in the pipe trench was inconsistent and was not in accordance with the Contract specifications and drawings. In some places, particularly under and around the 160mm diameter pipe, it was practically not possible to achieve the specified compaction because the pipe was too light and would move while the bedding was being tamped. These problems were not reported, however, and no action was taken to rectify the situation.

4.2 *Pipelaying*

The method specified on the drawing was to fill to above the level of the top of the pipe, excavate individual trenches for each of the two pipes, prepare the bedding cradles and backfill in specified sequences. Bedding cradles were to be as in SABS 1200LB, Section 5.3 for flexible pipes, but with impervious fill in place of granular material.

For convenience on site, neither the trench construction nor the bedding of the pipes was executed in accordance with the drawings and specification:

- A single wide trench was excavated into which both pipes were laid.
- The bedding cradles under both the pipes were omitted and the pipes were laid directly on the excavated and prepared flat surface.
- The pipe joints in the 160mm diameter pipe were not properly made.
- The three concrete collars around the 600mm diameter pipe were omitted.

5 POSTULATED CAUSE AND MECHANISM OF FAILURE

The failure is regarded as a typical piping failure through a dam wall. However, it could not be determined whether the failure was initiated along the 600mm or 160mm diameter pipe.

It is postulated that water flowed, albeit slowly, through inadequately compacted earthfill, probably under or around or along the outside walls of one or other of the 600mm diameter decant outlet pipe unobstructed by any collars or the 160mm diameter drainage outlet pipe. The easily erodible, possibly dispersive, poorly compacted fill material was washed away by the flowing water, which formed a "pipe" (a flow tube or channel) starting at the downstream end and eroding backwards into the dam. Once a connection was made with the water then standing inside the dam, the flow increased, the erosion accelerated, and the 2,5m high by 3,6m wide tunnel was quickly formed. The rush of water emptying the dam carried with it the concrete and plastic pipes (Figure 2).

The following construction deficiencies are considered most likely to have caused the failure:

a) The probable presence of a wedge of uncompacted soil at the point of contact between the pipes and trench floor.
b) The absence of the concrete collars on the 600mm diameter pipe to create a positive cutoff.
c) The poor compaction of the soil around the 160mm corrugated pipe.

6 RECONSTRUCTION

When the breached wall was reconstructed for the second time, at no cost to the Client, certain modifications were made to some of the details and precautions were taken to ensure that the construction was properly controlled.

6.1 *Earthworks*

The selected borrow material was subjected to the full range of indicator and compaction tests.

The floor of the excavation was tested with a handheld dropweight cone penetrometer (TMH6 Method ST6) to ensure that there were no loose patches underneath the compacted fill.

6.2 *Backfill around pipes*

Because of the practical difficulty of ensuring that the earthfill bedding cradle and blanket around the pipes would be adequately compacted to the required density, low permeability and absence of voids, soilcrete was used as backfill instead.

Figure 7. Outside of west wall where breach was repaired

6.3 *Sequence of filling, trenching and pipelaying*

1. All the loose material, the roof of the erosion tunnel and the sides of the breach were excavated to shape back down to original ground.
2. Earthfill was compacted in 150mm layers to the specified height over the top of the 160mm diameter pipe. A wacker was used to compact against the near vertical walls of the existing wide trench where the heavy roller could not reach.
3. A 450mm wide trench was excavated for the 160mm diameter pipe and the trench floor was compacted with a wacker. The recesses for the three concrete collars were excavated into the walls and floor of the pipe trench.
4. Pipe support saddles were formed out of stiff concrete, the pipes were laid and were tied down.
5. The concrete collars were cast against the ground except where shuttered inside the pipe trench itself.
6. Soilcrete was poured around the pipe in layers up to 300mm above the top of the pipe.
7. Earthfill was now compacted in 150mm layers to the specified height over the top of the 600mm diameter pipe, again making sure that the specified 93% mod. AASHTO compaction was achieved throughout including against the sides of the excavation.
8. A 1200mm wide trench was excavated for the 600mm diameter pipe, and the recesses for the three concrete collars were excavated.
9. Concrete pipe support saddles were formed and the pipes were laid.

10. The concrete collars were cast.
11. The pipe was surrounded by soilcrete.
12. Backfill was placed in 150mm layers compacted to 93% mod. AASHTO density, first to general floor level and then in the breached part of the wall up to final crest level. The sides of the breach were cut back to form steps. A wacker was used where the heavy roller could not reach to ensure adequate compaction throughout.
13. Density tests were carried out on every layer, using a Troxler nuclear meter, backed up by a sand replacement test for each day's production.

6.4 *Completion*

The embankment has now been reconstructed (Figures 7 and 8) and has performed without problems through the 1998-99 rainy season to the present.

7 LESSONS TO BE NOTED

This may seem a rather prosaic case history, but there are a number of lessons to be learnt so as to avoid such failures in the future:

1. Ensure that all the staff involved on the job are familiar with the standard specifications and the particular requirements of the job and that they are fully briefed about critical details.

2. Ensure that the design details are suitable for their required functions and that they can be constructed in practice.

3. Ensure that the specified quality control proce-

Figure 8. Inside of west wall with drop inlet overflow spillway outlet in the middle

dures are carried out, including the required test-ing and monitoring of test results.

8 ACKNOWLEDGEMENTS

The agreement of the Wastewater Subcluster of the Greater Johannesburg Metropolitan Council, Metro-politan Infrastructure and Technical Services, to the publication of this paper is gratefully acknowledged.

REFERENCES

- Gerber, FA & Harmse, HJ vM (1987): Proposed procedure for identification of dispersive soils by chemical testing. Civil Engineer in S Africa, Vol 29 No 10, pp 397-9
- National Institute for Transport and Road Re-search (1984): Technical methods for highways. Draft TMH6 – Special methods for testing roads. Committee of State Road Authorities of the Dept. of Transport, RSA (1984)

- South African Bureau of Standards (1983): Stan-dardised Specification for Civil Engineering Con-struction, SABS 1200 LB: Bedding (Pipes)
- USBR: Test designation 5405: Determining dis-persibility of clayey soils by the double hy-drometer test method. United States Dept of the Interior Bureau of Reclamation
- USBR: Test designation 5410: Determining dis-persibility of clayey soils by the pinhole test method. United States Dept of the Interior Bureau of Reclamation

Geotechnics for Developing Africa, Wardle, Blight & Fourie (eds) © 1999 Balkema, Rotterdam, ISBN 90 5809 082 5

Capacity of continuous flight auger piles in the Johannesburg CBD

K. Schwartz
Dura Piling (Pty) Limited, Johannesburg, South Africa

A.G. A'Bear
Africon Engineering International (Pty) Limited, Johannesburg, South Africa

J.H. Strydom
Africon Engineering International (Pty) Limited, Pretoria, South Africa

ABSTRACT: Continuous Flight Auger (CFA) piles were installed in residual igneous soil for a structure built in the Johannesburg CBD. The residual soils are low density, fine grained clayey silts derived from the weathering of diabase and andesite but have appreciable consistency below 12 m depth. The CFA piles, which relied on shaft adhesion capacity, were designed mainly on the basis of SPT related undrained shear strengths and employing an appropriate adhesion factor to reflect the anticipated degree of smear during the installation process. The piles were installed using conventional design and piling techniques. However, the failure of a test pile to meet performance specifications resulted in the downgrading of the pile capacity. This paper covers the geological setting of the site and the soil engineering properties. The design approach is discussed as well as the subsequent behaviour of the test piles and the possible reasons for the unexpected behaviour of the test piles. The assumptions made when designing CFA piles are critically examined and the value of pile tests in advance of pile construction are discussed.

1. INTRODUCTION

A new development in the centre of Johannesburg, known as ABSA Towers North, has been underway since the end of 1996. This R400-million development entails the construction of a six-storey building with three basement levels. The building covers two city blocks and is bounded by Commissioner, Main, Polly and Troye Streets. The portion of Fox Street previously located between the two city blocks has been included in the development. As a result of the wide column spacing used in the structure, loads are relatively high, reaching values of as much as 15,000 kN over the central part of the structure. Column loads at the perimeter of the structure are 2400 kN.

2. GEOTECHNICAL INVESTIGATION

The geotechnical investigation was conducted while demolition was in progress. Access to the site was thus restricted to a large extent by existing buildings, services and demolition work. Use was made of an earlier investigation carried out by Weber De Beer and Associates (1985) for proposed extensions to the old Rand Daily Mail building which were planned for the south eastern corner of the present development. In addition, the information available from the Carlton Centre was used to provide a basis for deciding on the relevance and accuracy of the data retrieved during the investigation. A total of ten trial holes and eight boreholes were drilled on the site. Standard penetration tests were carried out at regular intervals in the boreholes. Undisturbed samples retrieved from the sides of the trial holes were subjected to shear box and oedometer tests in the laboratory.

3. GEOLOGY

The geology of the site is described in detail in a companion paper (A'Bear et al, 1997) and only the salient geological features of the site will be discussed here.

According to the geological map of Johannesburg produced by De Beer (1965) the site is located over rocks of the Jeppestown Subgroup, Witwatersrand Supergroup. This Subgroup consists of alternating beds of arenaceous shales and argillaceous quartzites. In the vicinity of the site the beds are overturned and as a result dip steeply to the north. In this area the Jeppestown rocks have been

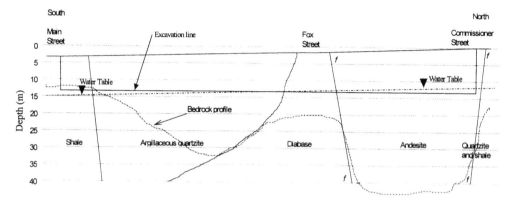

Figure 1 : Schematic geological cross section

intruded by a diabase dyke, known as the Fox Street dyke due to its association with that street. Figure 1 depicts a typical north-south cross section through the site. All contacts strike roughly east-west.

At the outset of the investigation it appeared that the dyke was notably wider at the ABSA Towers North site than indicated on De Beer's map, seeming in fact to occupy the entire northern half of the site. However, it was noted that within the dyke area shallow rock occurred in the vicinity of Fox Street, while the depth to rock further north was considerably deeper as shown on Figure 1. It was thus felt that two intrusives could be present on the site. This was further reinforced by the fact that, once exposed in the excavation, the residual soils over the shallow intrusive zone were noted to be speckled and showed relict grain sizes in the medium grain range. The soils overlying the deeper northern igneous body were uniform in colour and relict grains were not visible to the naked eye, that is the grain size of the parent rock was apparently very fine grained. Amygdales were also noted in places.

Quartz veins, breccia and slickensided planes, indicative of shearing and faulting, are present in the northern igneous body. These generally dip southwards at an angle of approximately 60°. This evidence of faulting taken with the nature of the rock, indicated that the northern intrusive may not be an intrusive but a wedge of andesite lava faulted in at that position. Further work involving thin section analysis was carried out and the rock was diagnosed as being a strongly altered andesite. It was noted that the original texture was still present and that the rock may have been a basaltic andesite. Thin section analysis carried out on the

diabase led to it being diagnosed as a uralitised and altered diabase typical of the older diabase sills of the Transvaal Sequence.

The investigation showed that the geology of the site is more complex than anticipated. The wide diabase "dyke" consists of both a dyke and an andesite wedge that had been faulted into the north of the dyke.

4.   GEOTECHNICAL PROPERTIES

As can be expected from the fairly complex geology which underlies the site, a range of geotechnical properties were encountered. A brief discussion of the various properties is given in the following sub-sections.

4.1   *Grading and indicator characteristics*

The results from indicator and grading tests show that the most of the materials on site are relatively inert in terms of potential heave. This accords well with the structure, or lack of it, found in the profile. Although difficult to perceive in the profile, there are slight differences between the residual argillaceous quartzite and residual soils derived from the igneous rocks. The residual quartzite is marginally coarser grained and has a grading modulus ranging from 0.33 to 0.42. The residual igneous soil, on the other hand, is predominantly a silt and has a grading modulus ranging from 0.01 to 0.27. The plasticity indices show little variation between the soil types and generally fall in the range of 9 to 15 %.

Table 1 : Comparison of dry densities and moisture content

Material type	Andesite		Diabase		Hillwash		Quartzite		Shale	
Number of samples	7		4		4		6		2	
Properties	DD*	MC**	DD*	MC**	DD*	MC**	DD*	MC**	DD*	MC**
Average	1342	30.7	1160	44.7	1428	17.4	1577	23.2	1695	18.7
Standard deviation	113	13.0	113	9.9	25	3.4	104	6.5	115	2.5
Maximum	1563	40.6	1278	54.3	1464	20.3	1762	28.5	1776	20.5
Minimum	1215	2.7	1051	32.6	1411	12.5	1450	11.2	1613	16.9

Legend: * Dry density (kg/m^3)
          ** Moisture content (%)

4.2   *In-situ dry density and moisture content*

The range of dry density and moisture content values given in Table 1 below are derived from the present and earlier investigations.

In general the andesite and diabase have the lowest dry densities with the former showing a range of values between 1200 and 1500 kg/m^3 and the latter showing an even lower range of between 1050 and 1300 kg/m^3.

The residual soils derived from the sedimentary rocks, on the other hand, show significantly higher densities with values ranging from 1400 to 1800 kg/m^3.

A plot of the dry density versus depth is given in Figure 2 and from this it can be seen that the in-situ density of the soils does not, as one would expect, increase with depth but rather remains fairly constant over the range of depths measured. Some

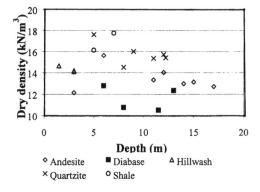

Figure 2 : Results of dry density versus depth

of the lowest values are recorded at a depth of approximately 17 m below surface, the maximum depth from which it was feasible to retrieve undisturbed samples.

The range of results found in the andesite body are similar to those reported for residual soils occurring within the Ventersdorp Lava 211 land pattern of the Johannesburg graben (Brink, 1979).

4.3   *Shear strength*

The results of shear box tests carried out on undisturbed samples taken from the various soil horizons are summarised in Table 2.

The main objective of the shear box tests was to obtain drained shear strength parameters (c' and ϕ') for the design of the lateral support system. Of concern in this regard are the high values of cohesion obtained from the tests. It is generally accepted that designers should be extremely cautious in deciding on a value of effective cohesion.   In this instance, an assessment of the results appears to indicate that the high values of cohesion measured are probably a function of the test procedure rather than representative of actual soil behaviour. This conclusion is supported by the low values of effective angle of shearing resistance measured in certain of the tests and the fact that these low values generally coincide with high measured values of cohesion. This indicates that the rate of stress used in the tests was probably too high to ensure fully drained conditions. These factors, as well as the relict jointing and slickensiding along certain joints within the residual soils, are important considerations in the choice of effective shear strength parameters.

Table 2 : Shear box test results

Material type		Andesite	Diabase	Quartzite	Shale
Number of tests		6	4	6	2
Depth range (m)		6 to 13	6 to 15	5 to 12	5 to 7
Cohesion (kPa)	Average	34	76	35	68
	Max	45	123	63	77
	Min	19.5	57	22	59
	Std deviation	8.2	31.7	14.4	12.7
Angle of Shearing resistance (degrees)	Average	26	17	27.6	28.3
	Max	29	20.5	29.5	29
	Min	24.5	11	26	28
	Std deviation	2.0	4.1	1.4	0.4

4.4 Compressibility

Figure 3 gives a plot of SPT N-values versus the modulus of compressibility (E) derived from oedometer test results. The oedometer modulus of compressibility was derived from the rebound curve after correcting the laboratory oedometer curves in accordance with the procedures described by Schmertmann (1955). Williams (1975) has shown that this procedure gives representative consolidation parameters for deeply weathered residual soils of low in-situ dry density. The SPT results have been related to depth by means of a linear regression and it has also been assumed that the SPT values found in the andesite are valid for the diabase as no results are available from this unit. A linear regression carried out on the results plotted in Figure 3 show that the ratio of E to N ranges between 300 and 550 for the andesite, diabase and argillaceous quartzite units. This is considerably lower than would normally be expected in stiff cohesive soils. Ratios of 1000 to 1500 would be more typical for soils with this plasticity (Stroud, 1974). Some doubt exists about the validity of the SPT results as there was limited head room where boreholes were drilled in basements. In addition it seems strange that the SPTs refused in soils profiled as firm to stiff and which exhibit dry densities as low as 1300 kg/m^3. It was thus deemed prudent to treat the SPT results with caution.

5. CFA PILE PERFORMANCE

5.1 Background

As severe collapse of the trial holes and boreholes was experienced during the investigation, underslurry auger piles were recommended as foundations for the building in all geological units except the shale. In the shale area conventional, cast in-situ auger piles were to be used. As there was a large variation in depth to bedrock and the strength of the in-situ soils was generally low, it was decided that all piles, which are generally heavily loaded, would be founded on and in rock. The successful contractor provided an alternative in which it was proposed to use cast in-situ auger piles, varying in diameter from 900 to 1800 mm, over most of the site and continuous flight auger (CFA) piles on the perimeter where columns carry lighter loads. The CFA piles were intended to shed load relying predominantly on side friction with limited end bearing contribution.

The design of CFA piles should be carried out taking into consideration of the construction technique and features that are specific to this particular file type. The pile is formed by screwing a long continuous flight auger with a hollow stem into the ground to the required depth. Grout is then pumped to the base of the auger through the hollow stem. Pumping of the grout under pressure is continued while the auger flight is slowly extracted. This continues until the auger flight is fully extracted and the pile shaft is completely filled with grout. Once the grouting operation is complete, the reinforcing cage is placed into the fresh grout to the specified depth.

The pile, however, has some negative features which need to be taken into consideration in evaluating the overall quality of the piling system when compared to other systems. Some of the features are as follows:

- The operation needs to be under the control of an experienced operator and site supervisor. This is necessary since the

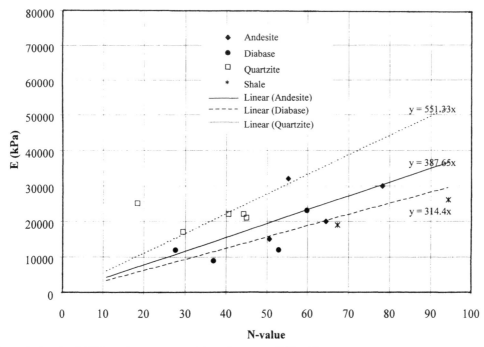

Figure 3: Graph of SPT N-value versus modulus of compressibility (E)

overall quality of the pile shaft is highly dependent on the installation procedure and in particular the monitoring of the grout flow during the extraction of the flight.

- The nature of the installation is such that it is not possible to inspect or evaluate the soil strength or the integrity of the shaft either during or after installation.

- The value of end bearing support needs to be treated with caution because of possible disturbance at the tip of the flight during drilling and grouting.

5.2   General design procedures for bored piles

The design procedures that should be used for CFA piles are dealt with extensively in various Codes of Practice and in the technical literature (SAA Piling Code 1978, Federation of Piling Specialists 1985, Derbyshire, 1985 and McAnally, 1981).

For CFA piles in cohesive soils (such as encountered on the site in question) it is recommended in the various codes and in the literature that the design procedures and soil parameters adopted should be those appropriate to normal bored (or auger) piles.

The ultimate pile capacity is obtained as follows :-

$$\text{End bearing} = A_b.N_c.C_u \tag{1}$$
$$\text{Shaft} \quad = A_s.\alpha.C_u \tag{2}$$
$$\text{Total} \quad = A_b.N_c.C_u + A_s.\alpha.C_u \tag{3}$$

where :

A_b = area of the base of the piles; A_s = area of the shaft of the piles; N_c = bearing capacity factor equal to 9 for deep piles; C_u = undrained shear strength of the soil either at the base or along the shaft; and α = adhesion factor.

A factor of safety needs to be introduced to obtain the allowable pile load.

The recommendations given by Burland et al (1966) are usually used in this regard.   This requires that an overall factor of safety of 2 is used or a minimum factor of safety on end bearing of 3. The allowable pile capacity is the lesser of the two expressions:

$$\frac{A_b.N_c\,C_u + A_s.\alpha.C_u}{2} \tag{4}$$

or

$$\frac{A_b.N_c\,C_u}{3} + A_s.\alpha.C_u \tag{5}$$

5.3   Site specific design

It is considered necessary to discuss three of the parameters used in the design in some detail. These are the undrained shear strength (C_u), the adhesion factor (α) and the factor of safety. These were derived as discussed in the following sections:

5.3.1   Undrained Shear Strength

Undrained shear strength (C_u) was determined from the results of the Index Property and Standard Penetration Tests (SPT) contained in the geotechnical investigation report provided at tender stage. The procedures recommended by Stroud (1974) were used in this regard as follows :

$$C_u = 6N \qquad (6)$$

Where C_u is in kPa and N is the SPT blowcount.

The SPT "N" value used in the design was consistent with the lower bound values obtained from the geotechnical investigation and these values and the associated C_u values are summarised in Table 3:

Table   3:   Summary   of   SPT   values   and corresponding undrained shear strenghts.

Depth (m)	Average "N" value	Cu (kPa)
Basement level to –11 m	60	360
-11 to -15 m (and below)	80	480

5.3.2.   Adhesion factor

Numerous tests on piles have shown that the full undrained shear strength (C_u) cannot be used in determining the ultimate shaft resistance of bored (or augered) piles in cohesive soils. This has led to the development of an empirical adhesion factor α to determine the ultimate shaft resistance. Research has shown that the value of α depends to a large extent on the type of pile, the undrained shear strength of the soil and the method of installation of the pile.

As indicated previously, research and the literature recommends that for CFA piles, the parameters adopted should be those appropriate for bored (or auger) piles. Much research has been carried out on the determination of α for bored (or auger) piles. This research has concentrated on the correlation of α with undrained shear strength. Typical values obtained from the literature for similar values of shear strength encountered at the site in question are tabulated below :

Table 4: Summary of adhesion factors from literature

Reference	Adhesion factor (α)
Weltman and Healy (1978)	0.4
Woodward Gardner and Greer (1972)	0.4
Skempton (1959)	0.45
Reese and O'Neill (1988)	0.55
Australian Code (1978)	0.3
Wright and Reese (1978)	0.5
Mohan (1988)	0.5
Fang (1991)	0.3
O'Riordan (1982)	0.4
Range	0.3 to 0.55
Average	0.42

There are no published values of adhesion factors for residual cohesive soils in South Africa. Tests carried out on residual Ventersdorp lava do however generally correlate well with the published literature and in some instances higher values than those published have been obtained.

Results of pull-out tests carried out on soil nails in residual andesite soils with SPT blow counts of up to 70 (or equivalent C_u of about 350 kPa) have been published by Heymann et al (1992). These results show that an α factor of 0.4 is appropriate to the determination of shaft friction. These results are considered to be particularly relevant to this design, since they have been carried out in similar soil conditions which occur at this site and are also applicable to a grout/soil interface as is the case for the CFA piles.

In the design of the CFA piles, consideration was given to all of the above factors and an adhesion factor α value of 0.4 was adopted.

5.3.3   Factors of safety

The procedures recommended by Burland et al (1966) were considered to be appropriate. The pile design was therefore computed both with an overall factor of safety of 2 and with a minimum factor of safety of 3 on end bearing.

5.3.4 Design comments

Based on the discussions presented above, the following comments are considered relevant with regard to the design:

- In carrying out the design an allowable end bearing of 1440 kPa, with a minimum factor of safety of 3, was obtained. In view of the construction procedures associated with the CFA piles this was considered to be too high and a value of 750 kPa was adopted.
- The design calculations indicated that a shaft length of 11 m would be sufficient for a 2500 kN working load. For convenience associated with the equipment used, this was increased to 15 m. If a back analysis is carried out assuming an allowable end bearing of 750 kPa and an overall factor of safety of 2, then the actual adhesion value (α) used in the design reduces to 0.34. This is equivalent to the lower bound values of adhesion as recommended in the literature.

5.4 Pile performance and backanalysis

A pile test on a 15 m long 750 mm diameter CFA pile showed that the ultimate load capacity of the pile was only about 2700 kN, while the required working load was 2500 kN. Due to the low ultimate capacity of the tested pile it was decided to derate the capacity of all the CFA piles to 1700 kN. The deficit in pile capacity was rectified by installing 900 mm diameter conventional augered piles to 12 m depth.

A second load test on a conventional augered pile of 11.8 m length and 485 mm diameter was performed for two reasons:

- to obtain site specific design parameters for the above-mentioned 900 mm auger piles, and
- to obtain a second set of test results on a bored pile for comparison with that obtained from the tests on the CFA pile.

As discussed earlier, although the CFA and augered piles represent different pile types, both are essentially bored piles which should behave similarly in terms of shaft capacity.

The 485 mm auger pile yielded an ultimate load of approximately 1930 kN. Backanalysis of the results of the two tested piles indicated that if an end bearing capacity of 750 kPa did apply, the values of α for the CFA and auger piles respectively are only 0.17 and 0.25. If no end bearing was allowed for, these values increased marginally to 0.19 and 0.27 respectively. Irrespective of the assumptions on end bearing, these values were both well below the initially assumed design value of 0.4 and the actual equivalent value of 0.34 finally used for the design.

Concern existed about the integrity of the pile shafts, given the low values of the adhesion factor. However, integrity tests both on the test piles and working piles indicated no deficiencies of this nature.

5.5 Reasons for pile behaviour and lessons learnt

An explanation should be sought for the low values measured for the adhesion factor or for some other reason that may have contributed to the pile behaviour. From the widely published results in the literature (see Table 3) the lowest values are 0.3 ie. about 50 % higher than the values derived from the test on the CFA pile.

It must be recognised that the adhesion factor is not only a function of the undrained shear strength, but certainly also the type of soil, installation procedure, type of pile (driven or bored) and time lapse between drilling and casting of the pile.

The possible reasons for the pile behaviour are not clear at this stage.

From the construction records no anomalous aspects could be traced that indicated inappropriate construction procedures. The piles were installed by experienced staff familiar with the requirements of CFA pile construction.

It should be noted that the adhesion factor is determined empirically from the undrained shear strength which in turn is derived empirically from the SPT results. Although there is scatter in the results which define the above-mentioned empirical relationships, the design procedure specifically incorporates appropriate safety factors to allow for variation (scatter) of this nature.

It may also be that the relationship between SPT results and the undrained shear strength of the low density, residual andesite found in the Johannesburg CBD is outside the norm. Further work would be required to examine this relationship.

The residual andesite in which the CFA piles were installed is a fine grained silt soil with a significant clay content. The soil type is not one in which CFA piles are commonly installed in South Africa and limited local experience has been obtained in this respect. It is possible that the properties of these residual soils are significantly different from the norm.

Stress relief below basement level could also possibly affect the shaft capacity of the piles, although this is not a problem commonly known to affect bored piles in fine grained soils.

It is worth noting though that CPT tests conducted after the basement had been excavated, yielded blow counts typically between 10 and 30 over the 5 m depth directly below basement level with fairly sudden refusals at depths of about 6 to 10 m. SPT results from the original geotechnical investigation indicated refusals at similar depths, although the blow counts over the corresponding 5 m depth were substantially higher, ie. between 50 and 90.

With the exact cause of the problem shrouded in uncertainty the emphasis shifts towards the lessons learnt. In this respect three important aspects come to mind:

- The adhesion factor of fine grained soils in the interior of South Africa seems to be substantially less than the values reported in the literature. This implies that the design of CFA piles in these soils should be treated with caution.
- The CFA piles were installed to depths of SPT refusal. Despite this and despite allowing for a low adhesion factor against the background of the pile test results, very little, if any, end bearing capacity was mobilized.
- It is important to perform a sufficient number of pile load tests in advance of construction or as early as possible during the construction phase. In view of the above mentioned uncertainty, the behaviour of prototype piles becomes an essential benchmark by which the design should be gauged.

6. CONCLUSIONS

During the initial construction phase of ABSA Towers North it was found that the site is geologically more complex than anticipated.

The performance of test piles indicate that the adhesion factors obtained were roughly only 50 to 70 % of those used as typically design values derived from in situ and laboratory tests, and values reported in the literature.

The reasons for the low adhesion factors are not clear. On the assumption of sound pile construction, reasons could include a departure from the empirical relationship between typical soil engineering parameters and the value of the adhesion factor for fine grained residual andesite, as well as little end bearing capacity in this material. Stress relaxation due to excavation and its possible effect on the adhesion factor is also unclear.

The results of the pile tests underline not only their value but the necessity of conducting them at the earliest opportunity.

7. ACKNOWLEDGEMENTS

ABSA Properties is kindly acknowledged for permission to publish this paper, as well as the valuable contributions made by Messrs. J Crous of Dura Piling and E Braithwaite of Franki on the properties of the soils encountered on this site.

REFERENCES

A'Bear, A.G. Strydom J.H. and Schwartz K, (1997). Geotechnical Considerations in the Construction of ABSA Towers North, Johannesburg CBD, *Proc. Conference on Geology for Engineering, Urban Planning and the Environment, Midrand, 13 – 14 November 1997.*

Brink, A.B.A. (1979). *Engineering Geology of Southern Africa, Volume 1.* Building Publications, Silverton, South Africa.

Burland, J.B. Butler F.G. and Dunican, P. (1966). The behaviour and design of large diameter bored piles in stiff clay. *Proceedings of the Symposium on large bored* piles. ICE, London.

De Beer, J.H. (1965). *Geological map of Johannesburg,* Department of Geology, University of the Witwatersrand, August 1965.

Derbyshire, P.H. (1984). Continuous Flight Auger Piling in the UK. *Piling and Ground Treatment.* Thomas Telford, London.

Fang, H.Y. (1991). *Foundation Engineering Handbook*, Second Edition. Van Nostrand Reinhold.

Federation of Piling Specialists (1985). *Specification for the Construction of Piles using Continuous Flight Augers and concrete or grout injection through the auger stem.* Published by *Piling Publications,* London.

Heyman, G.H. Rhode A.W. Schwartz K. and Friedlaender E (1992), Soil nail pull out ressistance in residual soils. *Proc. Int. symposium on Earth Reiforcement Practice.* Fuknoka, Japan.

McAnally, P.A. (1981). An analysis of load tests on Grout Injected Piles. *Proceedings of the Symposium on Grout Injected Piles.* Queensland Institute of Technology.

Mohan, D. (1988). *Pile Foundations,* A A Balkema, Rotterdam.

O'Riordan, N.J. (1982). The mobilisation of shaft adhesion down a bored cast in situ pile in the Woolwich and Reading beds. *Ground Engineering Vol 15, No 3.*

Reese, L.C. and O'Neill, M.W. (1988). Drilled Shafts: Construction Procedures and Design Methods. *US Department of Transportation.* FHA McLean, Virginia.

Schmertmann, J.M. (1955). The Undisturbed Consolidation of Clay. *Transactions ASCE Vol.* 120.

Skempton, A.W. (1959), Cast in situ Bore Piles in London Clay, *Geotechnique.*

Standards P Association of Australia (1978), Rules for the Design and Installation of Piling. *SAA Piling Code AS 2159 - 1978.*

Stroud, M.A. (1974). The Standard Penetration Test in insensitive clays and soft Rocks, Proc. *European Symposium on Penetration Testing (ESOPT 1),* PP. 367-375.

Weber De Beer and Associates (1985), Geotechnical Investigation. Proposed new SAAN building Fox, Polly and Main Streets, Johannesburg. *Report 1839.*

Weltmant, A.J. and Healy, P.R. (1978). Piling in boulder Clay and other glacial fills. *DOE/CIRIA Report PG 5,* London.

Wight, S.J. and Reese, L.C. (1979). Design of Large Diameter Bored Piles. *Ground Engineering Vol. 12, No 8.*

Williams, A.A.B. (1975). The settlement of three embankments on ancient residual soils. *Proceedings of the Sixth Regional Conference for Africa on Soil Mechanics and Foundation Engineering,* Durban.

Woodward, R.J. Gardner, W.S. and Gree, D.M. (9172). *Drilled Pier Foundations.* McGraw Hill.

Geotechnics for Developing Africa, Wardle, Blight & Fourie (eds) © 1999 Balkema, Rotterdam, ISBN 90 5809 082 5

Performance of a high road embankment over a deep zone of peat

D.L.Webb
D.L.Webb and Associates, Durban, South Africa

ABSTRACT: This paper deals with the construction of a high road embankment over soft peat up to 15m thick and describes the behavior of the embankment during placement and after opening the road to traffic. To ensure stability during construction it was necessary to determine the maximum safe rate of placement of fill and also the magnitude and rate of settlement due to consolidation of the peat. Comparisons are shown between vane shear strength, immediate undrained shear strength in the triaxial apparatus and point resistance in the Dutch cone penetration test. A relationship between water content and compressibility is established, and it is shown that, as to be expected, there is a very marked increase in undrained shear strength with increase in effective consolidation pressure.

1 INTRODUCTION

In the construction of the road works for National Route 2 across the Mhlatuze flood plain, near Richards Bay it was necessary to place an earth fill embankment up to 14m high over zones of peat varying in cumulative thickness up to 15m or more over a distance of 250m along the route.

A representative geological cross-section of the road embankment and the underlying subsoils is shown in Figure 1.

It was apparent that large settlements of the embankment would take place as a result of consolidation of the peat. The main emphasis of the site investigation was therefore to classify the peat material in terms of an accepted system and to investigate its shear strength and compressibility.

Shear strengths of the peat were determined from insitu vane tests and from laboratory tests on 100mm diameter tube samples taken in the boreholes, and from Standard Penetration Tests. Consolidometer tests were also carried out on tube samples.

2 UNDISTURBED SAMPLING

Undisturbed samples were obtained from each of four 150mm-diameter boreholes located approximately 1,5m away from the position of each of four-vane shear test holes. Initially, sampling was attempted using a standard U100 sampling tube driven into the soil. However, this method was unsuccessful as the peaty clay proved to be too soft, and the material tended to slip out of the sampling tube as it was extracted from the hole. It was then decided to abandon the U100 tube method in favour of a piston sampling technique. This consisted of a drive sampler with a retractable piston rod.

This piston sampler method proved successful and some 20 samples, each approximately 0,5m long, were obtained. All the tubes were sealed at either end with wax immediately after extraction so as to preserve the natural moisture content of the material. Special care was also taken in the transport of the tubes from the site to the laboratory in order to minimize disturbance.

3. LABORATORY TESTS AND INSITU TESTS

3.1 Grading Curves

The grading envelope for the peat is shown in Figure 2 below, from which it is apparent that the peat consists predominantly of silt and clay size particles with 5 to 25 percent by mass of fine sand size particles.

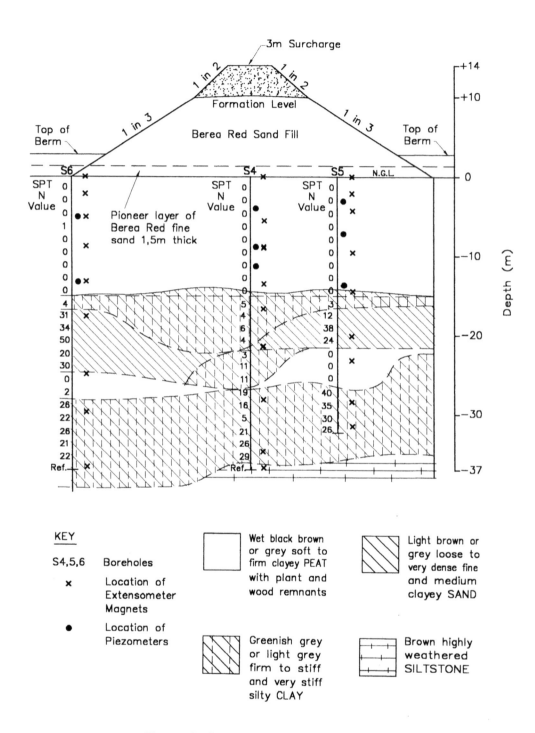

Figure 1. Representative Cross—Section

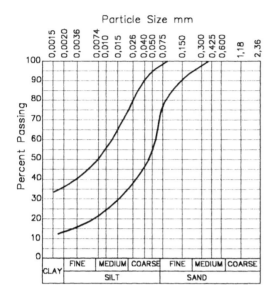

Figure 2. Grading Curve Envelope for Peat

The Symbols in Table 1

The Symbols in Table 1

P   Organic content

H   Humification

F   Fine Fibres

R   Coarse Fibres

W   Wood Remnants

The liquid limit ranged between 37 and 76 with an average of 55, and the Plasticity Index ranged from 13 to 36 with an average of 22.

Natural water contents in the peat ranged from slightly less than 250, to nearly 700 percent of the dry mass

3.2 Profile Descriptions

Representative profile descriptions of the peat are given in Table 1 in terms of the Modified Von Post classification (Landva, Pheeny, 1980) and the Standard MCCSSO classification (Jennings, Brink and Williams 1973)

3.3 Consolidation Tests

In a number of consolidation test on 75mm diameter specimens cut from 100mm diameter tube samples the void ratios ranged between 7,5 and 10 with corresponding water contents ranging from 321 to 396 percent.

The coefficient of volume compressibility m_v ranged between 1.45 and 1,64 m^2/MN. The rate of consolidation decreased markedly with increasing consolidation pressure.

TABLE 1: DESCRIPTION OF TUBE SAMPLES FROM A REPRESENTATIVE BOREHOLE

Depth Metres	Modified von post Classification	Standard MCCSSO Classification
4,5 – 5,0	P_{2-3}-H_3-F_2-R_1-W_0	Wet, black-brown peat with patches of red fine sand
6,0 – 6,5	P_3-H_{4-5}-F_1-R_1-W_{2-3}	Wet, black-brown soft clayey peat with much wood remnants
7,0 – 7,5	P_3-H_{3-4}-F_1-R_3-W_2	Wet, black-brown soft, clayey peat with plant remnants
8,0 – 8,5	P_3-H_3-F_1-R_3-W_{1-2}	Wet, black-brown soft, clayey peat with plant remnants
6,5 – 7,0	P_2-H_3-F_3-R_1-W_{2-3}	Wet, black-brown soft, clayey peat with wood remnants
7,5 – 8,0	P_3-H_3-F_2-R_3-W_2	Wet, black-brown soft, clayey peat with plant and wood remnants
10,5 – 11,0	P_1-H_{5-6}-F_2-R_0-W_0	Wet, grey black, soft, peaty clay

$11,0 - 11,5$	$P_2-H_{4-5}-F_1-R_2-W_{01}$	Wet, black-brown soft, clayey peat with plant remnants
$12,0 - 12,5$	$P_{1-2}-H_4-F_2-R_{0-1}-W_{0-1}$	Wet, black-brown soft, peaty clay
$12,5 - 13,0$	$P_{2-3}-H_{3-4}-F_2-R_1-W_1$	Wet, black-brown soft, peaty clay
$13,5 - 13,8$	$P_{2-3}-H_{3-4}-F_{1-2}-R_{0-1}-W_0$	Wet, black-brown soft, peaty clay
$14,5 - 14,8$	$P_1-H_6-F_{0-1}-R_0-W_0$	Wet, black, soft and pasty clay with patches of brown clayey sand
$16,0 - 14,8$	$P_1-H_7F_1-R_0-W_0$	Wet, black, soft, pasty peaty clay
$12,0 - 12,5$	$P_1-H_4-F_1-R_2-W_2$	Wet, black-brown, soft, clayey peat with wood and plant remnants
$13,0 \ 13,3$	$P_2-H_4-F_{0-1}-R_0-W_3$	Wet, black-brown soft decomposed wood remnants
$14,0 - 14,5$	$P_1-H_5-F_1-R_1-W_0$	Wet, black-brown, soft, pasty clayey, peaty clay
$12,0 - 12,5$	$P_1-H_4-F_1-R_2-W_2$	Wet, black-brown, soft, clayey peat with wood and plant remnants
$13,0 - 13,3$	$P_2-H_4-F_{0-1}-R_0-W_3$	Wet, black-brown, soft, decomposed wood remnants
$14,0 - 14,5$	$P_1-H_5-F_1-R_1-W_0$	Wet, black-brown, soft, pasty clayey peaty clay

3.4 Undrained Triaxial Compression Tests

The tests were carried out on 37mm diameter, 75mm high cylindrical specimens cut from the 100mm diameter piston samples. Typical results for a single borehole are plotted in Figure 3, and compared with the results of field vane tests, which exhibit a fairly close correlation. The vane tests were carried out by using a standard vane, 55mm x 110mm. An NX size hole was advanced by wash-boring to a predetermined depth. The bottom of the hole was then cleaned out and the jacketed vane inserted down to the bottom. The vane was then extended an additional 0,5m taking care not to rotate it as that would cause disturbance of the peat. The torque was then applied at a constant rate in the normal way.

3.5 Organic Content

The standard acid test was used for determining the organic content. The proportion by mass of organic material was found to average 79 percent in the range 72 to 94 percent. However, because of the low density of the peat fibres, their volume in the soil is greater than indicated by the percentages based on mass.

3.6 Dutch Cone Penetration Tests

The results of a typical Dutch cone penetration test is also shown in Figure 3. In some of the tests there were localized sharp increases in point resistance due, probably, to the presence of wood remnants.

It is seen that the ratio between point resistance and undrained strength or vane shear strength is about 38 or greater, in this series of tests.

4. CONSTRUCTION OF EMBANKMENT

Before placement of the embankment fill a pioneer layer of fine to medium sand, known locally as Berea red sand, was placed over the entire site of the embankment area to a depth of 1,5m. This was to provide access over the swampy ground for installation of instruments, and vertical sand wick drains required for acceleration of settlement. The dry density of the sand fill averaged approximately $1620kg/m^3$, and the moisture content about 14 percent. The instruments consisted of pneumatic piezometers, inclinometers, and magnetic settlement indicators.

Rate of placement of embankment fill and corresponding excess pore pressure in the peat and set-

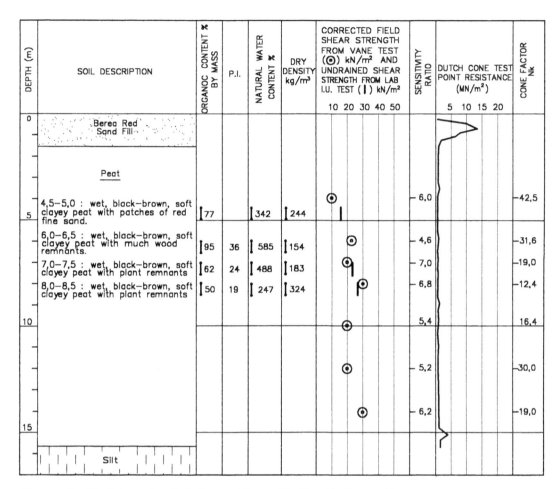

Figure 3. Variation in Soil Properties with Depth for a Representative Borehole

tlement of the original ground surface are shown for a representative cross section in Figure 4. The break in the fill height after 700 days was due to a flood which had an estimated return period of 1 in at least 30 years which washed away nearly the entire of the embankment. Most of the subsoil instrumentation, however, not was affected.

In the early stages of construction the rate of placement was just over 2m per year. The gain in shear strength and changes in pore pressure in the peat were continuously monitored to ensure stability of the embankment as its height increased. After the flood damage it became necessary to increase the rate of placement of fill to a maximum compatible with safety. As seen in Figure 4, the final rate of placement, up to the top of a 3m temporary surcharge was nearly 11m per year. Although there had been a marked increase in undrained shear strength of the peat as a result of consolidation during the

first 900 days, this rate of placement was excessive and localized slips began to occur on the shoulders of the embankment with heave of the natural ground beyond the 1,5m pioneer layer of sand. To stabilize the embankment it was therefore necessary to construct toe berms containing horizontal drains to remove the water reaching the pioneer sand layer from the vertical wick drains.

The drainage from the vertical sand wicks, which were on a grid of 2,5m x 2,5m to 3m x 3m, was impeded to some extent because, with a final settlement of nearly 7m, the top of the pioneer sand layer was over 5m below original ground level. The permeability of the slightly clayey Berea red sand comprising the fill was less than that of the pioneer sand layer.

From the vane shear tests it was apparent that there was a substantial gain in shear strength of the peat

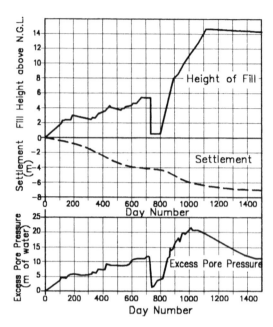

Figure 4. Observed Time Settlement Relationship

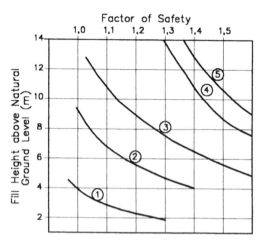

Figure 5. Estimated Factors of Safety of Embankment for Different Placement and Subsoil Conditions

during placement of the fill. It was shown that during construction the undrained shear strength of the peat could be expressed as 30 to 35 percent of the effective overburden pressure. Since the peat was still consolidating under the increasing mass of the embankment fill this ratio was expected to increase. Results of the consolidated undrained triaxial compression tests indicate a ratio between undrained strength and all round consolidation pressure of between 37 and 76 percent with an average of 55 percent, the higher ratios being associated with higher peat contents, or a lower clay content, of the samples.

The reason for the high ratios between consolidation pressure and undrained shear strength is possibly the relatively large angle of shearing resistance, ø, of the peaty material. In the lower stress ranges a value of ø in excess of 40 degrees was indicated for some of the peat samples. The average for all tests, including those on the more clayey peats, was 35 degrees.

The flood damage did not have any significant effects on stability of the embankment when reconstruction began. At that time the increment of loading applied to the peat was $97kN/m^2$. From the void ratio-pressure curves in the consolidation tests it is apparent that there was then already a marked decrease in void ratio. The rebound curve is very flat and, because of the low values of the coefficient of consolidation, any rebound would have occurred very slowly over a number of months. The washaway thus had little effect on the shear strength of the peat.

5. STABILITY AND SETTLEMENT OF THE EMBANKMENT

Variation in factors of safety with increasing height of embankment are shown in Figure 5. The procedure initially proposed for construction included placement of a berm and raising the height of the fill at the rate of 600mm per month, increasing later to 1000mm per month. This corresponds approximately to curve 5 in the diagram.

1. Initial constant strength and density of peat subsoil – rapid placement.

2. Shear strength increasing with depth according to empirical formulae, density equal to that of clay, rapid placement allowing for 3.5 metres of sand fill below original ground level, due to settlement.

3. Vane shear strength at 900 days and rapid placement, allowing for increase in density of clayey peat, due to consolidation.

4. Vane and laboratory shear strengths and measured densities of peat, with gain in strength during construction and with berm 25 metres wide, 2.5 metres high beyond toe of the embankment Placement rate 1000mm per month.

5. Vane and laboratory shear strength during con-

struction and with berm 25 metres wide, 2.5 metres high beyond toe, Placement rate 600mm per month.

The compressibility of the peat is indicated in Figure 6, in which the two parallel dashed lines represent the range of values of the compression index for similar peats found by the National Research Council of Canada (NRCC).

The ratio between time for 90% of consolidation settlement in the laboratory consolidometer tests and that in the field is

$$\frac{t_{90} \text{ Field}}{t_{90} \text{ Lab}} = \left[\frac{\text{H Field}}{\text{H Lab}} \right]^{i}$$

where t_{90} Field = the time for 90% consolidation of soil material in the field.

t_{90} Lab = the time for 90% consolidation of the laboratory test specimen.

H Field = the thickness of the soil layer in the field.

and   H Lab = the height of the test specimen in the laboratory.

In Terzaghi's classical unidimensional consolidation theory i is 2. However, in peat materials with very high initial water contents the value of i may be as

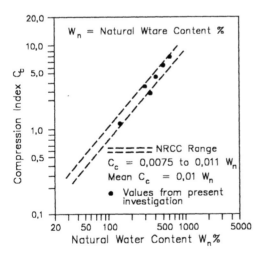

Figure 6.   Relationship Between Compression Index and Natural Water Content

low as 1, according to Lake, 1961. From all the data available from this site, the value of i. was found to be approximately 1.5 for the peat.

Similar values of the exponential, i, were determined by Samson and La Rochelle, 1972.

6.   LONGER TERM BEHAVIOR OF THE EMBANKMENT

After removal of the remaining surcharge fill, and constructing the asphalt premix surfacing, the road was opened as part of the motor way between Durban and Richards Bay.

Some relatively small settlements have continued at a reducing rate during the first three years of the road being in service.
These settlements, of about 100mm in places, have resulted in two parallel cracks across the entire width of the road where the embankment is underlain by the thickest zone of peat. At the surface the cracks were 3cm wide and 40cm apart.

Later, similar cracks also appeared after opening a section of motorway south of Durban, near Port Shepstone. Here the road was on an embankment 15m high over a variable deposit of soft normally consolidated clay. Vertical wick drains and temporary surcharge were also employed here to accelerate settlement. The cracks appeared, at the top of the formation level, before the road surfacing was placed.

The cracks are attributed to the inability of the rigid compacted fill to adjust to the relatively small but uneven secondary consolidation settlement of the underlying peat. Repairs have been carried out on a maintenance basis by excavating to depths of 1 to 2 metres, backfilling and re-establishing the road paving layers, then constructing asphalt premix surfacing over the width of the excavation. After a further two years the cracks had not re-opened.

7. EXPERIENCE GAINED

A number of features of this project are considered worth mentioning in relation to future similar works.

Firstly, to the knowledge of the author, such thick and extensive zones of classical peat have not previously been reported in relation to construction in South Africa.

Secondly, a particularly detailed subsoil investigation is necessary if, as in this case, the presence of peat is indicated anywhere along the proposed routes of roads which are to cross the wide flood plains of the rivers along the Kwa Zulu Natal coast.

Thirdly, possible cracking of a relatively rigid earth embankment due to differential settlement of underlying peat, or soft clays, must be taken into account. This mechanism has often been reported in relation to the design of earth embankment dams.
It is also observed, from Figure 4, that at 1400 days when settlement appeared to have virtually ceased the excess pore pressure had decreased by only 50%. Thus some continuing long term secondary consolidation is to be expected.

REFERENCES

Jennings J.E. 1973. *Revised guide to soil profiling for Civil Engineering purposes in Southern Africa.* Trans. S. Afr. Instn. Civ. Engns. Vol. 15, pp 3-12.

Lake, J.R. 1961, *Investigation of the problem of constructing roads on peat in Scotland.* Proc. Seventh Muskeg Res. Conf., NRC Technical Memorandum, No. 71, pp133-149.

Landva A.O, and Pheeny P.E. 1980, *Peat Fabric Structure.* Canadian Geotechnical Journal vol. 17, pp 416 – 435.

Samson, L., and La Rochelle, P. 1972. *Design and Performance of an Expressway Constructed over Peat, by Preloading.* Canadian Geotechnical Journal.

Author index
Index des auteurs